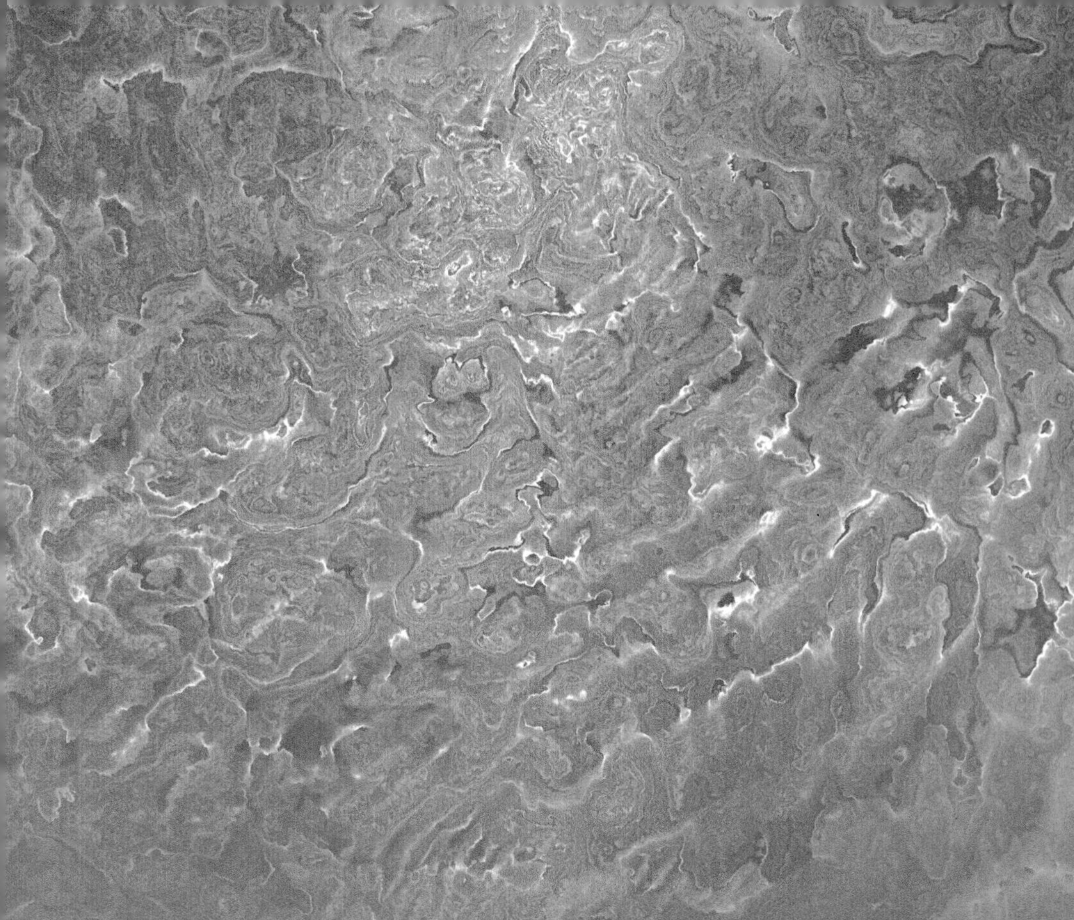

Mullite Formations

Mullite Formations

Analysis and Applications

Akshoy Kumar Chakraborty

Jenny Stanford
Publishing

Published by

Jenny Stanford Publishing Pte. Ltd.
Level 34, Centennial Tower
3 Temasek Avenue
Singapore 039190

Email: editorial@jennystanford.com
Web: www.jennystanford.com

British Library Cataloguing-in-Publication Data
A catalogue record for this book is available from the British Library.

Mullite Formations: Analysis and Applications

ISBN 978-981-4877-05-3 (Hardcover)
ISBN 978-1-003-03167-3 (eBook)

Dedicated to
the Revered Sadhana–Siddha Sadguru Swami Nigamananda,
renowned spiritual personality

Contents

Preface xxix

1. Introduction **1**
1.1 Mullite Research in Conventional Fields 1
1.1.1 The Phase Diagram 2
1.1.2 Forms of Mullite, Solid Solution, and Stoichiometry 2
1.1.3 Applications 4
1.2 Mullite Research in Advanced Fields and Its Importance 7
1.3 Existing Methodology 10
1.3.1 Sinter Process 10
1.3.2 Gelation Process 11
1.3.3 Coprecipitation Process 12
1.3.4 Colloidal Process 12
1.3.5 Codecomposition Process 12
1.4 Visualization of the Scope of the Mullite Book 14

Part I Review

2. Mullite Precursors Synthesized by the Monophasic Gelation Method **21**
2.1 Synthesis of Mullite by the Aqueous Sol-Gel Method 21
2.2 Synthesis of Mullite by the Monophasic Gel Method 27
2.3 Synthesis of Mullite by Polymeric and Colloidal Gel Methods 32
2.4 Summary 36

3. Mullite Precursors Synthesized by the Coprecipitation Method **43**
3.1 Precipitation of Water-Soluble Silicon and Aluminum Components in the Presence of NH_4OH 44

3.2 Hydrolysis and Subsequent Precipitation of the Organic-Silicon Component in the Presence of the Al Component (Basic Medium) 47
3.3 Hydrolysis and Subsequent Precipitation of Water-Soluble Silicon Component in the Presence of Organic Al Component (Purely Basic Medium) 49
3.4 Hydrolysis and Subsequent Coprecipitation of Organic Al and Si Components (Basic Medium) 49
3.5 Precipitation of Si and Al Components from Organic Sources by Using Urea 55
3.6 The Incorporation of Si in Al_2O_3 Structure: Case Studies by Different Researchers 57
3.7 Summary 62

4. Mullite Precursors Synthesized by the Colloidal Gelation Method 71
4.1 Synthesis of General Diphasic Gels 71
4.2 Synthesis of "in situ" Diphasic Gels 79
4.3 Synthesis of Other Diphasic Gels by Colloidal Methods 79
4.4 Synthesis of Solid-Gel-Coated Precursor by Microcomposite Methods 81
4.5 Summary 84

5. Mullite Precursors Synthesized by Codecomposition Processes 91
5.1 Introduction 91
5.2 Decomposition Method 92
5.2.1 Spray-Pyrolysis Route 92
5.2.2 Spray-Drying Route 93
5.2.3 Plasma Powder Process 98
5.2.4 Thermal Decomposition of Alkoxides 98
5.2.5 Flame Spraying 98
5.2.6 Freeze-Dry Process 100
5.2.7 Decomposition of an Aerosol 101
5.3 Summary 102

6. Mullite Precursors Synthesized from Other Miscellaneous Sources/by Other Processes 107

6.1 Mullite Using Components from the Same Source but with Different Processing Conditions 107

6.1.1 Producing Two Kinds of Mullite Precursors 107

6.1.2 Producing Three Kinds of Mullite Precursors 111

6.1.3 Producing Four Kinds of Mullite Precursors 112

6.1.4 Producing Six Kinds of Mullite Precursors 113

6.2 Synthesis of a Mullite Gel by the Organic Route 114

6.2.1 Citric Acid 114

6.2.2 Using ANN and TESPA by an Organic Gel–Assisted Processing Technique 115

6.2.3 Use of Organic Ligands 117

6.2.4 Ethyl Acetoacetate 117

6.2.5 Acrylamide 120

6.2.6 Urea 121

6.2.7 Aluminosiloxanes 122

6.3 Synthesis of Mullite Gel by Miscellaneous Methods 122

6.3.1 Synthesis of Mullite by the Hybrid Gel Route 122

6.3.2 Synthesis of Optically Clear Mullite Ceramic 123

6.4 Pilot-Scale Process of Mullite Synthesis by the Consecutive Precipitation Method 128

6.4.1 By the Aqueous Gel Route 128

6.4.2 By the Monophasic Gel Route 128

6.4.3 By Consecutive Precipitation and Oxide Mixtures 128

6.4.4 Colloidal Process 129

6.5 Summary 129

7. Phase Evolution Sequences of Mullite Precursors Synthesized by Different Methods Based on Components Used and Processing Variables 139

7.1 Monophasic Mullite Gel Prepared by the Aqueous Sol-Gel Method 139

7.1.1 Effect of pH on the Phase Transformation of Monophasic Mullite Gel Prepared by Aqueous Sol-Gel Method 139

7.1.2 Effect of Silica Sols from Different Sources on the Phase Transformation of Mullite Gel Prepared by Aqueous Sol-Gel Method 142

7.2 Monophasic Mullite Gel Prepared by the SH Gel Method and Variables Affecting Its Phase Evolution 145

7.2.1 Effect of pH on the Phase Transformation of This Monophasic Mullite Gel 145

7.2.1.1 Monophasic gel (SH) 145

7.2.1.2 Other types of monophasic gels (RH) 151

7.2.1.3 Development of diphasicity in a purely basic condition 153

7.2.2 Effect of Water on the Phase Transformation of Monophasic Mullite/SH Gel 161

7.2.3 Effect of Aging Time and/or Temperature of the Precursor Solution on the Phase Transformation of Monophasic SH Mullite Gel 168

7.2.4 Effect of the Drying Condition on the Mullite Gel 173

7.3 Synthesis of Polymeric Mullite Gel and Colloidal Mullite Gel 175

7.3.1 Effect of pH on the Phase Transformation of Polymeric and Colloidal Mullite Gels 182

7.3.2 Effect of Water on the Phase Transformation of Polymeric Colloidal Mullite Gel 183
7.3.3 Effect of Temperature on the Phase Transformation of Polymeric and Colloidal Mullite Gels 184
7.3.4 Effect of Washing and of Gel Surface on DTA 185
7.4 Mullite Gel Synthesized by the Diphasic Gelation Method 186
7.4.1 Effect of pH on the Phase Transformation of Mullite Gel Prepared by a Diphasic Gel 187
7.4.1.1 Possibility of formation of gels of a diphasic character in the basic pH range 193
7.4.1.2 Possibility of formation of a mixture of monophasic and diphasic gels 195
7.4.1.3 Importance of the reactivity of two components 195
7.4.2 Effect of Aging Silica 197
7.5 Mullite Precursor Synthesized by SD and SP Methods 197
7.5.1 Effect of Aging on the Phase Transformation of a Mullite Precursor Prepared by SP and SD methods 198
7.6 Verification of Transformation Processes of Mullite Gels Synthesized in Varying Processing Conditions from Components from Varying Sources vs. That from Components from Two Fixed Sources: A Case Study 199
7.7 Conclusion 201

8. Phase Evolution Studies of Various Mullite Gels/ Precursors by IR and Raman Spectral Techniques 211
8.1 Introduction 211
8.2 Basic Literature of IR Study 212
8.3 IR Study of Monophasic Gels 214

8.4 Comparison of IR Spectra of Monophasic Gels of Different Al_2O_3-SiO_2 Compositions on Heating 222
8.5 IR Study of Coprecipitated Gel and Diphasic Gel 225
8.6 Raman Spectral Study of a Monophasic Gel 228
8.7 Summary 230

9. Phase Evolution Studies of Various Mullite Precursors/Gels by MAS-NMR 235
9.1 Basic Literature 235
9.2 MAS-NMR Studies of Mullite Gels Synthesized by the Aqueous Sol-Gel Method 236
9.2.1 Mullite Gels: CM Sample 236
9.2.2 Coprecipitated Gels: Coprecipitated Sample 239
9.2.3 Mullite Gels Synthesized by Hydrolysis and Gelation 240
9.3 Monophasic and Diphasic Gels 240
9.4 Synthesis of Mullite Gels by Polymeric and Colloidal Methods 241
9.5 Synthesis of Mullite Gels by Coprecipitation Methods 242
9.6 Synthesis of Mullite Gels by the Diphasic Method 243
9.6.1 MAS-NMR Studies of Diphasic Gels 243
9.6.2 Phase Evolution of Mixed Gels by MAS-NMR of ^{29}Si 245
9.7 Synthesis of Mullite Gels by the Spray-Drying Method 246
9.8 Comparison of MAS-NMR Spectra of Mullite Precursors 246
9.8.1 CM and SGM 247
9.8.2 MI and MII Precursors 249
9.8.3 Aluminosilicate Gels and Silicon Aluminum Ester 253
9.8.4 Mullite Gel Synthesized by Hydrolysis and Gelation in Varying Amounts of Water/Alcohol 254
9.8.5 Mullite Precursors Synthesized by Polymeric, Colloidal, and Nitrate Gels 257

9.8.6 Mullite Precursors Synthesized by Polymeric Gel (Type I), Diphasic Gel (Type II), and Coprecipitated Gel (Type III) 259
9.8.7 Mullite Precursors Synthesized by Polymeric Gel, Coprecipitated Gel, and Diphasic Gel 262
9.8.8 Mullite Precursors of SP, CA, and PB Gels 264
9.8.9 Monophasic, Coprecipitated, Diphasic, and Spray-Pyrolyzed Precursors 265
9.8.10 Mullite Precursors Obtained from Components from the Same Source 271
9.9 Conclusions 273

10. Chemistry of Mullite Formation through the Sol-Gel Process 283
10.1 Mullite Gel by the Aqueous Sol-Gel Method Using Water-Soluble Components 283
10.2 Mullite Gel from a Monophasic Gel: Chemistry of Hydrolysis vis-à-vis Gelation-cum-Coprecipitation Reaction 287
10.2.1 DTA Studies of $Al(NO_3)_3 \cdot 9H_2O$ and TEOS and Their Interactions 298
10.3 Mullite Gel by Polymeric and Colloidal Methods Using TEOS and AlOBu 312
10.4 Mullite Gel by Coprecipitating Gel Methods 318
10.4.1 Both Water-Soluble Components Precipitated by NH_4OH 318
10.4.2 Water-Soluble Al Salt Precipitated by NH_4OH in the Presence of Organic Silica 318
10.4.3 Coprecipitation of Organic Al and Si Compounds 318
10.5 Mullite Gel by the Diphasic Method 319
10.5.1 General Synthesis of a Diphasic Gel 319
10.5.2 In situ Synthesis of a Diphasic Gel 319
10.5.2.1 NH_4OH added to a mixture of TEOS and ANN 319

10.5.2.2 Urea added to a mixture of ANN and TEOS 322
10.5.2.3 Gelation course in acidic vs. basic condition 323
10.6 Mullite Precursor Synthesis by Other Processes 323
10.7 Summary 325

11. Homogeneity of Al_2O_3-SiO_2 Gels/Precursors 335
11.1 Effect of Homogeneity 335
11.1.1 Views on Homogeneous Al_2O_3-SiO_2 Gels vs. Mullitization and Atomic-Scale Mixing of Al and Si Atoms 335
11.1.2 Mixing Scale vs. Mullitization Temperature 336
11.1.3 Intensity of Exotherm of Mullite Formation Related to Al-Si Bonding 337
11.1.4 Interrelationship between 980°C Exotherm and Mullite Formation 338
11.2 Methods of Achieving Homogeneity in the Alkoxide Method of Mullitization 340
11.2.1 First Approach: Prehydrolysis Technique 341
11.2.2 Second Approach: Reducing the Difference in the Hydrolysis Reaction Rates of Components 342
11.2.3 Third Approach: Matching the Hydrolysis Reaction Rates 343
11.3 Methods of Achieving Homogeneity in Both Water-Soluble Salt Methods of Mullitization 343
11.3.1 pH 343
11.3.2 Attempt to Produce Homogeneous Precipitation by Urea 348
11.4 Achieving Homogeneity by the Soluble Salt–Organic Silicon Method of Mullitization 348
11.4.1 Varying Use of Solvent 348
11.4.2 Amount of Water Used for Hydrolyzing Alkoxide 349
11.4.3 Method or Rate of Addition of Water 349
11.4.4 Time of Hydrolysis with or without Stirring Time 349

11.4.5 Rapid Solvent Removal during Spraying: An Important Factor? 350
11.4.6 Evaporation of Solvents in a Water Bath or on a Hot Plate 351
11.4.7 Role of the Alumina Component 351
11.5 Method of Achieving Homogeneity by Thermal Decomposition Process 353
11.5.1 Spray Drying 353
11.5.2 Codecomposition 355
11.5.3 Spray Pyrolysis 355
11.6 Temperature Scale of Mullitization 356
11.6.1 Homogeneity of Mullite Precursors vs. Crystallization Path of Mullite Formation 356
11.6.2 As Per Homogeneity of Mullite Precursors in MAS-NMR Study 357
11.7 Summary 359

12. Comparison of Thermal Transformation Processes of Six Mullite Precursors 367
12.1 Experimental Techniques for Phase Evolution Studies of Mullite Precursors 367
12.2 Variables Affecting Mullite Precursor Synthesis Methods and Phase Evolution 368
12.3 Transformation Behaviors of Six Mullite Precursors Studied by DTA, XRD, IR, and MAS-NMR 370
12.4 Transformation Processes of Precursors Synthesized under Varying Processing Conditions: A Few Examples 372
12.4.1 CM and SGM Gels 372
12.4.2 Precursors Synthesized from ANN and TEOS 373
12.4.3 Mullite Gels from SiO_2 and Al_2O_3 374
12.4.4 Mullite Gels Synthesized by Three Methods 377
12.4.5 Comparison of Monophasic Mullite Gels Made from TEOS and AIP by Different Techniques with Diphasic Gel 379

12.4.6 Comparison of Crystallization Sequences of Four Mullite Gels 382
12.4.7 Polymeric Gels Obtained by Changing Acidic Conditions to Basic Conditions 384
12.4.8 Mullite Powders Obtained by Four Methods at Two pH Conditions 385
12.4.9 Transformation of Al_2O_3-SiO_2 Precursors Prepared by Different Techniques 387
12.4.10 Comparison of Transformation of $3Al_2O_3$-SiO_2 Precursors with That of Oxide Mixtures 388
12.4.11 Comparison of Transformations of Mullite Precursors Obtained from the Same Components under Five Processing Variables 389
12.5 Various Gel-Mullite Transformation Processes: Problems and Questions 392
12.5.1 DTA and Densification Study 392
12.5.2 XRD Study 395
12.5.3 IR Study 397
12.5.4 NMR Study 397
12.5.5 Water Used for Hydrolysis 398
12.5.6 pH in the Gelation Process 399
12.6 Scientific Community: Controversies and Predictions 399
12.7 Predictions Made by the Scientific Community 400
12.7.1 Characterization of the Spinel Phase 401
12.7.2 Mullitization Reaction 402
12.7.3 Analysis of Previous Studies and Probable Solutions 402
12.8 Summary 403

Part II Identification: Characterization of Four Phases of Gel-to-Mullite Transformation Series

13. Nature and Its Characterization of the Noncrystalline Al_2O_3-SiO_2 Mullite Precursor Phase 419
13.1 Introduction 419
13.2 Chemical Techniques 420

13.3 IR Technique 421
13.4 MAS-NMR Technique 424
13.4.1 MAS-NMR Studies of the Hydrolysis/Gelation Process 424
13.4.2 MAS-NMR Studies of Monophasic Gels (Hydrolysis/Gelation of Four Aluminum Components) vs. Diphasic Gel 426
13.5 X-Ray Absorption Fine Structure Spectroscopy 431
13.6 RED Functions from LAXS Study 434
13.7 X-Ray Fluorescence Spectroscopy Study 438
13.8 Solid-State Reaction Study: A New Approach 438
13.9 Alkali Leaching Study 440
13.10 Physicochemical Changes of Raw Mullite Gel up to the Final Dehydroxylation 440
13.11 The Presence of –OH Groups in Noncrystalline Aluminosilicate Precursor Phase by FTIR and Its Role 443
13.12 Characteristics of Six Types of Precursor Phases 444
13.13 Requirements for the Exhibition of a 980°C Exotherm 448
13.14 Summary 451

14. Final Dehydroxylation of Noncrystalline Al_2O_3-SiO_2 Mullite Precursors 459

14.1 Intimacy of Mixing Al_2O_3/SiO_2 Powder, Homogeneity, and Agglomeration 459
14.2 Role of Protonic Species of the Noncrystalline Precursor Phase in Crystallization during a 980°C Exotherm 464
14.2.1 Occurrence of an Endothermic Dip Just before the 980°C Exotherm 465
14.2.2 Final Dehydroxylation at the Endothermic Dip 468
14.2.3 Phase Separation Prior to Crystallization of Mullite and/or Al-Si Spinel 471

14.2.4 Absence of the Metastable Phase Separation Process during the Final Dehydroxylation Process 472
14.3 Final Dehydroxylation of Noncrystalline Al_2O_3-SiO_2 Mullite Precursors 473
14.4 Conclusion 476

15. Spinel Phase: A Concise Review 481
15.1 Formation of the Spinel Phase in Various Gels Observed by Earlier Authors 481
15.2 Phase Transformation and Growth of Spinel Phase in "in situ" Diphasic Gels Observed by the Author 486
15.2.1 Synthesis and Transformation Behaviors of Diphasic Gels of Varying SiO_2 and Al_2O_3 Contents 486
15.2.1.1 Qualitative XRD study 486
15.2.1.2 DTA/DDTA studies of "in situ" diphasic gels 489
15.2.2 Qualitative XRD Study of Al-Si Spinel Phase Formation 492
15.3 Variations in the Evolution Processes of Diphasic Gels, Mixed Oxides, and Mixed Gels of Different Compositions: Examples 494
15.3.1 Comparison of Phase Transformation of Diphasic Gels with That of a Chemical Mixture of γ-Al_2O_3 and Silica (A) 494
15.3.1.1 Phase evolution of three diphasic gels 495
15.3.1.2 Phase evolution of mixed oxides 496
15.3.2 Mode of Crystallization Behavior of Mixed Gels 498
15.3.3 Crystallization of Al-Si Phase from Noncrystalline Aluminosilicate Precursor Phase during Heating of a Diphasic Gel 501

15.3.3.1 Amorphous bands of the noncrystalline precursor phase and the noncrystalline aluminosilicate residual phase 502
15.3.3.2 Al-Si spinel phase 503
15.4 Characterization of the Spinel Phase by a Solid-State Reaction Study with CaO 505
15.4.1 Characterization of Diphasic Gels and Precursors 505
15.4.2 Characterization of Monophasic and Diphasic Gels and Precursors 507
15.5 Characterization of the Spinel Phase by the Alkali Leaching Method of Heat-Treated "in situ" Diphasic Mullite Gels 513
15.5.1 Leached Residue Characterization by Different Researchers Using Different Techniques: Supplementary Case Studies 517
15.5.1.1 TEM/EDS and a few drawbacks 517
15.5.1.2 IR technique and a few drawbacks 520
15.5.1.3 MAS-NMR Study and a few drawbacks 522
15.6 Characterization of the Spinel Phase by QXRD Studies by Various Authors 523
15.6.1 Semiquantitative Amount of the Al-Si Spinel Phase by Earlier Studies 523
15.6.2 Semiquantitative Amount of the Al-Si Spinel Phase by the Author 525
15.6.2.1 Measurement of intensity of the Al-Si spinel phase 525
15.6.2.2 QXRD study of the spinel phase 528
15.6.3 Discussions of QXRD Results of the Al-Si Spinel Phase 528
15.6.4 Discussions on Al-Si Spinel Phase Composition First Suggested by Okada and Otsuka (1986b) 530

15.6.5 Discussions on XRD Analysis of the Spinel Phase by Wang and Thomson (1995) 532
15.6.6 Discussions on the Quantitative Data of the Spinel Phase by Wei and Halloran (1988) 533
15.7 Characterization of the Spinel Phase by Lattice Parameter Measurement 534
15.8 Interrelationship between Al-Si Spinel Phase and Mullite Formation: Case Studies by Different Researchers 536
15.9 Controlling Factor in Spinel Formation 538
15.9.1 Role of SiO_2 in Phase Evolution Processes of Mullite Precursors 540
15.9.2 Role of Aqueous Silica Sol in Spinel Formation 540
15.9.3 Role of Organic Silica Source in Spinel Formation 541
15.10 Summary 542

16. Mullite Phase: A Concise Review 559
16.1 Introduction on Mullite Formation in Different Cases of Gels 559
16.1.1 Monophasic Gels 560
16.1.2 Coprecipitated (CP) Gel 562
16.1.3 Colloidal Gel 563
16.1.4 Diphasic Gels 563
16.1.5 Spray Pyrolysis/Spray Dried Precursors 564
16.1.6 Chelated Monophasic Gel 564
16.1.7 Different Views of Mullite Formation 564
16.2 Mullite Formation Studies of Monophasic Gels 571
16.2.1 Quantitative XRD Study (QXRD) of Mullite Formation 571
16.2.2 Mullite Formation of Six Precursors Based on Qualitative and Quantitative Analysis: A Case Study 573
16.2.2.1 First case of a monophasic gel: Formation of mullite phase at first exotherm 574

16.2.2.2 Second case of a monophasic gel: Formation of mullite phase by two paths 576
16.2.2.3 Third case of a monophasic gel: Intense crystallization of Al-Si spinel phase 576
16.2.2.4 Fourth case of coprecipitated gel: Formation of mullite occurs at second exotherms 577
16.2.2.5 Fifth case of diphasic gel: Formation of mullite in diphasic gels 579
16.2.2.6 Sixth case in SD precursor: Formation of mullite in SD gels 580
16.2.3 Mullitization Behavior of Oxide Mixture in Comparison to Diphasic Gel 581
16.3 Mullite Formation of Diphasic Gels 583
16.3.1 Diphasic Gel Made Using NH_4OH at pH = 9 e.g., "in situ Gel" and It's Mullite Formation 584
16.3.2 Comparative Mullite Formation Process of Diphasic Gels in Stoichiometric vs. Nonstoichiometric Compositions 587
16.3.3 Transformation of Intermediate Al-Si Spinel Phase along with Θ-Al_2O_3 Formed in Mixed Gel to Mullite Phase 600
16.4 Lattice Constant Measurement Data of Mullite Phase, Relation of Unit Cell Parameter to Composition of Mullite/Solid Solution and Stoichiometry 604
16.4.1 Lattice Parameter Measurement of Mullite Obtained by Solid State Reaction 607
16.4.2 Lattice Parameter Measurement of Mullite Obtained by Monophasic Gel/Coprecipitated Gel Method 609

16.4.3 Changes in LC of Mullite Formed Out of Monophasic Gels of Different Compositions 613
16.4.4 Lattice Parameter Measurement of Mullite Obtained by Diphasic Gels Method 616
16.4.5 Composition of Alumina Rich Mullite 617
16.5 Changes in LC of Mullite Formed Out of Five Different Compositions of Diphasic Gels: A Case Study 617
16.6 Heat of Reaction/980°C Crystallization Enthalpy of Various Mullite Gels and Variation of Activation Energy for 980°C Crystallization 625
16.7 Summary 629

17. Nature of Residual Noncrystalline Aluminosilicate Phases Associated with Mullite Formation 649
17.1 Nature of Noncrystalline Al_2O_3-SiO_2 Mullite Precursor Phase Prior to 980°C Exotherm 649
17.2 Liberation of One Noncrystalline Aluminosilicate Phase & One Residual Phase during Final Dehydroxylation Process at 980°C Exotherm 653
17.3 Composition of Residual Noncrystalline Aluminosilicate Phase in the Light of MAS-NMR Study 656
17.4 Formation of Noncrystalline Aluminosilicate Phases during Mullite Formation Process in Six Cases of Precursors 657
17.5 Role of the Large Quantity of Noncrystalline Aluminosilicate Phase in Mullitization 658

Part III Critical Analysis

18. Critical Analysis and Characterization of Spinel Phase 663
18.1 Introduction 663
18.2 Identification and Characterization of Spinel Phase: A Summary 666
18.2.1 Indirect Evidences Al-Si Spinel 666
18.2.2 Direct Evidences of Al-Si Spinel 670

18.2.3 Characterization of Al-Si Spinel Phase Formed Out of Two Mullite Precursors in the Light of MAS-NMR Study: An Example 670
18.2.3.1 Al-Si spinel phase formation in case of monophasic gel 670
18.2.3.2 Al-Si spinel phase formation in case of diphasic gel 671
18.3 Evidence of Formation of Si Incorporated θ-Al_2O_3 as Coexisting Phase with Al-Si Spinel Phase during Transformation of Mixed Gels of Varying Compositions by Deconvolution of XRD Peaks and by Measurement of Peak Width 675
18.3.1 Precision Identification by Deconvolution of XRD Peak of Spinel Phase 675
18.3.2 Measurement of Peak Width 677
18.4 Stepwise Process of Crystallization of Al-Si Spinel Phase 680
18.4.1 Formation of Noncrystalline Aluminosilicate Phase during Dehydration and Decomposition of Diphasic Gel (400°C–600°C) 683
18.4.2 Crystallization of Noncrystalline Aluminosilicate Phase to Al-Si Spinel Phase and Interrelationship between Area of the Amorphous Hump at 22° 2ϑ to that of Peak at 67° 2ϑ of Al-Si Spinel Phase 684
18.4.3 Mechanism of Transformation of Alumina Phase Content of Al_2O_3-SiO_2 Gel 685
18.4.4 Critical Studies on Intermediate Noncrystalline Phase-to-Al-Si Spinel Phase-to-Mullite 686
18.4.4.1 Discussion on spinel formation studies of Ivankovic et al. (2003) 688

18.4.4.2 Discussion on microstructural studies of Wei and Halloran (1988) 690
18.4.4.3 Discussions on mullitization study of HB13 and sample B by Schneider et al. (1993) and Jaymes et al. (1996) 691
18.5 Variations in Quantity of Formations of Al-Si Spinel Phase from Six Cases of Precursor: A Case Study 691
18.6 Composition of Spinel Phase by Different Researchers: Supplementary Case Studies 693
18.6.1 TEM Study 693
18.6.2 IR Study 694
18.6.3 MAS-NMR Study 696
18.6.4 Quantitative X-Ray Diffraction Analysis (QXRD) 698
18.6.5 Polymorphic Transformation of Al-Si Spinel to Mullite 700
18.7 Solid Solution Range of Silica in γ-Alumina Phase to Form Al-Si Spinel Phase: Illustrative Examples 700
18.7.1 Introduction to Solid Solution Range of Silica in Al-Si Spinel Phase 700
18.7.2 Siliceous Side of Mullite Stoichiometry 703
18.7.2.1 Enthalpy data 706
18.7.3 Aluminous Side of Mullite Stoichiometry 706
18.7.3.1 Starting temperature of mullite formation 710
18.7.3.2 Crystallization temperature of mullite 712
18.7.3.3 Nucleation mechanism 716
18.8 Summary 717

19. Critical Analysis and Characterization of Mullite Phase 729
19.1 Introduction 729
19.2 Characterization of 980°C Mullite and Solid Solution: Different Views 731

19.3 Problems of Characterization of Weakly Crystalline Intermediate Mullite Phase vis-à-vis 980°C Reaction 734
19.3.1 Presence of Noncrystalline Aluminosilicate Residual Phase 735
19.3.2 Progressive Changes in Lattice Parameter of Mullite 735
19.3.3 Progressive Changes in Crystallinity of Mullite 736
19.3.4 Progressive Changes of AlO_4 to AlO_6 Ratios of Mullite 736
19.3.4.1 Observed by IR study 736
19.3.4.2 Observed by ^{27}Al NMR study 737
19.3.4.3 Alignment of ATEM and Lattice Parameter Curves 738
19.3.4.4 Progressive shift of Si resonance peak in ^{29}Si NMR study 738
19.3.5 Tentative Prediction of the Nature of 980°C Mullite Phase 738
19.4 Discussion on Earlier Results on Changes in Composition of Primary Mullite to Regular Mullite 741
19.4.1 Mullitization in First Type of Monophasic Gels e.g., SH Gel of Okada et al.; Type I of Schneider et al.; SHI of Haque and Chakraborty 741
19.4.2 Mullitization in Other Type of Monophasic Precursors e.g., RH Gel of Okada et al.; Type III of Schneider et al.; CP Gel and SHIII of Haque and Chakraborty 748
19.4.3 Discussions on Mullitization in Some Other Type of Monophasic Gel of Previous Authors 748
19.4.4 Effect of Heat on Changes in Mullite Type Phase at 1200°C–1600°C 753
19.4.5 Composition of Mullite Formed Out of Diphasic Gel 755
19.5 Summary 756

20. Emergence of Only Three Routes of Phase Transformation Sequences of Mullite Precursors and Critical Analysis of Intermediates **767**
20.1 Introduction 767
20.2 Mullite Precursors Synthesized by Acidic Medium as First Route of Mullitization and Their Phase Evolution 768
20.2.1 Phase Transformation of Aqueous Mullite Gels Showing Complete Mullitization Reaction in a Single Step at 980°C Exotherm 768
20.2.2 Phase Transformation of Monophasic Gels Showing Mullitization Reaction Mostly in a Single Step at 980°C Exotherm 769
20.2.3 Formation of Noncrystalline Aluminosilicate Phase is a Function of pH on in Monophasic Gels e.g., SHI, SHII, SHIII and SHIV 773
20.2.4 Formation of Noncrystalline Aluminosilicate Phase Based on Water of Hydrolysis Used in Monophasic Gels 774
20.2.5 Phase Transformation of (SHIII) Gel Out of Six Types of SH Mullite Gels: As a Representative Example 778
20.2.6 Phase Transformation of Polymeric Gels Showing Mullitization Reaction in a Single Step at 980°C Exotherm 779
20.3 Mullite Precursors Synthesized at Basic pH as Second Route of Mullitization and Their Phase Evolution 782
20.3.1 Mullite Formation Out of Diphasic Gel Synthesized by General Method Using Boehmite Sol Considering Alumina and TEOS, Ludox/Silica Sol as Source 782
20.4 Redesignation of Al_2O_3-SiO_2 Gels Synthesized by Earlier Researchers 782

20.4.1 Coprecipitated Gels Prepared by Using ANN and TEOS by NH_4OH at pH = 6 is to be Designated as Diphasic Gel: First Kind 782
20.4.2 Coprecipitated Gels Prepared by Using ANN and TEOS by NH_4OH at pH = 9 is to be Designated as "in situ Gel" Diphasic Gel: Second Kind 788
20.4.3 Coprecipitated Gels Prepared by Using ANN and TEOS by NH_4OH at pH = 9–14 is to be Designated as Diphasic Gel: Third Kind 789
20.4.4 Coprecipitated Gels Prepared Out of Large Silicon Hydroxide Colloids Considering Silica Component as Source is Designated as Diphasic Gel 790
20.4.5 Diphasic Gels Considering Alumina as Source Component is Designated as Diphasic Gel 791
20.4.6 Diphasic Gels Considering Use of Large Quantity of Hydrolysis Water 791
20.4.7 Probability of Formation of Mixed Monophasic and Diphasic Gel 792
20.4.8 Evolution Mixed Gel to Mullite Phase: Fourth Kind 793
20.5 Categorization of Diphasic Gels 793
20.6 Mullite Formation in AS-γ-Al_2O_3 Microcomposites Behaves as Like as Diphasic Route 801
20.7 Mullite Precursors Synthesized by Codecomposition Method as Third Route and Their Phase Evolution 804
20.8 Emergence of Composition of Intermediate Mullite and Spinel Phases 808
20.8.1 Composition of Weakly Crystalline Primary Mullite 808
20.8.2 Composition of Weakly Crystalline Spinel Phase 810

21. Critical Analysis of Classification Scheme of Mullite Gels 825
21.1 Introduction 825
21.2 Old Basis of Classification of Different Types of Mullite Gels 826
21.3 New Classification Scheme, Deviation from Normal Gelification Procedure of SH Gels, Colloidal Gels and Others 829

Author Index 841

Subject Index 847

Preface

The large area of application of mullite in refractories and in the ceramic field has led to a lot of interest among researchers/technologists. It possesses a defect structure and exists over a considerable range of solid solution and its behavior during melting give rise to a great controversy. Mullite occurs as a part of materials with a low-melting glassy phase and others in traditional ceramics, like structural clay products, art pottery, sanitary ware, and insulators.

Mullite has become increasingly important in the production of some conventional ceramics, such as technical porcelains, as a main constituent in refractories, etc.

Mullite is attractive in advanced structural and functional ceramics due its high temperature strength, good chemical stability, excellent creep resistance, low thermal expansion coefficient, and low thermal conductivity, the one problem area being its deficiency in terms of fracture toughness. To improve this characteristic of mullite materials, research is being carried out on composite development and property evolution vis-à-vis product development.

During the last 40 years, a lot of research has been conducted in many countries on various basic topics as well as in various fields of application. These findings are presented below, along with a list of books, workshops, symposiums on different topics of mullite, and published journals.

- The 1st International Conference was held in Tokyo, Japan, on November 9–10, 1987. The conference items included an Al_2O_3-SiO_2 phase diagram, mullite stoichiometry, solubility and solid solution in the mullite structure, powder preparation, densification, microstructure and property measurements, composite making and its property evolution, among others. The presented papers were published in *Ceramic Transactions*, Vol. 6. eds. S. Somiya, R. F. Davis, and J. A. Pask, American Ceramic Society, Westervikke, OH (1990).
- Progress on mullite research continued and the Symposium F of the 43rd Pacific Coast Regional Meeting of the American

Ceramic Society was held in Seattle, WA (1990). A significant number of papers covering different aspects of mullite, such as mullite processing, properties, and structures, including the current state of mullite research, remaining issues, different applications, and future directions, were published in the third topical issue of *J. Am. Ceram. Soc.* (1991).

- To assess further development on mullite research, the 2nd International Workshop Mullite 1994 was held in Irsee, Germany, and the presented papers were published in the special issue of *J. Euro. Ceram. Soc.*, **16**(2), (1996).
- The Mullite Workshop 2000 was held at the Oban Town Hall, Oban, England, on August 27–31, 2000.
- The International Workshop Mullite was held in 2006 and the presented papers published in *J. Euro. Ceram. Soc.*, **28**(2), 327–504 (2008). The text is available on www.sciencedirect.com.
- The 5th International Workshop on Mullite & Mullite-Type Materials took place at the Niemeyer Center in Avilés, Spain, in May 2011, after the previous mullite meeting, which was held in Vienna in 2006. There was considerable discussion with respect to the production, properties, and application of mullite ceramics, fibers, composites, and coatings, along with the advanced knowledge on the crystal chemistry and structure-related properties of mullite.

The origin of mullitization has been a subject of extensive and controversial investigations. A few books, as mentioned above, provide brief summaries of the results of earlier researchers as chapters covering a compilation of studies on the thermal effects of some precursors of mullite only. No comprehensive work covering the synthesis aspects of precursors versus their phase evolution has been documented so far. A close inspection of the basic literature of earlier mullite research activities reveals a lack of summing up. Most of the papers give the results of experiments, and often the statements of the different authors bring sharp contradictions to light.

My aim as the author is to mention a few topics, for example, powder preparation and phase evolution, only, considering the volume of the present monograph, instead of surveying all associated problems and applications of mullite. In the past few decades, the need for new forming methods—perhaps advanced chemical techniques—has become increasingly important. The vision of this book is to upgrade the knowledge of the synthesis of various kinds of mullite precursors by the chemical route (sol-gel route) and other techniques and their phase transformation processes with the view to establish the mechanisms of mullite formation. The topics dealt with are as follows:

- Various ways of synthesizing mullite precursors.
- Mullite formation by various methods, for example, gelation, coprecipitation, diphasic gelation, codecomposition, and microcomposite, using materials from different sources, which has also attracted the attention of a large number of researchers throughout the world. The extent of mullite research studies on the above topic is reflected in the vast literature (the reference list is given at the end of each chapter).
- Revelation of the evolution of phases on heating and the mechanism of formation of different phases, with the help of various physicochemical characterizing techniques, for example, XRD, DTA, IR, MASNMR, TEM, and EDS.
- Characterization of the intermediate Al-Si spinel phase and the weakly crystalline mullite phase.
- Mechanism of mullite formation with different mullite precursors. There are many points of disagreement regarding the phase transformation paths taken by precursors from different sources. However, there are some concurrent views on the formation of the Al-Si spinel phase other than γ-Al_2O_3.

The situation outlined above has suggested a comprehensive review work on the formation and characterization of intermediate phases, like Al-Si spinel and mullite. The book begins with a systematic survey of the experimentations done by various researchers and the results they obtained. These are examined in the light of various researchers' views and also with my own research studies involving different physicochemical methods. A concrete conclusion has been derived with reference to the exhibitions of three exothermic events observed in the high-temperature region of DTA of five mullite precursors. Finally, the book points out the extent to which and how some of the controversies are eliminated in the gel-to-mullite reaction series.

It is hoped that this book will be received by a large section of mullite researchers throughout the world with keen interest. The summary of each chapter of the book will serve as a text for both undergraduate and graduate students. Researchers and students belonging to the material science stream at the university level around the world are the primary target audience of this book. The new directions as presented in the book will be a source of inspiration and encouragement for further study and research for the teaching faculty and mullite researchers.

I must thank the publishers of various journals for their kind permission to use the figures and tables from their scientific journals, without which it would have been very difficult to publish this book. I have been fortunate enough to do my research studies in the clay-mullite field in the Central Glass and Ceramic Research Institute, Kolkata, India, taking advantage of the facilities it offered for my scientific endeavors. Moreover, I have used the library and administrative opportunities offered by the XRD and refractories divisions of the said institute for my continued research activities spanning 40 years.

My special thanks to my beloved wife, Mun Mun, and my family members for their support and continued encouragement to do research study and write on the mullite field of ceramics—a most disputed topic.

I am especially indebted to Jenny Stanford Publishing for their valuable suggestions and considerable assistance in improving the book as a whole and publishing it in a short span of time. Finally, it is appropriate to acknowledge the real affection and moral support accorded to me by my granddaughter and grandson in the planning and organizing of this book.

Akshoy Kumar Chakraborty

Chapter 1

Introduction

1.1 Mullite Research in Conventional Fields

Mullite is the only stable crystalline compound in the Al_2O_3-SiO_2 system under normal atmospheric pressure. It has a chemical composition range from $3Al_2O_3{\cdot}2SiO_2$ to $2Al_2O_3{\cdot}SiO_2$ approximately and crystallizes in the orthorhombic system most commonly in the form of elongated needle-shaped crystals, the exception being when it is prepared in the absence of a liquid phase. It is the most important crystalline phase in the fired products belonging to the Al_2O_3-SiO_2 system found in industrial ceramic products; however, it is rare in nature, so normally, mullite is synthesized for different purposes. The melting point of mullite is high above 1800°C, but values in the range of 1810°C–1900°C are reported. Controversy arises because mullite shows congruent melting in some cases and incongruent melting in others, and there is evidence in favor of both. In a silicate system, a purely eutectic (congruent) or a purely peritectic (incongruent) reaction seldom occurs. The formation of a double compound (solid) is usually favored. This, on melting, gives rise to a liquid phase of its own composition. This melting process is called eutectic (congruent) melting. In another melting process, the solid is split into a new solid phase and a new liquid phase, each of which is different in composition from the parent solid. This melting process is called peritectic (incongruent) melting.

Mullite Formations: Analysis and Applications
Akshoy Kumar Chakraborty

ISBN 978-981-4877-05-3 (Hardcover), 978-1-003-03167-3 (eBook)
www.jennystanford.com

1.1.1 The Phase Diagram

The diagram of Al_2O_3-SiO_2 is of significance to ceramic technology, being directly related to a number of ceramic materials, refractories in particular (silica, fireclay, high-alumina ceramics, sintered corundum, etc.). For this reason, the properties of the individual phases of this system are of great interest. The Al_2O_3-SiO_2 system also illustrates some of the problems in satisfactorily determining and evaluating the experimental data needed to assemble the phase diagram. Four phase equilibrium curves were obtained by different experimental and thermodynamic studies, conducted by Bowen and Greig (1924), Aramaki and Roy (1962), Aksay and Pask (1975), and Klug et al. (1987, 1990). Bowen and Greig first obtained a systematic phase diagram for this system, showing the compound mullite to melt incongruently at ~1828°C. Aramaki and Roy, on the other hand, proposed a revision that shows a congruently melting behavior for mullite at ~1850°C. The stable phase equilibrium diagram proposed by Aksay and Pask is in general agreement with Bowen and Greig's version. The only stable compound in the system Al_2O_3-SiO_2 is mullite containing 71.8 wt % of Al_2O_3 (theoretical composition). In fact, it is a solid solution with a composition range indicated by the phase diagram (60–63 mol % of Al_2O_3, up to 76 mol % for the metastable solid solution). Thus, it is a nonstoichiometric material. To evaluate the nature of the mullite crystallized just at 980°C, at 1250°C, and at 1300°C, exothermic peak temperatures in differential thermal analysis by the X-ray diffractometry technique on heating different types of mullite gels have been studied by various researchers.

1.1.2 Forms of Mullite, Solid Solution, and Stoichiometry

In an exploratory work, Wyckoff et al. (1926) and later Taylor (1928) noted a close structural relationship between sillimanite and mullite. Since the X-ray powder diffraction patterns are remarkably similar, on the basis of X-ray diffraction patterns, Wyckoff et al. inferred that mullite and sillimanite are almost indistinguishable. In sillimanite, continuous columns of octahedra (parallel to the *c* axis) are linked together by (SiO_4) and (AlO_4) groups; the (SiO_4) group is a regular tetrahedron, and the (AlO_4) tetrahedron is nearly regular. Taylor showed that the two minerals (sillimanite and mullite) have very similar structures and the unit cell of sillimanite contains $Al_8Si_4O_{20}$ and that of mullite contains $Al_9Si_3O_{19.5}$ but that mullite is less dense (3.15 gm/cc) than sillimanite (3.24 gm/cc), which suggests that the mullite structure could be derived from that of sillimanite by replacing the aluminum atom with silicon in some tetrahedral positions. This is the essential feature of the relation between mullite and sillimanite. They deduced a structural model of mullite out of the structure of sillimanite by substituting Si^{+4} by Al^{+3} and introducing oxygen vacancies (□). This resulted in a variation in the composition of mullite as per Eq. 1.1.

$$2\,Si^{+4}O^{-2} \rightarrow 2Al^{+3} + \square \qquad (1.1)$$

An increase in the Al content leads to a long-range disorder of Al and Si at tetrahedral sites. Several attempts were made to determine an average structure of it (Durovic, 1962, 1969; Sadanaga et al., 1962; Burnham, 1963, 1964; Nakajima et al., 1975; Saalfeld and Guse, 1981; Angel and Prewitt, 1986). The mullite structure likely consists of chains of edge-shared AlO_6 octahedra running parallel to the direction of the *c* axis. These chains are cross-linked by (Si,Al) O_4 tetrahedra forming double chains. which also run parallel to the direction of the *c* axis. In mullite, the chains of aluminum oxygen octahedra with common oxygen-oxygen edges run parallel to the crystallographic *c* axis. The octahedra are linked by double chains containing aluminum-oxygen and silicon-oxygen tetrahedra in a random sequence. The tetrahedral chains of mullite are distributed in a manner that some pairs of tetrahedra are misoriented, with the removal of some oxygen atoms from the structure. Thereby, two adjacent tetrahedral cations site are displaced. Combined with these displacements, substitutional replacement of tetrahedral silicon by aluminum takes place. The connection between octahedral and tetrahedral chains generates relatively wide structural channels running parallel to the *c* axis. Again the (AlO_4) and (SiO_4) groups are almost exactly superposed at the interval of one half the *c* axis. Since, the octahedron in the continuous chains is repeated at intervals of $c/2$, the whole structure would be repeated at intervals of $c/2$, and hence the true unit cell would be half the height of that actually observed. The structure is generally represented by the formula $Al_2(Al_{2+2x}\,Si_{2-2x})O_{10-x}$, where x denotes the number of missing oxygen atom relationships per average unit cell and varies from 0.17 to 0.5. According to Saalfeld and Guse (1981), due to the defect structure of mullite by the substitution effect and rise of vacancies, many different types of extra reflections in X-ray diffraction patterns are noted (Scholze, 1955; Nakajima et al. 1975; and Tokonami et al., 1980). The extra reflections in mullite obtained by melting and followed by the quenching process are classified into two groups: (i) satellite reflections and (ii) diffuse streaks in reciprocal space. In comparison, a disordered arrangement of oxygen vacancies of the D-mullite type are noted in mullite obtained by the solid-state reaction of oxides and from the phase evolution of kaolinite.

1.1.3 Applications

From the nature of phase equilibria in the Al_2O_3-SiO_2 system, it is clear that knowledge of phases in this system could be obtained for a wide variety of ceramic and glass products. For example, SiO_2 is the basis for the glass manufacturing industry whereas Al_2O_3, the other component of the phase diagram, is used in the production of electronic substrates, spark plugs, cutting tools, and lamp envelops. Mullite and many other compositions in this system are useful in various applications, as stated next.

In conventional ceramics: Mullite is the main constituent in white ware, low- and high-temperature (hard) porcelains, sanitary ware, hotel china, high-tension electrical porcelains, low-tension porcelains, insulators, structural clay products (pipes, tiles, etc.), architectural products, etc.

In refractories in metallurgical industries: The steel-making industry is the largest market for refractories, and mullite-based bricks are used as the lining in various furnaces, for example, smelting furnace, torpedo car, rolling mill, and regenerator, due to the high melting point (~1890°C). Mullite can accommodate a small excess of Al_2O_3 in a solid solution, as indicated in the diagram. Mixtures of mullite and Al_2O_3 form no liquid until heated to a temperature above 1840°C. Refractories used in steel making include silica brick, which usually contains less than 1 wt % of Al_2O_3. Bowen and Greig (1924) showed by the liquidus curve that a liquid phase begins to form at 1595°C and that a few percent of Al_2O_3 would markedly increase the proportion of liquid in the brick at temperatures between 1595°C and the melting point of SiO_2. The increase in the amount of liquid phase reduces the load-bearing capacity of the brick, and thus the maximum safe operating temperature is decreased by the presence of the Al_2O_3.

Mullite-based bricks are the main constituents of the lining of the upper parts of melting furnaces, hot blast stoves, hot iron runners, and continuous casting furnaces. They are also used to build the lower part of the combustion chamber of the blast furnace stove, where good thermal shock resistance is required. Classical uses of mullite include use in refractories in the metallurgical industries, in electric furnace roofs, in hot metal mixers, and in low-frequency induction furnaces.

Two types of mullite-based bricks have been distinguished: fireclay and high-alumina bricks. Fireclay refractories cover the range from about 40% Al_2O_3 to <50% Al_2O_3; the composition selected depends on the particular application. Mullite is one of the main phases of fireclay-based ceramic materials. In addition to a high melting point, it exhibits outstanding chemical resistance and high strength. It is volume-stable because it does not undergo any polymorphic change. Mullite has the shape of a needle, and interlocking of these needle-shaped crystals gives high mechanical strength and zero volume expansion. For these reasons, mullite is the most desirable phase in any refractory material. Fireclay bricks are manufactured using kaolin and/or pyrophyllite as the raw material. Therefore, these bricks are richer in SiO_2 than mullite and the microstructure is composed of mullite, cristobalite, and glassy phase. The high-alumina bricks have an Al_2O_3 content >50 wt % and are mainly manufactured from the aluminum silicate minerals kyanite, andalusite, and sillimanite; sinter-mullite; and bauxite plus silica. The microstructure of these bricks usually consists of well-developed mullite and/or α-Al_2O_3 grains interlocked with the glassy phase.

In refractories in glass melting industries: Mullite bricks are used, for example, in glass tank furnaces. Although most linings have been changed from fused-mullite bricks to alumina-zirconia-silica-fused bricks in recent years, mullite bricks are still used in the fore hearth of furnaces. Mullite bricks are mostly used in the form of blocks in the glass tank wall and base construction. In the glass industry, these refractories are employed in the upper structure of the tank in which the glass is melted and for constructing the drawing chambers.

In refractories in cement industries: Mullite refractory bricks are an important component for lining high-temperature tools in the cement industry, for example, in the calcination and cooler zones of rotary kilns.

In kiln furniture: Mullite shows little deformation under load due to good creep resistance. Thus, it is frequently used in furnace liners, kiln setting slabs and posts for firing ceramic ware, and in the lining of high-temperature reactors. Mullite shows a low coefficient of expansion, that is, it is a good thermal-shock-resistant material. As a consequence, it is used in making conveyor belts for continuous furnaces, replacing stainless-steel conveyor belts.

Production of spark plug: Mullite ceramic shows good strength both at low and high temperatures and good heat shock resistance.

As a corrosion-resistant material: Mullite shows good resistance to corrosion. In comparison to some ceramic materials, for example, alumina, magnesia, zirconia, silicon carbide, silicon nitride, and boron carbide, mullite ceramic shows better resistance to acidic solutions, molten salts, slag, and molten metal (Schneider et al., 1994, p. 237). The approximate limits of usefulness of mullite in air and vacuum are 1800°C and 1600°C, respectively. It is used in laboratory ware since it has excellent stability in acid and metal slags and is insoluble in most acids.

As a heat-resistant material: It is used in various types of crucibles. Mullite beads are used for cast-iron sand, placing sands, and sand for fluidized bed furnaces. It is used as a heat exchanger by replacing heat-resistant alloys because it shows good resistance to oxidation.

As a gas-impermeable material: Mullite shows good gas impermeability. Accordingly, it is used in pyrometer protection tubes, for example, thermocouple tubes.

As a break-lining material: Mullite has also found application as an ingredient in the friction-resistant materials used in the aircraft-braking system.

As a porous ceramic material: Porous mullite parts have been used in various applications, for example, breweries and wastewater disposal plants, diesel oil filters, water filters, sound absorbing materials, and catalyst carriers.

Because mullite materials are applicable in a large number of areas, a number of workshops and symposiums were held for discussing different areas (as shown earlier) and reviews were made by Davis and Pask (1971), Aksay and Wiederhorn (1991), Aksay et al. (1991), Okada et al. (1991), Somiya et al. (1991), Okada (1993), and Schneider and Komarneni (2005). It is possible to produce mullite precursors and mullite products for various applications, as stated above. Their production necessitates immense research activities on mullite. Many techniques have been developed to obtain mullite powders with high purity, high activity, and narrow size distribution. Mullite is generally synthesized by sintering mechanically mixed alumina and silica powders from different sources. Fenstermacher and Hummel (1961) used a mixture of α-Al_2O_3 and quartz powders in stoichiometric mullite composition for mullite synthesis in a laboratory. In this case, a considerable amount of time is required even on calcination at as high a temperature a >1700°C. According to Cho et al. (1998) densification of mullite ceramics proceeds above liquid formation temperature (≥1587 ± 10°C) and it develops a glassy phase during sintering, which reduces the credibility of ceramics, especially in terms of high-temperature strength.

1.2 Mullite Research in Advanced Fields and Its Importance

There is considerable interest among various researchers in the solution process, particularly the sol-gel and coprecipitation techniques and especially the use of metal alkoxide for making ultrahomogeneous glasses and ceramics for technological applications from nuclear fuel to ceramic fibers, window coatings, and abrasive grains and even in basic phase equilibrium studies (Roy & Osborn, 1954; Roy, 1956). The above methods utilize both organic and inorganic precursors. For mullitization studies, solution techniques are considered to be the most important because the aluminum- and silicon-containing precursors can be mixed in a very fine scale, allowing a low mullitization temperature to be achieved, and it is expected to be sintered at a lower temperature in comparison to conventional oxide sintering processes. Wheat (1977) stated that the key to the successful production of ceramic materials is the initial selection of high-quality raw materials and the different methods of producing reactive and homogeneous powders, for example, (i) precipitation techniques, (ii) colloid formation, (iii) codecomposition, and (iv) solution drying techniques. Various researchers have synthesized mullite by the chemical route and it is of great interest to them due to the physicochemical properties of the product, basically the high homogeneity and high purity that may be attained through this route in comparison to the oxide synthesis process.

Electronic packaging materials: Mullite is a stable crystalline material in the binary Al_2O_3-SiO_2 system, and it has attracted a great deal of interest in recent years due to its excellent high-temperature strength and creep resistance, good chemical and thermal stability, low thermal expansion coefficient, and good dielectric properties. Taking advantage of these favorable characteristics, mullite-based materials can be used as packaging materials and in semiconductor devices. Tummala (1991) reported that it is an important substrate material in the multilayer packaging industry, after studying other potential ceramic substrate materials for their electrical, thermal, and mechanical properties. Three important principal characteristics of mullite and mullite substrates are as follows:

- Generally, the transmission delay time is proportional to the root of the relative dielectric constant. The dielectric constant value ($\varepsilon = 6.7$) of mullite is less than that of alumina ($\varepsilon = 9.8$), so the signal transmission delay time is 17% lower for mullite in comparison to that for alumina.
- High ware density and low sintering temperature of mullite make it possible to achieve cosintering with metals, for example, copper. Mullite ceramic synthesized by the diphasic route can be densified at as low a temperature as 1250°C, which is comparatively lower than the sintering temperature of ~1600°C usually required for alumina ceramic.
- The area under solder connection must have thermal expansion match between silicon and the substrate. Mullite

has a low thermal expansion coefficient (40 × 10–7/°C in the range 20°C–200°C), which is very near to that of silicon. Thus, mullite stands out as a potential material for substrate making in packaging industries in addition to many refractory applications. Particularly, high-purity Al_2O_3 finds use in the electronic industry because of its excellent electrical characteristics when used as an insulator in power circuits or as a substrate material in hybrid circuits for computer chips and other modern applications.

As an optically transparent material: Mullite has been demonstrated in the spectrum from 3 to an infrared transparent window, especially at high temperatures.

As a solid-state laser activator: Polycrystalline fine-grained mullite is a good candidate for laser due to a low dielectric constant and its optical transmission property.

As transparent mullite: The quantum efficiencies in single-crystal mullite are close to 100%. The growth of single-crystal mullite is difficult. Mullite glass ceramic is an obvious choice. Its production requires temperatures as high as 2000°C for melting and prolonged annealing for controlled crystallization. An alternative choice is high-density, high-purity, and fine-grained transparent mullite. Hot pressing at 1500°C–1650°C at 30–50 MPa and/or vacuum hot pressing at 1500°C at 5 kpsi produces translucent mullite by an alkoxide route (Aksay et al., 1991).

Advanced structural and engineering ceramic materials: It may be used in many industrial fields, for example, as ceramic engine components, high-temperature protective coating, radomes, and gas turbines, due to high creep resistance and high flexural strength at elevated temperatures, and a number of ceramics have been considered for use in radome applications (Perry, 1973).

Composites: Mullite and mullite-containing composites have been used as a reinforced phase for making a variety of high-temperature structural components and high-performance applications. Mullite ceramics and/or mullite-cordierite composites have also been used for shelves, saggers, and props since these applications require little creep deformation, good thermal resistance, and good corrosion resistance during firing.

High-temperature protective coating: For information on this, refer to the book written by Schneider and Komarneni (2005).

Fibers: Mullite fibers have been used as heat-insulating materials for furnaces because of their good heat resistance and their favorable heat-insulating properties. Mullite fibers are also applied as heat-resistant fillers, seal materials, and packing materials.

As oxygen ionic conductors: Meng and Huggins (1983) pointed out that this special property of mullite depends on its composition—that is, oxygen ion conductivity occurs at high temperatures due to appreciable concentrations of oxygen vacancies in its structure—as per the following formula:

$(Al_{2+x}\ Si_{2-x})^{IV}\ Al_2{}^{VI}O_{10-x/2}\ V_{x/2}$, where V = oxygen vacancy, x varies from 0.25 to 0.4, and the Al_2O_3:SiO_2 ratio varies from 3:2 to 2:1.

In low-cost terrestrial solar cell development program: Mullite is used due to its low cost, ease of processing, and physical and chemical compatibility with silicon. Its thermal expansion at 800°C at the rate or 5×10^{-6}/°C is reasonably near to that of silicon (3×10^{-6}/°C). Even the slight difference in expansion coefficient is minimized by incorporating the required quantity of silica glass during processing of lower-expansion mullite bodies, as shown by Leipold and Sibold (1982).

Catalyst support: Amorphous solids prepared by the sol-gel method are chemically homogeneous and as such silicoaluminates are industrially important as a support material for catalysts like Pt due to high surface areas (Lopez et al., 1989). Other metallic cations, for example, Ru, Mn, and Sn, are incorporated during the polymerization process, as shown by Lopez et al. (1986) and Jones et al. (1984).

1.3 Existing Methodology

1.3.1 Sinter Process

Conventional mullite ceramic is generally processed by a solid-state diffusion process of solid reactants, which requires both a long time and a high temperature of >1650°C. Even though this process is economical, it hardly leads to a homogeneous microstructure. Retention of inhomogeneity is obvious. On the other hand, homogeneity of mullite ceramic depends on the initial uniform distribution of the reagents/components either in a solution or in a sol and in the oxide state. The retention of homogeneity from the initial reactant stage to the final mullite is greatly dependent on processing methods. The most advanced processing technique in the glass and ceramic field is the nonthermal way of synthesis through the sol-gel process, in contrast to the need of very high temperatures to develop these products. This route claims a high degree of reactivity between silica and alumina components. However, the sol-gel process shows two distinct boundaries between the polymerization method and the colloidal method of synthesis (Yoldas & Partlow, 1988). Silica and alumina component materials from different sources have been used for synthesizing mullite gels on the basis of their availability and choice of different researchers. Generally, mullite synthesis by different researchers have been divided into two different categories: gelation, where the solution is evaporated to concentrate the sol in order to set to the gel state, and precipitation, where the pH of the solution is changed to cause flocculation, etc. The sol-gel method of synthesizing mullite is considered excellent (Schneider et al., 1993). Probably, the homogeneity in the solution level will be achieved and mullitization will be lowered significantly in comparison to the formation of the same in Al_2O_3-SiO_2 oxide mixtures. Densification may occur at expectedly lower temperatures due to smaller particle size.

1.3.2 Gelation Process

In the sol-gel process, the first step of the silicon alkoxide–water reaction is hydrolysis. Polycondensation starts thereafter, which may lead to the gelation of monomers. The condensation reaction may not be complete although gelation is permitted to take a long time during the gel state. It may continue in the drying and heating stages. Dried gels finally transform into mullite on heating. It is still doubtful that homogeneous mixtures will arise from either organic precursor, because of differences in the hydrolysis rates of components. There is every possibility of the formation of heterogeneous mixtures.

1.3.3 Coprecipitation Process

In the precipitation process, both reactants, for example, $AlCl_3$ and $SiCl_4$, are highly soluble in water, that is, mixing occurs in the ionic stage. Both are coprecipitated (CP) as insoluble hydroxides in a homogeneously mixed way when ammonium hydroxide is used. Even in this technique, consecutive precipitation may occur that may lead to the segregation of the hydroxides of the two components unless precaution is taken.

1.3.4 Colloidal Process

In the colloidal process, the randomly dispersed component ions in a solution are clustered to form small colloidal particles 10–1000 Å in diameter, consisting of several molecular hydroxyls and finally stabilized. For example, Ludox is used as a commercial source of silica sol. For alumina sol, aluminum hydroxide is first precipitated from soluble aluminum salt with the addition of ammonia, then filtered and washed, and then dispersed with nitric acid. Two separate sols are mixed and gelled. Thus, the mixing homogeneity of the two components occurs in a nanometer range of particles in contrast to the purely coprecipitation process.

1.3.5 Codecomposition Process

In the codecomposition process, the two reagents are mixed in a true solution in aqueous and/or alcoholic solvents. The mixed solution is simultaneously dried in a conventional spray drier or calcined to form an oxide by atomizing or spraying onto a hot surface. It is hoped that spray-drying/roasting processes may retain the homogeneity of the oxide components as present in the solution stage.

It is possible to produce mullite products of specific properties for various applications as stated above, by keeping tight control on both the choice of component raw materials and processing, than to obtain mullite materials synthesized by oxide sinter process which shows varying properties.

Unfortunately, the researchers of mullite community are yet to solve some of the pertinent problems of various mullite precursors in their phase transformation processes during heating.

There are three main aspects to the problems, and these are of great concern to this scientific community:

- The composition of the Al-Si spinel phase formed at intermittent stage of thermal evolution process
- Composition of the weakly crystalline mullite formed at ~1000°C
- Mechanism of mullite formation by five different routes and the solid solution range of mullite

Solutions to these problems is the primary subject matter of this book.

Another pertinent problem is homogeneity. Both chemical purity and reproducibility may be maintained in sol-gel processes to achieve homogeneity, in contrast to conventional milling operation, which is usually practiced in an oxide mixture. It is necessary to re-examine how the starting components react on heating or the thermal evolution studies of mullite gels/precursors synthesized from the five techniques mentioned in the chapter and to what extent chemical homogeneity is maintained from the solution up to the mullite grain. It is also important to examine to what extent chemical homogeneity depends upon processing techniques and to what extent homogeneity of the gel has been achieved during gelation versus precipitation techniques. To reveal answers to these questions, the first approach undertaken is to elaborately study the wide variety of powder preparation methods of mullites mentioned in earlier studies, to present a thorough review, and to focus on the shortcomings of these research works. The stress is mainly on the phase transformation behaviors of different kinds of gels, that is, different ways of synthesis of various gels, their thermal evolutions, and characterization of different phases. The different types of synthesis techniques of mullite gels/precursors using different components followed by previous researchers are generalized.

The proper classification of methods of synthesis on the basis of phase transformation studies by various previous authors after synthesis of mullite gels or precipitates from varying water-soluble to organic Si and Al compounds will be presented. It is highly expected that sol-gel, CP, or spray-pyrolyzed mullite precursors may produce very homogeneous and dense mullite products of suitable microstructure to meet the properties for application in the high-performance ceramic domain. Literature shows that microstructure development and attainment of some of the outstanding thermal properties of sintered mullite depends upon many factors. The most important factor among them is precursor synthesis, that is, the choice of components and the technique of hydrolysis-cum-processing variables. It is noted that a minor change in the processing parameter plays havoc in the mullitization path. Thus, compete knowledge of the rate of hydrolysis of both organic silicon and aluminum alkoxides is necessary, since one may hydrolyze and polymerize faster than the other.

The chemistry of hydrolysis of components and the scale of homogeneity achieved during synthesis of mullite gels need to be discussed elaborately.

1.4 Visualization of the Scope of the Mullite Book

During the last few decades, many research studies have been conducted on various basic topics as well as on various fields of application. These findings are presented in the Review Chapters from 2–6 of the Book along with a list of references at the end of each chapter.

Synthesis of mullite by various authors are placed chronologically. Instead of presenting earlier studies yearwise, these are presented in terms of synthesis procedures in Chapters 2–12. Thereafter, the author elaborately addresses the various issues, for example, (i) the various problems faced earlier, (ii) difficulties encountered by distinguished authors, and (iii) predictions by the scientific community on the phase transformation of mullite gels, as stated above, in Chapters 13–17. The critical analyses are presented in Chapters 18–21, with proper cross-references from previous chapters.

The method of achieving homogeneity in different processing techniques is discussed more thoroughly after proper classification of mullite precursors.

The interrelationship between synthesis/processing conditions and intimacy/homogeneity of aluminum-silicon bonding in the gel network exhibiting a 980°C exotherm is shown.

The interrelationship between synthesis/processing conditions and the crystallization path to mullitization is also explained.

Accordingly, a few selected earlier studies have been cited and compiled chapterwise. But it is impractical to include each and every publication relevant to the subject of mullites, and the author apologizes for any inadvertent omissions of important topics.

Experimentally noted figures of several earlier studies of a large number of researchers are presented after obtaining their kind permission for reproduction, along with the inclusion of the author's own newer concepts and visualizations of the problems and mechanisms of thermal transformation reaction series. These results of various researchers and scientific personals in their individual investigative fields of study (Table 12.1) are comparable with different kinds of flowers blooming in various gardens. The present author has just tried to cite them from their gel-to-mullite research areas, like plucking colorful flowers and sorting them in terms of their shape, size, and color, for future investigation. Each chapter is analogous to assorted flowers kept in a tall vase. These are chronologically and elaborately presented in Chapters 2–12. The analysis of the literature of these chapters is presented in Chapters 13–21. This way of presenting earlier literature along with the new concepts and analysis of the present author in a chapterwise mode is done with a view to develop lucid pictures of mullitization processes, analogous to the conglomeration of flowers kept in different decorated colored ceramic vases, so that they are embraced globally by students, professors, researchers, and technocrats.

This presentation of review (Parts I and II) and critical analysis (Part III) is expected to evoke a melody in the minds of readers when they go through this book, the two parts being equivalent to imaginary two-flower bouquets or two colorful flower garlands.

The author humbly dedicates the two bouquets (Part I, Part II, and Part III of the book) to his revered Sadhan Siddha Sadguru Swami Nigamananda Paramahansa Deva.

References

1. N. Bowen and J. W. Greig, The system Al_2O_3-SiO_2. *J. Am. Ceram. Soc.*, **7**, 238–254 (1924).

2. S. Aramaki and R. Roy, Revised phase diagram for the system Al_2O_3-SiO_2. *J. Am. Ceram. Soc.*, **45**(5), 229–241 (1962).
3. I. A. Aksay and J. A. Pask, Stable and metastable equilibria in the system SiO_2-Al_2O_3. *J. Am. Ceram. Soc.*, **58**(11–12), 507–512 (1975).
4. F. J. Klug, S. Prochajka, and R. H. Doremus, Alumiosilicate phase diagram in mullite region. *J. Am. Ceram. Soc.*, **70**, 750–759 (1987).
5. F. J. Klug, S. Prochazka, and R. H. Doremus, Alumina-silica phase diagram in the mullite region, in *Ceramic Transactions*, Vol. 6, Mullite and Mullite Matrix Composities. eds. S. Somiya, R. F. Davis, and J. A. Pask, American Ceramic Society, Westervikke, OH, p. 15 (1990).
6. R. W. G. Wyckoff, J. W. Greig, and N. L. Bowen, The X-ray diffraction patterns of mullite and sillimanite. *Am. J. Sci.*, **11**, 459–472 (1926).
7. W. H. Tayor, The structure of sillimaite and mullite. *Z. Krist*, **68**, 503–521 (1928).
8. R. Sadanaga, M. Tokonami, and Y. Takeuchi, The structure of mullite, $2Al_2O_3.SiO_2$ and relationship with the structures of sillimaite and andalusite. *Acta Crystallogr.*, **15**, 65–68 (1962).
9. S. Durovic, A statistical model for the crystal structure of mullite. *Kristallografiya*, **7**, 339 (1962).
10. S. Durovic, Refinement of the crystal structure of mullite. *Chem. Zvesti*, **23**, 113–128 (1969).
11. C. W. Burnham, Crystal structure of mullite. *Carnegie Inst. Washington Year Book*, **62**, 158–165 (1963).
12. C. W. Burnham, Crystal structure of mullite. *Carnegie Inst. Washington Year Book*, **63**, 223–227 (1964).
13. H. Saalfeld and W. Guse, Structure refinement of 3:2-mullite ($3A_2O_3.2SiO_2$). *Neues Jahrb. Mineral. Monatsh.*, 145–150 (1981).
14. R. J. Angel and C. T. Prewitt, Crystal structure of mullite: a re-examination of the average mullitestructure. *Am. Mineral*, **71**, 1476–1482 (1986).
15. H. Scholze, Zum Sillimanit-Mullit problem. *Ber. Dtsch. Keram. Ges.*, **32**, 381–385 (1955).
16. M. Tokonami, Y. Nakajima, and N. Morimoto, Diffraction aspect and a structural model of mullite, $Al(Al_{1+2x}Si_{1-2x})O_{5-x}$. *Acta. Crystallogr.*, **A36**, 270–276 (1980).
17. Y. Nakajima, N. Morimoto, and E. Watanabe, Direct observation of oxygen vacancy in mullite $1.86Al_2O_3.SiO_2$ by high resolution electron microscopy. *Proc. Japan Acad.*, **51**, 173–148 (1975).

18. H. Schneider, K. Okada, and J. Pask (eds.), *Mullite and Mullite Ceramics*. John Wiley & Sons, Chichester, UK, p. 237 (1994).

19. R. F. Davis and J. A. Pask, Mullite in Refractory Material: A Series of Monographs, in *High Temperature Oxides*, Part 4, ed. A. M. Alper, Academic Press, NY, Chapter 3, pp. 37–76 (1971).

20. I. A. Aksay and S. M. Wiederhorn, Topical Issue: Symposium for mullite processing, structure and properties, *J. Am. Ceram. Soc.*, **74**, 2341–2342 (1991).

21. I. A. Aksay, D. M. Dabbs, and M. Sarikaya, Mullite for structural, electronic, and optical applications. *J. Am. Ceram. Soc.*, **74**, 2343–2358 (1991).

22. K. Okada, N. Otsuka, and S. Somiya, Review of mullite routes in Japan. *Ceram. Bull.*, **70**(10), 1633–1640 (1991).

23. S. Somiya and Y. Hirata, Mullite powder technology & application. *Am. Ceram. Soc. Bull.*, **70**, 1624–1632 (1991).

24. K. Okada, Mullite ceramics reviewed from raw materials to properties. *Taikabutsu Overseas*, **13**(3), 44–52 (1993).

25. H. Schneider and S. Komarneni, *Mullite*. WILEY-VCH, Weinheim, Germany (2005).

26. J. E. Fenstermacher and F. A. Hummel, High-temperature mechanical properties of ceramic materials: IV, Sintered mullite bodies. *J. Am. Ceram. Soc.*, **44**, 284–289 (1961).

27. Y. I. Cho, H. Kamiya, Y. Suzuki, M. Horio, and H. Suzuki, Processing of mullite ceramic alkoxide-derived silica and colloidal alumina with ultra-high cold isostatic pressing. *J. Euro. Ceram. Soc.*, **18**, 261–268 (1998).

28. R. Roy and E. F. Osborn. The system Al_2O_3-SiO_2-H_2O. *Am. Miner.*, **39**(11/12), 853–885 (1954).

29. R. Roy, Aids in hydrothermal experimentation: II, Methods of making mixtures for both "dry" and "wet" phase equilibrium studies. *J. Am. Ceram. Soc.*, **39**(4), 145–146 (1956).

30. T. A. Wheat, Techniques for producing reactive and homogeneous ceramic powders. *J. Can. Ceram. Soc.*, **46**, 11–18 (1977).

31. R. R. Tummala, Ceramic & glass ceramic packaging in 1990's. *J. Am. Ceram. Soc.*, **74**, 895–908 (1991).

32. G. S. Perry, Microwave dielectric properties of mullite. *Trans. J. Brit. Ceram. Soc.*, **72**, 279–283 (1973).

33. G.-Y., Meng and R. A. Ruggins, A new chemical method for preparation of both pure and doped mullite. *Mater. Res. Bull.*, **18**, 581–588 (1983).
34. M. H. Leipold and J. D. Sibod, Deveopment of low–thermal expansion mullite bodies. *J. Am. Ceram. Soc.*, **8**, C147–C148 (1982).
35. T. Lopez, A. Gaona, R. Gomez, and A. Campero, Preparación de catalizadores de rutenio por el método sol-gel. *Actas X Simp. Iberoam. Catal. Merida Ven.*, **2**, 722 (1986).
36. T. Lopez, M. Asomoza, L. Razo, and R. Gomez, Study of the formation of Silico aluminates by the sol-gel method by means of IR, DTA and TGA. *J. Non-Cryst. Solids*, **108**, 45–48 (1989).
37. K. Jones, H. G. Emblem, and H. M. Hafez, *J. Non-Cryst. Solids*, **63**, 201 (1984).
38. B. E. Yoldas and D. P. Partlow, Formation of mullite and other alumina-based ceramics via hydrolytic polycondensation of alkoxides and resultant ultra- and micro-structural effects. *J. Mater. Sci.*, **23**, 1895–1900 (1988).
39. H. Schneider, B. Saruhan, D. Voll, L. Merwin, and A. Sebald, Mullite precursor phases. *J. Euro. Ceram. Soc.*, **11**, 87–94 (1993).

Part I

Review

Chapter 2

Mullite Precursors Synthesized by the Monophasic Gelation Method

In Chapters 2–6, we will study mullite formation by various chemical methods and study their phase evolution behavior by differential thermal analysis (DTA) and X-ray diffractometry (XRD).

2.1 Synthesis of Mullite by the Aqueous Sol-Gel Method

Mullite is synthesized by the aqueous sol-gel method by the gelation of a water-soluble silica component in the presence of Al salt. The colloidal system of alumina-silica-water is synthesized on the basis of adsorption (chemisorption) of aluminum hydroxide on the silicon hydroxide gel surface. The first step of this process is the gelation of silica hydrosol in a definite condition. The water-glass solution (sodium silicate) is neutralized by the addition of dilute sulfuric acid when silicic acid sol is formed. It is positively charged at pH of ~2, which is the isoelectric point, and above this pH, the sol is negatively charged. The rate of condensation of silica sol is the highest at pH of ~7 and lowest at pH of ~2. In the second step, an aluminum sulfate solution is added to the silica sol at the latter pH value. In the third step, dilute ammonia solution is added to allow interaction

Mullite Formations: Analysis and Applications
Akshoy Kumar Chakraborty

ISBN 978-981-4877-05-3 (Hardcover), 978-1-003-03167-3 (eBook)
www.jennystanford.com

among them. Some of the most important parameters pointed out by Dessalces et al. (1992) are as follows:

- Decrease in the gelation time: A fast gelation rate produces silica gel with an open texture. Alumina incorporation depends on the porosity of the silica hydro gel.
- Influence of silica concentration: It also controls the openness of the gel structure.
- pH value and the concentration of aluminum solution: During the adsorption of aluminum on the silicic acid surface in the pH range of 3 to 4, the pH value decreases by a few tenths, for example, from 3.5 to 3.2 (de Boer et al., 1960).
- The amount of aluminum that can be adsorbed: This is also a function of the number of –OH groups on the surface of silica hydrosol.
- Molecular weight of silicic acid sol at pH of ~2: Direct silica-alumina combination takes place when a solution of aluminum sulfate is added to a low-molecular-weight silica solution, which shows spontaneous cogelation.

The spectroscopic measurement shows that alumina incorporation likely takes place to the extent of 30% in the silica framework. As a result, the surface properties of silica-alumina combinations differ to some extent. Tamele (1950) considered the generation of acidic sites as per the following reaction:

$$SiO_2 \text{ surface (OH) groups} + Al(OH)_3 \rightarrow$$
$$SiO_2 \text{ surface } Al^- - O^+H_2 + 2H_2O \qquad (2.1)$$

Incompletely polymerized silica hydrosol reacts or condenses with hydrolyzed aluminum ions. Cloos et al. (1969) reviewed the earlier studies of relation of acidity to activity of alumina-silica cracking catalysts by Milliken et al. (1950) and Oblad et al. (1951). They considered that condensation of the hydroxyls at the interface occurs between silica and alumina particles, which leads to interbonding among them, with the local formation of an aluminate structure with fourfold-coordinated aluminium. This aluminate structure is stabilized by strongly positive H^+ ions. This process is continued with the addition of Al ions to silica sol to the extent of 30 wt % of Al^{+3} ions. The amount of aluminate decreases steadily with the further increase of alumina content. An XRD study showed that, thereafter, excess alumina crystallizes into a mixture of boehmite and bayerite and suggested the probable structure of silico alumina. Leonard et al. (1971) synthesized two series of silico-alumina gel over the entire range of compositions using ethyl orthosilicate and aluminum isopropoxide (A series) and aluminum nitrate nonahydrate (ANN, B series) by the cohydrolysis method and homogenized the first in acetic acid solution and then in ammonia solution at pH 7. The gels were calcined at 500°C for 24 h and finally studied for radial electron distribution (RED). The first profile decomposes maximum into three vectors for different ratios of Al_2O_3 and Al_2O_3+SiO_2. For lower amounts of Al_2O_3 content, the deconvolution process indicates the formation of a tetrahedral network in which silicon and a fraction of aluminum are intimately interchanged. This explains the origin of the negative lattice charge*/acidity and catalytic cracking activity and increase of the cation-exchange capacity value. This synthetic silico-alumina shows a large surface area and high surface acidity. An explanation of the catalytic activity of this compound is needed. According to Charles (1949), a catalyst is a chemical compound associated with AlO_4 groups in combination with SiO_4 groups and acidic hydrogen ions. During calcination of such Al/SiO_2 gel, extra Al^{+3} may be introduced by rupturing stable Si–O–Si bridges at a temperature of around 350°C. More information regarding phase evolution at much higher temperatures is given later in this book.

Insley and Ewell (1935) synthesized five gels with approximate Al_2O_3:SiO_2 ratios of 1:4, 1:2, 1:1, 2:1, and 4:1 using sodium silicate and aluminum sulfate as raw materials and sodium hydroxide as the neutralizing agent. The soda content of the dried, coprecipitated (CP) 1:2 gel was ~0.65%. The phase transformation sequences in two cases, namely clays and CP gels synthesized by them, are analogous as follows:

- Both are amorphous to X-ray. CP gels exhibit a 980°C exotherm, which resembles that of dehydrated kaolinite.
- After 980°C, kaolin shows the γ-Al_2O_3 phase. Al_2O_3-2SiO_2 gel shows only mullite. Gels high in Al_2O_3 show γ-Al_2O_3 in addition to mullite. They concluded that γ-Al_2O_3 formation out of the alumina (A) present in the metakaolin may be the cause of the exothermic heat effect in the case of kaolinite. On the other hand, mullite formation may be the cause of the same effect in the case of gels.

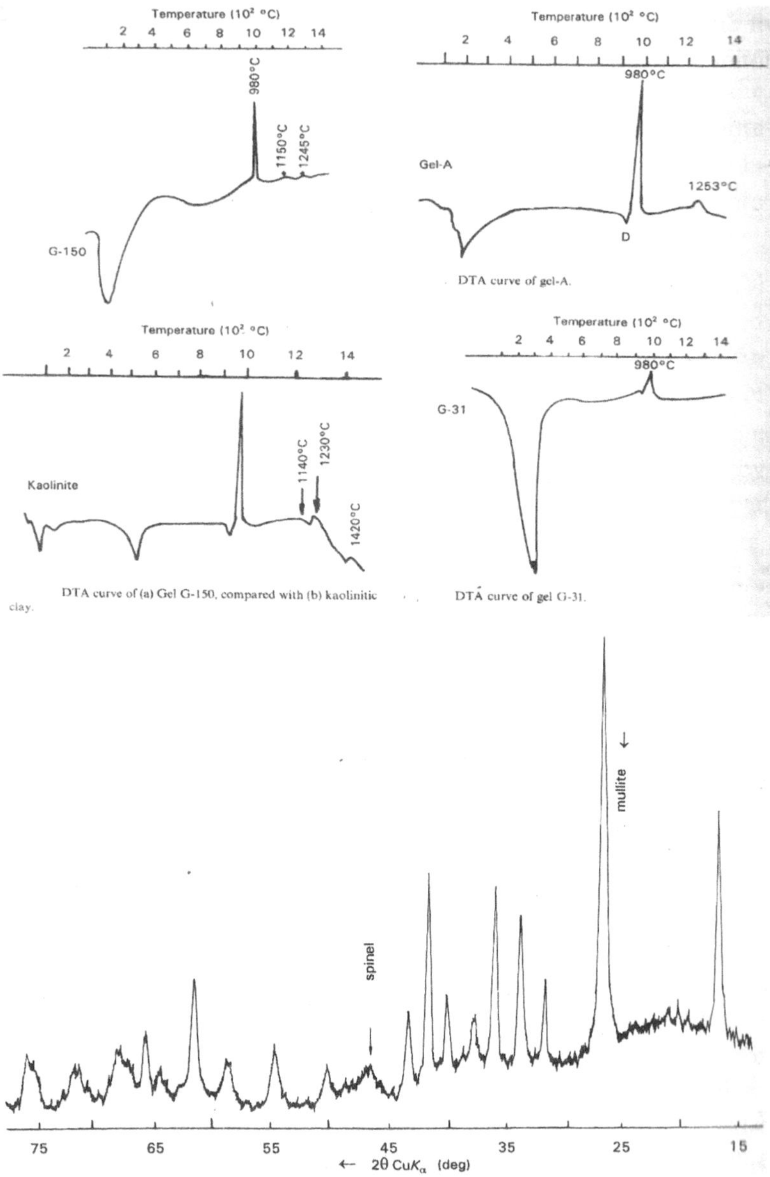

Figure 2.2 Characterization of Gel-A by DTA and XRD compared with others (Chakraborty, 1994b).

2.2 Synthesis of Mullite by the Monophasic Gel Method

Mullite was synthesized by the monophasic gel method by the hydrolysis and gelation of the organic silicon component in the presence of a water-soluble component in absolute alcohol and an acidic medium. Hoffman et al. (1984) first synthesized two monophasic mullite gels using ANN and tetraethyl orthosilicate (TEOS) in absolute alcohol as the medium of hydrolysis in two different quantities for a few days in a water bath condition at ~60°C. The phase transformation behavior of these nanostructure xerogels made by them by using less ethanol was studied by DTA and X-ray diffraction methods and is given in Table 2.1.

Table 2.1 Phase transformation of single phase xerogels prepared by using TEOS and ANN in absolute alcohol; gelling at 60°C in a water bath for several days

Temperature	Phase
150°C–950°C	Amorphous to X-ray
At 960°C	Sharp exotherm visible in DTA; a diffraction pattern of poorly crystalline mullite-type phase at 1050°C; an XRD peak visible at ~46° 2θ, which corresponds to the γ-Al_2O_3-type spinel phase, in addition to the broad peaks of weakly crystalline mullite, on careful observation by Chakraborty and Ghosh (1986)
At a higher temperature	Well-crystalized mullite

Hoffman et al. (1984) pointed out that both single-phase gel and kaolinite exhibit a 980°C exotherm and according to them, this close similarity may predict the amorphous structure of the xerogel having a unit cell mixing of silica and alumina. Al-Jarsha et al. (1985) chose $AlCl_3$ and TEOS for the synthesis of a mullite precursor by the gelation process. They noted mullite formation at >1200°C.

Okada and Otsuka (1986) repeated the process of synthesizing monophasic mullite gel the way Hoffman et al. (1984) did. The reactants TEOS and ANN were dissolved in absolute ethanol and slowly heated in an air oven at 60°C for 1 to 2 weeks. The gelation in this case proceeded slowly and gradually and thus this gel was named slow hydrolysis (SH) gel and is known as the first type of monophasic gel. The phase evolution behavior of the SH xerogel synthesized by them is given in Table 2.2.

Table 2.2 The phase evolution behavior of the SH xerogel

Temperature	Phase
150°C–950°C	SH xerogel is amorphous.
At 980°C	Exothermic peak is the sharpest and highest at the composition $3Al_2O_3$-$2SiO_2$. The peak intensity decreases as the composition deviates from the composition above. XRD showed the formation of a mullite-type phase along with γ-Al_2O_3-type spinel (small amount). The specimen richer in alumina than 3:2 mullite showed a spinel phase in addition to mullite and only spinel phase is noticed in the specimen rich in alumina than 5.4 $Al_2O_3 \cdot 2SiO_2$.
900°C–1400°C	Mullite starts forming at around 900°C and rapidly increases in amount until about 1000°C and then its formation ceases in the temperature range between 1000°C–1100°C. Thereafter, it again increases gradually from 1100°C–1400°C and develops into a well-crystallized variety.

Skorodumova et al. (1989) used $AlCl_3 \cdot 6H_2O$ and TEOS for the synthesis of mullite gel. A DTA study of $AlCl_3$ showed its complete dissociation at 823°K. Thereafter, it remained amorphous in the temperature range 823°K–1023°K. It crystallized into γ-Al_2O_3 at a 1073°K exotherm, and it finally crystalized to α-Al_2O_3 with the occurrence of an exotherm at 1443°K. The gel obtained by the hydrolysis of ethyl silicate and aluminum chloride in an acidic medium crystallized into mullite and a spinel phase on heating at 1273°K. The authors assumed that amorphous silica did not react completely at this temperature and the yield of mullite was low. High-heat treatment was necessary to increase its yield.

Chakraborty and Ghosh (1988) resynthesized mullite gels with TEOS and ANN as the components at different water and alcohol contents. There phase transformations show differences as shown in Table 2.3.

Table 2.3 DTA and XRD analysis of different mullite gels

DTA/XRD		Amount of water			Amount of alcohol				Amount of absolute alcohol	
Analysis	Volume of gel	20 mL G-143	4 mL G-144	2 mL G-145	60 mL G-139	40 mL G-138	20 mL G-137	10 mL G-142	10 mL G-141	5 mL anhydrous alumina G-31
Appearance of 980°C DTA		Very small	A[a]	A	Prominent peak[b]	Most prominent peak	More prominent peak	P	P	P, broad
	-	Qualitative	XRD	Analysis	of	980°C heat treated	Gels	-		
Presence of γ-Al_2O_3-type spinel phase		P	P	P	P	P	P	P	P	A
Presence of orthorhombic mullite		A	A	A	A	A	A	A	P	P

[a]A: Absent
[b]P: Prominent peak

It was shown that the concentration of water and/or alcohol plays a vital role in the development of gels of different characters, which in turn determines the development of either Al-Si spinel or mullite or both at a 980°C exotherm (Table 2.3). The extent of Al-Si spinel formation depends on the concentration of alcohol used during gelation. For example, for the gel marked (G-26) during the synthesis process, 10 mL of ordinary alcohol and ANN were used (it is assumed that water is available from two sources, some amount of free water from alcohol and some from the water of crystallization present in ANN). It subsequently crystallizes into the spinel phase. When 10 mL of ordinary alcohol is replaced by 10 mL of absolute alcohol, then the water of crystallization (nine molecules) is the only water available. Even then the spinel phase is formed in association with some quantity of mullite phase, as observed in the DTA of the gel marked (Gel A). On the basis of these two observations, it was concluded that the nature of phase development depends on (i) pH and (ii) the quantity of solvent used during the processing of gels.

Hyatt and Bansal (1990) repeated the SH gel procedure for the synthesis of a monophasic gel. They prepared the monophasic gel by comixing ANN and TEOS in absolute alcohol without using any additional water as water present in aluminum nitrate gave a water:TEOS mol ratio of ~2.7. The resultant gel showed a strong exotherm at ~975°C, which was attributed to the Al-Si spinel. The enthalpy change, $-\Delta H$, corresponding to this exotherm was calculated from the peak area out of a run of a precalcined sample at 490°C and the value was determined to be −166 J/g.

Li and Thomson (1991) studied the effect of the hydrolysis condition based particularly on the time of hydrolysis on mullite formation reaction from a number of single-phase gels, a colloidal gel made from the same components (TEOS and ANN), and a diphasic gel by using high-temperature dynamic X-ray diffraction (DXRD) as well as DTA. In monophasic gel preparation, ANN and TEOS were used, whereas in the diphasic gel preparation, boehmite sol and TEOS were taken. Characteristics of the gel preparation conditions are as follows:

- 2H gel was prepared in absolute ethanol and gelled in an open container on a hot plate at 75°C for 2 h; nitrate/ethanol (N/E) = 1.3.

- 1D gel was prepared in absolute ethanol and gelled in a covered container on a hot plate at 60°C for 1 day; N/E = 1.3.
- 2D gel was prepared in absolute ethanol and gelled in a sealed container on a hot plate at 60°C for 2 days; N/E = 1.3.
- 2W gel was prepared in absolute ethanol and gelled in a sealed container on a hot plate at 60°C for 2 weeks; N/E = 0.4.
- 2WC gel was prepared in absolute ethanol and gelled in a sealed container on a hot plate at 60°C for 2 weeks; N/E = 0.4; it was then calcined at 500°C for 6 h.
- CG-TEOS was first hydrolyzed in deionized water at pH 2, then ANN was added and gelled for 1 day in an oven at 60°C.
- DG-TEOS was hydrolyzed in boehmite dispersion at pH ≈ 2 and gelled in an oven at 60°C for 12 h.

A DXRD scan of integrated intensities of the 120 and 210 double peak of mullite showed two plateaus for 2WC, 2W, 2D, 1D, and 2H gels and suggested a two-stage transformation of mullite. The first step of transformation corresponded to the first exotherm, at ~980°C. The second step of mullite transformation likely occurred at the second exotherm, at 1200°C. CG gel did not show any mullite formation; it occurred rapidly at 1200°C and matched with the second exotherm. DG developed mullite suddenly at ~1300°C. The gelation time is directly correlated with the extent of mullite formation at a 980°C exotherm.

Chakraborty (1994a) studied the phase transformation of Al_2O_3-SiO_2 gel prepared in a highly acidic condition (pH ≤ 1) by mixing aqueous silica sol with aluminum sulfate in the stoichiometric ratio (3:2) of mullite. At pH ≈ 1, the solution was gellified within 10 min. after mixing. During DTA study, the dried gel exhibits a small exotherm at 980°C. XRD of this heated gel shows only the formation of traces of mullite, along with a large concentration of the spinel phase. At ~1200°C, the gel develops a mixture of corundum and β-cristobalite. The intensities of these two phases increase with the rise of temperature to 1300°C. At ~1400°C, mullite formation takes place with marked diminution in the intensities of both α-Al_2O_3 (0.348 nm peak) and β-cristobalite (0.249 nm peak) phases. The crystallization behaviors of these phases show that the spinel phase formed on heating this gel before the formation of corundum is γ-Al_2O_3 and as a result there is no exothermic peak in the vicinity of 980°C in the DTA. Therefore, pH plays a vital role in the development of the character of the gel, which in turn determines the subsequent thermal sequence. This gel, which was prepared at pH ≤ 1, forms mullite via intermediate γ-Al_2O_3 formation.

Saha and Pramanik (1994) synthesized gels with the composition of mullite ($3Al_2O_3 \cdot 2SiO_2$) from an aqueous solution of aluminum oxalate and tetraethoxysilane by varying the temperature of the formation of gels and gelling time. They noted broad XRD peaks of the spinel phase and showed two high-temperature exotherms, at 990°C and 1250°C, respectively. High-purity stoichiometric mullite powders were produced by calcination of the gel powders with a small exothermic event at 1250°C. The chemical and structural evolution were characterized by DTA and thermogravimetric analysis of the gel materials and XRD and infrared of the fired powders. They concluded that the composition of mullite varied with the temperature of the heat treatment.

2.3 Synthesis of Mullite by Polymeric and Colloidal Gel Methods

Mullite can be synthesized by the polymeric and colloidal gel methods involving hydrolysis and subsequent cogelation of organic Al and Si compounds in a neutral to a slightly acidic medium. Synthesis of ceramic materials from metal-alkoxide precursors is a promising process, providing some attractive properties, for example, high purity, submicron particle size (typically 20–50 Å), molecular-level homogeneity among component materials, and high reactivity due to high surface area and these lead to lower processing temperatures to produce sintered products than those required by classical methods. Ceramic bonding between silica and alumina, previously achievable by thermal means during the interdiffusion process, likely takes place by the chemical polymerization method. Yoldas (1977, 1979) introduced an emerging process in the ceramic field based on the nonthermal formation of ceramic substructures by chemical polymerization. Metal alkoxides are promising materials in the ceramic forming process. They are generally converted to metal oxides either by thermal decomposition or by hydrolysis followed by dehydration. In the presence of low and insufficient quantity of water, silicon alkoxide may partially hydrolyze to form silanol, which is soluble in ethanol.

$$Si(OR)_4 * H_2O \rightarrow (RO)_3Si - OH + ROH \quad (2.2)$$

When Al(OR′) is added to this silanol solution, a reaction takes place leading to the formation of a aluminosiloxane derivative, as follows:

$$(RO)_3Si - OH + Al(OR')_3 \rightarrow (RO)_3Si - O - Al\,(OR')_2 + R'OH \quad (2.3)$$

The reaction product contains Si and Al connected by oxygen. This process is called chemical polymerization. The addition of more water and alkoxide to the above aluminosiloxane derivative solution causes further polymerization, cross-linking, and entrapping of solvent/solution and finally leads to gelation.

Yoldas (1980) synthesized a monolithic Al_2O_3-SiO_2 precursor by a colloidal-based method using AlOBu and TEOS. First, a colloidal aluminum precursor sol was prepared by the hydrolysis of AlOBu with water by the following reaction:

$$Al(OC_4H_9)_3 + 2H_2O \rightarrow AlO(OH) + 3C_4H_9(OH) \quad (2.4)$$

The aqueous alumina sol was peptized with 0.07 mol of nitric acid/mol of alkoxide to form a clear liquid. In the second step, stoichiometric quantity of silicon ester was added. Immiscibility of $Si(OR)_4$ was overcome by the addition of alcohol. The resulting mixture was reacted while stirring to produce a gel of mullite composition.

Yoldas and Partlow (1988) were of the opinion that hydrolytic polycondensation rates of aluminum and silicon alkoxides are substantially different. They conjectured that the polycondensation of a mixture of $Al(OR)_3$ and $Si(OR)_4$ and water gives gross inhomogeneities in the oxide networks due to large differences in the reaction rates. The bond formation that occurs in two cases of gels, as shown above, may be feasible in a 1:1 molar proportion, but it is not known whether complete bond formation between silicon and aluminum would occur when 1 mol of silicon alkoxide is allowed to react with 3 mol of aluminum alkoxide to form mullite gel. There is every possibility that Al–O–Al bond development should be predominant because of the following two reasons: (i) the reactivity between two active reactants obviously decreases after the monomeric species of silicon alkoxide reacts with aluminum alkoxide, and (ii) the hydrolysis and polycondensation rate of aluminum alkoxide is faster than that of silicon alkoxide. A thick white slurry is generally obtained instantaneously when a water/ethanol mixture was added. Due to these reasons, the growth of alumina colloids is more rapid than the interaction between two reactants. Preferentially, self-condensation of alumina colloids takes place rather than bond formation with $Si(OR)_4$.

Yoldas and Partlow (1988) and thereafter Yoldas (1992) discussed and engineered partial hydrolysis methods to overcome the problem of inhomogeneity that occurred during hydrolysis and condensation processes. They synthesized three ways of gel preparation in the Al_2O_3-SiO_2 binary:

- The first method, described earlier, was a colloidal process. In this process, an aluminum hydroxyl bond was created from $Al(OC_4H_9)_3$ without significant self-condensation and the condensate was peptized in a clear aqueous sol containing AlO(OH) colloidal particles, which was then reacted with alkoxy bonds of Si to produce bridging oxygen bonds between silicon and aluminum. The resultant aluminosilicate gel was termed by them as "colloidal gel."

$$\equiv \mathrm{Si-OR} + \underset{\text{(Aq. colloidal AlO(OH)sol)}}{\mathrm{HO-[AlO(OH)}_{(n-1)/n}]_n} + \underset{\text{(Mutual solvent)}}{\text{Alcohol}} \longrightarrow \underset{\text{(Colloidal gel)}}{\equiv \mathrm{Si-O-[-AlO(OH)}_{(n-1)/n]n}} + \mathrm{ROH} \quad (2.5)$$

  They assumed that a chemical bond of hydrolyzed silica shell forms around and between AlO(OH) colloids. Thus the ultrastructure is a nanocomposite and the homogeneity is extended up to the dimension of AlO(OH) colloids. They expected the homogeneity is up to the size of alumina colloids, that is, 10–30 nm in size.

- The second method is based on the synthesis of polyorganoaluminum condensate polymer from 1 mol of aluminum alkoxide by hydrolysis with 0.5–1 mol of water in dry ethanol. At the first stage of hydrolysis, a thick white slurry was formed, which on vigorous stirring at 50°C for 1–2 h transformed into a clear liquid. After the liquid became clear, TEOS was added as per Al_2O_3:SiO_2 in the ratio 3:2

(stoichiometric mullite), along with 0.5 mol of water and 0.5 g of HCl in alcohol at a time during the shaking process. The following important observations were made: (i) gelification occurred immediately after the addition of TEOS, water, and catalyst, (ii) this gel transformed into a clear sol within 5–10 min., (iii) and when another 1–2 mol of water in ethanol was added, the sol irreversibly transformed into a gel, termed as "polymeric gel."

- To attain a higher degree of homogeneity, they first preferentially hydrolyzed $Si(OR)_4$ partially and then reacted it with unhydrolyzed aluminum alkoxide, as per the following reaction:

$$\underset{\text{(Partially hydrolyzed)}}{(RO)_3\,Si-OH} + RO-Al(OR)_2 \longrightarrow \underset{\text{(Polymeric gel)}}{(RO)_3\,Si—O—Al(OR)_2} + ROH \qquad (2.6)$$

The aluminosiloxane derivative on aging in water leads to the subsequent polymerization between silicon and aluminum through oxygen. These reactions finally lead to the formation of a gel. Partial hydrolysis was done with a view that the hydrolysis and self-condensation rate of $Si(OR)_4$ is much slower than that of $Al(OR)_3$. The phase transformation of these colloidal and polymeric gels will be reported in Chapter 7. The two methods of synthesis of colloidal and polymeric gels affect the following:

- A difference in the ultrastructure (see Yoldas, 1992, Fig. 2) and crystallization of intermediate phases (see Pask et al., 1987) is expected.
- The polymeric gel produced a mixture of tetragonal mullite, or t-mullite, (mullite-type phase, major) and a γ-Al_2O_3-type spinel phase (minor); in contrast the colloidal gel didn't develop t-mullite at ~1000°C.
- The densification/sintering rate (see Yoldas, 1988, Fig. 6) of the polymeric gel showed ~35% linear shrinkage at ~980°C. In comparison, the gel made out of prehydrolyzed $Si(OC_2H_5)_4$ and AlOBu showed a gradual shrinkage up to ~900°C and a more rapid shrinkage between 900°C and 1100°C.

Hirata et al. (1985, 1989) hydrolyzed a mixture of $Al(OPr)_3$ and TEOS by refluxing at ~80°C for 4 h by stirring in benzine with the addition of a mixture of water with isopropyl alcohol and by maintaining it at pH 5.5. The hydrolysis product was evaporated on a hot plate to develop powders. The produced powder ($2.94Al_2O_3{\cdot}2SiO_2$) was amorphous and exhibited an exotherm at 984°C in DTA. The X-ray diffractogram pattern of the phases formed after the exotherm indicated a mullite-type phase with a small amount of γ-Al_2O_3-type phase. The latter phase existed in the temperature range of 1000°C–1100°C. On heating the powder to a temperature beyond 1200°C, complete mullitization was noted.

2.4 Summary

The chemisorption of aluminum hydroxide on the surface of the silicon hydroxide gel was shown by Tamele (1950), de Boer et al. (1960), Cloos et al. (1969), Dessalces et al. (1992), and others. These authors reported (i) a decrease in the pH and (ii) changes in the surface properties during such interaction. The incorporation of alumina in the silica network structure was also shown by the deconvolution of the RED curve of heat-treated gels by Leonard et al. (1971) and conclusively showed the formation of a tetrahedral network containing interchanged Si and Al, which explains the origin of the negative lattice charge. Several gels have been synthesized by various researchers in the mullite field, and the thermal evolution of these gels will be shown. Three processes of synthesis of mullite precursors are described: monophasic, CP, and colloidal. A DTA of gels synthesized by the three methods invariably show the first exotherm and then crystallization of the mullite. The synthesis of mullite by the aqueous sol-gel route is claimed to be the best process among the three methods (Insley & Ewell, 1935; Ossaka, 1962; Chakraborty & Ghosh, 1988; Jaymes, 1994), since it does not develop an Al-Si spinel as a secondary phase along with mullite, even in a minor quantity.

- The aqueous sol-gel method: In this method, the gel made from silica sol and soluble aluminum salt showed crystallization exotherm. The gel synthesized from silica sol/gel and ANN or Al-sulfate did not crystallize when mixed in its own crystallization paths (Chakraborty & Ghosh, 1988).

A DTA study showed no exothermic peak at either 1060°C or at 1130°C due to crystallization peaks of β-cristobalite and α-alumina but showed crystallization of mullite at 980°C. It is further established that the intensity of the 980°C exotherm is related to the XRD intensity of orthorhombic mullite (o-mullite). Direct mullite formation during the exhibition of a 980°C exotherm was observed by Insley and Ewell (1935), Ossaka (1961), Chakraborty and Ghosh (1988), and Schneider et al. (1992) in their CM samples, as shown in this chapter. The tentative thermal reaction is as follows:

$$\underset{\text{(Amorphous to X-ray)}}{Al_2O_3 \cdot SiO_2 \text{ gel}} \xrightarrow[\text{Exotherm}]{1^{st}} \text{Mullite type phase} \tag{2.7}$$

- The monophasic/SH gel method: In this method, the silica sol was replaced by silica from an organic source, for example, TEOS. Both mullite (major quantity) and the spinel phase (minor quantity) during exotherm were noted by Hoffman et al. (1984) and Okada et al. (1986). The tentative thermal reaction of this mullite gel noted by them is as follows:

$$\underset{\text{Monophasic/SH gel}}{Al_2O_3 \cdot SiO_2 \text{ gel}} \xrightarrow[\text{Exotherm}]{1^{st}} \underset{\text{(major)}}{\text{Mullite type phase}} + \underset{\text{(minor)}}{\gamma\text{-}Al_2O_3 \text{ type-spinel phase}} \tag{2.8}$$

Significantly, mullite gels synthesized from TEOS and ANN at (i) different water and/or alcohol contents, (ii) of different pH values, and (iii) with silica sol of different characters showed variations. Phase transformations in these cases also showed differences (Chakraborty & Ghosh, 1988). The hydrolysis condition affected both the intensity of the 980°C peak as well as the ratio of the Al-Si spinel to mullite crystallization and likely the steps of mullitization (Li & Thomson, 1991). A vital role was played by the pH, which determined the sequence of phase development of the Al_2O_3-SiO_2 gel. The gel prepared at pH ≤ 1 formed γ-alumina instead of mullite (Chakraborty, 1994a). Besides the pH and the water content used, other factors, like the temperature of the formation of gels and the gelling time, also affect the phase transformation (Saha & Pramanik, 1994),

- The polymeric and colloidal gel method: Al_2O_3-SiO_2 precursor gels were synthesized by using AlOBu and TEOS through the chemical polymerization process by Yoldas (1979, 1980) in a neutral medium. Significantly, three ways of synthesis process were shown by Yoldas and Partlow (1988) and Yoldas (1992) for obtaining a colloidal gel, a polymeric gel, and a polymeric gel by the partial hydrolysis method. Thermal transformations of the above two gels by DTA and XRD studies show differences. While the polymeric gel produced a mixture of t-mullite (major) and Al-Si spinel (minor) during a 980°C exotherm, the colloidal gel neither exhibited the first exotherm nor developed t-mullite at ~1000°C.

  A mixture of TEOS and $Al(OPr)_3$ instead of AlOBu as done by Yoldas (1992) was hydrolyzed at pH 5.5 by Hirata et al. (1989). It indicated mullite with a small amount of Al-Si spinel phase at a 984°C exotherm. The latter phase existed in the temperature range of 1000°C–1100°C. Complete mullitization occurred on heating to a temperature beyond 1200°C. The thermal reaction of this polymeric mullite gel noted by them is tentatively similar to that shown in the case of monophasic/SH gels.

Problems

1. What should be the probable structure whose theoretical composition belonging to mullite lies in the composition range between $(0.6–0.67)Al_2O_3$ + $(0.4–0.33)SiO_2$?

   Does some quantity of alumina hydroxide remain outside because quantity of silica is low? Is there any possibility of interaction between $Si(OH)_4$ with external $Al(OH)_3$ during dehydroxylation on heating?
2. What will happen when this silica-alumina hydrogel is heat-treated further?
3. γ-Al_2O_3 formation is absent in $Al_2O_3{\cdot}2SiO_2$ gel with low Al_2O_3 content synthesized by Insley and Ewell (1935) and heat-treated at 980°C. Explain, why does kaolinite form γ-Al_2O_3 type spinel phase in addition to mullite during the exhibition of 980°C exotherm?

4. Why does the problem of inhomogeneity occur during the synthesis of mullite by polymeric and colloidal gels?
   Can the partial hydrolysis methods overcome the problem of inhomogeneity that occurs during hydrolysis and condensation processes?
5. Describe the hydrolysis processes of organic silicon and aluminum compounds in neutral conditions as well as in acidic-basic conditions in an aqueous medium.

References

1. G. Dessalces, I. Biay, F. Kolenda, J. F. Quinson, and J. P. Reymond, Study of porous texture of silica and silica-alumina hydrogels using experimental design theory. *J. Non-Cryst. Solids*, **147–148**, 141–145 (1992).
2. J. H. de Boer, B. G. Linsen, and C. Okkerse, *Proc. Kon. Akad. B*, **63**, 360 (1960).
3. M. W. Tamele, Chemistry of the surface and the activity of alumina silica cracking catalysts. *Disc. Faraday Soc.*, **8**, 270 (1950).
4. P. Cloos, A. J. Leonard, J. P. Moreau, A. Herbillon, and J. J. Fripiat, Structural organization in amorphous silico-aluminas. *Clays Clay Miner.*, **17**, 279–287 (1969).
5. A. J. Leonard, P. Ratnasamy, F. D. Declerck, and J. J. Fripiat, Structure and properties of amorphous silico-alumnous. Part 5. Nature and poperties of silico-auminous surfaces. *Disc. Faraday Soc.*, **52**, 98–108 (1971).
6. I. Thomas Charles, Chemistry of cracking catalysts. *Ind. Eng. Chem.*, **41**, 2564–2571 (1949).
7. H. Insley and R. H. Ewell, Thermal behavior of the kaolin minerals. *J. Res. Natl. Bur. Stand.*, **14**(5), 615–627 (1935).
8. A. K. Chakraborty and D. K. Ghosh, Synthesis and 980°C phase development of some mullite gels. *J. Am. Ceram. Soc.*, **71**(11), 978–987 (1988).
9. A. K. Chakraborty, Effect of pH on 980°C spinel phase–mullite formation of Al_2O_3-SiO_2 gels. *J. Mater. Sci.*, **29**, 1558–1568 (1994a).
10. A. K. Chakraborty, Role of hydrolysis water-alcohol mixture on mullitization of Al_2O_3-SiO_2 monophasic gels. *J. Mater. Sci.*, **29**, 6131–6138 (1994b).

11. H. Schneider, I. Merwin, and A. Sebald, Mullite formation from non-crystalline precursors. *J. Mater. Sci.*, **29**, 805–812 (1992).
12. G. P. Thim, C. A. Bertran, V. E. Barlette, M. I. F. Macedo, and M. A. S. Oliveira, Experimental and monte carlo–simulation: the role of urea in mullite synthesis. *J. Euro. Ceram. Soc.*, **21**, 759–763 (2001).
13. D. W. Hoffman, R. Roy, and S. Komarneni, Diphasic xerogels, a new class of materials: phases in the system Al_2O_3-SiO_2. *J. Am. Ceram. Soc.*, **67**, 468–471 (1984).
14. A. K. Chakraborty and D. K. Ghosh, Comment on "Diphasic xerogels, a new class of materials phases in the system Al_2O_3-SiO_2." *J. Am. Ceram. Soc.*, **69**(8), C-202–C-203 (1986). Reply by S. Komarneni and R. Roy, ibid., **69**(8), C-204 (1986).
15. Y. M. M. Al-Jarsha, K. D. Biddle, and A. K. Das, Mullite formation from ethyl silicate and aluminum chlorides. *J. Mater. Sci.*, **20**, 1773–1781 (1985).
16. K. Okada and N. Otsuka, Characterization of the spinel phase from SiO_2-Al_2O_3 xerogels and the formation process of mullite. *J. Am. Ceram. Soc.*, **69**(9), 652–656 (1986).
17. O. B. Skorodumova, Crystallization of mullite in mixture obtained using sol-gel technology. *Glass Ceram.*, **46**(7–8), 340–342 (1989).
18. M. J. Hyatt and N. P. Bansal, Phase transformations in xerogels of mullite composition. *J. Mater. Sci.*, **25**, 2815–2821 (1990).
19. D. X. Li and W. J. Thomson, Effects of hydrolysis on the kinetics of high temperatture transformations in aluminosilicate gels. *J. Am. Ceram. Soc.*, 74574–74578 (1991).
20. S. K. Saha and P. Pramanik, Aqueous sol-gel synthesis of mullite powder by using aluminum oxalate and tetraethoxysilane. *J. Mater. Sci.*, **29**, 3425–3429 (1994).
21. B. E. Yoldas, Preparation of glasses and ceramics from metal-organic compounds. *J. Mater. Sci.*, **12**, 1203–1208 (1977).
22. B. E. Yoldas, Monoithic glass formation by chemical polymerization. *J. Mater. Sci.*, **14**, 1843–1849 (1979).
23. B. E. Yoldas, Microstructure of monolithic materials formed by heat treatment of chemically polymerized precursors in the Al_2O_3-SiO_2 binary. *Ceram. Bull.*, **59**(4), 479–483 (1980).
24. B. E. Yoldas and D. P. Partlow, Formation of mullite and other alumina-based ceramics via hydrolytic polycondensation of alkoxides and resultant ultra- and micro-structural effects. *J. Mater. Sci.*, **23**, 1895–1900 (1988).

25. Y. Hirata, K. Sakeda, Y. Matsushita, and K. Shimada, Preparation of fine SiO_2-Al_2O_3 powders by hydrolysis of mixed alkoxides. *Yogo Kyokai Shi*, **93**(9), 577 (1985).

26. Y. Hirata, K. Sakeda, Y. Matsushita, K. Shimada, and Y. Ishihara, Characterization and sintering behavior of alkoxide-derived aluminosilicate powders. *J. Am. Ceram. Soc.*, **72**(6), 995–1002 (1989).

27. B. E. Yoldas, Effect of ultrastructure on crystallization of mullite. *J. Mater. Sci.*, **27**(24), 6667–6672 (1992).

28. J. A. Pask, X. W. Zhang, A. P. Tomsia, and B. E. Yoldas, Effect of sol-gel mixing on mullite microstructure and phase equilibria in the α-Al_2O_3-SiO_2 system. *J. Am. Ceram. Soc.*, **70**(10), 704–707 (1987).

29. J. Ossaka, Tetragonal mullite type phase from co precipitated gels. *Nature (London)*, **19**, 1000–1001 (1961).

Additional References

1. T. H. Milliken, Jr., G. A. Mills, and A. G. Oblad, The chemical characteristics and structure of cracking catalysts. *Discuss. Faraday Soc.*, **8**, 279–290 (1950).

2. W. Lewenistin, The distribution of aluminum in the tetrahedra of silicates and aluminates. *Am. Mineral.*, **39**(1–2), 92–96 (1954).

3. G. C. Bye and G. T. Simpkin, Influence of Cr and Fe on Formation of α-Al_2O_3 from γ-Al_2O_3. *J. Am. Ceram. Soc.*, **57**, 367–371 (1974).

4. D. L. Trimm, Catalytic combustion (review). *Appl. Catal.*, **7**, 249–282 (1983).

5. E. Lippmaa, A. Samoson, and M. Magi, High-resolution aluminum-27 NMR of aluminosilicates. *J. Am. Chem. Soc.*, **108**, 1730–1735 (1986).

6. O. Sakurai, N. Mizutani, and M. Kato, Preparation of mullite powders from metal alkoxides by ultrasonic spray pyrolysis. *J. Ceram. Soc. Jpn.*, **96**, 639–645 (1988).

7. J. A. Pask and A. P. Tomsia, Formation of mullite from sol-gel mixtures and kaolinite. *J. Am. Ceram. Soc.*, **74**, 2367–2373 (1991).

8. L. A. Xue and I. Chen, Influence of additives on the γ-to-α transformation of alumina. *J. Mater. Sci. Lett.*, **11**, 443–445 (1992).

9. S. K. Pradhan, A. Chatterjee, A. Datta, M. De, and D. Chakravorty, Low temperature mullite formation from sol-gel precursors by hot pressing, *J. Mater. Res.*, **9**(10), 2474–2475 (1994).

10. D. R. Treadwell, D. M. Dabbs, and I. A. Aksay, Mullite ($3Al_2O_3$-$2SiO_2$) synthesis with aluminosiloxanes. *Chem. Mater.*, **8**, 2056–2060 (1996).

11. C. W. Won and B. Siffert, Preparation by solgel method of SiO_2 and mullite ($3Al_2O_3.2SiO_2$) powders and study of their surface characteristics by inverse gas chromatography and zetametry. *Colloids Surf., A*, **131**, 161–172 (1998).

12. K. Okada, S. Yasohama, S. Hayashi, and Yasumori, Sol-gel synthesis of mullite long fibres from water solvent systems. *J. Eur. Ceram. Soc.*, **18**, 1879 (1998).

13. K. C. Song, Preparation of mullite fibers from aluminium isopropoxide-aluminium nitrate-tetraethyl orthosilicate solutions by sol-gel method. *Mater. Lett.*, **35**, 290 (1998).

14. Dj. Janaćković, V. Jokanović, Lj. Kostić-Gvozdenović, and Lj. Živković, Synthesis, morphology, and formation mechanism of mullite particles produced by ultrasonic spray pyrolysis. *J. Mater. Res.*, **11**(7), 1706–1716 (1996).

15. F. Mizukami, K. Maeda, M. Toba, T. Sano, S.-I. Niwa, M. Miyazaki, and K. Kojima, Effect of organic ligands used in sol-gel process on the formation of mullite. *J. Sol-Gel. Sci. Technol.*, **8**, 101–106 (1997).

Chapter 3

Mullite Precursors Synthesized by the Coprecipitation Method

Generally, ammonia solution is used as a precipitant in the mixed solution of two components. On its addition, the pH of the medium changes from acidic to neutral to basic. As a result, the solubility of the aluminum component decreases to initiate the formation of gelatinous aluminum hydroxide colloidal precipitates. The ammoniated hydroxide later accompanies the silica component to form silicon hydroxide, and it is assumed that the two components are coprecipitated. The precipitant is used in such a quantity so as to show the pH of the resulting slurry as either less than or greater than 7. The pair of components for the synthesis of a mullite gel may be divided into three categories: (i) both water-soluble salts, (ii) an organic silicon compound and aluminum nitrate nonahydrate (ANN), and (iii) both organic compounds. Urea or hexamethylenetetramine may also be used for precipitation. These precipitants are added to the solution beforehand, and subsequent precipitation of components occurs by decomposition either by heating or by a change in the pH of the solution. This process is called homogeneous precipitation.

Mullite Formations: Analysis and Applications
Akshoy Kumar Chakraborty

ISBN 978-981-4877-05-3 (Hardcover), 978-1-003-03167-3 (eBook)
www.jennystanford.com

3.1 Precipitation of Water-Soluble Silicon and Aluminum Components in the Presence of NH_4OH

Horte and Wiegman (1956) prepared aluminosilicate precursors (overall compositions in the range of 20–85 wt % of Al_2O_3) from mixed aluminum and silicon chloride solutions. Gelatinous precipitates were obtained by using ammoniated water. The gels were subsequently washed until ammonium chloride was free. X-ray diffractometry (XRD) results indicated that the gels were initially amorphous. Differential thermal analysis (DTA) showed a sharp exothermic peak in the range of ~980°C, but the temperature of mullite formation was not specified.

Demediuk and Cole (1960) prepared gels by mixing $SiCl_4$ and $AlCl_3{\cdot}6H_2O$ followed by cohydrolysis with the help of dilute NH_4OH solution. A DTA of these gels showed a 980°C exotherm, and an XRD analysis showed a diffuse pattern of orthorhombic mullite. The exothermic peak area was the highest for gels of composition 3:2 mullite. These mullite gels were put to the moisture expansion test. To perform it, Demediuk and Cole (1960) synthesized coprecipitated gels of alumina and silica with SiO_2:Al_2O_3 molecular ratios of 4, 2, 1.5, 1, 0.8, 0.6, and 0.3 by using the above components in a controlled pH environment. The dried gels were heat-treated at the rate of 100°C/h, accompanied by soaking for 4 h at each at the following temperatures: 800°C, 850°C, 900°C, 950°C, 1000°C, 1050°C, 1100°C, and 1200°C. Then the gels were cooled in desiccators. Specimens were autoclaved at 200°C for 24 h, and finally the change in dimension was measured with a micrometer. The effect on the moisture expansion of the heated-autoclaved specimens of different compositions with the temperature of firing was plotted, which showed that for each firing temperature, the expansion drops to a minimum at a SiO_2:Al_2O_3 ratio of 0.67, the composition of mullite. At this composition, the development of mullite is maximum. The crystalline mullite showed a small specific surface and low surface energy and thus contributes the least in moisture expansion.

Ossaka (1961) dissolved sodium silicate and potassium aluminum sulfate—$KAl(SO_4)_2{\cdot}12H_2O$—in dilute aqueous sulfuric acid. The mixture was coprecipitated with excess of hexamethylenetetramine instead of ammonia. The pH value during the precipitation process was not mentioned. The precipitates were filtered and washed thoroughly to remove alkali metal ions and amine components. The phase transformation of the gel synthesized by him is given in Table 3.1.

Table 3.1 Phase transformation of aqueous mullite gel obtained by coprecipitation of water soluble components mixture by amine solution

Temperature	Phase
At 910°C	XRD patterns showed a mullite-type phase, called t-mullite.
1100°C/5 h	Spinel formation did not occur in this temperature range. A DTA showed a sharp 980°C exothermic peak.
At 1250°C/5 h	XRD patterns showed well-crystallized mullite.

Judging from the mullitization behavior, Ossaka (1961) considered the silica and alumina components to be in a molecularly mixed state. It has to be seen whether during homogeneous precipitating conditions, as thought of, the composition of the starting solution and the composition of the final precipitates are alike because of the less solubility difference of the components in solution. He conceived that as this mullite shows considerably broad peaks, it may be tetragonal in nature.

Croft and Marshall (1967) used aluminum and silicon chloride solutions to prepare aluminosilicate precursors with compositions in the range 56.5–60.9 mol % of Al_2O_3. To minimize selective hydrolysis, they initially dissolved the $AlCl_3$ and $SiCl_4$ in diethyl ether. Hydrolysis/precipitation was carried out by injecting the mixed salt solution into a vigorously stirred aqueous ammonia solution that had been heated to 50°C in order to "flash" off the ether. The precipitated powder was collected by centrifugation and subsequently filtered, dried, and calcined. Despite the turbulent conditions during the "flash hydrolysis" process, segregation occurred to some extent during precipitation. XRD results indicated that the dried powder was amorphous. However, calcined powders showed γ-Al_2O_3 at 800°C and α-alumina at 1200°C. Mullite formation was observed at temperatures between 1000°C and 1200°C, although the reaction was not complete after 2 h at 1200°C, but after 2 h at 1400°C, there is no further trace of γ-Al_2O_3.

McGee and Wirkus (1972) prepared mullite from aluminum chloride and silicon tetrachloride that were initially dissolved in absolute methanol. The mixed salt solution was coprecipitated using ammonium hydroxide solution. The gels were dried at 150°C to produce a free-flowing powder and fired in air for 40 h at a constant temperature in the range of 900°C–1500°C. XRD results showed that the dried and calcined powders with stoichiometric composition (~72 wt % of Al_2O_3) were amorphous up to temperatures of 1050°C (40 h). Crystallization to mullite was observed at 1100°C (40 h), although the temperature for the complete reaction was not specified. They reported that the mullitization was low at 1050°C. However, these chlorides, especially silicon chloride, are unstable and control of the chemical composition of the precipitates was difficult. They concluded that the molecular-scale mixing in a nonaqueous solution prior to coprecipitation permitted complete conversion of aluminosilicates to high-purity mullite at ~1150°C.

The effect of pH in the coprecipitation process was shown by Lee et al. (2002, 2003, 2004) and Kim et al. (2003). They synthesized coprecipitated gels from ANN and colloidal silica in three pH conditions (acidic, intermediate, and basic). Only the rates of gelation in these cases were different. In acidic pH, it was relatively slow and the gel was translucent. In comparison, the other two gels were opaque and the gelation process was fast due to the addition of ammonia solution. A DTA of the acidic sample showed two exotherms, at 910°C and 965°C, and showed the absence of any exotherm in the basic sample. The mullitization temperature of the acidic gel sample was lower than those of intermediate and basic gel samples. Cristobalite formation was seen in three samples in the 1200°C–1300°C temperature range. The peak intensity of cristobalite increased from 1200°C to the maximum at 1300°C. Thereafter, it changed to glassy phase and reabsorbed to form mullite at >1400°C. XRD results of acidic samples showed calcination at 1200°C, indicating the (i) coexistence of mullite and cristobalite in the silica-rich composition, (ii) existence of mullite, cristobalite, and metastable alumina in the stoichiometric composition, and (iii) existence of mullite and corundum in the alumina-rich composition. The XRD results of the heat-treated intermediate and basic samples were very similar, indicating that the precursor's pH and sintering temperature did not affect the developed phases. In the intermediate and basic samples, θ-Al_2O_3 and α-Al_2O_3 coexisted at 1400°C and showed complete mullitization at 1500°C.

3.2 Hydrolysis and Subsequent Precipitation of the Organic-Silicon Component in the Presence of the Al Component (Basic Medium)

Okada and Otsuka (1986) synthesized rapid hydrolysis (RH) gel by using ANN and tetraethyl orthosilicate (TEOS) that coprecipitated rapidly (hence the name) on using ammonia solution. However, the prevailing pH during the precipitation process and the XRD pattern of this raw RH gel were unknown. The phase transformation process of RH gel according to them is shown in Table 3.2.

Table 3.2 Phase transformation of (RH) mullite gel prepared by coprecipitation of water–ethanol soluble component mixture by ammonia solution

Temperature	Phase
150°C–950°C	Amorphous.
At 980°C	In comparison to SH gel, the 980°C exothermic peak in RH gel was small in magnitude. A specimen of the composition Al_2O_3 2.8 SiO_2 showed the highest peak intensity. XRD shows a γ-Al_2O_3-type spinel phase.
1100°C–1400°C	In this case, mullite formation started abruptly at ~1100°C and formation increased up to 1150°C. Thereafter, it increased gradually up to 1400°C.

Chakraborty and Ghosh (1988) synthesized mullite gels from ANN and TEOS in a basic condition, that is, by using ammonia solution as per the procedure adopted by Grofesic and Vago (1961). The gel marked (G-133) in DTA showed a small exotherm at 980°C and formed a γ-Al_2O_3-type spinel phase (major) with mullite (trace). Ammonia likely plays a dual role; it hydrolyses both ANN and TEOS. Aluminum hydroxide colloids are produced in the aqueous stage in the former case. But hydrolysis of the latter component is much more complex in a basic condition than in an acidic condition.

Hsi et al. (1989) synthesized a mullite gel in a basic condition by using TEOS and ANN. They showed that the formation of the crystalline phase depends on the pH value.

The effect of pH on phase transformation was also shown by Huang et al. (1997), who used TEOS and $AlCl_3 \cdot 6H_2O$ as sources of silica and alumina. To prepare mullite precursors by prehydrolysis of TEOS under refluxing conditions, a clear solution was first prepared. To this solution, ammonia solution was added dropwise while stirring the mixed solution at pH 9. The resultant milky-white coprecipitate was washed and divided into six portions. These were separately treated with suitable volumes of 3N nitric acid and ammonia solution while they were kept in air for 8 h, to adjust pH values to 1, 3, 5, 7, and 11. The effect of the pH prevailing during gelation on temperature-induced phase transformations was studied to ascertain the influence of pH on the mixing scale of the components on mullitization. The XRD patterns of MI (strong acidic condition) heated at 950°C showed a spinel phase. The intensity of the spinel phase decreased but remained still at 1200°C, when much of mullite crystallization started. The latter phase increased at 1350°C. The phase development of MII showed some differences. The spinel phase crystallizes gradually on heating and develops into a major quantity at ~1200°C, with some quantity of θ-Al_2O_3 phase. It, thereafter, transforms to mullite at 1350°C. Thus, the stability range of the spinel phase differs in two regions of pH. Huang et al. suggested that an improved mixing scale was observed at pH 1.5 and according to that, a sharp 980°C exotherm occurred and mullitization took place at ~1200°C, with the beginning of the decrease in the intensity of the spinel phase. At pH 11.5, a heterogeneity likely developed due to the formation of four phases—bayerite, gibbsite, boehmite, and silica (A)—this resulted in the absence of exotherm but the spinel phase started developing at a temperature as low as 450°C and continued up to 1200°C; thereafter mullitization took place. According to Huang et al., the appearance of a 980°C exothermic peak could be taken as evidence of better homogeneity in the degree of mixing of aluminum and silicon components in the precursor system. But the nature of the spinel phase formed in acidic pH is likely different than that noted in basic pH.

3.3 Hydrolysis and Subsequent Precipitation of Water-Soluble Silicon Component in the Presence of Organic Al Component (Purely Basic Medium)

Jaymes et al. (1994) used a new combination, of silicon tetrachloride and aluminum isopropoxide (AIP), as the source components. These are dissolved in tetrahydrofuran and hydrolyzed by ^{17}O-enriched water. This powder showed only mullite after a 980°C exotherm. They claimed this powder as the most homogeneous. The enthalpy of this peak, measured by differential calorimetry, is 121 kJ/mole. A similar result is obtained with precursor synthesized by $AlCl_3$ and TEOS. They concluded that an exchange reaction probably takes place between alkoxides and chlorides, which was previously reported as the preparation of monolithic binary oxide gels by a nonhydrolytic sol-gel process.

3.4 Hydrolysis and Subsequent Coprecipitation of Organic Al and Si Components (Basic Medium)

Mazdiyasni and Brown (1972) prepared a mullite precursor having Al_2O_3:SiO_2 in the mol ratio of 3:2 by using AIP, or $Al(OC_3H_7)_3$, and silicon isopropoxide, or $Si(OC_3H_7)_4$. The alkoxides were mixed and refluxed in excess isopropyl alcohol prior to hydrolysis/condensation reactions. The latter part was carried out by slowly adding the alkoxide solutions to ammoniated water. The precipitate obtained was subsequently washed repeatedly, dried, and calcined. This mullite precursor after being vacuum dried (60°C for 16 h) and calcined at 60°C for 1 h decreased in the specific surface area from 550 m^2/g to 280 m^2/g. XRD showed that the powder was amorphous up to ~1080°C. Extensive mullite formation was observed above ~1185°C, although higher temperatures were required for the completion of the reaction.

Prochazka and Klug (1983) dissolved TEOS and AIP in cyclohexane separately and mixed them so as to have the overall compositions of 72.3 and 76 wt % of Al_2O_3, respectively. The system was hydrolyzed by slowly adding excess water than that needed for the hydrolysis of the alkoxide. The resulting gels were filtered, dried, and calcined at temperatures in the range of 850°C–1200°C. The powders were amorphous to X-rays below ~850°C. Spinel phase formation proceeded after mullitization. At ~950°C, the characteristic three broad X-ray peaks appeared at 0.24, 0.195, and 0.138 nm. Mullite formation required a temperature higher than 1150°C.

Somiya (1985) mixed AIP and TEOS and then dissolved the mixture in benzene and refluxed it for 5 h at ~80°C in a nitrogen atmosphere. Hydrolysis/condensation of the mixed alkoxide was carried out hydrothermally at temperatures in the range 300°C–600°C and pressures in the range of 10–100 MPa. The product consisted of 1–5 μm agglomerates composed of acicular (~100 nm long) primary particles. Conventional calcination at 1000°C–1200°C (for 1 h) resulted in the formation of a weakly crystalline Al-Si spinel phase, while mullite formation occurred at 1300°C (1 h). Small amounts of Al-Si spinel and mullite formed when hydrothermal treatment was carried out at 500°C–600°C and 50–100 MPa pressures. In contrast to the results obtained with the colloidal sols, the increase in the hydrothermal treatment temperature did not enhance mullitization during subsequent calcination at 1200°C (1 h).

Hirata et al. (1985) synthesized SiO_2-Al_2O_3 powder from TEOS and AIP at different pH values with the help of ammonia in the temperature range of 20°C–80°C. Mullite formation started at ~1200°C and the rapidity of formation increased ~1300°C onward. At 1600°C, mullite became well crystallized but was associated with some quantities of θ- and α-Al_2O_3 phases. This view was supported by electron microscopy. It was observed that the average particle size of SiO_2-Al_2O_3 powders decreased slightly from 0.08 to 0.05 mm up to 1300°C. On further rise of temperature, the size decreased abruptly to 0.5 nm at 1600°C owing to rapid grain growth of mullite.

Hamano et al. (1985) prepared mullite powder from AIP and tetramethyl orthosilicate by the coprecipitation technique using a dilute ammonium hydroxide solution in such a way that the powder was amorphous in nature. Thus, this process seems to be analogous to the process of Mazdiyasni and Brown (1972) and Hirata et al. (1985). They prepared two sets of coprecipitates:

- Coprecipitate A was obtained when a mixture of aluminum alkoxide and methyl silicate diluted in isopropanol was added dropwise to a dilute ammonia solution at 60°C followed by centrifugation.
- Coprecipitate B was collected when an isopropanol and water mixture was added to a mixed alkoxide solution for hydrolysis.

Both coprecipitates were amorphous, and both exhibited a 980°C exotherm. However, the phase evolution in the two cases were different. Coprecipitate A developed an Al-Si spinel phase.

Hamano et al (1986) first synthesized the coprecipitated powder as above from AIP and methyl silicate by using ammonia solution and examined the properties and microstructure of the fired bodies. XRD of the coprecipitated material showed it to be amorphous up to ~950°C. It crystallized at a 980°C exotherm to the spinel phase exclusively. Thereafter, this crystalline spinel phase was present up to < 1250°C and further transformed at the second exotherm at ~1250°C to the mullite phase. When mullite was fired at 1780°C, it showed a high bulk density of 3.14 gm/cc. From 1500°C to 1690°C, the lattice parameters $\mathbf{a_0}$ and $\mathbf{b_0}$ decreased, but $\mathbf{a_0}$ increased at 1780°C. At about 1780°C specimen obtained by coprecipitation showed smallest amount of glass phase than that by sol mixture (SM) and oxide mixtures (OM) in scanning electron microscope (SEM). Thermal changes of CP, SM and OM materials on firing at various temperatures have been studied by XRD which revels that SM is diphasic in character.

Sakurai et al. (1988) prepared mullite powders from TEOS and $Al(O^iPr)_3$. Initially, TEOS was dissolved in a solution prepared by mixing dehydrated ethanol, water, and a catalyst. It was prehydrolyzed by refluxing at 75°C for 6 h in a nitrogen atmosphere. The prehydrolyzed TEOS was dissolved in the ethanol-water solution. Then, $Al(O^iPr)_3$ was dissolved in this solution and it was refluxed again under the same conditions to cohydrolyze both alkoxides. The experimental conditions and crystalline phases detected for corresponding materials are listed.

Low and McPherson (1989) synthesized three aluminosilicate gels, containing 63%, 72%, and 80% alumina, respectively. They hydrolyzed commercial purity silicon tetraethoxide and AIP for their gel preparation. The gels were heat-treated in a Rigaku DTA furnace at a heating rate of 10°C per min. and phases formed were analyzed by X-ray. The phase transformation of 72 wt % mullite gel is shown in Table 3.3.

Table 3.3 Phase transformation of 72 wt % mullite gel obtained out of two organic components during heating

Temperature	Phase
At 1150°C	A γ-Al_2O_3-type spinel phase was the only crystalline phase formed. Its particle size lies in the range of 1 to 5 nm, which coarsened rapidly with an increase in temperature.
At 1200°C	Both γ-Al_2O_3-type spinel and mullite-type phases coexisted. Mullite peaks at (120) and (210) lines as also peaks at (250) and (520) lines were not distinctly split; rather they were broad and diffuse. Onset of splitting occurred with the disappearance of the spinel phase at ~1300°C.

Low and McPherson (1989) measured the unit cell parameter of mullite formed at different temperatures and plotted unit cell edge versus composition. The variation in cell volume versus cell edge finally showed the composition of mullite as a function of temperature.

Suzuki et al. (1990) also thought the problem of inhomogeneity was due to the difference in the hydrolysis rates of Si and Al alkoxides, and for that they attempted to synthesize ultrafine mullite powder using the partial hydrolysis method of TEOS in acidic water before mixing it with AIP. Finally, they hydrolyzed the mixed mullite precursor solution with dropwise addition of distilled water. The heat-treated precursor showed an amorphous pattern up to 900°C, and it crystallized to Al-Si spinel at 1000°C, which was seen at 1100°C.

Schneider et al. (1992) synthesized mullite precursors by the sol-gel method marked, where TEOS and aluminum butyrate were mixed with isopropanol for homogenization. A white gel was produced by adding water dropwise. The phase transformation of these two gels have been studied by heat treatment, followed by the characterization of the phases produced by XRD, magic angle spinning–nuclear magnetic resonance spectroscopy (MAS-NMR),

DTA, etc. The phase development sequences monitored by DTA, XRD, and MAS-NMR are given later on.

Taylor and Holland (1993) tried to synthesize a mullite precursor using a water-free sol-gel approach. The two samples were made from TEOS and AIP in the presence of dry butan-1-ol and isopropyl alcohol as the medium and stirred for ~20 days and more. GYW contained H_2O/-OR in a ratio of >5, while in GNW, external water was not used.

A study of the effect of pH on coprecipitation was done by Schneider et al. (1993). They first tried to synthesize three different mullite precursors, namely Type I, Type II, and Type III, by different synthesis methods from the same starting materials, TEOS and AlOBu, by varying the amount of water, the velocity of the hydrolysis process, and the pH of the solvent. For Type I, no prehydrolysis was done and only the components were mixed in isopropanol; for Type II, TEOS and AlOBu were fully hydrolyzed at pH 13 to develop large alumina colloids particles of Al-oxyhydrate mixed with silicon hydroxides; and for Type III too, the components were fully hydrolyzed, the only difference from the Type II gel being the pH of the medium, which was reduced from 13 to 10 so that boehmite does not developed.

The Type I sample was designated as NH (no prehydrolysis), and this process is in accordance with the synthesis of polymeric gel mark C of Yoldas and Partlow (1988) vide the following reaction:

In humid air

$$\underset{\text{(AlOBu)}}{Al(OR')_3} + \underset{\text{(TMOS or TEOS)}}{Si(OR)_4} + SLV + H_2O \rightarrow (OR')_2\,Al\text{–}O\text{–}Si\,(OR)_3 + R'OH + ROH \quad (3.1)$$

where $Al(OR')_3 = Al(OBu)_3$, $Si(OR)_4$ may be TMOS = tetra methyl oxysilane or TEOS = tetra ethyl ortho silicate; SLV = solvent (alcohol); R, R′ = organic groups.

The mixed alkoxides were allowed to be in contact with air for 14 days to develop homogeneous network linkages. The sequential phase evolution was studied in the temperature range of 150°C to 1150°C by XRD measurement, infrared (IR), etc. Schneider et al. (1993) tried to determine the constitution of the resultant γ-alumina spinel phase on heat treatment of the Type II mullite precursor synthesized from TEOS and AlOBu.

In the Type II precursor, in the first step, TEOS was fully prehydrolyzed with a portion of isopropanol and water under a strong basic condition (pH 13), as per the following reaction. It was conceived that hydrolysis yields monomeric $Si(OH)_4$.

At pH ≥ 10

$$Si(OR)_4 + \text{excess } H_2O \rightarrow Si(OH)_4 + H_2O \quad (3.2)$$

In the second step, hydrolyzed silica sol was mixed with AlOBu solution. The excess water present in the sol immediately hydrolyzed AlOBu in a basic condition and formed large colloids of pseudoboehmite. By the XRD line broadening technique, their size was estimated to ≥100 Å to ≤1000 Å. The combined reaction is as follows:

At pH = 13

$$Si(OH)_4 + Al(OR') + \text{SLV (solvent)} + H_2O \text{ (excess)} \rightarrow AlO(OH) + Si(OH)_4 \quad (3.3)$$

According to Schneider et al. (1993), the character of Type II precursor is that large colloidal particles of oxyhydrate coexist with monomeric silicon hydroxide. The mullite precursor, heat-treated at 350°C, contained pseudoboehmite and some small amounts of transition alumina (γ-Al_2O_3). Higher annealing temperatures, of up to 1150°C, yielded XRD peaks of γ-Al_2O_3 spinel only. With increasing temperature, the XRD peaks of the latter phase became sharper and more intense and the reflections tended to split, which indicated that the phase was not cubic but rather tetragonal or orthorhombic (δ-phase).

Geradin et al. (1994) synthesized three types of mullite precursors to examine their crystallization paths. The M1 precursor was synthesized by using prehydrolyzed TEOS in an aqueous acid solution at pH 1 and AIP, with continuous stirring at room temperature. M2 was synthesized by mixing TEOS and AIP in 2-proponal at 60°C, followed simultaneously with hydrolysis by the addition of water. M3 was made by adding TEOS directly to an aqueous dispersion of boehmite and stirring for 24 h. DTA scans of the three precursors annealed at 800°C for 6 h will be given in Chapter 12. It will be shown that by varying the hydrolysis conditions, but using components from the same source, TEOS and AIP, two types of mullite precursors were formed, which on examination during heating showed different crystallization paths.

Chakraborty (1996) studied the effect of water on the phase transformation of coprecipitated gel. The author synthesized gels from the same components with three different amounts of water and studied the DTA and XRD to reveal the mechanism of their transformation leading to the formation of mullite. The first gel, marked G-152 (Al:Si = 3:1), was synthesized by using a large volume of water (50 mL) for prehydrolysis of AlOBu (7 mL) and then mixed with 2 mL of TEOS in 10 mL of alcohol followed by 2 drops of diluted HNO_3. The resultant gel forms pseudoboehmite and is diphasic in character. The second and third gels, marked G-152(i) and G-152(ii), were synthesized as per the above procedure except that the amount of water was reduced to 10 mL and in the last case hydrolysis was carried out in a humid atmosphere. Both these gels are monophasic in nature but show different behaviors in mullitization.

Sales and Alarcon (1996) synthesized three gels, marked A, B, and C. Sample A (in 36.1 wt % of Al_2O_3 and 63.9 wt % of SiO_2), sample B (in 71.8 wt % and 28.2 wt % of SiO_2), and sample C (in 88.4 wt % and 11.6 wt % of SiO_2) were prepared from TEOS and AlOBu by the prehydrolysis technique as followed by Yoldas (1990).

3.5 Precipitation of Si and Al Components from Organic Sources by Using Urea

In this case, generally urea or hexamethylenetetramine was used as the precipitant and added into the mixed solution of the components beforehand. It decomposed during heating and ammonia evolved, changing the pH of the solution and causing homogeneous precipitation of the components.

Yamada and Kimura (1962) prepared coprecipitated gels of different compositions from the Al_2O_3-SiO_2 system by hydrolyzing silicon ethylester and aluminum ethylester in an alcoholic solution followed by coprecipitation of alumina silica gel at pH values from 5 to 8 with the help of decomposing urea at 70°C. The phase transformation behaviors of the three coprecipitated gels (Al_2O_3:SiO_2 = 1:1, 3:2, and 2:1, marked as specimens A,B, and C) synthesized by Yamada and Kimura, along with two component gels and a mechanical mixture of gel, have been studied with help of DTA, IR, and XRD. The results are summarized in Table 3.4.

Table 3.4 Phase transformation of coprecipitated gels obtained by hydrolyzing of double alkoxides by using urea during precipitation process

Temperature	Phase
At 100°C–200°C	An endothermic peak was observed in all cases of gels, and this peak was due to the dehydration of the adsorbed water in the gels.
At a temperature below the first exotherm	The gels were amorphous.
At the temperature of the first exotherm	A γ-Al_2O_3-type spinel phase was detected by the XRD technique, wherein only two broad diffraction peaks, at (400) and (440), were noted. It was stable in the temperature range of 1000°C to 1200°C.
At the temperature of the second exotherm	The spinel phase thermally decomposed and mullite appeared.

It was considered that the second exothermic peak is due to the evolution of heat energy liberated during the transition of the spinel-type phase crystal, which is the metastable stage to the formation of a stable crystal of mullite. In the temperature range of 1000°C–1200°C, mullite exhibits 2D grain growth. At temperatures higher than 1200°C, both the velocity of mullite formation and 3D grain growth of mullite rapidly increase.

There are many studies on the combination of silicon and aluminum alkoxides used for the synthesis of mullite precursors. According to Jaymes et al. (1995), the experimental conditions in this case of synthesis are to be carefully controlled to promote the simultaneous reaction and copolymerization of the two components. This is perhaps doubtful since inhomogeneities occur frequently. They showed that the crystallization of mullite at 980°C is achieved easily when aluminum alkoxide is replaced by ANN.

Shiga et al. (1991) synthesized a mullite precursor by the homogeneous precipitation technique by using urea as the precipitant from a mixture of colloidal silica and aluminum sulfate solution in reflux condition. The resultant precipitate was filtered out by suction and dried. It produced a large quantity of spinel phase and a minor amount of mullite at 1000°C. Extensive mullitization on further heating occurred at and below 1400°C.

Jaymes et al. (1995) synthesized mullite gels in a homogeneous way in three steps, as follows: In the first step, a stoichiometric amount of TEOS was added to a solution of ANN, which was stirred for 0.5 h, or until a clear solution was obtained. To it, 1.8 mol of urea per mol of ANN was added and stored. After a few hours, a gel was formed. In the second step, when it was stored for 3 days, a clear sol was obtained. In the third step, this sol turned into a gel again after 1 week. The thermal evolution of this gel will be shown later on.

3.6 The Incorporation of Si in Al_2O_3 Structure: Case Studies by Different Researchers

Yoldas (1975) produced a new active alumina by the chemical polymerization of alumina sol into a monolithic and transparent bulk form at a temperature as low as 500°C. It contained pores that were extremely small in size (4 nm). During conversion to α-Al_2O_3, it lost its transparency. Yoldas (1976) subsequently stabilized this active alumina form by the incorporation of 6 wt % of SiO_2 in the gel network by way of addition of TEOS or partially hydrolyzed TEOS to alumina sol. This monolith during moderate heating stabilized the active alumina form into the silica-substituted structure of δ-Al_2O_3. Consequently, the transformation temperature of δ-Al_2O_3 was increased from 1200°C to 1400°C and likely increased the transformation temperature of α-Al_2O_3 crystallization.

Yoldas (1980) first synthesized a monolithic Al_2O_3-SiO_2 precursor by using the seq. AlOBu and TEOS. The phase transformation of undoped alumina and gel of the composition 63% Al_2O_3 and 37% SiO_2 has been studied by DTA and XRD.

Undoped active alumina shows a sharp exotherm at ~1200°C due to alpha alumina formation. During this transformation, the pores are virtually eliminated and material shrinks and turns opaque. The crystalline phases noted by XRD before and after the occurrence of this peak are δ-Al_2O_3 and α-Al_2O_3.

As the active alumina is doped with silica, it is introduced in the lattice structure of the former phase at very low temperatures by the chemical polymerization technique. The DTA peak making α-Al_2O_3 transformation at 1200°C for pure Al_2O_3 occurs at higher and higher temperatures.

Alumina doped with 6 wt % of silica shows an increase in exotherm at ~1400°C when analyzed by DTA. Both surface area and porosity are found to be minimum at this concentration of silica. Yoldas concluded that a silica concentration of up to 6 wt % stabilizes the δ-Al_2O_3 phase structure and this δ-phase is extremely resistant to sintering as well as pore migration and thus used as a catalyst.

At ~20% silica concentration most of the porosity is eliminated and mullite is formed. At ~35% silica, porosity is eliminated and a fine textured matrix is noted in electron microscopic studies. With the increase of this silica by ~20%, crisscrossing plates of mullite crystals are seen under microscopy. The crystalline phases detected by XRD on calcination of materials in the entire range of Al_2O_3-SiO_2 are also shown.

Chakraborty and Ghosh (1987) studied the crystallization behavior of Al_2O_3 in the presence of SiO_2 during the heating process. They investigated three questions related to gel synthesized by the coprecipitation method.

Question 1: What happens when $Al(OH)_3$ gel is heated in the presence of increasing amounts of $Si(OH)_4$ silicic acid gel?

When heated at 1000°C for 1 h, aluminum hydroxide gel G-68 (obtained by adding NH_4OH to an aqueous solution of ANN in the absence of silica) formed γ-Al_2O_3 (as identified by a 0.139 nm peak) and α-Al_2O_3 (0.140 nm and 0.137 nm peaks) (Fig. 3.1). With the addition of 5 wt % of SiO_2, γ-Al_2O_3 formed and corundum formation ceased. Thus, SiO_2 hinders the crystallization of the alumina component into α-Al_2O_3 near 1000°C and increases its crystallization to ~1200°C. This behavior continued until 72% Al_2O_3 composition in the mixed gel. At 72% and below, α-Al_2O_3 did not form at all, even when the gel was heated to a high temperature. The phase that developed near 1000°C from a gel had ~72 wt % of Al_2O_3, as designated earlier by Chakraborty and Ghosh (1988) as Al-Si spinel instead of simple γ-Al_2O_3. The composition of this cubic spinel was the same as that of 3:2 mullite.

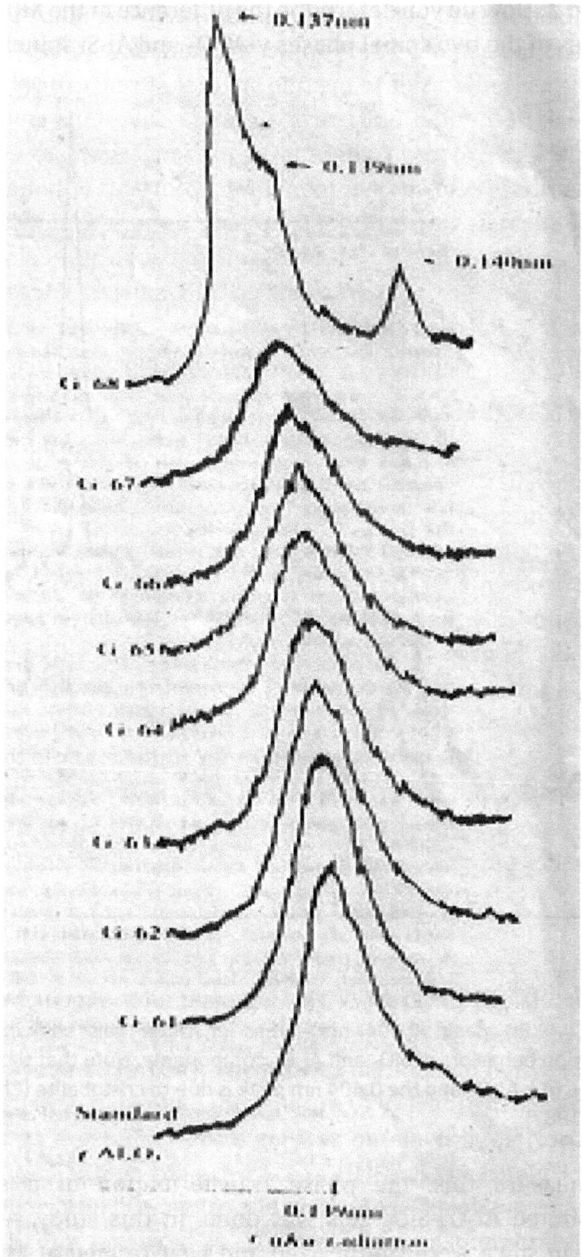

Figure 3.1 XRD (0.137 nm) peaks of gels heated to 1000°C for 1 h compared with that of standard γ-Al_2O_3 (Chakraborty & Ghosh, 1987).

alumina xerogel shows the presence of monophasic pseudoboehmite in a dried state, the presence of γ-Al_2O_3 when heated to 500°C, and finally the presence of α-Al_2O_3 when heated to 1100°C. Alumina gel doped with 1% molar silica remarkably changes the sequence of transformation behavior. At 500°C, it forms γ-Al_2O_3 and then forms δ-Al_2O_3 at 1100°C. They believed that a small amount of silicon atoms in the alumina network leads to the replacement of some atoms with silicon atoms and the formation of punctual defects of transitional alumina phases. These defects help in the stabilization of the δ-Al_2O_3 tetragonal structure and increasing the temperature of the phase transformation to the α-Al_2O_3 phase. Introduction of 1% molar of alumina in xerogels of silica changed the sequence of phase transformations in relation to pure silica xerogels. Alumina-doped silica xerogel crystallizes to β-cristobalite on heat treatment at 1100°C, while it shows an amorphous structure in undoped gel. The practical examples given in the chapter conform to the introduction of Si in the Al_2O_3 structure and vice versa.

3.7 Summary

Monophasic gel (aqueous solution of two components by the coprecipitation method)

With a view to synthesize uniform and homogeneous coprecipitates of $Al(OH)_3$ and $Si(OH)_4$ gels, aqueous solutions of $SiCl_4$ (a likely hazardous chemical) along with $AlCl_3$ were used in earlier days by Horte and Wiegman (1956), Demediuk and Cole (1960), Crofts and Marshall (1967), and McGee and Wirkus (1972). Mullite formation is the main crystalline phase in the thermal evolution processes of this coprecipitated gel during exhibition of a 980°C exotherm in DTA.

Monophasic gel (organic-silicon component and Al salts by the coprecipitation method)

In this method, a mixture of ANN and TEOS solution was rapidly hydrolyzed by using ammonia solution, called the RH method by Okada and Otsuka (1986). The 980°C exothermic peak in this case is small in magnitude due to the spinel phase instead of mullite formation. Chakraborty and Ghosh (1988) repeated this process of synthesis and also showed a small 980°C exotherm. The tentative thermal reaction of this RH mullite gel by using ammonia solution is given in Eq. 3.4 and looks quite different from that of SH gel as shown in Eq. 2.8 in Chapter 2.

$$\underset{\text{Monophasic/RH gel}}{Al_2O_3\text{-}SiO_2 \text{ gel}} \xrightarrow[\text{Exotherm}]{1^{st}} \underset{\text{(major)}}{\gamma\text{-}Al_2O_3} \text{ type spinel phase} \qquad (3.4)$$

Monophasic gel (organic Al and Si components by the coprecipitation method)

Coprecipitation of both organic Al and Si esters was first performed by Mazdiyasni and Brown (1972) using ammoniated water. Thereafter, this technique was repeated by several authors, as given below. They used component esters of their own choice based on availability, dissolving solvents are of various kinds, and finally hydrolysis/condensation techniques used are also different. pH adjustment was done in several cases; in some cases, it was made neutral by addition of water, whereas in other cases, it was alkaline.

- Prochazka and Klug (1983) mixed TEOS and AIP in cyclohexane separately and hydrolyzed this mixture by only adding water. Spinel phase formation preceded mullitization. At ~950°C, three broad characteristic X-ray peaks of the spinel phase at 0.24, 0.195, and 0.138 nm appeared. Mullite formation required a temperature above 1150°C.
- Somiya (1985) carried out a hydrothermal process for the hydrolysis/condensation of mixed alkoxides. The product on conventional calcination at ~1000°C showed an Al-Si spinel phase.

In conclusion, spinel formation is the predominating phase in the phase transformation monophasic gel synthesized from double alkoxide used as components. The thermal reaction of this monophasic gel crystallized to mullite phase via spinel as an intermediary phase is as follows:

$$\underset{\text{Monophasic}}{Al_2O_3\text{-}SiO_2 \text{ gel}} \xrightarrow{\text{At} \sim 950°\text{C}} \gamma\text{-}Al_2O_3\text{-type spinel phase} \xrightarrow{\text{At } 1150°\text{C}-1250°\text{C}} \text{Mullite type phase} \qquad (3.5)$$

The effect of pH: Coprecipitates were synthesized at different pH values by various authors (Hirata et al., 1985; Hamano et al., 1985, 1986) earlier. By adjusting the pH of TEOS and AIP, three kinds of powders were synthesized by Hirata et al. (1985). The powder prepared at pH 8.2 was amorphous, the same at pH > 10 contained boehmite, and at pH 11.9 it showed aluminum trihydroxide. The intermediate spinel phase was characterized by them as Al-Si spinel, The disappearance of it coincided with the formation of a large quantity of mullite.

With components from the same source, AIP ($Al(Oi-C_3H_7)_3$) and TMOS ($Si(OCH_3)_4$), a mullite gel was synthesized by Mitachi et al. (1990). In the A series, hydrolysis was done at pH 7 simply with distilled water. In the B series, hydrolysis was done at pH 10 with aqueous ammonia solution and the large difference in the crystallization behaviors of the two series of powders was noted. The ratio of spinel to mullite formed at the first exotherm was different in the two series.

Like Yoldas (1980), Paulick et al. (1987) also considered the difference in the rate of hydrolysis of silicon and aluminum alkoxides resulting in differential precipitation of component hydroxides during hydrolysis. Accordingly, urea was used instead of direct ammonium hydroxide since the former evolves ammonia in a slow and controlled way. Coprecipitated gels of compositions different from the Al_2O_3-SiO_2 system were prepared by Yamada and Kimura (1962) on hydrolysis of silicon ethylester and aluminum ethylester in an alcoholic solution followed by the coprecipitation of alumina silica gel at pH values from 5 to 8 with the help of decomposing urea at 70°C. This monophasic gel invariably developed Al-Si spinel during the 980°C exotherm and crystallized to mullite at the second exotherm. Urea was also used for coprecipitation of a water-soluble mixture of colloidal silica and aluminum sulfate by Shiga et al. (1991). The precipitate crystallized to a spinel phase, as shown by Yamada and Kimura (1962).

Four mullite gels were synthesized by Hsi et al. (1989) by dissolving ANN and TEOS in absolute alcohol at four different pH values, 8.3, 9.5 10.1, and up to 10.4, and proceed differently in their thermal transformations. These mullite gels synthesized in basic pH conditions may develop diphasicity.

The effect of water: A list of transformation processes of coprecipitated mullite gels synthesized in the presence of NH_4OH or simple water has been discussed above. The following monophasic gels were synthesized by hydrolysis and subsequent gelation of organic silicon and aluminum component by various authors using large quantities of water.

- Colloidal gel synthesized using the same components but with the use of an excess amount of water. Synthesis and phase change studies of Pask et al. (1987).
- Type II (prepared using high water content at pH > 10 where hydrolysis was very fast when boehmite was precipitated out.
- Two precursors, GTW and GNW, with difference in the water content, shown by Taylor and Holland (1993).

Problems

1. Why does segregation occur during the coprecipitation process of aqueous solutions of $AlCl_3$ and $SiCl_4$ in ammoniacal condition? How can you minimize this effect?
2. Is the exothermic peak observed in DTA related to the Al_2O_3:SiO_2 ratio in mullite coprecipitates?
3. Are the starting composition of Al to Si in a solution of sodium silicate and potassium aluminum sulfate and the composition of the final precipitates after amine addition alike?
4. To what extent does homogeneous precipitation really occur during molecular-scale mixing of aluminum and silicon chloride solutions? Aqueous or nonaqueous solution might have a role prior to coprecipitation.
5. The reasons behind the formation of the spinel phase in both cases of RH gel preparation by Okada et al (1986) and the same in the cases of Chakraborty and Ghosh (1988), Hsi et al, (1989), and Huang et al. (1997) require to be explained. The question is, what happens in the gelation process when the ammonia solution is used as a precipitant?
6. Coprecipitated gels obtained by hydrolysis and subsequent

coprecipitation of both organic Al and Si components in a basic medium or even by using urea do not crystallize to mullite directly. Is it due to a variation in the homogeneity of the precursor gel?

References

1. C. H. Horte and J. Wiegmann, Reaction between amorphous SiO_2 and Ai_2O_3. *Naturwiss*, **43**, 9 (1956).
2. T. Demediuk and W. F. Cole, Exothermic reaction of metal sols in between 950°C–1000°C. *Nature*, **181**, 1400 (1958).
3. J. Ossaka, Tetragonal mullite type phase from co precipitated gels. *Nature (London)*, **19**, 1000–1001 (1961).
4. D. Croft and W. W. Marshall, A novel synthesis of alumino-silicates and similar materials. *Trans. Br. Ceram. Soc.*, **64**(3), 121–126 (1967).
5. T. D. McGee and C. D. Wirkus, Mullitization of aluminum-silicate gels. *Am. Ceram. Soc. Bull.*, **51**, 577–581 (1972).
6. E. Lee, J.-W. Kim, and Y.-G. Jung, Effect of precursor pH and sintering temperature on synthesizing and morphology of sol-gel processed mullite. *Ceram. Int.*, **28**, 935–940 (2002).
7. E. Lee, J.-W. Kim, and Y.-G. Jung, Effect of precursor pH and composition on the grain morphology and size of mullite in aqueous system. *Mater. Lett.*, **57**(21), 3239–3244 (2003).
8. E. Lee, J.-W. Kim, and Y.-G. Jung, Mullite precursor synthesis aqueous conditions: dependence of mullite crystallization and grain size and morphology on solution pH and precursor salt. *J. Mater. Res.*, **19**(4), 1133–1138 (2004).
9. J. W. Kim, J. E. Lee, Y. G. Jung, C. Y. Jo, J. H. Lee, and U. Paik, Synthesis behavior and grain morphology in mullite ceramics *with* precursor pH and sintering temperature. *J. Mater. Res.*, **18**(1), 81–87 (2003).
10. K. Okada and N. Otsuka, Characterization of the spinel phase from SiO_2-Al_2O_3 xerogels and the formation process of mullite. *J. Am. Ceram. Soc.*, **69**(9), 652–656 (1986).
11. A. K. Chakraborty and D. K. Ghosh, Synthesis and 980°C phase development of some mullite gels. *J. Am. Ceram. Soc.*, **71**(11), 978–987 (1988).
12. J. Grofesik and E. Vago, in *Mullite, Its Structure, Formation and Signeficance*, eds. J. Grofesik and F. Tamas, Publishing House of the Hungarian Academy of Sciences, Budapest, Hungary, p. 96 (1961).

13. C. S. Hsi, H. Y. Lu, and F. S. Yen, Thermal behavior of aluminosilica xerogels during calcinations. *J. Am. Ceram. Soc.*, **72**, 2208–2210 (1989).

14. Y. X. Huang, A. M. R. Senos, J. Rocha, and J. L. Baptista, Gel formation in mullite precursor obtained via TEOS prehydrolysis. *J. Mater. Sci.*, **32**, 105–110 (1997).

15. I. Jaymes, A. Douy, P. Florian, D. Massiot, and J. P. Coutures, New synthesis of mullite. Structural evolution study by ^{17}O, ^{27}Al and ^{29}Si MAS NMR spectroscopy. *J. Sol-Gel Sci. Technol.*, **2**, 367–370 (1994).

16. K. S. Mazdiyasni and L. M. Brown, Synthesis and mechanical properties of stoichiometric aluminum silicate (mullite). *J. Am. Ceram. Soc.*, **55**(11), 548–552 (1972).

17. S. Prochazka and F. J. Klug, Infrared-transparent mullite. *J. Am. Ceram. Soc.*, **66**(12), 874–880 (1983).

18. S. Somiya, ed. *Mullite*. Uchida Rokakuho Publishing Co., Tokyo, Japan (1985).

19. Y. Hirata, H. Minamizono, and K. Shimada, Property of SiO_2-Al_2O_3 powders prepared from metal alkoxide. *Yogo Kyokai Shi*, **93**(1), 36–54 (1985).

20. K. Hamano, Z. Nakagawa, G. Cun-Ji, and T. Sato, in *Mullite*, ed. S. Somiya, Uchida Rokakuho Publishing Co., Tokyo, Japan, p. 37 (1985).

21. K. Hamano, T. Sato, and Z. Nakagawa, Properties of mullite prepared by coprecipitation and microstructure of fired bodies. *Yogo Kyokai Shi*, **94**(8), 818–822 (1986).

22. O. Sakurai, N. Mizutani, and M. Kato, Preparation of mullite powders from metal alkoxides by ultrasonic spray pyrolysis. *J. Ceram. Soc. Jpn.*, **96**, 639–645 (1988).

23. M. Low and R. McPherson, The origins of mullite formation. *J. Mater. Sci.*, **24**, 926 (1989).

24. H. Suzuki, H. Saito, Y. Tomokiyo, and Y. Suyama, Processing of ultrafine mullite through alkoxide route, in *Ceramic Transactions*, Vol. 6, Mullite and Mullite Matrix Composites. eds. S. Somiya, R. F. Davis, and J. A. Pask, American Ceramic Society, Westervikke, OH, p. 263 (1987).

25. H. Schneider, I. Merwin, and A. Sebald, Mullite formation from non-crystalline precursors. *J. Mater. Sci.*, **29**, 805–812 (1992).

26. A. Taylor and D. Holland, The chemical synthesis and characterization sequence of mullite. *J. Non-Cryst. Solids*, **152**, 1–17 (1993).

27. H. Schneider, B. Saruhan, D. Voll, L. Merwin, and A. Sebald, Mullite precursor phases. *J. Euro. Ceram. Soc.*, **11**, 87–94 (1993).

28. B. E. Yoldas and D. P. Partlow, Formation of mullite and other alumina-based ceramics via hydrolytic polycondensation of alkoxides and resultant ultra- and micro-structural effects. *J. Mater. Sci.*, **23**, 1895–1900 (1988).

29. C. Gerardin, S. Sundaresan, J. Benziger, and A. Navrotsky, Structural investigation and energetics of mullite formation from sol-gel precursors. *Chem. Mater.*, **6**, 160–170 (1994).

30. A. K. Chakraborty, DTA characterisation of three types of Al_2O_3-SiO_2 gels made from TEOS-Al(OBu)$_3$ mixture with variation of water. *Ceram. Int.*, **22**, 463–469 (1996).

31. M. Sales and J. Alarcon, Synthesis and phase transformations of mullite obtained from SiO_2-Al_2O_3 gels. *J. Euro. Ceram. Soc.*, **16**, 781–789 (1996).

32. B. E. Yoldas, Mullite formation from aluminum and silicon alkoxides. *Ceram. Trans.*, Vol. 6, edited by S. Somiya, R. F. Davis and J. A. Pask, *Am. Ceram. Soc.*, Westerville, OH (1990), p. 255.

33. H. Yamada and S. Kimura, Studies on co-precipitates of alumina and silica gels and its transformations at higher temperatures. *Yogo Kyokai Shi*, **70**, 87–93 (1962).

34. H. Shiga, M. G. M. U. Ismail, and K. Katayama, Sintering of Zr02 toughened mullite ceramics and its microstructure. *J. Ceram. Soc. Jpn.*, **99**, 798–802 (1991).

35. I. Jaymes, A. Douy, and D. Massiot, Synthesis of a mullite precursor from aluminum nitrate and tetraethoxysilane via aqueous homogeneous precipitation: an ^{27}Al and ^{29}Si liquid- and solid-state NMR spectroscopic study. *J. Am. Ceram. Soc.*, **78**(10), 2648–2654 (1995).

36. B. E. Yoldas, A transparent porous alumina. *Ceram. Bull.*, **54**(3), 286–288 (1975).

37. B. E. Yoldas, Thermal stabilization of an active alumina and effect of dopants on the surface area. *J. Mater. Sci.*, **11**, 465–470 (1976).

38. B. E. Yoldas, Microstructure of monolithic materials formed by heat treatment of chemically polymerized precursors in the Al_2O_3-SiO_2 binary. *Ceram. Bull.*, **59**(4), 479–483 (1980).

39. A. K. Chakraborty and D. K. Ghosh, Crystallization behavior of Al_2O_3 in the presence of SiO_2. *J. Am. Ceram. Soc.*, **79**(3), C-46–C-48 (1987).

40. Ph. Colomban, Structure of oxide gels and glasses by infrared and Raman scattering part 1 alumina. *J. Mater. Sci.*, **24**, 3002–3010 (1989).

41. R.-L. Orefice and W.-L. Vasconcelos, Sol-gel transition and structural evolution on multicomponent gels derived from the alumina-silica system. *J. Sol-Gel Sci. Technol.*, **9**, 239–249 (1997).

42. S. Mitachi, M. Matsuzawa, K. Kaneko, S. Kanzaki, and Y. Tabata, Characterization of SiO_2-Al_2O_3 powdes prepared from metal alkoxides, in *Ceram. Trans.*, Vol. 6 (1990), pp. 275–286.

43. L. A. Paulick, Y.-F. Yu, and T.-I. Mah, Ceramic powders from metal alkoxide precursors, in *Advances in Ceramics*, Vol. 21: Ceramic Powder Science (1987), pp. 121–129.

44. J. A. Pask, X. W. Zhang, A. P. Tomsia, and B. E. Yoldas, Effect of sol-gel mixing on mullite microstructure and phase equilibria in the α-Al_2O_3-SiO_2 system. *J. Am. Ceram. Soc.*, **70**(10), 704–707 (1987).

Chapter 4

Mullite Precursors Synthesized by the Colloidal Gelation Method

4.1 Synthesis of General Diphasic Gels

Hoffman and Komarneni (1984) first synthesized a diphasic gel, consisting of two discrete silica (silica rich) and alumina (alumina rich) phases, each noncrystalline in nature in some cases and one component pseudocrystalline (as in boehmite) in other cases. It was synthesized by mixing aqueous silica sol with boehmite sol dispersed in ethanol. These gels yielded no differential thermal analysis (DTA) exotherm at 960°C but a slight peak at about 1250°C. X-ray diffraction data for the samples showed the presence of boehmite in the gel, which formed a spinel phase above 1000°C. Mullitization was first achieved at ~1300°C. In comparison to monophasic gel, the diphasic gel behaved differently in its phase transformation sequence. The phase transformation process of the diphasic gel as per them is shown in Table 4.1.

Hamano et al. (1985, 1986) mixed boehmite and silica sols to prepare mullite powders so that the stoichiometric composition of mullite was maintained. The boehmite sol was prepared from aluminum ethoxide by hydrolysis/condensation reactions, using HCl as a catalyst. The silica sol was prepared by hydrolysis/condensation of methyl silicate. The two sols were then mixed and

Mullite Formations: Analysis and Applications
Akshoy Kumar Chakraborty

ISBN 978-981-4877-05-3 (Hardcover), 978-1-003-03167-3 (eBook)
www.jennystanford.com

subsequently dried by slowly adding the mixture to a hot (160°C) petroleum bath. X-ray diffractometry (XRD) results indicated the presence of a weakly crystalline Al-Si spinel phase in powders calcined in the 1000°C–1250°C range and the initial formation of mullite began at about 1250°C. However, this process apparently did not result in homogeneous mixing of the alumina and silica, since XRD results showed that calcined powders contained cristobalite in the temperature range of 1250°C–1600°C and α-Al_2O_3 in the temperature range of 1300°C–1700°C.

Table 4.1 Phase evolution of diphasic mullite gel prepared out of two discrete sols in aqueous medium (pH unknown) on heating

Temperature	**Phase**
Diphasic xerogels prepared by mixing aqueous silica sol and boehmite sol followed by evaporation at room temperature	X-ray showed boehmite and a band pertaining to silica (A).
At 400°C	DTA showed an endotherm. Decomposition of boehmite occurred.
At 1010°C	No trace of exotherm was observed. The XRD pattern, according to Hoffman and Komarneni (1984), contained γ-alumina and cristobalite. They claimed that two discrete phases are reacting independently up to ~1000°C. But in reality, the pattern did not contain any peaks relating to cristobalite (see comments).
At 1330°C	A broad exotherm was noted at 1200°C; X-ray showed well-crystallized mullite with traces of corundum.

Ismail et al. (1986, 1987) mixed aqueous silica (~130 m^2/g specific surface area) and boehmite (aluminum monohydroxide) sols under acidic conditions (pH ~1.8), followed by gelling through solvent evaporation. The boehmite sol was prepared by hydrolysis of γ-Al_2O_3 (~68 m^2/g specific surface area) using a vigorously agitated nitric acid solution at 95°C. The finer particle size of the boehmite was indicated by the high specific surface area (277m^2/g) of the dried gel mixture. Thereafter, the gel was evaporated to a state of dryness and powdered. DTA of this diphasic gel showed an exotherm at 1296°C due to mullite formation. After calcination to 1400°C for 1 h, they found the specific surface area to have decreased from 277 m^2/gm to 1.7 m^2/gm. The mullitization routes for these samples were essentially the same as for those mullite-type phases prepared with about 74 wt % of Al_2O_3 and 26 wt % of SiO_2 from tetraethyl orthosilicate (TEOS) and aluminum butylate between 950°C and 1000°C. The most significant characteristic of the X-ray patterns was the lack of angular separation of the reflection pairs 120/210, 240/420, 041/401, and 250/520. The lack of reflection splitting of the doublets means that the symmetry was tetragonal instead of orthorhombic. The pseudotetragonal mullite developed when highly reactive metal organic compounds were used. The metastable tetragonal phase gradually transformed into orthorhombic mullite at a temperature above 1000°C.

Shinohara et al. (1986) mixed a colloidal boehmite suspension and a partially hydrolyzed TEOS solution at pH 2.5 for ~4 days. The mixture was then heated at 60°C and cast into a plastic mold just prior to gelation. Samples were sintered at temperatures in the 1100°C–1600°C range. Samples with a stoichiometric mullite composition sintered to ~94% relative density after 1 h at 1250°C, but no additional densification was observed with longer sintering times (up to 8 h). The cessation of densification indicated the formation of mullite. XRD showed that a sample sintered at 1200°C for 2 h was primarily θ-Al_2O_3, with small amounts of mullite and cristobalite, whereas only mullite appeared when the sample was sintered at 1250°C for 2 h. These results clearly illustrate the detrimental effect of mullitization on the densification process. Even after sintering at 1600°C, the relative density increased only to ~97%.

Sonuparlak (1988) used colloidal boehmite suspensions and TEOS to prepare mullite by the sol-gel process. Boehmite powders were initially dispersed in water under acidic conditions (pH < 2). TEOS was then added to the sol in the mullite compositional ratio. The sol was converted to gel at 25°C in air. The gel monoliths were air-dried very slowly for a considerable length of time (more than

2 months) at room temperature in covered molds with pinhole vents to prevent cracking. Dried samples when sintered at 1250°C for 1 h showed 96% relative density, but no additional densification occurred when the sample was heated up to a temperature of 1700°C. He also prepared a gel with the same procedure but under pH 6–7 using ammonium hydroxide. This resulted in destabilization of the sol, as large flocks were observed to form immediately. The relative density of the sample after sintering (at 1250°C for 1 h) increased with increasing compaction pressure, reaching a maximum value of ~97%.

Huling and Messing (1989, 1990) synthesized two diphasic gels, marked Gel A and Gel B, using identical boehmite and silica sols but different gelation procedures. They designed to promote heterocoagulation between boehmite powder and silica sol and maintained the pH at 3 by adding HNO_3 to obtain maximum homogeneity of the mixed gel marked Gel A. Gel B was designed to develop heterogeneous precursor mixing by controlling the pH. In the first stage, colloidal silica in the acidic stabilization form was prepared from Ludox by changing the pH to nearly 2 by adding HNO_3. The acidic sol was added to boehmite sol at pH 3 for gelation. In this method of heterocoagulation, silica particles are not discrete, as in Gel A, but in small clusters. Such cluster formation needs to be checked. The crystalline phases detected by them on heating are shown in Table 4.2.

Heterocoagulation between boehmite and silica particles yielded homogeneous Gel A that transformed to mullite between 1270°C and 1300°C. In contrast, the heterogeneous precursor mixing that occurred in Gel B when the pH was varied led to preferential crystallization of components prior to mullitization. But the quantity of each phase versus temperature was not reported. According to Huling and Messing (1989, 1990), the time and temperature required to complete mullitization were dependent upon the gel's homogeneity. Gel A transformed fully to mullite between 1270°C and 1300°C without component crystallization. In Gel B, containing silica clusters, individual crystallization of components as well as mullite formation occurred between 1270°C and 1350°C. The variation in the sequence of phase changes showed a difference in the microstructural development.

Table 4.2 Comparative phase transformations behavior of homogeneous diphasic mullite gel (Gel A) and hetero coagulated gel (Gel B) on heating

	Gel A
Temperature	**Phase**
Below 1250°C	It contained δ- and θ-Al_2O_3 stabilized by silica.
At 1270°C	Mullitization occurred. DTA revealed a small exothermic peak at 1300°C–1350°C. Complete mullite formation occurred at and above 1300°C. The smooth transition of phases seemed to be consistent with the homogeneous mixture of colloidal precursors on the nanometer scale.
	Gel B
Temperature	**Phase**
At ~1250°C	Different to the phase transformation in Gel A, cristobalite formation took place in addition to the formation of δ- and θ-Al_2O_3.
At 1270°C	Crystallization of corundum took place in addition to the maximum amount of cristobalite. Mullite formation started. Some amount of δ- and θ-Al_2O_3 still remained. DTA showed exotherm due to mullite formation, which was 20°C higher than what occurred in Gel A.
At 1300°C	Trace amounts of θ-Al_2O_3, δ-Al_2O_3, and cristobalite were present. The formation of both mullite and corundum increased.
At 1350°C	θ-Al_2O_3 and δ-Al_2O_3 disappeared. The maximum amount of corundum formed. Mullitization proceeded.
At 1550°C	Complete mullitization occurred.

Klaussen et al. (1990) synthesized a mullite gel using pseudoboehmite in the form of a powder and colloidal silica in the form of a suspension. The former component was first peptized in water at a pH between 1 and 2 and to it, colloidal silica was added and finally the pH was adjusted to obtain a gel. It was dried at 100°C to make a powder. DTA of pseudoboehmite showed an exothermic peak at 1216°C. The DTA curve of the gel powder ran at 5°C/min.

and exhibited three exothermic peaks, 1188°C due to δ-alumina, 1254°C due to θ-alumina, and 1328°C due to the crystallization of mullite. However, the DTA curve of the gel powder ran at 15°C /min. and exhibited one exothermic peak, due to mullite at 1354°C. All other crystallization peaks due to transitional alumina had disappeared. The sequential development of the three phases were analyzed by XRD analysis of the gel powder heated to 1100°C, 1200°C, 1250°C, and 1300°C, with different holding times. δ-alumina formed at 1100°C. θ-alumina was identified at 1200°C and existed with δ-alumina even at 1300°C. On the contrary, Wei and Halloran (1988) noted the formation of δ-alumina and θ-alumina at lower temperatures (850°C and 1150°C, respectively). The results of these two researchers show that crystallization temperatures of two alumina modifications have definitely increased. Mullite started forming at 1250°C, and it showed a major jump at 1300°C, which nearly coincides with results of the DTA experiment.

Wei and Halloran (1988) synthesized a translucent white-type diphasic gel from TEOS and colloidal pseudoboehmite sol in acidic condition at 50°C and aged for 24 h. The transformation sequence in the presence of amorphous silica is as follows:

1050°C–1200°C

$$\gamma\text{-}Al_2O_3^- \xrightarrow{1050°C-1200°C} \delta\text{-}Al_2O_3 \xrightarrow{1050°C-1200°C} \alpha\text{-}Al_2O_3 \quad (4.1)$$

The silica component did not crystallize even at 1200°C. According to them, mullite forming by a solid-state reaction was conjectured by Hoffman et al. (1984).

1300°C

$$\gamma\text{-}Al_2O_3 + SiO_2(A) \xrightarrow{1300°C} \text{Mullite} \quad (4.2)$$

Ismail et al. (1986) did not observed cristobalite formation when the gel was prepared from boehmite sol and colloidal silicon dioxide at pH 1.8. Klaussen (1990) prepared a diphasic gel at pH between 1 and 2 from pseudoboehmite powder and colloidal SiO_2. The gel showed exotherms at 1188°C and 1254°C due to the transition of δ-Al_2O_3 and θ-Al_2O_3 and thereafter formed mullite at 1328°C. The diphasic gel marked Gel A prepared by Huling and Messing (1989) also transformed as per the sequence shown by Wei and Halloran (1988).

Hyatt and Bansal (1990) prepared a diphasic gel by using AlO(OH) and Ludox in the presence of HNO_3. The DTA curve shows two endotherms due to the removal of physically adsorbed water and decomposition of boehmite and one exotherm between 1300°C and 1350°C due to mullite formation. According to them, the phase transformation of this diphasic gel occurred as per the following XRD data: It is observed from the transformation equations that a fraction of component oxides were independently crystallized. This might be due to the use of HNO_3 during the gelation process. As a result, their gel was opaque in appearance (see process chart). At a high temperature, mullite was formed between α-Al_2O_3 and β-cristobalite. A similar observation was made by Hamano et al. (1986). They noted β-cristobalite between 1250°C and 1600°C and α-Al_2O_3 between 1300°C and 1700°C and thus confirm the true diphasic route in their crystallization behavior. However, the mode of phase transformation of this diphasic gel is completely different from the modes described by Wei and Halloran (1988) and Ismail et al. (1986). At pH 2, on the other hand, it formed mullite at ~1300°C. The cause of the variation in the phase transformation behaviors of the diphasic gel made by various researchers was not explained as yet. It may have been due to a variation in pH during gelation in a wide range (2–7).

$$\text{Diphasic gel} \xrightarrow{400^\circ\text{C}} \begin{matrix}\text{Al}_2\text{O}_3\text{(A)}\\ +\ \text{SiO}_2\text{(A)}\end{matrix} \xrightarrow{900^\circ\text{C}-1350^\circ\text{C}} \begin{matrix}\alpha\text{-Al}_2\text{O}_3\text{(via }\delta\text{-Al}_2\text{O}_3\\ \&\ \theta\text{-Al}_2\text{O}_3)\\ +\ \beta\text{-crystobalite}\end{matrix} \xrightarrow{1600^\circ\text{C}} \begin{matrix}\text{Mullite (major)}\\ \alpha\text{-Al}_2\text{O}_3\text{ (minor)}\\ \beta\text{-cristobalite}\\ \text{(disappeared)}\end{matrix} \quad (4.3)$$

On the contrary, Sonuparlak (1988) showed quite different phase transformation behaviors when the diphasic gel was processed under acidic versus basic conditions, as follows:

Acidic pH

$$\text{Diphasic gel} \xrightarrow[\text{at} \sim 1250°\text{C}]{\text{Densified}} \gamma\text{-Al}_2\text{O}_3 + \text{Glassy phase} \xrightarrow[\text{1 h}]{1300°\text{C}} \text{Spinel} + \text{Mullite (minor)} \quad (4.4)$$

(prepared at ~1250°C + 1 h + acidic pH) Amorphous phase + Mullite (minor)

Basic pH

$$\text{Diphasic gel} \xrightarrow[\text{at} \sim 1250°\text{C}]{\text{Densified}} \gamma\text{-Al}_2\text{O}_3 + \text{Glassy phase} \xrightarrow[\text{1 h}]{1300°\text{C}} \text{Mullite} \quad (4.5)$$

(prepared at ~1250°C + 1 h + basic pH) Amorphous phase

This reaction shows that densification occurred before mullitization. Wei and Halloran (1988) noted densification at ~1190°C after 72 h, Komarneni et al. (1985), of course, first observed a similar phenomenon. It was explained that this early densification mechanism is due to viscous phase development, as noted by scanning electron microscope (SEM) study. If that is true, then the question arises about how mullitization occurs. No one has considered the role of viscous phase formation before mullitization. The glassy phase should not be pure silica phase alone. It may be an aluminosilicate when the solid-state reaction between amorphous silica and alumina is practically feasible because of the possibility of interdiffusion of two cations during heating. It might not be ruled out that Si would substitute Al in the γ-Al_2O_3 structure during its formation. The striking factor is that if the above possibility at all materializes, the resultant spinel phases (may be a mixture of γ-alumina and Si-substituted spinel, depending upon the time and the temperature of heating) could not be differentiated by the XRD technique due to identical fingerprints.

Pach et al. (1995) synthesized diphasic $3Al_2O_3$-$2SiO_2$ mullite gel using boehmite (10 nm) and silica sol (12 nm). The sol mix was peptized by mixing with HNO_3 (pH 2) at 55°C. The phase composition, relative density, and microstructures were studied by different techniques. Intensive sintering followed by crystallization of mullite occurred in a narrow temperature range for diphasic gels. Here, SiO_2 particles lost their identities and formed SiO_2 continuum. This behavior took place in unpressed gels at 1240°C and in pressed (1.5 GPa) ones at 1220°C. If the annealing temperature of the gel was about 20°C lower, sintering and crystallization of mullite were significantly retarded and mullite crystals were not detectable even after annealing for 4 h at 1220°C and 1180°C in the case of unpressed and pressed gels, respectively. Nucleation of mullite appeared to occur at points of contact (interface) of SiO_2/γ-Al_2O_3 particles.

Wu et al (1993) synthesized an alumina-silica gel of stoichiometric mullite ($3Al_2O_3$-$2SiO_2$) composition by using submicrometer alumina powder and Ludox colloidal silica sol as precursors. The pH of Ludox sol was adjusted from 9 to 2–3 using dilute HCl (1 M). Alumina powder was dispersed in the sol by magnetic stirring. Dilute HCl was again added to maintain the pH at 5–6 so that gelation occurred in 30 min. The bulk gel was slowly dried at room temperature, then calcined at 500°C for 2 h, and finally sintered. Bulk alumina-silica gels were almost fully densified (over 98% of the theoretical) density at 1300°C and almost completely mullitized at 1600°C. The low sintering temperature according to them was due to the viscous sintering of silica in the presence of alumina. Unidirectional fiber composites were synthesized by single-stage infiltration of carbon fibers with the alumina-silica sol. The composite was hot-pressed to over 97% of full theoretical density at 1400°C and exhibited nonbrittle failure with a mean flexural strength of 766 MPa.

4.2 Synthesis of "in situ" Diphasic Gels

Instead of using boehmite sol and TEOS or Ludox, as used by several earlier researchers, for example, Hoffman et al. (1984), a new technique of synthesizing diphasic gel was used by the author (1979). Boehmite was developed "in situ" during the gelification process itself from Ludox and aluminum nitrate nonahydrate (ANN) in an ammonia solution. It was heat-treated at 1000°C and leached with alkali to solubilize free silica content. The phase transformation process of this gel will be discussed in Chapter 15.

4.3 Synthesis of Other Diphasic Gels by Colloidal Methods

McAtee and Milligan (1950) synthesized two series of mullite gels. They first prepared the respective gels by a solution precipitation technique and thereafter ground the gel mixtures for intimate mixing. In the case of series I, aluminum hydroxide gel was precipitated from the ANN solution by the use of ammonium hydroxide solution. Silica gel was prepared from sodium metasilicate with the help hydrochloride acid. Both gels were washed to remove soluble impurities by centrifugation. In the case of series II, alumina gel and silica gel were prepared from aluminum isopropoxide and silicon ethoxide by a hydrolysis/condensation process. Mixed gels of stoichiometric mullite composition (i.e., 60 mol % of Al_2O_3/40 mol % of SiO_2; ~71.8 wt % of Al_2O_3/28.2 wt % of SiO_2) on heating at 1750°C for 2 h indicated the presence of mullite with unconverted/unreacted α-Al_2O_3 and β-cristobalite phases.

Ghate et al. (1973) prepared alumina sol by dispersing fume alumina γ-Al_2O_3 (~100 m^2/g specific surface area) particles in hydrochloric acid solution. Silica sol (~230 m^2/g specific surface area) was added slowly to this suspension, and the pH was adjusted to 6–7 with NH_4OH to produce powders having a stoichiometric mullite composition (i.e., Al_2O_3:SiO_2 in the mol ratio of 3:2). In this pH range, the surfaces of alumina particles are positively charged whereas those of silica particles are negatively charged. These studies suggest that heterocoagulation resulted from a decrease in stability (against flocculation) of both the alumina and silica suspensions in the pH range of 6–7. The gel pieces, when dried (120°C), crushed, and sieved, crystallized into mullite on heating to 1200°C. Complete mullitization took place at 1400°C after 20 h. Corundum did not appear at any stage of heating. At an early stage of heating, cristobalite was detected and that too disappeared on heating for long. They conjectured that fume alumina, γ-Al_2O_3 structure, accepted silica by the diffusion process and subsequently rearranged to an intermediate mullite-like phase and steadily transformed to well-crystallized mullite at ~1400°C. The particles of mullite were submicron in size, globular (as against the common, needlelike morphology), and of 99.9% pure.

Sacks and Pask (1982a, 1982b) mixed the aqueous suspensions of γ-Al_2O_3 and Ludox by vigorous mechanical agitation and by electrostatic attraction of the oppositely charged alumina and silica particles. They synthesized the gel by pH adjustment and finally evaporated the excess moisture by heating. For complete mullitization, they heat-treated this gel to a temperature of 1450°C for 24 h.

Transformation of mullite powder prepared from aluminum sulfate and fumed silica is given by Mizuno and Sato (1989). DTA and a thermogravimetric analysis showed a large endotherm at around 860°C, corresponding to the decomposition of aluminum salt, as per Eq. 4.6.

Dehydration Decomposition

$$Al_2(SO_4)_3{\cdot}18H_2O\ Al_2(SO_4)_3 \xrightarrow{\text{At } 860°C} \gamma\text{-}Al_2O_3 \quad (4.6)$$

This fine γ-Al_2O_3 may start a solid-state reaction with SiO_2. DTA trace beyond 1200°C showed an upward shift, indicating the evolution of heat energy with crystallization. XRD analysis showed the start of formation of poorly crystalline mullite at 1200°C in the presence of unconverted γ-Al_2O_3. Predominant mullitization occurred at 1250°C. On continued heating up to 1700°C, the XRD peak of the mullite peak sharpened. The formation of mullite, which occurred at around 1250°C, agrees with the finding of Ghate et al. (1973). The mullite powder developed at 1350°C is agglomerated which on attrition milling ground to fine powder composed of uniform equiaxial particles 0.1–0.2 µm in size, as shown in the transmission electron microscopy (TEM) photograph in the work by Sacks et al. (1990, p. 168).

4.4 Synthesis of Solid-Gel-Coated Precursor by Microcomposite Methods

- This may either be pure alumina core (α-alumina) or pure γ-Al_2O_3 core, upon which coating was done. Sacks et al. (1991) prepared a coated precursor by using fine α-Al_2O_3 as the core material coated with hydrated silica obtained from the hydrolysis of TEOS. Later on, microcomposites of varying alumina/silica ratios [α-Al_2O_3/silica (A)] were synthesized and monitored by DTA, XRD, and SEM technique by Sacks et al. (1997). The DTA plots of microcomposite powders having overall alumina:silica weight ratios of ~66:34, ~74:26, and ~83:17 are shown in Fig. 2 of Sacks et al. (1997). The endothermic peak temperatures of the composites were 1518°C, 1499°C, and 1489°C, respectively. This result showed that the temperature of mullitization reaction decreased as the alumina:silica weight ratio increased.

Question 1: Why is there a decrease in crystallization temperatures?

Sacks and coworkers (1991, 1995, 1997) showed that the microstructure of the sintered microcomposite at ~1300°C consisted of alumina particles distributed in a dense, continuous siliceous matrix. Mullite was seen to rapidly form during heating between 1425°C and 1450°C in the first step and rather slowly in the last stage, between 1500°C and 1600°C. Microstructure evolution studies indicated that the first step of mullite formation occurred primarily by nucleation and growth within the siliceous phase. The remaining alumina particles may have reacted with residual silica through interdiffusion.

Question 2: Why is there rapid mullitization between 1425°C and 1450°C?

Sacks et al. (1995) showed that with the incorporation of 2 wt % of mullite seed in a green powder microcomposite, extensive mullitization occurred at ~1400°C and the reaction was essentially completed. Sacks et al. (1997) investigated the effect of the alumina:silica ratio of silica-coated α-alumina composite powder on the mullitization behavior—rather the determination of the reaction rate. Low-temperature mullite formation in AS-γA nanocomposites was explained as controlled by dissolution-precipitation reactions. A similar mechanism was valid for mullitization in the diphasic aluminosilicate gel system as proposed by Sundaresan and Aksay (1991). According to these authors, alumina species dissolved in the coexisting SiO_2 liquid until a critical alumina concentration was reached. At this stage, random mullite nucleation ensued in the alumina-rich silica phase. Thus, the dissolution velocity of alumina into the SiO_2 matrix may be the rate-controlling step for mullite nucleation and crystal growth.

- Bartsch et al. (1999) made experimentations the results of which given in Table 4.3.

Table 4.3 Comparative mullite formation and densification studies of As-γ-Al_2O_3 and As-α-Al_2O_3 nanocomposites

Nanocomposite (AS-γA)	
As received	γ-Al_2O_3 as crystalline phase (~10 nm)/High amorphous background due to SiO_2(A) thinner coating (1–2 nm).
At 1275°C	XRD pattern consists of γ-Al_2O_3/some θ-Al_2O_3/ amorphous band.
At 1285°C	Major intensity of mullite + traces of γ-Al_2O_3.
At 1300°C	Completion of mullite formation likely.
At 1360°C	DTA exhibits exotherm.
Microcomposite (AS-αA)	
As received	α-Al_2O_3 (500 nm)/Amorphous background due to SiO_2(A) coating (25–30 nm thicker) as per TEM observation.
At 1400°C	Cristobalite formation starts, mullite does not form. Compact sample shows high-density >90% TD by TVS mechanism as per Sacks et al. (1991).
1425°C–1475°C	Extensive cristobalite formation occurs. Mullite formation starts.
At 1575°C	The intensities of both cristobalite and α-Al_2O_3 decrease sharply. In lieu, the intensity of mullite increases to a large extent and complete mullitization occurs at 1600°C.

- Monolithic disc samples of the mullite composition using Ludox and α-alumina were made. The powders were synthesized by processing at the final pH of 5.5 by Wu et al. (1993, 1996).

Table 4.4 Phase transformation of composite powder prepared at acidic pH and studied by TEM

Temperature	Phase
At the first stage at ~1300°C	A high level of densification was reported by density measurement and by TEM study.
At the second stage, in the 1300°C–1400°C range	Amorphous colloidal silica component of the composite crystallized.
At 1500°C	A significant amount of mullite crystallized at the expense of α-alumina present already in the composite and with newly formed cristobalite. Unreacted components were fully reacted to developed dominant crystallization of mullite, as observed in both XRD and TEM studies.

Thus the XRD pattern of 74 wt % of alumina/26 wt % of silica powder compact showed corundum when sintered at 1300°C for 2 h. An SEM micrograph of a polished and unetched sample showed almost full densification. At 1400°C, partial cristobalite formation occurred without any crystallization of mullite. At 1500°C, sharp, extensive mullite formation took place. Complete mullitization occurred at and above 1600°C.

4.5 Summary

Hoffman et al. (1984) termed a colloidal gel prepared from aqueous silica sol/Ludox and boehmite sol a "diphasic gel." Since this study, many researchers have turned their focus to this topic of research. A critical review of the thermal sequence presented above has shown significant variations. The reaction process of diphasic gel to noncrystalline aluminosilicate precursor phase to spinel to mullite showed two significant aspects. The phase transformation sequence up to mullitization constitutes one aspect of research. The crystallization of phases and the densification characteristics of the diphasic gel are another aspect of study. The latter phenomenon is more important in the diphasic gel in view of obtaining sintered mullite ceramic. Phase transformations of diphasic gels synthesized by various researchers by choosing components from different sources and using different techniques vary to a great extent.

- On the basis of the sequence of transformation of two components: It was shown that a fraction of boehmite component crystallized following its own path. It first developed into γ-alumina, then into δ-alumina, then into θ-alumina, and finally into its last polymorph. Hoffman et al. (1984), Hyatt and Bansal (1990), Hamano et al. (1985, 1986), and Huling and Messing (1989) in their Gel B noted a few polymorphs of alumina prior to mullite formation. Contrary to monophasic (single phasic) gel, diphasic gel did not exhibit DTA exotherm at 960°C but formed the spinel phase, which was assumed by Hoffman et al. (1984) to be γ-alumina. According to them, the latter phase transformed into mullite by a solid-state reaction with cristobalite at ~1300°C following a different path, as noted in monophasic gel. XRD results of Hamano et al. (1985) noted the crystallization of components of their diphasic gels, such as cristobalite, in the temperature range of 1250°C–1600°C and α-alumina at 1300°C–1700°C, Hyatt and Bansal (1990) showed one exotherm in the 1300°C–1350°C range and explained it as due to mullite formation in their diphasic gel. Prior to mullitization, they observed partial crystallization of three polymorphs of the alumina component and crystallization of β-cristobalite of the silica component. Mullite crystallization started at ~1350°C and continued all along with the crystal phases of the two components up to 1600°C.
- On the basis of the development of a significant amount of glassy phase prior to mullitization: The following researchers noted early densification and subsequent crystallization of mullite: Komarneni et al. (1985), Shinohara et al. (1986), Sonuparlak (1988), Pach et al. (1995), and Wu et al. (1993, 1996). Komarneni et al. (1985) and Shinohara et al. (1986) noted the densification of their diphasic gel samples at ~1250°C when heated for 1 h. Further densification did not occur even with an increase in the sintering time, which was explained by them as due to crystallization of a small quantity

of mullite. Sonuparlak (1988) obtained 96% of relative density on sintering the sample at 1250°C. However, he showed quite different phase transformation behaviors when the diphasic gel was processed in an acidic condition and when it was processed in a basic condition. The alumina transformation sequence in the presence of amorphous silica is delayed. Pach et al. (1995) noted intensive densification followed by mullitization of their diphasic gel in a narrow temperature range. The Unpressed gel showed sintering at 1240°C. In comparison, the pressed gel (1.5 GPa) densified at 1220°C. Wu et al. (1993), of course, noted a theoretical density of 98% of their diphasic gel at ~1300°C, due to viscous sintering of silica in the presence of alumina.

- Contrary to the thoughts of Hoffman et al. (1984), during the phase transformation of synthesized diphasic gel, its components did not crystallize following their own path. For example, cristobalite did not crystallize at all. Traces of corundum did crystallize far above its independent path of formation. Two questions arise: Why does cristobalite not form? And why does corundum not crystallize to its full extent at 900°C–1200°C?
- Incorporation by diffusion of SiO_2 into alumina structure during heating is a universal process and verified by the pioneering studies of Yoldas (1976) and others. The consequences of the substitution phenomenon and the reasons of noncrystallization of the components of the diphase gel in some instances will be explained in later chapters.

Problems

1. What might be reason for viscous phase development during the heating of a diphasic gel system?
2. The transformation processes of diphasic gels synthesized by earlier researchers, like Hoffman et al. (1984) and Wei and Halloran (1988), vary to a great extent. The reason might be changes in reaction conditions, particularly, the pH of the gelation process and the role of sizes of component particles. Explain.

3. What is the cause of early densification, and what is the mechanism of densification? Is it due to the viscous flow of amorphous silica?
4. Compare the densification process of a microcomposite powder with that of the diphasic gel precursor shown by Sonuparlak (1988). Show qualitative data of mullite formation versus the temperature of heat treatment of the diphasic gel.
5. Explain the differences in mechanisms of mullitization in a diphasic gel and a microcomposite powder.
6. Are the formation of siliceous glassy phase and its composition depend upon alumina to silica ratio? Will the amount glassy phase change with the rise in temperature of heat treatment?

References

1. D. W. Hoffman, R. Roy, and S. Komarneni, Diphasic xerogels, a new class of materials: phases in the system Al_2O_3-SiO_2. *J. Am. Ceram. Soc.*, **67**, 468–471 (1984).
2. K. Hamano, Z. Nakagawa, G. Cun-Ji, and T. Sato, in *Mullite*, ed. S. Somiya, Uchida Rokakuho Publishing Co., Tokyo, Japan, p. 37 (1985).
3. K. Hamano, T. Sato, and Z. Nakagawa, Properties of mullite prepared by coprecipitation and microstructure of fired bodies. *Yogo Kyokai Shi*, **94**(8), 818–822 (1986).
4. K. Okada, Comment on diphasic xerogels, a new class of materials: phases in the system Al_2O_3-SiO_2. *J. Am. Ceram. Soc.*, **68**(3), C-85 (1985).
5. A. K. Chakraborty and D. K. Ghosh, Comment on diphasic xerogels, a new class of materials: phases in the system Al_2O_3-SiO_2. *J. Am. Ceram. Soc.*, **69**(8), C-202–C-203 (1986). Reply by S. Komarneni and R. Roy, ibid., **69**(8), C-204 (1986).
6. M. G. M. U. Ismail, Z. Nakai, K. Minegishi, and S. Somiya, Synthesis of mullite powder and its characteristics. *Int. J. High Technol. Ceram.*, **2**, 123–134 (1986).
7. M. G. M. U. Ismail, Z. Nakai, and S. Somiya, Microstructure and mechanical properties of mullite prepared by the sol-gel method. *J. Am. Ceram. Soc.*, **70**(1), C-7–C-8 (1987).
8. N. Shinohara, D. M. Dabbs, and I. A. Aksay, Infrared transparent mullite through densification of monolithic gels at 1250°C. *Proc SPIE*, **0683**, 19–24 (1986).

9. B. Sonuparlak, Sol-gel processing of infrared transparent mullite. *Adv. Ceram. Mater.*, **3**(3), 263–267 (1988).
10. J. C. Huling and G. L. Messing, Hybrid gels for homoepitactic nucleation of mullite. *J. Am. Ceram. Soc.*, **72**(9), 1725–1729 (1989).
11. J. C. Huling and G. Messing, Surface chemistry effects on homogeneity and crystallization of colloidal mullite gels, in *Ceram. Trans.* vol. 6, ed. S. Somiya, R. F. Davis, and J. A. Pask, p. 221 (1990).
12. G. Klaussen, Microstructural evolution of sol-gel mullite. *Ceram. Eng. Sci. Proc.*, **11**, 1087–1093 (1992). K. Okada, Comment on diphasic xerogels, a new class of materials: phases in the system Al_2O_3-SiO_2. *J. Am. Ceram. Soc.*, **68**(3), C-85 (1985).
13. W.-C. Wei and J. W. Halloran, Phase transformation of diphasic aluminosilicate gels. *J. Am. Ceram. Soc.*, **71**(3), 166–172 (1988).
14. J. Hyatt and N. P. Bansal, Phase transformations in xerogels of mullite composition. *J. Mater. Sci.*, **25**, 2815–2821 (1990).
15. S. Komarneni, R. Roy, C. A. Fyfe, and G. J. Kennedy, Preliminary characterization of gel precursors and their high- temperature products by 27 Al magic-angle spinning NMR. *J. Am. Ceram. Soc.*, **68**(9), C-243–C-245 (1985).
16. L. Pach, A. Iratni, Z. Hrabe, S. Svetik, and S. Komarneni, Sintering and crystallization of mullite in diphasic gels. *J. Mater. Sci.*, **30**, 5490–5494 (1995).
17. J. Wu, M. Chen, F. R. Jons, and P. F. James, Mullite and alumina-silica matrices for composites by modified sol-gel processing. *J. Non-Cryst. Solids*, **162**, 197–200 (1993).
18. A. K. Chakraborty, Formation of silicon-aluminium spinel. *J. Am. Ceram. Soc.*, **62**, 120 (1979).
19. J. L. McAtee and W. O. Milligan, X-ray diffraction examination of synthetic mullite. *Tex. J. Sci.*, **2**, 200–205 (1950).
20. B. B. Ghate, D. P. H. Hasselman, and R. M. Spriggs, Synthesis and characterization of high purity, fine grained mullite. *Ceram. Bull.*, **52**(9), 670–672 (1973).
21. M. D. Sacks and J. A. Pask, Sintering of mullite containing materials: I, Effect of composition. *J. Am. Ceram. Soc.*, **65**, 65–70 (1982a).
22. M. D. Sacks and J. A. Pask, Sintering of mullite containing materials: I, Effect of agglomeration. *J. Am. Ceram. Soc.*, **65**, 70–77 (1982b).
23. M. Mizuno and H. Saito, Preparation of highly pure fine mullite powder. *J. Am. Ceram. Soc.*, **72**, 377–382 (1989).

24. M. D. Sacks, H. Lee, and J. A. Pask, A review of powder preparation methods and densification procedures for fabricating high density mullite. *Ceram. Trans.*, Vol. 6, ed. S. Somiya, R. F. Davis, and J. A. Pask, p. 167 (1990).

25. M. D. Sacks, N. Bozkurt, and G. W. Scheiffele, Fabrication of mullite and mullite-matrix composites by transient viscous sintering of composite powders. *J. Am. Ceram. Soc.*, **74**(10), 2428–2437 (1991).

26. M. D. Sacks, Y.-J. Lin, G. W. Scheiffele, K. Wang, and N. Bozkurt, Effect of seeding on phase development, densification behavior, and microstructure evolution in mullite fabricated from microcomposite particles. *J. Am. Ceram. Soc.*, **78**(11), 2897–2906 (1995).

27. M. D. Sacks, K. Wang, G. W. Scheiffele, and N. Bozkurt, Effect of composition of mullitization behavior of α-alumina/silica microcomposite powders. *J. Am. Ceram. Soc.*, **80**(3), 663–672 (1997).

28. S. Sundaresan and I. A. Aksay, Mullitization of diphasic aluminosilicate gels. *J. Am. Ceram. Soc.*, **74**(10), 2388–2392 (1991).

29. M. Bartsch, B. Saruhan, M. Schmucker, and H. Schneider, Novel low-temperature processing route of dense mullite ceramics by reaction sintering of amorphous SiO_2-coated γ-Al_2O_3 particle nanocomposites. *J. Am. Ceram. Soc.*, **82**(6), 1388–1392 (1999).

30. J. Wu, M. Chen, F. R. Jones, and P. F. James, Mullite and alumina-silica matrices for composites by modified sol-gel processing. *J. Non-Cryst. Solids*, **162**, 197–200 (1993).

31. J. Wu, M. Chen, F. R. Jones, and P. F. James, Characterization of sol-gel derived alumina-silica matrices for continuous fibre reinforced composites. *J. Euro. Ceram. Soc.*, **16**, 619–626 (1996).

32. B. E. Yoldas, Thermal stabilization of an active alumina and effect of dopants on the surface area. *J. Mater. Sci.*, **11**, 465–470 (1976).

Chapter 5

Mullite Precursors Synthesized by Codecomposition Processes

5.1 Introduction

With alkoxides as starting component sources, a large number of investigations have been done by several researchers to synthesize mullite gels, as described in Chapters 2–4. They faced difficulties in synthesizing chemically homogeneous high-purity mullite precursors. Very stringent processing parameters in terms of, for example, pH, the ratio of water to alkoxide, and the use of an acid or a base as a catalyst, have to be followed. Achieving the control to promote simultaneous hydrolysis of two alkoxides and homogeneous copolymerization is a rather difficult task. How does one avoid localized high water concentration? With finding the answer to this question as the goal, a group of researchers has pursued another method, where the group has replaced aluminum alkoxide with water-soluble aluminum nitrate nonahydrate (ANN). The controlled addition of water is not important; one can use excess water for the hydrolysis of tetraethyl orthosilicate (TEOS) as an additional advantage.

Mullite Formations: Analysis and Applications
Akshoy Kumar Chakraborty

ISBN 978-981-4877-05-3 (Hardcover), 978-1-003-03167-3 (eBook)
www.jennystanford.com

5.2 Decomposition Method

5.2.1 Spray-Pyrolysis Route

Kanzaki et al. (1985) synthesized a mullite powder by the spray-pyrolysis (SP) method in the following way: ANN and TEOS were dissolved in a 1:1 water:methanol solution. The mixed solution was sprayed with compressed air through a glass nozzle into a preheated quartz reaction tube at a temperature of 350°C–650°C, when the solution decomposed to form an oxide powder.

A thermal transformation study of the SP precursor was done by the usual X-ray diffractometry (XRD) and differential thermal analysis (DTA) methods. The following important observations were made:

- Direct formation of mullite from an amorphous precursor: The sprayed powder synthesized by Kanzaki et al. (1985) was amorphous to X-ray. DTA showed a very sharp exothermic peak at 970°C. XRD analysis of the quenched SP powder heated previously at 980°C showed significant crystallization of mullite.
- Direct relation between the mullitization process and the change in the surface area: The spray-pyrolyzed powder was composed of spherical hollow particles with a large surface area, and it was dependent upon the spraying temperature. The surface area data of 17 m^2/g (obtained during pyrolysis at 450°C) dropped sharply to 5 m^2/g after completion of the exotherm at 1000°C.
- Mullitization temperature unaffected by the variation in composition: Kanzaki et al. (1990) and Kumazawa et al. (1986) showed that a single sharp exotherm was observed at 980°C in all three samples: 68A (67.8% Al_2O_3), STD (72.1% Al_2O_3) and 78A (77.8% Al_2O_3). The former two samples showed mullite only above 980°C. However, the latter sample showed corundum in addition to mullite. Apparently, the height of the sharp exotherm is found to be the highest in the STD sample and it falls on the other side of the STD sample.

Sato et al. (1998) took aluminum sulfate and TEOS, and the spray-pyrolyzed powder showed spinel formation at 1000°C. Suttor et al. (1997) synthesized monolithic 3:2 mullite by pyrolysis of metal organic polymer compounds such as siloxane with Al_2O_3 or Al in stoichiometric proportions.

Sakurai et al. (1988) spray-pyrolyzed a mixture of TEOS and aluminum alkoxide by an ultrasonicator. The resultant powder showed direct mullitization at 1000°C.

5.2.2 Spray-Drying Route

Hamano et al. (1986) spray-dried an etheral mixture of ANN and TEOS in a furnace at 250°C. DTA studies of this spray-dried powder showed a strong 980°C exotherm and mullite at the same temperature. But the spray-dried powder of aluminum sulfate with TEOS showed reprecipitation of aluminum sulfate. As a result, a weak 980°C exotherm was noted.

Jaymes and Douy (1992) synthesized mullite precursor powders from aqueous solutions or sols by a spray-drying (SD) method using TEOS and ANN as components. The phase transformations of the three spray-dried precursors obtained from a solution not aged and solutions aged for different time periods—marked A (without aging), B (aged for 3 h), and C (aged for 10 days)—were different. Precursor A showed a single, sharp exotherm at 975°C (enthalpy of reaction = 240 J/g). The XRD pattern showed complete crystallization into pseudotetragonal mullite at 1000°C. Precursors obtained by the aging procedure exhibited three exotherms. Precursor B showed double exotherms—the first peak at 924°C (enthalpy = 46 J/g), followed by the second exotherm at 925°C (enthalpy = 137 J/g)—with crystallization of spinel and some amount of mullite. Thereafter, it also showed a weak exotherm at 1275°C, when a single-phase orthorhombic mullite (o-mullite) was obtained. Precursor C also showed a double exotherm, the first one at 893°C (enthalpy = 58 J/g) and the second, broad exotherm at 975°C (enthalpy = 111 J/g). After the second exotherm, strong crystallization of the transient spinel phase was noted, along with a minor amount of mullite. The third peak, which occurred at 1275°C, corresponded to the o-mullite.

Thus, the time of aging shows an inverse relationship with mullite phase development at around 1000°C. The peak height of mullite (as shown in XRD patterns) formed in precursor A (formed with no aging time) is the highest, compared to precursor B, and it is insignificant in the case of precursor C, obtained with 10 days' aging time. Secondly, a similar indirect relationship is noted between the time of aging and the height of a DTA exotherm. It is conceivable that during aging of an aqueous, hydrolyzed solution of TEOS and aluminum nitrate, the silicic acid component went into polymerization. With time, the mixed solution became opalescent and at a later stage underwent gelation and grew into silica colloids. This may be the reason for the spontaneous spinel formation during the 970°C exotherm, as in the case of precursor C. This spinel phase in due course of heating transformed into mullite at ~1275°C, which is in contrast to the mullite formation behavior of precursor A, which formed mullite only at the occurrence of a single exotherm at 970°C.

Jaymes and Douy (1995) pointed out rightly that the powder obtained by spraying an aqueous solution mixture made by hydrolyzing TEOS and aqueous ANN is a homogeneous alumina-silicate precursor. They also pointed out that this chemical homogeneity is difficult to preserve and problematic during the synthesis of a solid powder precursor by the removal of water from the solution. They engineered a coprecipitation technique by raising the pH rapidly by using salts by the isopropanol route/aqueous route or raising it slowly by using urea.

When the powder was sprayed into a solution of ammonia in isopropanol, first neutralization of the acidic solution of aluminum nitrate occurred and then it was hydrolyzed and finally condensation of aluminum hydroxide colloids continued in the presence of silicic acid. The latter component, in the ammonia environment, started condensing simultaneously and grew into silica gel. These colloidal mixed precipitates in an isopropanol suspension were spray-dried to develop a solid precursor. This precursor (Mull A) exhibited a strong exotherm at 990°C, with an enthalpy of 80.5 kJ/mol (Fig. 5.1).

However, XRD analysis of the powder after the completion of this peak showed the crystallization of both pseudotetragonal mullite and a small quantity of the spinel phase. This observation definitely concludes that this technique failed to produce a purely chemically homogeneous mullite precursor (Fig. 5.2).

When it was sprayed into an aqueous solution of ammonia or ammonium carbonate, a colloidal precipitate was obtained, which was later spray-dried as usual. A thermogravimetric analysis curve shows a larger weight loss (51%) than 42%, which occurred in Mull A. This precursor (Mull B) exhibited two weak exotherms, one at 988°C (31.85 kJ/mol), with the formation of a weakly crystallized spinel phase. The second exotherm, at 1260°C (8.1 kJ/mol) corresponded to o-mullite. So this spray-dried precursor is too far from chemical homogeneity. Likely, by the rapid precipitation in an aqueous medium, the precursor tends to be diphasic character.

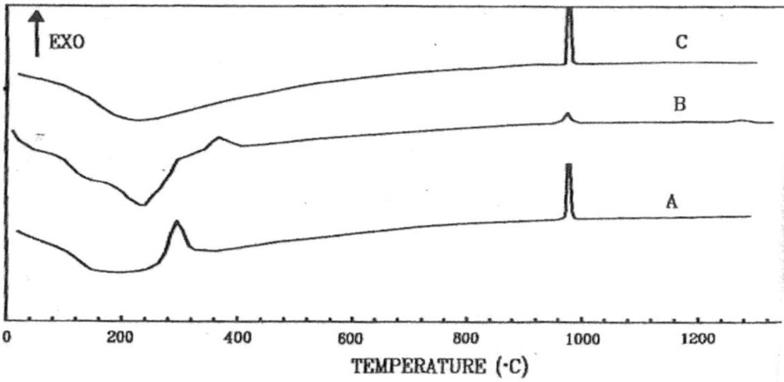

Figure 5.1 DTA traces of three powders (Jaymes & Douy, 1995).

A dried xerogel (Mull C) was synthesized by the slow coprecipitation of two components of the said solution mixture during in situ generation of ammonia by the thermal hydrolysis of urea, followed by the filtration and washing of the ammonium nitrate formed during neutralization of nitrate ions with ammonia. This precursor showed less weight loss (34%) compared to the other two precursors. It exhibited only one sharp exotherm, at 975°C, in DTA, with the evolution of a larger enthalpy of crystallization (109 kJ/mol) than in Mull A. The resultant pseudotetragonal mullite phase development is also quite large, as revealed by comparing the XRD intensities of the mullite formed in the two cases. It is to be claimed that a slow and homogeneous coprecipitation technique produces a homogeneous mullite precursor. It is shown that during the synthesis of the precursor, the most important parameters are the solvent used and the time of precipitation (whether it is slow or fast).

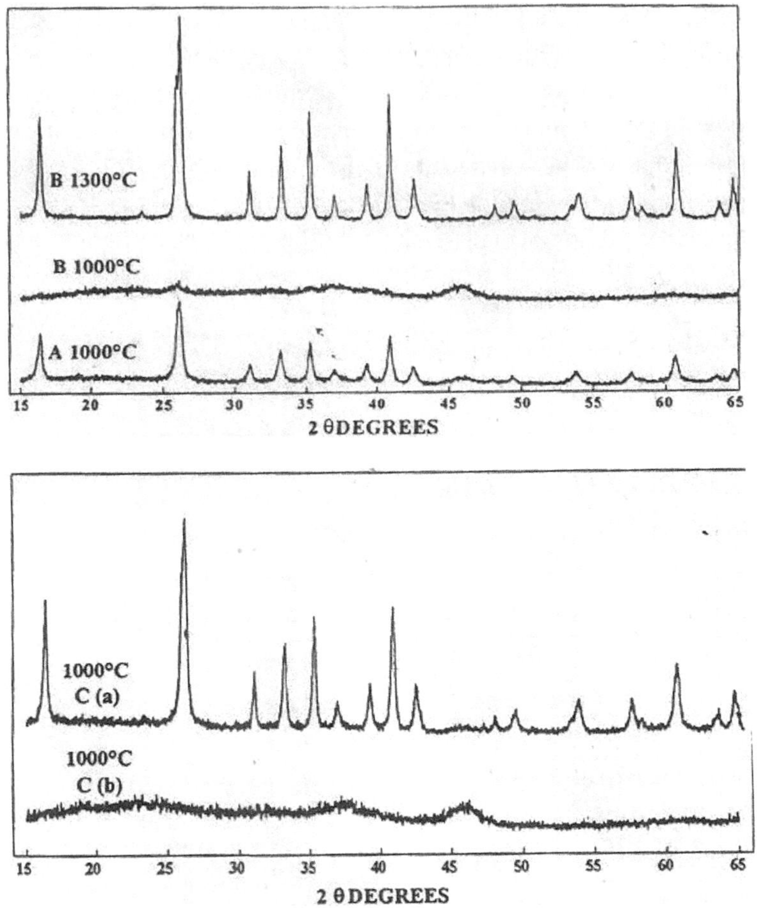

Figure 5.2 XRD diagrams of a precursor powder calcined for 1 h at different temperatures. Mull A at 1000°C; Mull B at 1000°C and at 1300°C; and Mull C at 1000°C (a) by eliminating NH_4NO_3 by washing and (b) without eliminating NH_4NO_3 by washing.

Bhattacharya et al. (1996) claimed that the homogeneity of the alumina-silica gel could be achieved by the SD process. To reach the above conclusion, the following experimentations were done in a stepwise manner. They chose commercially available silica sol and four varieties of inorganic alumina sols for making mullite gel: "Degussa" aluminum oxide C, Condea "dispersal" pseudoboehmite powder, Hoerscht aluminum chlorohydrate "Locron P," and an inorganic alumina sol prepared by extracting NO_3^{-1} using tertiary alkyl amine from an aluminum nitrate solution. The Degussa-based colloidal gel did not form mullite at 1000°C. The Condea-based mixture showed a strong reflection of the spinel phase. The chlorohydrate-based gels exhibited two exotherms, one broad, which occurred at 800°C and formed mullite along with a spinel phase and amorphous silica, and one small but sharp, which occurred at ~1000°C and showed 20% mullite and a strong spinel phase. A nuclear magnetic resonance (NMR) study of this chlorohydrate sol indicated three peaks. The first one showed the presence of six coordinated Al at 0 ppm, indicative of the monomeric hexa-aquo Al^{+3} ions. The other two peaks were due to a five-coordinated species at 11 ppm and a four-coordinated species at 63 ppm. These two peaks are characteristic of the Keggin ion structure. The method of production influences the proportion of monomeric, polymeric (Al_{13}), aggregated polymeric (Al_{13}), and colloidal species in aluminum hydrate solutions. Thus, the chlorohydrate sol might have contained aggregated polymeric species whose are unreactive toward silica sol. A partial quantity of mullite is formed at ~ 800°C exotherm, but a high temperature is needed for completion.

In comparison, a nitrate-based sol in an NMR study showed a spectrum that consists of two distinct six-coordinated species at 0.6 ppm (major) and 4.9 ppm (minor). The monomeric hexa-aquo Al^{+3} is a reactive species and does not have any polymeric ions. The resultant gel on firing at 1000°C, of course, developed equal concentrations of mullite, spinel, and cristobalite, thus predicting increased reactivity. The formation of the latter two phases might be due to inhomogeneity of the bulk gel, which occurred due to segregation of the silica and the alumina during the drying process. To increase the homogeneity, the mixed sol was spray-dried, which effectively froze the sol and the product showed a very sharp exotherm at 920°C and formed only mullite, indicating complete mullitization in a single step. In these ways, Bhattacharya et al. (1996) arrived at the conclusion that a spray-dried gel is the most homogeneous and thus substantiated the fact that the homogeneity of a given reactive aluminum species is of paramount importance in the crystallization of mullite.

During the synthesis of a mullite precursor by the SP method, a variation in the phase crystallization at 980°C is observed due to (i) use of fume silica or tetramethyl orthosilicate (TMOS) and/or silicic acid sol instead of TEOS as a source of silica component and (ii) the technique of spraying. For example, Moore et al. (1992) noted the spinel phase. Ocana et al. (1993) noted the formation of both spinel and mullite phases when an aerosol of hydrolyzed Al *sec*-butoxide and TMOS was spray-dried. Janackovic et al. (1996) observed only the spinel phase by the ultrasonic SP of a mixture of ANN and silicic acid at the decomposition temperature of ~900°C.

5.2.3 Plasma Powder Process

Gani and McPherson (1977) synthesized an alumina-silica powder by plasma-spraying a mixture of Al_2Br_3 and $SiCl_4$ in an argon stream by it injecting into the tail flame of the oxygen-argon high-frequency plasma torch and the powder was collected by passing the exhaust gases through an electrostatic precipitator.

5.2.4 Thermal Decomposition of Alkoxides

Roy et al. (1977) synthesized fine oxide particles by evaporative decomposition. Lewis (1991) obtained a mullite powder by direct oxidation of a mixture of organometallic compounds. Inoue et al. (1996) decomposed a mixture of aluminum isopropoxide (AIP) and TEOS in the presence of toluene in an autoclave at 300°C. The question is, what sorts of reactions occur during autoclaving? AIP alone in an inert solvent at 300°C for 2 h during autoclaving produced x-alumina. In the same conditions, TEOS formed a clear solution instead of decomposing. A mixture of the two during autoclaving in the same conditions inhibited the formation of x-alumina but facilitated the decomposition of TEOS and did not show any precipitation when water was added to the supernatant of the reacted AIP and TEOS mixture. Thermogravimetric and differential thermal analysis (TG-DTA) profiles of the product obtained by the thermal decomposition of AIP and tetraethoxysilane in toluene at 300°C for 2 h exhibited an exotherm at ~1000°C.

5.2.5 Flame Spraying

Gani and McPherson (1977) used an amorphous aluminosilicate powder for densification. Chung et al. (1992) synthesized a mullite powder by the combustion of $SiCl_4$ and $AlCl_3$ using a counterflow diffusion flame. The oxidant (O_2 diluted with N_2) flows downward. The fuel (H_2 diluted with N_2) flows upward. The particles generated in the flame were collected at the burner exit for material characterization. The powder synthesized at a low temperature of 2000°C was amorphous agglomerated and that consisted of primary particles 20–30 nm in size. However, when the powder was synthesized using a high-temperature flame, >2250°C, the particles were spherical agglomerates ~150 nm in size and showed poorly crystallized mullite and spinel structure under XRD.

Baranwal et al. (2001) attempted an alternative approach in chemical processing for the synthesis of a mullite precursor. They developed a simple method of producing a ceramic powder with ultrafine and nanosized particles by subjecting the molecular and polymer precursor to flame spray pyrolysis (FSP). They first synthesized (TEA Al)$_4$ and (TEA-Si-egH), where TEA stands for triethanol amine, as follows: $Al(OH)_3{\cdot}xH_2O$, triethanolamine $(TEAH)_3$, and ethylene glycol $(EGH)_2$ were reacted by stirring and heating to 200°C in a flask for 3–4 h in a N_2 atmosphere. Fume silica, $TEAH_3$, and EGH_2 were also reacted in the same manner. This (TEA-Si-egH) solution was added to the previously synthesized (TEA Al)$_4$, stirred, heated to 200°C under a N_2 atmosphere, cooled, and fed into an FSP apparatus for powder preparation. The reactions are as follows:

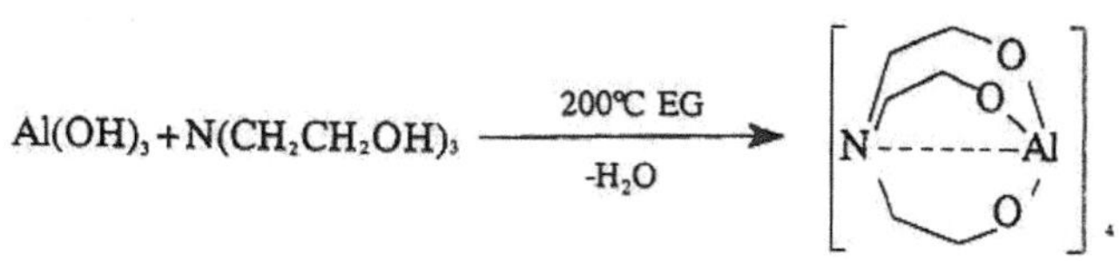

(5.1)

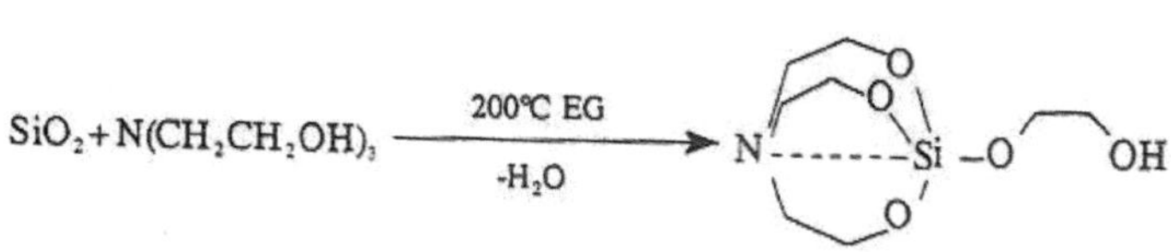

(5.2)

Baranwal et al. (2001) discussed that the production of a homogeneous material is often difficult if the gaseous species have different condensation rates. The ultimate result will be nonstoichiometric, inhomogeneous powder, even for nanoparticles.

The DTA profile of the as-processed powder revealed only one event: an exotherm at 980°C. Both XRD and Fourier-transform infrared studies suggested the formation of tetragonal mullite (t-mullite) only. X-ray analysis showed the splitting of the (120) (210) peaks at 26.15° 2θ in t-mullite into 25.97° and 26.26° 2θ in o-mullite during the heating process. The splitting of this peak was used to estimate the percent conversion of t-mullite into o-mullite. Roughly, at 1000°C, t-mullite is 100% (since no splitting occurs). At 1200°C, o-mullite is >90%. At 1400°C also, it is >90%. At 1600°C, o-mullite is 100%. This result corroborates the phase transformation of the FSP powder with the XRD result.

5.2.6 Freeze-Dry Process

Wheat et al. (1979) attempted to synthesized mullite by THE freeze-dry process, wherein a mixture of commercial components was obtained. A thermal transformation of the freeze-dried gel on heat treatment and on X-ray analysis shows the following (Table 5.1):

Table 5.1 Sequential transformation of spinel-to-mullite phase on heating of freeze-dried gel

Temperature	Phase
Freeze-dried powder	Weak X-ray peaks occurred.
At 1000°C	According to Wheat et al. (1979), an X-ray showed the powder to be amorphous. In fact, on careful observation, although a 2θ value was not given, it was found that it contained broad peaks due to a spinel phase and an amorphous band due to the presence of amorphous silica as in the dried powder. A DTA study of this powder showed a very small exothermic peak, preceded by a large endotherm at ~870°C due to the decomposition of aluminum sulfate
At 1200°C	The bulk of the mullite crystallized together with a trace of α-alumina. This free alumina did not transform or consume to develop a single-phase mullite even on continued heating for 24 h at the same temperature.
At 1400°C	Single-phase mullite developed without giving any soaking at that temperature.

According to Wheat et al. (1979), in the sol-gel process, gelation may occur slowly and thus can also lead to a segregation of the components. They are of the opinion that only in rare cases can genuine coprecipitation occurs. On the contrary, it is assumed that the freeze-dry process is most desirable as it avoids the possibility of segregation. In solution, rapid immobilization of ions may occur. Of course, this process actually hinders the recrystallization of aluminum sulfate but an intimate mixing of the two component oxides is not achieved. Since single-stage mullitization is not found to take place followed by a sharp exotherm at ~1000°C in the DTA study, it may be assumed that amorphous alumina liberated after the decomposition of aluminum sulfate at a high temperature is a highly active material, which readily reacts in the solid state at ~1200°C with another active silica phase to crystallize mullite. Therefore, the reaction sequence in this case is not analogous to the two-stage transformation of kaolinite to mullite, as assumed by them.

Guo et al. (1986) synthesized a sol mixture with a mullite composition from aluminum sulfate and ethyl silicate both by freeze drying and SD methods. They showed the DTA and TG curves of the dried materials. Aluminum sulfate showed decomposition at ~850°C. but there is an absence of the 980°C exotherm in both cases. These observations indicate that Si–O–Al bond formation did not occur either during the freeze-drying process or during the SP operation of this sol mixture.

5.2.7 Decomposition of an Aerosol

Moore et al. (1992) synthesized high-purity mullite by the aerosol decomposition technique by using an aqueous solution of aluminum nitrate and fumed silica in a reactor by the SP technique by passing it through a heated tube reactor at 900°C. By scanning electron microscope (SEM) micrograph study and by a particle size distribution analysis by the sedimentation technique, they claimed that the powder particles have a mean diameter on the order of 0.6 μm. All particles were roughly spherical in shape and in the 0.1–2 μm size range and there was no evidence of hard agglomerates and hollow shell particle morphology, as observed in the case of the spray-pyrolyzed powder of Mizuno and Saito (1989). This precursor, however, produced a spinel phase and remained stable at as high a temperature as 1400°C, and at 1600°C, it transformed into mullite. Control over the final microstructure and the grain size of mullite is affected by the choice of processing route, component sources (e.g., salts, alkoxide, or colloids), and processing conditions.

West and Gray (1958) chose metal organic compounds of silicon and aluminum, and the mixed solution was evaporated till it dehydrated and dried at 450°C. The thermogram of this SiO_2-Al_2O_3 powder is similar to that of kaolinite. XRD analysis of the powder heated to 980°C showed the presence of both spinel and mullite.

Ogihara et al. (1994) synthesized a monodispersed spherical fine mullite powder by the hydrolysis of Al(*sec*-$OC_4H_9)_3$ and TEOS in alcohol/acetonitrile as a dipolar aprotic solvent. Synthesis factors such as reagent concentration, volume ratio of the solvent, and reaction temperature influenced the formation of monodispersed mullite fines. XRD analysis showed that the as-prepared spherical powders were amorphous and crystallized to mullite at ~1000°C, which corroborated the DTA exotherm. The morphology was retained after crystallization, as shown in an SEM photograph. The monodispersed spherical mullite precursor powder transformed into cracked spherical particles after being calcined at 1000°C.

5.3 Summary

Precursors are synthesized by the decomposition method, using either the SP route or the SD process. By the SP method, Kanzaki and Tabata (1985) were the first to synthesize a mullite precursor. It showed the direct formation of mullite from the amorphous precursor, with the exhibition of a very sharp exothermic peak at 970°C and without any exotherm at a higher temperature. Kanzaki et al. (1990) and Kumazawa et al. (1986) showed that the single sharp exotherm was the highest for STD (71.2 wt % Al_2O_3) and it fell on either side of this composition. Direct mullitization at 1000°C was noted on the thermal reaction of the mullite precursor in the spray-pyrolyzed method in the studies by Kanzaki et al. (1990) and Kumazawa et al. (1986), and the tentative reaction is as follows:

$$\underset{\text{(S P powder)}}{Al_2O_3{\cdot}SiO_2} \xrightarrow[\text{Exotherm}]{1^{st}} \text{Mullite-type phase} \qquad (5.3)$$

The powder synthesized by the SP method by Kumazawa et al. (1991) exhibited a large exotherm at 980°C. A mixture of TEOS and aluminum alkoxide was spray-pyrolyzed by an ultrasonicator by Sakurai et al. (1988). On the contrary, when aluminum sulfate replaced ANN (Mizuno et al., 1990) and was spray-pyrolyzed with fume silica, the powder showed spinel formation at 1000°C.

References

1. S. Kanzaki, H. Tabata, T. Kumazawa, and S. Ohta, Sintering and mechanical properties of stoichiometric mullite. *J. Am. Ceram. Soc.*, **68**(1), C-6–C-7 (1985).
2. S. Kanzaki, H. Tabata, and T. Kumazawa, Sintering and mechanical properties of mullite derived via spray pyrolysis. *Ceram. Trans.*, **6**, 339–351 (1990).
3. T. Kumazawa, S. Kanzaki, J. Asaumi, O. Abe, and H. Tabata, Sinterability of SiO_2-Al_2O_3 powders prepared by spray pyrolysis. *Yogo Kyokai Shi*, **94**(5), 485–490 (1986).
4. Y. Sato, T. Takei, S. Hayashi, A. Yasumori, and K. Okada, Effects of amorphous and crystalline SiO_2 additives on γ-Al_2O_3-to-alpha-Al_2O_3 phase transitions. *J. Am. Ceram. Soc.*, **81**, 2197–2200 (1998).
5. D. Suttor, H.-J. Kleebe, and G. Zieger, Formation of mullite from filled siloxanes. *J. Am. Ceram. Soc.*, **80**(10), 2541–2548 (1997).
6. O. Sakurai, N. Mizutani, and M. Kato, Preparation of mullite powders from metal alkoxides by ultrasonic spray pyrolysis. *J. Ceram. Soc. Jpn.*, **96**, 639–645 (1988).
7. K. Hamano, T. Sato, and Z. Nakagawa, Properties of mullite prepared by co-precipitation and microstructure of fired bodies. *Yogo Kyokai Shi*, **94**(8), 818–822 (1986).
8. I. Jaymes and A. Douy, Homogeneous mullite-forming powders from spray-drying aqueous solutions. *J. Am. Ceram. Soc.*, **75**(11), 3154–3156 (1992).
9. I. Jaymes and A. Douy, Homogeneous precipitation of mullite precursors. *J. Sol-Gel Sci. Technol.*, **4**, 7–13 (1995).
10. A. K. Bhattcharya, A. Hartridge, and K. K. Mallick, Inorganic aluminium precursors in the synthesis of mullite: an investigation. *J. Mater. Sci.*, **31**, 5551–5554 (1996).

11. K. A. Moore, J. Cesarano III, D. M. Smith, and T. T. Kodas, Synthesis of submicrometer powder via high temperature aerosol decomposition. *J. Am. Ceram. Soc.*, **75**, 213–215 (1992).
12. M. Ocana, J. Sanz, T. Gonzales-Carreno, and S. Serna, Spherical mullite particles prepared by hydrolysis of Aerosols. *J. Am. Ceram. Soc.*, **76**, 2081–2085 (1993).
13. Dj. Janaćković, V. Jokanović, Lj. Kostić-Gvozdenović, Lj. Živković, and D. Uskoković, Synthesis, morphology, and formation mechanism of mullite particles produced by ultrasonic spray pyrolysis. *J. Mater. Res.*, **11**(7), 1706–1716 (1996).
14. M. S. J. Gani and R. McPherson, Glass formation and phase transformations in plasma prepared Al_2O_3-SiO_2 powders. *J. Mater. Sci.*, **12**, 999–1009 (1977).
15. D. M. Roy, P. R. Neurgaonkar, T. P. O. Holleran, and R. Roy, Preparation of fine oxide powders by evaporative decomposition of solutions. *Am. Ceram. Soc. Bull.*, **56**(11), 1023–1024 (1977).
16. D. J. Lewis, Techniques for production mull & other mixed oxide system. *J. Am. Ceram. Soc.*, **74**, 2410–2413 (1991).
17. M. Inoue, H. Kominami, and T. Inui, Thermal decomposition of alkoxides in an inert organic solvent: novel method for the synthesis of homogeneous mullite precursor. *J. Am. Ceram. Soc.*, **79**(3), 793–795 (1996).
18. S.-L. Chung, Yu-C. Sheu, and M.-S. Tsai, Formation of SiO_2, Al_2O_3, and $3Al_2O_3.2SiO_2$ particles in a counterflow diffusion flame. *J. Am. Ceram. Soc.*, **75**(1), 11–23 (1992).
19. R. Baranwal, M. P. Villar, R. Garcia, and R. M. Laine, Flame spray pyrolysis of precursors as a route to ano-mullite powder: powder characterization ad sintering behavior. *J. Am. Ceram. Soc.*, **84**(5), 951–961 (2001).
20. T. A. Wheat, E. M. H. Sallam, and A. C. D. Chaklader, Synthesis of mullite by a freeze-dry process. *Ceramurgia Int.*, **5**(10), 42–44 (1979).
21. C.-J. Guo, Z.-e Nakagawa, and K. Hamano, Effect of drying method on mullite ceramics prepared from sol mixture. *Yogo Kyokai Shi*, **94**(6), 583–589 (1986).
22. M. Mizuno and H. Saito, Preparation of highly pure fine mullite powder. *J. Am. Ceram. Soc.*, **72**, 377–382 (1989).
23. R. R. West and T. J. Gray, Reactions in silica-alumina mixtures. *J. Am. Ceram. Soc.*, **41**(4), 132–136 (1958).

24. T. Ogihara, T. Yanagawa, N. Ogata, K. Yoshida, M. Iguchi, N. Nagata, and K. Ogawa, Synthesis of monodispersed, spherical fine mullite powders by alkoxide method, *J. Ceram. Soc. Jpn.*, **102**(8), 778–784 (1994).
25. T. Kumazawa, S. Ohta, S. Kanzaki, and H. Tabata, Influence of powder characteristics on microstructural and mechanical properties of mullite ceramics (74 wt. % Al_2O_3). *J. Jpn. Ceram. Soc.*, **99**, 1228–1233 (1991).
26. M. Mizuno, M. Shiraishi, and H. Sato, Microstructure and bending strength of highly pure mullite ceramics. *Ceram. Trans.*, Vol. 6, ed. S. Somiya, R. F. Davis, and J. A. Pask, p. 413 (1990).

Chapter 6

Mullite Precursors Synthesized from Other Miscellaneous Sources/by Other Processes

In this chapter, we will look at the synthesis of mullite using components from the same source or varying sources and either changing the processing conditions or keeping them constant.

6.1 Mullite Using Components from the Same Source but with Different Processing Conditions

Synthesis of mullite using components from the same source but with different processing conditions produces gels or precursors of different characteristics.

6.1.1 Producing Two Kinds of Mullite Precursors

Only two different kinds of mullite gels can be produced when components from the same source are used.

- Two components (tetraethyl orthosilicate [TEOS] and aluminum nitrate nonahydrate [ANN]) were used and

Mullite Formations: Analysis and Applications
Akshoy Kumar Chakraborty

ISBN 978-981-4877-05-3 (Hardcover), 978-1-003-03167-3 (eBook)
www.jennystanford.com

processed at two different pH values, which led to the development of two different kinds of mullite gels (Okada & Otsuka, 1986; Chakraborty & Ghosh, 1988)—see Chapters 2 and 7.

- A mixture of two components (AlOBu and TEOS) was processed using two different processing conditions—varying pH and varying the water content—which led to the formation of two kinds of mullite gels, polymeric and colloidal (Pask et al., 1987). The phase transformation sequences of the two gels are given in Tables 6.1 and 6.2 (Yoldas & Partlow, 1988; Yoldas, 1992)—see Chapter 2.

Table 6.1 Phase transformation of polymeric gel (X-ray figure not provided) during heating

Temperature	Phase
Beginning	The gel was amorphous to X-ray.
At 950°C	It was still amorphous.
At 1050°C	A crystalline mullite-type phase along with diffuse XRD peaks at 46° 2θ and 67° 2θ appeared, assigned to the γ-Al_2O_3-type spinel phase or the trace amount of transitional alumina. The crystallization temperature of mullite is pinpointed with a strong exothermic peak.

Table 6.2 Thermal transformation of colloidal gel (X-ray figure not provided) during heating

Temperature	Phase
Beginning	The gel was amorphous to X-ray.
At 950°C	Diffuse peaks appeared at 46° 2θ and 67° 2θ, which may have been due to γ-Al_2O_3-type spinel or other aluminous phases.
At 1050°C	There was neither a hint of mullite crystallization nor the occurrence of a 980°C exotherm.
At 1200°C	The some amount of mullite developed; untransformed spinel and/or transition alumina remained. A diffuse DTA peak with one-tenth the intensity of the 980°C peak of the polymeric gel system occurred between 1200°C and 1300°C.
At 1300°C	Complete mullite formation occurred.

- Two coprecipitated (CP) mullite powders (A and B) were synthesized from aluminum isopropoxide (AIP) and tetramethyl orthosilicate by the coprecipitation technique using a dilute ammonium hydroxide solution by Hamano et al. (1986)—see Chapter 7.
- Colloidal boehmite suspensions and TEOS were used to prepare mullite by the sol-gel process at two pH conditions (Sonuparlak, 1988). The first one was prepared under acidic conditions (pH < 2). The second one was prepared by the same procedure but under pH 6–7 using ammonium hydroxide (see Chapter 8).
- Two different methods were followed by Colomban (1989) for the synthesis of precursors of two different characters. The first one was a slow hydrolysis (SH) technique (monolith synthesis) and the other was a rapid hydrolysis technique (powder synthesis), where the mixture was hydrolyzed rapidly with the use of a large amount of water under vigorous stirring to obtain a gel powder. The mullite monolith obtained by the slow route exhibited an intense exotherm between 1020°C and 1030°C, which is monophasic in character. In comparison, the powder obtained by the rapid route contained boehmite and also showed bayerite as the crystalline phase of less magnitude, which is diphasic in character. The powder obtained using silicon butoxide showed an endothermic peak, like glass transition anomaly, just before the 980°C exotherm. And the precursor obtained from ethoxide and ANN showed double exotherms, at 920°C and 1015°C. Colomban (1989) questioned the local structure of mullite gels and glasses and the influence of synthesis on the structure.
- Using aluminum silicate powder with a hydrolyzing mixed solution of AIP ($Al(Oi\text{-}C_3H_7)_3$) and TMOS ($Si(OCH_3)_4$) in benzene, Mitachi et al. (1990) synthesized two CP powders in two different processing conditions. In A series, hydrolysis was done at pH 7 simply with distilled water. In B series, hydrolysis was done at pH 10 with aqueous ammonia solution.
- Using identical boehmite and silica sols but different in gelation procedures, two diphasic gels, marked Gel A and Gel B, were synthesized by Huling and Messing (1990)—see Chapter 8.

- From $Al(OPr)_3$ and $Si(OMe)_4$, Kumazawa et al. (1990) synthesized mullite precursors by two different methods: gelation by hydrolysis with H_2O and gelation by hydrolysis with H_2O and NH_4OH (see Chapter 7).
- Ge et al. (1992) used the sol-gel method with silicon tetraethoxide and ANN in the presence of three catalysts; two mullite powders were synthesized (see Chapter 12).
- Two mullite precursor samples were made from TEOS and AIP by Taylor and Holland (1993), one by using a water-free approach in the presence of dry butan-1-ol and isopropyl alcohol as a medium, with the mixture stirred for ~20 days, and the other by the addition of external water. GYW contained a ratio of H_2O/–OR >5, while in GNW, external water was not used (see Chapter 7). Phase transformations of these two were different.
- Geradin et al. (1994) made two mullite precursors from TEOS and AIP but in different hydrolysis conditions—particularly varying the amount of water added (see Chapter 7).
- Huang et al. (1997) used components from the same source (TEOS and aluminum chloride) for gelation (monophasic gel as per Chapter 3), with a variation in pH from the acidic to the basic range (see Chapter 7). They pointed out that mullitization temperature is an important criterion to assess the mixing scale or the degree of aluminosilicate (–Si–O–Al–) bonds present in the precursor system and the pH prevailing during gelation also influences the mixing scale of the two silicon aluminum components. The phase transformations of two gels, one made at pH 1.5 (marked as M I) and the other made at pH 11.5 (marked as MII), were shown to be different. They conceived two different gel structures at the two extreme pH values. Finally, they conjectured that acidic gels are homogeneous monophasic in nature, in contrast to basic gels, which are heterogeneous and diphasic in nature. A mullite gel prepared from TEOS and aluminum chloride at pH 11 showed spinel and other alumina polymorphs at a temperature as high as 1200°C, which may be regarded as diphasic as the dried xerogel contained gibbsite, boehmite, and bayerite and amorphous silica. They conceived two different gel structures at the two extreme pH values.

- Components from the same source (TEOS and ANN) at two different pH values led to the development of mullite gels different in terms of thermal character (Fonseca et al., 1997). Alumina-silica gels and powders of two compositions, namely $60Al_2O_3{\cdot}40SiO_2$ and $64.3Al_2O_3{\cdot}35.7\ SiO_2$, were synthesized from TEOS and ANN in varying conditions, and these were marked as MG (monophasic) and MAG/MP, MAPIF, MAPIR (drying on a hot plate), and MPP (diphasic) in a basic condition (see Chapter 12).
  MG gel prepared in an acidic condition and dried slowly at 60°C for 7 days showed a strong 975°C exotherm and formed mullite as an identifiable phase at 1000°C. MAG gel synthesized via the SH gel route formed the mullite phase at 1000°C in spite of the excess alumina present in this precursor. According to Fonseca et al. (1997), these gels might have high chemical homogeneity. On the contrary, MPP powder showed a very small exotherm and formed only the spinel phase at 1000°C. They suggested that MG and MAG behave as single-phase precursors while MPP is clearly diphasic in character.
- The effect of heating on the phase and structural evolution of two kinds of mullite precursors obtained by varying the method of preparation, for example, polymeric gels PA and PB made in acidic and basic conditions (Yoldas, 1990; Yoldas & Partlow, 1988), were investigated by Nieto et al. (1998).

6.1.2 Producing Three Kinds of Mullite Precursors

Using ANN and TEOS but changing both the pH and the water for processing produces three kinds of mullite gels:

- Three precursors, Type I, Type II, and Type III, were synthesized by Schneider et al. (1993) (Chapter 3) from components from the same source but hydrolyzed with varying quantities of water at two different pH levels, leading to a change in the hydrolysis rates of the components. These showed wide variation in the intensity of the exotherm (980°C–1250°C), with variation in the crystallization of either mullite or spinel. For example, Type I (prepared in very low water content so the hydrolysis was very slow) exhibited a sharp 980°C exotherm, with the formation of tetragonal mullite (t-mullite, major quantity) in addition to crystallization of the spinel phase (minor amount). Thus, Type 1 is designated as a monophasic gel. Type III gel (prepared using high water content at pH < 10, where the hydrolysis was fast) exhibited the first exotherm of a lower intensity than that of Type I and crystallized only to spinel instead of t-mullite (major amount), unlike what was noted in Type I. This intermediate spinel further transformed to mullite at the second exotherm. Thus, Type III, synthesized by Schneider et al. (1993), may be designated as a diphasic gel as the spinel is crystallized. On the other hand, Type II (prepared using high water content at pH > 10, where the hydrolysis was very fast when boehmite was precipitated out) followed a completely different path to mullite formation compared to the other two cases of gels. It is purely diphasic in character.

- Chakraborty (1996) synthesized gels from the same components but with three different amounts of water and did their DTA and X-ray diffractometry (XRD) to reveal the mechanisms of their transformation into mullite. The first gel, marked G-152 (Al:Si = 3:1), was synthesized by using a large volume of water (50 mL) for the prehydrolysis of AlOBu and then mixed with 2 mL of TEOS in 10 mL of alcohol, followed by 2 drops of diluted HNO_3. The resultant gel formed pseudoboehmite and was diphasic in character. The other two gels, marked G-152(i) and G-152(ii), were synthesized by the same procedure but the amount of water was reduced to 10 mL and 7 mL, respectively (see Chapter 7).
- Precursors synthesized by Rajendran et al. (1990) are of two characters. Mull A, Mull B, and Mull C were synthesized from the same source by Jaymes and Douy (1995) but are of different characters.

6.1.3 Producing Four Kinds of Mullite Precursors

Four kinds of mullite precursors can be produced using ANN and TEOS but changing the processing conditions. Precursor B was synthesized via the homogeneous precipitation method with TEOS and ANN using urea (Jaymes & Douy, 1995), and the precipitation occurred within the pores of the organic gel. Precursor E was obtained by the same homogeneous precipitation technique using TEOS and ANN but without an organic gel. Urea was added to a solution mixture of ANN and silicic acid (Jaymes & Douy, 1995). Two other precursors (C and D) were synthesized by the aqueous spray-drying (SD) technique by Jaymes and Douy (1992). They spray-dried a clear aqueous solution of ANN mixed with hydrolyzed TEOS in a laboratory spray drier with compressed air preheated to 200°C. The resultant powder (precursor C) was heat-treated at various temperatures and analyzed by XRD and DTA techniques to reveal the crystallization scheme. Precursor D was obtained after aging a diluted clear solution of the components as above at 100°C for 3 days prior to SD (see Chapter 12). These four precursors, B, E, C, and D, showed differences in crystallization behavior, as revealed by both Al and Si magic angle spinning–nuclear magnetic resonance spectroscopy (MAS-NMR) and by XRD (Jaymes et al., 1996).

6.1.4 Producing Six Kinds of Mullite Precursors

Haque (2000) synthesized three types of SH gels and two types of CP gels at two different pH values by utilizing the same components, TEOS and ANN. Later on, as many as six types of mullite precursors (four monophasic, one diphasic, and one spray-dried precursor) were synthesized by changing the pH and the water content. 20 g of ANN and 4 mL of TEOS were taken, and precursors were prepared in the following ways:

In the case of SHI, ANN was repeatedly heated on a hot plate at ~120°C to remove its water of crystallization. Finally, it was dissolved in absolute alcohol. To it, a measured quantity of TEOS was added to make a stoichiometric mullite gel. The solution was swirled and its pH measured and it was allowed to set to a gel in an air oven at ~60°C. In SHII, weighed quantities of both ANN and TEOS were simply mixed in absolute alcohol and the mixture was gelled as above. In SHIII, 10 mL of distilled water was added to an ANN and TEOS mixture, as in SHII, and the mixture was gelled. In SHIV, required amounts of ANN and TEOS were mixed in a Pyrex beaker containing 200 mL of distilled water. To the mixture, a 1:1 ammonium hydroxide solution was added dropwise and the pH of the solution was raised to 6, when gelatinous precipitation occurred.

The solution was filtered and dried. To make a diphasic gel, only the pH of the solution, as in SHIV, was raised to 9, when boehmite was precipitated in situ. An SD precursor was made by spraying an alcoholic solution of ANN and TEOS in a dryer at ~120°C. See Chapters 12 and 15 for variations in their transformation paths.

6.2 Synthesis of a Mullite Gel by the Organic Route

West and Gray (1958) synthesized a series of intimate mixtures of silica and alumina from metal-organic compounds by vacuum-dehydrating the mixtures at 450°C. The thermograms showed the progress from pure silica to pure alumina. Three samples, with 49.63%, 33.32%, and 25.47% silica (Al_2O_3:SiO_2 = 1.0152, 2.002, and 2.927) showed large exotherms at slightly below 1000°C. The crystallization of the spinel as well as some quantity of mullite were identified. Later on, these samples exhibited a broad exothermic peak from 1200°C to 1400°C due to further mullitization. Other organic compounds influence the powder characteristics of the precursors, as shown by different authors.

6.2.1 Citric Acid

Marcilly et al. (1970) synthesized a highly dispersed mixed oxide by the amorphous citrate process on pyrolysis. Douy and Odier (1989) obtained a mullite powder by the gelification of a mixture of Al citrate and silica hydro sol prepared from the hydrolysis of TEOS by using polyacrylamide followed by calcining to 800°C. Thereafter, Douy (1991) extended the use of the citric acid complexation process of metal ions but pointed out that silicon did not make any complex. Thus, the homogeneity of the mixed oxide is complicated. He used TEOS and 3-(triethoxysilyl)-propyl-amine (TESPA) as the silica source, Al salt (nitrate, citrate, chloride, or sulfate) as the source of the alumina component, and polyacrylamide for the synthesis of various mullite gels. The following materials were used for organic gel preparation: a monofunctional monomer (acrylamide or 2-hydroxyethymethacrylate), a bifunctional monomer (cross linking)-*N,N'*-methylene diacrylamide, a radical polymerization initiator (AIBN or ammonium peroxodisulfate), and a radical transfer agent (Temed).

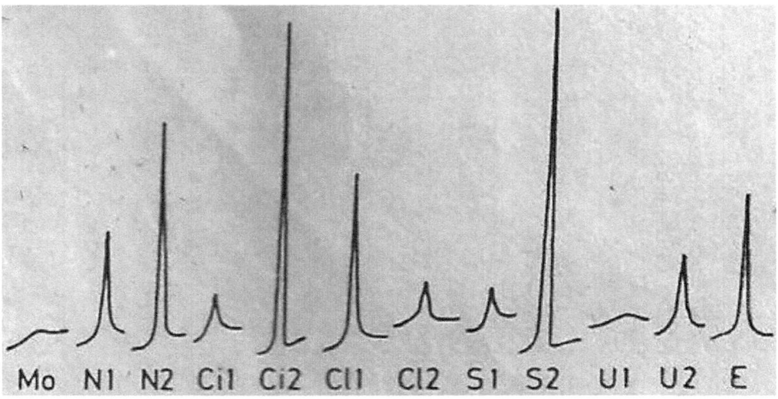

Figure 6.1 970°C DTA exotherms of the powders studied (Douy, 1991).

Portions of 980°C exotherms of DTA-analyzed organic gels are shown in Fig. 6.1 and a summary of the characteristics is shown in Table 6.3. Douy (1991) predicted that there was no direct relation between the intensity of the 970°C exotherm and the crystallization of the powder precursor into mullite. The DTA signal intensity must reflect something else at the microscopic level. The chemical homogeneity likely changed during the synthesis process of the precursor. The intensity of the 980°C exotherm and the nature and intensity of the crystalline phase or phases developed were dependent on the source of the aluminum salt used and type of silicon alkoxide chosen as the source.

6.2.2 Using ANN and TESPA by an Organic Gel–Assisted Processing Technique

Precursor A was made by an organic gel–assisted method (Douy, 1991) wherein the reactivity of aluminum alkoxide was reduced by chelating it with citric acid in the presence of TESPA followed by the addition of ammonia solution at pH $\approx$ 2. Secondly, the solution was gelled by the in situ formation of a polyacrylamide network.

Table 6.3 Powders prepared and some of their characteristics

Powder	Al salt	Si alkoxide	Organic gel	Solvent	Specific surface areas in m^2g^{-1} (750°C, 5 h)	Crystalline phases[a] (950°C, 2 h)
M0[b]	Citrate	TEOS	polyacrylamide	H_2O	155	S
N1	Nitrate	TEOS	polyacrylamide	H_2O	45	S, m
N2	Nitrate	TESPA	polyacrylamide	H_2O	58	S, m
Ci1	Citrate	TEOS	polyacrylamide	H_2O	195	S
Ci2	Citrate	TESPA	polyacrylamide	H_2O	235	M
Cl1	Chloride	TEOS	polyacrylamide	H_2O	360	S
Cl2	Chloride	TESPA	polyacrylamide	H_2O	350	s, m
S1	Sulphate	TEOS	polyacrylamide	H_2O	310	S, m
S2	Sulphate	TESPA	polyacrylamide	H_2O	300	s, M
U1	Nitrate–urea	TEOS	polyacrylamide	H_2O	300	S
U2	Nitrate–urea	TESPA	polyacrylamide	H_2O	310	S
E	Nitrate	TEOS	pHEMA[c]	Ethanol	60	s, M

Source: (Douy, 1991)

6.2.3 Use of Organic Ligands

Hydrolysis-cum-polymerization of metal ions can generally be modified by the use of complexing agents. Mizukami et al. (1990) and Mizukami et al. (1997) tried to report a relationship between the complexing ability of various ligands during the sol-gel process of the synthesis of a precursor and the formation of mullite. They synthesized two series of samples by choosing two different sources of the Al component—A series from TEOS and ANN and B series from TEOS and $Al(OBu)_2(AcAcEt)$ (dibutoxy ethyl acetoacetate aluminum)—using the following organic ligands of varying bridging coordination abilities: ethylene glycol (EG), 1,3-propanediol (PD), BD, 2-methyl-2,4-pentanediol (MPD), and diethylene glycol monoethyl ether (DEME). At ~1000°C, mixed oxides produced mullite and spinel phase of varying intensities. In the first series of mixed oxides, mullite was apt to appear in the order EG > PD > MPD, which was consistent with that of ligands in complexing ability with original ligands of aluminum complexes. ANN gives the aquo. complex $Al(H_2O)_6^{+3}$. Water ligands are easy to replace with EG and develop an $Al–OCH_2CH_2O–Si$ bond, as shown by the NMR study, in a mixed oxide precursor. Thus bonding means good mixing of alumina and silica. On the contrary, in B series, original ligands are firmly bonded to aluminum and with such aluminum complexes it is difficult to produce heteronuclear polymers. As a result, the alumina is apt to turn into the spinel phase due to bad miscibility between the components.

6.2.4 Ethyl Acetoacetate

Heinrich et al. (1991) synthesized an ultrafine and homogeneous mullite powder by the controlled hydrolysis and condensation of AlOBu and TEOS.

An alcoholic solution (2-propanol) of acetyl acetone was first added to ALOBu while stirring, which chelated the alkoxide and thus decreased its reactivity. TEOS solvated in 2-propanol was also added. The molar ratio of AlOBu:(acetyl acetone):TEOS was 3:1.5:1

to achieve a 3:2 mullite composition. Heat treatment of the gel below 1263°K showed an amorphous phase and the same above 1263°K led to crystallization and a spinel phase (cubic mullite or c-mullite) with additional orthorhombic mullite (o-mullite). It converted to o-mullite at 1470°K. Heinrich and Raether (1992) further synthesized single-phase gels from two starting compounds: aluminum-*sec*-butylate chelated with β-diketone and tetraalkoxysilane in isopropanol in a 0.1 N HCl medium. The second system was ANN-ethanol/TEOS and water. The crystallization behaviors of gel of four categories are shown in Table 6.4.

Samples were prepared from aluminum-sec-butylate, chelated with acetylacetone and TEOS. Exceptions are: (I) β-diketone: methylacetoacetate; (II) β-diketone: ethylacetoacetate; (III) silicon source: $Si(OEt)_{3.4}(O^iPr)_{0.6}$; (IV) silicon source: $Si(OEt)_{2.6}$ $(O^iPr)_{1.4}$; (V) silicon source: TMOS; (VI) silicon source: TPOS; (VII) aluminum source: aluminum nitrate nonahydrate.

R_C, molar ratio β-diketone to aluminum-*sec*-butylate; C_H, molar concentration of water; C_A, molar concentration of silicon alkoxide; C_{tM}, weight percent of tM.

Chen and Vilminot (1994) synthesized four mullite gels using TEOS and aluminum tri-*sec*-butoxide chemically modified by ethyl acetoacetate as precursors. The function of ethyl acetoacetate is to substitute one (–OBu) group in aluminum alkoxide. The mixed components were hydrolyzed at three different water:alkoxide ratios in a neutral condition, marked as MUL1, MUL2, and MUL3. A second gel, marked MUL4, was prepared in a basic condition. All samples exhibited the first exotherm and formed t-mullite, depending upon the hydrolysis ratio. Besides t-mullite, MUL1 showed partial crystallization of corundum, which persisted even after annealing at 1250°C. This suggests that heterogeneity might be present to some extent in this xerogel. All gels except MUL1 exhibited a second exotherm. Thus hydrolysis conditions, for example, the water:alkoxide ratio and neutral or basic condition, show noticeable influence on the phase evolution of gels.

Table 6.4 List of samples and results

Sample no.	R_c	c_H (mol⁻/l)	c_A (mol/l)	c_{tM}
A1	0.3	3.3	0.26	0.10
A2	0.5	3.3	0.26	0.23
A3	0.7	3.3	0.26	0.37
A4(I)	0.5	3.3	0.26	0.11
A5(I)	0.75	3.3	0.26	0.20
A6(I)	1.0	3.3	0.26	0.42
A7(II)	0.5	3.3	0.26	0.05
A8(II)	1.0	3.3	0.26	0.16
A9(II)	1.5	3.3	0.26	0.68
S1	0.5	3.3	0.26	0.38
S2(III)	0.5	3.3	0.26	0.21
S3(IV)	0.5	3.3	0.26	0.06
S4(V)	0.5	3.3	0.26	0.44
S5	0.5	3.3	0.26	0.23
S6(VI)	0.5	3.3	0.26	0.14
R1	0.5	3.3	0.26	0.45
R2	0.5	3.3	0.18	0.23
R3	0.5	3.3	0.09	0.21
R4	0.5	10.0	0.26	0.04
R5	0.5	10.0	0.18	0.03
R6	0.5	10.0	0.09	0.09
R7	0.5	16.6	0.26	0.05
R8	0.5	16.6	0.18	0.09
R9	0.5	16.6	0.09	0.09
N1(VII)	–	8.3	0.31	1.0
N2(VII)	–	11.4	0.29	0.39
N3(VII)	–	18.9	0.24	0.10

Source: (Heinrich & Raether, 1992)

6.2.5 Acrylamide

Sin et al. (2001) adopted acrylamide polymerizations. They synthesized mullite powders for three different applications. They used Al powder and granular silica and dissolved them in HNO_3 and 40% HF solutions, respectively. Al^{+3} is complexed with ethylenediaminetetraacetic acid and added to a Si solution. To this mixed solution, acrylamide, *bis*-acrylamide monomers, and a thermal initiator for polymerization were also added and the solution was heated to 80°C to set to gel.

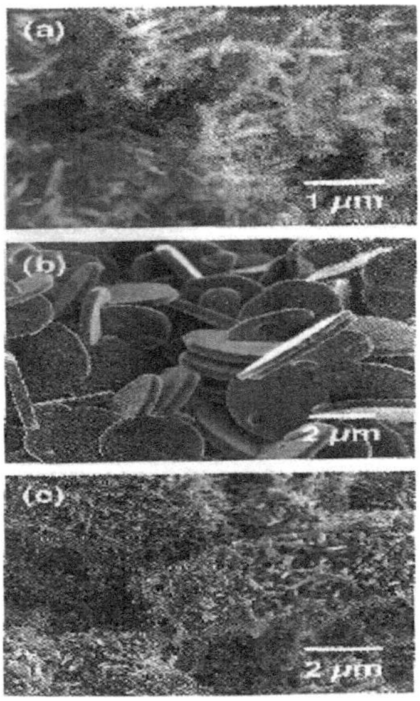

Figure 6.2 SEM photomicrographs of (a) mullite heated first in low-pressure O_2 and then in air to 1300°C for 6 h, (b) mullite with 20% starch heated in air to 1315°C for 6 h, and (c) the same gel and conditions as in (b) but without starch (Sin et al., 2001).

The dried gel was first heated to 450°C in flowing oxygen and then reheated in air in a conventional way to 1100°C–1350°C. The resulting mullite powder was acicular, and the average particle size was less than 1 μm. During combustion of the gel with the addition of rice starch in a fraction of 0.25 of the dry mass to 1315°C for 6 h, finer particles and more easily separated well-defined platelets are produced. In the absence of starch, the dried gel produced a ceramic foam that was composed of struts that consisted of intergrown rods (Fig. 6.2a–c). Two basic forms of mullite shape showed different utilities. For example, the fine powder may be suitable for the fabrication of monoliths. The platelets may be suitable for toughened composites.

6.2.6 Urea

Thim et al. (2001) and others cited acetyl acetone, carboxylates, carboxylic acids, and amides as compounds used to control the Al^{+3} hydrolysis rate by chelating it. This reduces the fraction of hydrated aluminum ions in aqueous solution as the reactant, and these chelating agents reduce the rates of the olation and oxolation reactions and drying by acting as dry chemical control agents (DCCAs). This reduces the solvent interfacial tension effect. To explore the role of urea, they did a Monte Carlo simulation to investigate the site-site correlation functions for ion-urea $g(r_{\text{IOu}})$ and ion-water $g(r_{\text{IOW}})$ interactions and liquid ^{27}Al NMR spectroscopy studies.

Like those agents, Thim et al. (2001) chose a high concentration of urea to synthesize mullite from an aqueous solution of silicic acid and aluminum nitrate, as stated earlier. They showed two sharp peaks at 2 Å and two more broad peaks around 4 Å and 6 Å due to water and urea distance distributions around aluminum ions. They also showed the coordination numbers between Al^{+3} and the oxygen sites of the water molecule (O_w) and between Al^{+3} and the oxygen sites of the urea molecule (O_u). It indicates the presence of four water and two urea molecules in the first coordination shell of the Al^{+3} ion and another ion coordination pattern for urea and water in the outside shells. With these results, they concluded that urea takes part in the ion coordination shell and even in the outside shells.

An ^{27}Al NMR spectrum for this aqueous solution showed five peaks—at 0.4, –2.3, –4.2, –5.8, and –7.4 ppm—which indicated the existence of five different types of solvated Al^{+3} ions. The peaks at –2.3 and –4.2 ppm were attributed to $[Al(H_2O)_5\cdot(urea)]^{+3}$ and $[Al(H_2O)_5\cdot(urea)_2]^{+3}$. They predicted from the NMR result that urea is part of the Al^{+3} coordination shell with the replacement of water molecules in the ion salvation shell.

Even with the extensive use of urea, a small fraction of spinel still crystallized along with the t-mullite at ~1050°C. They also calculated the activation energy, 770 KJ/mol, for this crystallization.

6.2.7 Aluminosiloxanes

Aluminosiloxanes were synthesized by Mehrotra and Pant (1963). Pouxviel et al. (1986) synthesized a mullite precursor with the help aluminosiloxanes other than separate alkoxides for a homogeneous aluminosilicate gel. They chose aluminosiloxane ($Al(OPr)_2OSiMe_3$) and silicon aluminum ester $(EtO)_3SiO\ Al(OBu)_4$ as starting materials for the synthesis of mullite. The latter source developed mullite at 1000°C.

6.3 Synthesis of Mullite Gel by Miscellaneous Methods

6.3.1 Synthesis of Mullite by the Hybrid Gel Route

Huling and Messing (1989) described a technique of processing seeded sol-gel mullite. The processing approach was a hybrid of the individual polymeric and colloidal methods. They studied the crystallization sequences of hybrid gels spanning the two ends of the spectrum—polymeric/molecular and colloidal/diphasic. A hybrid gel is a combination of polymeric/molecularly mixed alumina-silica sol and colloidal sol.

- Aqueous polymeric sol (seeding sol) was added to the colloidal sol to produce hybrid gels containing 2.5–30 wt % of polymeric mullite sol. These gels were denoted as 2.5S–30S hybrid gels. A DTA study of a polymeric mullite gel revealed a strong exotherm at 1017°C and of a colloidal gel exhibited a less intense exotherm at a significantly higher temperature of 1364°C. With an increase in the amount of seed polymeric sol, the hybrid gel exhibited a decrease in the DTA peak temperature. The 30% polymeric gel showed the start of the first exotherm in addition to the usual 1335°C exotherm. Transformation temperatures vis-à-vis phase development during heating indicated that the polymeric gel first crystallizes to mullite in the hybrid gel during heating and this in situ mullite acts as the seed crystals for nucleation during the subsequent transformation of the major colloidal part of the hybrid gel to mullite.

- Hybrid gels were also prepared by combining molecular sol with colloidal sol to the extent of 0 wt % to 10 wt% to 25 wt %, and these were denoted as 100M, 90M, 75M, etc. They showed that the molecular gel 100M crystallized directly and fully to mullite, with a sharp and intense exotherm. With the addition of 10 wt% of colloidal gel, the resultant hybrid gel (90M1) crystallized to a trace quantity of mullite with a less intense exotherm. A substantial amount of spinel formed in lieu of mullite. This indicated that the associated spinel phase impeded mullite formation and changed the independent crystallization behavior of the molecular gel. This may be the reason for the noncrystallization of ~90% of the mullite as theoretically expected on the basis of the composition of the 90% polymeric gel content. With further increase of the content of colloidal gel of the hybrid gel, 75M1 crystallized to spinel only and the associated exothermic peak intensity was substantially reduced.

Besides, Huling and Messing (1991) observed that the impediment of mullite also happened when the molecular gel was mixed with simple boehmite and was analyzed for DTA and XRD. It developed in situ γ-alumina. With these results, they proposed that γ-alumina formed in either of the cases served as the substrate for the nucleation of the spinel in the hybrid gel. They also stressed that such epitactic nucleation of spinel on γ-alumina is favorable as both are isostructural.

6.3.2 Synthesis of Optically Clear Mullite Ceramic

Yoldas (1980) synthesized optically clear mullite (OCM) ceramic with ~63% Al_2O_3 by high-temperature sintering. Colomban and Mazerolles (1990) developed OCM ceramic by the sol-gel processing method using similar component mixtures but with different hydrolysis conditions. It was a slow route. They took ALOBu as the alumina component and varying types of silicon alkoxides, for example, methoxide, ethoxide, *sec*-butoxide, and isopropoxide. The component mixture in a nonaqueous solvent like propanol and in nonprotonic and nonpolar media, like hexane and cyclohexane, was slowly hydrolyzed with atmospheric moisture (relative humidity 60%) for many weeks or even 4–6 months. The water vapor met the solution surface through gas invasion at the interstices through the small pinholes in the cover. The phase transformation of this slow monolith was studied by energy-dispersive X-ray in order to obtain the complete view of the structural change at both local and microscopic levels. Colomban and Mazerolles (1990) showed a strong correlation between some crystallochemical changes. Departure of the OH^{-1} defects present in the polymeric glassy phase (0.5% in weight), intense DTA exotherm at ~1000°C, strong macroscopic shrinkage (≥8%), and simultaneous crystallization of the γ-Al_2O-like spinel phase were noted (Figs. 6.3a and b). The micrograph remained characteristic of a glassy phase, with particle sizes less than 5 nm and with ceramic-like fractures.

On thermal treatment of OCM at 1100°C, a few c-mullite crystals 20–50 nm in size grew at the surface. Even on annealing for 3 weeks, there was no significant grain growth. Colomban and Mazerolles (1990) was of the opinion that the crystallinity of mullite (peak intensity modification, narrowing) increased with temperature but not with the annealing time (hours, days, or weeks). The X-ray pattern of the sample heated to 1100°C for 4 days still showed an amorphous band at 0.4 nm. Furthermore, he noted optical clarity even up to 1480°C. With these two observations, he opined that no segregation likely takes place during the transformation of the gel to o-mullite and a progressive ordering at the local scale may take place.

At 1400°C, a glass-like phase still remained. A fracture micrograph of it showed a mirror habit. Thus, the slow-hydrolyzed monolith remained optically clear up to 1400°C and showed glass-like mechanical property. At 1450°C and above, a homogeneous fine microstructure was observed, which showed an intergranular ceramic-like structure. In an alumina-rich composition ($x > 0.4$), Colomban and Mazerolles (1990) observed high-temperature form of the disordered spinel with a composition close to $0.75Al_2O_3$-$0.25SiO_2$.

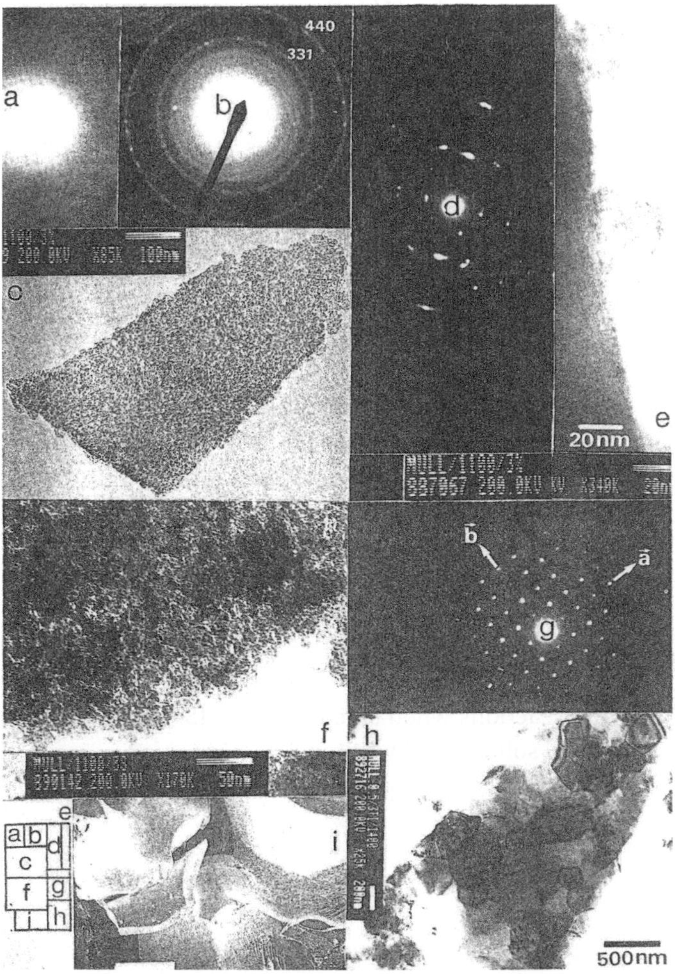

Figure 6.3 Evolution of mullite from the glassy state to the well-crystallized state: (a) Electron diffraction patterns of mullite porous glass heated at 800°C for a week. (b) Electron diffraction pattern of an optically clear monolith heated for 20 h at 1100°C; the rings correspond to the spinel γ-Al_2O_3-like structure. The corresponding electron diffraction is given in (c). (d) At the surface of a particle, small cubic crystals of disordered mullite can be obtained. After 3 weeks at 1100°C, (f) the amorphous aspect of mullite is always observed but (g) well-crystallized particles are also visible. (i) Finally, after heating the monolith at 1400°C, a homogeneous fine microstructure is observed, whereas the microscopic aspect of the monolith fracture remains glass-like (Colomban & Mazerolles, 1990).

Orefice and Vasconcelos (1997) synthesized two types of mullite gels of 40 mol % silica: (i) a transparent monolithic xerogel, which was obtained by hydrolyzing a mixture of prehydrolyzed TEOS (water:TEOS molar ratio 2) and hydrolyzed AIP in the pH range 2.5–5 at room temperature. This gel was found to be monophasic in character, (ii) a non-transparent gel with macroscopic defects produced by using same two sources but processed at a higher temperature (90°C) instead at room temperature.

Chemical method: Reynen and Faizullah (1974) reported a wet chemical method of mullite synthesis in which they used TEOS and aluminum sulfate as the precursors and the hot petroleum drying technique. The process involves boiling the precursor in refluxing kerosene, filtering it, and drying it in the presence of $Al(OH)_3$.

Meng and Huggins (1983) chose three component mixtures: (i) silicon acetate and ANN in a water medium, (ii) silicon acetate and aluminum sulfate in methanol, and (iii) silicon acetate and ANN in methanol. The mixed suspensions were gelled by drying on a hot plate. In these case, well-defined mullite was noted on heating at a temperature >1300°C.

In the wet chemical method, Meng and Huggins (1983) stated the following reactions when silicon acetate [$Si(C_2H_3O_2)_4$] was put into a methanol solution of hydrated ANN.

$$\begin{array}{c} Si(C_2H_3O_2)_4 + 4CH_3OH \rightarrow Si(OH)_4 + 4CH_3C{=}O(CH_3) \\ \downarrow \\ SiO_2{\cdot}2H_2O \end{array} \tag{6.1}$$

The reaction in the methanol solution was gentle and formed hydrated silica and that was dispersed in in the solution. It was conceived that silica gel absorbed aluminum nitrate and formed an intimate mixture during the drying process. On heating at 150°C–200°C, the gel dehydrated and aluminum nitrate decomposed and produced a fine precursor phase.

$$\begin{array}{c} Si(C_2H_3O_2)_4 + 4H_2O \rightarrow Si(OH)_4 + 4CH_3C{=}O(OH) \\ \downarrow \\ SiO_2{\cdot}2H_2O \end{array} \tag{6.2}$$

In an aqueous system, however this reaction was very violent because a lot of heat was generated in this reaction.

Kansal and Laine (1997) described a low-cost route to an easily processed aluminosilicate precursor by directly reacting SiO_2 with $Al(OH)_3$ in EG and triethanol amine (TEA), called the oxide one-pot synthesis (OOPS) process. A mullite precursor was also synthesized by reacting the individual components (TEA-Al and TEA-Si-EG) in a 3:1 ratio. This precursor behaved identically to precursors synthesized by other methods, like SH gels. Moreover, this can be made within a few hours, in contrast to the sol-gel processing of mullite materials, which may require a few weeks.

Suttor et al. (1997) followed an alternative strategy to synthesize mullite from Al_2O_3-filled siloxanes. They used a commercially available thermosetting solid meltable polymethyl siloxane as the silica precursor and α-alumina (0.8 μm) as the filler component. These are homogenized in a solvent like ethanol or acetone and subsequently evaporated to obtain granules, which are then cross-linked in a conventional die press at 200°C and a pressure of 15 MPa into dense granules and finally calcined for oxidation and mullitization.

Iwahiro et al. (2001) prepared mullite by the geopolymer technique to obtain nanometer-sized mullite polycrystals. They reacted alkaline silicate, that is, water glasses directly with ANN at room temperature in view of the polycondensation mechanism of $[SiO_4]$-monomers of the water-soluble component. Mullite crystallization occurred at around 1000°C, with the exhibition of an exotherm, when metasilicate was used, while it occurred at around 1100°C when disilicate was used.

Huizhong et al. (2003) synthesized mullite powder with nanosized particles using AIP and TEOS as starting materials through the sol-gel method followed by a supercritical fluid drying process, as per the process shown in Fig. 6.4.

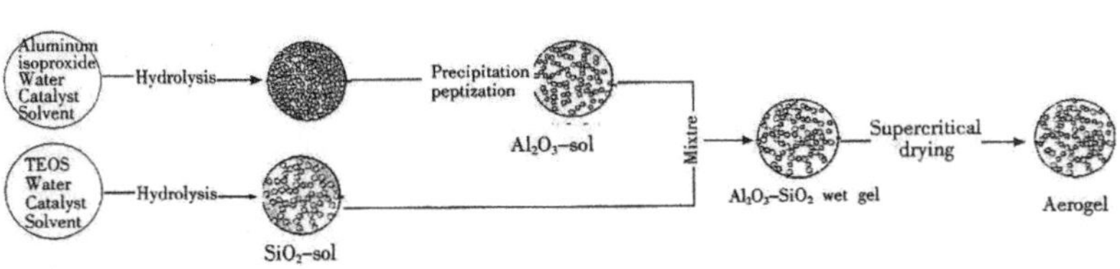

Figure 6.4 Schematic process (Huizhong et al., 2003).

XRD patterns of this aerogel on heat treatment at increasing temperatures showed first the formation of a spinel phase at ~1020°C, which transformed into mullite at ~1200°C. The size of the mullite is 50 nm and the surface area is 95.81 m^2/g.

6.4 Pilot-Scale Process of Mullite Synthesis by the Consecutive Precipitation Method

6.4.1 By the Aqueous Gel Route

Insley and Ewell (1935) synthesized three mixed gels, marked Gel A (precipitated Al_2O_3 and SiO_2 gels were mixed and filtered), Gel B (dried Al_2O_3 and SiO_2 gels were mixed and ground thoroughly), and Gel C (where silica gel was precipitated around alumina particle), giving a voluminous precipitate and finally compared them for their heat effects. These mixed gels too gave an exothermic effect at the same temperature as observed in the case of CP gels, shown above.

6.4.2 By the Monophasic Gel Route

Perry (1973) hydrolyzed $(EtO)_4Si$ with an ammonia solution in the presence of $Al(OH)_3$ synthesized from the hydrolysis of $AlCl_3$. The precipitate was filtered, washed with water, dried, and calcined to 1400°C for conversion into mullite.

6.4.3 By Consecutive Precipitation and Oxide Mixtures

Kumazawa et al. (1991) synthesized two precursors, as follows. Precursor D was made by the sequential hydrolysis of two components $Al(OPr)_3$ and $Si(OMe)_4$ with H_2O and NH_4OH and subsequent mixing. Precursor E was synthesized by the sequential hydrolysis of the same two components as in D, but the hydrolyzed powder was calcined and subsequently the two oxides were mixed. Samples marked D (assumed to be diphasic in character) and E (oxide mixture) showed the absence of a 980°C exotherm. D exhibited one broad exotherm in the temperature range of 1200°C to 1250°C and then a broad exotherm at ~1280°C, whereas E showed an exotherm at ~1340°C due to mullitization. The DTA exothermic behaviors of these precursors are different. The crystalline phases formed are shown in Chapter 12.

6.4.4 Colloidal Process

Metcalfe and Sant (1975) used a colloidal process in which the mixed sols were dried (instead of being chemically gelled and named sequential or consecutive precipitation). Aluminum hydroxide was first formed by mixing an NH_4OH solution with an $AlCl_3$ solution. TEOS was subsequently added, and hydrolysis/condensation reactions were carried out by heating the mixture to 95°C for 4 h. The Al_2O_3 content was ~71.8 wt % in the gels. The dry-milled powders were calcined for times up to 144 h. The ground powders were classified by the size of the particles. Powders were dry-pressed at 155 MPa and compacts were subsequently sintered at 1700°C. The sintered density increased with a decrease in the size of the particles from 4 μm. Compacts could be sintered to ~98% relative density in 1 h if all the powder particles were <4 μm. XRD results showed that mullite formation occurred after calcination at 1200°C but complete mullite formation required heat treatment at 1400°C for 4 h.

Colloidal methods shown by Ghate et al. (1973), review regarding the production process of mullite by Okada et al. (1991) in Japan, and finally a survey of mullite synthesis by Sacks et al. (1990) are most useful in such context.

6.5 Summary

- Mullite synthesized using the same source components but differing in processing conditions produces gels/precursors of different characters.
- Mullite gels can be synthesized by the organic route as follows: A metal-organic silica and alumina mixture was dehydrated by vacuum at 450°C by West and Gray (1958) to synthesize a precursor powder. Three samples, with an Al_2O_3:SiO_2 ratio of 1.0152, 2.002, and 2.927, exhibited intense 980°C exotherms with the crystallization of spinel and mullite and thereafter exhibited a broad exothermic peak at temperatures ranging from 1200°C to 1400°C due to the transformation of spinel

to mullite. To minimize differential hydrolysis of components and to avoid inhomogeneity during gelation processes, synthesis of mullite gel via organic processes was thought of. In the organic method, metal ions can generally be modified by using some organic substances prior to the gelation process. Some organic compounds deployed in mullite synthesis are as follows:

- A complexation process of metal ions was introduced by Douy (1991) extending the use of citric acid to obtain Al citrate as a source of Al cation. TESPA was used as the source of the silica component by Douy (1991). Secondly, the solution was gelled by the in situ formation of a polyacrylamide network.
- Complexing agents generally modify hydrolysis-cum-polymerization of metal ions. Mizukami et al. (1997) tried to report a relationship between the complexing ability of Al cation and various ligands. They synthesized two series of samples by choosing two different sources of the Al component—A series from TEOS and ANN and B series from TEOS and dibutoxy ethyl acetoacetate aluminum—using the following organic ligands of varying bridging coordination abilities: EG, PD, BD, MPD, and DEME.
- To control the hydrolysis and condensation of AlOBu and TEOS, Heinrich et al. (1991) used an alcoholic solution (2-propanol) of acetyl acetone, which acts as a chelating compound and thus decreases the reactivity of the alkoxide. The molar ratio of AlOBu:(acetyl acetone):TEOS was 3:1.5:1 to synthesize a mullite powder with ultrafine and homogeneous particles. They further synthesized single-phase gels from aluminum-*sec*-butylate chelated with β-diketone and tetraalkoxysilane in isopropanol and 0.1 N HCl medium. With the use of two chelating compounds, the hydrolysis of AlOBu was chemically modified (Heinrich et al., 1991; Chen & Vilminot, 1994). The function of ethyl acetoacetate was to substitute one (–OBu) group in aluminum alkoxide.
- Acrylamide, *bis*-acrylamide monomers, and a thermal initiator are used for polymerization (Sin et al., 2001).

- Several compounds were used in the synthesis of mullite gels by Thim et al. (2001). (i) Acetyl acetone, carboxylates, carboxylic acids, and amides were the compounds used to control the Al^{+3} hydrolysis rate by chelating it. This reduced the fraction of hydrated aluminum ions in the aqueous solution as the reactant, and these chelating agents reduced the rates of the olation and oxolation reactions. (ii) The drying process was controlled by using a DCCA. This reduced the solvent interfacial tension effect.

To explore the role of urea, Thim et al. (2001) did a Monte Carlo simulation to investigate the site-site correlation functions for ion-urea $g(r_{\text{IOu}})$ and ion-water $g(r_{\text{IOW}})$ interactions and liquid ^{27}Al NMR spectroscopy studies. It was concluded that urea takes part in the ion coordination shell due to the presence of four water and two urea molecules in the first coordination shell of the Al^{+3} ion and another ion coordination pattern for urea and water in the outside shells.
Several reagents have been used for the synthesis of mullite gel by the organic route:

- Alcohol/acetonitrile as a dipolar aprotic solvent was used to obtain monodispersed spherical particles. Fine mullite powder was synthesized by Ogihara et al. (1994).
- Exchange reactions between alkoxides and chlorides have been done.
- The OOPS process was followed by Kansal and Laine (1997).
- The organic route was used an alternative strategy to synthesize mullite from Al_2O_3-filled siloxanes by Suttor et al. (2001).
- Mullite gel was synthesized by the decomposition of aluminosiloxanes by Pouxviel et al. (1986).

- Mullite gel can be synthesized by other miscellaneous methods:
  - Synthesis of mullite by the hybrid gel route: Originally, a strong exotherm at 1017°C was noted in the polymeric gel and a comparatively less intense exotherm at a significantly higher temperature at 1364°C was noted in the colloidal gel (Huling & Messing, 1989). Both

transformation temperatures as observed in DTA as well as phase developments during heating as monitored by XRD can be modified by adding a polymeric sol to the colloidal sol and vice versa. For example, with a gradual increase in the amount of polymeric sol to colloidal sol, the DTA exothermic peak temperature of the latter sol decreased and the intensity of mullite increased. In the reverse case, with an increase in the colloidal gel content, the intensity of the 980°C peak drastically decreased. The intensity of the formation of mullite and the spinel phase decreased.

 - An OCM monolith was developed by Colomban and Mazerolles (1990) by the sol-gel processing route in the presence of different solvents. Instead of oven drying or water bath drying of the wet gel, Huizhong et al. (2003) followed a new technique to obtain the aerogel—they used the sol-gel method but with the supercritical fluid drying process.

- A pilot-scale process of mullite synthesis done by the consecutive precipitation method and others are cited in the chapter.

References

1. K. Okada and N. Otsuka, Characterization of the spinel phase from SiO_2-Al_2O_3 xerogels and the formation process of mullite. *J. Am. Ceram. Soc.*, **69**(9), 652–656 (1986).
2. A. K. Chakraborty and D. K. Ghosh, Synthesis and 980°C phase development of some mullite gels. *J. Am. Ceram. Soc.*, **71**(11), 978–987 (1988).
3. J. A. Pask, X. W. Zhang, A. P. Tomsia, and B. E. Yoldas, Effect of sol-gel mixing on mullite microstructure and phase equilibria in the α-Al_2O_3-SiO_2 system. *J. Am. Ceram. Soc.*, **70**(10), 704–707 (1987).
4. B. E. Yoldas and D. P. Partlow, Formation of mullite and other alumina-based ceramics via hydrolytic polycondensation of alkoxides and resultant ultra- and micro-structural effects. *J. Mater. Sci.*, **23**, 1895–1900 (1988).
5. B. E. Yoldas, Effect of ultrastructure on crystallization of mullite. *J. Mater. Sci.*, **27**(24), 6667–6672 (1992).

6. K. Hamano, T. Sato, and Z. Nakagawa, Properties of mullite prepared by coprecipitation and microstructure of fired bodies. *Yogo Kyokai Shi*, **94**(8), 818–822 (1986).
7. B. Sonuparlak, Sol-gel processing of infrared transparent mullite. *Adv. Ceram. Mater.*, **3**(3), 263–267 (1988).
8. Ph. Colomban, Structure of oxide gels and glasses by infrared and Raman scattering part 2 mullite. *J. Mater. Sci.*, **24**, 3011–3020 (1989).
9. S. Mitachi, M. Matsuzawa, K. Kaneko, S. Kanzaki, and Y. Tabata, Characterization of SiO_2-Al_2O_3 powders prepared from metal alkoxides. *Ceram. Trans.*, **6**, 275–286 (1990).
10. J. C. Huling and G. L. Messing, Surface chemistry effects on homogeneity and crystallization of colloidal mullite sol-gels. *Ceram. Trans.*, **6**, 221–229 (1990).
11. T. Kumazawa, S. Ohta, S. Kanzaki, and H. Tabata, Influence of powder characteristics on microstructural and mechanical properties of mullite ceramics. *Ceram. Trans.*, **6**, 401–411 (1990).
12. M. Ge, H. Yang, Z. Jiang. Y. Wang, and F. Zhang, Ultrafine pure mullite powder by sol-gel method. *J. Non-Cryst. Solids*, **147–148**, 565–568 (1992).
13. A. Taylor and D. Holland, The chemical synthesis and characterization sequence of mullite. *J. Non-Cryst. Solids*, **152**, 1–17 (1993).
14. C. Geradin, S. Sundaresan, J. Benziger, and A. Navrotsky, Structural investigation and energetics of mullite formation from sol-gel precursors. *Chem. Mater.*, **6**, 160–170 (1994).
15. Y. X. Huang, A. M. R. Senos, J. Rocha, and J. L. Baptista, Gel formation in mullite precursor obtained via TEOS prehydrolysis. *J. Mater. Sci.*, **32**, 105–110 (1997).
16. A. M. L. M. Fonseca, J. M. F. Ferreira, I. M. M. Salvado, and J. L. Baptista, Mullite based compositions prepared by sol-gel techniques. *J. Sol-Gel Sci. Technol.*, **8**, 403–407 (1997).
17. B. E. Yoldas, Mullite formation from aluminium and silicon alkoxides. *Ceram. Trans.*, vol. 6, ed. S. Somiya, R. F. Davis, and J. A. Pask, *Am. Ceram. Soc.*, Westerville, OH, p. 255 (1990).
18. M. I. Nieto, G. Urretavizcaya, A. L. Cavalieri, and P. Rana, Structural changes in colloidal and polymeric aluminosilicate gels with mullite composition. *Br. Ceram. Trans.*, **97**(1), 17–23 (1998).
19. H. Schneider, B. Saruhan, D. Voll, L. Merwin, and A. Sebald, Mullite precursor phases. *J. Euro. Ceram. Soc.*, **11**, 87–94 (1993).

20. A. K. Chakraborty, DTA characterisation of three types of Al_2O_3-SiO_2 gels made from TEOS-Al$(OBu)_3$ mixture with variation of water. *Ceram. Int.*, **22**, 463–469 (1996).
21. S. Rajendran, H. J. Rossell, and J. V. Sanders, Crystallization of a coprecipitated mullite precursor during heat treatment. *J. Mater. Sci.*, **25**, 4462–4471 (1990).
22. I. Jaymes and A. Douy, Homogeneous precipitation of mullite precursors. *J. Sol-Gel Sci. Technol.*, **4**, 7–13 (1995).
23. I. Jaymes and A. Douy, Homogeneous mullite-forming powders from spray-drying aqueous solutions. *J. Am. Ceram. Soc.*, **75**(11), 3154–3156 (1992).
24. I. Jaymes, A. Douy, D. Massiot, and J. P. Coutures, Characterization of mono- and diphasic mullite precursor powders prepared by aqueous routes, ^{27}Al and ^{29}Si MAS-NMR spectroscopy. *J. Mater. Sci.*, **31**, 4581–4589 (1996).
25. M. Haque, Thesis, University of Kolkata (2000).
26. R. R. West and T. J. Gray, Reactions in silica-alumina mixtures. *J. Am. Ceram. Soc.*, **41**(4), 132–136 (1958).
27. C. Marcilly, P. Courty, and B. Delmon, Preparation of highly dispersed mixed oxides and oxide solid solutions by pyrolysis of amorphous organic precursors. *J. Am. Ceram. Soc.*, **53**(1), 56–57 (1970).
28. A. Douy and P. Odier, The polyacrylamide gel: a novel route to ceramic and glassy oxide powders. *Mater. Res. Bull.*, **24**, 1119–1126 (1989).
29. A. Douy, Organic gels preparation of silica-alumina powders, I, mullite. *J. Euro. Ceram. Soc.*, **7**, 117–123 (1991).
30. F. Mizukami, K. Maeda, and M. Toba, Effect of organic ligands on the formation of mullite, in *Abstracts of the 3rd Autuman Symposium of Ceramics*. Paper No. 6-1A01, Ceramic Society of Japan, Tokyo (1990).
31. F. Mizukami K. Maeda, M. Toba, T. Sano, S.-I. Niwa, M. Miyazaki, and K. Kohma, Effect of organic ligands used in sol-gel process on the formation of mullite. *J. Sol-Gel Sci. Technol.*, **8**, 101–106 (1997).
32. T. Heinrich, F. Raether, O. Spormann, and J. Fricke, SAXS measurements of condensation in mullite precursors. *J. Appl. Cryst.*, **24**, 788–793 (1991).
33. T. Heinrich and F. Raether, Structural characterization and phase development of sol-gel-derived mullite and its precursors. *J. Non-Cryst. Solids*, **147–148**, 152–156 (1992).

34. Y.-F. Chen and S. Vilminot, Thermal evolution of TEOS-Al(*sec*BuO)$_3$-ETAC mullite gels. *J. Sol-Gel Sci. Technol.*, **2**, 399–402 (1994).

35. A. Sin, J. J. Picciolo, R. H. Lee, F. Gutierrez-Mora, and K. C. Goretta, Synthesis of mullite powders by acrylamide polymerization. *J. Mater. Sci. Lett.*, **20**, 1639–1641 (2001).

36. G. P. Thim, C. A. Bertran, V. E. Barlette, M. I. F. Macedo, and M. A. S. Oliveira, Experimental and Monte Carlo-simulation: the role of urea in mullite synthesis. *J. Euro. Ceram. Soc.*, **21**, 759–763 (2001).

37. R. C. Mehrotra and B. C. Pant, *Ind. J. Appl. Chem.*, **26**, 109–111 (1963).

38. J. C. Pouxviel, J. P. Boilot, A. Daugerr, and L. Huker, Chemical route to aluminosilicate gels, glasses and ceramics, in *Better Ceramics Through Chemistry II*, ed. C. J. Brinker, D. E. Clark, and D. R. Ulrich, Materials Research Society, Pittsburgh, pp. 269–274 (1986).

39. J. C. Huling and G. L. Messing, Hybrid gel for homoepitactic nucleation of mullite. *J. Am. Ceram. Soc.*, **72**, 1725–1729 (1989).

40. J. C. Huling and G. L. Messing, Epitactic nucleation of spinel in aluminosilicate gels and its effect on mullite crystallization. *J. Am. Ceram. Soc.*, **74**(10), 2374–2381 (1991).

41. B. E. Yoldas, Microstructure of monolithic materials formed by heat treatment of chemically polymerized precursors in the Al_2O_3-SiO_2 binary. *Ceram. Bull.*, **59**(4), 479–483 (1980).

42. Ph. Colomban and L. Mazerolles, SiO_2-Al_2O_3 phase diagram and mullite non-stoichiometry of sol-gel prepared monoliths: influence on mechanical properties. *J. Mater. Sci. Lett.*, **9**, 1077–1079 (1990).

43. R.-L. Orefice and W.-L. Vasconcelos, Sol-gel transition and structural evolution on multicomponent gels derived from the alumina-silica system. *J. Sol-Gel Sci. Technol.*, **9**, 239–249 (1997).

44. P. Reynen and M. Faizullah, Sintering of mullite, in *Materials Science Monographs: 4. Sintering—New Developments*, ed. M. M. Ristic, Elsevier Science, Amsterdam-Oxford-New York (1974).

45. G.-Y., Meng and R. A. Huggins, A new chemical method for preparation of both pure and doped mullite. *Mater. Res. Bull.*, **18**, 581–588 (1983).

46. P. Kansal and R. M. Laine, A processable mullite precursor prepared by reacting silica and aluminum hydroxide with triethanolamine in ethylene glycol: structural evolution on pyrolysis. *J. Am. Ceram. Soc.*, **80**(10), 2597–2606 (1997).

47. D. Suttor, H.-J. Kleebe, and G. Zieger, Formation of mullite from filled siloxanes. *J. Am. Ceram. Soc.*, **80**(10), 2541–2548 (1997).

48. D. Suttor, H.-J. Kleebe, and G. Zieger, Low shrinkage mullite derived from filled polymeric siloxanes. *Key Eng. Mater.*, **132–136**, 448–451 (1997).

49. T. Iwahiro, Y. Nakamura, R. Komatsu, and K. Ikeda, Crystallization behavior and characteristics of mullites formed from alumina-silica gels prepared by the geopolymer technique in acidic conditions. *J. Euro. Ceram. Soc.*, **21**, 2515–2519 (2001).

50. Z. Huizhong, J. I. Daojun, L. E. I. Zhongxig, W. Houzhi, and Z. Wenjie, Synthesis and characterization of nanosized mullite powder. *J. Chin. Ceram. Soc.*, **31**(12), 1216–1220 (2003).

51. H. Insley and R. H. Ewell, Thermal behavior of the kaolin minerals. *J. Res. Nat. Bur. Stand.*, **14**(5), 615–627 (1935).

52. G. S. Perry, Microwave dielectric properties of mullite. *Trans. J. Br. Ceram. Soc.*, **72**, 279–283 (1973).

53. T. Kumazawa, S. Ohta, S. Kanzaki, and H. Tabata, Influence of powder characteristics on microstructural and mechanical properties of mullite ceramics (74 wt% Al_2O_3). *J. Jpn. Ceram. Soc.*, **99**, 1228–1233 (1991).

54. B. L. Metcalfe and J. H. Sant, The synthesis, microstructure and physical properties of high purity mullite. *Trans. J. Br. Ceram. Soc.*, **74**, 193–201 (1975).

55. B. B. Ghate, D. P. H. Hasselman, and R. M. Spriggs, Synthesis and characterization of high purity, fine grained mullite. *Ceram. Bull.*, **52**(9), 670–672 (1973).

56. K. Okada, H. Otsuka, and S. Somiya, Review of mullite synthesis routes in Japan. *Ceram. Bull.*, **70**(10), 1633–1640 (1991).

57. M. D. Sacks, H.-W. Lee, and A. Pask, A review of powder preparation methods and densification procedures for fabricating high density mullite. *Ceram. Trans.*, **6**, 167–207 (1990).

58. T. Ogihara, T. Yanagawa, N. Ogata, K. Yoshida, M. Iguchi, N. Nagata, and K. Ogawa, Synthesis of mono dispersed, spherical fine mullite powders by alkoxide method. *J. Ceram. Soc. Jpn.*, **102**(8), 778–784 (1994).

Additional References

1. J. Parmentier and S. Vilminot, Influence of synthesis and composition on mullite crystallization. *Chem. Mater.*, **9**(5), 1134–1137 (1997).

2. A. M. R. Senos, J. Rochaj, and L. Baptista, Gel formation in mullite precursors obtained via tetraethylorthosilicate (TEOS) pre-hydrolysis. *J. Mater. Sci.*, **33**(17), 4435–4446 (1998).
3. K. Sinkó, and R. Mezei, Preparation effects on sol-gel aluminosilicate gels. *J. Non-Cryst. Solids,* **231**, 1–9 (1998).
4. D. Voll, A. Beran, and H. Schneider, Temperature-dependent dehydration of sol-gel-derived mullite precursors: an FTIR spectroscopic study. *J. Eur. Ceram. Soc.*, **18**, 1101–1106 (1998).
5. E. Tkalcec, D. Hoebbel, R. Nass, and H. Schmidt, Structural changes of mullite precursors in presence of polyethyleneimine. *J. Non-Cryst. Solids,* **243**, 233 (1999).
6. U.-Y. Hwang, J.-W. Lee, J.-H. Choi, H.-S. Park, S.-J. Yoo, H.-s. Yoon, and Y.-R. Kim, Synthesis of spherical pre-mullite particles by sol-gel method and mullitization mechanism of pre-mullite, *Hwahak Konghak*, **38**(5), 669–675 (Journal of the Korean Institute of Chemical Engineers) (2000).
7. V. V. Vol'khin, I. L. Kazakova, P. Pongratz, and E. Halwax, Mullite formation from highly homogeneous mixtures of Al_2O_3 and SiO_2. *Inorg. Mater.,* **36**, 375–379 (2000).
8. H. Shin, C.-S. Kim, and S.-N. Chag, Mullitization from a multicomponent oxide system in the temperature range 1200°–1500°C. *J. Am. Ceram. Soc.*, **83**(5), 1237–1240 (2000).
9. P. Padmaja, G. M. Anilkumar, P. Mukundan, G. Aruldhas, and K. G. K. Warrier, Characterisation of stoichiometric sol-gel mullite by Fourier transform infrared spectroscopy. *Int. J. Inorg. Mater.*, **3**(7), 693–698 (2001).
10. S. Satoshi, C. Contreras, H. Juárez, A. Aguilera, and J. Serrato, Homogeneous precipitation and thermal phase transformation of mullite ceramic precursor. *Int. J. Inorg. Mater.*, **3**(7), 625–632 (2001).
11. K. Sinkó, R. Mezei, and M. Zrínyi, Gelation of aluminosilicate systems under different chemical conditions. *J. Sol-Gel Sci. Technol.*, **21**, 147–156 (2001).
12. Y.-f. Tang, A.-d Li, H.-q. Ling, Y.-j. Wang, Q.-y. Shao, Y.-n. Lu, and Z.-d. Ling, Fabrication of ultrafine mullite powders. *Mater. Chem. Phys.*, **75**, 301–304 (2002).
13. U. Y. Hwang, J. U. Lee, H. S. Park, Y. R. Kim, and G. G. Gu, The formation mechanism of pre-mullite particles by the mixed solvents method. *Korean Chem. Eng. Res.*, **41**(4), 509–509 (2003).

14. H. Zhao, K. Hiragushi, and Y. Mizota, Mullite formation of colloidal matrix hybrid aluminosilicate gel. *J. Sol-Gel Sci. Technol.*, **27**, 287–291 (2003).

15. U.-Y. Hwangn, J.-W. Lee, H.-S. Park, Y.-R. Kim, and K.-K. Koo, The formation mechanism of pre-mullite particles by the mixed solvents method. *Hwahak Konghak*, **41**(4), 509–516 (2003).

16. J. Macan, H. Ivanković, and E. Tkalčec, Crystallization behavior of hybrid premullite powder synthesized by sol-gel method. *Conference on Materials, Processes, Friction and Wear MATRIB'04*, Vela Luka (2004).

17. Y. F. Chen, M. C. Wang, and M. H. Hon, Phase transformation and growth of mullite in kaolin ceramics. *J. Am. Ceram. Soc.*, **24**, 2389–2397 (2004).

18. A. C. Osawa and C. A. Beltran, Mullite formation from mixtures of alumina and silica sols: mechanism and pH effect. *J. Braz. Chem. Soc.*, **16**(2), 251 (2005).

19. E. R. Sola, F. Estevan, F. J. Torres, and J. Alarcón, Effect of thermal treatment on the structural evolution of 3:2 and 2:1 mullite monophasic gels. *J. Non-Cryst. Solids,* **351**, 1202–1209 (2005).

20. J. Anggono, Mullite ceramics: its properties, structure, and synthesis. *Mater. Sci., Jurnal Teknik Mesin*, **7**(1), 1–10 (2005).

Chapter 7

Phase Evolution Sequences of Mullite Precursors Synthesized by Different Methods Based on Components Used and Processing Variables

7.1 Monophasic Mullite Gel Prepared by the Aqueous Sol-Gel Method

A monophasic gel was first synthesized by Insley and Ewell (1935) and then by Ossaka (1961), who coprecipitated (CP) $KAl(SO_4)_2$ and sodium silicate hexaethylenetetraamine. Later, Chakraborty and Ghosh (1988) synthesized a monophasic gel from silica sol and aluminum sulfate; then Schneider et al. (1992) synthesized one by the aqueous sol-gel method. It exhibited only one exotherm, with the formation of mullite as the major phase and occasionally traces of the spinel phase. The phase evolution processes of these monophasic gels are already given in Chapter 2.

7.1.1 Effect of pH on the Phase Transformation of Monophasic Mullite Gel Prepared by Aqueous Sol-Gel Method

Example 1: A monophasic gel prepared from silica sol and aluminum sulfate formed mullite only at the 980°C exotherm (Chakraborty &

Mullite Formations: Analysis and Applications
Akshoy Kumar Chakraborty

ISBN 978-981-4877-05-3 (Hardcover), 978-1-003-03167-3 (eBook)
www.jennystanford.com

Ghosh, 1988). The amount of mullite produced was dependent on the pH maintained during the gelation processes. It was further noted by X-ray diffractometry (XRD) that the intensity of the 980°C exotherm and the quantity of mullite formed were directly proportional. Both the differential thermal analysis (DTA) intensity and the XRD intensity of mullite increased with an increase in the pH of the mixed solutions from 2.5 to 4, and the intensities of both events were found to be maximum at ~4.5 pH.

Example 2: Al_2O_3-SiO_2 xerogel marked 7030(N) prepared from aqueous silica sol and aluminum sulfate at pH ≤ 1 did not exhibit either exotherm or a spinel phase but developed γ-Al_2O_3 (Table 7.1). On heating at >1200°C, it developed corundum and cristobalite. The intensities of these two phases first increased with an increase in temperature to 1300°C. At ~1400°C, mullite formed, with marked diminution in the intensities of both reactant phases. The crystallization behaviors of these three phases are shown in Fig. 7.1, and the mullitization behavior of Al_2O_3-SiO_2 xerogel marked 7030(N) was compared with that of monophasic gel G-104. Both the DTA study of the dried gel and the XRD studies of the heat-treated gel 7030(N) were distinctly different from those of monophasic gel G-104 (Fig. 7.2). These variations are found to be due to variations

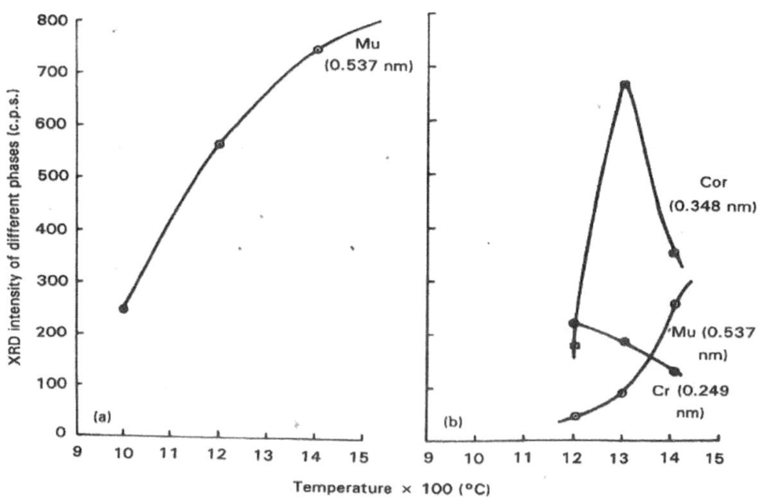

Figure 7.1 XRD intensities of phases formed against the firing temperature of gels (a) G-104 and (b) 7030(N) (Chakraborty, 1994a).

Table 7.1 Summarized results of DTA and XRD analyses of monophasic mullite gels to 980°C

Gel mark	Starting material	pH of the salt–sol mixture	DTA analysis	XRD analysis (980°C-heated gel sample)
G-104	Silica sol made by ion exchange of sodium silicate and aluminium sulphate hydrate	~4.5	Very strong 980°C exotherm	Mullite
G-96	As above	4.0 (adjusted by use of NH_4OH solution)	Strong 980°C exotherm	Mullite
G-107	As above	3.8	Medium-strong 980°C exotherm	Mullite
7030 (N)	Same silica sol as above and aluminium sulphate hydrate of (S, M)	~1.0	Very insignificant 980°C exotherm	γ-Al_2O_3

Source: (Chakraborty, 1994a)

in the pH of the mixed solution produced by the use of aluminum sulfate hydrate from two different sources (Table 7.1). Therefore, pH plays a vital role in the development of gels of different characters, which in turn determines the differences in crystallization processes that occur during subsequent thermal sequences. In the pH range of 3–4.5, Al_2O_3-SiO_2 xerogel transforms directly into mullite at the 980°C exotherm, whereas the gel prepared at pH ≤ 1 forms mullite via intermediate γ-Al_2O_3 formation. Accordingly, it is conceived that there are two different mechanisms of gelation reaction. The phase transformation of Al_2O_3-SiO_2 xerogels in the entire pH range of 1 to 7 to 14 were shown by Chakraborty (1994a).

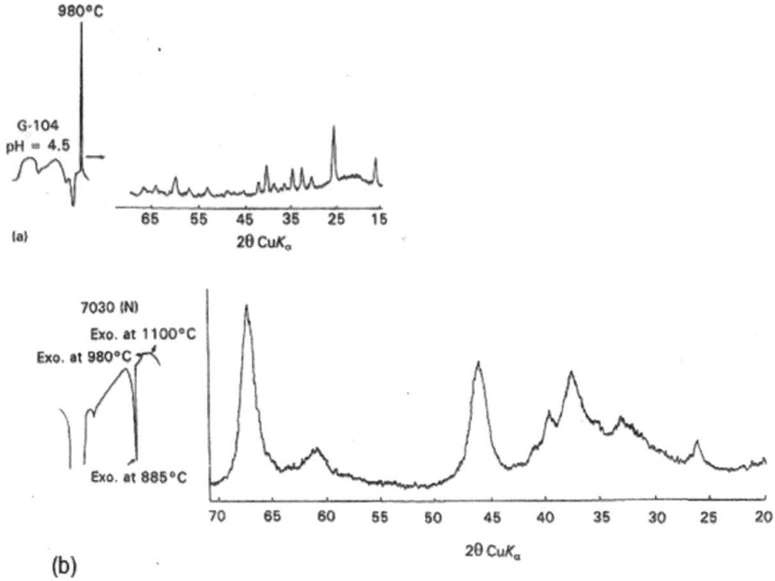

Figure 7.2 Comparative DTA and XRD intensities of phases formed against the firing temperature of gels (a) G-104 and (b) 7030(N) (Chakraborty, 1994a).

7.1.2 Effect of Silica Sols from Different Sources on the Phase Transformation of Mullite Gel Prepared by Aqueous Sol-Gel Method

Example 1: The character of the aqueous silica sol is a factor for determining the extent of orthorhombic mullite (o-mullite) formed at the 980°C exotherm and its exhibition. Table 7.2 shows the nature of exotherm vis-à-vis phases crystallized from different Al_2O_3-SiO_2 gels at 980°C. These events are a function of the starting silica sol as the source material. The formation of any form of mullite, o-mullite, or cubic mullite (Al-Si spinel, also called c-mullite) either alone or as a mixture is primarily responsible for the choice of starting components and for the process of gelation. Now we need to ask two questions, (i) what are the roles of the individual silica and alumina reacting components derived from different sources and (ii) how do the reactive species interact or function during the process of gelation? As far as interactions of silica and alumina component sols are concerned, Iller (1979) demonstrated the basic reaction in which the two sols chemically react and form aluminosilicic acid with the liberation of 1 equivalent of acid.

Table 7.2 The 980°C DTA peak and formation of phase from Al_2O_3-SiO_2 gel, which are dependent

No.	Source of silica sol	Nature of exotherm	Phase developed during exotherm
1.	Sodium silicate obtained from a solution of SiO_2(A) in hot NaOH Sol made by the ion-exchange method Al_2O_3-SiO_2 gel obtained at pH 3 using ANN	Very sharp	Mullite-type phase
2.	Commercial sodium silicate (1:2) Similar process of gel making	Sharp	γ-Al_2O_3-type spinel phase + o-mullite
3.	Sodium silicate made by the carbonate fusion method, having Na_2O:SiO_2 = 1:3.3 Similar process of gel making	Sharp	Mullite-type phase + γ-Al_2O_3 type spinel
4.	Ludox used as the source Similar process of gel making	Small peak	γ-Al_2O_3-type spinel

Source: (Chakraborty & Ghosh, 1988)

The details are given in Chapter 10. The reactivity and extent of the condensation reaction above depend upon the active sites of the reacting components. It is believed that silica sol should be the primary (–Si–OH) species instead of polymerized units, since the reactivity decreases with an increase in polymerized $Si(OH)_4$ units. It is presumed that the nature of silica sol obtained from the four sources may vary. The molecular weight of the silica sol is to be determined to find out the cause of this variation. An example is cited now for presenting a comparative behavior with a study when a larger particle of silica is used as a source.

Example 2 (the role of monosilicic acid sol versus silicic acid sol from fumed silica): Two different processes were carried out for mullite formation and examined by Nishu et al. (1989). The first process was based on combining the $Al(OH)_3$ and the monosilicic solution. They prepared alumina sol $Al(OH)_3$ by adjusting the pH of $AlCl_3$ solution to 8 by adding aqueous ammonia solution to it. The monosilicic acid solution (mostly monomer) was prepared by dissolving and breaking silica units of silica gel under boiling conditions with water adjusted to pH 9 by the addition of aqueous ammonia. The alumina sol and silica solution were mixed at pH 9 and stirred for 5 h at room temperature to form a gel, which formed tetragonal mullite (t-mullite) directly on heating to 980°C. In the second process, the silica sol obtained by dissolving fumed silica (particle size < nanometers) was combined with aluminum salt, for example, nitrate, sulfate, or chloride. Spinel formation preceded mullite formation, that is, spinel formed at 1000°C and transformed to mullite on further firing to a temperature above 1200°C. The large differences in phase development and mullite formation paths of these mullite gels depend upon the development of the polymeric source of silica component. The formation of spinel prior to the formation of mullite is due to the choice of silica source as fumed silica of agglomerated particles of silica were used for the synthesis of SiO_2 sol. Thus the amount of polymeric silicic acid units plays a major role in phase evolution.

Example 3: Colloidal silica along with usual aluminum nitrate nonahydrate (ANN) were used by Kim et al. (2003) for the synthesis of a mullite precursor by the coprecipitation process at acidic pH. This precursor developed cristobalite during the phase evolution process. Its peak intensity increased from 1200°C to the maximum at 1300°C. Thereafter, it decreased and finally the precursor crystallized to mullite at >1400°C. It is presumed that the crystallization of the silica component to cristobalite prior to the formation of mullite may be due to the choice of highly polymerized colloidal silica source instead of tetraethyl orthosilicate (TEOS) as the commonly used component. Likely, the large colloidal silica particles did not penetrate smoothly into the alumina network the during gelation process and even during the heating process, as shown by Fukuoka et al. (1993).

Thus, a monophasic gel synthesized by the aqueous sol-gel route, as per the method shown in Chapter 2, showed direct mullite formation during the exhibition of 980°C exotherm, as shown by Insley and Ewell (1935), Ossaka (1961), and by Chakraborty and Ghosh (1988). The nature of the exotherm and phase development are dependent on the pH of the solution, as shown in Table 7.2. The tentative thermal reaction was shown in Chapter 2.

7.2 Monophasic Mullite Gel Prepared by the SH Gel Method and Variables Affecting Its Phase Evolution

7.2.1 Effect of pH on the Phase Transformation of This Monophasic Mullite Gel

7.2.1.1 Monophasic gel (SH)

Monophasic gel was first synthesized by Hoffman et al. (1984); thereafter, Okada and Otsuka (1986) synthesized it and designated it as slow hydrolysis (SH) gel. These authors showed the formation of mullite and spinel (minor) phases after a sharp first exotherm. Of all the parameters of hydrolysis, the role of pH on phase development is one of the important subject matters and accordingly, the relevant works of earlier researchers are cited next and it is shown that with components from the same source but with differing values of pH of processing, there will be variation in phase evolution.

In acidic pH

Example 1: A few monophasic gels were synthesized by Chakraborty and Ghosh (1988) in acidic conditions by using different quantities of alcohol, which led to changes in the pH values, as in gels marked G-139 to G-142, to finally attain during gelation an alcoholic solution of pH value 3.5–4.5. These gels showed differences in 980°C phase developments. Thus, when an acid was used externally as a hydrolyzing catalyst, as in the case of coprecipitation or cohydrolysis reactions of silica (TEOS) and alumina components (ANN), or when it was liberated by the hydrolysis of acidic salts, such as aluminum nitrate, aluminum sulfate, or aluminum chloride, the resultant gels exhibited 980°C exotherm and formed predominantly mullite, along with a minor amount of Al-Si spinel phase (Okada & Otsuka, 1986; Chakraborty & Ghosh, 1988).

Example 2: Li and Thomson (1991) varied the hydrolysis condition of some monophasic gels synthesized from components from the same source, example, TEOS and ANN (see Chapter 2). A dynamic X-ray diffraction (DXRD) scan of integrated intensities of the 120 and 210 double peak of mullite showed two plateaus for 2WC, 2W, 2D, 1D, and 2H gels and suggested a two-stage transformation of mullite. The first step of transformation corresponded to the first exotherm at ~980°C. The second step of mullite transformation likely occurred at the second exotherm at 1200°C. But the colloidal gel did not show any mullite formation; it occurred rapidly at 1200°C and matched with the second exotherm. The phase transformation of the several gels synthesized from TEOS and ANN at different gelation times was studied by Li and Thomson (1991). Varying degrees of Al-Si-spinel-cum-t-mullite formation during the heating process at ~980°C was shown by them (Table 7.3).

Table 7.3 Apparent variation of 980°C exothermic peak and formation of mullite with processing of monophasic gels

980°C reaction	The exothermic peak was detected in 2H (insignificant), 1D (highest), 2D (medium), 2WC (less medium), and 2W (small).
Mullite formation	Two-stage mullite formation were seen in DXRD versus temperature plots of different gels.

The first stage of transformation took place at 980°C for all single-phase gels although the extent of mullite formation at this temperature varied among gels. Regarding the nature of transformation, it was noted that the rapid transformation, which started around 1000°C, soon levelled off before reaching full conversion. Thus, 980°C DTA for single-phase gels was due to mullite formation. The second stage of mullite transformation, again for single-phase gels, took place at the very weak second exothermic peak at ~1200°C in DTA and thereafter, it proceeded gradually with a rise in temperature.

The diphasic gel transformed into a significant amount of mullite at a temperature of ~1300°C, which is higher than 1200°C, as had happened in other cases of single-phase gels and colloidal gels.

The following was concluded: (i) A slower hydrolysis gel tends to produce a gel with comparatively better molecular mixing. (ii) The so-called single-phase alumina-silicate gel is not purely single phase. It may be a mixture of a single-phase gel and a diphasic gel because of the fact that it exhibits a two-stage conversion of mullite. (iii) The formation of both spinel and mullite results in an exothermic peak at 980°C according to Li and Thomson (1991). However, it is noted that the intensities of DTA exotherms of 2H, 1D, 2D 2W, and 2WC as shown by them do not validate the corresponding X-ray intensity of the mullite formed in those noted gels. Secondly, no correlation is noticed between the gelation time for the synthesis of monophasic gels with 980°C exotherms. Gels were levelled according to the length of gelation time by Li and Thomson (1991). They correlated them with the gels of previous authors.

- Considering the time of gelation, they tried to compare a 2H gel with a rapid hydrolysis (RH) gel and a 2W gel with the SH gel by Okada and Otsuka (1986). This comparison does not support the 980°C peak height data given by the authors.
- According to nitrate-to-ethanol weight ratios used by Hoffman et al. (1984), for their "more ethanol" and "less ethanol" gel samples, they again correlated those with 2W and 2D gels, respectively. However, both 2W and "more ethanol" samples exhibited small exotherms. Comparatively, the nature of exotherm as shown by 2W is more significant than that shown by Hoffman et al. (1984).
- Furthermore, they synthesized a gel by introducing ANN into a completely hydrolyzed TEOS solution, which is more or less analogous to the procedure as per the concept of Yoldas (1990) and marked as a colloidal gel. Moreover, the phase evaluation process of these gels showed itself to be very different from the phase evaluation processes of gels synthesized by previous authors.

The 2H gel showed formation of a partial mullite-type phase whereas the RH gel did not form mullite but formed a γ-Al_2O_3-type spinel phase.

The 2W gel exhibited a less intense DTA exotherm and lesser extent of mullite formation in comparison to the SH gel of Okada and

Otsuka (1986) but corresponded with the "less ethanol" gel sample. The exothermic peak heights tentatively showed the following order: SH > "less ethanol" gel > 2D gel > 2W gel samples. Mullite formation at a 980°C exotherm showed the following order: SH > 2WC > 2W > 2D > 1D > 2H. Mullite formation at ~1250°C showed the reverse order. Mullite formation data of "less ethanol" and "more ethanol" are not available. Although synthesized at acidic pH, some monophasic gels in exceptional cases transform to mullite in a different way. One such example is colloidal gel synthesized by Li and Thomson (1991) and marked as CG; gels made by Ge et al. (1992) at different catalytic conditions and gels made by Huang et al. (1997) at different pH.

The colloidal gel exhibited a small 980°C exotherm and formed spinel and not analogous to colloidal gel prepared by Yoldas (1990), rather it behaved as RH gel and showed significant mullite formation at second exothermic region as revealed in DXRD analysis.

Example A: Ge et al. (1992) synthesized mullite powders by the sol-gel method using silicon tetraethoxide and ANN in the presence of three catalysts. Different gelation catalysts modified the formation temperature for pure mullite. It was the lowest (980°C) in the case of the acid catalyst process. It was intermediate (1200°C) in the case of the base catalyst process. However, it was as high as 1300°C when ammonium acetate (CH_3COONH_4) was used as the catalyst. These catalysts also changed the phase transformation processes of those precursors. In the case of the acid catalyst, mullite formed directly; in the basic medium, mullite formed via an intermediate spinel phase. Contrarily, in the presence of ammonium acetate, the formation of alumina polymorphs was predominant in the intermediate stages during heating (γ-Al_2O_3 formed at ~980°C and then α-Al_2O_3 formed at ~1200°C) prior to mullite formation at 1300°C, with the complete disappearance of α-Al_2O_3 (Fig. 7.3). XRD patterns of the spinel phase formed in the case of the ammonia-catalyzed sample and γ-Al_2O_3 formed in the case of the ammonium acetate–catalyzed sample showed significant differences. Some of the d Å peaks of γ-Al_2O_3 were absent and some were of reduced intensities in the case of the Al-Si spinel formed in the intermediate stage.

Example B: Comprehensive data on the effects of pH in the entire range on the thermal transformations of different Al_2O_3-SiO_2 gels were presented by Chakraborty (1994a).

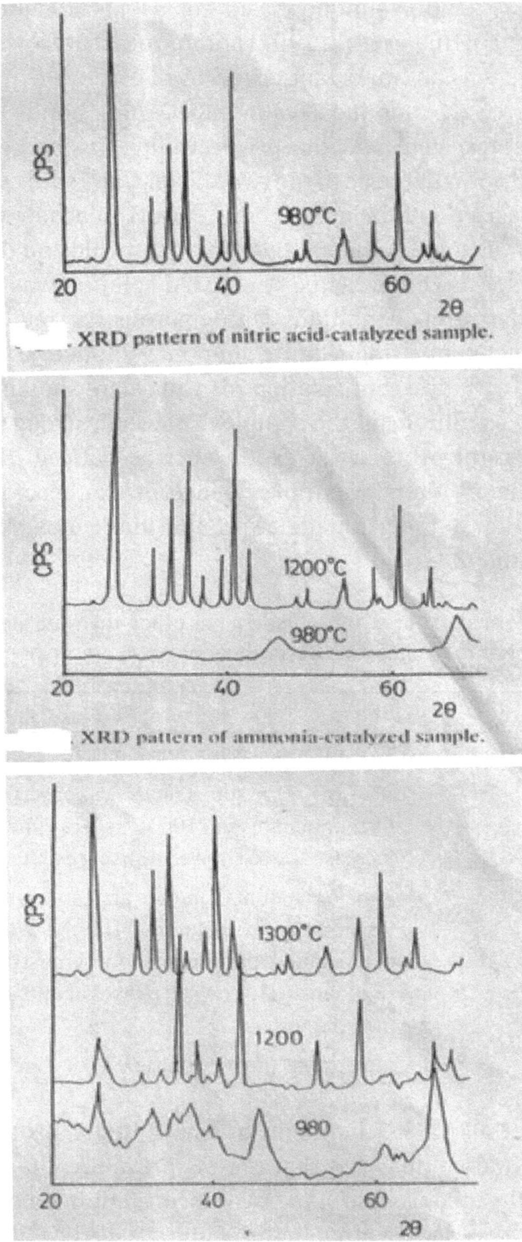

Figure 7.3 XRD pattern of an ammonium acetate–catalyzed sample (Ge et al., 1992).

Example C (gelation—monophasic gel as per Chapter 2—with variation of pH): Huang et al. (1997) pointed out that the mullitization temperature was an important criterion to assess the mixing scale or the degree of aluminosilicate (–Si–O–Al–) bonds present in the precursor system and the pH prevailing during gelation also influences the mixing scale of the two silicon aluminum components. They synthesized mullite gels at different pH values by using TEOS dissolved in absolute ethanol and aluminum chloride dissolved in distilled water as components. The mixed solution was hydrolyzed at 70°C under reflux conditions with vigorous stirring for 8 h. The pH was adjusted by using a dilute ammonia solution. DTA curves of all samples with different gelation pH values are shown in Fig. 7.4. In an acidic condition, all the samples showed a strong exothermic peak. At neutral pH, a weak exotherm was noticed. However, in basic pH region, there was no evidence of exotherm. The phase transformation of the gel made at pH 1.5 (marked as MI) has been shown in Table 7.4.

Table 7.4 Formation of γ-Al_2O_3-type spinel other than mullite type phase prior to mullitization during heating of MI precursor

Temperature	Phase
At room temperature	The precursor was amorphous.
At 950°C	A γ-Al_2O_3-type spinel phase was observed. On further heating, to 1100°C, its X-ray intensities increased. DTA showed a strong exothermic peak.
At 1200°C	A major amount of mullite appeared with a trace amount of θ-alumina. The crystallinity of the mullite pattern was low because the (120) and (210) crystal phases of mullite were indistinguishable.
At 1350°C	Mullite was the only phase.

Huang et al. (1997) pointed out that a 980°C exotherm is pH dependent and emphasized that the pH prevailing during gelation influenced the mixing scale of the two silicon aluminum components. Specifically, they showed that in the acidic range, the temperature of exotherm decreases from 980°C at pH 1.5 to 971°C at pH 3 to 961°C at pH 5 (Fig. 7.4). However, the gel that is prepared at pH 11 shows spinel and other alumina polymorphs at a temperature as high as

1200°C. This gel may be regarded as diphasic as the dried xerogel containing gibbsite, boehmite, bayerite, and silica (A) (Table 7.5). They conceived two different gel structures at two extreme pH ranges. Finally, they conjectured that acidic gels are homogeneous in contrast to basic gels, which are heterogeneous.

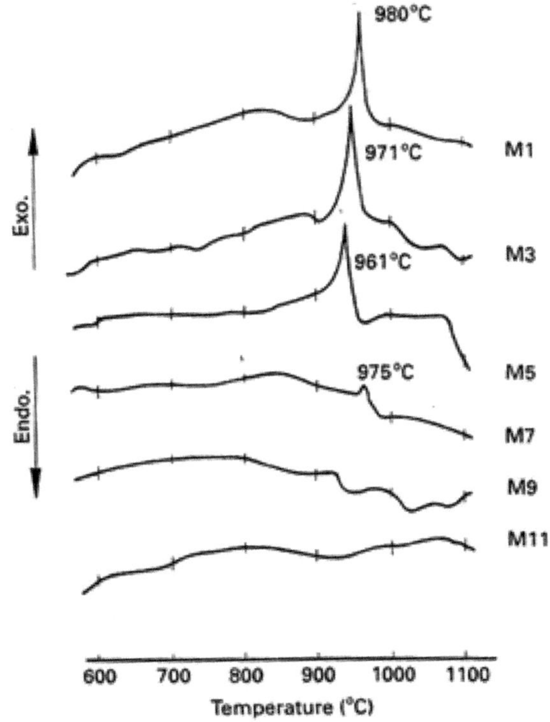

Figure 7.4 DTA curves of all samples with different gelation pH values (Huang et al., 1997).

7.2.1.2 Other types of monophasic gels (RH)

These types of gels were prepared by a rapid coprecipitation process by the hydrolysis and subsequent gelation of the organic silicon component in the presence of a water-soluble Al component in a basic medium (monophasic gel as per Chapter 3 versus monophasic gel as per Chapter 2) and the process was named as the rapid hydrolysis (RH) method by Okada and Otsuka (1986).

Table 7.5 Phase transformation of MII precursor on heating process

Temperature	Phase
At room temperature	The xerogel showed four phases: bayerite, gibbsite, boehmite and amorphous silica.
At 450°C	All crystalline phases of aluminum hydroxides were decomposed.
At 950°C	Only γ-alumina was present. DTA exhibited no crystallization peak.
At 1200°C	γ-Alumina, which was formed earlier, continued its rise in X-ray intensities. Some of it transformed into θ-alumina.
At 1350°C	Mullite was the only phase.

In basic pH

Example 1: When NH_4OH was used as a hydrolyzing catalyst, both silicon and aluminum components were uniformly precipitated by the OH concentration. But the prevailing pH was not known. Both DTA and the phase evolution process of (RH) mullite gels synthesized were compared with those of monophasic (SH) gel. The resultant gels sometimes exhibited a weak 980°C exotherm and formed a spinel phase (Okada & Otsuka, 1986). Observing the exhibition of 980°C exotherm, even in reduced size, this gel is termed as the other kind of monophasic gel. Thus, by changing the condition of the hydrolyzing medium from acidic to basic, both the intensity of the 980°C exotherm and the phase crystallization process alter.

Example 2: Subsequently, other kinds of monophasic gels were synthesized by Dafadar et al. (1998), at pH 6–7, and marked as CP6 (colloidal gel) by Li and Thomson (1991).

Example 3: CP gels was prepared after Yamada and Kimura (1962). They hydrolyzed silicon ethylester and aluminum ethylester in an alcoholic solution followed by coprecipitating alumina silica gel at a pH value of 5–8 with the help of decomposing urea at 70°C. At basic pH, when NH_4OH was used as a hydrolyzing catalyst and generated in situ, both silicon aluminum components were uniformly precipitated by the OH concentration and the resultant gels exhibited moderate 980°C exotherm and formed an Al-Si spinel phase.

Example 4: Mullite precursors were synthesized by Schneider et al. (1993), namely Type III, by using TEOS and AlOBu at pH 10. The XRD of this gel did not shown the presence of boehmite.

Example 5: A CP powder marked C was obtained by Kumazawa et al. (1990) by adding ammonia solution to a mixture of ANN and ethyl silicate. Two powders, Ha and Hb, from $Al(OPr)_3$ and $Si(OMe)_4$ (at pH 7 and at pH 10, respectively) by using ammonia solution but different processing conditions were also synthesized by them. All powders in the dried state were amorphous to XRD. Ha powder showed a single exothermic peak at 980°C and formed an Al-Si spinel in addition to predominant mullite. On the other hand, two exotherms were observed, at ~980°C and at 1300°C, for both Hb and C.

7.2.1.3 Development of diphasicity in a purely basic condition

Example 1: Some mullite gels were synthesized by Hsi et al. (1989) on dissolving ANN and TEOS in absolute alcohol at four different pH values, 8.3, 9.5, 10.1, and 10.4, by titrating with ammonia solution. The pH of the reaction mixture was found to be a factor that influenced the morphology of the crystalline phase in the xerogels. The development of the crystalline phase depended on the pH value. At a lower pH value, of 8.3, fibrous pseudoboehmite was formed, whereas at pH greater than 9.5, bayerite appeared and it became predominant at pH of 10.4. It was noted that precipitation pH in the mixed Al_2O_3-H_2O and SiO_2-H_2O was not similar to that of the pure Al_2O_3-H_2O. As a result, the pH at which boehmite and bayerite formed changed. Generally, the shift to the higher values may be due to the preservation of the SiO_2-H_2O system. The phase morphology/transformation during calcination studied by transmission electron microscopy (TEM) and XRD methods is shown in Fig. 7.5.

Example 2: Four xerogels of (0.84–2.2) $Al_2O_3 \cdot 1SiO_2$ by the chemical coprecipitation of ANN and TEOS were synthesized by Su Yen et al. (1991) in an extension of the previous study. Three gels, those that were precipitated at pH ~8.3 by using ammonia solution, showed pseudoboehmite in the initial gels. The fourth gel was precipitated at pH ~10.4 and showed bayerite. Molar ratios of Al_2O_3:SiO_2 for the xerogels were 0.84:1 (Al-poor), 1.4:1 (stoichiometric mullite), and 2.2:1 (Al-rich).

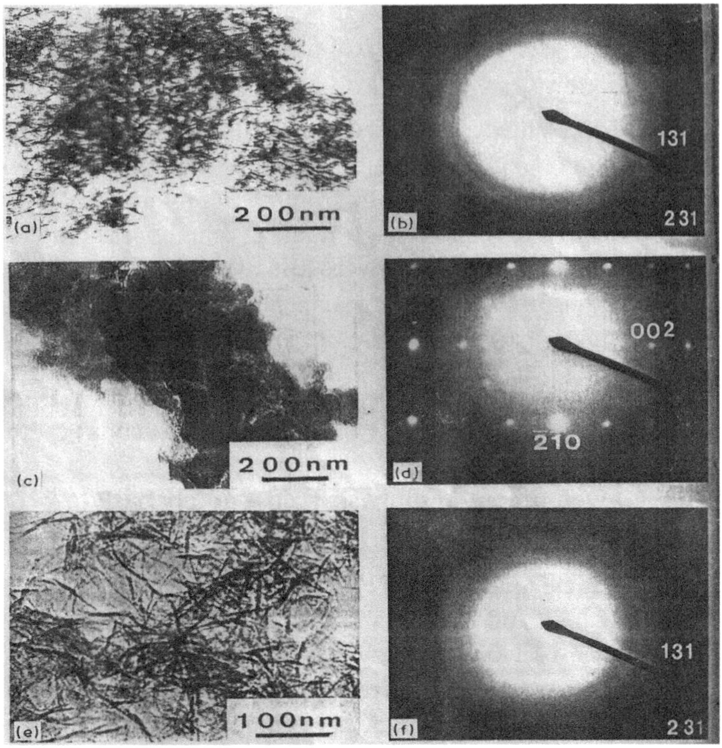

Figure 7.5 The TEM photographs of particles of Al_2O_3-SiO_2 xerogels at different pH values. (a) at pH 8.3 at a temperature of 25°C; (b) SADP; (c) at pH 10.4 at a temperature of 25°C; (d) SAD particles; (e) at pH 9.5 at a temperature of 60°C (Hsi et al., 1989).

On the basis of DTA and XRD quantitative studies of mullite formation and its crystallite size determination by Sherrer equation, Su Yen et al. (1991) suggested that gel samples were mixtures of monophasic and diphasic xerogels. In this case, the spinel phase was observed in TEM studies (Fig. 7.6). They predicted three thermal reaction paths for mullite formation: from an Al-poor and stoichiometric samples (containing pseudoboehmite) by Path A, from an Al-rich sample (containing pseudoboehmite) by Path B; and from an Al-rich sample (containing bayerite) by Path C. All four xerogels are basically diphasic in nature because they were either predominantly a mixture of pseudoboehmite and silica or a mixture containing predominantly bayerite and silica with traces of pseudoboehmite. This does not agree with the author's claim that the former mixture was a single-phase xerogel. On the basis of the XRD patterns of the two extreme gel samples, the transformation sequences of the two diphasic gels, one containing predominantly pseudoboehmite and the other containing bayerite, are given as follows:

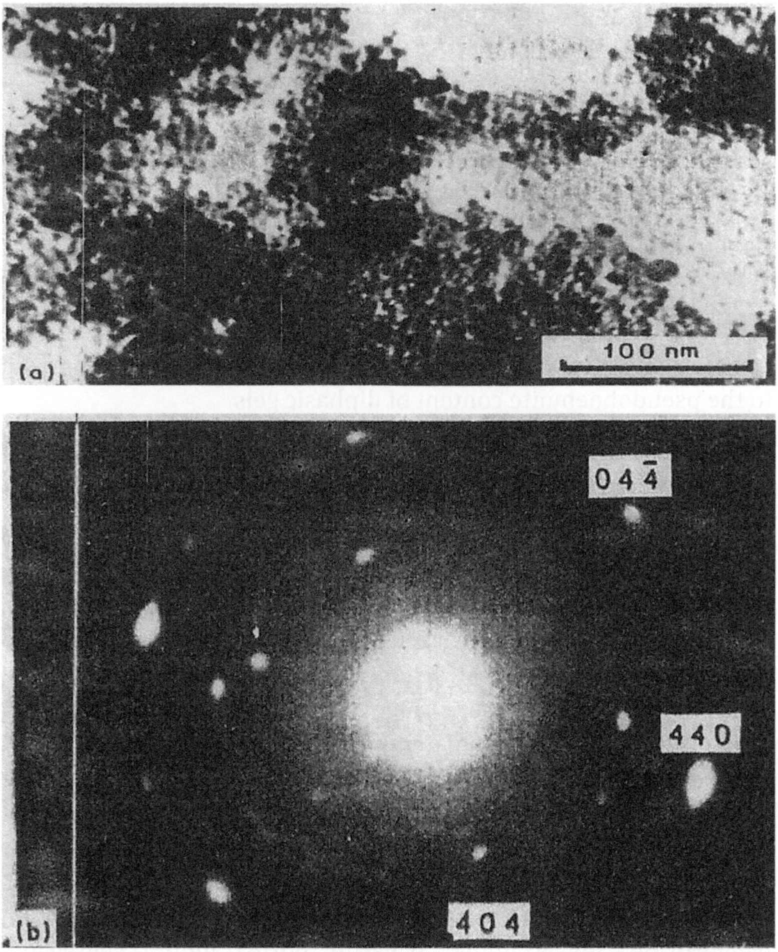

Figure 7.6 (a) Spinel phase formed during calcination at 600°C shows aggregates in microscopy. (b) The corresponding SADP reflection shows bright spots of (440) and (404) planes of the spinel phase (Su Yen et al., 1991).

$$\text{Pseudoboehmite} + SiO_2 \xrightarrow{600°C} \text{Amorphous} + \gamma\text{-}Al_2O_3 \xrightarrow{960°C}$$
$$\gamma\text{-}Al_2O_3\text{-type spinel phase} \xrightarrow{1250°C}$$
$$\text{Mullite-type phase by Path A} \qquad (7.1)$$

The γ-Al_2O_3 phase appeared initially at ~ 600°C after the decomposition of xerogel, which transformed into a spinel-like phase on heating at ~1000°C. It appeared as aggregates of fine particles (Fig. 7.6) in both TEM and diffraction studies.

The thermal transformation path for mullite formation of xerogel containing bayerite phase is given in Eq. 7.2.

$$\text{Bayerite} + SiO_2 \xrightarrow{300°C} \varepsilon\text{-}Al_2O_3 + Si(OH)_4 \xrightarrow{600°C} \varepsilon\text{-}Al_2O_3 +$$
$$(\text{Amorphous } SiO_2) \xrightarrow{960°C} \theta\text{-}Al_2O_3 + (\text{Amorphous } SiO_2) \xrightarrow{1300°C}$$
$$\text{Mullite by Path C} \qquad (7.2)$$

Thus, a higher temperature is required to achieve the transformation of a diphasic gel containing bayerite in comparison to the pseudoboehmite content of diphasic gels.

Example 3: Two types of mullite gels, one in an acidic condition and the other in a purely basic condition, were made by Huang et al. (1997), gels marked MI (Table 7.4) and MII (Table 7.5), made at pH 1.5 and 11.5, respectively.

Example 4: Hirata et al. (1985) studied the effect of pH on coprecipitation. They synthesized SiO_2-Al_2O_3 powder from TEOS and aluminum isopropoxide (AIP) at different pH values with the help of ammonia in the temperature range of 20°C–80°C. In the first case of preparation, excess ammoniated water (pH > 7) was added slowly to the solution mixture with constant stirring. The powder sample marked 2 (~76.3% Al_2O_3) prepared at pH 8.2 was amorphous. At higher pH values (>10), segregation took place since crystalline aluminum hydroxide, for example, aluminum monohydroxide (boehmite), precipitated. At still higher pH, of 11.9, aluminum trihydroxide, for example, bayerite nordstrandite, segregated out. The sample marked 2 developed a spinel phase on heating at ~1100°C, which was found to exist up to temperatures as high as 1200°C. From an infrared (IR) analysis, they interpreted the spinel phase as the Al-Si spinel structure where the $Si(Al)O_4$ tetrahedron and the AlO_6 octahedron are bonded at random by sharing oxygen atoms. The most interesting observation was that it developed a large quantity of mullite at ~1300°C, with the disappearance of the spinel phase. This result is comparable with the mullitization behavior of G-150 made by Chakraborty (1994a).

Example 5: Mitachi et al. (1990) studied the effect of pH on coprecipitation (Table 7.6). They synthesized aluminum silicate powder by hydrolyzing a mixed solution of AIP ($Al(Oi\text{-}C_3H_7)_3$) and TMOS ($Si(OCH_3)_4$) in benzene in two ways, as follows: In A series, hydrolysis was done at pH 7 simply with distilled water. In B series, hydrolysis was done at pH 10 with aqueous ammonia solution.

Table 7.6 Phase transformation of powder A series and powder B series examined by DTA and XRD on heating process

	Powder A series
Beginning	The powder was amorphous.
At 980°C	It exhibited a sharp exotherm.
	The calcined powder heated at the rate of 10°C/min. developed mullite (major) with traces of the spinel phase (marked as δ on two peaks on the XRD chart) at 1000°C. There was a sharp drop in the specific surface area during this transformation.
	Powder B series
Beginning	The powder was amorphous.
At 980°C	The exotherm was much less intense than that noted in A series.
At 1000°C	It formed an Al-Si spinel phase.
At 1250°C	The exothermic peak (this was the second one) was exhibited, which was also weak in nature. The XRD pattern showed mullite formation, and this was attributed by them to the conversion of Al-Si spinel to mullite. Accordingly, a drop in the specific surface area occurred twice.

To account for the large difference in crystallization behaviors of the two series of powders, Mitachi et al. (1990) suggested the following mechanism. In the A series powder, fairly large chains and networks of siloxanes bonds containing Al (–O–Si–O–Al–O–Si–) were formed by partial hydrolysis. In comparison, in the B series powder, hydrolysis was accelerated by ammonia, that is, by hydroxyl groups and resulted in the formation of short chains of siloxane bonds. There was also a possibility of the formation of some hydroxides partially. They are of the opinion that Si and Al elements were distributed homogeneously on an atomic level in the case of the A series and, thus, this precursor transformed directly into crystalline mullite on heating. On the contrary, in B series, the distribution of Si and Al was less homogeneous than that in A series and the precursor transformed into the spinel phase followed by the formation of mullite on further heating.

Example 6: Paulick et al. (1987) studied the effect of pH on coprecipitation. They carried out hydrolysis and condensation reactions using TEOS and $Al(OC_3H_7)_3$ in the entire range from acidic to alkaline conditions. They considered the differences in the rates of hydrolysis of silicon and aluminum alkoxides resulting in the differential precipitation of component hydroxides during hydrolysis. They hydrolytically decomposed the alkoxide mixture in various conditions, for example, at various pH values, in varying temperatures, and in different reaction times. Finally, it was concluded that the proposed decomposition was controlled mainly by the rate of TEOS decomposition, since $Al(OPr)_3$ can be readily hydrolyze in water without any catalyst.

- Nucleophilic-hydrolysis-cum-condensation of TEOS in a basic solution generally produces highly cross-linked species and with incomplete hydrolysis. At the alkaline condition of pH 10, TEOS is partially hydrolyzed even after heating at 70°C for 20 h.
- Aluminum trisisopropoxide hydrolyzes at a faster rate than TEOS. The hydrolysis is aggravated in the basic pH condition of ~10, which leads to differential precipitation during hydrolysis. Furthermore, during the heating process, it remains unreacted and appears as corundum, in addition to the usual mullite formation. This shows that differential hydrolysis rates of two alkoxides may be the reason stoichiometric homogeneous powders are not formed.
- In contrast, in electrophilic hydrolysis, the condensation mechanism may produce weakly cross-linked species, which tend probably to complete hydrolysis. At an acidic

condition of pH 2, they noted that hydrolytic decomposition was completed within 7 h at room temperature. The heated product is stoichiometric mullite.

- At a neutral condition, pH 6.5, the degree of hydrolysis was moderately affected by temperature and not by time. The oxide powder obtained was mainly alumina, with a minor amount of mullite.

Example 7: Schneider et al. (1993) studied the effect of pH on coprecipitation. They synthesized two different mullite precursors, Type II and Type III, by different synthesis methods from the same starting materials, TEOS and AlOBu, varying the amount of water, the velocity of the hydrolysis process, and the pH of the solvent, besides Type I, where no prehydrolysis had been done and only the components were mixed in isopropanol. In Type II, full hydrolysis of TEOS and AlOBu took place at pH 13, which led to the development of large alumina colloidal particles of Al-oxyhydrate mixed with silicon hydroxides; and in Type III, full hydrolysis of the components also took place, the only difference from the Type II gel being that the pH of the medium was reduced from 13 to 10 so that boehmite did not develop.

Example 8: Rajendran et al. (1990) prepared a mullite gel from TEOS and ANN by the coprecipitation technique using ammonia solution at pH 9 accompanied by vigorous stirring. The dried gel consisted of a well-crystallized bayerite. Thus, in the hydrolysis-cum-coprecipitation process, the pH is found to play a vital role in producing a homogeneous aluminum hydroxide–silicon hydroxide mixed gel. There is a problem of segregation when there is a local increase in the pH due to inefficient stirring or an increase in the pH of the solution as a whole. Phase development and exhibition of both 980°C and 1250°C DTA peaks in the case of the CP gel depend on the pH of the coprecipitation medium and the agent, for example, ammonia, urea, or amine, used.

- The CP gel prepared in a close-to-neutral condition (pH 6–7) exhibited a small and broad 980°C exotherm with the formation of an Al-Si spinel phase, as in the cases of Hsi et al. (1989), Mitachi et al. (1990), and others.

- The CP gel prepared at pH ≥ 9 did not exhibit a 980°C exotherm and might develop an Al-Si spinel over a wide temperature range (900°C–1300°C) with the exhibition of a broad exotherm (T_m = 1260°C) due to the slow crystallization of the γ-Al_2O_3-type spinel phase. Thereafter, it exhibited an exotherm at 1270°C–1300°C with the formation of mullite (Chakraborty, 1996b).
- Instead of ammonia, when urea was used as the precipitating agent, the precipitate obtained from the two esters showed Al-Si spinel formation at the first exotherm. It transformed, thereafter, into mullite at the second exotherm (Yamada & Kimura, 1962). As the 980°C phase development is different in the sol-gel process compared to that in the coprecipitation process, the next course, that is, mullitization processes of these gels likely show considerable variations among them. We come to the following conclusions:
    - Two types of mullitization routes were observed in mullite gels synthesized from raw materials from the same source, that is, TEOS and ANN, but with a change in the pH (Okada & Otsuka, 1986). Variations in the intensity of the 980°C DTA peak as well as the nature of the two phases in acidic against basic conditions have been demonstrated by Hirata et al. (1985), Hamano et al. (1986), Paulick et al. (1987), and Yamene et al. (1984). It was suggested by the latter authors that the hydrolysis and subsequent polymeric reactions of TEOS were influenced by the pH of the solution, for example, in the acid catalyzed reaction, it was thought that a homogeneous precursor was probably formed and thus produced mullite directly at 1000°C.
    - When materials from the same source were used but the hydrolysis conditions differed, particularly the pH, the synthesized gels still showed variations in the intensity of the 980°C DTA peak as well as in the nature of the phases, for example, the Al-Si spinel (c-mullite) or the (t-mullite). Such variations in developments in acidic against basic conditions have been demonstrated by Hirata et al. (1985) and Hamano et al. (1985).

7.2.2 Effect of Water on the Phase Transformation of Monophasic Mullite/SH Gel

The amount of hydrolysis agent used is an important factor influencing the gel/powder characteristics. Some of the investigations are compiled next.

Example 1: Hoffman et al. (1984) first synthesized Al_2O_3-SiO_2 xerogel using 10.2 g of ANN and 2 mL of TEOS with varying amounts of absolute alcohol. The Al_2O_3-SiO_2 xerogel exhibited a sharp and intense 980°C exothermic peak when 10 mL of ethanol, that is, "less ethanol," was used. But a similarly sharp peak with a much less intense peak occurred when 30 mL of ethanol, that is, "more ethanol," was used.

Example 2: The above gels were resynthesized by Li and Thomson (1991), who termed them 2D and 2W, respectively. Both gels exhibited 980°C exotherms, analogous to those of Hoffman et al., and but formed mullite, with an additional quantity of spinel—more so in 2D than in 1W. At the first exothermic region, a high-temperature DXRD study showed that the intensity of the mullite peak in 2W was more than that in 2D. At the second exothermic peak, the spinel phase disappeared and further mullitization occurred. With this evidence, Li and Thomson suggested a two-stage conversion in the mullitization process. However, the reasons behind the variations in the 980°C phase generation using different components are not explained fully. The extent of mullitization during the first exotherm was assumed to be due to the homogeneous mixing of the $Al_2O_3$and SiO_2 components, that is, the highest degree of Al–O–Si bonding in the precursor system (Hoffman et al., 1984; Okada & Otsuka, 1986; Pask et al., 1987). However, complete mullitization is a problem. Under almost all preparative conditions, for example, the monolithic gel route, the SH gel method, the polymeric method, or the hybrid gelation technique, a small amount of spinel phase is always formed together with the regular formation of mullite. The cause of formation of the said spinel phase was assumed to be segregation in the Al_2O_3-SiO_2 system arising from the dissimilar rates of hydrolysis of the components in water, which actually aggravated the homogeneity in practice. Epitactic nucleation by γ-Al_2O_3 may be an additional cause for promoting spinel formation (Huling & Messing, 1991).

Example 3: A process of synthesizing only pure mullite at 980°C was demonstrated by Chakraborty and Ghosh (1988) and further indicated that the use of varying amounts of water-alcohol was one of the four major factors behind the variation in the 980°C phase development. Using the same Al_2O_3 and SiO_2 components, ANN and TEOS, and varying the quantity of the hydrolyzing agent used during the gelation process, Chakraborty (1994b) prepared three gels: G-150, Gel-A, and G-31.

- Synthesis of three monophasic gels differences in water content during processing:

G-150: 10 g of $Al(NO_3)_3{\cdot}9H_2O$ was taken in a beaker, to which 60 mL of ethyl alcohol was added. The mixture was swirled gently to dissolve the salt. Then 2 mL of TEOS was added. It was warmed in a hot water bath at 60°C–65°C until gelation was completed and then it was dried at 110°C in an air oven.

Gel-A: It was prepared by using the same components in the same proportion. Instead of using a large volume of alcohol, 10 mL of absolute alcohol was added as per the synthesis of the monophasic gel by Hoffman et al. (1984) or the SH method as adopted by Okada and Otsuka (1986).

G-31: 10 g of $Al(NO_3)_3{\cdot}9H_2O$ was taken in a beaker and heated very gently on a hot plate in a temperature range of 100°C–105°C until dry. The heating was stopped when nitrous fumes began emerging, and the compound was cooled to room temperature. The beaker containing a white solid residual was dissolved in 5 mL of absolute alcohol. To it, 2 mL of TEOS was added. The mixture was swirled, gelled as per the usual procedure, and finally dried.

- Variation in DTA exothermic reaction behaviors of three monophasic gels prepared at different water contents:

The exothermic phenomenon and crystalline phases of the preceding three gels synthesized by gradually decreasing the alcohol content as studied by DTA & XRD are shown in Table 7.7.

G-150: The thermo gram of G-150 shows the usual 980°C exotherm as noted by several researchers of the mullite community. It exhibited two more exotherms, one at $T_m \approx 1150$°C and the other at $T_m \approx 1245$°C (Fig. 7.7).

Table 7.7 Phase transformation of different types of gel compared with kaolinite

		Transformation temperature		
Sample	**Water/TEOS ratio**	**980°C**	**1100–1300°C**	**1250°C**
Kaolinite	—	Exhibits a sharp exotherm and forms cubic mullite and aluminosilicate phase	Exhibits a broad exotherm. Aluminosilicate phase crystallizes to mullite	Exhibits a less-sharp exotherm. Cubic mullite polymorphically transforms into orthorhombic mullite
G-150	4:1	As above	Exhibits a broad exotherm at ~1150°C owing to crystallization of mullite from aluminosilicate phase	As above
Gel A	External water addition was not used, water of crystalization of $Al(NO_3)_3$ is the source of water	Exhibits a sharp exotherm and forms a major amount of tetragonal mullite and a minor amount of cubic mullite	No exotherm	As above
G-31	Nil	Exhibits a sharp exotherm and transforms into tetragonal mullite	No exotherm	—

Source: (Chakraborty, 1994b)

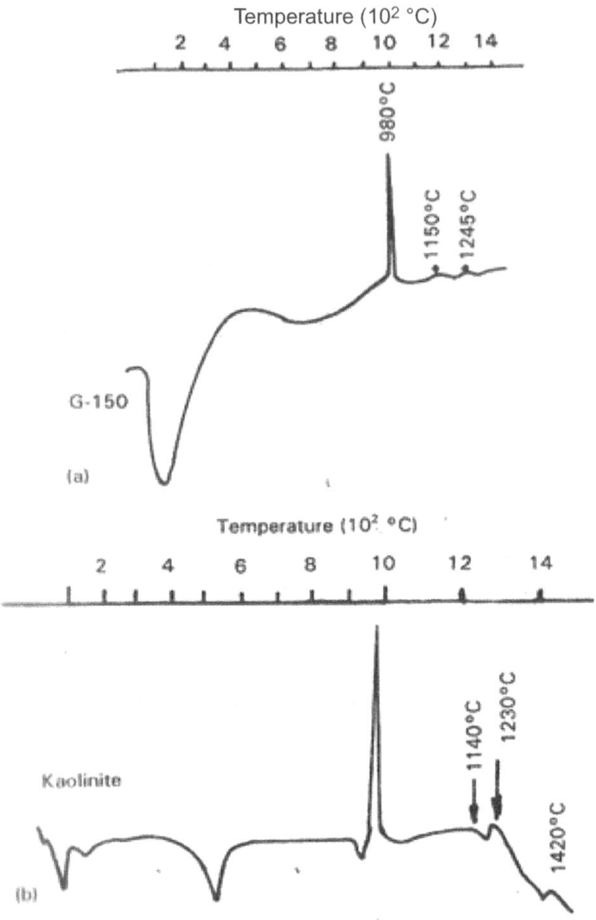

Figure 7.7 A comparison of the DTA curves of (a) G-150 and (b) kaolinite clay (Chakraborty, 1994b).

These three exothermic events correspond with those occurring in the DTA trace of a kaolinite clay (Fig. 2.2) but differ in their temperatures of occurrence. Therefore, it would not be erroneous to conclude that the thermal transformation phenomenon of G-150 bears a similarity to the high-temperature reaction of kaolinite.

Gel-A: The dried Gel-A decomposed below 980°C, with the elimination of residual –OH groups at the temperature marked "D" in the DTA trace, and then exhibited a very strong 980°C exotherm

(Fig. 7.8) with the formation of both Al-Si spinel and maybe t-mullite. The former phase on further heating exhibited a broad peak of a lower magnitude at ~1253°C with the formation of o-mullite. These results confirmed the mullitization behavior of SH gel (Okada & Otsuka, 1986) and DXRD data for 2H, 1D, and 2D gels (Li & Thomson, 1991).

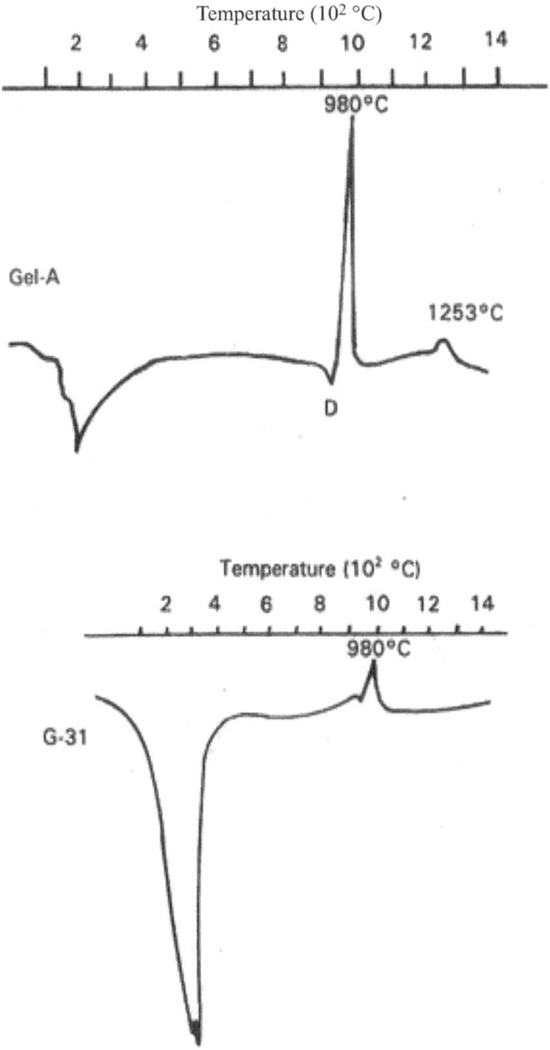

Figure 7.8 DTA curves of Gel-A and G-31 (Chakraborty, 1994b).

G-31: The thermo gram of gel G-31 showed a broad 980°C exotherm proceeded by a very small endothermic dip (Fig. 7.8).

- Comparative mullitization behaviors of three monophasic gels prepared at different water contents as studied by semi quantitative XRD analysis:

G-150: It was shown that kaolinite at 980°C exotherm formed 20–25 wt % cubic mullite (Al-Si spinel) and left ~ 40 wt % noncrystalline aluminosilicate phase as residue (Chakraborty & Ghosh, 1991). Interestingly, Chakraborty (1994b) showed that these two phases exhibited two exotherms with abundant crystallization of o-mullite. Like kaolinite, G-150 first formed c-mullite at the first exotherm and then transformed to o-mullite at 1250°C. Secondly, the growth of mullite in G-150 consisted of two stages (Fig. 7.9). In the first stage, mullite formation occurred rapidly at 1200°C–1250°C and thereafter in the second stage it increased gradually with rise of temperature.

The occurrence of the DTA exotherm at 1245°C coincided with the formation of the o-mullite from complete transformation of c-mullite as per X-ray analysis. Thus, the mullitization in G-150 was analogous to kaolinite and it followed two reaction paths : (i) c-mullite formed at the first exotherm further transformed polymorphically to o-mullite at 1245°C, constituting path II and (ii) the exhibition of the exotherm at 1150°C could be explained as due to mullitization in the noncrystalline aluminosilicate phase, as in path III, similar to kaolinite. This process proceeds at a comparatively slower rate than path II. The initial step portion accounts for mullite formation by path II, and the subsequent gradual rise is due to mullitization out of noncrystalline aluminosilicate phase by nucleation and subsequent crystallization by path III, as given below.

Gel-A and G-31: The growth of mullite was found to be very sudden (Fig. 7.9) and the transformation was found to be completed in one step, as the growth curve did not increase with increasing temperature. Except for the first exotherm, G-31 did not exhibit any other high-temperature peak between 1200° and 1300°C, such as shown in Gel-A and G-150. The variations in the 980°C phase development in the two types of monophasic gel G-31 and G-150 and the variations in their mullitization mechanism as outlined above are in reality due to two different types of monophasic aluminosilicate gels had been formed during gelling due to the varying water-alcohol contents, even though both gels were synthesized from materials from the same sources. These two gels likely followed two separate mechanisms during the 980°C reaction, e.g., (i) the gel marked G-31 formed exclusively o- or t-mullite and exhibited 980°C exotherm and (ii) the gel marked G-150 formed only Al-Si spinel during said exotherm.

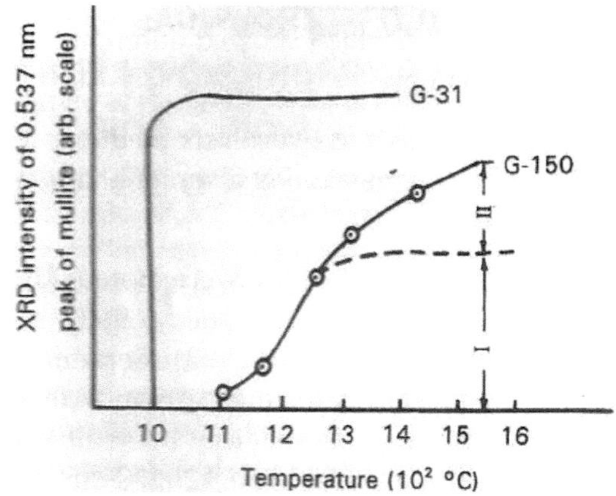

Figure 7.9 XRD intensity of o-mullite versus the firing temperature of two kinds of gels (Chakraborty, 1994b).

Thus, the main processing parameter allowing direct control over the gel structure was influenced by a variation in the use of water and/or alcohol during the gelation process. The interrelationship between the use of water-alcohol and the formation of the Al-Si spinel has been substantiated. G-150 invariably developed an appreciable amount of noncrystalline aluminosilicate phase, which is unwanted in the present-day requirements for achieving mullite ceramics in so far as improved hot modulus of rupture and high creep resistance properties are concerned. Even when Gel-A was prepared by eliminating the addition of external water and instead using ordinary alcohol, with 9 mol of water of crystallization of ANN the only source of water, the objective was not fulfilled because a minor amount of spinel phase was always formed, although t-mullite developed in a major quantity. The question is, what type of methodology is to be adopted for obtaining absolutely pure t-mullite/o-mullite, free from any liquid phase, as thought by Pask et al. (1987)? A comparison of the phase transformation behaviors of G-150, Gel-A, and G-31 vide Table 7.7 suggests that the water used for the hydration and hydrolysis of the components, for example,

TEOS, should be minimized so that extensive polymerization would be drastically eliminated. For example, G-31 transformed absolutely to o-/t-mullite, without any intermediary spinel phase and residual noncrystalline aluminosilicate phase.

Example 4: Two mullite gels from TEOS and $Al(OPr)_3$ based on the content of water were synthesized by Taylor and Holland (1993) and named GTW and GNW, which showed differences in phase and exothermic peak formations.

Example 5: Using water as the hydrolysis medium, it was shown that mullite gels invariably developed a spinel phase (Prochazka & Klug, 1983; Suzuki et al., 1990).

Other examples: Similar behaviors were noted for the sol-gel material synthesized by Schneider et al. (1993) and GYW gel synthesized by Taylor and Holland (1993). Geradin et al. (1994) synthesized the gel named M2 by mixing TEOS and AIP in 2-proponal at 60°C and hydrolyzing it simultaneously by adding water. It also exhibited two exotherms, at 987°C and 1252°C. At 1000°C, poorly crystalline spinel was noted. In addition, it developed characteristic patterns of δ- and θ-alumina phases. Thus, the M2 precursor likely behaves like other kinds of monophasic gels.

7.2.3 Effect of Aging Time and/or Temperature of the Precursor Solution on the Phase Transformation of Monophasic SH Mullite Gel

Example 1: Using the dip-coating method, Okada and Otsuka (1990) prepared transparent mullite films from the coating solution prepared by dissolving TEOS and ANN in absolute ethanol. They showed that the formation of mullite on heating coated films at 1050°C increased with aging time and showed a maximum at around 30 days and then decreased as the aging time increased further, with the crystalline phase in the film finally changing from mullite to the spinel phase. The formation of the spinel phase with longer aging time was considered by them to be caused by the polymerization of the silica component. They also noted the orientation of (hk0) of the mullite phase formed during heating in the 30-day-aged specimen. Accordingly, they calculated the Lotgering index. It increased rapidly with aging time and became 1, which corresponds to perfectly oriented (hko) reflections. They considered that structural change might have occurred during the aging process. On aging for a longer period, the silica hydrosol formed from TEOS polymerized, which caused spinel formation in due course of heating. With a view to explain the cause of mullite formation in preference to spinel when the dip-coated sample was heated, they remarked that a small amount of the precursor solution rapidly evaporated from the solvent.

Chakraborty (1994a) believed that the freshly prepared silica sol will not remain in true equilibrium. Condensation of two silanol groups might occur easily with the formation of siloxane linkages. Consequently, on standing, the silica sol became colloidal, with continued rise in viscosity as well as molecular weight. Therefore, on mixing aluminum salt with silica sol, two reactions likely took place simultaneously: polymerization of the silicic acid itself and the formation of aluminosilicate hydrate. The polymerized silica sol units will react with the Al^{+3} ions later. The effect of polymerized silica sol on the 980°C reaction of different Al_2O_3-SiO_2 gels was studied and is shown in Fig. 7.10a. The intensity of the 980°C DTA peak (height) and the quantity of mullite formed (XRD intensity of the 0.537 nm peak) at the same temperature (Fig. 7.10b) are both significant in the case of gel prepared from aluminum sulfate (BDH) and freshly prepared sol (i.e., with no aging time). Gels prepared with silica sol with increased aging times show that (i) the intensity of the 980°C exotherm decreased rapidly from the highest value and then proceeded very slowly and (ii) the amount of mullite also decreased rapidly and thereafter the formation of Al-Si spinel commenced and then increased gradually. Thus, the existence of an interrelationship between the 980°C exothermic peak in DTA and the nature/amount of 980°C phase development of silica sol aged for increasing durations is established. This suggests that the essential condition for obtaining pure mullite is that the silica component may either be monomeric or in the form of small aggregated colloidal particles.

Example 2: Okada et al. (1996) dissolved TEOS and ANN in ethanol solution to synthesize a mullite precursor. The solution was aged for 5 days at various temperatures to study the character of the mullite precursor structure by using liquid-state nuclear magnetic resonance (NMR), small-angle X-ray scattering (SAXS), and XRD studies.

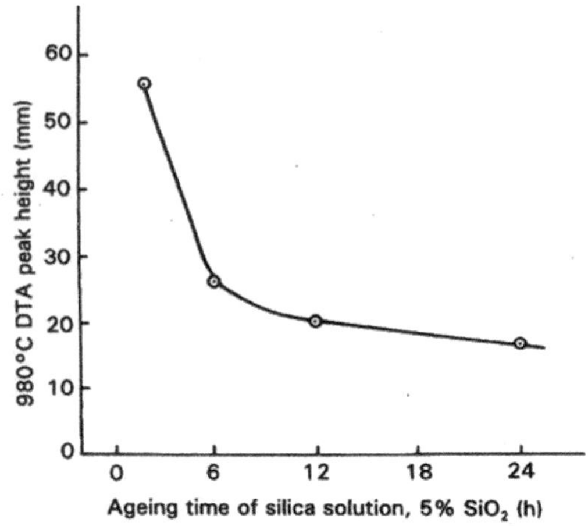

(a)

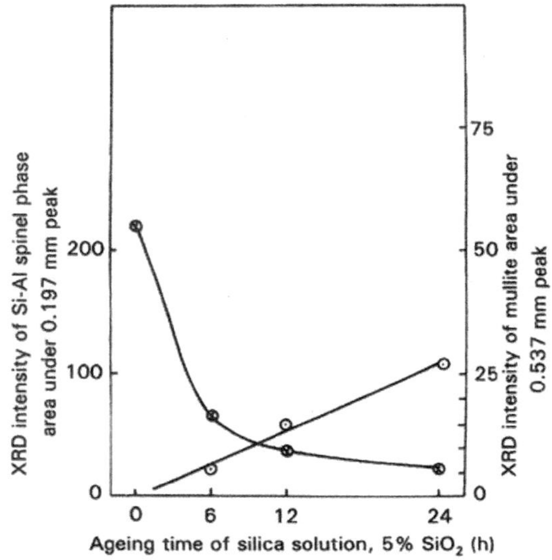

(b)

Figure 7.10 (a) 980°C DTA peak height (mm) against aging time of silica sol used for synthesizing different Al_2O_3-SiO_2 gels (Chakraborty, 1994a). (b) Intensity of Θ mullite and ⊙ Al-Si spinel formed on heating at 1000°C of 0 h against the aging time of silica sol used for synthesizing different Al_2O_3-SiO_2 gels (Chakraborty, 1994a).

With an increase in the aging temperature, the following changes take place:

Liquid-state ^{29}Si NMR spectroscopy showed that three peaks, at around −91, −100, and −107 ppm, assigned to Q^2, Q^3, and Q^4 structures, respectively, decreased with temperature. At 60°C, all three peaks disappeared and the spectrum related to the hydrolyzed product of TEOS was a 3D silica framework structure.

SAXS study showed that the size of the sol particles increased by gradual coalescence of silica colloids. The solution aged at 50°C showed particles of two size distributions, 7 nm and 14 nm. When the solution was aged at 60°C, the sizes were 9 nm and 26 nm.

On heating the aged samples at 1000°C, it was noted that the X-ray intensity of the spinel increased on aging from 25°C to 40°C to 50°C and then decreased at 60°C and formed mullite predominantly (Fig. 7.11). Okada et al. (1996) also showed that no dominant mullitization was found in all samples aged for various lengths of time below 60°C. At 60°C, the 1-day aged sample showed little mullite formation, and it increased with an increase in the aging time and then attained a constant on being aged for over 5 days. The questions are:

- Why is spinel formed from samples aged below 60°C, and why is mullite formed predominantly at the cost of spinel when the aged sample is heated at 60°C?
- What is the state of silica gel at stage of solution at 60°C? At 60°C, the mixed solution contained a silica gel network structure trapping a high concentration of ANN solution. Since ANN showed great solubility at that temperature and TEOS hydrolyzed to a 3D form, they conceived this state of affair was analogous to that observed by Hoffman et al. (1988).
- What is the state of the dried precursor, and what is happening during drying?

All these questions remain to be answered.

A liquid-state ^{29}Si NMR precursor solution aged for 5 days at 60°C showed spectra free from Q^2, Q^3, and Q^4 structural bonds. It is assumed that hydrolysis and thereafter polymerization of TEOS occurred with the formation of 3D polymerized silica sol particles. An SAXS measurement showed two particle size distributions, which indicated that silica sol particles coalesce into bigger silica gel particles. What will happen when silica gel with its network structure impregnated with ANN solution is dehydrated in a water bath or instantly dried at 500°C? The following conclusions are drawn about the hydrolysis-cum-gelation method of synthesis of mullite:

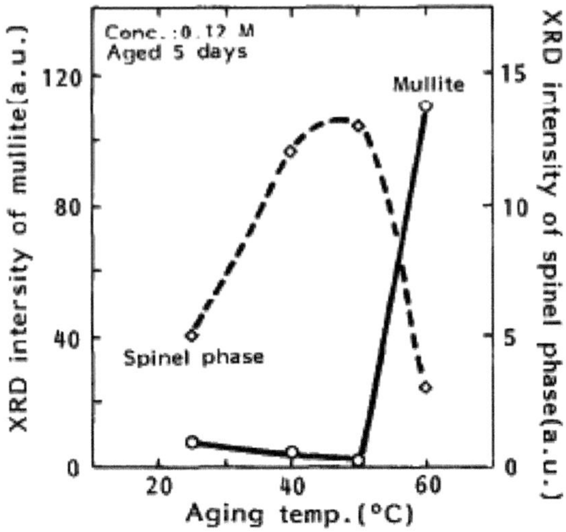

Figure 7.11 XRD intensity of crystalline phases in the samples fired at 1000°C for 6 h and aged at various temperatures (Okada et al., 1996).

When organic silica was used instead of aqueous silica as the silica source and the other component was water soluble, the resultant gel exhibited the first exotherm and formed mullite (major extent) with some small quantity of spinel phase. Regarding the nature of the exotherm, the peak was found to be the highest at the composition equivalent to the composition of 3:2 mullite. Rapid mullitization was noted at or beyond 1000°C. The extent of spinel formation was found roughly dependent on the concentration of the solvent, that is, alcohol/water present during the gelation process. It was shown that 980°C DTA characteristic and mullitization path varied with the hydrolyzing agent. Spinel formation decreased with the water content; on the other hand, the geometry of the exotherm and the intensity of mullite formation increased. Besides the role of the hydrolyzing agent, the temperature of gelation and the aging of the solution mixture affect spinel-phase-to-mullite-phase ratio. Other conditions of hydrolysis, for example, the temperature during gelation in the range 25°C to 40°C to 50°C to 60°C and aging for 1 day to 5 days, largely affected the exhibition of exotherm and mullite formation thereof.

7.2.4 Effect of the Drying Condition on the Mullite Gel

Example 1: A single-phase gel of mullite composition and a diphasic mullite gel were synthesized by Komarneni and Rustiser (1996). For the monophasic gel, ANN was mixed with 95% ethyl alcohol and the mixture was stirred for 1 h. TEOS was added, and the sol was poured into tubes and kept at 60°C until gelation. Diphasic mullite gels were prepared from colloidal silica and boehmite using concentrated nitric acid to maintain the pH at 3. After gelation, gels were exchanged in methanol for 4–12 days more. Xerogels were made by drying at 60°C and aerogels were made by critical point drying. The critical point drying conditions were a temperature between 260°C and 270°C and pressure of 8.27 MPa in a 50 mL Hastalloy C autoclave. Both single-phase and diphasic mullite gels were dried simultaneously so they would have identical drying conditions. The single-phase xerogels showed an intense exothermic peak at about 980°C, while their counterparts, aerogels (both monophasic and diphasic), did not show any detectable exotherm in DTA. These results suggested that the structure of single-phase gels were changed during critical point drying. Both single-phase and diphasic aerogels of mullite composition exhibited high surface areas after sintering in the temperature range of 1000°C–1400°C. Segregation of the alumina component occurred during critical point drying of single-phase gels of mullite composition. Sintered aerogels of mullite composition might be useful for high-temperature catalytic applications. Thus, the drying condition of mullite gels affects both exothermic events and phase evolution.

Example 2: Fonseca et al. (1997) synthesized alumina-silica gels and powders of two compositions (in mol %), $60Al_2O_3$:$40SiO_2$ and $64.3Al_2O_3$:$35.7SiO_2$ from TEOS and ANN at varying conditions and tried to correlate the exothermicity of the 980°C peak with mullite formation with the following variables: (i) acidic condition, (ii) basic condition, (iii) water content, (iv) alcohol content, and (v) method of preparation of powder (whether the sol mixture was dried rapidly by dropping the sol on a hot plate). It was observed that the thermal evolution and the phase transition are strongly influenced by the synthesis route adopted.

The DTA and phase transformation behaviors of gels prepared for the composition $60Al_2O_3$:$40SiO_2$ are as follows:

- MG gel, prepared in an acidic condition and dried slowly at 60°C for 7 days, showed a strong 975°C exotherm and formed mullite as the identifiable phase at 1000°C. This is monophasic in nature.
- The moisture of the MP powder prepared in an acidic condition evaporated rapidly when the sol was dropped on a hot plate. The intensity of the exotherm in DTA decreased abruptly and formed a spinel phase together with less well-defined mullite. This powder is like other kinds of gel.
- MPP powder–mixed sol was precipitated by using dilute ammonia solution and subsequently evaporated on a hot plate. The resulting powder showed a very small exotherm and formed only a spinel phase at 1000°C. Fonseca et al. (1997) suggested that MG behaves like a single-phase precursor while the MPP is clearly diphasic in character, where the MP represents an intermediary situation. According to them, these results confirmed the earlier model of Okada and Otsuka (1986), which suggested that a decreasing segregation scale favored mullite crystallization.

The phase transformation behaviors of gels of composition 64.3Al_2O_3:35.7 SiO_2 are as follows:

- MAG gel was synthesized as per the SH gel route and formed only the mullite phase at 1000°C, in spite of the excess alumina present in this precursor. According to Fonseca et al. (1997), this gel might attribute high chemical homogeneity among gel.
- MAPIF, prepared by dropping the sol on a stationary or fixed hot plate, showed spinel and mullite of less intensity than that observed in MAG at 1000°C.
- MAPIR, prepared by dropping the sol on a rotating hot plate, showed spinel and also more quantity of mullite than what occurred in MAPIF.

The rotation of the hot plate may expose new hot zones to the falling drops of sol mixtures. This may be the reason for the increase of mullite content at the cost of a decrease in the spinel phase. By the spraying technique on the hot plate, Jaymes et al. (1996) achieved more mullite content. It was thought further that the size of the droplets was reduced in the spraying process. It is concluded that the phase transition is strongly influenced by the synthesis route adopted and one of the synthesis parameters is the drying condition. Jaymes et al. (1996) applied slow drying versus fast drying to very fast drying techniques in their gel preparation processes. When the MG gel dried slowly at 60°C for 7 days, it showed a strong 975°C exotherm and formed mullite. When the moisture of the MP gel was evaporated very rapidly by dropping it on a stationary hot plate, the DTA peak decreased with the formation of more spinel phase and less mullite. Dropping the MAPIR sol on a rotating hot plate instead of a stationary hot plate showed that the spinel formation quantity of mullite crystallization increased in comparison to that in MP. Thus the rotation of the hot plate and/or spraying may be helpful in achieving more mullite content.

7.3 Synthesis of Polymeric Mullite Gel and Colloidal Mullite Gel

Let us look at the hydrolysis and subsequent cogelation of organic Al and silicon compounds in a neutral medium, as already discussed in Section 2.3, Chapter 2.

Example 1: By sol-gel methods, polymeric gel and colloidal gel powders containing 72 wt % of Al_2O_3 and 28 wt % of SiO_2 were synthesized by Yoldas and Partlow (1988) and Yoldas (1992), and the same were resynthesized by Pask et al. (1987), as shown in Chapter 2. The DTA curve for ground 1A specimen (colloidal) showed the absence of a 980°C exothermic peak but exhibited a sharp exothermic peak at ~1267°C. Broad diffraction peaks for the spinel phase were observed by XRD in a specimen heated to 1000°C and the formation of mullite was seen at the temperature of exotherm. On the contrary, the DTA curve for specimen 2A (polymeric) indicated a sharp exothermic peak at 985°C and XRD after 1000°C showed mullite peaks with faint peaks of the spinel phase. Thereafter, a second small exothermic peak was noted at 1235°C; XRD showed only mullite crystallization at this temperature.

Example 2: Three kinds of Al_2O_3-SiO_2 gels were resynthesized from AlOBu and TEOS systematically by Chakraborty (1996a), as follows:

The first kind of gel, marked as G-152, was synthesized by using a large volume of water using the following steps: A mixture of ALOBu, water, and alcohol in the ratio of 7mL:50 mL:10 mL followed by the addition of 2 drops of 1:10 dilute HNO_3 was hydrolyzed by warming. To this mixed solution, 2 mL of TEOS was added and the mixture was heated in a constant water bath at 65°C, with occasional shaking, until gelation took place. Finally, the gel was dried at 110°C in an air oven.

The second kind of gel, marked as G-152(i), was synthesized following the same procedure except that the amount of water added was only 10 mL.

The third kind of gel, marked G-152(ii), was synthesized again following the same procedure except that external water was not used directly for the hydrolysis of ester. Gelation was performed by keeping the mixed reactants in a humid atmospheric condition (room temperature 27°C; air humidity 80%–90%) for 7 days. It was assumed that under these conditions, hydrolysis of both components took place at a very low rate only due to the moisture available from air.

The causes of the occurrence of the exothermic events of these gels in DTA traces are revealed by subsequent X-ray studies. The corroborative results of these studies are summarized and shown in Table 7.8. In the presence of a large amount of water, as in gel G-152, X-ray analysis indicated the formation of pseudoboehmite as the crystalline phase and an amorphous band of $Si(OH)_4$ derived from the hydrolysis of ALOBu and TEOS separately. Obviously, this gel was characterized as a diphasic gel and it showed crystallization of the Al-Si spinel on gradual heating in the temperature range of 800°C–1200°C. Accordingly, it didn't exhibit exotherm at 980°C due to crystallization of the Al-Si spinel and/or mullite and did not exhibit exotherm at 1250°C due to the formation of o-mullite, as revealed by XRD. This may be the reason why Yoldas and Partlow (1988) did not notice an exotherm in the DTA of their colloidal marked A. The amorphous SiO_2 component of the gel did not crystallize to β-cristobalite as reported by Hyatt and Bansal (1990). Only at ~1320°C was an exotherm noted, and it may be explained as due to the polymorphic transformation of accumulated c-mullite to o-mullite.

Table 7.8 DTA and X-ray characteristics of three different Al_2O_3-SiO_2 gels compared to kaolinite

Samples	Water used during gelation	Nature of the gel formed	DTA results	X-ray characterisation of crystalline phases formed in DTA analysed sample
G-152	Excess amount than that required for hydrolysis of mixed esters	Diphasic	At 980°C no exotherm At 1320°C small exotherm	Formation c-mullite continued[30] Forms o-mullite
G-152(i)	Moderate amount	Monophasic	At 980°C sharp exotherm At 1150°C broad exotherm At 1250°C less sharp exotherm	Forms major c-mullite Forms o-mullite Forms additional c-mullite
G-152(ii)	May be very small (not quantified)	Monophasic	At 980°C very sharp exotherm At 1150°C no exotherm At 1270°C very small exotherm	Forms major t-mullite and minor o-mullite Forms o-mullite
Kaolinitic clay[32]	—	—	At 980°C very sharp exotherm At 1100–1400°C broad exotherm At 1250°C sharp exotherm within the above broad exotherm	Forms c-mullite' Forms o-mullite Forms o-mullite

Source: (Chakraborty, 1996a)

When the water amount was reduced from 50 mL to 10 mL, as in the case of gel G-152(i), colloidal hydrolyzed species of silicon and aluminum interacted mutually and developed an X-ray amorphous monophasic aluminosilicate hydrate gel. The DTA of this gel showed a 980°C exotherm of reduced intensity, which is similar to that noted by Yoldas and Partlow (1988) in their sample marked B. X-ray analysis of the heated gel revealed c-mullite (major) and mullite (minor) as crystalline phases. Besides the exhibition of 980°C exotherm, it showed two other exotherms, at 1150°C and 1250°C. These exotherms resemble the peaks exhibited in the DTA trace of a kaolinitic clay. Comparing these DTA phenomena, it is concluded that the thermal transformation sequence of G-152(i) bears a similarity to the high-temperature reaction sequences of kaolinite and the gel marked G-150(i), as shown above. Therefore, two high-temperature reaction paths of mullite formation, namely path I and path III, as shown in the cases of kaolinite and G-150, reveal the origin of the two separate exotherms in the DTA curve of G-152(i) (Figs. 7.12A and 7.12B).

Similar phenomena in the three cases suggest to us that some amount of the noncrystalline aluminosilicate phase likely remains as a residual phase in the 980°C reaction of G-152(i).

When water was not directly available for use, the obtained gel marked G-152(ii) crystallized sharply to mullite, with the exhibition of a pronounced exotherm at 980°C, which conformed with the findings of the molecular gel marked C made by Yoldas and Partlow (1988). In addition, the molecular gel exhibited a small exotherm at 1270°C. To ascertain the cause of this exotherm, XRD analysis of the 980°C heated gel was carried out and showed the crystallization of both c-mullite and mullite, which Yoldas and Partlow did not analyze. Thus, the exhibition of 980°C was related to both forms of mullite and the 1270°C exotherms was explained as due to the subsequent polymorphic transition of c- to o-mullite. The absence of any exotherm in the vicinity of 1200°C, as in the case of G-152(i), suggests that noncrystalline aluminosilicate phase formation did not occur. On comparing the mullitization behaviors of these gels, the following inferences are made:

- The large amount of water used in hydrolysis is responsible for the development of gels that are diphasic in character.

- Even with the use of a small quantity of water for the hydrolysis of two components of the gel, the unwanted noncrystalline aluminosilicate phase invariably forms.
- The formation of both c-mullite and the noncrystalline aluminosilicate phase could be minimized directly by avoiding or reducing the use of water.

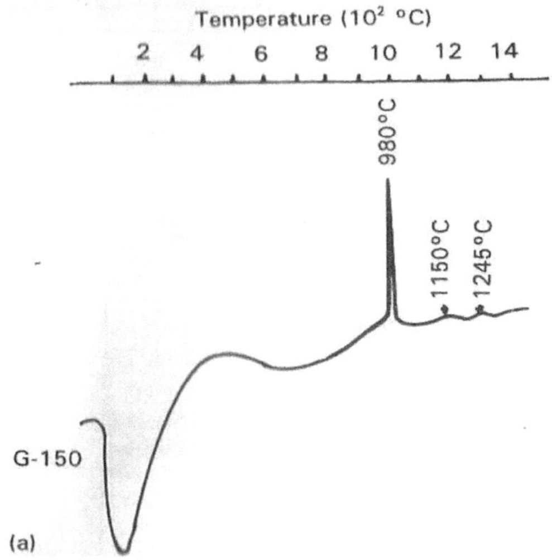

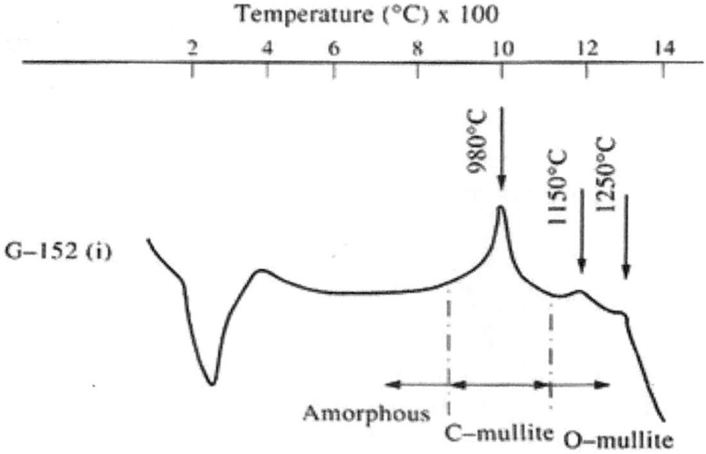

DTA curve of G-152 (i).

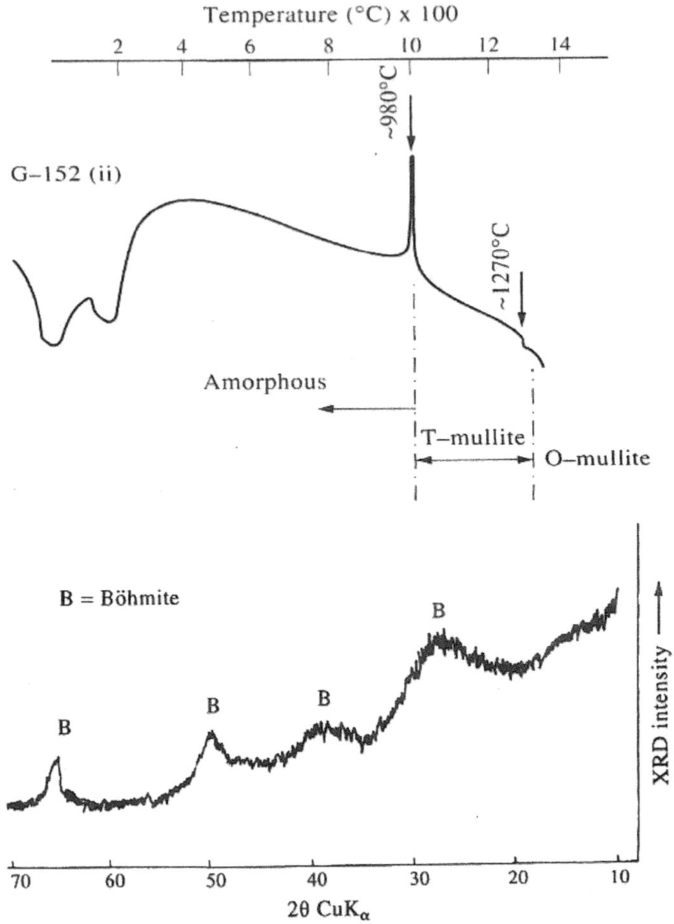

Figure 7.12 (A) DTA curves of G-152 (upper), G-152(i), and G-152(ii) and the XRD pattern of dried G-152 showing boehmite (Chakraborty, 1996a).

Other examples:

A Type III gel was synthesized using a high water content at the prevailing pH of 10 by Schneider et al. (1993). It showed differently from precursors made using low water content. It exhibited a weak exotherm due to spinel and another weak exotherm due to mullite formation. Both exotherms in DTA were also noted by Geradin et al. (1994) in their gel marked MII, which was synthesized by hydrolyzing components by the addition of water.

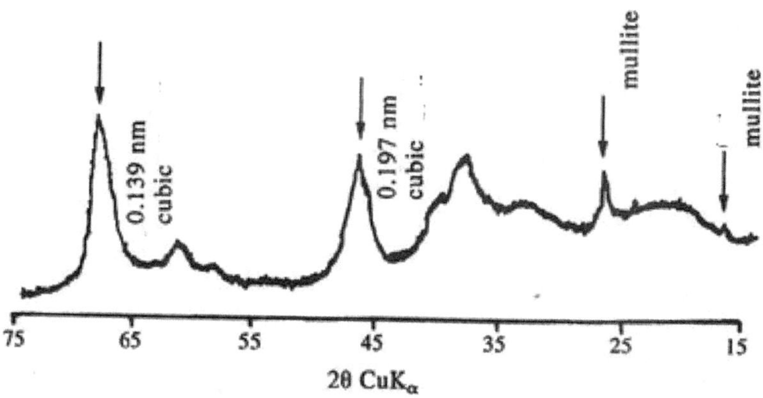

Figure 7.12 (B) XRD pattern of G-152(i) heated in a DTA cell at a 980°C exotherm, showing cubic mullite (major) and o-mullite (minor) (Chakraborty, 1996a).

Three types of precursors with TEOS and AlOBu as the starting material and using low water content with slow hydrolysis and high water content with fast hydrolysis by controlling pH values were also prepared by Ruscher et al. (1996), and these showed differences in phase evolution.

Five mullite precursors from $Al(OPr)_3$ and $Si(OMe)_4$, hydrolyzed either with water or with ammonia or using the sequential hydrolysis-cum-mixing technique, were synthesized by Kumazawa et al. (1991). The obtained gels exhibited differences in DTA traces and in the formation of thermal crystalline phases.

Mullite precursor synthesis generally involves hydrolysis and condensation of two alkoxides in the presence of solvents and a catalyst. The gelation mechanisms by using SAXS and ^{29}Si NMR spectroscopy techniques were studied by Heinrich et al. (1991). These two processes were found to be reaction-limited cluster-cluster aggregation. Primary particles are formed immediately after hydrolysis, aggregating exponentially in time. The parameters influencing the gelation behavior were dilution with water and aging effect. The high-temperature phase development was related to two important process parameters (Heinrich & Raether, 1992): (i) the reactivity difference between silicon and aluminum alkoxide (dealt with in Chapter 11) and (ii) the water content of the precursor solution. Both parameters influenced the homogeneity of the resultant gel. They opined that both have to be minimized to avoid local enrichment of alumina, which could act as nuclei for spinel crystallization. They synthesized single-phase gels from two starting compounds: (i) aluminum-*sec*-butylate chelated with β-diketone and tetraalkoxysilane in isopropanol and 0.1 N HCl medium and (ii) ANN-ethanol/TEOS and water. The crystallization behavior of gels of four categories of samples are shown in Table 6.4. The gelation time was correlated with the water content used. The precursor powder derived by them from gels with a water concentration ≥ 10 mol/L formed traces of t-mullite in samples R4–R9. Powders derived from gels with 3.3 mol/L developed 30 wt % of t-mullite after calcination (R1–R3). Secondly, three mullite gels were synthesized from the ANN and TEOS system with different water contents. Nitrate powders (N1–N3) showed that with increasing water content of the precursor solution, the quantity of t- mullite decreased on calcination. At ~1273°K crystallization of the spinel-like phase occurred.

7.3.1 Effect of pH on the Phase Transformation of Polymeric and Colloidal Mullite Gels

A polymeric gel (PB2) was prepared by Sanz et al. (1991) in a basic condition, but the pH was not mentioned. Two other polymeric gels, PA and PB, were also made by Nieto et al. (1998) in acidic and basic conditions (as per Yoldas, 1992). Both these gels showed the usual 980°C exothermic peak and an exotherm at 1230°C–1250°C. However, the intensity of the first exotherm for PA was more than that for PB. The total weight loss in the colloidal sample (10.3%) occurred below 1000°C, and in the acid and basic polymeric samples (13.6% and 11.5%, respectively) below 800°C, and this may be related to the amount of –OR and –OH present in the desiccated gels. In both PA and PB gels, sintering occurred in two steps, which is analogous to two exotherms in DTA. An XRD examination showed the formation of a small amount of spinel below 1000°C in the case of the colloidal gel but it formed at 1000°C for both polymeric gels, which is in agreement with the corresponding exothermic peaks noted in DTA. However, the intensities of the spinel phase in the three cases were not given.

7.3.2 Effect of Water on the Phase Transformation of Polymeric Colloidal Mullite Gel

Besides the choice of ANN and TEOS as popular sources of components, various authors used ALOBu and TEOS as the basic raw materials for the synthesis of mullite by hydrolyzing them either in an acidic or in an ammonia medium. By gradually reducing water/moisture conditions for hydrolysis and from the choice of stoichiometric mixture of those two reagents, three different types of mullite gels were synthesized first by Yoldas and Partlow (1988). They first diluted the alkoxides in a nonaqueous solvent and then introduced water in small quantities as per the following equation:

$$x\text{Al(OR)}_3, y\text{Si(OR)}_4, z\text{SLV} + w\text{H}_2\text{O} \rightarrow \text{Al–O–Si}\equiv +$$
$$\text{ROH}, z\text{SLV (nonaqueous solvent or dry ethyl alcohol)}, \quad (7.3)$$

where $w << x$ and $y << z$.

In this case, a stoichiometric amount of the two components was mixed with dry ethyl alcohol and divided into three portions.

The first portion, which was mixed directly with excess water, showed no characteristic exotherm in a DTA study (Fig. 7.13).

The second portion was spread thinly over a tray and was allowed to hydrolyze and dry in humid air and showed a 980°C exotherm of a small magnitude.

The third portion was put in an open flask and stirred for 3 months and assuming very slow and restricted hydrolysis, showed a significantly large exotherm.

The resultant precursors exhibited variations in the intensities of the exothermic peaks. The following observations were made: (i) The intensity of the 980°C exothermic peak followed an inverse relation to the quantity of water used during the gelation process. (ii) The exhibition of 980°C might be used as an index for predicting the gel structure. Yoldas and Partlow (1988) were of the opinion that the occurrence and size of the 980°C peak are closely related to the intimacy or homogeneity of aluminum and silicon atoms through oxygen in the precursor. However, the XRD patterns of the DTA-analyzed samples indicating the presence or absence of phases associated with the exhibition of the 980°C exotherm were not shown.

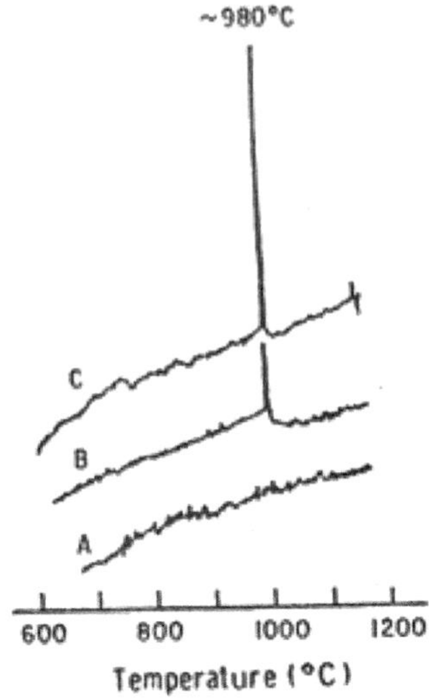

Figure 7.13 The occurrence and intensity of the ~980°C DTA peak of $Al_6Si_2O_{13}$ condensates is a function of the rate of hydrolysis of the base solution prepared from a stoichiometric mixture of $Al(OC_4H_9)_3$ and $Si(OC_2H_5)_4$ in alcohol. (A) The solution was hydrolyzed vigorously with excess water; (B) a thin layer of the solution was poured over a tray and hydrolyzed in humid air within hours; (C) the solution was hydrolyzed in humid air over 90 days while stirring (Yoldas & Partlow, 1988).

The effect of condensation reactions during the chemical polymerization of alkoxides in the Al_2O_3-SiO_2 system versus the crystallization behavior on heating three kinds of precursors were suggested. Yoldas and Partlow (1988) are of the opinion that the alkoxide mixtures formed a more homogeneous oxide network when the chemical encounter rates with water were reduced.

7.3.3 Effect of Temperature on the Phase Transformation of Polymeric and Colloidal Mullite Gels

The influence of the temperature of hydrolysis on the correlation of the structural evolution was studied by Orefice and Vasconcelos (1997). They synthesized two types of mullite gels with 40 mol % of silica: (i) A transparent monolithic xerogel that was obtained by hydrolyzing a mixture of prehydrolyzed TEOS (water/TEOS molar ratio = 2) and hydrolyzed AIP in the pH range 2.5–5 at room temperature. This gel is found to be monophasic in character. (ii) A gel hydrolyzed rapidly at 90°C, which developed pseudoboehmite; thus the resultant gel was nontransparent and was diphasic in nature. These two gels exhibited different sequences of phase transformation. The monophasic gel transformed to mullite only at ~1100°C. The diphasic gel on heat treatment at 500°C showed the presence of typical IR resonance peaks of γ-Al_2O_3. On further heating to 1100°C, it formed Al-Si spinel, which supports the experimental observation of Chakraborty (1996b).

7.3.4 Effect of Washing and of Gel Surface on DTA

Three sets of aluminosilicate gel containing 47 wt % of Al_2O_3 were prepared by Fahrenholtz et al. (1991). These gels were subjected to different washing treatments to probe the structure of a gel as a function of heat treatment. The first one was a nitrate powder, which was prepared from a stoichiometric solution of TEOS and ANN by gelling at pH ~6 using ammonia solution. The second was ASB powder, and it was prepared from an alcoholic solution of TEOS and ALOBu mixture at pH 6. Stoichiometric amounts of the two components were added to ethanol to produce 47 wt % of an alumina precursor solution. HCl dissolved in ethanol was added to the precursor solution for pH adjustment to hydrolyze those. The molar ratio of aluminum to silicon to HCl was ~1:1:1. The mixture was refluxed for 24 h and subsequently gelled by the addition of 100 mol of water per mol of silicon in solution. The accessible surface area measured after drying at 100°C for 24 h was 1 m^2/g. After washing with an ethanol sample marked as ASB WASHED, it showed a large surface area, on the order of 500 m^2/g. The results showed that the phase evolution process during heating was influenced greatly by the physical structure, specially the pore structure, of these gels.

DTA showed significant differences in the apparent peak positions and heights. The nitrate gel exhibited a small peak at 1050°C, while the ASB gel showed two peaks, one at 980°C and another at 1025°C. The magnitudes of both were small. On the other hand, the high-surface-area ASB powder exhibited a large exothermic peak at ~1025°C. At 1100°C, the nitrate powder still remained amorphous, while traces of mullite were noted in the case of ASB. Only a minor quantity of mullite developed in the washed ASB powder.

A variation in the phase development in these three cases was observed, and according to Fahrenholtz et al. (1991), it was attributable to the difference in mixing among the precursors, and magic angle spinning–nuclear magnetic resonance spectroscopy data addressed this point. ^{27}Al resonance showed a difference in the reactivity of the nitrate compound to the alkoxide-derived gel. The nitrate gel showed a narrower Si resonance and shifted to the high field side in comparison to ASB and the ASB-washed gel, which suggests that less aluminum was incorporated into the second coordination sphere of silicon, that is, molecular mixing is less in this case in comparison to that in alkoxide gels.

Relative changes in the surface areas of raw xerogels may be due to differences among the physical structures of aluminosilicate of gels. On heat treatment, profound changes in surface area curves were noted. ASB gels showed a gradual increase in the area with a rise in temperature, which reached a maximum at ~1000°C and decreased gradually up to 1400°C. The observed increase in the surface area may be due to a fracture in the gel surface during heating and marked openings to increase the accessibility of fluids. In comparison, the ASB WASHED gel showed a sharp decrease in the surface area in two steps. The first step coincided with the occurrence of the DTA exotherm at ~980°C due to the formation of a minor quantity of mullite. The second step occurred over a temperature range of 1250°C to 1400°C, where a large quantity of mullite crystallized. Concurrently, a dilatometry study at this temperature range showed a higher shrinkage in this gel in comparison to ASB gel. But how it developed is difficult to predict due to the unavailability of XRD patterns at different temperatures.

7.4 Mullite Gel Synthesized by the Diphasic Gelation Method

There are two general sources of synthesis of diphasic gels: (i) aqueous alumina sol [AlO(OH)] and aqueous silica sol/Ludox (silica sol) and (ii) boehmite sol and TEOS. In both cases, no exotherm occurred at 980°C and the spinel phase crystallized below 900°C remained stable far above 1250°C. In the 1270°C–1300°C temperature range, the spinel phase suddenly disappeared and mullite rapidly crystallized with the occurrence of an exotherm. A variation in the crystallization path noted in the diphasic gel method of mullitization was primarily dependent upon the reactivity of the two components chosen: (i) in some cases, independent crystallization of two component oxides crystallized partially to corundum and cristobalite prior to mullite formation and (ii) in others, spinel was noted as the intermediate phase prior to mullitization. The cause of variations in the phase transformation behaviors may be due to variation in the pH during the gelation process.

7.4.1 Effect of pH on the Phase Transformation of Mullite Gel Prepared by a Diphasic Gel

The true diphasic route of the crystallization behavior of a mullite gel was reported by Hamano et al. (1986). They noted β-cristobalite formation between 1250°C and 1600°C and α-Al_2O_3 formation between 1300°C and 1700°C and thus confirmed the true diphasic route in its crystallization behavior. According to these authors, the mullitization reaction was due to a solid-state reaction between α-Al_2O_3 and β-cristobalite at 1600°C apart from the γ-Al_2O_3/amorphous SiO_2 reaction. Besides, the importance of pH and its role are highlighted as follows:

- At pH 1.8 a diphasic gel was synthesized by Ismail et al. (1987) using boehmite sol and silica sol.
- Sonuparlak (1988) prepared a diphasic mullite gel using colloidal boehmite suspensions and TEOS in acidic conditions (pH < 2) by the sol-gel process. An XRD study showed the presence of an amorphous phase. A spinel phase with partial mullite formation occurred in a sample sintered at 1250°C for 1 h.
- Sonuparlak (1988) also prepared a gel by the same procedure but at pH 6–7 and using ammonium hydroxide. An XRD study showed that extensive mullitization occurred in the previously sintered sample after further heat treatment at 1300°C for 1 h. Thus, utilizing components from the same source, that is, colloidal boehmite suspensions and TEOS, Sonuparlak (1988)

showed quite different phase transformation behaviors of a diphasic gel processed under acidic versus basic conditions. The diphasic gel prepared in both conditions densified at ~1250°C, with the development of a γ-Al_2O_3 spinel phase. When the gel was further heated to 1300°C for 1 h, only mullite was observed in the gel synthesized in the acidic condition.

- Gels differing in homogeneity: Huling and Messing (1990) synthesized two diphasic gels using identical boehmite and silica sols as a model case. The phase transformation behaviors and the microstructure developments of the two diphasic gels were noted to be different. Various researchers too showed different mullitization behaviors of diphasic gels prepared from TEOS and colloidal pseudoboehmite sol in acidic conditions.
- Wei and Halloran (1988) noted that on heating, the alumina component crystallized to γ-Al_2O_3 and thence to δ-Al_2O_3 and some amount of θ-Al_2O_3 but the silica component did not crystallize to β-cristobalite. With these observations, they believed that mullite formed by a solid-state reaction of δ-Al_2O_3 and amorphous silica during heating. But they didn't explain the reason behind the noncrystallization of cristobalite.
- At pH between 1 and 2, Klaussen (1990) synthesized another diphasic gel mullite gel using pseudoboehmite in the form of a powder and colloidal silica. The silica component did not crystallize but exothermic peaks were noted at 1188°C and 1254°C and claimed as due to the formation of δ-Al_2O_3 and θ-Al_2O_3 transition phases prior to mullite formation at a 1328°C exotherm.
- A diphasic gel was prepared by Hyatt and Bansal (1990) by using AlO(OH) and Ludox in the presence of HNO_3. They observed the crystallization of all transitional alumina phases from the two chosen components. They showed mullitization besides partial crystallization of both cristobalite and corundum on heating to 1350°C. These fractions of independent crystallization of component oxides might be due to the use of an acid during the gelation process, as a result of which inhomogeneity developed since the gel was opaque in appearance.

- For a gel prepared even at pH 2, mullite formation was noted at 1300°C instead of the usual very high temperature of 1600°C, where both α-Al_2O_3 and β-cristobalite were crystallized at the intermediate stage of firing (Li & Thomson, 1991).
- In a pH range of 0.84–2.2, four mullite xerogels of the composition $Al_2O_3 \cdot SiO_2$ were synthesized by Su Yen et al. (1991) by the chemical coprecipitation of ANN and TEOS. The gel precipitated at pH ~8.3 by using ammonia solution showed pseudoboehmite in the initial stage. On the other hand, the gel precipitated at pH ~10.4 showed bayerite. It was suggested that the gel samples were mixtures of monophasic and diphasic xerogels. These earlier studies show variations both in the phase transformation behaviors and in mullitization reaction paths. The cause of the variations is to be ascertained.
- In varying pH conditions, a series of diphasic gels were synthesized by Chakraborty (2003) and subsequently studied for thermal behavior by the XRD technique. Figures 7.14–7.17 depict that the pH had a paramount influence on the hydrolysis of TEOS and the resultant silica particles ultimately guide the course of the mullitization reaction.

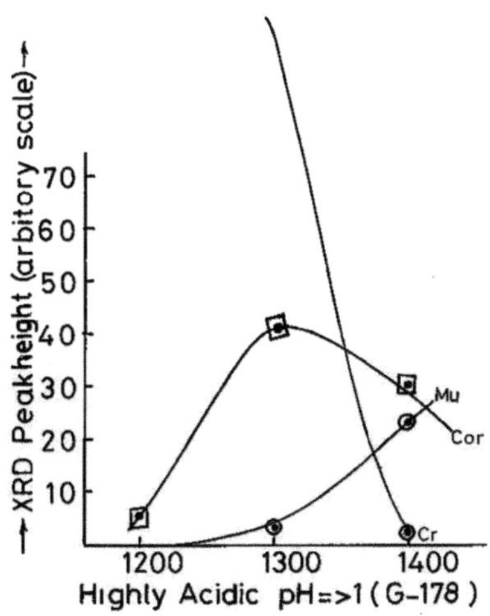

(a)

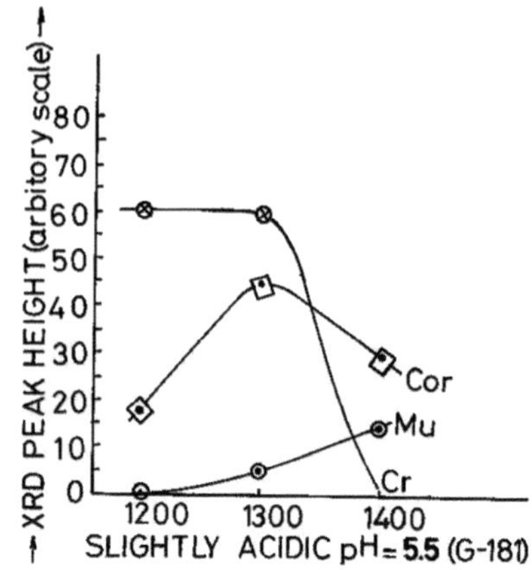

(b)

Figure 7.14 The formation of mullite, cristobalite, and corundum phases versus the temperature of heat treatment (°C) of (a) G-178 and (b) G-181.

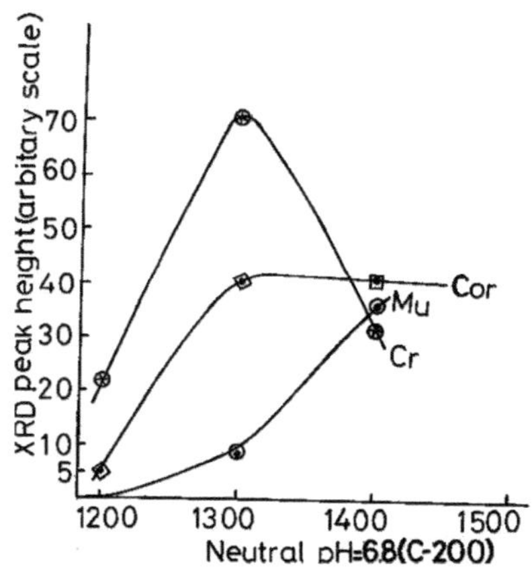

Figure 7.15 The formation of mullite, cristobalite, and corundum phases versus the temperature of heat treatment (°C) of G-200.

The figures showed the crystallization of components of diphasic gels to cristobalite and corundum. The formation of the noncrystalline aluminosilicate phase was sensitive to the pH of the gelation process. This was also true for the next step, that is, the mullitization process. In a highly acidic pH range, a solid-state reaction occurred between corundum and cristobalite, leading to the development of the liquid phase for mullite nucleation. At a moderately acidic pH, nucleation and crystallization were most operative for nucleation in the aluminosilicate matrix (Fig. 7.18). In a highly basic pH region, the Al-Si spinel phase developed by the incorporation of Si in the aluminous phase as an intermediary phase. Its polymorphic transformation may be the cause of sudden mullitization (Fig. 7.17). The changes in the crystallization sequence in the three distinct processes may be due to a variation in the nature and size of the silicic acid particles formed by the hydrolysis and condensation of TEOS at different pH values. Contrarily, Chakraborty (2003) noted intermediate Al-Si spinel phase formation in the diphasic $Al_2O_3{\cdot}SiO_2$ gel prepared in a basic condition before mullite formation. Thus, earlier studies showed variations both in the phase transformation behavior and in the mullitization reaction path. The cause of the variations is not explained as yet. The comprehensive picture depicts that the pH not only influences the hydrolysis of TEOS and the formation of the resultant silica particles but also controls the course of the mullitization reaction.

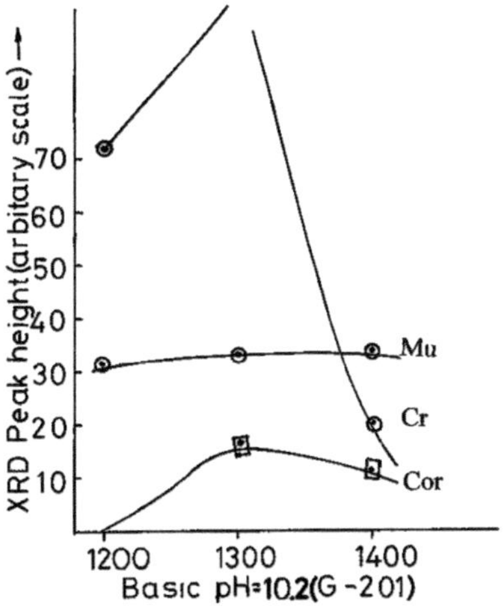

Figure 7.16 The formation of mullite, cristobalite, and corundum phases versus the temperature of heat treatment (°C) of G-201.

Diphasic gels at three different pH conditions 3.5, 5.5, and at 8 were synthesized from TEOS and freshly peptized boehmite sol using ammonia solution, followed by stirring for 4 h for homogenization. The effect of the precursor's pH on the phase formation characteristics of diphasic gels showed a variation in the development of the mullite phase as follows:

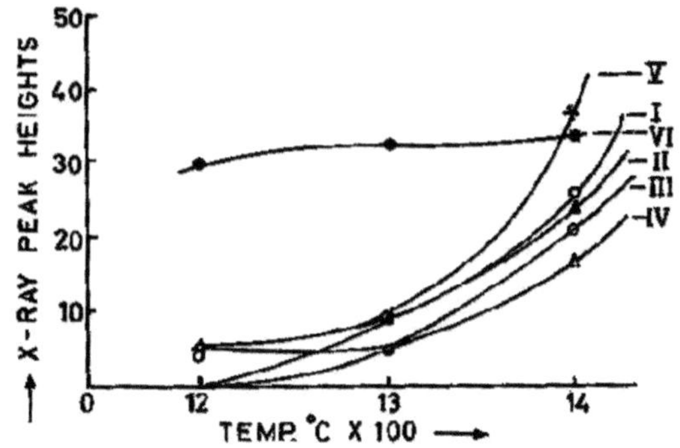

Figure 7.17 XRD peak height of mullite versus calcination temperature (°C) of diphasic gels. (I) G-177, (II) G-178, (III) G-181, (IV) G-181, (V) G-200, and (VI) G-201 (Chakraborty, 2003).

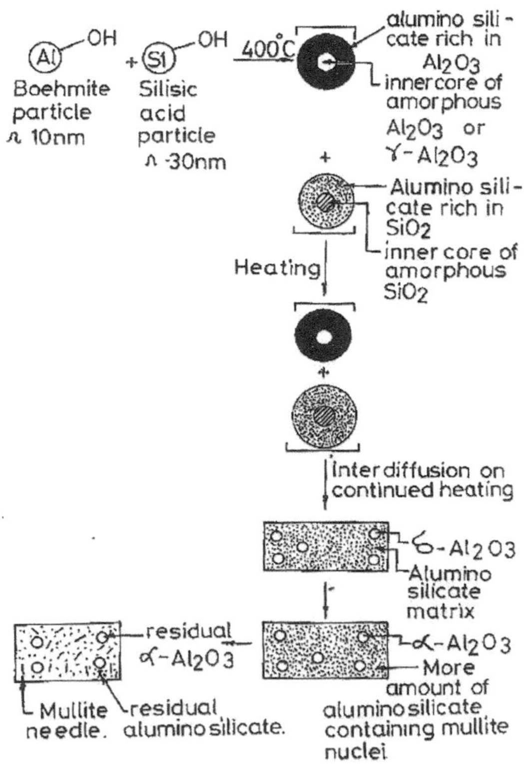

Figure 7.18 A reaction of two discrete phases of the diphasic gel system (Chakraborty, 2003).

A sharp reactivity in G-101 and in the gel marked VI is observed to develop mullite among others.

On heating at 1225°C, the precursor prepared at pH 3.5 showed a transition alumina phase while the precursor prepared at a higher pH, of 5.5, predominantly developed an α-Al_2O_3 phase along with transitional alumina phases. When the precursor's pH was further raised to 8, the peak intensities of both corundum and transitional alumina were reduced.

On further heating, to 1250°C, the sample prepared at pH 3.5 transformed to o-mullite while the samples prepared at higher pH showed corundum as the major phase with some mullite. The former phase still existed even when the firing temperature was raised to 1500°C.

7.4.1.1 Possibility of formation of gels of a diphasic character in the basic pH range

- By dissolving ANN and TEOS in absolute alcohol at four different pH values, 8.3, 9.5, 10.1, and 10.4, Hsi et al. (1989) synthesized diphasic gels. The phase evolutions in the four cases were, however, different. Al_2O_3-SiO_2 xerogels were either predominantly monophasic, a mixture of monophasic and diphasic, or purely diphasic, depending upon the amount of bayerite formed. Hsi et al. showed that in a pH range of 10–14, the resultant gel behaved as diphasic in character and formed an Al-Si spinel phase prior to mullitization at ~1300°C exotherm (Table 7.9).
- A gel of the composition of mullite was prepared by Rajendran et al. (1990) from TEOS and ANN by a coprecipitation technique using ammonia solution at pH 9, along with vigorous stirring. The product was kept at room temperature for 0.5 h, filtered and washed with deionized water until free from nitrate, dispersed twice ultrasonically in acetone, and then dried in a vacuum oven at 75°C for 24 h. DTA, electron microscopy, and XRD analyses were performed. The dried gel consisted of well-crystallized bayerite ($Al(OH)_3$), that is, it was diphasic in nature. On heating the gel to 300°C, the bayerite decomposed and the pattern became amorphous. They showed that on calcination below 1100°C, the CP gel developed γ-Al_2O_3 and

perhaps cristobalite crystallites within the basic grains, whose morphology was otherwise invariant with temperature. When mullite formed above 1100°C by a reaction between γ-Al_2O_3 and SiO_2 crystallites, the grain morphology changed markedly. Small exothermic events occurred at 1000°C and the small exothermic peak at 1280°C appeared due to the formation of mullite. Yoldas (1979) prepared a polymeric mullite precursor from TEOS and AliOP by the usual polymeric technique. It exhibited a large exothermic peak at 1000°C, which could be associated with the decomposition of the amorphous aluminosilicate precursor phase to crystalline γ-Al_2O_3 and SiO_2, and a small exothermic event at 1250°C due to mullite formation. In comparison, the CP gel might be diphasic (Fig. 7.5) while the polymeric gel is monophasic.

Table 7.9 Quantity of crystalline and noncrystalline phases formed on heating diphasic gels at 1300°C and 1400°C

Gel mark	pH	Amount of cristobalite 1300°C/ 1400°C	Amount of corundum 1300°C/ 1400°C	Amount of mullite 1300°C/ 1400°C	Amount of non-crystalline phase 1300°C/ 1400°C
G-177 I	<1	22.2/10.7	47.7/27.2	11.8/34.2	18.2/27.7
G-178 II	1.5	17.7/1.4	51.1/34	10.5/31.5	20.5/22.8
G-179 III	4.5	15.5/1.1	54.5/30.6	6.5/28.9	23.3/39.2
G-181 IV	5.5	12.2/nil	48.8/35.2	6.5/21	32.3/44.7
G-200 V	6.8	10/4.9	50/45.5	11.8/48.6	28.1/1
G-201 VI	10.2	11.1/0.8	18.1/12.5	43.4/44.7	27.2/49.9

- Alumina-silica gels named MPP synthesized in a basic condition exhibited a very small exotherm and formed a spinel phase at 1000°C because it acquired the diphasic characteristic of the powder (Fonseca et al., 1997).

7.4.1.2 Possibility of formation of a mixture of monophasic and diphasic gels

Four xerogels of (0.84–2.2) $Al_2O_3 \cdot SiO_2$ were prepared by Su Yen et al. (1991) by the chemical coprecipitation technique using ANN and TEOS. Three of the gels, precipitated at pH ~8.3 by using ammonia solution, showed pseudoboehmite in the initial gels, whereas the fourth gel was precipitated at pH ~10.4 and showed bayerite. Su Yen et al. performed DTA and XRD quantitative studies of the four xerogels and determined the mullite formation and the crystallite size of the mullite by Scherrer equation. They suggested that during pH adjustment, gel samples produced mixtures of monophasic and diphasic xerogels and accordingly they predicted three thermal reaction paths for mullite formation.

7.4.1.3 Importance of the reactivity of two components

Sugita et al. (1998) prepared a mullite powder by a process similar to the one used by Mizuno et al. (1990). A mixture of fume silica (Aerosil 200) in an aqueous solution of aluminum sulfate with ammonium bisulfate was heated, and a homogeneous precipitation of basic aluminum sulfate salt precipitated out and coated the fume silica surface. DTA showed an endotherm at ~ 850°C due to the decomposition of aluminum sulfate. Only a very broad exothermic peak was noticed at ~1150°C. Mullite was produced at a temperature above 1250°C. The final product was an agglomerate consisting of mullite particles 100–200 nm in size. They depicted three routes of phase transformation depending on mixture homogeneity of the mullite precursors (Fig. 7.19).

Two diphasic mullite precursors prepared from TEOS and boehmite with and without dehydroxylation at 400°C were studied by Padmaja et al. (1998). They noted a large reactivity difference in the phase evolution processes. α-Al_2O_3 formed at a temperature as low as 1000°C in a precursor gel made with dehydroxylated boehmite as a component, and it remained till as high a temperature as 1500°C. This result indicated the absence of any reaction between dehydroxylated boehmite and silica hydrogel at the early drying temperature. This supports the previous findings of Chakraborty (2004). Precursor gel (1) showed weight loss amounting to 17%, in thermogravimetric analysis, due to the dehydroxylation of the prepared from boehmite calcined at 400°C and TEOS in the temperature range of 400°C–500°C. On the other hand, precursor gel (2), prepared from boehmite without calcination and TEOS, showed a higher quantity of weight loss, of ~29%. Both precursors showed differential shrinkage peaks at ~1250°C. Precursor gel (1) showed earlier mullite formation, at 1225°C, than precursor gel (2), at ~1250°C, as indicated by XRD patterns.

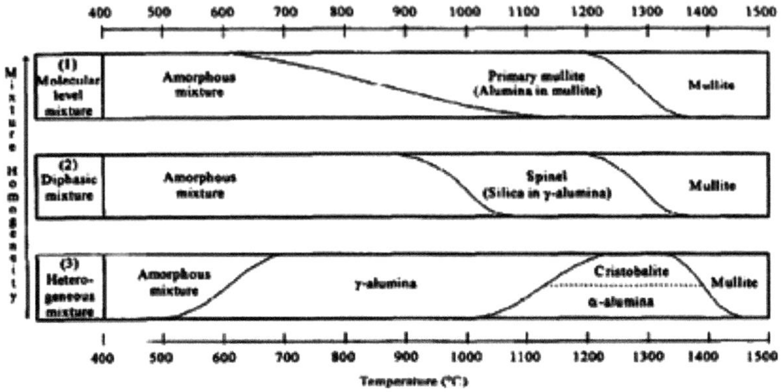

Figure 7.19 Phase transformation sequences of mullite precursors depending on the mixture's homogeneity (Sugita et al., 1998).

The slight reduction in the transformation temperature of mullite was interpreted by Sugita et al. (1998) to be probably due to the availability of the alumina phase, which acts as seed particles. However, there is a large reactivity difference in the phase evolution process of the two precursor gels. The formation of α-Al_2O_3 occurred at as low as 1000°C in precursor gel (1) and it remained till as high a temperature as 1500°C.

TEOS and $ACl_3{\cdot}6H_20$ were used by Jin et al. (2002) as starting materials for the synthesis of a diphasic gel. The former component was prehydrolyzed overnight at room temperature in an alcohol water solution. To this solution, the second component was added. The mixture was stirred vigorously and kept at 70°C and finally ammoniated by slowly adding 4N ammoniated water till complete precipitation. The precipitate was washed free of Cl^- and finally dried at 110°C and marked as a hydrolysis-coprecipitated powder (HCP). For comparison, they also prepared a mixed alumina-silica powder (ASP) from separately prepared dried alumina and silica gels from the above sources followed by mixing and grinding in an agate mortar. The structural evolutions of these on heat treatment were monitored by XRD and IR techniques. Results showed that the phase transformation of HCP and ASP powders proceed in different ways.

7.4.2 Effect of Aging Silica

Stoichiometric mullite gels, using the diphasic route, with silica particles of varying sizes were synthesized by Fahrenholtz et al. (1993) and compared with the gel obtained from the solution precursor. They tailored the phase transformation behavior of diphasic gels by varying the particle size of the silica particles while keeping a fixed source of the boehmite component (average particle size after dispersion in deionized water was 50 nm, with a range of 40–100 nm). The size of the silica spheres with varying diameters—20, 130, and 250 nm—was controlled by the TEOS:ethanol:NH_4OH ratio used in the hydrolysis process. Thereafter, particulate gels were made by mixing aqueous components followed by gelation using NH_4OH at pH 6. They conceived that the variation in the size of the silica particles likely changed the structure of the gel. For example, the DP (P) gel (DP here stands for "diphasic") prepared with fine silica particles should have a connected silica phase that is completely separated by the boehmite particles. This gel transformed directly to mullite at 1300°C, while the gels prepared with larger silica particles, as in DP (20)–DP (130), have a connected boehmite phase separated by silica particles. All these gels showed crystallization of a fraction of the cristobalite at 1300°C and α-alumina at 1400°C. When the size of the silica particle was much bigger, as in DP (250), both cristobalite and α-alumina crystallized at a still lower temperature (1300°C). Complete mullite formation requires a temperature greater than 1500°C in these three cases.

7.5 Mullite Precursor Synthesized by SD and SP Methods

- Mullite precursor by spray pyrolysis (SP): In this process, ANN and TEOS were dissolved in a 1:1 water:methanol solution. The mixed solution was sprayed with compressed air through a glass nozzle into a preheated quartz reaction tube at a temperature of 350°C–650°C when decomposition of the solution took place to form an oxide powder that was amorphous to X-ray (Kanzaki et al., 1990). DTA showed a very sharp exothermic peak at 970°C (Fig. 1), with complete transformation of the powder to mullite.

- Mullite precursor by spray drying (SD): Amorphous spray-dried precursors were synthesized by Douy (2006) in the whole range of composition in the Al_2O_3-SiO_2 system, who then studied the crystallization behaviors by XRD and differential scanning calorimetry (DSC) techniques. It was noted that for up to 70 mol % of Al_2O_3, only mullite crystallizes around 980°C–1000°C; between 70 mol % and 80 mol % of Al_2O_3, mullite and spinel crystallize together; and for more than 80 mol % of Al_2O_3, only spinel is formed. However, the nature of this latter spinel phase was not properly identified in terms of whether it is Al-Si spinel or γ-Al_2O_3 or a mixture of the two.

7.5.1 Effect of Aging on the Phase Transformation of a Mullite Precursor Prepared by SP and SD methods

- Phase transformation studies of the three spray-dried precursors obtained from a solution not aged and aged for different time periods—marked A (without aging), B (aged for 3 h), and C (aged for 10 days)—were done by Jaymes and Douy (1992). According to them, aging affects the heterogeneity of the powder. In powder A, mullite formation was observed through a strong 980°C exothermic peak, as shown in the DSC curves. They decided that this precursor was the most homogeneous and assumed that mixing might have occurred at the atomic level. In powder B, exothermicity was partly reduced and crystallization of the spinel ensued, in addition to some quantity of mullite, as in powder A. They inferred that the chemical homogeneity of mullite was reduced on aging the solution for 3 h. In powder C, the exotherm at 975°C was drastically reduced and the spinel phase was crystallized at 1000°C in a major quantity, which requires heating at a high temperature, on the order of 1275°C, for complete mullitization (Jaymes & Douy, 1992, Fig. 3). This precursor, obtained after aging the solution for 10 days prior to SD, was designated as the least homogeneous. They inferred that homogeneity could be controlled by varying the aging process. The phase transformation of these three spray-dried precursors obtained from a solution aged for different time periods showed variation in the chemical homogeneity of the precursor powders.

- Three other spray-dried precursors, marked Mull A, Mull B, and Mull C, were synthesized by Jaymes et al. (1995). In this process, the mixture of TEOS and ANN produced a homogeneous alumina-silica powder, which in turn exhibited the first exothermic peak and formed t-mullite. Spinel formation was sometimes noted when the mixed solution was aged for a time. A similar behavior was noted in the case of aging of SH gels. Precursor D was obtained after aging a diluted clear solution of the components at above 100°C for 3 days prior to SD by Jaymes et al. (1996). According to them, powder D showed two exotherms, at 924°C and 975°C. For the two combined peaks in place of one exothermic peak, the enthalpy of crystallization was 170 J/g. At 1275°C, this precursor showed one weakly crystalline peak due to orthorhombic mullitization and the enthalpy of crystallization was on the order of 19 J/ g. They are of the opinion that the process of aging leads to heterogeneity of the two oxide systems. It is probable that the reaction between alumina and silica does not occur; on the contrary, polycondensation of silicic acid may take place.

7.6 Verification of Transformation Processes of Mullite Gels Synthesized in Varying Processing Conditions from Components from Varying Sources vs. That from Components from Two Fixed Sources: A Case Study

The thermal behaviors of the following six representative mullite precursors chemically synthesized by various techniques and characterized by DTA and XRD by different researchers were shown in Chapters 2–6, and these have been compared in this chapter. Using components from the same source and varying the processing parameters, some researchers synthesized up to three different mullite precursors and these follow one of the five different routes of mullitization on heating, as mentioned above. Choosing two major variable parameters, such as pH and water, and using different source materials, they have shown the importance of processing techniques, as narrated chronologically and shown above.

Interestingly, using components from the same source with five variables in processing conditions, three kinds monophasic mullite gels (hydrolysis and subsequent gelation of organic silicon component in the presence of water-soluble Al component at different pH values, as per Chapter 3) and two other kinds of CP gels and an SD precursor were synthesized by Haque (2000). All these experiments were repeated later by Chakraborty (2008, unpublished data). The high-temperature portions of the DTA run of all six precursors are shown in Fig. 7.20.

The characteristic peaks observed are summarized as follows:

SHI exhibited only one exothermic peak, at ~1000°C (called 1st exo), followed by a small endotherm, marked as D, at ~965°C.

SHII showed a broad first exotherm at ~980°C, preceded by an endothermic dip at ~960°C, as noted in SHI. It further exhibited a small and broad second exotherm at 1250°C.

SHIII showed three exotherms. The first exotherm was broad and occurred at ~990°C. A new exotherm occurred at 1150°C (called 3rd exo). An as usual second exotherm occurred at 1250°C. Both later peaks were small in magnitude.

SHIV also exhibited two exotherms. The first one was exhibited at 980°C. The new exotherm, of a small intensity, called 3rd exo, occurred at 1180°C. An as usual second, broad exotherm, of a moderate magnitude, occurred at 1200°C.

The diphasic gel exhibited a completely different thermal curve than all the previous monophasic SH gels. It showed a broad exotherm over a wide range of temperatures, 1000°C–1300°C, and exhibited an exotherm of a small magnitude at ~1320°C.

Spray-dried gels exhibited 980°C exotherms, analogous to that exhibited by the SHI precursor. However, the nature of the peak was very sharp compared to that of SH gels.

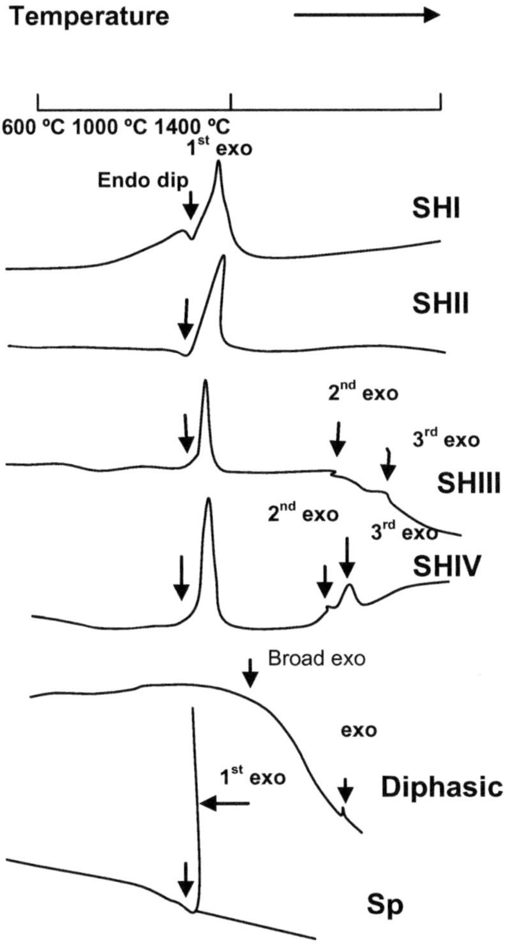

Figure 7.20 DTA traces of different types of Al_2O_3-SiO_2 gels.

7.7 Conclusion

Six different methods of synthesis of mullite precursors from different silica and alumina components have been discussed.

Accordingly, the six ways of phase transformations of these are enumerated.

In the aqueous sol-gel method, both pH and gelation period are the factors in developing direct mullitization.

In the SH gel method, acidic pH and a very low amount or the absence of water are good criteria for abundant mullitization at an earlier stage of firing. However, it is difficult to avoid the partial formation of the spinel phase.

Even in the polymeric gel method, both the water content used and the prevailing pH play a significant role in the formation of a mixture of both mullite and the spinel phase. The favored processing parameter is found to be low amount of water.

The most promising method in the crystallization of pure mullite is, of course, the SP method. The spraying must be done without soaking.

Finally, during the synthesis of a mullite gel, it has been observed that the use of a large quantity of water or an ammoniacal medium turns the gel diphasic.

Therefore, the water used for the hydrolysis of both TEOS and Al alkoxides and soaking during gelation processes must be restricted in order to obtain the predominating mullite phase, for example, Yoldas (1992) polymeric gel, Schneider et al.'s (1993) Type I, and Taylor and Holland's (1993) Type I. Even in the presence of water at pH 7, a series of powders by Mitachi et al. (1990) showed t-mullite formation.

The following methods of precursor synthesis of mullite are described in this chapter: aqueous sol-gel method, monophasic SH gel method, polymeric gel method, diphasic gelation method, and SD and SP methods.

The phase evolution processes of all these mullite gels prepared by the above methods are affected by the following processing variables: (i) pH, (ii) water/alcohol used for cohydrolysis, (iii) the source of silica sols and its aging, (iv) the use of large-sized aluminum hydroxide colloids, and (v) the drying condition of the final gel. Finally, all five processes are substantiated by the author for the same component choice, only with a variation in processing. Results of the five transformation paths of the mullite precursors have been interpreted.

The following observations are made regarding various earlier studies:

- The phase development of the Al_2O_3-SiO_2 gel at 980°C is a function of the two starting components from different sources chosen by various authors.
- The SiO_2 and Al_2O_3 components react differently with a variation in the pH, in the acidic as well as the basic range. A few results are highlighted in this book. The reasons behind these results are to be ascertained. A drastic difference is observed (SH to RH) when the pH of gelation is changed from acidic to basic. Raw gels of both are amorphous; boehmite is not crystallized even in a basic condition. The phase evolutions and geometries of the 980°C exotherm are different. The cause for this is unknown. These two types of monophasic gels are designated as monophasic gel of the first type and monophasic gel of the other type. The heat-treated precursor of the monophasic gel of the other type transforms first to the spinel phase and then to mullite. The thermal transformation of this gel and kaolinite are analogous and likely the explanations for the 980°C peak in both cases will also be analogous. The results of phase transformations of polymeric, diphasic, colloidal, and SD/SP precursors synthesized by various authors are elaborated in the form of examples in this chapter, but the reasons for the results are not provided by said authors.
- The reason for the 980°C exotherm: An interrelationship exists between the crystallization of weakly crystalline mullite and the spinel phase in terms of the geometry of the 980°C peak. Therefore, the specific formation of a phase or phases in the exothermic region is responsible for the occurrence of the first exotherm.

Problems

1. What is the function of water in hydrolysis? Explain the role of using a restricted amount of water versus a large quantity of water for the gelation of organic esters.

2. Explain the role of silica sols and the effect of their aging on the phase transformation of mullite gels. Explain the mechanism of its action on the Si–O–Al bond formation.

References

1. H. Insley and R. H. Ewell, Thermal behavior of the kaolin minerals. *J. Res. Natl. Bur. Stand.*, **14**(5), 615–627 (1935).
2. J. Ossaka, Tetragonal mullite type phase from co precipitated gels. *Nature (London)*, **19**, 1000–1001 (1961).
3. A. K. Chakraborty and D. K. Ghosh, Synthesis and 980°C phase development of some mullite gels. *J. Am. Ceram. Soc.*, **71**(11), 978–987 (1988).
4. H. Schneider, I. Merwin, and A. Sebald, Mullite formation from non-crystalline precursors. *J. Mater. Sci.*, **29**, 805–812 (1992).
5. A. K. Chakraborty, Effect of pH on 980°C spinel phase-mullite formation of Al_2O_3-SiO_2 gels. *J. Mater. Sci.*, **29**, 1558–1568 (1994a).
6. R. K. Iller, *The Chemistry of Silica*. Wiley-Inter Science, London, N.Y. (1979).
7. K. Nishu, T. Yokoyama, T. Watanabe, and T. Tarutani, Characterization of amorphous aluminosilicate formed by adsorption of silicic acid on aluminium hydroxide, in *Abstracts of the 27th Symposium on the Basic Science of Ceramics*. Paper No. IB08. Ceramic Society of Japan, Tokyo (1989).
8. J. W. Kim, J. E. Lee, Y. G. Jung, C. Y. Jo, J. H. Lee, and U. Paik, Synthesis behavior and grain morphology in mullite ceramics with precursor pH and sintering temperature. *J. Mater. Res.*, **18**(1), 81–87 (2003).
9. M. Fukuoka, Y. Onoda, S. Inoue, K. Wada, A. Nukui, and A. Makishima, The role of precursors in the structure of SiO_2-Al_2O_3 sols and gels by the sol-gel process. *J. Sol-Gel Sci. Technol.*, **1**, 47–56 (1993).
10. D. W. Hoffman, R. Roy, and S. Komarneni, Diphasic xerogels, a new class of materials: phases in the system Al_2O_3-SiO_2. *J. Am. Ceram. Soc.*, **67**, 468–471 (1984).
11. K. Okada and N. Otsuka, Characterization of the spinel phase from SiO_2-Al_2O_3 xerogels and the formation process of mullite. *J. Am. Ceram. Soc.*, **69**(9), 652–656 (1986).
12. D. X. Li and W. J. Thomson, Effects of hydrolysis on the kinetics of high-temperature transformations in aluminosilicate gels. *J. Am. Ceram. Soc.*, **74**, 574–578 (1991).

13. B. E. Yoldas, Mullite formation from aluminum and silicon alkoxides. *Ceram. Trans.*, vol. 6, ed. S. Somiya, R. F. Davis, and J. A. Pask, *Am. Ceram. Soc.*, Westerville, OH, p. 255 (1990).

14. M. Ge, H. Yang, Z. Jiang, Y. Wang, and F. Zhang, Ultrafine pure mullite powder by sol-gel method. *J. Non-Cryst. Solids*, **147–148**, 565–568 (1992).

15. Y. X. Huang, A. M. R. Senos, J. Rocha, and J. L. Baptista, Gel formation in mullite precursor obtained via TEOS prehydrolysis. *J. Mater. Sci.*, **32**, 105–110 (1997).

16. M. H. Dafadar, S. Das, and A. K. Chakraborty, Effect of heat on Al_2O_3-SiO_2 gel co-precipitated at different pH. *Trans. Ind. Ceram. Soc.*, **57**(4), 100–102 (1998).

17. H. Yamada and S. Kimura, Studies on co-precipitates of alumina and silica gels and its transformations at higher temperatures. *Yogo Kyokai Shi*, **70**, 87–83 (1962).

18. H. Schneider, B. Saruhan, D. Voll, L. Merwin, and A. Sebald, Mullite precursor phases. *J. Euro. Ceram. Soc.*, **11**, 87–94 (1993).

19. T. Kumazawa, S. Ohta, S. Kanzaki, and H. Tabata, Influence of powder characteristics on microstructural and mechanical properties of mullite ceramics. *Ceram. Trans.*, **6**, 401–411 (1990).

20. C. S. Hsi, F. S. Yen, and Y. H. Chang, Characterization of co-precipitated Al_2O_3-SiO_2 gels. *J. Mater. Sci.*, **24**, 2041–2046 (1989).

21. B. Fu-Su Yen, C. S. Hsi, Y. H. Chang, and H. Y. Lu, Mullite formation from xerogels of (0.84-2.2) Al_2O_3.1SiO_2. *J. Mater. Sci.*, **26**, 2150–2156 (1991).

22. Y. Hirata, H. Minamizono, and K. Shimada, Property of SiO_2-Al_2O_3 powders prepared from metal alkoxide. *Yogo Kyokai Shi*, **93**(1), 36–54 (1985).

23. S. Mitachi, M. Matsuzawa, K. Kaneko, S. Kanzaki, and Y. Tabata, Characterization of SiO_2-Al_2O_3 powders prepared from metal alkoxides. *Ceram. Trans.*, **6**, 275–286 (1990).

24. L. A. Paulick, Y.-F. Yu, and T.-I. Mah, Ceramic powders from metal alkoxide precursors. *Advances in Ceramics: Ceramic Powder Science*, **21**, 121–129 (1987).

25. S. Rajendran, H. J. Rossell, and J. V. Sanders, Crystallization of a coprecipitated mullite precursor during heat treatment. *J. Mater. Sci.*, **25**, 4462–4471 (1990).

26. A. K. Chakraborty, New data on thermal analysis of diphasic mullite gels. *J. Therm. Anal.*, **46**, 1413–1419 (1996b).

27. M. Yamane, S. Inoue, and A. Yasumori, Sol-gel transition in the hydrolysis of silicon methoxide. *J. Non-Cryst. Solids*, **63**, 13–21 (1984).
28. K. Hamano, Z. Nakagawa, G. Cun-Ji, and T. Sato, in *Mullite*, ed. S. Somiya, Uchida Rokakuho Publishing Co., Tokyo, Japan, p. 37 (1985).
29. J. A. Pask, X. W. Zhang, A. P. Tomsia, and B. E. Yoldas, Effect of sol-gel mixing on mullite microstructure and phase equilibria in the α-Al_2O_3-SiO_2 system. *J. Am. Ceram. Soc.*, **70**(10), 704–707 (1987).
30. J. C. Huling and G. L. Messing, Epitactic nucleation of spinel in aluminosilicate gels and its effect on mullite crystallization. *J. Am. Ceram. Soc.*, **74**(10), 2374–2381 (1991).
31. A. K. Chakraborty, Role of hydrolysis water-alcohol mixture on mullitization of Al_2O_3-SiO_2 monophasic gels. *J. Mater. Sci.*, **29**, 6131–6138 (1994b).
32. A. K. Chakraborty and D. K. Ghosh, Kaolinite-mullite reaction series. The development and significance of a binary aluminosilicate phase. *J. Am. Ceram. Soc.*, **74**, 1401–1406 (1991).
33. A. Taylor and D. Holland, The chemical synthesis and characterization sequence of mullite. *J. Non-Cryst. Solids*, **152**, 1–17 (1993).
34. S. Prochazka and F. J. Klug, Infrared-transparent mullite. *J. Am. Ceram. Soc.*, **66**(12), 874–880 (1983).
35. H. Suzuki, H. Saito, Y. Tomokiyo, and Y. Suyama, Processing of ultrafine mullite powder through alkoxide route, in *Ceramic Transactions*, Vol. 6, Mullite and Mullite Matrix Composites. eds. S. Somiya, R. F. Davis, and J. A. Pask, American Ceramic Society, Westervikke, OH, pp. 263–274 (1990).
36. C. Geradin, S. Sundaresan, J. Benziger, and A. Navrotsky, Structural investigation and energetics of mullite formation from sol-gel precursors. *Chem. Mater.*, **6**, 160–170 (1994).
37. K. Okada and N. Otsuka, Preparation of transparent mullite films by dip coating method. *Ceram. Trans.*, **6**, 425 (1990).
38. K. Okada, C. Aoki, T. Ban, S. Hayashi, and A. Yasumori, Effect of aging temperature on the structure of mullite precursor prepared from tetraethoxysilane and aluminium nitrate in ethanol solution. *J. Euro. Ceram. Soc.*, **16**, 149–153 (1996).
39. S. Komerneni and C. Ruscher, Single-phase and diphasic aerogels and xerogels of mullite: preparation and characterization. *J. Euro. Ceram. Soc.*, **16**, 143–147 (1996).

40. A. M. L. M. Fonseca, J. M. F. Ferreira, I. M. M. Salvado, and J. L. Baptista, Mullite based compositions prepared by sol-gel techniques. *J. Sol-Gel Sci. Technol.*, **8**, 403–407 (1997).

41. I. Jaymes, A. Douy, D. Massiot, and J. P. Coutures, Characterization of mono- and diphasic mullite precursor powders prepared by aqueous routes, ^{27}Al and ^{29}Si MAS-NMR spectroscopy. *J. Mater. Sci.*, **31**, 4581–4589 (1996).

42. B. E. Yoldas and D. P. Partlow, Formation of mullite and other alumina-based ceramics via hydrolytic polycondensation of alkoxides and resultant ultra- and micro-structural effects. *J. Mater. Sci.*, **23**, 1895–1900 (1988).

43. B. E. Yoldas, Effect of ultrastructure on crystallization of mullite. *J. Mater. Sci.*, **27**(24), 6667–6672 (1992).

44. A. K. Chakraborty, DTA characterisation of three types of Al_2O_3-SiO_2 gels made from TEOS-Al(OBu)$_3$ mixture with variation of water. *Ceram. Int.*, **22**, 463–469 (1996a).

45. C. H. Ruscher, G. Schrader, and M. Gotte, Infra-red spectroscopic investigation in the mullite field of composition: $Al_2(Al_{2+2x}Si_{2-2x})O_{10-x}$ with 0.55>x>0.25. *J. Euro. Ceram. Soc.*, **16**, 169–175 (1996).

46. T. Kumazawa, S. Ohta, S. Kanzaki, and H. Tabata, Influence of powder characteristics on microstructural and mechanical properties of mullite ceramics (74 wt% Al_2O_3). *J. Jpn. Ceram. Soc.*, **99**, 1228–1233 (1991).

47. T. Heinrich, F. Raether, O. Spormann, and J. Fricke, SAXS measurements of condensation in mullite precursors. *J. Appl. Cryst.*, **24**, 788–793 (1991).

48. T. Heinrich and F. Raether, Structural characterization and phase development of sol-gel–derived mullite and its precursors. *J. Non-Cryst. Solids*, **147–148**, 152–156 (1992).

49. J. Sanz, I. Sobrados, A. L. Cavalieri, P. Pena, S. de. Aza, and J. S. Moya, Structural changes induced on mullite precursors by thermal treatment: a ^{27}Al MAS-NMR investigation. *J. Am. Ceram. Soc.*, **74**(10) 2398–2403 (1991).

50. M. I. Nieto, G. Urretavizcaya, A. L. Cavalieri, and P. Rana, Structural changes in colloidal and polymeric aluminosilicate gels with mullite composition. *Br. Ceram. Trans.*, **97**(1), 17–23 (1998).

51. R.-L. Orefice and W.-L. Vasconcelos, Sol-gel transition and structural evolution on multicomponent gels derived from the alumina-silica system. *J. Sol-Gel Sci. Technol.*, **9**, 239–249 (1997).

52. W. G. Fahrenholtz, S. L. Hietala, P. Newcomer, N. R. Dando, D. M. Smith, and C. J. Brinker, Effect of physical structure on the phase development of aluminosilicate gels. *J. Am. Ceram. Soc.*, **74**(10), 2393–2397 (1991).

53. K. Hamano, T. Sato, and Z. Nakagawa, Properties of mullite prepared by co-precipitation and microstructure of fired bodies. *Yogo Kyokai Shi*, **94**(8), 818 -822 (1986).

54. M. G. M. U. Ismail, Z. Nakai, and S. Somiya, Microstructure and mechanical properties of mullite prepared by the sol-gel method. *J. Am. Ceram. Soc.*, **70**(1), C-7–C-8 (1987).

55. B. Sonuparlak, Sol-gel processing of infrared transparent mullite. *Adv. Ceram. Mater.*, **3**(3), 263–267 (1988).

56. J. C. Huling and G. L. Messing, Surface chemistry effects on homogeneity and crystallization of colloidal mullite sol-gel, in *Ceramic Transactions*, Vol. 6, Mullite and Mullite Matrix Composites. eds. S. Somiya, R. F. Davis, and J. A. Pask, American Ceramic Society, Westervikke, OH, pp. 221–229 (1990).

57. W.-C. Wei and J. W. Halloran, Phase transformation of diphasic aluminosilicate gels. *J. Am. Ceram. Soc.*, **71**(3), 166–172 (1988).

58. G. Klaussen, Microstructural evolution of sol-gel mullite. *Ceram. Eng. Sci. Proc.*, **11**, 1087–1093 (1992).

59. M. J. Hyatt and N. P. Bansal, Phase transformations in xerogels of mullite composition. *J. Mater. Sci.*, **25**, 2815–2821 (1990).

60. A. K. Chakraborty, Preliminary study on the effect of pH on thermal transformation of some diphasic Al_2O_3-SiO_2 gels. *J. Sol-Gel Sci. Technol.*, **28**, 87–95 (2003).

61. S. Rajendran, H. J. Rossell, and J. V. Sanders, Crystallization of a coprecipitated mullite precursor during heat treatment. *J. Mater. Sci.*, **25**, 4462–4471 (1990).

62. B. E. Yoldas, Monolithic glass formation by chemical polymerization. *J. Mater. Sci.*, **14**, 1843–1849 (1979).

63. S. S. Sueyoshi and C. A. Contreras Soto, Fine pure mullite powder by homogeneous precipitation. *J. Euro. Ceram. Soc.*, **18**, 1145–1152 (1998).

64. P. Padmaja, G. M. Anilkumar, P. Krishna Pillai, A. D. Damodaran, and K. G. K. Warrier, Formation characteristics and densification behavior of diphasic mullite gels under various pH conditions. *Br. Ceram. Trans.*, **97**(5), 232–235 (1998).

65. A. K. Chakraborty, Characterization of monophasic and diphasic mullite precursors by solid state reaction study. *Br. Ceram. Trans.*, **103**(1), 33–36 (2004).

66. X.-H. Jin, L. Gao, and J.-K. Guo, The structural change of diphasic mullite Ge studied by XRD and IR spectrum analysis. *J. Euro. Ceram. Soc.*, **22**, 1307–1311 (2002).

67. W. G. Fahrenholtz, D. M. Smith, and J. Cesarano, Effect of precursor particle size on the densification and crystallization behaviour of mullite. *J. Am. Ceram. Soc.*, **76**(2), 433–437 (1993).

68. S. Kanzaki and H. Tabata, Sintering and mechanical properties of stoichiometric mullite. *J. Am. Ceram. Soc.*, **68**(1), C-6–C-7 (1985).

69. A. Douy, Crystallization of amorphous spray-dried precursors in the Al_2O_3-SiO_2 system. *J. Euro. Ceram. Soc.*, **26**, 1447–1454 (2006).

70. I. Jaymes and A. Douy, Homogeneous mullite-forming powders from spray-drying aqueous solutions. *J. Am. Ceram. Soc.*, **75**(11), 3154–3156 (1992).

71. I. Jaymes, A. Douy, and D. Massiot, Synthesis of a mullite precursor from aluminum nitrate and tetraethoxysilane via aqueous homogeneous precipitation: an ^{27}Al and ^{29}Si liquid- and solid-state NMR spectroscopic study. *J. Am. Ceram. Soc.*, **78**(10), 2648–2654 (1995).

72. M. Haque, Thesis, University of Kolkata (2000).

73. A. K. Chakraborty, An analysis of the phase evolution of six types of mullite gels. Unpublished data (2008).

Additional References

1. Y.-W. Kim, H.-D. Kim, and C. B. Park, Processing of microcellular mullite. *J. Am. Ceram. Soc.*, **88**(12), 3311–3315 (2005).

2. E. Tkalcec, S. Kurajica, and H. Ivankovic, Diphasic aluminosilicate gels with two stage mullitization in temperature range of 1200–1300°C. *J. Eur. Ceram. Soc.*, **25**, 613 (2005).

3. Z. Chen, L. Zhang, L. Cheng, et al., Novel method of adding seeds for preparation of mullite. *J. Mater. Process. Technol.*, **166**(2), 183–187 (2005).

4. C. C. Osawa, and C. A. Bertran, Mullite formation from mixtures of alumina and silica sols: mechanism and pH effect. *J. Braz. Chem. Soc.*, **16**(2), 251–258 (2005).

5. J. Leivo, M. Lindén, C. V. Teixeira, and J. Puputti, Sol-gel synthesis of a nanoparticulate aluminosilicate precursor for homogeneous mullite ceramics. J. Mater. Res., **21**(5), 1279–1285 (2006).
6. E. R. Sola, F. J. Torres, and J. Alarcón, Thermal evolution and structural study of 2:1 mullite from monophasic gels. *J. Eur. Ceram. Soc.*, **26**, 2279–2284 (2006).
7. O. Burgos-Montes and R. Moreno, Colloidal behaviour of mullite powders produced by combustion synthesis. *J. Eur. Ceram. Soc.*, **27**(16), 4751–4757 (2007).
8. L. S. Cividanes, T. M. B. Campos, L. A. Rodrigues, D. D. Brunelli, and G. P. Thim, Review of mullite synthesis routes by sol-gel method. *J. Sol-Gel Sci. Technol.*, **55**(1), 111–125 (2010).
9. K. Yoshida, H. Hyuga, N. Kondo, and H. Kita, Synthesis of precursor for fibrous mullite powder by alkoxide hydrolysis method. *Mater. Sci. Eng. B*, **173**(1–3), 66–71 (2010).
10. Buljan, C. Kosanovic, and D. Kralj, A novel synthesis of nano-sized mullite from aluminosilicate precursors. *J. Alloys Compd.*, **509**(32), 8256–8261 (2011).
11. T. Hongbin, Preparation of mullite fibers by sol-gel process and study of their morphology. *Mater. Manuf. Processes*, **24**, 374–1377 (2011).
12. F. W. Fernandes, T. M. B. Campos, L. S. Cividanes, J. P. B. Machado, E. A. N. Simonetti, and G. P. Thim, Influence of ethylene glycol on the mullite crystallization processes analyzed by Rietveld refinement. *J. Aerosp. Technol. Manag.*, **5**(4), 431–438 (2013).
13. L. T. Jurado, R. M. A. Hernández, and E. Rocha-Rangel, Sol-gel synthesis of mullite starting from different inorganic precursors. *J. Powder Technol.*, **2013**, Article ID 268070 (2013).
14. M. Fukushima and Y. I. Yoshizawa, Fabrication and morphology control of highly porous mullite thermal insulators prepared by gelation freezing route. *J. Eur. Ceram. Soc.*, **36**(12), 2947–2953 (2016).
15. S. Bhattacharyya and R. Singh, Effect of solution pH on mullite phase formation from a diphasic precursor powder. *J. Aust. Ceram. Soc.*, **52**(2), 20–31 (2016).

Chapter 8

Phase Evolution Studies of Various Mullite Gels/Precursors by IR and Raman Spectral Techniques

8.1 Introduction

In Chapters 2–6, mullite precursors synthesized by gelation, coprecipitation techniques, diphasic gelation method, and codecomposition process were studied for the evolution of phases jointly by X-ray diffractometry (XRD) and differential thermal analysis (DTA). Comparisons of the phase evolution processes of mullite precursors synthesized by different processing techniques were shown. Even precursors prepared from the same two component sources but with varying gelation or coprecipitation parameters were presented. The brief results so far are as follows:

- Monophasic gels/polymeric gels/Type I significantly formed poorly crystalline mullite (major) and also weakly crystalline spinel (minor) at the first exotherm.
- Some other monophasic gels (colloidal gels)/coprecipitated gels/Type III formed exclusively spinel phases at the first exotherm and then mullite at the second exotherm.
- Diphasic/colloidal gels/Type II exhibited neither the first nor the second exotherm but formed spinel phases over a

Mullite Formations: Analysis and Applications
Akshoy Kumar Chakraborty

ISBN 978-981-4877-05-3 (Hardcover), 978-1-003-03167-3 (eBook)
www.jennystanford.com

long range of temperatures. These precursors crystallized to mullite at the third exotherm, at ~1300°C.

- The precursor synthesized by the spray-pyrolyzed method crystallized into pure mullite at the first exotherm only.
- In addition to these four precursors, the author synthesized two more and made an inroad into the subject and the findings will be discussed in later chapters. Now the question is to characterize the two 980°C heat-treated phases: mullite and spinel. In the following two chapters, we will show how these phases were analyzed by infrared (IR) and by magic angle spinning–nuclear magnetic resonance spectroscopy techniques and how the five thermal transformation processes could be explained in the right ways.

8.2 Basic Literature of IR Study

Three basic reference studies were commonly used for the interpretation of phase transformation of mullite precursors in IR applications by Percival et al. (1974), Mackenzie (1972), and Cameron (1977). Assignment of mullite bands was presented by Percival et al. (1974). IR frequency calculation for ideal mullite and its spectrum was shown by Mackenzie (1972). Raman frequencies of mullite is shown by McMillan and Piriou (1982). A comparison of the IR frequencies is shown in Table 8.1. In the spectral range of 1100–1200 cm^{-1}, Mackenzie showed the superposition of two peaks, one of high intensity and one of low intensity, and assigned these to AlO_4 (1165 cm^{-1}) and SiO_4 (1125 cm^{-1}) species, respectively.

Some mullite samples varying in chemical composition and calcination temperature were taken by Cameron (1977) with a view to investigate the field of solid solutions of mullite bearing the empirical formula Al_2(VI) [$Al_{2+2x}Si_{2-2x}$(IV)] $O_{10-x}V$. It depends first on the accommodation of oxygen vacancies. The second view of investigation is to development of the lattice constant (LC) value of mullite with variation of *x*. The **a** lattice parameter of orthorhombic mullite (o-mullite) increases linearly from **a** = 0.754 nm of 3:2 mullite (sinter variety where $x = 0.25$) to **a** = 0.757 nm for 2:1 mullite (melt variety where $x = 0.4$). He emphasized the first technique to estimate the composition of mullite based on lattice parameter measurement by XRD method.

Table 8.1 IR and Raman frequencies of mullite

Number of bands	Percival et al. (1972) Al(vi–O) cm^{-1}	Al(iv–O) cm^{-1}	Si–O cm^{-1}	Mackenzie (1972)	McMillan and Piriou (1982)
1			1172 s	A 1165–AlO_4	1160 m
2			1120 ms, sh	B 1125–SiO_4	1040 sh
3			{960 ms, s 927 s, br 901 s, sh}	C 950–SiO_4	960 s
4		832 s, br		D {860–AlO_4	880 m
5				E 814} br-AlO_6	
6		740		F 730–AlO_4	
					710
7	613 s, br			G [589–AlO_6]	610 m
8	567 s			H [548–AlO_4]	
			{542 s, sh		
9			500 m sh 445 m, sh} 362 w, sh	I 498–SiO_4	410 s 310 s 150 m

Source: (Percival et al., 1972; Mackenzie, 1972; McMillan & Piriou, 1982).

Mackenzie (1972) also investigated the possibility of changes in local ordering under IR spectroscopy in the 3:2 to 2:1 mullite composition range. He noted a systematic change in the line profile of the IR spectrum in the 1100–1200 cm^{-1} range as a function of the alumina content of mullite. Accordingly, he introduced this parameter as a second technique to estimate the composition of mullite. It was noted that one peak was much more intense than the other. He reported the relation between the absorbances of 1130 cm^{-1} (A_{1130}) and 1170 cm^{-1} (A_{1170}) and chemical composition. When A_{1130} is stronger than A_{1170}, the mullite is richer in alumina.

When A_{1170} becomes much stronger than A_{1130}, the chemical composition approaches mullite of 3:2 stoichiometric composition (~60 mol % of alumina). As the alumina content of mullite increases, the peak moves to the lower wave numbers and decreases in intensity relative to the other peak at still lower frequencies. The peak positions and their sharpness vary smoothly over the 66 mol % to 71 mol % range of Al_2O_3.

8.3 IR Study of Monophasic Gels

Example 1 (one type of monophasic gel, SH, when mullite is a major phase at 1000°C)

The IR-spectra heat-treated SH gel as synthesized (Chapter 2) in the vicinity of 1000 cm^{-1} to 1200 cm^{-1} was measured by Okada, Hoshi, and Otsuka (1986a) and Okada and Otsuka (1986b) and then compared with data published by Cameron (1977). The following observations were made:

Amorphous silica showed an absorption band at 1100 cm^{-1}, assigned to the stretching vibration of Si–O–Si bonds. It shifted to a lower wave number (1030 cm^{-1}) when heat-treated at ~900°C. They considered this shift to be caused by the substitution of alumina by silica in the amorphous solid and further predicted the presence of a discrete amorphous substance to be dependent on the composition of the gel.

With an increase in the firing temperature to around 900°C, mullite-type phase formation occurred in the slow hydrolysis (SH) xerogel and it rapidly increased in amount until 1000°C. Mullitization ceased and showed a plateau between 1000°C and 1100°C.

Mullitization again increased slowly with a rise in temperature. Okada, Hoshi, and Otsuka (1986a) and Okada and Otsuka (1986b) considered two stages for mullite formation in the SH gel. The first stage corresponds to the first step of the mullite formation curve between 900°C and 1000°C. Second stage of mullite formation starts at 1100°C and increases slowly and gradually up to ~1300°C (see Fig. 16.1).

The IR absorption spectra of the heat-treated SH gel at 1000°C–1600°C showed changes in its A_{1130} and A_{1170} absorbance as follows: from $A_{1130} > A_{1170}$ to $A_{1130} = A_{1170}$ and then to $A_{1130} < A_{1170}$ (Fig. 8.1). They correlated the changes with Cameron's (1977) data and predicted such changes with the composition of mullite changing from 70 mol % to 60 mol % of Al_2O_3. Thus the IR result was used to interpret the mullitization process of SH, which showed major mullite crystallization during the first exotherm. According to Okada, Hoshi, and Otsuka (1986a) and Okada and Otsuka (1986b), this mullite contained more than 60 mol % alumina, unlike the data shared by Cameron (1977). On continued heating at a higher temperature, the data might correspond to ~60 mol % of alumina, which coincides with the bulk composition of the SH gel.

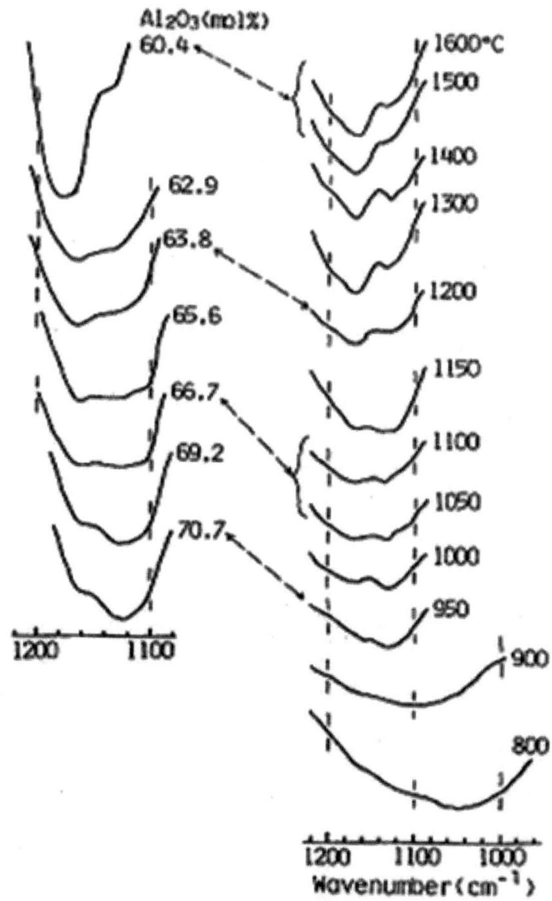

Figure 8.1 Infrared spectra around 1000 cm^{-1} to 1200 cm^{-1} for SH xerogels (Okada, Hoshi, & Otsuka, 1986a). The spectra and chemical composition data on the left are from Cameron (1977).

Ruscher et al. (1996) reproduced three types of precursors with tetraethyl orthosilicate and aluminum-*sec*-butylate as the starting materials as per Schneider et al. (1993). The products were formulated with the stoichiometric composition $3Al_2O_3{:}2SiO_2$. Type I had low water content and thus underwent slow hydrolysis. Types II and III had high water contents and so underwent fast hydrolysis, with pH values of >10 and <10, respectively. These are categorized as per Schneider et al. (1993), as shown in Chapter 3. The materials were calcined at 350°C and given further heat treatments at temperatures between 80°C and 1650°C. The IR absorption of 2:1 and 3:2 mullites and a series of heat-treated mullite were investigated in the spectral range of 400–1400 cm^{-1} using the KBr powder method. It was shown that the intensity variation of the absorption band in the spectral range of 1100–1200 cm^{-1} would provide a useful empirical scale for the determination of mullite composition.

This absorption feature contained the superposition of three peaks, which are related to vibrations of the mullite-specific tetrahedral units $[SiO_4]$, $[AlO_4]$, and $[Al^*O_4]$, with the frequency maxima at about 1165 cm^{-1}, 1130 cm^{-1}, and 1108 cm^{-1}, respectively. Ruscher et al. (1996) opined that the chemical variation of mullite $Al_2(VI)\ [Al_2(IV)Al_{2x}\ Si_{2-2x}(IV)]O_{10-x}$ should lead to a dependence of the intensity ratio $I(1108\ cm^{-1})/I(1165\ cm^{-1})$ as $2x/(2-2x)$, assuming that there is no frequency shift and no change in the absorption cross section of each species as a function of x. Secondly, for the ratio $I(1130\ cm^{-1})/I(1165\ cm^{-1})$, a dependence according to $2/(2-2x)$ is expected that increases more smoothly as a function of x than does the former. Their findings are as follows:

- The IR absorption spectra for Type I precursor heat-treated at higher temperatures showed a significant change in the 1100 cm^{-1} to 1200 cm^{-1} absorption profiles. Similar changes were reported by Cameron (1977) for mullite of different $AI_2O_3{:}SiO_2$ ratios ($0.6 < x < 0.25$)—see Fig. 16.6. Ruscher et al. (1996) roughly determined the intensities of the 1130 cm^{-1} and 1170 cm^{-1} peaks and plotted the ratios of intensities as a function of firing temperatures. They noted the variations and predicted that the composition of the mullite changes gradually from Al-rich mullite to bulk 3:2 mullite as a function of increasing temperature of the firing process. This suggestion is supported by the behavior of the **a** lattice parameter, which also showed a similar functional dependence (first technique) to estimate the composition of mullite. Accordingly, they further correlated the intensity ratio $I(1130\ cm^{-1})/I(1170\ cm^{-1})$ as a function of **a** lattice constants (Å) for mullite composition. It is concluded that the proposed intensity ratio follows a linear relationship with the composition of mullite and the ratio might be used as an empirical scale for the determination of mullite composition $Al_2^{vi}(Al_{2+2x}Si_{2-2x})^{iv}O_{10-x}$.

- Absorption ratios $I(1130\ cm^{-1})/I(1170\ cm^{-1})$ for mullite products of the Type II and II precursors would indicate $x = 0.25$, that is, there is no change in the composition as a function of firing temperature between 1200°C and 1600°C. Some quantity of noncrystalline silica or noncrystalline aluminosilicate phase is expected to be present during the final dehydroxylation of the SH gel or Type I mullite gel at ~1000°C, which is in association with the weakly crystalline mullite phase. These phases also contribute to the IR absorption in the 1000–1100 cm^{-1} spectral range. The intensities of absorption peaks of these decrease on heating due to further mullitization step. It is likely that the decrease in the intensity of 1130 cm^{-1} on gradually heating Type I gel from 1200°C to 1400°C may be related to the decrease of noncrystalline substances, leading to slow mullite formation.

Example 2 (other type of monophasic gel when the Al-Si Spinel Is a major phase at 1000°C)

The phase change of the Al_2O_3-SiO_2 gel system during heating was studied by Hirata et al. (1985). Al_2O_3-SiO_2 powder was prepared from mixtures of metal oxides by hydrolyzing Al_2O_3-SiO_2 with ammoniated water at pH below ~9. In the temperature range of 20°C to 80°C, it was amorphous and consisted of spherical particles. A DTA study exhibited an exotherm at 1005°C, with the formation of the Al-Si spinel phase. At and above 1200°C, mullite formation proceeded rapidly, with the transformation of the intermediate Al-Si spine phase as revealed by XRD analysis. Besides XRD and DTA studies, the phase evolution of Al_2O_3-SiO_2 was also monitored by IR in comparison with IR spectrums of silica (A) and boehmite gel. Later on, Hirata et al. (1989) substantiated the IR study of Al_2O_3-

SiO_2 gels of different compositions. The transformation processes consisted of the following steps:

1. They opined that the gel heated to 900°C was not a mere mixture of silica and aluminum hydroxide but had Si–O–Al bonds. The absorption band at 1100 cm^{-1} for SiO_2 shifted first to a lower wave number (e.g., 1010 cm^{-1}), maybe due to the introduction of Al_2O_3 in the SiO_2 network and the formation of Si–O–Al bonds. But the question is, how much alumina connected with silica and formed noncrystalline alumina silicate? This may not be ascertained during the heating process by noting the IR shifting of the silica band either (i) before the commencement of crystallization of the cubic spinel phase or (ii) after the crystallization of the spinel phase during the exothermic peak temperature. Thus, the constitution of spinel may also be difficult to ascertain.
2. The absorption band in the 500–800 cm^{-1} region would be associated with tetrahedral and octahedral Al–O bonds of mullite. The wave numbers for θ_{Si-O} shift abruptly from 1012 cm^{-1} to 1134 cm^{-1}, with structural change at 980°C exotherm and the formation of a γ-Al_2O_3-type spinel phase (major amount) and mullite (minor) at 1000°C. They summarized the relation between heating temperature and wave number. The two absorption bands, marked (I) and (B), varied with the Al_2O_3:SiO_2 ratio. These are relatively weak with increasing Al_2O_3 content. They also tried to correlate the amount of SiO_2 that does not form mullite. The IR spectra of the monophasic precursor of composition $2.95Al_2O_3 \cdot 2SiO_2$ synthesized (Chapter 3) and heated at different temperatures are shown by Hirata et al. (1989). On the basis of IR changes, the phase developments of alumina silica powder heated to different temperatures are given in Table 8.2.

The question is, is the mullite phase that crystallized from the other kind of monophasic gel at 1200°C instead at 1000°C alumina rich of ~70 mol % of Al_2O_3 content, as conjectured? On the basis of Fig. 3 and Table 1 in the article "Characterization and sintering behavior of alkoxide – derived aluminosilicate powders," by Hirata et al. (1989), it was observed that mullite formed substantially at 1200°C. The IR pattern showed clear absorption bands at 1170 cm^{-1}

and 1130 cm^{-1}. This band ratio of course changed further on heating to 1400°C.

Table 8.2 Changes in the IR frequency shift during phase transformation of SiO_2-Al_2O_3 powder of ~70 mol % Al_2O_3 content

Temperature	Phase
As-produced powder	The powder was amorphous. First shift: The absorption band of silicic (1100 cm^{-1}) shifted to a lower wave number on the introduction of Al into the silica network. The IR spectra showed a tetrahedral Al–O–Si absorption band at 1010–1020 cm^{-1}.
At 1000°C	A structural change occurred. XRD showed the formation of a γ-Al_2O_3-type spinel phase (major amount) and a mullite-type phase (minor). Second shift: The absorption band at 1020 cm^{-1} shifted abruptly to higher values at 1134 cm^{-1} due to the formation of the latter two phases. Two peaks relating to AlO_6, at ~598 or 600 cm^{-1} and AlO_4 at ~548 or 550 cm^{-1}, were just discernable. Another absorption band (weak), at 740 cm^{-1}, due to the AlO_4 group was also observed.
1200°C–1400°C	Absorption bands due to mullite were clearly observed when more mullite was produced at and above 1200°C. Third shift: The Al–O–Si band further shifted to the higher-energy field on gradual heating of the powders from 1000°C to 1200°C, when the absorption band at 1165 cm^{-1} and with a shoulder at 1130 cm^{-1} associated with AlO_4 and SiO_4 tetrahedral vibrations was distinctly observed. The band ratio of 1165 cm^{-1} to 1130 cm^{-1} also became more distinct. Besides, two octahedral peaks, at 600 cm^{-1} and 550 cm^{-1}, and one tetrahedral peak, at 740 cm^{-1}, became more prominent. This IR result corroborates the formation of well-crystallized mullite, as observed in XRD patterns.

Example 3 (another type of monophasic gel)

Suzuki et al. (1987) reported changes in the IR spectra of amorphous mullite precursor powders (synthesized as shown in Chapter 3) on heat treatment. XRD showed that on heating the precursor powder, a mullite-type phase formed from the γ-Al_2O_3-type spinel phase at 1100°C. But the IR spectrum of the said powder did not coincide with that of ideal mullite calculated by Mackenzie et al. (1972). Two peaks, 1165 cm^{-1} and 1130 cm^{-1}, between 1100 cm^{-1} and 1200 cm^{-1} varied much with the composition of mullite sensitivity. As the alumina content of mullite increased, the first peak, of the higher wave number, moved to a lower wave number and decreased in intensity compared with the second peak. They noted that when the powder calcined to 1300°C, the peak at 1165 cm^{-1} showed a higher intensity compared with the second one, which coincided with that of ideal mullite of a stoichiometric composition. On calcination of the powder to 1500°C and up to 1700°C, the ratio of the two peaks did not change.

Example 4 (another type of monophasic gel)
The IR spectrum of a monophasic gel (Chapter 2) vis-à-vis that of a diphasic (Chapter 4) gel heated to 1350°C and 1550°C for 1 h were compared by Hyatt and Bansal (1990). They noted the difference in heating temperatures from 1250°C for 1 h to 1530°C for 1 h with changes in the IR spectrum of the diphasic gel and particularly in relation to the differences in the absorbances in the 1130 cm^{-1} and 1170 cm^{-1} region. With an increase in the temperature, the 1170 cm^{-1} band increased in intensity relative to the 1130 cm^{-1} band at a lower temperature. In the case of monophasic gel in the temperature range of 1150°C–1530°C, the IR absorbances of 1130 cm^{-1} to 1170 cm^{-1} also showed similar changes. Although changes in the absorption ratio are similar in nature, the temperature of occurrence of this phenomenon is not the same. Mullite, however, crystallized in this case on the transformation of the Al-Si spinel phase on heating at 1250°C for 1 h as per the XRD chart. Besides the formation of mullite at ~1300°C, the solid-state reaction between weakly crystalline, alumina-rich mullite and the noncrystalline, silica-rich phase in the temperature range of 1150°C–1530°C is also the cause of mullitization. Although the mechanisms of mullite formation in the two cases are different, both showed changes of a similar nature in the IR band in the said region at two different temperatures.

Example 5 (another type of monophasic gel)
Three different aluminosilicate gels of mullite composition were synthesized by Nieto et al. (1998) according to colloidal and polymeric methods (Yoldas, 1992). The differential thermal

analysis–thermogravimetric analysis (DTA-TGA) curves of a colloidal gel, marked as CA, and two polymeric gels, PA and PB, synthesized in acid and base conditions, were shown. CA exhibited only one exotherm, at ~1285°C, due to mullitization, whereas the two polymeric gels showed two exotherms. Substantiating Yoldas's DTA work, Nieto et al. (1998) showed that the peak intensities of both exotherms were weaker in the gel prepared in the basic condition than the gel prepared in the acidic condition. They also studied the thermal evolutions by IR spectroscopic studies of the three types of gels. A broad band and a weak band between 1000 cm^{-1} and 1100 cm^{-1} were related to Si–O bonds. The absorption band at 1010 cm^{-1} corresponded to Al(IV)–O–Si bonds. However, this band falls in the region of the Si–O stretching vibration band, and the detection of such a band is difficult. The band at 580 cm^{-1} was shown at 500°C, that is, at a lower temperature in the CA sample than in the PA and PB samples (heated to 1000°C and 1100°C, respectively). The colloidal sample did not change until the crystallization of mullite, up to 1200°C. On the other hand, PA and PB showed spectral changes at temperatures as low as 900°C and 1100°C, respectively, where spinel crystallization started and further changes followed during mullitization.

Absorption bands of the three precursors in Fourier-transform infrared spectra are different in the following regions: 500–800 cm^{-1}, 800–950 cm^{-1}, and 1000–1200 cm^{-1}. The first spectral change occurred on heating PA and PB from 900°C to 1000°C, showing a band at 580 cm^{-1}, assigned to γ-Al_2O_3. The second spectral change occurred on heating PA and PB from 1200°C to 1300°C at 550 cm^{-1} and 730 cm^{-1}, assigned to vibrations of octahedral and tetrahedral bonds of mullite. Spinel formation (γ-Al_2O_3) was noted at as low as 500°C in the case of colloidal gel and then a major spectral change occurred during mullite formation at ~1200°C. PA was another type of monophasic gel. It showed mullite formation and an exotherm at ~1200°C and accordingly showed both absorption bands at 1137 cm^{-1} and 1170 cm^{-1}, respectively. The absorption band between 1000 cm^{-1} and 1200 cm^{-1} became narrow and the ratio of the two peaks changed and finally the band at 1170 cm^{-1} was higher than the peak at 1137 cm^{-1}.

8.4 Comparison of IR Spectra of Monophasic Gels of Different Al_2O_3-SiO_2 Compositions on Heating

Example 1: The first precursor powder ($1.03Al_2O_3 \cdot 2SiO_2$) by Hirata et al. (1989) was high in silica, that is, there was an excess amount of silica besides 3:2 mullite formation. Consequently, the shift of Si–O–Al from 1020 cm^{-1} to higher energies (1165 cm^{-1}), with a small shoulder at 1130 cm^{-1}, on heating to 1000°C, with the formation of mullite, did not show a perceptible change. The major portion of silica was left at 1100 cm^{-1}. That's why the shift showed a minimum change instead of an abrupt change in wave numbers, as noted in other cases. Hirata et al. (1989) showed the IR spectra of four precursors of varying compositions heat-treated at 1400°C. Two points are noted.

- The contour of the absorption bands between 1100 cm^{-1} and 1200 cm^{-1} seems to be a function of the chemical composition of the precursors.
- Two absorption bands related to Si–O vibrations at 1125 cm^{-1} and 498 cm^{-1}, designated (B) and (I), were shown by MacKenzie (1972). The intensities of the bands associated with free silica (Si–O–Si) observed at 1130 cm^{-1} and 440 cm^{-1} showed a change from a high-silica to a high-alumina composition.

Example 2: Sales and Alarcon (1996) synthesized three precursors (in mol %) A (25 Al_2O_3 : 75 SiO_2); B (71.8 Al_2O_3 : 28.2 SiO_2); C (88.4 Al_2O_3 : 18.2 SiO_2) by prehydrolysis technique (see Chapter 3). The IR study observations of these three heat-treated samples are shown in Fig. 8.2, 8.3 and 8.4, respectively.

- The characteristic vibration modes associated with Si–O bonds were not clearly distinguished in samples B and C.
- In the 500–800 cm^{-1} region, the bands related to tetrahedral and octahedral Al–O coordination were clearly observed in sample B and specially in sample C, which is rich in alumina.
- A broad band at 875 cm^{-1} increased in intensity with a rise in temperature, maybe due to Si–O–Al bond vibration.
- The peak at 1020 cm^{-1} was due to Si–O–Al bond vibrations. It was derived from Si–O stretching vibrations by the

introduction of tetrahedral Al from its shift from the original vibration at 1100 cm^{-1}.

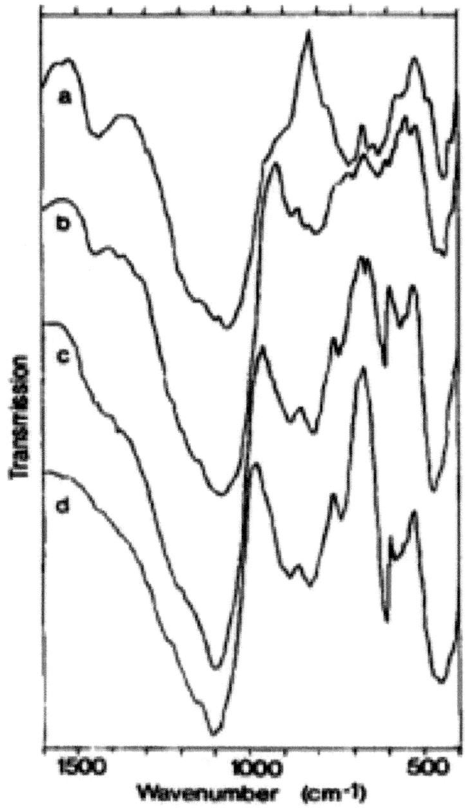

Figure 8.2 IR absorption spectra of sample A after several heat treatments; (a) dried gel; (b) heat-treated at 750°C for 3 h; (c) heat-treated at 1000°C for 3 h; (d) heat-treated at 1300°C for 3 h (Sales & Alarcon, 1996).

- This Si–O–Al bond vibration shifted to higher energies when the gel powder was heated to a higher temperature. At 1300°C, the band was centered at 1165 cm^{-1}, with a shoulder at 1130 cm^{-1}, associated with SiO_4 and AlO_4 tetrahedral vibration modes, respectively.

Sales and Alarcon (1996) tried to correlate the intensity of the peak ratio of 1170–1130 cm^{-1} peaks in samples A, B, and C heated at 1300°C, where o-mullite is fully developed, and again heated at

1000°C for 15 h, in which case a mullite-type phase was observed only in sample +B. Finally, they tried to predict the composition of the mullite using Cameron's (1977) relation.

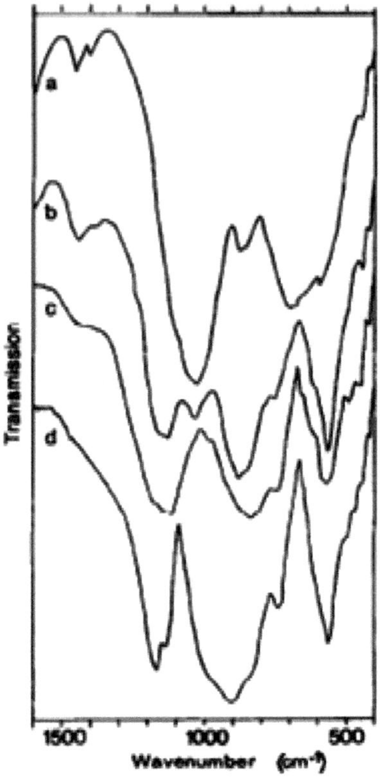

Figure 8.3 IR absorption spectra of sample B after several heat treatments; (a) dried gel; (b) heat-treated at 750°C for 3 h; (c) heat-treated at 1000°C for 3 h; (d) heat-treated at 1300°C for 3 h (Sales & Alarcon, 1996).

According to Sales and Alarcon (1996), at the first stage of mullite formation in sample B, heated at ~1000°C for 15 h, the IR absorption band at 1130 cm^{-1} showed to be stronger than the 1165 cm^{-1} band. The IR absorbance band for sample C at 1300°C at 1170 cm^{-1} was stronger than the Si–O–Al bond vibrations at 1130 cm^{-1}. However, the IR absorbance band for sample A at 1300°C at 1170 cm^{-1} was always stronger during heating at 1000°C–1300°C, because it was richer in silica composition. But the peak at 1170 cm^{-1} was observable.

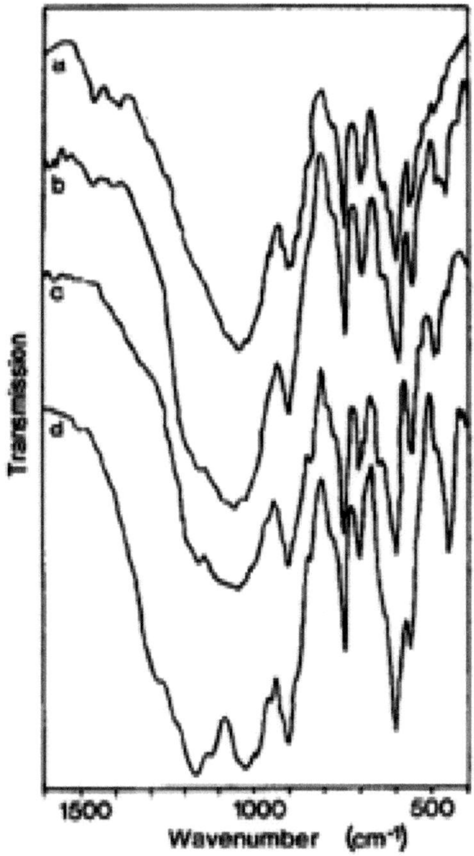

Figure 8.4 IR absorption spectra of sample C after several heat treatments; (a) dried gel; (b) heat-treated at 750°C for 3 h; (c) heat-treated at 1000°C for 3 h; (d) heat-treated at 1300°C for 3 h (Sales & Alarcon, 1996).

8.5 IR Study of Coprecipitated Gel and Diphasic Gel

Diphasic gel

The IR spectra of the Type II gel by Schneider et al. (1993) heat-treated at 900°C exhibited strong bands at ~1100 cm^{-1} and 470 cm^{-1} and a broad absorption area of ~820 cm^{-1} and 580 cm^{-1}. The strong absorption peaks at ~1100 cm^{-1} and 470 cm^{-1} are attributed to Si–O–Si stretching and O–Si–O bending vibrations, and the absorption region between ~820 cm^{-1} and ~580 cm^{-1} may be due to the spinel phase. The structural evolutions of hydrolysis-coprecipitated powder (HCP) and alumina-silica powder (ASP) on heating were monitored by XRD and IR techniques by Jin et al. (2002). The results are shown in Fig. 8.5 and in Tables 8.3 and 8.4.

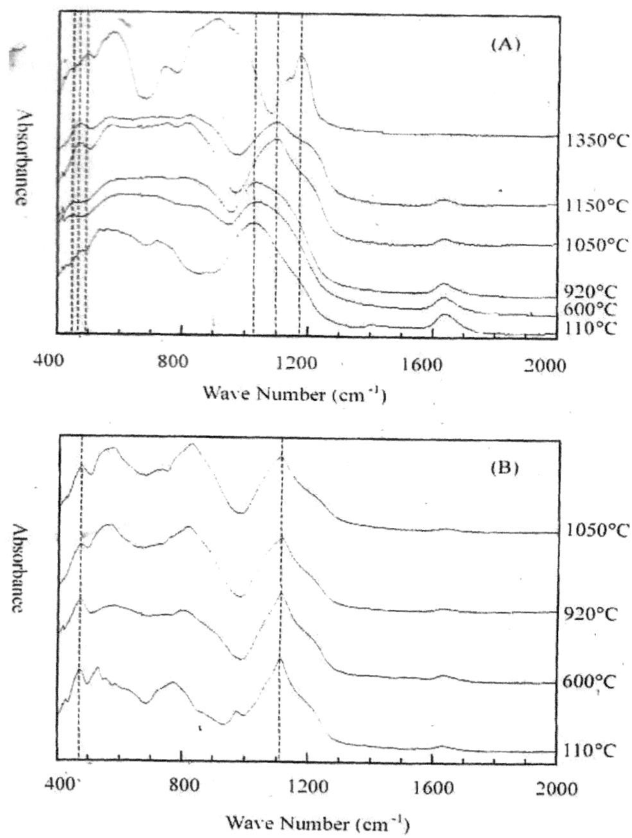

Figure 8.5 IR spectra of (A) diphasic gel (HCP) and (B) mixed gel (ASP) powders heat-treated at various temperatures (Jin et al., 2002).

Orefice et al. (1997) heat-treated two varieties of mullite precursors and did XRD and IR to characterize these. The phase transformations are shown in preceding chapters.

Table 8.3 IR spectral studies of HCP gel

Variation in Al–O bands		
As received HCP gel (contains bayerite)	During crystallization of the spinel at 920°C	Subsequent transformation to mullite at 1350°C
AlO_6 band (strong) at 560 cm^{-1} (due to hexacoordinated Al ion)	AlO_4 appeared at a lower temperature, becoming comparable to and now overlapping AlO_6. The broad brand, stretching from 500 $^{-1}$ to 900 cm^{-1}, remained.	Three peaks were observed: one due to AlO_6 at 560 cm^{-1} and two due to AlO_4 groups, one at 740 cm^{-1} and one at 830 cm^{-1}.
Variation in Si–O bands		
	The ~1100 cm^{-1} Si–O band shifted to a higher wave number, and the 470 cm^{-1} band also shifted slightly to a higher wave number.	

Table 8.4 IR Spectral studies of ASP gels and others

Variation in Al–O bands	
As-received ASP gel (contains bayerite of strong XRD intensities)	In the temperature range of 600°C to 1050°C
AlO_6 band (strong) at 560 cm^{-1}	XRD showed crystallization of γ-Al_2O_3 and θ-Al_2O_3. The intensities of the AlO_4 band at 830 cm^{-1} and 740 cm^{-1} overlapped into a broad band stretching from 500 cm^{-1} to 900 cm^{-1}, gradually increasing from 600°C to 920°C, and at 1050°C the broad band depicted the appearance of one more distinct broad peak.
Variation in Si–O bands	
	Two Si–O bands, one at ~1100 cm^{-1} and one at 470 cm^{-1}, remained apparently unchanged during the heating process up to 1050°C.

8.6 Raman Spectral Study of a Monophasic Gel

Colomban (1989) attempted to study the local structure of mullite gels and glasses and the influence of synthesis on this structure. DTA traces of typical mullite monoliths and powders synthesized are shown in Chapter 3. An intense exothermic peak was observed between 1020°C and 1330°C. A second exotherm was noticeable in some cases between 1230°C and 1425°C. Nitrate gel exhibited a weak double exotherm around 920°C–1015°C. According to him, XRD patterns showed an amorphous structure up to 900°C and then a progressive transformation into a γ-alumina-like pattern was observed and the pores closed. Above the exothermic peak, the mullite phase was observed. Its crystallinity (peak intensity, modification, narrowing) increased progressively with temperature.

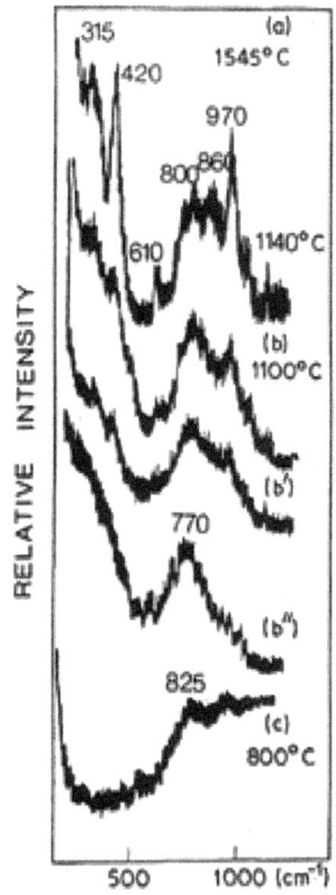

Figure 8.6 Micro Raman spectra of mullite (x = 0.25); optically clear samples produced from SH in hexane using silicon propoxide (a), silicon methoxide (b′), and silicon ethoxide (b″ and c) and heated at various temperatures. A spectrum of a monolith prepared using $Al(NO_3)_3$ solution is given in (b) (Colomban, 1989).

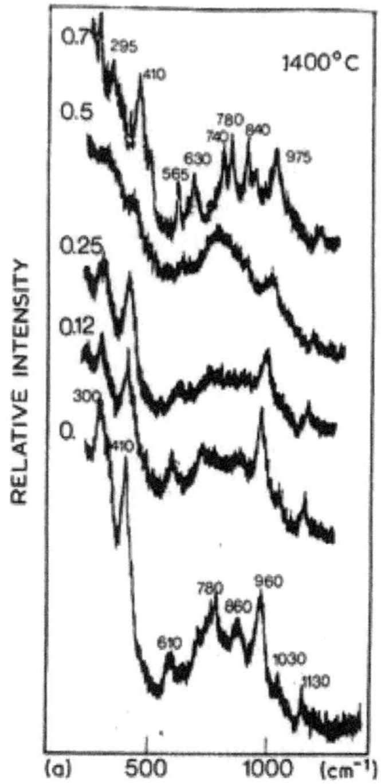

Figure 8.7 Micro Raman spectra of mullite monoliths of various compositions ($0 \leq x \leq 0.7$) prepared by slow hydrolysis of AlOBu and silicon methoxide in hexane and heated for 1 h at various temperatures between 800°C and 1650°C. Monoliths with $x = 0.5$, 0.25, and 0.12 are optically clear (Colomban, 1989).

In Raman spectra (Fig. 8.6), the formation of crystalline mullite was observed by the new bands, at 320 cm^{-1}, 430 cm^{-1}, and 970 cm^{-1}. The degree of modification depended upon the synthesis method. Band narrowing continued up to 1500°C. During the evolution process, the monoliths remained optically clear and the evolution of the spectra was progressive. With these indications, Colomban (1989) predicted that the transformation of the gel first to glass then to spinel, and then to mullite is continuous. According to Colomban (1989), no segregation takes place during the transformation of the gel to o-mullite. A progressive ordering at the local scale takes place. The X-ray powder patterns and DTA traces indicated a strong modification in the 1000°C–1100°C temperature range. The decrease in orientational disorder can explain the narrowing of the SiO_4 mode at 970 cm^{-1} (above 1000°C) and AlO_4 (above 1300°C) stretching modes. The counterparts of Raman spectra are the condensation of the bending modes below 500 cm^{-1}, the frequency shift of the Al–O "stretching mode" (from 820 cm^{-1} to 770 cm^{-1}) indicating a considerable strengthening of the bond at the glass → spinel transition. The mean Al–O frequency increased when the mullite phase occurred according to unit cell dilation. In the IR spectra (Fig. 8.7), the spinel-like phase is characterized by the 350 cm^{-1} shoulder and the appearance of the 550 cm^{-1} band. The shift of the mean Si–O and Al–O stretching modes was straightforward.

8.7 Summary

Researchers Okada, Hoshi, and Otsuka (1986a), Hirata et al. (1985, 1989), Colomban (1989), Li and Thomson (1991), Hyatt and Bansal (1990), Suzuki et al. (1987), Sales and Alarcon (1996), Ruscher et al. (1996), and Nieto et al. (1998) applied the IR technique as a tool to reveal the sequential phase evolution process of the monophasic/Type I gel.

1. This step of the transformation involves the dehydration/decomposition of residual alkoxy groups and the development of Si–O–Al bonds between two components by substitutions as per DTA-TGA and IR observation. Several researchers (Okada, Hoshi, & Otsuka, 1986a; Hirata et al., 1985, 1989) noted the shift of Si–O–Si bonds at ~1100 cm^{-1} to a lower frequency (~1020 cm^{-1}) in their IR investigations of the gel undergoing heating. The tentative reaction on heating the mullite gel at <950°C would be a noncrystalline aluminosilicate hydrate phase, called the mullite precursor. This process continued on heating up to nearly the commencement of the 980°C exotherm.
2. Crystallization of the mullite precursor phase occurs at the first exotherm. At the first exothermic step, the aluminosilicate hydrate precursor phase crystallized into a poorly crystalline mullite-type phase (major) and a weakly crystalline γ-Al_2O_3-type spinel (minor) phase. Although the XRD intensity peaks of mullite are predominantly high, the IR absorption profile of the mullite precursor heated at ~1000°C is very weak. Hirata

et al. (1989) and Li and Thomson (1991) made the following observations during the exothermic process:

- A sudden jump of absorption from 1020 cm^{-1} to higher values (1134 cm^{-1})
- Shift of the contour from the 1200 cm^{-1} region to the 1100 cm^{-1} region
- Appearance of faint IR absorption peaks of AlO_6 groups at 589 cm^{-1} and 548 cm^{-1} and the same for AlO_4 groups at 850 cm^{-1} and 730 cm^{-1}

3. With continued heating, at ~1400°C, Si–O–Al bonds shift to a large extent to high-energy field, to 1170 cm^{-1}, with a shoulder at ~1130 cm^{-1}, corresponding to AlO_4 and SiO_4 tetrahedra of mullite. Thus, mullite becomes well crystallized on heating at 1400°C. The observed IR absorption bands correspond well with the calculated absorption band for ideal mullite analyzed by Mackenzie (1977). A peak at 1170 cm^{-1} shows a higher intensity compared to the one at 1130 cm^{-1}, which coincides with that of stoichiometric mullite. However, the relative IR intensities of the two peaks change gradually from ~1000°C to 1600°C (Okada, Hoshi, and Otsuka, 1986a).

Changes in the IR pattern and the relative resolution of the Al(4)/Al(6) ratio during heat treatment of mullite precursors: Experimental results

Okada, Hoshi, and Otsuka (1986a) noted that initially $A_{1130} > A_{1170}$, then $A_{1130} \pm A_{1170}$, and finally $A_{1130} < A_{1170}$ for SH xerogel in Fig. 8.1, along with the IR spectra and chemical composition data, citing from Cameron (1977). Subsequent researchers, for example, Hirata et al. (1989), first showed the Al–O–Si absorption band of the monophasic precursor at ~1020 cm^{-1}, which shifts abruptly to higher values, from 1020 cm^{-1} to 1134 cm^{-1}, due to the formation of a mullite-type phase. It further shifts to a higher-energy field, at 1165 cm^{-1}, with a shoulder at 1130 cm^{-1}, associated with AlO_4 and SiO_4 tetrahedral vibrations. In the first stage of mullite formation in sample B, at ~1000°C, (Sales & Alarcon, 1996), the IR absorbance band at 1130 cm^{-1} is stronger than the one at 1170 cm^{-1}. Hyatt and Bansal (1990) noted that with an increase in temperature, the 1170 cm^{-1} band increases in intensity relative to the 1130 cm^{-1} band.

Ruscher et al. (1996) opined that the absorption feature contained the superposition of three peaks, which are related to vibrations of the mullite-specific tetrahedral units [SiO_4], [AlO_4], and [Al^*O_4], with the frequency maxima at about 1165 cm^{-1}, 1130 cm^{-1}, and 1108 cm^{-1}, respectively. They plotted the ratios of intensities of the 1130 cm^{-1} and 1170 cm^{-1} peaks as a function of the firing temperature of the Type I precursor.

Conclusions based on the referred documents

- The relation between the IR absorbances of 1130 cm^{-1} and 1170 cm^{-1} and the chemical composition of mullite over the range of 66 mol %–71 mol % of Al_2O_3 was first reported by Cameron (1977). After comparing with Cameron's (1977) chemical composition the IR data in the 1100–1200 wave number (cm^{-1}) region, Okada, Hoshi, and Otsuka (1986a) predicted the formation of two types of mullite during the heating process of monophasic gel. When the 1130 cm^{-1} peak is stronger than the 1170 cm^{-1} peak, the chemical composition of mullite is rich in Al_2O_3. In the first stage of mullite formation, at ~980°C exotherm, the band 1130 cm^{-1} seems stronger. Accordingly, it is also suggested after the initial formation of Al_2O_3-rich mullite (~70 mol %), the Al_2O_3 content drops to 60 mol % when the mullite is heated gradually to a higher temperature in the second stage. This observation that the 1170 cm^{-1} band increases in intensity relative to the 1130 cm^{-1} band with heat treatment conforms to the earlier report of Cameron (1977). Sales and Alarcon (1996) tried to predict the composition of the mullite using Cameron's (1977) relation. Ruscher et al. (1996) correlated that the composition of the mullite changes gradually from Al-rich mullite to bulk 3:2 mullite as a function of increasing temperature of the firing process. In the second correlation, they indicated the intensity ratio $I(1130\ cm^{-1})/I(1170\ cm^{-1})$ as a function of **a** lattice constants (Å) for mullite composition.
- **Identification of the spinel phase:** Neito et al. (1998) conjectured that the band at 580 cm^{-1}, as shown at a lower temperature, at 500°C, in the CA sample is γ-alumina rather than the spinel formed, in comparison to PA and PB samples

heated at 1000°C and 1100°C. Distinction between spinel other than γ-alumina is required to be shown. Regarding the characterization of the spinel phase, nothing much has been observed in several IR studies.

References

1. H. J. Percival, J. F. Duncan, and P. K. Foster, Interpretation of the kaolinite-mullite reaction sequence from infrared absorption spectra. *J. Am. Ceram. Soc.*, **57**, 57–61 (1974).
2. K. J. D. Mackenzie, Infrared frequency calculation for ideal Mullite ($3Al_2O_3.2SiO_2$). *J. Am. Ceram. Soc.*, **55**(2), 68–71 (1972).
3. W. E. Cameron, Composition and cell dimensions of mullite. *Am. Ceram. Soc. Bull.*, **56**(11), 1003–10011 (1977).
4. P. McMillan and B. Piriou, The structures and vibrational spectra of crystals and glasses in the silica-alumina system. *J. Non-Cryst. Solids*, **53**, 279–298 (1982).
5. K. Okada, Y. Hoshi, and N. Otsuka, Formation reaction of mullite from SiO_2-Al_2O_3 xerogels. *J. Mater. Sci. Lett.*, **5**, 1315–1318 (1986a).
6. K. Okada and N. Otsuka, Characterization of the spinel phase from SiO_2-Al_2O_3 xerogels and the formation process of mullite. *J. Am. Ceram. Soc.*, **69**(9), 652–656 (1986b).
7. C. H. Ruscher, G. Schrader, and M. Gotte, Infra-red spectroscopic investigation in the mullite field of composition: $Al_2(Al_{2+2x}Si_{2-2x})O_{10-x}$ with 0.55>x>0.25. *J. Euro. Ceram. Soc.*, **16**, 169–175 (1996).
8. H. Schneider, B. Saruhan, D. Voll, L. Merwin, and A. Sebald, Mullite precursor phases. *J. Euro. Ceram. Soc.*, **11**, 87–94 (1993).
9. Y. Hirata, H. Minamizono, K. Shimada, Property of SiO_2-Al_2O_3 powders prepared from metal alkoxide. *Yogo Kyokai Shi*, **93**(1), 36–54 (1985).
10. Y. Hirata, K. Sakeda, Y. Matsushita, K. Shimada, and Y. Ishihara, Characterization and sintering behavior of alkoxide-derived aluminosilicate powders. *J. Am. Ceram. Soc.*, **72**(6), 995–1002 (1989).
11. H. Suzuki, H. Saito, Y. Tomokiyo, and Y. Suyama, Processing of ultrafine mullite through alkoxide route, in *Ceramic Transactions*, Vol. 6, Mullite and Mullite Matrix Composities. eds. S. Somiya, R. F. Davis, and J. A. Pask, American Ceramic Society, Westervikke, OH, p. 263 (1987).
12. M. J. Hyatt and N. P. Bansal, Phase transformations in xerogels of mullite composition. *J. Mater. Sci.*, **25**, 2815–2821 (1990).

13. M. I. Nieto, G. Urretavizcaya, A. L. Cavalieri, and P. Rana, Structural changes in colloidal and polymeric aluminosilicate gels with mullite composition. *Br. Ceram. Trans.*, **97**(1), 17–23 (1998).

14. B. E. Yoldas, Effect of ultrastructure on crystallization of mullite. *J. Mater. Sci.*, **27**(24), 6667–6672 (1992).

15. M. Sales and J. Alarcon, Synthesis and phase transformations of mullite obtained from SiO_2-Al_2O_3 gels. *J. Euro. Ceram. Soc.*, **16**, 781–789 (1996).

16. X.-H. Jin, L. Gao, and J.-K. Guo, The structural change of diphasic mullite Ge studied by XRD ad IR spectrum analysis. *J. Euro. Ceram. Soc.*, **22**, 1307–1311 (2002).

17. R.-L. Orefice and W.-L. Vasconcelos, Sol-gel transition and structural evolution on multicomponent gels derived from the alumina-silica system. *J. Sol-Gel Sci. Technol.*, **9**, 239–249 (1997).

18. Ph. Colomban, Structure of oxide gels and glasses by infrared and Raman scattering part 2 mullite. *J. Mater. Sci.*, **24**, 3011–3020 (1989).

19. D. X. Li and W. J. Thomson, Effects of hydrolysis on the kinetics of high temperature transformations in aluminosilicate gels. *J. Am. Ceram. Soc.*, **74**, 574–578 (1991).

Chapter 9

Phase Evolution Studies of Various Mullite Precursors/Gels by MAS-NMR

Synthesis of precursors like monophasic gels/polymeric gels/Type I, colloidal gels/coprecipitated gels/Type III (also monophasic gels), diphasic gels/colloidal gels/Type II, and spray-pyrolyzed samples was shown in the previous chapters. The phase evolution processes of these precursors were also shown by differential thermal analysis (DTA)/differential scanning calorimetry, X-ray diffractometry (XRD), and infrared IR methods. This chapter will show how the intermediate mullite-type phase and γ-Al_2O_3-like spinel phase, which crystallize on heat treatment at 980°C, are analyzed/characterized by magic angle spinning–nuclear magnetic resonance spectroscopy (MAS-NMR) techniques and secondly focus on the phase transformations of these gels/precursors to mullite reaction series.

9.1 Basic Literature

According to Engelhadt and Michel (1987), the silica tetrahedra in a solid silicate have been classified into five species on the basis of the number of bridging oxygen atoms forming Si–O–Si bonds. In aluminosilicate, the silica tetrahedra with four bridging oxygen atoms are also classified into five species on the basis of the number of adjacent aluminum atoms bonded to oxygen (Table 9.1).

Mullite Formations: Analysis and Applications
Akshoy Kumar Chakraborty

ISBN 978-981-4877-05-3 (Hardcover), 978-1-003-03167-3 (eBook)
www.jennystanford.com

Table 9.1 Classification of tetrahedral silica in solid silicates

Q_n, i.e., $Si(OSi)_nO_{4-n}$	Range of chemical shift with respect to tetramethylsilane	$Q_4(mAl)$, i.e., $Si(OSi)_{4-m}(OAl)m$	Range of chemical shift
Q_0	−60 to −82 ppm	$Q_4(4Al)$	−81 to −91 ppm
Q_1	−68 to −83 ppm	$Q_4(3Al)$	−85 to −94 ppm
Q_2	−74 to −93 ppm	$Q_4(2Al)$	−91 to −100 ppm
Q_3	−91 to −101 ppm	$Q_4(1Al)$	−97 to −107 ppm
Q_4	−106 to −120 ppm	$Q_4(0Al)$	−101 to −116 ppm

Risbud et al. (1987) presented ^{29}Si MAS-NMR spectra of silica glasses with a wide range of compositions showing a wide range of chemical shifts and demonstrating a range of local structural environments.

9.2 MAS-NMR Studies of Mullite Gels Synthesized by the Aqueous Sol-Gel Method

9.2.1 Mullite Gels: CM Sample

MAS-NMR studies of the coprecipitated material (CM) sample of mullite gel (gelation of water-soluble silica component in the presence of Al salt as per Chapter 2) were conducted by Schneider et al. (1992). Their phase transformations were studied by using NMR spectroscopy, which is given in Table 9.2 on the basis of Figs. 9.1 and 9.2.

This shows that mullite formation in the CM precursor synthesized by the aqueous gel method is analogous to that of slow hydrolysis (SH) gel prepared by the monophasic gel method. Mullite formation in both cases occurs in two steps. In the first step, a mullite-type phase and a γ-Al_2O_3-type spinel are formed. In the second step, both these phases react with the noncrystalline phase to crystallize to mullite. It is seen that although nearly complete mullitization occurs in the 1100°C to 1200°C range (as per ^{29}Si resonance data), mullite is not attained in a stable state. To attain stable mullite, calcination needs to be done far above 1100°C.

Table 9.2 Phase transformation of CM studied by ^{29}Si and ^{27}Al MAS NMR spectroscopy

Temperature	Phase
600°C–800°C	An XRD showed an amorphous pattern. A ^{29}Si NMR study showed an asymmetrical shape of the ^{29}Si resonance peak, with a shoulder toward a more negative chemical shift at ~95 ppm. ^{27}Al resonance showed a characteristic triplet.
At 900°C	Si resonance started splitting at 800°C. It became pronounced when the gel was heated to 900°C. An XRD showed a mullite-type phase (major) with a γ-Al_2O_3-type spinel (minor) phase. DTA showed a broad exotherm. Al resonance still showed a 30 ppm peak, due to a penta-coordinated Al–O bond.
At 1000°C	The Si resonance peak separated well into –90 and –110 resonance peaks. A shoulder at –80 ppm was noted. The intensity of the mullite increased. Al spectra are apparently similar to that of mullite, but the signals are very broad.
At 1100°C	The –110 ppm peak decreased. In the 1000°C–1100°C range, the intensity of the –110 resonance peak changed significantly and more mullitization occurred. The ratio of AlO_4 to AlO_6 started changing. The AlO_4 peak tended to increase. The ratio of AlO_4 to AlO_6 did not correspond to that of standard mullite.
At 1200°C	The DTA exhibited no exotherm. In the 1100°C–1650°C range, the ratio of AlO_4 to AlO_6 changed remarkably. For example, AlO_4 increased at the cost of AlO_6. Moreover, AlO_4 split. The Si resonance peak at –110 ppm vanished on firing before 1400°C, with crystallization of pure mullite only. The Si resonance peaks around –90 ppm (–87, –90, and –94) became increasingly sharper and narrower.
At 1400°C	An XRD showed mullite only. Both Si and Al spectra concur with that of ideal mullite.

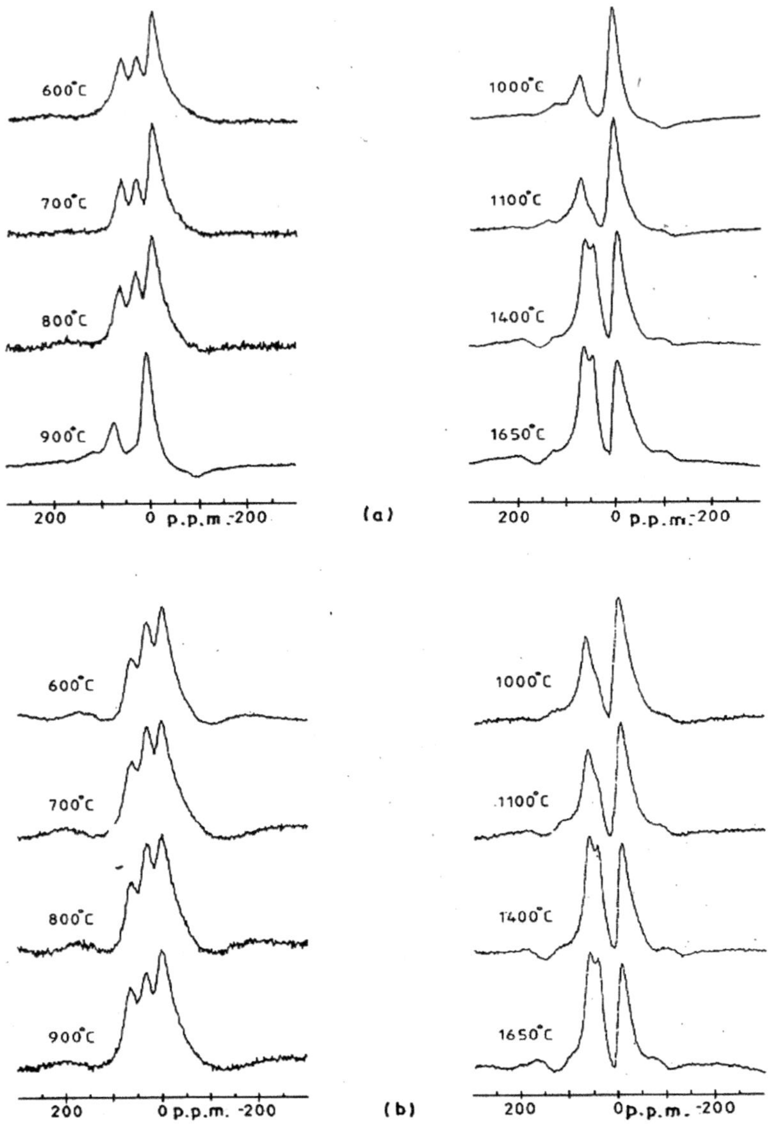

Figure 9.1 ^{27}Al MAS-NMR spectra of mullite precursor materials heat-treated at different temperatures: (a) Sol-gel material (SGM) and (b) coprecipitated material (CM) (Schneider et al., 1992).

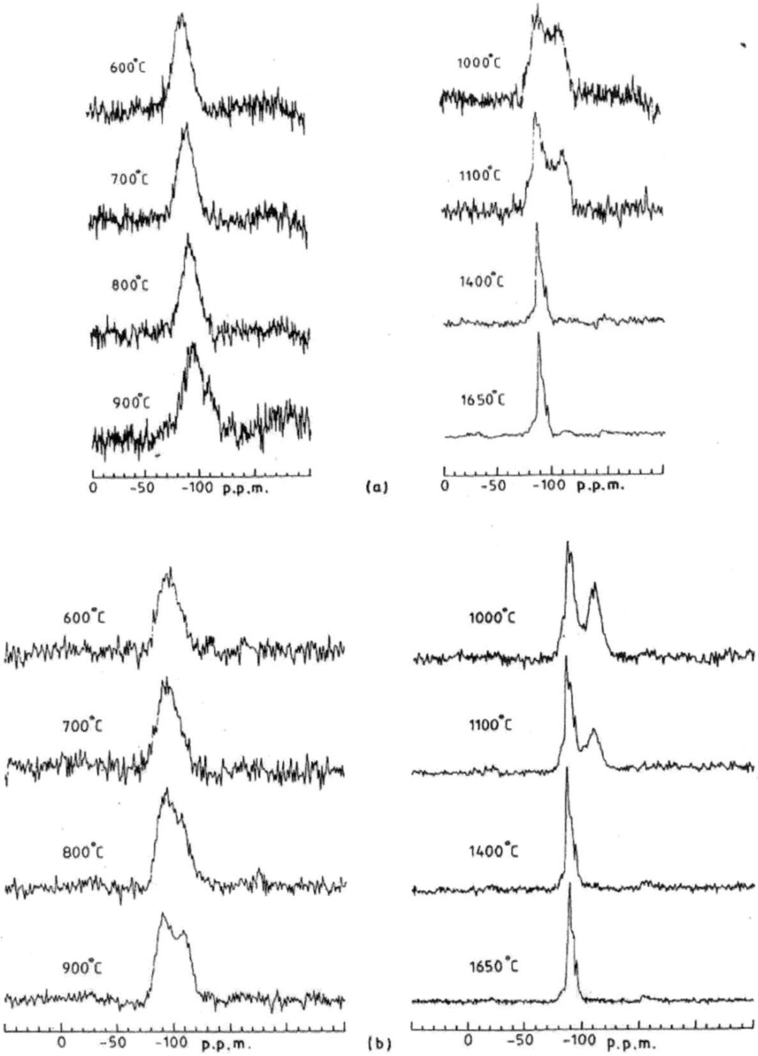

Figure 9.2 ^{29}Si MAS-NMR spectra of mullite precursor materials heat-treated at different temperatures: (a) Sol-gel material (SGM) and (b) coprecipitated material (CM) (Schneider et al., 1992).

9.2.2 Coprecipitated Gels: Coprecipitated Sample

MAS-NMR studies of the coprecipitated sample of mullite gel were done by Nishu et al. (1989). Mullite gel was synthesized by the coprecipitation of two components, aluminum chloride and silica gel. Aluminum chloride solution was adjusted to pH 8 by adding ammonia solution. Silica gel was turned into a sol by boiling with water and the pH was increased to 9. The mixture of the two sols was stirred for 5 h at 25°C at pH 9 to form a precipitate. It was filtered and dried. A DTA study showed mullitization at 980°C with the absence of the spinel phase. A ^{27}MAS-NMR study showed the occurrence of both fourfold- and sixfold-coordinated Al in this precipitate.

9.2.3 Mullite Gels Synthesized by Hydrolysis and Gelation

Jaymes et al. (1994) did MAS-NMR studies of mullite gels synthesized by hydrolysis and gelation in the presence of water-soluble silica content. Mullite gel was synthesized by the hydrolysis of an organic aluminum compound, for example, aluminum isopropoxide, in the presence of a water-soluble silica component, for example, $SiCl_4$, in tetrahydrofuran as the solvent. The mixture was subsequently hydrolyzed with ^{17}O-enriched water. The ^{29}Si NMR spectrum of the dried precursor at 100°C showed a large peak centered at –86 ppm only, which according to them reflects a good homogeneity, and even a small peak at –110 ppm. On further heating the precursor to 850°C, the Si peak shifted to –92 ppm. The cause of the shifting is not known. They made differentiation between tetragonal mullite (t-mullite) and orthorhombic mullite (o-mullite) forms observed in XRD as also in NMR by the apparent ratio "Al(IV)/Al(VI)," which increases on heating the precursor from 1000°C to 1300°C.

9.3 Monophasic and Diphasic Gels

Komarneni et al. (1985) conducted MAS-NMR studies of mullite gel (hydrolysis and subsequent gelation of the organic silicon component in the presence of water-soluble Al component-acidic medium as per Chapter 3) and diphasic gel (as per Chapter 4). The phase evolution of monophasic and diphasic gels synthesized earlier (Chapter 2) was studied by ^{27}Al MAS-NMR spectroscopy, wherein Al coordination in the precursor and their crystallization products were determined. The changes in the coordination result from phase change were monitored by the study. The Al spectra of the two gels were strikingly different. The single-phase gel showed a narrow octahedral resonance, indicative of high order, whereas the diphasic gel showed a broad resonance, indicative of some disorder.

Monophasic gel: At 1075°C, the single-phase mullite gel showed the transformation of some octahedral Al to tetrahedral and both peaks became broad, indicating the rise of some disorder in the structure. X-ray diffraction analysis revealed the formation of some poorly crystalline mullite, while DTA showed a clear exotherm. Computer deconvolution of the spectra showed that the ratio of tetrahedral to octahedral Al is 0.48. This value does not concur either with the ratio 0.71 for the theoretical value of tetrahedral to octahedral Al of 2:1 mullite or with the ratio 1.25 for the same for 3:1 mullite.

Diphasic gel: At 1010°C, the diphasic gel showed the formation of a spinel phase while DTA showed the absence of an exotherm. Komarneni and Roy (1986) measured the ^{27}Al MAS-NMR spectra of the diphasic mullite gel heated to 1010°C and showed that the tetrahedral to octahedral ratio of aluminum was 0.45. This value is very different from the corresponding value of 37.5/62.5 = 0.6 for γ-Al_2O_3. According to them, the overall spectrum was similar to that of γ-alumina.

At 1300°C, both gels showed extensive conversion of octahedral Al to two broad tetrahedral coordination values, at 61.2 ppm and 47.1 ppm, in mullite. The ratio of tetrahedrally to octahedrally bonded Al approached 1.25, which is the ratio of 3:1 mullite. X-ray analysis showed strong reflections of mullite in contrast to the broad resonances observed in the Al spectrum. MAS-NMR studies of these diphasic gels will be discussed in the following paragraphs.

9.4 Synthesis of Mullite Gels by Polymeric and Colloidal Methods

Yoldas (1992) conducted MAS-NMR studies of a polymeric mullite gel made from tetraethyl orthosilicate (TEOS)/AlOBu (hydrolysis and subsequent cogelation of organic Al and silicon compounds in a neutral medium as per Section 2.1.3, Chapter 2. ^{27}Al MAS-NMR spectra of this gel on heating to 500°C, 950°C, and 1050°C are shown in Fig. 9.3.

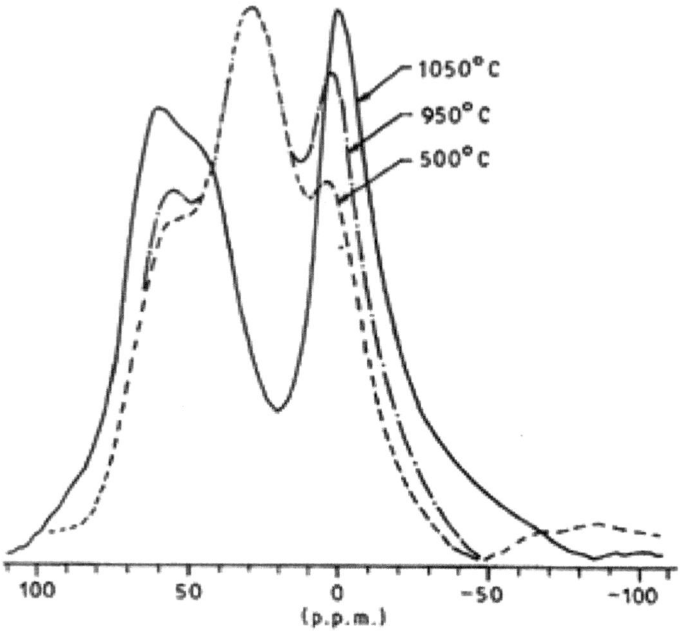

Figure 9.3 ^{27}Al MAS-NMR indicating an entirely different aluminum environment in polymeric $Al_6Si_2O_{13}$ gels (Yoldas, 1992).

The spectra in Fig. 9.3 show that a fundamental transformation has occurred between 950°C and 1050°C. The aluminum environment was pentahedraly coordinated. During the fundamental transformation from 950°C to 1050°C, 5-coordinated aluminum sites were eliminated and it changed to broad 4- and 6-coordinated sites, attributed to the characteristic of mullite MAS-NMR spectra.

9.5 Synthesis of Mullite Gels by Coprecipitation Methods

Jaymes and Douy (1995) performed MAS-NMR studies of gels synthesized by hydrolysis and subsequent gelation of an organic silicon component in the presence of a water-soluble Al component, using urea as per Chapter 5. A mullite precursor from aluminum nitrate nonahydrate (ANN) and tetraethoxysilane via the aqueous homogeneous precipitation technique using urea was synthesized. The thermal treatment of this coprecipitated gel showed a drastic modification in the local environments of the silicon and aluminum atoms in their spectroscopic study. They noted that on heat treatment of the gel, a regular shift of the ^{29}Si NMR band occurred. Al NMR showed a shift toward the tetrahedral and pentahedral coordination Al apparently on heating. At 980°C, DTA showed a strong exotherm and formed pseudotetragonal mullite. There were three sites of Al, one octahedral and two tetrahedral, ascribed as T and T* sites of the mullite structure, on heating at ~1000°C. Concurrently, Si NMR showed the replacement of the broad resonance at –90 ppm by –86, –90, –94, and –80 ppm and a broad signal at –110 ppm. The main resonance, at –86 ppm, is interpreted as being Si atoms in a sillimanite-type structure (Al_2SiO_5), while the peak at –94 ppm and shoulders at –90 and –80 ppm are due to Si atoms in mullite-type Al/Si ordering.

9.6 Synthesis of Mullite Gels by the Diphasic Method

9.6.1 MAS-NMR Studies of Diphasic Gels

Chakraborty (2005) conducted MAS-NMR studies of diphasic gels of three different compositions. The spinel phases derived from three different diphasic mullite compositions were reinvestigated by MAS-NMR study. Raw DG72 might be a homogeneous mixture of pseudoboehmite and silica gel. The MAS-NMR study of raw gel showed no AlO_4 groups. At ~900°C the ^{27}Al spectrum showed penta- and hexacoordinated aluminum in addition to tetrahedrally coordinated aluminum in the usual triplex nature of the spectrum (Fig. 9.4). This result indicated that a noncrystalline aluminosilicate phase may develop at this stage of heating. At ~1000°C, when Al-Si spinel is formed, AlO_5 groups disappear and the amounts of AlO_4 and AlO_6 increase to 31% and 69%, respectively. On subsequent heating to 1300°C, when the Al-Si spinel completely transforms to mullite, AlO_4 further increases to 50% at the cost of AlO_6 groups (Table 9.3).

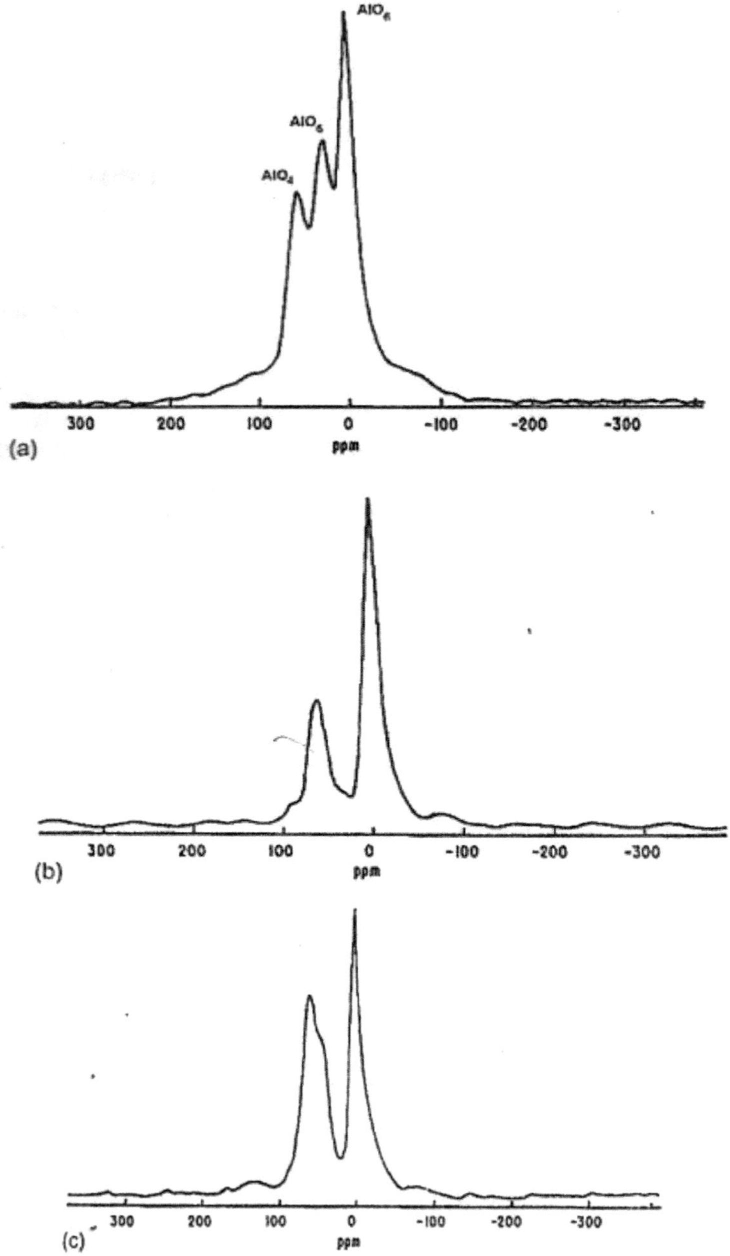

Figure 9.4 ^{27}Al MAS-NMR spectra of DG72 at (a) 900°C, (b) 1000°C, and (c) 1300°C.

Table 9.3 Contents of tetra-, penta-, and octahedral aluminum in heat-treated DG72 derived from MAS-NMR of ^{27}Al

	Diphasic gel			
Heated to	**AlO_4 ppm%**	**AlO_6 ppm%**	**AlO_5 ppm%**	**AlO_4/AlO_6 ppm%**
900°C	25.8	58–46	30.7–29	
1000°C	63.7–31	68–69	Nil	0.44
1300°C	66.1–50	41–50	Nil	

In view of the variation in solid solution data as mentioned earlier, the maximum wt % of silica that could be accommodated in the γ-Al_2O_3 phase prior to its transformation to α-Al_2O_3 was re-examined by Chakraborty (2006). Accordingly, the solid solution range between γ-Al_2O_3 and amorphous SiO_2 could be found out.

9.6.2 Phase Evolution of Mixed Gels by MAS-NMR of ^{29}Si

Figure 9.5 shows changes in the Si spectrum of MG72 during the heating process. At ~900°C, the spectrum showed a silicon resonance peak at about –95 ppm, which is indicative of the development of a noncrystalline aluminosilicate precursor phase. Thereafter, at 1000°C, two peaks appeared. The first one occurred at the earlier –95 ppm position, and the other one occurred at –110 ppm. The latter peak is due to the formation of a noncrystalline silica-rich alumina phase. Thus phase separation occurred while heating the gel at 1000°C. When the gel was heated further, these two peaks disappeared and the gel formed o-mullite. The Si spectrum resembles that of mullite. Figure 9.6 shows the Si spectrum of different gels of varying ratios of silica to alumina heat-treated to 1000°C. The spectrum of each gel consists of two peaks, one at about –95 ppm and other at –110 ppm. The characteristic feature was that as the silica content of the gel decreased, the latter peak also decreased and showed asymmetry. This result demonstrates that the noncrystalline silica-rich alumina phase formed on heating high-alumina mixed gels decreased with decreasing concentration of the silica component. It also indicates that as phase separation occurred during the heating of the gel with the higher alumina content (>72 wt % alumina) to 1000°C, some minor quantity of noncrystalline silica-rich alumina phase invariably persisted.

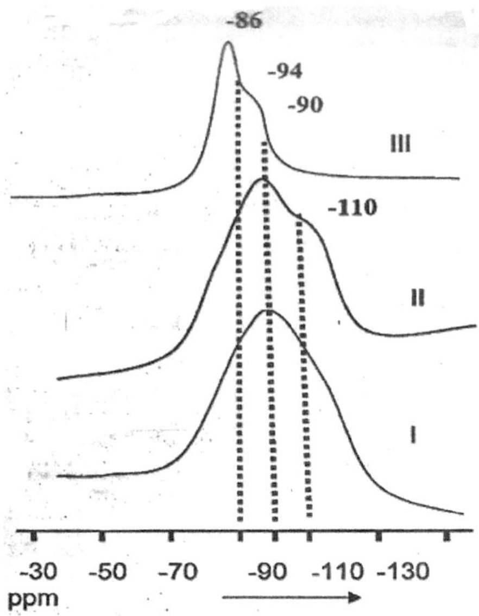

Figure 9.5 Changes in the ^{29}Si MAS-NMR spectrum of MG72 mixed gel on heat treatment at different temperatures. (I) MG40; (II) MG55, (III) MG72, and (IV) MG80.

9.7 Synthesis of Mullite Gels by the Spray-Drying Method

Jaymes et al. (1996) performed MAS NMR study of a spray dried precursor powder marked C which is prepared out of TEOS and ANN. The phase transformation process of it is compared with other precursors and is given in Section 9.8.9 below.

9.8 Comparison of MAS-NMR Spectra of Mullite Precursors

In this section, comparisons are made of MAS-NMR spectra of mullite precursors synthesized by different methods under different processing conditions by various authors, along with illustrative examples.

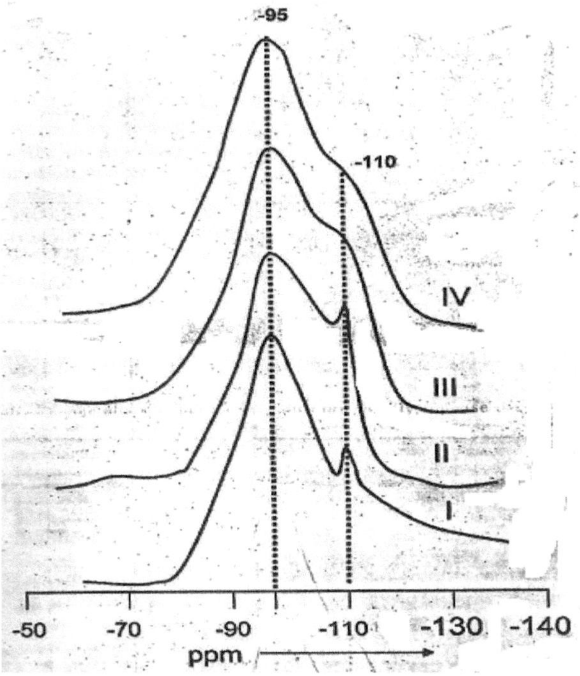

Figure 9.6 ^{29}Si MAS-NMR spectra of different mixed gels on heat treatment at 1000°C.

9.8.1 CM and SGM

Schneider et al. (1992) compared the MAS-NMR studies of CM, synthesized by the aqueous sol-gel technique and the gelation of water-soluble silica component in the presence of Al salt as per Chapter 2, with those of polymeric gel marked SGM, synthesized by hydrolysis and subsequent cogelation of organic Al and silicon compounds in a neutral medium as per Chapter 2.

The sequences of phase development of SGMs studied by using NMR spectroscopy are given in Table 9.4 on the basis of Figs. 9.1 and 9.2.

Table 9.4 Phase transformation of SGM precursor studied by ^{29}Si and ^{27}Al MAS NMR spectroscopy

Temperature	Phase
600°C–800°C	XRD showed an amorphous pattern. A ^{29}Si NMR study showed symmetrical ^{29}Si resonance at ~92 ppm. ^{27}Al NMR showed Al resonances at ~2, 30, and 58–60 ppm. These are due to six, five, and two tetrahedrally coordinated Al.
Above 900°C	Si resonance split into two peaks at –90 and –110 ppm. These are indicative of separation into silica-poorer (about –90 ppm) and silica-richer domains (about –110 ppm). The formation of a silica-rich phase occurred earlier in the CM sample than in the SGM sample.
At 900°C	XRD showed spinel (major) and mullite formations (small). At ~988°C, a DTA trace showed an exothermic peak. Al resonance showed predominant octahedral resonance and tetrahedral (minor) resonances. The 30 ppm peak due to penta-coordinated Al disappeared. Thus, amorphous resonance collapsed into an intermediate reaction–type resonance.
At 1000°C	A ^{29}Si resonance peak at –90 and –110 ppm peaks were clumped together. Si resonance also displayed a big shoulder of low intensity at –80 ppm. The octahedral peak shifted to 8 ppm, and the tetrahedral peak moved to 70 ppm. This spectrum shows great similarity to that of kaolinite related at 1000°C. The intensity of the tetrahedral peak is less than that noted in CM. In the 900°C–1100°C range, the ratio of AlO_4 to AlO_6 did not change significantly.
At 1100°C	The –90 ppm peak became sharp. Al resonance showed three broad peaks, at –5, 45, and 62 ppm, which are characteristic of typical mullite resonances. In the 1100°C–1650°C range, profound changes occur in the ratio of AlO_4 to AlO_6. For example, AlO_4 increases at the cost of AlO_6. Moreover, AlO_4 splits.
At 1200°C	DTA showed a small exothermic peak.
At and above 1400°C	Single-phase mullite developed. Si resonance displayed sharp peaks near –87 and 94 ppm and shoulders near –90 ppm, which are typical of mullite. Both Si and Al spectra concur with that of 3:2 mullite.

The SGM sample is greatly different from the CM sample in terms of DTA, TGA, and MAS-NMR characteristics.

- When the SGM exhibited two exotherms, the CM showed only one exotherm. Both, however, formed mullite and spinel phases.
- As a major amount of mullite formation took place at the first exotherm, the tetrahedral Al resonance peak of the CM sample was more in the sample heated at 1000°C. Contrarily, spinel phase formation was major in the SGM sample and due to that the intensity of the tetrahedral Al resonance was found to be less as expected for the suggested theory of Chakraborty's spinel model analogous to 3:2 composition of mullite (Chakraborty, 1979 and Chakraborty & Das 2003b). The substantial amount of this spinel phase later on transformed into mullite at a 1253°C exotherm Thus, the Al spectra of two types of successive heated gels are dramatically different. The DTA patterns in the high-temperature region for both SGM and kaolinite are similar.

The CM sample is greatly different from the SGM sample in terms of its DTA and TGA profiles. Mullite formation in the SGM takes place via the intermediary spinel phase in one step at the second exotherm. It was observed that mullitization is complete at 1100°C–1250°C (as per ^{29}Si resonance data), but this mullite is not attained in a stable state. Calcination needs to be done at a temperature far above 1100°C.

9.8.2 MI and MII Precursors

Huang et al. (1997) compared MAS-NMR studies of the monophasic precursor MI (as per the procedure in Chapter 2) and the diphasic precursor MII, obtained at a higher pH. At 600°C–800°C, the phase transformation behaviors of MI and MII at different pH values and the ^{29}Si and ^{27}Al resonances MAS-NMR studies of MI and MII (Figs. 9.7 and 9.8) are given in Tables 9.5 and 9.6.

- The MI (amorphous) precursor showed different behavior than the MII (crystalline) precursor, and this was explained as related to their amorphous or crystalline nature.
- MI showed three signals, at 5–6, 32, and 63 ppm, whereas MII showed only two signals, one at 6 and the other at 64 ppm.

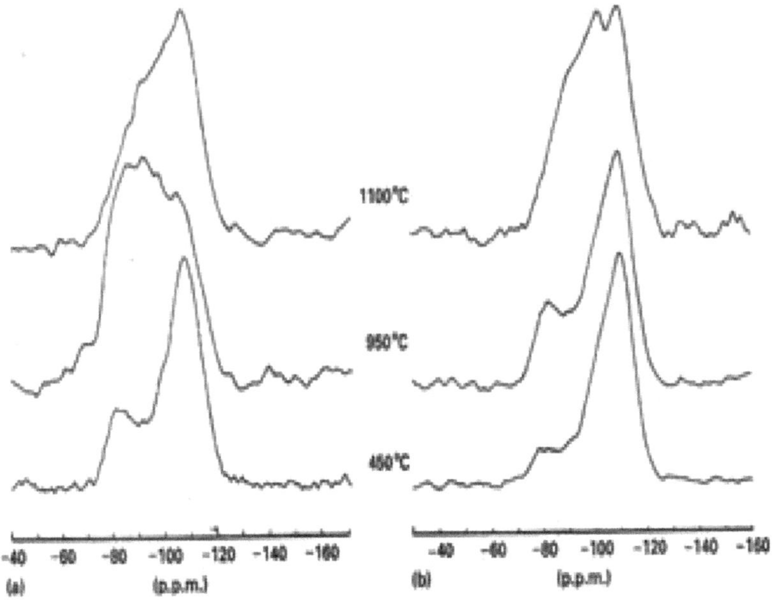

Figure 9.7 ^{29}Si MAS-NMR spectra of (A) MI and (B) MII powders heated at 450°C, 950°C, and 1100°C for 1 h (Huang et al., 1997).

- There was another difference in the Si spectra of the two precursors heated to 950°C: MI showed a rather broad peak in the 80–100 ppm range whereas MII showed a distinctive peak at –83 ppm, although X-ray showed a spinel phase in both precursors.

Table 9.5 Phase transformation of MI precursor in the light of ^{27}Al MAS NMR spectroscopic study

Temperature	Phase
At 450°C	Two main resonances appeared, one at –80 ppm and one at –108 ppm. The –80 ppm peak was assigned by Lippmaa et al. (1980) to silicon atoms connected to four aluminum atoms via oxygen, that is [Q_o(4Al) sites] and the –108 ppm resonance was assigned to [Q_4(0Al) sites].
At 950°C	The overlap of signals by different $Q_n(m\text{Al})$ sites are assumed as a new observation.
At 100°C	Three signals appeared: at –90, –101, and –109 ppm.

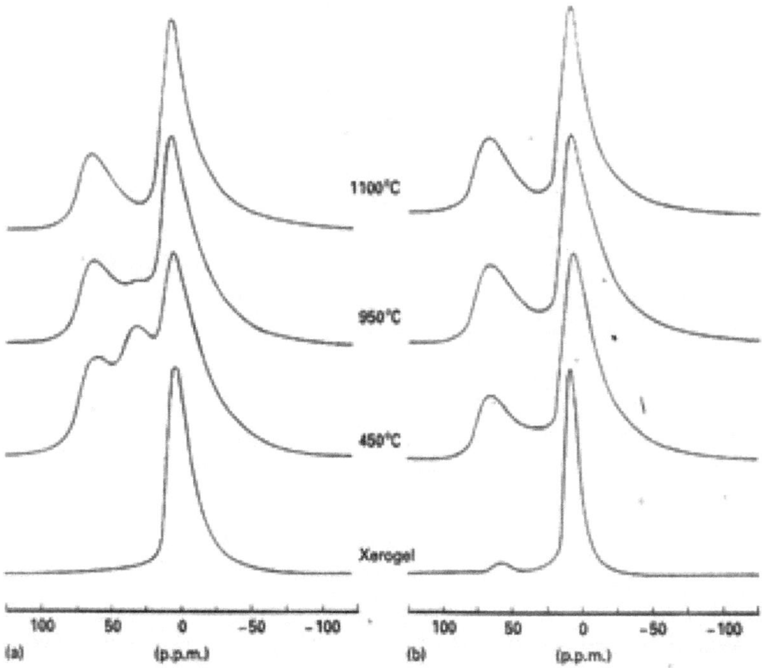

Figure 9.8 ^{27}Al MAS-NMR spectra of (A) MI and (B) MII powders heated at 450°C, 950°C, and 1100°C for 1 h (Huang et al., 1997).

Table 9.6 Phase transformation of MII precursor in the light of ^{27}Al MAS NMR spectroscopic study

Temperature	Phase
At 450°C	The spectrum of the MI xerogel displayed a signal at 4.4 ppm, whereas the MII xerogel showed a strong signal at 7.5 ppm and a weak signal at 58 ppm.
At 950°C	MI showed three signals, at 5–6, 32, and 63 ppm, whereas MII showed only two signals, one at 64 ppm and the other at 6 ppm.
At 100°C	MI and MII showed only two peaks.

These authors suggested that the ^{29}Si resonance at about –80 ppm might be due to (i) the resonance of silicon in an Al_2O_3-rich noncrystalline phase or (ii) the incorporation of silicon in γ-alumina. They further showed that the precursor heated at 1100°C showed peaks at –80 and – 110 ppm. Thus, the character of the two gels is a function of the processing parameters.

When MI precursor was synthesized at pH 1: Two main Si resonances, at –80 and –108 ppm, were observed at the first stage, on heating the MI precursor at 450°C. The corresponding Al resonance showed the signals at 5–6, 32, and 63 ppm. These results indicate the start of bond formation and development of AlO_5 and AlO_4 groups in the aluminosilicate phase from the siliceous phase (showing intense resonance at –108 ppm) with the aluminous phase (showing intense resonance at 5–6 ppm) present in the xerogels.

During the exhibition of the 980°C exotherm, spinel formation was noted to crystallize at the second stage. Overlap of signals by different $Q_n(m\mathrm{Al})$ sites was seen. As a consequence, the Si spectrum of MI was rather broad in the ppm range of –80 to –100. On the other hand, tetrahedral Al resonance became prominent relative to the octahedral peak. ^{29}Si NMR resonance peaks of the spinel phase and the residual silica-rich aluminous phase formed at the earlier stage may be situated in the same prominent broad resonance. XRD revealed the existence of spinel at 1200°C and mullite formation started above 1200°C at the third stage of transformation.

When MII precursor was synthesized at pH 11: The bayerite phase was crystallized at pH 11 during the addition of ammonia in this xerogel. The main Si resonance, at –108 ppm, was observed on heating the MII precursor to 450°C at the first stage. The corresponding Al resonance showed a strong Al signal due to AlO_6 and a weak signal due to AlO_4 at 58 ppm.

Spinel formation was noted to crystallize at the second stage in the temperature range of 450°C–1200°C but it did not exhibit a 980°C exotherm. The intensity of AlO_4 increased on heating at 950°C due to the gradual development of a Si-incorporated spinel phase.

MII at 950°C showed a sharp ^{29}Si resonance peak at –83 ppm, unlike the display of a broad resonance in the case of MI. This is indicative of the presence of noncrystalline aluminosilicate or a Si environment at the Al-Si spinel phase over the presence of a broad resonance (in the ppm range of –80 to –100). The –80 ppm peak became broader while heating MII at 1100°C, due to the formation of a larger quantity of former phases.

There is another difference in the two precursors heated at 950°C: the Si spectrum of MI is rather broad in the –80 to –100 ppm range whereas the same in MII showed a distinctive peak at –83 ppm, although X-ray showed a spinel phase in both precursors. It is assumed that the spinel phases in the two cases are of different characteristics. XRD revealed the existence of this spinel phase at 1200°C and the crystallization of mullite at a temperature >1350°C at the third stage of transformation, which is 100°C above the formation temperature of mullite in the case of the MI sample. Thus, where mixing homogeneity is concerned, the MI (amorphous) precursor shows a different behavior from that of the MII (crystalline) precursor and this is related to their amorphous or crystalline nature. It was conjectured that acidic gels are homogeneous in contrast to basic gels, which are heterogeneous in nature.

9.8.3 Aluminosilicate Gels and Silicon Aluminum Ester

Yasumori et al. (1990) synthesized aluminosilicate gel from silicon aluminum ester (SAE) and tetramethyloxysilane and studied its evolution on heating by NMR spectroscopy and finally deconvoluted it into Gaussian curves to estimate the peak area of the NMR spectrum of each curve.

Al spectra: The S-20 xerogel showed mostly fourfold-coordinated Al and was similar to the SAE starting material. During heating, sixfold-coordinated Al at 0 ppm decreased whereas the fourfold-coordinated Al at 55 ppm increased. At 950°C, fivefold-coordinated Al at 30 ppm appeared. At 1000°C, fivefold-coordinated Al disappeared and fourfold-coordination peaks of Al at 47 and 65 ppm appeared. On continued heating of the xerogel, the heights of the two peaks for Al at the fourfold coordination reversed and corresponded to that of mullite.

Si spectra: The spectrum of mullite showed a sharp peak in the vicinity of –85 ppm, which is attributed to SiO_4 tetrahedra surrounded by bridging oxygens bonding to Al ions, that is, Q4(4Al). The spectrum of S20 showed a broad peak in the range –70 to –120 ppm, whose chemical shift gradually moved to lower frequencies and the peak height decreased gradually on heat treatment. A drastically sharp decrease in the Si resonance peak was noted. At 1000°C, it became –111 ppm, which may be attributed to Q_4(1Al) or Q_4(0Al).

Yasumori et al. (1990) plotted the changes in the ratios of the estimated peak areas of different Al sites during heat treatment. Finally, they compared the peak areas of different Al sites in samples after heat treatment estimated from NMR with those of various mullite crystals estimated from XRD.

On heating the mullite at 400°C, the peak position for Al in the midfrequency range, at around 55 ppm (4MC), shifted slightly to the high-frequency side. Al in 5C (fivefold coordination at around 35 ppm) and Al in 6C (sixfold coordination states at ~0 ppm) remained unchanged.

On heating the mullite to temperatures between 900°C and 1000°C, when a drastic increase in the bulk density of the xerogel occurred and complete sintering took place because of mullite formation, 5C disappeared and two fourfold Al peaks, one at around 65 ppm (4HC) and another on the low-frequency side, at around 45 ppm (4LC), along with 6C were noted.

On isothermal soaking of the mullite at 1000°C, the proportion of Al ions in the various sites of the gel derived from the sample marked S20 changed directly and became close to that of mullite. Yasumori et al. (1990) noted that the ratio of Al ions in sixfold coordination in S20 decreased on isothermal treatment. The configuration of Al ions came close to that of mullite, which is a tendency opposite to that exhibited by any metastable phase separation.

9.8.4 Mullite Gel Synthesized by Hydrolysis and Gelation in Varying Amounts of Water/Alcohol

Taylor and Holland (1993) studied and compared the crystallization sequences of two samples, marked GNW and GYW, by ^{27}Al MAS-NMR spectroscopy (Figs. 9.9, 9.10; Table 9.7).

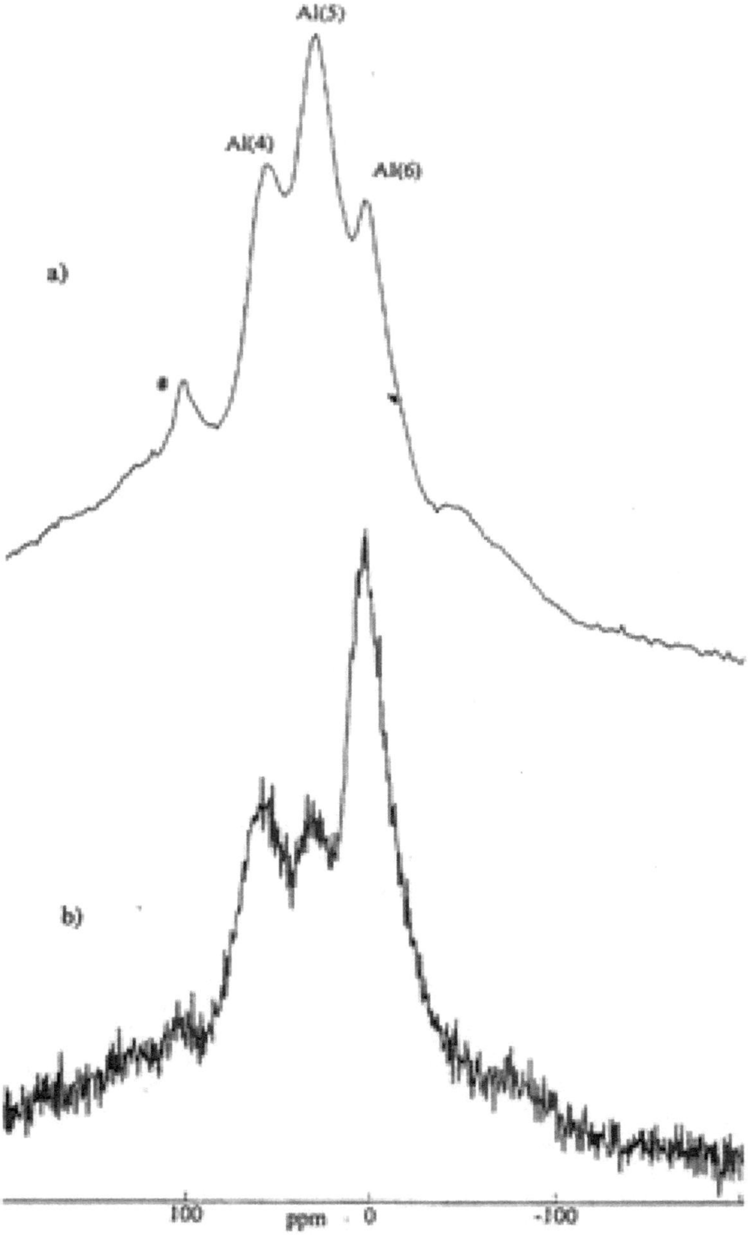

Figure 9.9 ^{27}Al MAS-NMR spectra of the two xerogels (a) GNW and (b) GYW (Taylor & Holland, 1993).

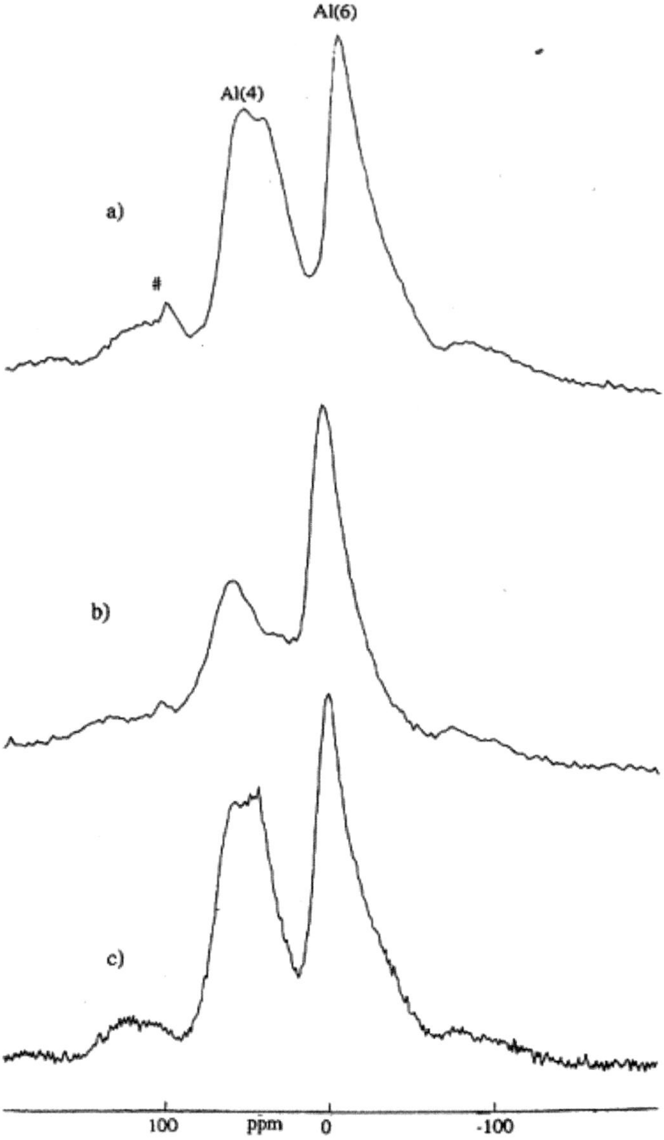

Figure 9.10 ^{27}Al MAS-NMR of (a) GNW and (b) GYW after completion of the exotherm at 980°C and (C) GYW after heating to 1100°C for 18 h (Taylor & Holland, 1993).

Taylor and Holland (1993) were of the opinion that the structure of the amorphous xerogel appeared to play a central role in determining the nature and relative amounts of crystalline phases produced. NMR spectroscopy gives an idea about the local structure, and it also indicates a relationship between the first coordination sphere of aluminum in the xerogel and the nature of the crystal phase formed and the completion of 980°C. They inferred that GNW prepared without the use of water was more homogeneous than GYW, where water was added. They indicated that the structure of amorphous xerogel appeared to play a key role in determining the nature and relative amounts of crystalline phases formed. For example, when the octahedral coordination was the preferred environment in the xerogel, spinel was the major crystalline phase. When aluminum was preferentially penta-coordinated, mullite was the resultant crystalline phase.

Table 9.7 Phase transformation of GYW and GNW precursors in the light of ^{27}Al MAS NMR spectroscopic study

	GYW
Temperature	**Phase**
	The xerogel showed varying amounts of three peaks, relating to Al(4), Al(5), and Al(6), of which octahedral Al was the preferred site.
At 980°C	The xerogel crystallized to the spinel phase. It showed significantly more octahedral coordinated aluminum.
At 1100°C for 18 h	Orthorhombic mullite developed. Al(4) content increased.
	GNW
Temperature	**Phase**
	The xerogel showed predominantly pentahedral Al.
At 980°C	The xerogel crystallized to mullite. It showed two types of tetrahedral aluminum, as shown in the Al spectrum.

9.8.5 Mullite Precursors Synthesized by Polymeric, Colloidal, and Nitrate Gels

Yoldas (1992) conducted ^{27}Al MAS-NMR study to compare the thermal transformation behaviors of polymeric, colloidal, and nitrate gels (Fig. 9.11; Table 9.8).

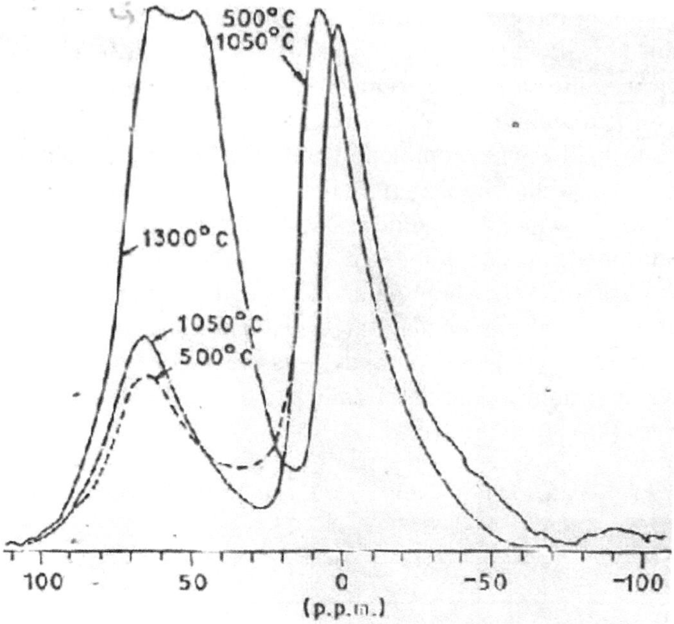

Figure 9.11 Aluminum environment in a colloidal aluminosilicate gel as indicated by ^{27}Al MAS-NMR heated to 500°C, 1050°C, and 1300°C (Yoldas, 1992).

Table 9.8 Thermal transformation behaviors of polymeric, colloidal, and nitrate-derived gels in the light of ^{27}Al MAS NMR spectroscopic study

Polymeric gel	
Temperature	**Phase**
	Major Al sites of the gel heated to 500°C were small penta-coordinated peaks of tetra- and octahedral Al.
At 950°C	A slight increase in tetrahedral and octahedral peaks occurred. X-ray showed an amorphous phase.
At 1050°C	A drastic change in Al coordination occurred. Penta-coordinated Al sites disappeared. A broad Al spectrum characteristic of mullite appeared, which conformed to X-ray and DTA observations.

Colloidal gel	
Temperature	**Phase**
	The Al spectrum showed tetrahedral and octahedral peaks at 64.5 and 7.1 ppm, respectively.
At 1050°C	The Al environment altered. The AlO_6 peak was much larger than the AlO_4 peak. XRD at 500°C–1050°C showed a cubic spinel phase only.
At 1300°C	Mullite was developed fully. The Al spectrum resembled that of mullite.

Nitrate-derived gel	
Temperature	**Phase**
	The Al environment was analogous to that of polymeric gel, for example, the major site at 500°C was penta-coordinated.
At 1050°C	The Al spectrum resembled that of colloidal gel heated to the same temperature, that is, it showed an octahedral peak (major) at 2 ppm, which arose from Al in octahedral coordination resonance; a frequency resonance of fivefold-coordinated Al near 30 ppm (minor); and a tetrahedral peak at the 58–60 ppm resonance peak. XRD showed a cubic spinel (major) phase and mullite (minor).

9.8.6 Mullite Precursors Synthesized by Polymeric Gel (Type I), Diphasic Gel (Type II), and Coprecipitated Gel (Type III)

Schneider et al. (1993) made a comparison of the MAS-NMR spectra of the mullite precursors synthesized by polymeric gel (Type I) as per sample C of Yoldas and Partlow (1988, Chapter 3), diphasic gel (Type II) (Chapter 3), and coprecipitated gel (Type III).

^{29}Si and ^{27}Al NMR spectroscopy and analytical transmission electron microscopy (ATEM) studies were performed and are given in Fig. 9.12 and Tables 9.9–9.11.

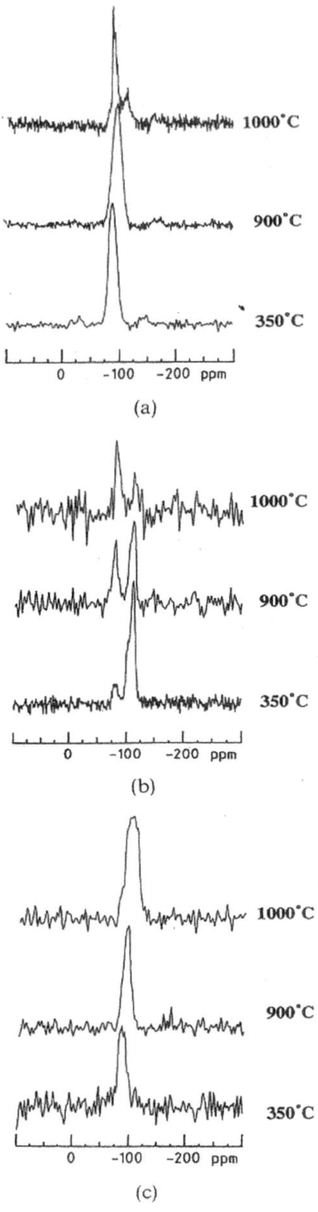

Figure 9.12 ^{29}Si MAS-NMR spectra of mullite precursors heat-treated at 350°C, 900°C, and 1000°C for 15 h in each case: (a) Type I (H), (b) Type II (HB13), and (c) Type III (HB10) (Schneider et al., 1993).

Table 9.9 Phase transformation of NH/Type I MI precursor in the light of XRD and ^{29}Si MAS NMR spectroscopic studies

Temperature	Phase
As-received gel	The gel was amorphous to X-ray.
At 350°C	Si resonance showed an intense and sharp symmetric peak at –85 ppm.
At 900°C	The Si resonance peak shifted from –85 to –90, that is, to a more negative value.
At 1000°C	DTA exhibited a sharp and strong exotherm at 980°C and no peak at 1250°C. X-ray showed a poorly crystalline mullite (major) phase together with a spinel (minor) phase. The Si resonance showed two maxima, one at –90 ppm, considered to be a characteristic of mullite, and the second at –110 ppm.

Table 9.10 Phase transformation of HB13 precursor in the light of XRD and ^{29}Si MAS NMR spectroscopic studies

Temperature	Phase
As-received gel	X-ray showed pseudoboehmite only.
At 350°C	Si resonance consisted of two components, one at –80 ppm and the other at –110 ppm.
At 1000°C	DTA exhibited no exotherm up to a temperature of 1400°C. γ-alumina formed after dehydration at the endotherm. The intensity and sharpness of the X-ray peaks were temperature dependent. With an increase in temperature, both ordering and growth of γ-alumina crystallites occurred. The phase was stable up to 1250°C. Mullite occurred thereafter. The Si resonance peak increased in intensity, whereas the –110 ppm peak decreased. The –80 ppm peak is ascribed to the presence of silica either in the alumina-rich noncrystalline phase or in γ-alumina. An increase in temperature indicated an increase in the amount of silicon in both phases.

Table 9.11 Phase transformation of HB10 precursor in the light of XRD and ^{29}Si MAS NMR spectroscopic studies

Temperature	Phase
As-received gel	The gel was amorphous.
At 350°C	Si resonance showed peak maxima at −85 ppm.
At 900°C	The Si resonance peak became sharper and shifted to −95 ppm.
At 1000°C	DTA exhibited a sharp exotherm at 980°C, but it was of a lower intensity than that of the Type I mullite precursor. It also exhibited another broad exotherm of low intensity at 1250°C. X-ray showed only a spinel phase. The Si pattern showed three maxima: at −80, −100, and −110 ppm.

According to Schneider et al. (1993), the Type I precursor transformed directly to alumina-rich mullite, whereas Type II and Type III transformed to mullite in multistep reactions with the intermediate formation of γ-alumina and silica.

9.8.7 Mullite Precursors Synthesized by Polymeric Gel, Coprecipitated Gel, and Diphasic Gel

Geradin et al. (1994) compared the MAS-NMR spectra of mullite precursors synthesized by a polymeric gel (as per Yoldas, 1980, using prehydrolyzed TEOS), a coprecipitated gel (in the presence of water), and a diphasic gel and investigated their structural evolutions. They also noted the predominant five coordinated environments just before mullite crystallization in their M1 precursor. After its crystallization, three types of environments that are characteristic of mullite (AlIV, AlIV*, and AlVI) were noted, but in proportions different from those found in the completely crystallized state. The ratio of tetrahedra Al to octahedra Al was increased from the precursor calcined at 950°C to the value corresponding to well-crystallized mullite formed at 1200°C. ^{29}Si MAS-NMR spectra of those fired at different temperatures are shown in Fig. 9.13.

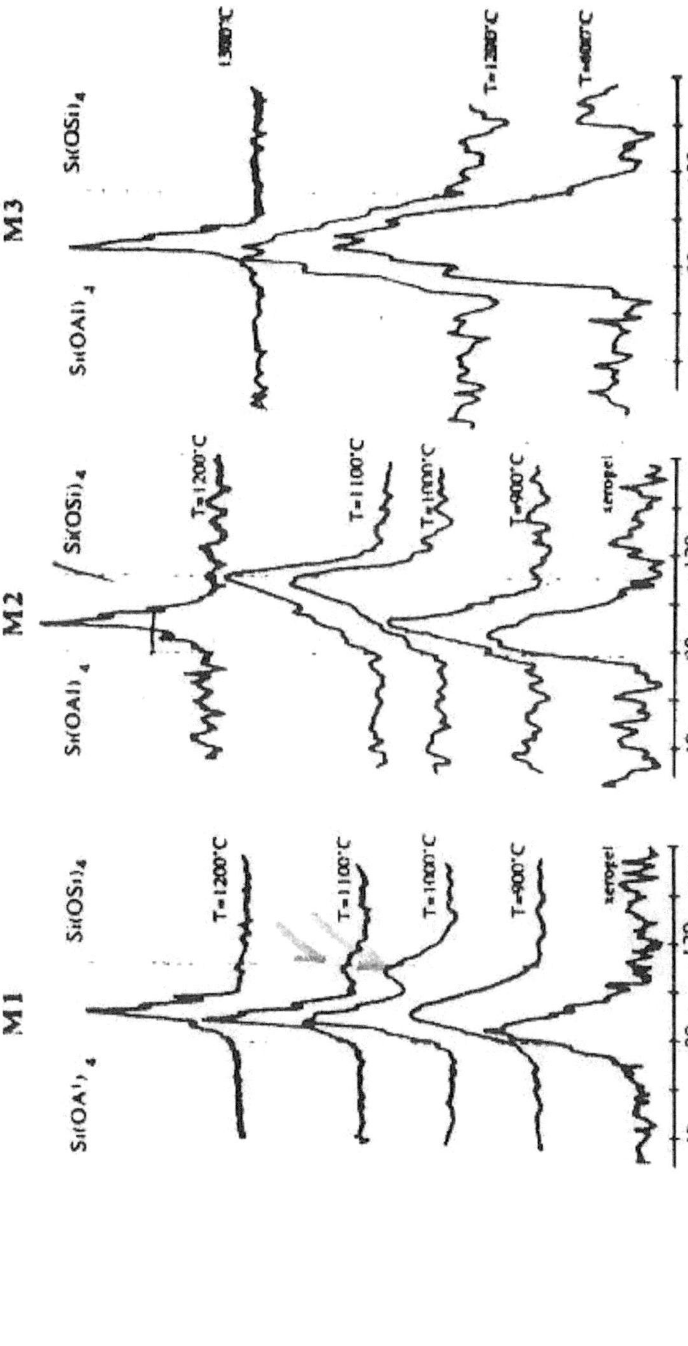

Figure 9.13 ^{29}Si MAS-NMR spectra of samples (A) M1, (B) M2, and (C) M3 heated at various temperatures (Geradin et al., 1994).

According to Geradin et al. (1994), Si spectra below 900°C were characteristic of amorphous aluminosilicate structures with a random distribution of silicon and aluminum. The local structure was described as Si(OAl)n(OSi)4–n, where n varies between 1 and 4.

9.8.8 Mullite Precursors of SP, CA, and PB Gels

Sanz et al. (1991) compared the MAS-NMR spectral studies of mullite precursors of SP, CA, and PB gels. NMR spectroscopy study was done to determine the local structure of Al_2O_3-SiO_2 precursors and also to characterize the intermediate phases produced during the thermal treatment of the precursors. These studies were complemented with XRD analysis of the colloidal sample and the polymeric sample made in acidic conditions, marked CA and PA, and the same prepared in basic conditions, marked PB and PB2.

SP precursor: At 400°C, the precursor was amorphous to X-ray whereas Si and Al were diluted. An Al NMR showed three peaks, at 65, 35, and 6 ppm, which corresponded to tetrahedral, pentahedral, and octahedral Al configuration. Among them, the pentahedral Al peak was predominant. The next predominant peak was the tetracoordinated Al, and the least was the octahedral Al. When the precursor was further heated, up to 900°C, the octahedral component decreased further. This data strongly favored the development of chemical linkages between the two oxides in this precursor.

At 1000°C, the mullite crystallized, exhibiting a sharp 980°C exotherm and an expansive effect detected at the same temperature. These two phenomena are interrelated. NMR showed complete elimination of penta-coordinated Al with the formation of a new tetrahedral peak at 42 ppm and octahedral peak at 2 ppm. The nature and intensities of the peaks of the new spectra nearly corresponded with 3:2 mullite.

PB2 and CA precursors: XRD patterns of polymeric gels were mostly amorphous up to 900°C, whereas colloidal gel in the same temperature range showed four XRD peaks, at 38, 46, 61, and 66.5° 2θ, which are characteristics of the cubic spinel phase. ^{27}Al NMR spectra of the colloidal gel heated up to 1000°C, with peaks at ~65 and ~6 ppm, are similar to those of the spinel phase. However, in addition to tetrahedral and octahedral peaks, the polymeric sample showed a pentahedral peak at ~32 ppm. According to Sanz et al. (1991), the relative intensities of tetra- and octahedral peaks in polymeric samples were not very different from those detected in the colloidal sample, indicating that the local structural arrangement of Al may have similarities in both samples. The polymeric sample showed two small DTA exotherms, one at 985°C and the other at 1250°C. These two exotherms were related to two shrinkage steps (two expansive effects, one at 980°C and one at 1250°C) noted in TMA. Thus, PB2 transformed into mullite in two steps via spinel as the intermediate phase. The colloidal gel, however, showed one exotherm, at 1285°C, which was certainly of a different character than the exotherm noted at 1250°C, and reasonably only one sharp shrinkage step was noted in TMA. Thus, the phase evolution processes of PB2 and CA are dissimilar. The intermediate spinel phase developed in CA at ~600°C showed broad reflections in the XRD pattern. The peaks of the spinel phase narrowed with continued heating up to 1250°C. The stability of this phase even at such a high temperature precludes the γ-Al_2O_3 concept. Incorporation of Si in Al-rich domains occurred in the cases of CA and PB2 in the heating range of 1100°C–1300°C. Moreover, two correlated effects, one DTA at 1285°C and one expansive effect in dilatometry, indicated a rapid thermal change. XRD confirmed the crystallization of mullite from the spinel phase. These three complementary observations indicate a sudden structural transformation of the precursor CA at 1285°C.

9.8.9 Monophasic, Coprecipitated, Diphasic, and Spray-Pyrolyzed Precursors

Jaymes et al. (1996) compared the NMR studies of four different types of precursors: monophasic, coprecipitated, diphasic, and spray pyrolyzed. The NMR spectra of the phase evolutions of five different precursors synthesized from ANN ad TEOS by different techniques in an aqueous solution are analyzed as follows:

Precursor A: A symmetrical resonance peak centered at –89 ppm was shown for the sample heated at 750°C. On further heating the sample at 900°C, the peak maxima shifted to the high field region to the –95 ppm position. At 1000°C, this resonance divided into two resonance bands, one relative to mullite centered at –89 ppm and the other at –109 ppm (Fig. 9.14).

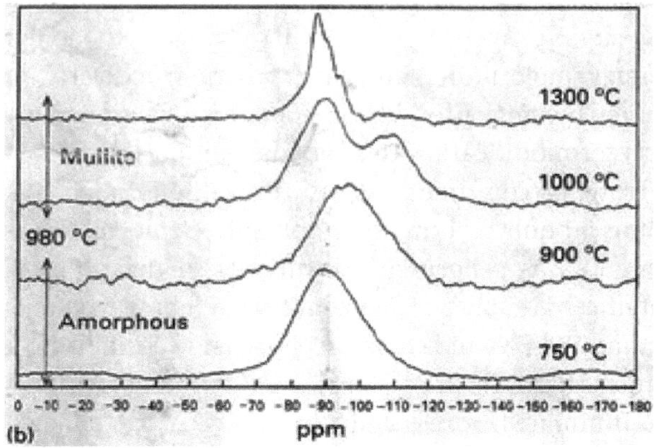

Figure 9.14 ^{29}Si resonance spectra of precursor A heat-treated at different temperatures (Jaymes et al., 1996).

Precursor B: It showed one large asymmetric peak centered at –90 ppm due to the aluminosilicate network and one small peak at –110 ppm. The –90 ppm peak decreased drastically at 1000°C, and in lieu of it, the peak at –110 ppm increased largely and became narrower and more symmetric on further heating, to 1200°C (Fig. 9.15).

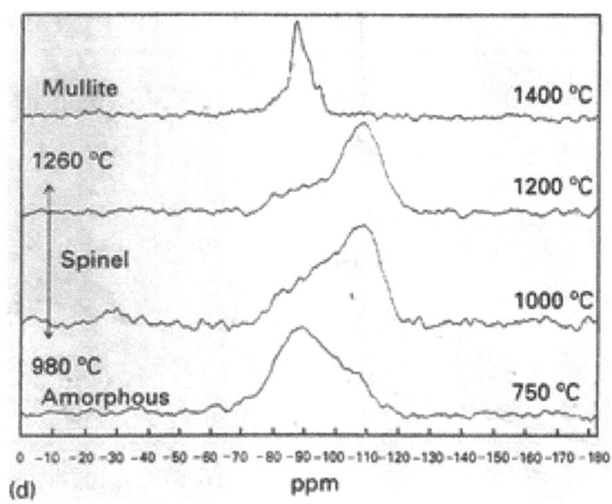

Figure 9.15 ^{29}Si resonance spectra of precursor B heat-treated at different temperatures (Jaymes et al., 1996).

Precursor D: A large peak is centered at –110 ppm and shows a tail starting from –70 ppm. On simulation, there might be two bands, centered at –93 and –108 ppm, respectively. At 840°C, a peak at ~90 ppm was perceptible from the tail portion of the spectrum.

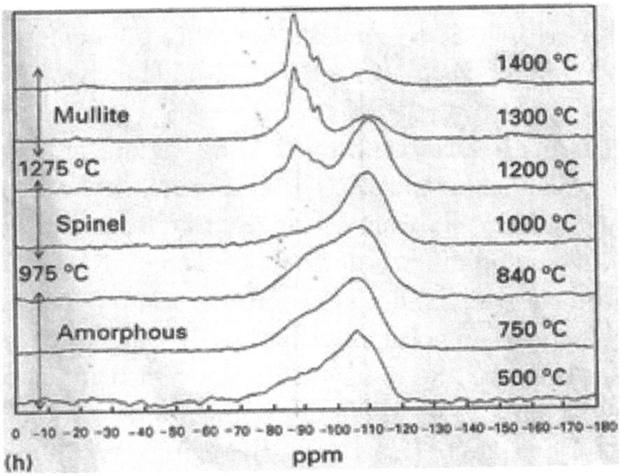

Figure 9.16 ^{29}Si resonance spectra of precursor D heat-treated at different temperatures (Jaymes et al., 1996).

The main peak, at –110, became sharp at 1000°C. On further heating the precursor to 1200°C, the –110 ppm resonance was sharper and more symmetrical and a new peak centered, at –88 ppm, appeared due to mullitization. However, the –110 ppm resonance decreased at 1300°C, with further mullitization. Even then, a small resonance, at –110, persisted at 1400°C, which is certainly a new observation and requires explanation (Fig. 9.16).

In precursor D and powder B synthesized by Jaymes and Douy (1992), the aged silica sol, that is, the condensed Si-O-Si units likely produced more silica-rich phase and as a result the Si NMR shifted gradually to the right of the spectrum, that is, toward an increasingly negative value of resonance. Simultaneously, some portion of the siliceous units may diffuse into the aluminous group and produce aluminosilicate, as seen by the appearance of a Si NMR peak at –90 ppm of precursor D heated-treated at 1200°C (Jaymes et al., 1996, Fig. 2h). This result is analogous to changes in the Si resonance pattern of the Type II precursor prepared at pH 13 by Schneider et al., wherein the –80 ppm peak increased at the cost of the –110 ppm peak of silica components on heating from 350°C to 900°C to 1000°C. This leads us to assume that the silica and alumina components likely interact on heating of precursor Type II of precursor D. This interaction, that is, interdiffusion continues with an increase in heating. It is not that two amorphous alumina and silica components remain idle with heat treatment. And this phenomenon should occur also in monophasic gel, diphasic gel, and the amorphous oxide mixture.

In precursor B, spinel developed. Similarly, M2 of Geradin et al. (1994) showed a spinel phase. The Si resonance of both precursors showed symmetrical Si resonance at 750°C in B and the same in M2 at 900°C. On heating the precursor further, the Si NMR peak shifted toward more negative values, up to 1100°C–1200°C. It signifies that preferably the Si–O–Si bond units, that is, the siliceous glassy phase grows more than the aluminosilicates. Thus, instead of an increase in the homogeneity, more and more segregation occurs when these precursors are heated. The silica-rich region, the aluminosilicate region, and a portion of the amorphous pure alumina region remain side by side on heating. The *Si spectra of precursor B by Jaymes et al. (1996), Type II by Schneider et al. (1993), and M2 precursor by Geradin et al. (1994) did not show a double peak in the samples heated at ~1000°C, where the spinel crystallized. They showed a single broad peak around –110 ppm.

Precursor B exhibited two DTA exotherms in the mullite formation region between 1200°C and 1280°C. What is the reason for the broad exotherm before the occurrence of exotherm due to mullite formation at 1260°C from the spinel phase formed earlier, at the 980°C exotherm? On the basis of its Si MAS-NMR spectrum, it is explained that the broad Si resonance at –110 ppm, which is noted as the accumulation of aluminosilicate, may nucleate and crystallize at the occurrence of the broad exotherm prior to the exotherm peak at 1260°C. The exhibition of this broad peak indirectly indicates the generation of the binary aluminosilicate phase earlier by the interaction of two amorphous components in the heat treatment process.

Precursor E: At 80°C, the resonance was very sharp, at –80 ppm. At 300°C, it became very broad. The question is, why does the center of this peak gradually shift toward the more negative values, say

−95 ppm, when the temperature rises to 900°C? At 1000°C, this peak dissociates into mullite resonance and usual resonance to −110 ppm (Fig. 9.17).

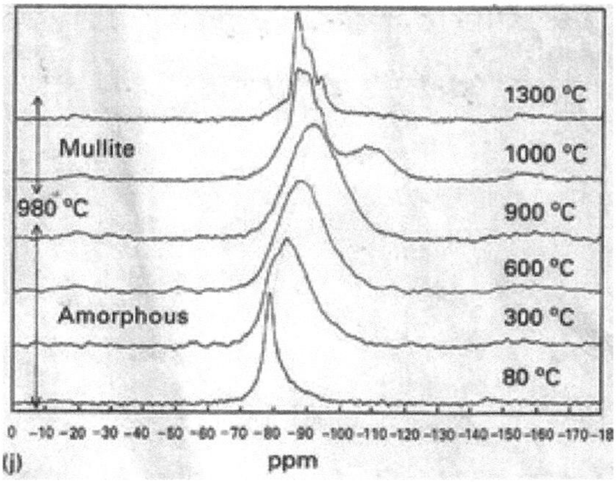

Figure 9.17 ^{29}Si resonance spectra of precursor E heat-treated at different temperature (Jaymes et al., 1996).

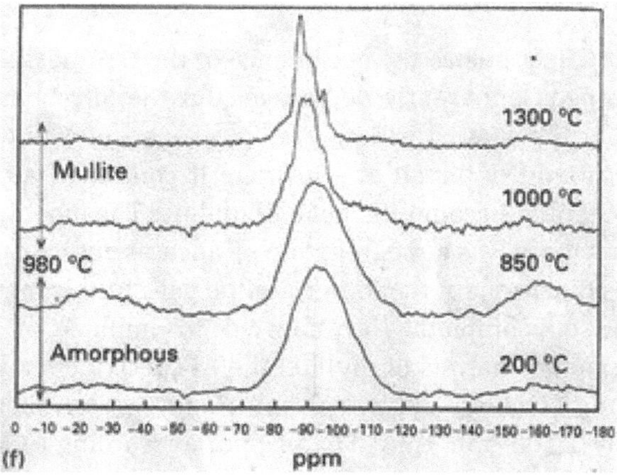

Figure 9.18 ^{29}Si resonance spectra of precursor C heat-treated at different temperatures (Jaymes et al., 1996).

Precursor C: The ^{29}Si spectra of precursor C showed a broad resonance at about −90 ppm (Fig. 9.18). Its ANN component decomposed at ~150°C. ^{27}Al spectra showed that in the spray-dried precursor, Al is present as Al(VI), which indicated that ANN was not fully decomposed during the rapid evaporation of the solvent. With a rise in the heat treatment temperature, the bands corresponding to Al(IV) and A(V) increased while the one corresponding to Al(VI) decreased (Fig. 9.19).

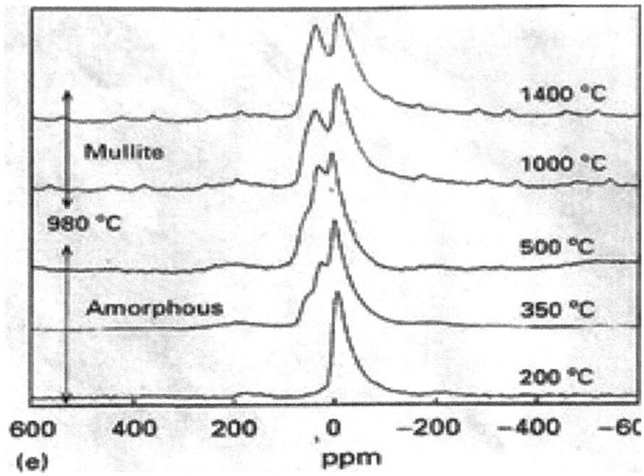

Figure 9.19 ^{27}Al resonance spectra of precursor C heat-treated at different temperatures (Jaymes et al., 1996).

At 970°C, just before the occurrence of the DTA peak, the Al(VI) resonance peak apparently decreased. But the Si resonance peak remained at the same position. The Si resonance peak was highly symmetrical and occurred at −90 ppm. It continued at 850°C. At 1000°C, −90 ppm became the peak of mullite. The most interesting point is that there was no segregation of silica, as noted in the case of the phase transformation of other mullite gels. At this temperature, the precursor completely crystallized to mullite. In terms of crystallographic analysis of mullite, the Al spectrum showed three Al sites, one octahedral site (~0 ppm) and two tetrahedral sites, denoted by T (60 ppm) and T* (44 ppm), according to Merwin et al. (1991) and subsequently according to Schneider et al. (1992). At this temperature, the precursor completely crystallized to mullite.

The Si spectra showed one resonance band and a shoulder near −94 ppm due to mullite and absolutely no band at about −110 ppm. At 1400°C, there was a slight increase in the ratio of Al(IV)/Al(VI) due to the development of well-crystallized mullite. Si resonance bands resemble characteristic o-mullite. The spectrum consisted of one main resonance, at about –80, analogous to sillimanite-type SiO_4 tetrahedra, one peak at about –90, and finally a shoulder at about –94 ppm. This ^{29}Si MAS-NMR pattern corresponded to the structure analysis of mullite by Ban and Okada (1993).

However, the precursors marked B and D are not true diphasic gels, as they claimed. As per DTA of precursor B, especially at the high-temperature side, there were two exotherms (one near 1200°C and one at 1260°C as usual), observed due to mullitization. Thus, mullite formation occurred in two paths as noted by the author in the case of G-150 gel (Fig. 7.9). Precursor D showed two exotherms in the vicinity of the first exotherm, one at 924°C and one at 980°C due to spinel and mullite formation, respectively (Fig. 12.1).

9.8.10 Mullite Precursors Obtained from Components from the Same Source

Comparison of NMR studies of mullite precursors obtained from components from the same source with three variables in processing conditions (by gelation of the organic silicon component in the presence of the water-soluble Al component at different pH values, as per Chapter 3) shows differences in the MAS-NMR patterns (Haque, 2000; Chakraborty, 2008).

^{29}Si MAS-NMR studies of mullite precursors of two types of SH gels and a diphasic gel are as follows:

SHII gel: At 900°C, ^{29}Si MAS-NMR spectra displayed symmetrical resonance at about –95 ppm. On heating the gel to ~1000°C, the resonance became asymmetric and showed two new maxima, one narrow peak centered at –96 ppm and a broad peak centered at –10 ppm. (Fig. 9.20). An XRD study of this sample showed Al-Si spinel (major) and mullite (minor). At ~1300°C, it showed three resonance maxima corresponding to the resonances of o-mullite.

SHIV: At 900°C, ^{29} Si MAS-NMR spectra showed resonance at ~94 ppm, as shown by SHIII precursor. But when the gel was heated to 1000°C, the resonance pattern completely changed. It showed two peaks, one asymmetric broad peak centered at about –88 ppm and one peak centered at –104 ppm. This showed that a profound segregation into noncrystalline aluminosilicate phase took place that might be equivalent to decomposition into two separate phases.

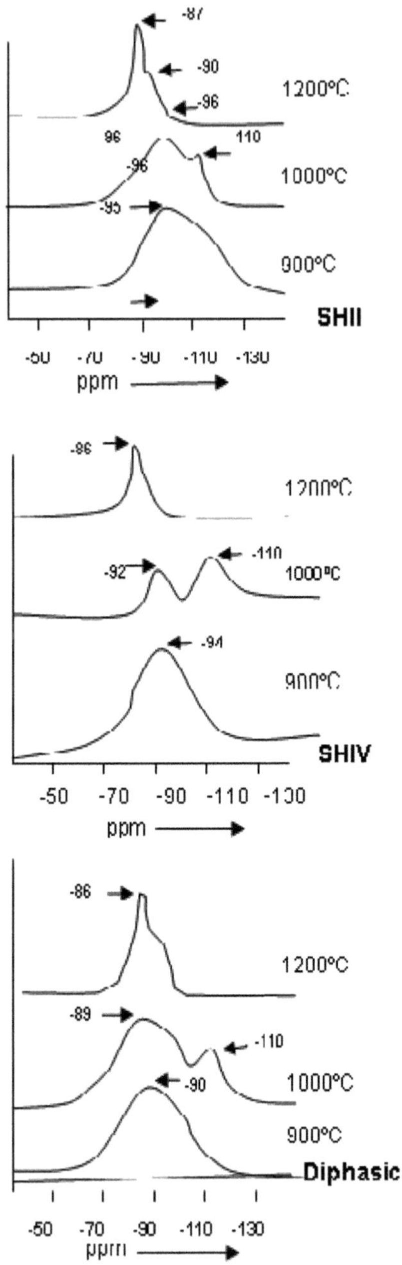

Figure 9.20 ^{29}Si MAS-NMR spectra of SHII, SHIV, and diphasic gels calcined at different temperatures (Haque, 2000; Chakraborty, 2008).

However, XRD showed a predominantly spinel phase at 1000°C. This spinel transformed rapidly at ~1200°C and formed mullite. The ^{29}Si NMR spectrum of this sample heated to 1300°C concurred with that of o-mullite. At the intermediate stage, the Al-Si spinel phase and noncrystalline silica-rich alumina phase (A) might be accounted for by the –88 ppm and –104 ppm resonance peaks, respectively. However, the latter phase is not free silica (A) since free silica was not detected during the alkali treatment process. The former Al-Si spinel phase transforms polymorphically to mullite at ~1200°C. Thereafter, a solid-state reaction occurs between newly formed mullite (may be alumina-rich mullite) and the proposed silica-rich alumina phase. It is expected that when the gel is further heated to a high temperature, Si from the silica-rich phase gradually diffuses into the mullite grain and develops ordered o-mullite.

Diphasic gel: The boehmite component first decomposed and formed alumina (A). It reacted immediately with free reactive silica derived from the dehydration of the silica gel component and developed noncrystalline aluminosilicate, as shown by the author earlier. A ^{29}Si MAS-NMR study showed a broad peak at –90 ppm when the gel was heated to 900°C. On further heating the gel to 1000°C, the Si spectrum showed a broad peak at about –89 ppm and a small shoulder at –109 ppm. At a high temperature of ~1300°C, the broad ^{29}Si resonance at –89 ppm slightly shifted to a most intense narrow peak at –86.6 ppm. The other peak, at –94.7 ppm, also corresponded well with a minor peak reported between –94.1 and –94.3 ppm. The shoulder that was fitted to a peak at –90.3 ppm was also reported at –90 ppm. Lastly, the low-intensity peak at –91.9 ppm was also due to well-crystallized mullite.

9.9 Conclusions

Here are some conclusions drawn about the phase transformations of six mullite precursors at first, second, and third stages as revealed by MAS-NMR studies. The mullite precursors (i) monophasic gel by the aqueous sol-gel method as per Chapter 2, (ii) monophasic gel by the sol-gel method in Chapter 3/SH gel, (iii) polymeric gel, (iv) coprecipitated gel, (v) diphasic gel, and (vi) mullite precursor by spray-drying or spray-pyrolysis (SP) method as per the current chapter usually showed three steps of transformation.

- Thermal transformation of aqueous mullite gel/precursors by MAS-NMR spectroscopy: The monophasic mullite gel (CM) prepared by the aqueous sol-gel method studied by Schneider et al. (1992) exhibited an exotherm at 988°C and the absence of any exotherm at ~1250°C. A broad symmetrical ^{29}Si resonance peak at about –90 or –92 ppm was noted at the first stage of transformation of this mullite precursor. It became asymmetrical and remained so on heating from 600°C to 900°C. The Si resonance collapsed into a crystalline resonance at the second stage of transformation, with the formation of mullite (major) and spinel (minor). The asymmetrical ^{29}Si resonance peak at ~92 ppm divided into two broad peaks, one at –90 ppm and one at –110 ppm. Al spectra was not similar to those of 3:2 mullite, but the signals were broad, which conformed with XRD data apparently. The –110 ppm peak decreased, and the spinel phase disappeared at the third stage of transformation, with the formation of only mullite.
- Thermal transformation of monophasic mullite gel/precursors by MAS-NMR spectroscopy: Komarneni et al. (1985) mentioned that the single-phase gel showed a narrow octahedral resonance by ^{27}Al MAS-NMR spectroscopy, indicative of high order at the first stage, on heating the gel at ~65°C. During the transformation at the second stage, some portion of the octahedral Al resonance peak changed to tetrahedral and both peaks became broad, indicating some rise in disorder in the structure. On monitoring the changes with DTA, an exotherm was noted at 980°C and the formation of poorly crystalline mullite was observed at 1075°C in XRD study. Extensive conversion of octahedral Al to two broad tetrahedral coordination, one at 61.2 ppm and one at 47.1 ppm in mullite, was noted further at the third stage of heating, at ~1300°C. The corresponding XRD study confirmed mullite formation.
- Thermal transformation of polymeric mullite gel/precursors by MAS-NMR spectroscopy: Two main Al resonances were observed on heating at 500°C at the first stage of

the polymeric gel/precursor (Yoldas, 1992). The first, small peak corresponded to AlO_4 groups, and, in addition, pronounced AlO_5 groups confirming bond formation were noted in the 27Al MAS-NMR spectral study. This pattern remained stable even when the gel was heated to 950°C. A fundamental transformation occurred at the second stage, between 950°C–1050°C. At this stage, a profound change in Al coordination occurred during mullitization after the exhibition of the first exotherm, as noted in DTA. Five-coordinated aluminum sites were eliminated, and it changed to broad four- and six-coordinated sites, attributed nearly to the characteristic of mullite MAS-NMR spectra.

- Thermal transformation of coprecipitated gel/mullite gel/precursors by MAS-NMR spectroscopy: Thermal treatment of the coprecipitated gel showed a drastic modification in the local environments of the silicon and aluminum atoms in their spectroscopic study (Jaymes & Douy, 1995). At the first stage of heat treatment, a regular shift of ^{29}Si NMR band occurred. Al NMR showed a shift toward tetrahedral and pentahedral coordination. Penta-coordinated Al was apparent on heating the precursor to 900°C. At the second stage, at 980°C, DTA showed a strong exotherm and formed pseudotetragonal mullite. There were three sites of Al, one octahedral and two tetrahedral, ascribed as T and T* sites of mullite structure on heating the gel at ~1000°C. Concurrently, Si NMR showed the replacement of the broad resonance at – 90 ppm by –86, –90, –94, and –80 ppm and a broad signal at –110 ppm. The main resonance, at –86 ppm, was interpreted as being Si atoms in a sillimanite-type structure (Al_2SiO_5), while the peak at –94 ppm and shoulders at –90 ppm and –80 ppm were due to Si atoms in a mullite-type Al/Si ordering. At the third stage of heating, at 1300°C, it crystallized into mullite. Pentahedral Al disappeared. One octahedral and two tetrahedral peaks resembling mullite appeared at 1300°C. The Si resonance pattern showed concurrence with mullite.
- Thermal transformation of diphasic gel/mullite gel/precursors by MAS-NMR spectroscopy: (i) By ^{27}Al MAS-NMR spectroscopy, the diphasic gel showed a broad resonance,

indicative of some disorder (Komarneni et al., 1985). At the second stage, at 1010°C, the diphasic gel showed the formation of a spinel phase while DTA showed the absence of any exotherm. The measured value of tetrahedral to octahedral Al ratio is 0.45. At the third stage, at 1300°C, it showed extensive conversion of octahedral Al to two broad bands of tetrahedral coordination at and around 61.2 ppm and 47.1 ppm in mullite. There was no peak separation of two tetrahedral resonances in the diphasic gel. (ii) Diphasic gel marked "insitu" by Chakraborty (2008) at the first stage of heating, at ~ 900°C, showed penta- and hexa-coordinated aluminum in addition to tetrahedrally coordinated aluminum in the usual triplex nature of the ^{27}Al spectrum. In the ^{29}Si spectrum, the characteristic feature showed that a noncrystalline aluminosilicate phase might develop at this stage and as the silica content of the gel decreased, the –110 ppm peak also decreased and showed an asymmetry. This result further demonstrated that the content of noncrystalline silica–rich alumina phase formed on heating high alumina mixed gels decreased with decreasing concentration of the silica component. Secondly, it indicated that as phase separation occurred on heating the gel with a higher alumina content (>72 wt % alumina) at 1000°C, some minor quantity of noncrystalline silica-rich alumina phase invariably persisted. At the second stage of heating, at ~1000°C, when Al-Si spinel formed, AlO_5 groups disappeared and the quantum of AlO_4 and AlO_6 increased to 31% and 69%, respectively. At the third stage of heating, at 1300°C, when Al-Si spinel transformed to mullite, AlO_4 further increased to 50% at the cost of AlO_6 groups.

- Thermal transformation of spray-dried/SP mullite precursors by MAS-NMR spectroscopy: Alumina-silica gels by the SP were resynthesized by Sanz et al. (1991) as per the process followed earlier by Kanzaki et al. (1985).

At the first stage of transformation, on being heated at 400°C, the precursor was amorphous to X-ray where Si and Al were diluted. Al NMR showed three peaks, at 65, 35, and 6 ppm, which corresponded to tetra -, penta-, and octahedral Al configuration. Among them, the pentahedral Al peak was predominant. The next predominant peak was that of tetracoordinated Al and the least was that of octahedral Al.

At the second stage, on heating the precursor to 900°C, the octahedral component decreased further. This data strongly favored the development of chemical linkages between the two oxides in this precursor. At ~1000°C, mullite crystallized with the exhibition of a sharp 980°C exotherm in DTA and sharp expansive effect of TMA. The NMR peak related to AlO_5 disappeared, a new peak appeared at 42 ppm, and the profile approached that of mullite. At the third stage, during crystallization of mullite, two tetrahedral peaks and one octahedral peak were observed. With continued heating to 1200°C, the Al peak at 42 ppm became higher than the peak at 65 ppm. This observation supports the finding of Yasumori et al. (1990) and the spectra resembles the spectra of mullite.

The most important MAS-NMR study was done by Jaymes et al. (1996). The ^{29}Si spectra of precursor C showed a broad resonance at about –90 ppm at the first stage of heating. With a rise in temperature, the bands corresponding to Al(IV) and A(V) increased while that corresponding to Al(VI) decreased. At the second stage of heating, at ~970°C, just before the occurrence of DTA peak, the Al(VI) resonance peak apparently decreased. The Si resonance peak was highly symmetrical and occurred at –90 ppm. At 1000°C, –90 ppm became the peak of mullite in XRD. At the third stage of heating, between 1000°C and 1400°C, it was realized that the AlO_4 group increased on calcination but the mullite formation was still completed, which predicts regularization of attaining the structure of mullite without any segregation at the intermediate stage.

Problems

1. In the CM sample (Schneider et al., 1992) heated at 600°C–800°C, what could be the reason for (i) the asymmetrical shape of the ^{29}Si resonance peak and (ii) the shoulder with a more negative chemical shift?
2. Why did the –110 resonance peaks seem more intense in the SGM sample than in the CM sample? What is the role of the peak in mullitization?
3. Give reasons for the following speculations made by researchers: (i) the –110 ppm resonance peak may be due to

the noncrystalline phase in the CM sample containing nearly pure SiO_2, (ii) intermediate spinel formation may be due to γ-alumina-type phase, and (iii) mullite formation occurs in a multistep reaction.

4. Why is the 30 ppm peak due to the penta-coordinated Al-O bond still noted when mullite has already occurred in a major quantity in the CM sample heat-treated at 900°C?
5. If silica is introduced into the γ-alumina structure, why does its AlO_4 content increase?
6. According to Schneider et al. (1992), why does phase separation occur due to silica-poorer (–90 ppm) and silica-richer domains (–110 ppm) when –90 ppm Si resonance is itself due to the crystallization of mullite?
7. Why did both octahedral and tetrahedral Al peaks in the MAS-NMR spectrum as shown by the study of Komarneni et al. (1985) of the monophasic sample become broad on heating the sample at 1075°C? What type of disorder was present in the newly formed mullite?
8. Why are DTA observations of MI and MII precursors largely different? Huang et al. (1997) conjectured that the MI (amorphous) precursor behaved differently from the MII (crystalline) precursor and this was related to their amorphous or crystalline nature. There was another difference in the Si spectra of the two precursors heated at 950°C: the MI spectrum was rather broad, in the 80–100 ppm range, whereas MII showed a distinctive peak at –83 ppm, although X-ray showed a spinel phase in both precursors. Give reasons for these observations.
9. The Si resonance of MII (Huang et al., 1997) shows a pronounced broad resonance at 1100°C. Explain this phenomenon.
10. How can one account for the formation of the spinel phase at 950°C along with mullite from the ^{27}Al spectrum of Yoldas (1992)?
11. Why are the Al spectra of the heated CM and SGM gels (Schneider et al., 1992) dramatically different?
12. What is the role of water during gel synthesis? What are the differences in the structures of the two amorphous xerogels GNW and GYW (Taylor & Holland, 1993)?

13. Why does ANN used in the nitrate gel as synthesized by Yoldas (1992) show a spinel phase (major)? Generally, it shows t-mullite (see SH gel's result, Ban & Okada, 1993).
14. Although the Al spectrum of the nitrate gel resembles that of the colloidal gel of Yoldas (1992) when both are heated to 1050°C, are the phase evolution paths in the two cases similar?
15. The small quantity of the spinel phase could not be characterized by spectral studies of Si or Al resonances of Type I mullite gel. This is true even in the case where spinel is the major crystalline phase, for example, in the case of HB13 or HB10 precursors. How can proper identification of its sharp and gradual crystallization be made? What are the mechanisms of mullite formation in the two cases? What are the multistep reactions that occur on the intermediate formation of γ-alumina and with silica on heating Type II and Type III precursors?
16. Why does the Si resonance peak gradually broaden on heating Type III (HB10) gel to 1100°C?
17. Why does Si resonance peak gradually broaden and incline toward the region richer in silica on during M2 (Geradin et al., 1994) to 1100°C? How is mullite crystallized from this broad matrix?
18. Why does the Si resonance peak remain broad with heating even M3 (Geradin et al. 1994) to 1100°C? What is the nature of this broad matrix containing aluminosilicate?
19. Explain the difference between the two Al spectra of the heat-treated M1 (Geradin et al., 1994) sample at 950°C and at 1200°C with respect to the ratios of Al(iv), Al(iv*)m and Al(vi). Show that the mullite formed at 980°C exotherm is not well crystallized.
20. Compare the thermal transformations of the mullite precursors of SP, CA, and PB gels synthesized by Sanz et al. (1991) and answer the following:
    (i) Why is the temperature of mullite formation higher in colloidal gels than in polymeric gel?
    (ii) Why is the intensity of 980°C maximum in the case of spray-pyrolyzed powder, minimum in polymeric gel (PB2), and absent in colloidal gel (C)?

21. What could be the reason for the apparent similarities in the local structural arrangements of Al in PB2 and CA precursors up to 900°C?
22. Why are the phase evolution processes of PB2 and CA dissimilar?
23. What is the reason for the drastically sharp decrease in the Si resonance peak on heating S-20 sample (Yasumori et al., 1990)?
24. Even on using chelating compounds during the synthesis of precursor A, why does phase separation still occur on heating the gel at 1000°C? Si NMR shows a –110 ppm resonance peak. Powder A is the same as that in the Ci2 sample synthesized by Douy (1991).
25. Explain the following thermal effects:
    (i) The nature of transformation of precursor B seems analogous to the powder U1 sample synthesized by Douy (1991).
    (ii) The nature of transformation of precursor C seems analogous to the spray-pyrolyzed sample by Kanzaki et al. (1985) and the powder A sample synthesized by Jaymes and Douy (1992).
    (iii) The nature of transformation of precursor D seems analogous to the powder B sample synthesized by Jaymes and Douy (1992).
    (iv) The nature of transformation of precursor E seems analogous to the powder Mul C sample synthesized by Jaymes and Douy (1992) or by Jaymes et al. (1995).
26. Explain the occurrence of the double exotherm in the vicinity of the 980°C exotherm in precursor D?

References

1. G. Engelhadt and D. Michel, *High Resolution Solid State NMR of Silicates and Zeolites.* John Wiley and Sons, New York, p. 54 (1987).

2. S. H. Risbud, R. J. Kirpatrick, A. P. Taglialavore, and B. Montez, Solid state NMR evidence of 4-, 5-, 76- fold aluminium sites in Roller-quenched SiO_2-Al_2O_3 glasses. *J. Am. Ceram. Soc.*, **70**, C10–C12 (1987).

3. E. Lippmaa, M. Magi, A. Samoson, G. Engelhardt, and A. R. Grimmer, Structural studies of silicates by solid-state high resolution ^{29}Si NMR. *J. Am. Chem. Soc.*, **102**(15), 4889–4893 (1980).

4. H. Schneider, I. Merwin, and A. Sebald, Mullite formation from non-crystalline precursors. *J. Mater. Sci.*, **29**, 805–812 (1992).

5. K. Nishu, T, Yokuyama, T. Wataabe, and T. Tarutani, Characterization of amorphous aluminosilicate formed by absorption of silicic acid on aluminum hydroxide, in *Abstracts of the 24th Symposium of Basic Science of Ceramics*. Paper No. IB08. Ceramic Society of Japan, Tokyo (1989).

6. I. Jaymes, A. Douy, P. Florian, D. Massiot, and J. P. Coutures, New synthesis of mullite. Structural evolution study by ^{17}O, ^{27}Al and ^{29}Si MAS-NMR spectroscopy. *J. Sol-Gel Sci. Technol.*, **2**, 367–370 (1994).

7. S. Komarneni, R. Roy, C. A. Fyfe, and G. J. Kennedy, Preliminary characterization of gel precursors and their high- temperature products by ^{27}Al magic-angle spinning NMR. *J. Am. Ceram. Soc.*, **68**(9), C-243–C-245 (1985).

8. S. Komarneni and R. Roy, Solid-state ^{27}Al and ^{29}Si magic-angle spinning NMR of aluminosilicate gels. *J. Am. Ceram. Soc.*, **69**(3), C-42–C-44 (1986).

9. B. E. Yoldas, Effect of ultrastructure on crystallization of mullite. *J. Mater. Sci.*, **27**(24), 6667–6672 (1992).

10. I. Jaymes and A. Douy, Homogeneous precipitation of mullite precursors. *J. Sol-Gel Sci. Technol.*, **4**, 7–13 (1995).

11. A. K. Chakraborty, Reinvestigation of Al-Si spinel phase in diphasic Al_2O_3-SiO_2 gel. *J. Am. Ceram. Soc.*, **88**(1), 134–140 (2005).

12. A. K. Chakraborty, Range of solid solutions of silica in spinel type phase. *Adv. Appl. Ceram.* **105**(6), 297–303 (2006).

13. I. Jaymes, A. Douy, D. Massiot, and J. P. Coutures, Characterization of mono- and diphasic mullite precursor powders prepared by aqueous routes, 27 Al and 29 Si MAS-NMR spectroscopy. *J. Mater. Sci.*, **31**, 4581–4589 (1996).

14. A. K. Chakraborty, Formation of silicon aluminium spinel. *J. Am. Ceram. Soc.*, **62**(3–4), 120–24 (1979).

15. A. K. Chakraborty and S. Das, Al–Si spinel phase formation in diphasic mullite gels. *Ceram. Int.*, **29**, 27–33 (2003).

16. Y. X. Huang, A. M. R. Senos, J. Rocha, and J. L. Baptista, Gel formation in mullite precursor obtained via TEOS prehydrolysis. *J. Mater. Sci.*, **32**, 105–110 (1997).

17. A. Yasumori, M. Iwasaki, H. Kawazoe, M. Yamane, and Y. Nakamura, Nuclear magnetic resonance study of the structure of aluminosilicate gel and glass. *Phys. Chem. Glasses*, **31**(1), 1–9 (1990).

18. A. Taylor and D. Holland, The chemical synthesis and characterization sequence of mullite. *J. Non-Cryst. Solids*, **152**, 1–17 (1993).

19. H. Schneider, B. Saruhan, D. Voll, L. Merwin, and A. Sebald, Mullite precursor phases. *J. Euro. Ceram. Soc.*, **11**, 87–94 (1993).

20. B. E. Yoldas and D. P. Partlow, Formation of mullite and other alumina-based ceramics via hydrolytic polycondensation of alkoxides and resultant ultra- and micro-structural effects. *J. Mater. Sci.*, **23**, 1895–1900 (1988).

21. C. Geradin, S. Sundaresan, J. Benziger, and A. Navrotsky, Structural investigation and energetics of mullite formation from sol-gel precursors. *Chem. Mater.*, **6**, 160–170 (1994).

22. J. Sanz, I. Sobrados, A. L. Cavalieri, P. Pena, S. de. Aza, and J. S. Moya, Structural changes induced on mullite precursors by thermal treatment: a ^{27}Al MAS-NMR investigation. *J. Am. Ceram. Soc.*, **74**(10), 2398–2403 (1991).

23. I. Jaymes and A. Douy, Homogeneous mullite-forming powders from spray-drying aqueous solutions. *J. Am. Ceram. Soc.*, 75(11), 3154–3156 (1992).

24. B. E. Yoldas, Microstructure of monolithic materials formed by heat treatment of chemically polymerized precursors in the Al_2O_3-SiO_2 binary. *Ceram. Bull.*, **59**(4), 479–483 (1980).

25. L. H. Merwin, A. Sebald, R. Rager, and H. Schneider, ^{29}Si and ^{27}Al MAS NMR spectroscopy of mullite. *Phys. Chem. Minerals*, **18**, 47–52 (1991).

26. T. Ban and K. Okada, Analysis of local cation arrangement in mullite using ^{29}Si MAS NMR resonance spectra. *J. Am. Ceram. Soc.*, **76**(10), 2491–2496 (1993).

27. M. Haque, Thesis, University of Kolkata (2000).

28. A. K. Chakraborty, An analysis of the phase evolution of six types of mullite gels. Unpublished data (2008).

29. S. Kanzaki, H. Tabata, T. Kumazawa, and S. Ohta, Sintering and mechanical properties of stoichiometric mullite. *J. Am. Ceram. Soc.*, **68**(1), C-6–C-7 (1985).

30. A. Douy, Organic gels preparation of silica-alumina powders. I. Mullite. *J. Euro. Ceram. Soc.*, **7**, 117–123 (1991).

Chapter 10

Chemistry of Mullite Formation through the Sol-Gel Process

In this chapter, we will discuss the predicted gelation reactions of mullite gels synthesized by various methods.

10.1 Mullite Gel by the Aqueous Sol-Gel Method Using Water-Soluble Components

During the gelation processes as used by Insley and Ewell (1935), as discussed in Chapter 2, NaOH solution was added to a mixture of aluminum sulfate and sodium silicate. Sodium aluminosilicate sulfate hydrate precipitation is expected to have occurred, which on dialysis formed aluminosilicic acid hydrate with the removal of sodium ion. A similar reaction might have occurred in the case of the neutralization reaction of a mixture of sodium silicate and potash alum on using hexaethlyamine instead of sodium hydroxide, as done by Ossaka (1961). Instead of using any base as the neutralizing agent and eliminating the tedious process of sodium removal, Chakraborty and Ghosh (1988) took silica sol obtained from the same silica source as before and freed the sodium ion by exchanging it with H^+ ion by using a cation exchanger.

Mullite Formations: Analysis and Applications
Akshoy Kumar Chakraborty

ISBN 978-981-4877-05-3 (Hardcover), 978-1-003-03167-3 (eBook)
www.jennystanford.com

Gel containing hydroxyl alumina-silicate sulfate: In this case, it was conceived that silica sol might have reacted with aluminum sulfate to form a chemical compound, likely hydroxyl alumina-silicate sulfate. The gelation mechanism of aluminum sulfate/silica sol is given below (Eq. 10.1). The monosilicic acid molecule $Si(OH)_4$ is very unstable due to incomplete coordination of silicon (since in hydrated form coordination number of Si^{+4} is 6) with respect to OH^{-1}, and thus condensation of two silanol (–Si–OH) groups may take place readily, leading to the formation of siloxane (–Si–O–Si–) bonds with the elimination of the water molecules. As a result, silica sol did not remain in true equilibrium. There was room for irreversible interaction of the hydroxyl group with other $Si(OH)_4$ molecules to form an increasing number of siloxane linkages. Consequently, on letting the sol stand for sufficient duration, it became colloidal with continued rise in viscosity as well as molecular weight. Therefore, when aluminum salt is mixed with silicic acid sol, two reactions likely take place simultaneously: polymerization of silicic acid itself and the formation of aluminosilicate hydrate. The polymerized silica sol units will react with Al^{+3} ion later. The effect of polymerized silica sol on the 980°C reaction of different Al_2O_3-SiO_2 gels was reported (Table 7.1). The intensity of the 980°C differential thermal analysis (DTA) peak (height) and the quantity of mullite formed (X-ray diffractometry [XRD] intensity of the 0.537 nm peak) were both significant in the case of the gel prepared from aluminum sulfate (BDH) and freshly prepared silica sol (i.e., with no aging time). Gels prepared with silica sol with increased aging times showed the following:

- The intensity of the 980°C exotherm decreased rapidly from the highest value and then proceeded very slowly (Fig. 7.10a).
- The amount of mullite also decreased rapidly, and thereafter the formation of Al-Si spinel commenced. It increased gradually with the aging period of silica sol (Fig. 7.10b).

Thus, the existence of an interrelationship between the 980°C exothermic peak in DTA and the nature and amount of formation of the 980°C phase versus silica sols aged for increasing amounts of time was shown for the synthesis of mullite gels. The essential condition for obtaining pure mullite was based primarily on the silica component, which will be monomeric or contain small aggregated colloidal particles. The mechanism of the reaction between silicic acid sol and Al^{+3} for the formation of simple units of aluminum sulfate hydrate is suggested to consist of two sequential reactions. Initially, one silicic acid molecule reacts with three aluminum atoms to form a simple unit of aluminosilicate complex. Two such polymeric units polymerize in the next step, as shown in Eqs. 10.1 and 10.2.

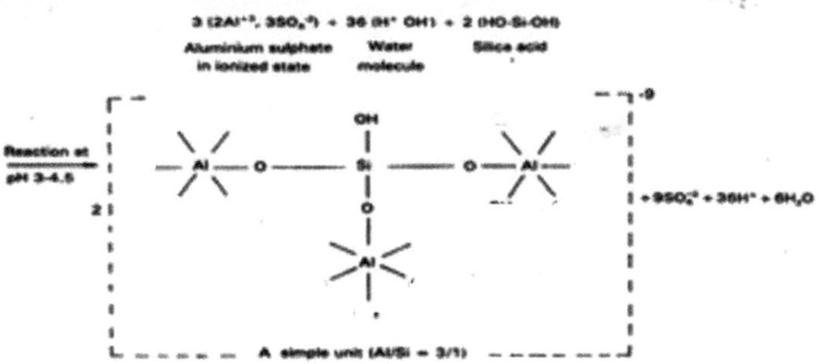

(10.1)

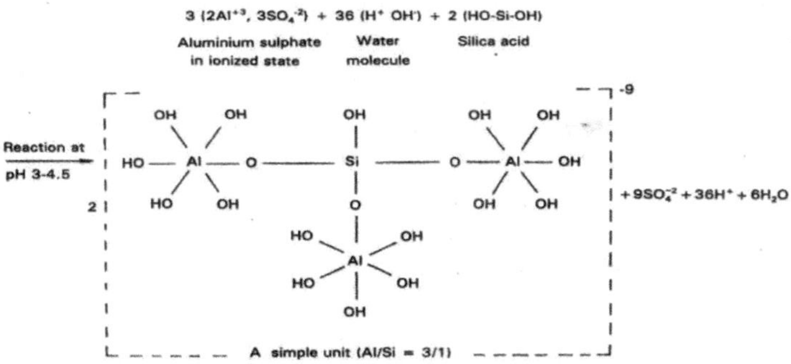

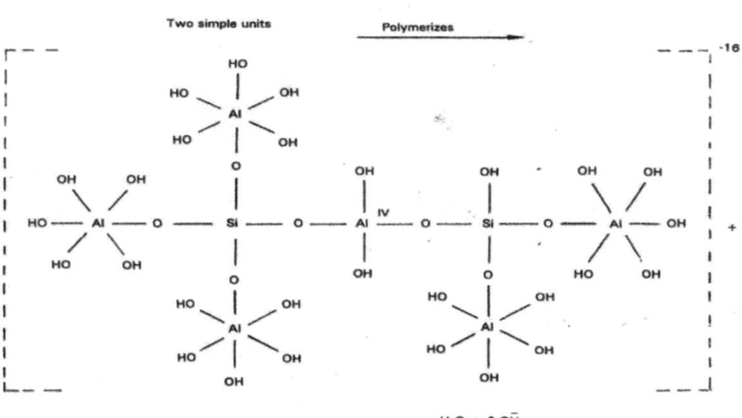

(10.2)

Equations 10.1 and 10.2 schematically represent the 1st models of the gel made by the aqueous sol-gel method showing the formation of two complex masses (Chakraborty 1994a).

Gel containing unidirectional chain: The formation of the above complex mass can now be imagined as the homogeneous mixing of silica and alumina components as in mullite stoichiometry (Si:Al = 1:3). Direct mullitization was observed at the 980°C exotherm. It was suggested that the chemical composition of the gel powder will remain consistent up to mullitization. Furthermore, the polymerized unit of the silica-alumina complex will appear to be a feasible model as it explains the existence of tetracovalent aluminum in addition to some Al-O groups, as noted by infrared (IR) studies (Hirata et al., 1985) and magic angle spinning–nuclear magnetic resonance spectroscopy (MAS-NMR) studies (Komarneni et al., 1985). However, inhomogeneity may likely arise as a side-reaction during gelation, and as a result, mullite might contain an Al-Si spinel phase (minor quantity) at ~1000°C. As stated above, silicic acid micelles generally grow preferentially into colloidal particles of different sizes (Carman, 1940; Kruyt, 1952). Carman (1940) and Kruyt (1952) suggested the interior structure of the particle, which showed each silicon surrounded on all sides by oxygen atoms in an orderly 3D state. The surface silicon will complete its tetrahedral coordination by the adsorption of water/hydroxyl groups. It is obvious that interactions of such variously sized charged silica particles with aluminum sulfate may produce gel particles of different Al:Si ratios, leading to gross inhomogeneity. For example, gels prepared by using aluminum sulfate and silica sol aged for 24 h would form a mixture of both mullite and Al-Si spinel. This may explain the findings of many earlier researchers that pure mullite does not form at 980°C exotherm but what forms is an admixture of Al-Si spinel. In this connection, during gelation, it is recalled that kaolinite/metakaolinite contained tetrahedral sheets proceed in the *a-b* direction. It also formed mullite with Al-Si spinel phases analogous to some Al_2O_3-SiO_2 gels, as shown above. The presence of silica components in the form of a cluster or a polymeric chain of silica may be responsible for the crystallization of the Al-Si spinel phase at the 980°C exotherm (Chakraborty, 1994b).

Kubota and Takagi (1986) and Mizuno et al. (1990) synthesized gel from a mixture of silica sol (obtained from fume silica) and aluminum salts (e.g., sulfate, nitrate, and chloride). In this reaction model, the $Si(OH)_4$ sol obtained from condensed 3D –Si–O–Si– unites with fume silica and likely contains polymerized units to form a 3D SiO_2 framework structure in the starting material. And as such, the gel formed mullite at ~1200°C via spinel formation at ~1000°C. Nishu et al. (1989) prepared a gel from a mixture of monosilicic acid solution (obtained by digesting silica gel at pH 9 by aqueous ammonia) and $Al(OH)_3$ sol (obtained by adding aqueous ammonia to an aluminum chloride solution at pH 8). This mixed sol was stirred at room temperature to form a gel. It showed direct mullitization at 980°C. The cause of the early mullitization was explained after measuring the polymerization state of the Si ion in the gelation state by the absorption method and trimethylsilylation. Most of the silicic acid was found to be a monomer. Nishu et al. (1989) also compared the coordination states of Al in pseudoboehmite sol and in mixed gel by ^{27}Al MAS-NMR. It was noted that only sixfold-coordinated aluminum was present in the boehmite gel whereas both fourfold and sixfold-coordinated aluminum was noticed in the mixed gel. As such it was considered that monosilicic ion likely entered into the reaction with the boehmite particle, resulting in the development of Si–O–Al(IV) linkages during gelation and in a drying process and may avoid polymerization itself to develop Si–O–Si bonds.

A similar observation was made by Komarneni and Roy (1986) in their single-phase gel by MAS-NMR studies. According to Okada et al. (1991), the degree of polymerization was certainly a factor and the preparation method resembled that of Ossaka (1961). Schematic structure models of the starting materials (i) mixed gel, (ii) coprecipitated gel, (iii) kaolin mineral, and (iv) glassy material were suggested by Okada and Otsuka (1986). Furthermore, it was predicted that the transformation of the mixed gel may be similar in nature to that of G-104, G-96, and G-10 synthesized by Chakraborty and Ghosh (1988).

10.2 Mullite Gel from a Monophasic Gel: Chemistry of Hydrolysis vis-à-vis Gelation-cum-Coprecipitation Reaction

Before understanding this problem, introduced in Chapter 3, it is necessary to critically review first the hydrolysis and condensation process of tetraethyl orthosilicate (TEOS). Thereafter, one can visualize any sort of physicochemical interaction between the two constituents. The hydrolysis and gelation process of TEOS consists of four stages: (i) hydrolysis of the alkoxy group of TEOS to form silanols, (ii) polymerization of the hydrolyzed species to form particles; (iii) growth of particles and linking of particles together to form chains and then 3D networks, and (iv) setting to form gel. The final morphology, structure, and properties of the gel are influenced by the gel preparation conditions. The different conditions and procedures attributed to the kinetics of hydrolysis and subsequent polymerization processes are outlined below. It is absolutely necessary to present first the chemistry of hydrolysis of silica and alumina components separately. Thereafter, coagulation and coprecipitation processes occur at the start of mullite synthesis, as presented in Chapter 2, and will be discussed for a thorough understanding prior to mullite evolution.

Hydrolysis of TEOS: Some of the parameters that guide the hydrolysis of TEOS to form silanols are as follows: the nature of the hydrolysis reaction, steps of hydrolysis usually practiced (single step or double step), the rate of hydrolysis (slow or fast), the order of the reaction, the mechanism (electrophilic or nucleophilic; comparison of acid- versus base-catalyzed reaction).

Polymerization/Condensation/Gelation: Some of the parameters that guide the polymerization reactions are as follows: the nature or character of the polymer species (size/type of polymerized molecule), the rate of condensation, the structure of the wet gel, the –OH content in the gel, the rate of hydrolysis versus the rate of condensation, and modification of the structure of the wet gel during drying.

Hydrolysis and gelation reaction of TEOS: The basic parameters of hydrolysis and gelation reaction are as follows: the pH, the water concentration, gelation time, and temperature. The effect of the condition of the hydrolysis and subsequent spontaneous polycondensation of $Si(OC_2H_5)_4$ was reported by Nogami and Moriya (1980), and these are dependent on the amount of water, the nature of the catalyst (e.g., HCl and NH_4OH solutions taken), and the temperature of hydrolysis. They also reported the time-dependent rise of viscosity of hydrolyzed sol in various conditions. In the initial stage, the viscosity increases gradually with time and then it increases strikingly so the sol sets and becomes a gel. It was shown that the time required for gel formation is shorter with increasing temperature. In the latter case, the gel particles are spherical (~100 Å in diameter) under an electron microscope. The gelling time of the solution hydrolyzed with HCl is longer than that in the previous case, and no spherical particles are observed.

On the basis of small-angle X-ray scattering (SAXS) observation, Yamane et al. (1979) noted the formation of colloidal particles at the beginning of hydrolysis of $Si(OCH_3)_4$ with ammoniated water. As the reaction progresses, aggregation occurs to form big and larger particles. When $Si(OC_2H^5)_4$ is hydrolyzed with NH_4OH, monodispersed colloidal SiO_2 spherulites are observed. The sol-gel transition occurring during the hydrolysis and polycondensation of TEOS was investigated by Sakka and Kamiya (1982), and the hydrolyzing processes of TEOS on changing the amount of water (i.e., water:alkoxide ratio) in the presence of either an acid or a base as a catalyst were described in detail. The acid-catalyzed solution is clear after hydrolysis and polycondensation, which results in a transparent gel. On the contrary, the base-catalyzed solution shows slightly cloudy precipitates as the reaction progresses and the resultant gel is not clear. What should be the size of the hydrolyzed silica particles when TEOS is catalyzed by ammonia? Near the gelling point, an abrupt increase in viscosity (η) was noted both by Sakka and Kamiya (1982) and Nogami and Moriya (1980). This phenomenon was ascribed by them to the aggregation of discrete spherical particles, which led to a 3D network structure. Many researchers have shown the formation of monodispersed colloidal silica spherulites during the hydrolysis of TEOS by ammonia. For the complete hydrolysis-condensation of 1 mole of TEOS into SiO_2 of a 3D network structure, 2 mol of water are required, as follows:

Hydrolysis reaction

$$n\text{Si(OC}_2\text{H}^5)_4 + 4n\text{H}_2\text{O} \rightarrow n\text{Si(OH)}_4 + 4n\text{C}_2\text{H}_5\text{OH} \qquad (10.3)$$

Polycondensation reaction

$$n\text{Si(H)}_4 \rightarrow n\text{SiO}_2 + 2n\text{H}_2\text{O} \qquad (10.4)$$

Net reaction

$$n\text{Si(OC}_2\text{H}^5)_4 + 2\text{H}_2\text{O} \rightarrow n\text{SiO}_2 + 4n\text{C}_2\text{H}_5\text{OH} \qquad (10.5)$$

They (Nogami and Moriya, 1980; Sakka and Kamiya, 1982) suggested that the addition of 1 mole of water to 1 mole of TEOS is required to produce a chain-like polysiloxane polymer, as per the following hydrolysis-polycondensation reaction:

$$\begin{array}{cccccccc}
\text{OR} & & \text{OR} & & \text{OR} & & \text{OR} & \\
\text{I} & & \text{I} & & \text{I} & & \text{I} & \\
n\text{Si(OR)}_4 & + n\text{H}_2\text{O} \rightarrow \text{RO–} & \text{Si} & \text{–(–O–} & \text{Si} & \text{---)}_{n-2}\text{–O---} & \text{Si} & \text{–OH} + (2n-1)\text{ROH} \\
\text{I} & & \text{I} & & \text{I} & & \text{I} & \\
\text{OR} & & \text{OR} & & \text{OR} & & \text{OR} &
\end{array} \qquad (10.6)$$

They (Sakka and Kamiya, 1982) are of the opinion that the solutions containing linear polymers are more viscous than those containing spherical polymers, as shown in the case of ammonia-catalyzed solution or when a large quantity of water is used. On the other hand, using a large quantity of water with a metal alkoxide favors the formation of monolithic bulk gel. Their major observations are as follows:

- First group: When a small amount of water was used and HCl was used as the catalyst, the solution showed linear polymers and spinnability.
- Second group: When a large amount of water was used and HCl or NH_4OH was used as the catalyst, the solution showed nonlinear polymers and prohibited the occurrence of spinnability before gelling and the solution solidified into bulk elastic gels.

The degree of polymerization according to Konard et al. (1867, 1929) is guided by the following equation:

$$\text{SP} = n/(n - m),$$

where p = number of silicon atoms in the polymer, n = mol of TEOS, and m = mol of water. Thus, they noted that every desired degree of polymerization may be chosen by adding the calculated amount of water. In the case of under-stoichiometric water addition, unhydrolyzed groups are present and likely some linear types of oligomers are formed, unlike in the case of excess water, with total hydrolysis of $Si(OR)_4$ and preferential formation of a 3D cross-linking.

The nature or character of polymer species or particles, for example, the size/type of polymerized molecule, is dependent upon the type of catalyst chosen. Whether it is a chain-like polymer, a 3D network, or a spherical particle is also dependent upon the type of catalyst used. The pH during hydrolysis strongly influences the properties of the product, for example, the size of the particles depends upon the concentration of the acid as well as ammonia used (Gossink, 1975). Concentrated acids give conglomerates of smaller particles. Dilute acids give particles of several nanometers. At low concentrations of NH_4OH, hydrolysis leads to fine precipitates of silica. With concentrated ammonia, the particles are dense/nonporous and spherical. The specific surface area value of the particle is highly porous. The size is ~0.5 micron.

The gel formation for acid- and base-catalyzed systems was illustrated schematically by Schaefer et al. (1984) and Brinker et al. (1982). Under acid-catalyzed conditions and especially with the addition of less water (e.g., $H_2O:Si \leq 5$), primarily linear or randomly branched polymers are formed that entangle and form additional branches, resulting in gelation. Under basic conditions and/or with addition of a larger amount of water, more highly branched clusters are formed, which do not interpenetrate prior to gelation and thus behave as discrete species. The gel formation time in the case of the acid-catalyzed process is 10 times more than in the case of the ammonia-catalyzed process (Nogami & Moriya, 1980). They also showed that a base-catalyzed gel forms spherical particles of bulk density whereas the acid-catalyzed gel does not show spherical particles.

Rate of hydrolysis (slow vs. fast): Schmidt et al. (1984) investigated the hydrolysis and condensation reaction by monitoring the H_2O content of the reaction system. They showed that proton-catalyzed hydrolysis of esters of silicic acid in ethanol solution generally displays a first-order reaction and a linear reaction rate dependence on proton concentration. They noted that the course of base-catalyzed hydrolysis of TEOS is not first order, indicating no dependence on only one concentration of a starting material. However, they also noted no second-order dependence. Aelion et al. (1950) noted higher orders for NH_4OH-catalyzed hydrolysis of tetraalkoxysilanes.

Nature of polymeric gel: Generally, hydrolysis and condensation reactions occur almost concurrently. As the rates of (i) hydrolysis of alkoxide groups to develop silanols, (ii) condensation of silanols to form polymeric linkages, and (iii) growth of polymers to set into gel are different, it is possible to vary the relative rates of the three processes by monitoring conditions by use of water-to-TEOS ratios and pH. To relate the nature of the gelation processes of TEOS to the properties of the dried gel, Brinker et al. (1982) used a two-step hydrolysis process for preparing solutions of different compositions. In the first step, TEOS, alcohol (ethanol, *n*-propanol, or isopropanol), water, and acid (HCl) in a molar ratio of 1:3:1:0.0007 are mixed. Initially, a quarter the stoichiometric amount of water is added to fully hydrolyze TEOS over a period of 1.5 h to monosilicic acid. In the second step, an additional amount of water, acid, or NH_4OH is added, and the solution is maintained at 40°C until gelification. They showed that the polymer species are different in nature in acid- and base-catalyzed gelation process.

First case: When the hydrolysis reaction proceeds much more rapidly than the condensation reaction, the basic mechanism for the condensation reaction will be base catalyzed and the nucleophilic attack on silicon is as per the following reactions:

$$(RO)_3SiOH + OH^{-1} \rightarrow (RO)_3SiO^{-1} + H_2O \text{ (fast)} \quad (10.7)$$

$$(RO)_3SiO^{-1} + (RO)_3SiOH \rightarrow (RO)_3SiOSi(OR)_3 + OH^{-1} \quad (10.8)$$

The net result is that the larger and highly condensed polymers tend to react with feebly acidic monomers to form even more highly condensed species, that is, the resulting sol will be a relatively large (50–200 Å) polymer due to extensive cross-linking. Thus, when hydrolysis was fast, larger and more highly condensed polymers were formed during gelation. This gel dried to a low-density coarse-textured material.

Second case: When the rate of condensation of silanols (either by dehydration or by dealcoholation) is much greater than their rate of production by hydrolysis. This occurs during acid-catalyzed hydrolysis at low water and acid concentrations. The form of the polymers is governed by the mechanisms of hydrolysis reaction and not by condensation reaction. The acid-catalyzed hydrolysis proceeds by a mechanism that involves an electrophilic attack on an alkoxide oxygen as per Aelion et al. (1950). Conversely, when hydrolysis is slow, smaller and less highly condensed polymers are formed. Microstructural analysis by scanning electron microscope (SEM) and transmission electron microscopy (TEM) showed extremely fine microstructural features, which is consistent with SAXS results. This gel dries into a high-density fine-textured gel. In basic and acidic conditions, with the use of a large amount of water for hydrolysis, the gel appears to consist of course (~900 Å) particles. Thus, the form of the resulting polymers is governed by the relative rates of hydrolysis and condensation. The important aspects of polymer growth prior to gelation were further discussed by Brinker et al. (1984). In the acid-catalyzed hydrolysis process, a wide distribution of hydrolyzed and unhydrolyzed monomers, dimers, and chains is formed instead of colloidal particles. In the second stage, hydrolysis seems to be complete, resulting in linear or branched polymers. In a base-catalyzed hydrolysis process, the resulting polymers are highly condensed clusters instead of colloidal. Generally, the hydrolysis is incomplete, unlike in the acid-catalyzed reaction, where hydrolysis may be complete in the second stage. For example, when condensation is rapid compared to hydrolysis, polysiloxane chains or rings result, while in the reverse case, more extensively cross-linked polymeric clusters are formed.

Different mechanisms for employing catalysts in sol-gel processing were proposed by Pope and Mackenzie (1986). The gelation processes for metal alkoxides are composed of two reactions, hydrolysis and condensation, and the gelation time is the result of both reaction processes. The slowest of these reactions is the rate-determining step in the total gelation process. The total rate of gelation is a function of both hydrolysis and polymerization reactions. Acid-catalyzed hydrolysis is an electrophilic reaction as follows:

$$(RO)_3SiOR + H_3O^+ \rightarrow (RO)_3SiOH + H^+ \quad (10.9)$$

$$H^+ + H_2O \rightarrow H_3O^+ \quad (10.10)$$

Here, the reaction rate is governed by the concentration of hydronium in the solution. The lower the pH, the faster is the rate at which the reaction proceeds. In the base-catalyzed hydrolysis, a nucleophilic substitution of hydroxyl ions for OR groups occurs as follows:

$$(RO)_3SiOR + OH^- \rightarrow (RO)_3SiOH + OR^- \quad (10.11)$$

$$OR^- + H_2O \rightarrow HOR + OH^- \quad (10.12)$$

Analogous to the acid-catalyzed reaction, the rate of base-catalyzed hydrolysis is a function of the hydroxyl concentration in the solution (Aeilon et al., 1950). On the basis of studies of later authors and Iller (1979), they further proposed a generalized inverse rate of reaction versus pH of the solution model. The inverse reaction rates for general acid- and base-catalyzed hydrolysis and polymerization were observed by Pope and Mackenzie (1986). The total rate of gelation is a function of both hydrolysis and polymerization reactions. Aeilon et al. (1950) have shown that the hydrolysis rate of TEOS is linearly proportional to the concentration of HCl and NH_4OH. Iller (1979) demonstrated that the polycondensation rate of silicic acid in an aqueous solution exhibits more or less a sinusoidal shape. The dashed lines in Fig. 10.1 of Iller represent the rate of reaction without impurities and the reaction rate with impurity is marked by the solid line. Figure 10.2 shown by him demonstrates the gelation time as a function of pH value below 5 with HCl as the catalyst. They explained the microstructural differences between acid- and base-catalyzed gels by considering and combining Iller's result for the effect of pH on the polymerization reaction with Aelion's relationship for the effect of catalyst concentration on hydrolysis.

For pH values above the isoelectric point (~2), increasing the OH concentration increases the rate of polymerization. Accordingly, a continuous decline in gelation time is noted and the size of the particles correspondingly increases. The gelation time is a function of pH for HCl-catalyzed solutions (Pope & Mackenzie, 1986). The shape becomes spherical in a high pH–catalyzed solution. In a low-pH region, preferably the chain length seems to be linear. The transition between linear chain growth and colloidal particle growth is affected by a change in the pH and also affects the apparent density, porosity, and gelation time.

Formation of intermediates: Hydrolysis is an indispensable step, as observed, and it requires water. Therefore, it should be possible to isolate intermediates by restricting the amount of water to be used during hydrolysis. Then, a part of the =Si–OR groups is able to react with =Si–OH and condense to siloxane bonds.

The previous works on the hydrolysis-condensation reaction of alkoxysilane was reviewed by Schmidt (1984), who demonstrated that many different intermediates are possible in an alkoxysilane-water reaction and these reaction paths are complicated. Schmidt (1984) is of the opinion that it is very difficult to give an exact thermodynamical description of what might be possible or not in these complicated reaction paths. It is difficult to separate hydrolysis from condensation leading to gelation.

$$Si(OR)_4 + H_2O = HO\ Si(OR)_3 + ROH \qquad (10.13)$$

$$HO\ Si(OR)_3 + H_2O = (HO)_2Si(OR)_2 + ROH \qquad (10.14)$$

$$(HO)_3Si\text{-}OR + H_2O = Si(OH)_4 + ROH \qquad (10.15)$$

$$=Si–OR + HO–Si= \ = \ =Si–O–Si= + ROH– \qquad (10.16)$$

$$=Si–OH + HO–Si= \ = \ =Si–O–Si= + HOH \qquad (10.17)$$

In most cases, reactions according to Eqs. 10.16 and 10.17 will start soon after Eq. 10.13 and if the degree of condensation has reached a sufficient state, gelation occurs. Of course, it should be possible to isolate intermediates by restricting the amount of water used in the hydrolysis process. A monomeric silicic acid sol method is emphasized by introducing $Si(OR)_4$ or $SiCl_4$ vapor into water. In some preparation conditions, monomeric and dimeric acids are stable for hours. Secondly, a minimum of the condensation rate at a pH of ~3 was established. Iller (1979) also measured that the gelling time of aqueous monosilicic acid solution was at a minimum pH of ~6. The polycondensation process occurs in stages. The process can be monitored by the partial hydrolysis of the gel in alcoholic solutions.

Growth and form of polymer prior to gelation: Comparison of the rate of hydrolysis and the rate of condensation. When the rate of hydrolysis is more than rate of condensation, what happens? The relative rates of hydrolysis of TEOS and condensation predict the final polymer.

Different natures of polymer species in acid- and base-catalyzed gelation processes: This difference is explained by Brinker et al.

(1982). The relative rates of hydrolysis of the alkoxide group of TEOS, condensation of the silanols, and linking of the polymers to form a gel depend upon three factors: concentration of water, concentration of TEOS, and the pH of the mixed solution.

First case: When the hydrolysis reaction proceeds much more rapidly than the condensation reaction, the basic mechanism for the condensation reaction will be a base-catalyzed nucleophilic attack on silicon as per the following reactions:

$$(RO)_3SiOH + OH^{-1} \rightarrow (RO)_3SiO^{-1} + H_2O \text{ (fast)} \quad (10.18)$$

$$(RO)_3SiO^{-1} + (RO)_3SiOH \rightarrow (RO)_3SiOSi(OR)_3 + OH^{-1} \quad (10.19)$$

The net result is that the larger and highly condensed polymers tend to react with feebly acidic monomers to form even more highly condensed species, that is, the resulting sol will be a relatively large (50–200 Å) polymer due to extensive cross-linking.

Second case: The rate of condensation of silanols (by dehydration or dealcoholation) is much greater than their rate of production by hydrolysis. Two-step hydrolysis processes of TEOS are discussed, along with the important aspects of polymer growth prior to gelation (Brinker et al., 1984). In an acid-catalyzed hydrolysis process, a wide distribution of hydrolyzed and unhydrolyzed monomers, dimers, and chains is formed instead of colloidal particles. In the second stage, hydrolysis seems to be complete, resulting in linear or branched polymers. In the base-catalyzed hydrolysis process, the resulting polymer is made of highly condensed clusters instead of being colloidal in nature. Generally, the hydrolysis is incomplete in comparison to an acid-catalyzed reaction, where hydrolysis may lead to completion in the second stage.

Structure of the wet gel (OH content in the gel): Yoldas (1988) correlated the amount of water used during gelation with the hydroxyl content of the gel. He proposed different structural models with different degrees of network connectivity.

Mechanisms (electrophilic or nucleophilic): Comparison of acid- vs. base-catalyzed reaction: Both acid and base hydrolysis behavior of silicon alkoxide were explained by Uhlmann et al (1984). An acid may increase the electrophilicity of the electrophilic silicon atom and transforms –OR into the better-leaving group –ROH. A base may increase the nucleophilicity of H_2O by producing OH^- and suggests two possible mechanisms. The hydrolysis reaction of TEOS in an acidic solution proceeds by an electrophilic mechanism. Alkaline-catalyzed hydrolysis of TEOS takes place by a nucleophilic substitution mechanism (Aelion et al., 1950). In an acid-catalyzed reaction, the rate of hydrolysis of a silicate tetrahedron tends to decrease as the alkoxide groups are removed. This favors the production of silanol sites at the end of chains, thus generating linear polymers. The electrophilic reaction mechanism tends to produce weakly cross-linked species, which tend to completely hydrolyze in solution (Keefer, 1984). In base-catalyzed reactions, it is argued that each subsequent hydrolysis of a tetrahedron should proceed more rapidly than the previous one, producing numerous branch points that are the preferred site for condensation. Nucleophilic hydrolysis and condensation in alkaline solutions tend to produce highly cross-linked species. Hydrolysis may be incomplete.

NMR study applied in the hydrolysis of silicon alkoxide: The hydrolysis-polymerization/condensation process of silicon alkoxides in the presence of water and/or alcohol is strongly dependent upon catalytic conditions employed. Irwin et al. (1987) used ^{29}Si NMR spectroscopy to monitor the extent of the polycondensation process during the heat treatment of raw gel. Recent resolution and sensitivity enhancement techniques like MAS and cross polymerization have the ability to investigate the chemical structure of the amorphous gel sample while heating the environment. Irwin et al. (1987) showed the ^{29}Si MAS-NMR spectra of uncatalyzed and base-catalyzed gels synthesized from tetramethyl orthosilicate (TMOS) with methanol and water under constant stirring. They noted three peaks, at about –91, –101, and –110 ppm, and assigned them as due to the environment of the silicon nucleus occupying Q^2, Q^3, and Q^4 sites, respectively. The relative percentage of each site on heating the silica gel was calculated from the integral of its peak. With an increase in the heating temperature, both Q^2 and Q^3 decrease and Q^4 increases in both samples. The spectra of these samples using the cross polymerization–MAS technique also confirm these results. Irwin et al. (1987) indicated the following:

- The base-catalyzed gel showed $Q^4 > Q^2$ groups, that is, more highly condensed than the uncatalyzed gel at room temperature. Schaefer and Keefer (1984) showed the formation of a highly branched polymer in base-catalyzed conditions but a predominantly linear polymer in acid-catalyzed gels. Linear chains were also noted by Sakka and Kamiya (1982) during spinning of fibers from acid-catalyzed sols.
- The structural evolution of the two gels on heating showed marked differences. The condensation of the uncatalyzed gel increases continuously at each stage of heating. Q^4 grows at the cost of Q^2 and Q^3 groups. The catalyzed gel behaves differently in the temperature range of 150°C–450°C and up to 800°C. After that, Q^4 distribution is similar to that of the uncatalyzed gel.

Hydrolysis and gelation reaction of aluminum alkoxide: For details, see the papers of Yoldas (1974, 1975).

10.2.1 DTA Studies of $Al(NO_3)_3{\cdot}9H_2O$ and TEOS and Their Interactions

What will happen when a system containing two pure components—aluminum nitrate nonahydrate (ANN) and TEOS—and a small amount of water and a mineral acid is heated in a water bath? To reveal any sort of interaction between ANN and TEOS, a series of DTA studies of two pure components under different conditions is given chronologically:

1. Pure TEOS exhibited an exothermic peak at 275°C (Fig. 10.1A) due to the decomposition of ethoxy groups.
2. Pure silica gel obtained from the ion-exchange process showed only one endothermic peak, at ~300°C (Fig. 10.1B), due to the loss of the physically adsorbed water.
3. Pure $Al(NO_3)_3{\cdot}9H_2O$ exhibited two endotherms (Fig. 10.1), a small one at ~95°C due to the elimination of its water of crystallization and one relatively broad and large in magnitude at ~170°C due to the decomposition of nitrate.

4. To detect the temperature of occurrence of the reaction of the ANN and TEOS, Chakraborty (2008b) carried out a series of DTA studies, which are discussed next.

The DTA analysis of pure ANN (curve 1) showed a small endotherm at ~95°C, maybe due to the elimination of free moisture, and thereafter a broad endotherm, due to the removal of nine molecules of water of crystallization and decomposition of nitrate.

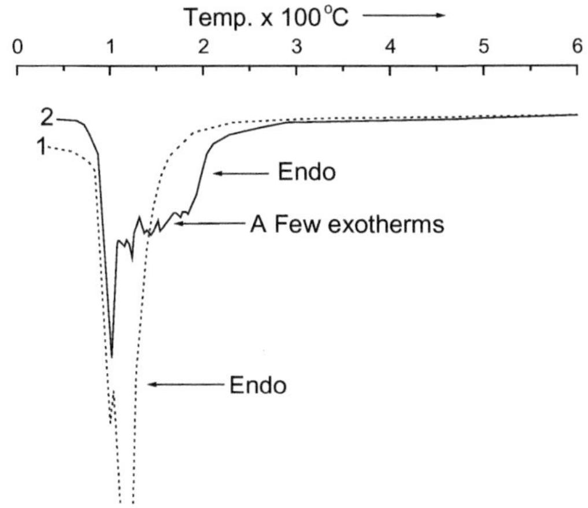

Figure 10.1 (curve A) DTA trace of pure ANN and (curve B) DTA trace of a mixture of ANN and TEOS (Chakraborty, 2008).

A DTA run of a mixture of ANN and TEOS in the case of ISG72 (curve 2) showed a few exothermic peaks of small magnitude within the broad endotherm of ANN. The exothermic nature of the reaction is also ascertained during the DTA run of the mixture of the two components (curve 2) when compared with the same for pure ANN (curve 1). The exhibition of a few exotherm peaks of small magnitudes indicates the ongoing reaction of the two components with a dynamic rise in heating during the DTA run.

Chakraborty (2008) used a constant temperature water bath at 80 ± 2°C to isothermally heat the mixture of the two components in different SiO_2:Al_2O_3 ratios. The gel marked as ISG40 (containing ~40 wt % of alumina) and other mixtures, namely ISG65, ISG66, ISG72, ISG76, and ISG80, were synthesized. For the isothermal study, 4 mL of TEOS was taken in a round-bottomed flask and was placed in the water bath until it attained the temperature of the bath. To this fixed amount of TEOS, different amounts of ANN were added to produce mixed gels of different SiO_2:Al_2O_3 ratios, as shown in Table 10.1.

Table 10.1 Batch composition of ANN and TEOS taken in the isothermal heating process

Gel mark	Approx. wt % Al_2O_3:SiO_2	Amount of TEOS (mL)	Amount of ANN (g)
ISG 40	40:60	4	5
ISG 56	56.25:43.75	4	10
ISG 66	66.66:33.33	4	15
ISG 72	72:28	4	20
ISG 76	76.27:23.72	4	25
ISG 80	79.41:20.58	4	30

Source: (Chakraborty, 2008)

A weighed amount of ANN crystals was taken in a separate beaker, and its temperature was raised gradually to ~80°C, when the solid melted first into a solution and then became concentrated on continued heating. After a constant temperature was attained, the concentrated ANN solution was added to the flask containing TEOS at the same temperature ~80°C. The temperature of the mixed solution was found to rise slowly at the beginning and then it rose rapidly immediately after the flask was swirled, with simultaneous effervescence of brown fumes of NO_2. During this heat evolution process, the temperature of the solution was noted at regular intervals using a thermometer, from the beginning of the experiment to the end of it. In the case of ISG72 gel, the temperature of the reaction mixture finally reached a maximum of 89°C. Thereafter, it decreased slowly with time, until the temperature subsided to ~80°C. Gelation of the reaction mixture started during this period. On continued heating, the gel gradually dehydrated and decrepitated inside the flask. The product was collected and stored for X-ray analysis. The experiment was repeated for successive batches. The evolved exothermic energy as a measure of rise in the temperatures of the reaction batches versus time period was plotted. The resultant gels were heated to 500°C and analyzed for phases by XRD.

While heating TEOS and ANN at a constant temperature of ~80°C in the water bath, it was observed that the temperature of the reaction medium suddenly increased and NO_2 fumes emerged vigorously. The two phenomena indicate the occurrence of an exothermic reaction during heating of the two reactants. Figure 10.2 shows that the temperature of the reaction bath increased with the Al_2O_3:SiO_2 ratio of the ISG gels. With an increase in the wt % of Al_2O_3 in the ISG gels under study, the apparent area under the time-temperature curve during heat evolution in the temperature range of 80°C–100°C increased slowly from ISG40 to ISG56 to ISG66, reaching its highest in ISG72. Even with an increase in the wt % of Al_2O_3, the said area decreased, as noted in the case of ISG76, and it further decreased in ISG80.

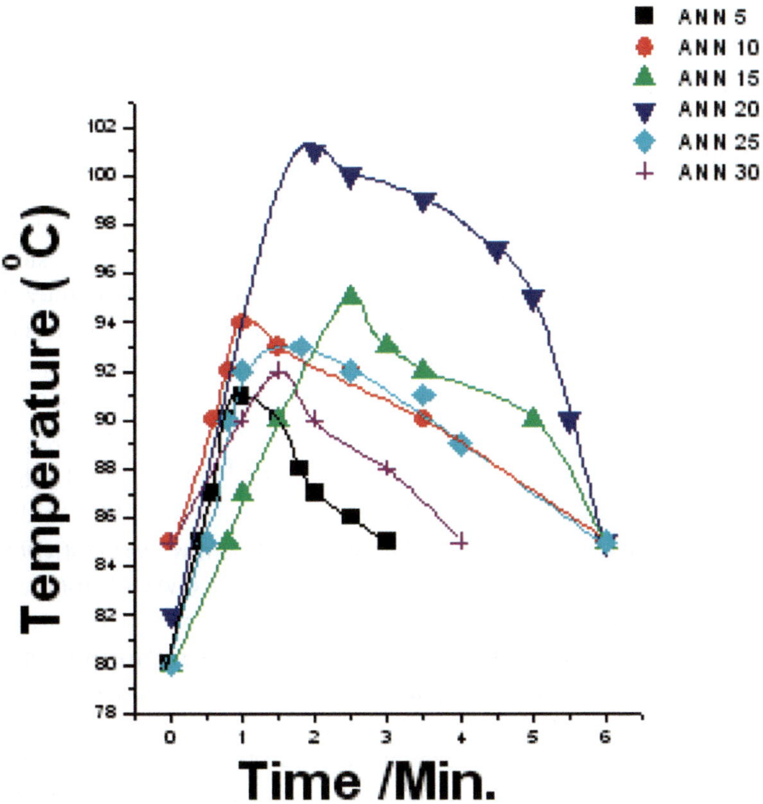

Figure 10.2 Time-temperature plot of various mixtures of TEOS-ANN at ~80°C.

In isothermal reaction conditions, various mixtures of TEOS and ANN reacted vigorously at ~80°C in a water bath, with the evolution of exothermic energy as heat, and subsequently formed a noncrystalline aluminosilicate precursor phase.

Jones and Fischbach (1988) noted that no reaction was apparent when a mixture of water and TEOS was kept for several hours. But when it was acidified with HNO_3, the reaction started showing an exotherm for about 15 sec. and thereafter the solution became clear. The exothermicity depended upon the effects of the water:TEOS ratio and the HNO_3:TEOS ratio used in the reaction. In this study, the maximum heat evolution occurs in the gel marked ISG72, which corresponds to the 3:2 mullite composition. Slow hydrolysis (SH) gels also formed intermediate noncrystalline aluminosilicate precursor phases during heating in the temperature range of 400°C–900°C other than being present as an intimate mixture of amorphous silicon and aluminum oxides. These phases exhibited 980°C exotherm during DTA study. The peak height depends on the alumina content of the gel, and it was maximum for the gel of composition Al:Si = 3:1 (Fig. 10.3).

Thus, the observations in the two cases of synthesis of gels are analogous. Accordingly, the aluminosilicate phase analogous in composition to mullite is most stable. Therefore, the results derived from the isothermal heating experiment and DTA study of the mixture of two components concur and definitely suggest the formation of a binary noncrystalline aluminosilicate phase.

When an organic silicon compound is used in the presence of Al salt (Chapter 3), the predicted gelation reactions in three categories of mullite precursor forming processes where TEOS was used in the presence of ANN are given next.

Gelation reaction: DTA study of a TEOS-anhydrous $Al(NO_3)_3$ mixture showed that TEOS might have reacted with aluminum nitrate at ~80°C (Fig. 10.3). This reaction is exothermic in nature. These data corroborate the observation of multiple small exothermic peaks within the large endotherm in DTA. With an increase in the wt % of Al_2O_3 in Al_2O_3-SiO_2 gels under study, the area of the time-temperature curve during heat evolution in the temperature range of 80°C–90°C first increased slowly in batches 2 to 3 and then attained more or less a constant temperature of 80°C in batches 4 and 5. Finally, it decreased even with an increase in the wt % of Al_2O_3. This result indicates that the interaction between TEOS (liquid) and melted aluminum nitrate crystals likely shows the maximum heat evolution with the Al_2O_3:SiO_2 ratio in the range of 3:2 to 2:1 due to the maximum increase in temperature.

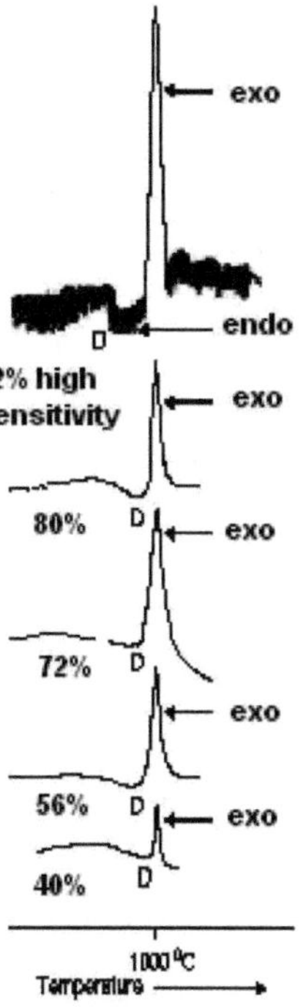

Figure 10.3 DTA traces of mullite gels of different alumina contents.

In this case of gel synthesis, TEOS is not allowed to hydrolyze; instead, a warm concentrated solution of aluminum nitrate is added to it at ~80°C. So the possibility of hydrolysis of aluminum nitrate into $[Al(H_2O)_6]^{+3}$ is less likely to occur. Most probably, silanol groups are not formed and likely a 2D polymeric chain or a 3D Si–O–Si network is also not generated. Therefore, no question Al^{3+} substitution occurs due to want of sufficient time. Considering that large heat evolution started just after mixing of the two reactants around 80°C, it may be conceived that the concentrated form of nitrate radicals of aluminum nitrate oxidizes ethoxy groups of TEOS with the liberation of nitrous fumes and generates linkages between Si and Al atoms. The heat of reaction of TEOS and ANN may be responsible for release of the large amount of energy. Therefore, this dried precursor is a noncrystalline aluminosilicate other than free oxide mixture, as assumed earlier. This study indicates that bonding between two components occurs during the gelation process. As already discussed above, Si–O–Al(IV) bond formation gradually occurs with increasing temperature while the xerogel is heated.

Komarneni and Roy (1986) showed by an MAS-NMR study that considerable substitution of Al for Si occurs in the Si–O–Si chain in the single-phase gel derived from TEOS and ANN. They noted resonances around 55 and 0 ppm from $Al(H_2O)_6^{+3}$ in the single-phase gel and the absence of the same in diphasic gels. They further showed that the tetrahedral coordination of Al increased regularly with a decrease in the Al:Si atom ratio from 4:1 to 1:4 (Komarneni & Roy, 1986, Fig. 1) and confirmed the homogeneity of the two oxides.

Yasumori et al. (1990) investigated a derived material (S20) of the composition $20Al_2O_3 \cdot 80SiO_2$ (mol %) prepared from silicon ethylester and a melt-quenched glass (M 20). Al spectra of those samples showed peaks between 40 and 70 ppm and were assigned to the tetrahedral network structure and the peak at 0 ppm was assigned to octahedral Al. It is interesting to note that the major portion of Al in the xerogel was found to have fourfold coordination. A similar observation was noted in M 20 quenched glass, where in addition to the tetrahedral peak, a pentahedral Al peak was noted along with an insignificant octahedral peak. This observation is obvious for the already made glass phase. This proves that Al–O–Si bonding takes place during gelation itself in the case of the S20 sample and some other examples.

Example 1: Gelation of anhydrous ANN and TEOS

In the case of G-31 synthesized by Chakraborty and Ghosh (1988), external water was not used and moreover a major part of nine molecules of water of crystallization of ANN was removed by a repeated solution-evaporation technique. In such a case, where both external water and the water of crystallization of ANN are almost absent, the pH of the alcoholic medium is found to be higher. Secondly, a major amount of NO_3 groups is eliminated. According to an NMR study by Bhattacharya et al. (1996), it is assumed that monomeric hexa-aquo Al^{+3} might be formed during the repeated solution-evaporation process of ANN and then partially hydrolyzed dimeric species $[Al_2(mu\text{-}OH)_2(OH)n(H_2O)8\text{-}n]$ $(4\text{-}n)+$ might be generated. Accordingly, the ionic reaction between $Al[(H_2O)_6]^{+3}$ with silicon hydroxide monomers should be more predominant due to an increase in reactivity. As a consequence, the probability of formation of a spinel phase would be less.

In the case of gel G-31, it is expected that owing to this restricted rate of hydrolysis of TEOS, large clusters of $Si(OH)_4$ are not formed by extensive polymerization nor chain-like polymeric species. The degree of polymerization is reduced to a large extent due to the absence of added water of hydrolysis and also due to minimization of water of crystallization of ANN. Accordingly, no such linear or spherical polymers of $Si(OH)_4$ are generated by the addition of absolute alcohol only, as conjectured by Sakka and Kamiya (1982). Instead, simple molecules of ethoxy silicate hydroxyls–aluminum hydroxyl complex may form during interactions with TEOS and anhydrous ANN in alcoholic solution at ~80°C in an isothermal heat treatment process, which on further heating completely transformed into mullite at the 980°C exotherm.

The 1st models of the gel have been given schematically earlier in the chapter, as Eqs. 10.1 and 10.2. The 2nd model of the gel particle consists of an ethoxy silicate–$(OH)^{-1}$–aluminum–$(OH)^{-1}$ complex (10.20).

Example 2: Gelation of ANN and TEOS

For monophasic gels, we have SHII of Haque (2000) ($9H_2O$ was the source of water and there was no external water source; hydrolysis was done in absolute alcohol), the single-phase gel by Hoffman et al. (1984), SH gel of Okada and Otsuka (1986), and 2WC and 2W precursors of Li and Thomson (1991). A mullite gel was synthesized by Hsi et al. (1989) by dissolving ANN and TEOS and we have Gel A of Chakraborty (1994b).

When SH of TEOS is carried out in a water bath for a long amount of time, TEOS mostly hydrolyzes to monomeric species. Substitution of Si^{+4} for Al^{+3} likely occurs, resulting in the gelation of aluminosilicate hydrate oven during drying. A fraction of monomeric silicic acid may get the chance to polymerize to develop some small chains due to the presence of water of hydroxyls that comes from the $9H_2O$ molecules of ANN.

The effect of pH when 9 mol of water of crystallization is the only source of water: The first step of the gelation reaction will be the hydrolysis of TEOS in the presence of a hydrolyzing agent/acid as the catalyst. The second step is the polymerization and/or condensation of $Si(OH)_4$ monomers and/or partially hydrolyzed TEOS. These two reactions may be dependent on the pH and the quantity of water used during gelation.

When the pH of the solution is maintained between 3 and 4.5 and excess of water is used, possibly a cluster of 3D chain of Si–O–Si may develop. However, in the absence of free moisture, when the water of crystallization of aluminum nitrate is the only source and hydrolysis of TEOS is restricted, as conjectured by Konard et al. (1929), a 2D chain of Si–O–Si might develop. Now, when an aluminum nitrate solution is added to the TEOS solution in either of the two conditions above, there may be some interaction between the two and that may lead to the substitution of Al^{3+} in the Si–O–Si network. The resulting gel was termed as "slow hydrolysis gel" by Okada and Otsuka (1990), "Gel A" by Chakraborty (1994), and "SHII" by Haque (2000).

The question is, to what extent Al^{3+} replaces Si^{4+} in the tetrahedral position. The usual rule is that Al_2O_3 to the extent of 30% by weight of SiO_2 may be substituted in the network and form aluminosilicate. The amount of substitution, of course, is affected by the acidity of the reaction mixture. It would be presumed that residual Al_2O_3 is present a in free state and mixed uniformly with aluminosilicate. Various researchers called this state of mixture as homogeneous mixing of two oxide components. There is a possibility that amorphous Al_2O_3 might diffuse into the aluminosilicate network during the drying process and on continued heating.

In this environment, it may be suggested that during evaporation on a water bath at ~60°C for maybe 1–2 weeks, ANN may actually interact very slowly to some extent with monomers like $Si(OEt)_3OH$, $Si(OEt)_2(OH)_2$, and $Si(OH)_4$ particles formed just after the hydrolysis of TEOS, with the formation of a complex salt by the substitution of at least some of tetrahedral Si^{+4} ions by Al^{+3} of aluminum nitrate. It is further presumed that a complex salt of aluminum nitrate–hydroxy/ethoxy silicate hydrate is formed, in which the nitrate $(NO_3)^{-1}$ group may be present as counterions. The dried complex gradually eliminates hydroxyl, then ethoxy groups through oxidation during heating, and lastly nitrate groups to form a noncrystalline aluminosilicate precursor phase. These observations have been supported by DTA studies wherein two endotherms have been seen. The IR study of Fukuoka et al. (1993), Tsuchiya (1989), Li and Thomson (1991), and Haque (2000) showed the gradual removal of both water and nitrate. This aluminosilicate finally decomposes before the temperature reaches 980°C, with the elimination of residual –OH groups at the temperature marked "D" in DTA, and then exhibits a very strong 980°C exotherm with the formation of both spinel (as minor) phase and weakly crystalline mullite (as major) phase, as evidenced earlier by X-ray analysis. The spinel phase, thereafter, transforms into orthorhombic mullite (o-mullite) at 1250°C.

Therefore, the exhibition of 980°C exotherm indirectly confirms that the complex salt formation between two components might have occurred during the gelation process, which subsequently decomposes during the ongoing heating process to form an aluminosilicate precursor phase. The prerequisites of TEOS hydrolysis is a suitable solvent, for example, ethyl alcohol and sufficient time and temperature so as to form a complex of aluminum nitrate–hydroxy/ethoxy silicate hydrate.

MAS-NMR data of some researchers are cited here to resolve the state of mixing of two oxides. Schneider et al. (1993) showed the shifting of Si NMR from –85 to –90 ppm during heating of their Type 1 precursor. This is the silica-poorer region of the aluminosilicate precursor phase. Similar data are also been observed in the NMR spectra of the samples marked A and E by Jaymes et al. (1996).

Fukuoka et al. (1993) conducted IR and MAS-NMR studies jointly to reveal the behavior of aluminum in the formation of a glass network. They showed that the change in the IR absorption intensity ratio of NO_3^{-1} to Si–O–Si bond showed a similar tendency to that in the number of aluminum atoms excluded from the gel made from TEOS and ANN into water, as previously done by Tsuchiya et al. (1989). Both ratios decreased with an increase in temperature (Fig. 13.2). The decrease of both was closely related to the increase in the Al(IV) ratio. They conjectured that active aluminum is generated by the decomposition of ANN at 300°C–400°C and that aluminum is incorporated into the silica gel structure and bonding develops.

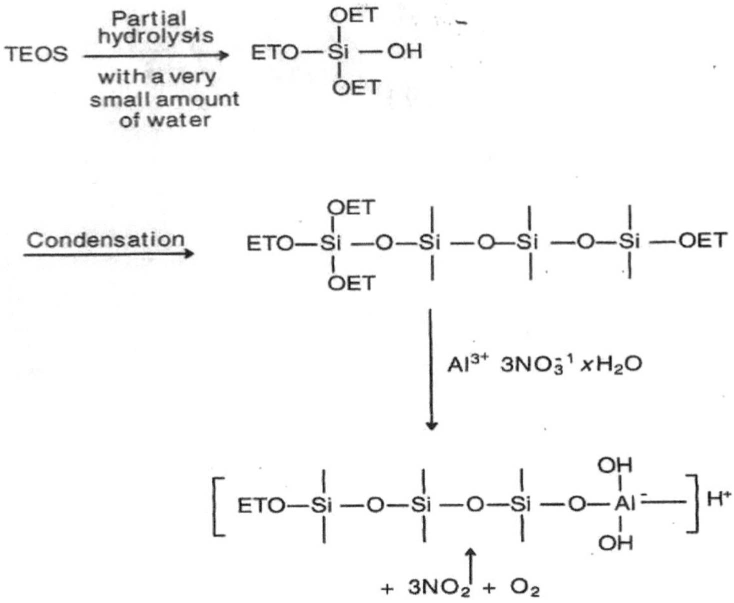

(10.21)

Equation 10.21 schematically represents the 3rd model of the gel, called Gel A, containing unidirectional short chain (Chakraborty, 1994b). The 1st and 2nd models are shown earlier in the chapter.

Gel A showed two types of mullite formation at the 980°C exotherm. Therefore, Gel A may be similar to the polymeric gel synthesized by Yoldas and Partlow (1988). They observed two exotherms, one at 980°C and one at 1250°C. It is argued that the appearance of the latter peak is indicative of the formation of the Al- Si spinel in addition to the usual tetragonal mullite (t-mullite) formation at the 980°C exotherm. It is also presumed that the small amount of water available during the gelling of Gel A and in the preparation of the polymeric gel by Pask et al. (1987) was responsible for the crystallization of the spinel phase. So segregation may eventually occur during heating, and the mullitization took place from the two-phase system by a two-stage conversion process, as shown by Li and Thomson (1991) in their dynamic X-ray diffraction diagram. Thus, when a small amount of water is present, the gel powder has the equivalent oxide composition of mullite and is not essentially rearranged into a mullite lattice. There may be some possibility of interdiffusion into phase-separated phases prior to mullitization. Thus the resultant hydrolyzed product of TEOS in this case where the availability of moisture content is low will form a chain-like polymer instead of a giant 3D network of $Si(OH)_4$. The interacted product of the polymeric chain with Al^{+3} may be as follows:

The formation of a chain-like polymer is also suggested by Sakka and Kamiya (1982). Schmidt et al. (1984) suggested the formation of many types of intermediate forms of compounds in the alkoxysilane-water hydrolysis process that may interact with ANN in various ways and develop Si–O–Al linkages. The schematic model of Gel A shows that only a few aluminum atoms are tetracoordinated, which is supported by the data of the coordination measurement by Leonard et al. (1971) in their X-ray fluorescence spectroscopy and radial electron density distribution studies of amorphous silicoaluminous.

Example 3: Gelation of ANN in excess water and TEOS

SHIII gel by Haque (2000), G-150 by Chakraborty (1988), and gel made with more ethanol by Hoffman et al. (1984) are cited. A possible explanation of the ANN-TEOS reaction on heating in a water bath with excess of water is first presented.

TEOS is first hydrolyzed to develop silanol groups as per reaction 10.22, which may be because Sakka and Kamiya (1982) used a large quantity of water and this may increase the hydroxyl content of the gel.

$$\begin{aligned} Si(OR)_4 &\xrightarrow{H_2O} \underset{+\,ROH}{Si(OR)_3(OH)} \xrightarrow{H_2O} \underset{+\,ROH}{Si(OR)_2(OH)_2} \xrightarrow{H_2O} \underset{+\,ROH}{Si(OR)(OH)_3} - \\ &\longrightarrow \underset{+ROH}{Si(OH)_4} \longrightarrow \text{Bulk } Si(OH)_4 \text{ gel} \qquad (10.22) \end{aligned}$$

These monomeric species may react with other silicon hydroxyl (HO–Si=) groups and thereafter polymerize or interact with another TEOS molecule to develop Si–O–Si linkages. Thus, when TEOS is used as the SiO_2 source, there is every possibility that hydrolysis reactions and self-polymerization reactions will be more rapid as the concentration of water increases, as discussed in Chapter 2. However, such type of bonding is unwanted in the context of direct mullitization. Aluminum nitrate may react with a monohydroxide, that is, $Si(OR)_3(OH)$, dihydroxide, trihydroxide, or $Si(OH)_4$ molecules. It is assumed that the reactivity between them gradually decreases as the extent of TEOS hydrolysis increases. Therefore, in the presence of water/alcohol, different types of intermediate products may form, for example, $[(OR)_3Si–O–Si(OR)_3]$, $[(OR)_3Si–O–Al(NO_3)_2]$, hydroxides of –Si–O–Si–O–, hydroxides of –Si–O–Si–O–Al–O–, 2D or 3D chains, 3D cross-linked polymeric clusters, etc. These colloidal particles ultimately develop into a gelatinous mass when allowed to set. Again, when the conditions of the intermediate reaction are changed and especially the temperature is increased from 60°C to 80°C, then either the interaction between the newly formed $(OR)_3Si(OH)$ and $Al(NO_3)_3$ is increased or a direct reaction between the free components may take place in the partially acidic medium, as follows:

$$\underset{\text{ANN}}{Si(OET)_4 + Al(NO_3)\cdot 9H_2O} \xrightarrow{\sim 80^\circ C} \underset{+ NO_2 + CO_2 + H_2O}{(ETO)_3\text{–}Si\text{–}O\text{–}Al(NO_3)_2} \xrightarrow[\text{reaction}]{\text{Further}}$$

$$\underset{+ NO_2 + CO_2 + H_2O}{(ETO)_2Si\text{–}[O\text{–}Al(NO_3)_2]_2} \xrightarrow[\text{reaction}]{\text{Further}} \underset{+ NO_2 + CO_2 + H_2O}{(ETO)Si\text{–}[O\text{–}Al(NO_3)_2]_3} \xrightarrow[\text{reaction}]{\text{Further}}$$

$$Si\text{–}[O\text{–}Al(NO_3)_2]_4 + NO_2 + CO_2 + H_2O \qquad (10.23)$$

In this case, it is probable that aging accelerates the polymerization of the silica unit to a 3D colloidal gel, resulting in crystallization to spinel of more quantity at the cost of mullitization at 1000°C. An alternative explanation may also be visualized as follows: It may be assumed that colloidal polymeric molecules of $Si(OH)_4$ derived from the hydrolysis of TEOS in the environment of excess water absorb Al^{+3} ions from the aqueous solution. In this case, a partial substitution absorbing Al^{+3} ions from the solution of aluminum in the siloxane chain may occur, which will cause the formation of aluminosilicate hydrate and possibly a more complex aluminosilicate hydroxy polymeric gel particle. The model of this particle could be contemplated as a 3D chain of the type shown in reaction 10.24.

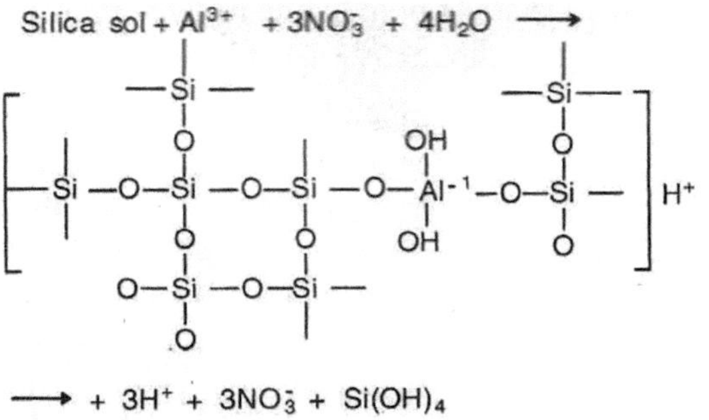

(10.24)

Equation 10.24 schematically represents the 4th model of the gel, called G-150, containing a 3D cross-linked chain (Chakraborty, 1994b)

SAXS, light scattering, and molecular weight determination indicated the steps of polymerization. In the early stage, the polymerization showed linear or slightly branched chains as the polymeric species. In the second stage, cross-linking of these chains occurred through diffusion and collisions between chains, which eventually led to gel formation (Bechtold, 1955; Brinker et al., 1984).

Sinko and Mezei (2001) gave special attention to preparing high-alumina-content aluminosilicate gel by the energy-consuming sol-gel method, which is more than that achieved by the traditional melting process. Al alkoxides are quite difficult to handle owing to their rapid hydrolysis and low solubility in water and alcohol, which may lead to phase separation. Inorganic aluminum salts are inexpensive, but anions of the salts need to be removed. The sol-formation step is time consuming. Several solvents are used, for example, ethanol, methanol, 1-propanol, 2-propanol, 1-butanol, and ethylene glycol.

Al incorporation: The amount of bonded Al atoms was considerably lower in the chloride-containing gel than in the nitrate-containing gel. Nitrate ions escaped during the gelation process at 80°C and thereafter on heat treatment. Contrarily, chloride ions remained in the gel. ^{27}Al MAS-NMR spectra of the gel prepared by $AlCl_3$ showed only one nonbonded sharp peak at −1.2 ppm. It is characteristic of octahedral Al ions. The area under this peak amounts to 98%–99%. The spectra from the gel prepared from aluminum nitrate showed three peaks. The broad peak, at 0 ppm, is characteristic of AlO_6 octahedra, attributed to the hydrated Al^{+3}; the area amounts to 67.5%. Al resonance at ~4 ppm is due to dimers or polymeric Al species; the area amounts to 25.5%. The third peak, at 58 ppm (7% of the alumina content), belongs to AlO_4 groups, indicating Al–O–Si bond formation.

- In SAXS, as shown in the strong and broad distortion of the scattering curve B (chloride) at ~0.17 $Å^{-1}$, is due to nonbonded Al content and shows the beginning of phase separation. Contrarily, the scattering curve A (nitrate) is slightly distorted and shows no phase separation (see Sinko & Mezei, 2001, Fig. 2).
- Effect of the solvent: The SAXS curve showed that the gel prepared in 1-propanol is most suitable for gelation and Al incorporation. The highest aluminum incorporation is found to be 0.53 mol ratio of bonded Al/(Si+Al).

10.3 Mullite Gel by Polymeric and Colloidal Methods Using TEOS and AlOBu

In these cases, both components organic compounds (Chapter 8). Water/alcohol is used for the hydrolysis of AlOBu. Partially hydrolyzed TEOS is used, which also contains water to some extent (Yoldas 1992, 1980). The gelation process of AlOBu/$Al(O^iPr)_3$ is as follows:

Hydrolysis

$$\text{Aluminum alkoxide} \rightarrow \text{AlO(OH)} \quad (10.25)$$

(aluminum monohydroxide)

This monohydroxide further tends to convert into trihydroxide on addition of excess water, as follows:

$$\text{AO (OR)}_3 + 2\text{H}_2\text{O} \rightarrow \text{AlO(OH)} + 3\text{R(OH)}$$

$$\text{AlO(OH)} + \text{H}_2\text{O} \rightarrow \text{Al(OH)}_3 \quad (10.26)$$

Other reactions

An aluminum hydroxyl bond (in a dilute condition, self-condensation is mostly prevented) is first synthesized and then allowed to react with alkoxy bonds of silicon as follows:

$$\text{Aqueous alumina sol} + [\text{Si}(\text{OC}_2\text{H}_5)_4] \xrightarrow[\text{polymerization}]{\text{Chemical}} \text{---Al---O---Si---} \tag{10.27}$$

Alkyl groups of silicon polymerization

$$[\text{Si}(\text{OC}_2\text{H}_5)_4]$$

Partially hydrolyzed $Si(OR)_4$ is synthesized first and then allowed to react with unhydrolyzed aluminum alkoxide.

$$(\text{RO})_2\,\text{Al---OR} + \text{HO---Si---(OR)}_3 \rightarrow$$
$$(\text{RO})_2\,\text{Al---O---Si---(OR)}_3 + \text{R(OH)} \tag{10.28}$$

In the following process, an aqueous colloidal AlO(OH) sol is first prepared and allowed to react with $Si(OR)_4$ to create chemically bonded silica around and between the alumina colloids.

$$(\text{RO})_3\,\text{Si–OR} + \text{HO---\{---AlO(OH)}_{(n-1)n}\}_n \rightarrow$$
$$(\text{RO})_3\,\text{Si---O---\{---AlO(OH)}_{(n-1)n})_n + \text{ROH} \tag{10.29}$$

Polycondensation reactions may also take place as side-reactions.

$$(\text{RO})_2\,\text{Al---OH} + (\text{RO})\,\text{Al–(OR)}_2 \rightarrow (\text{RO})_2\,\text{Al---O---Al–(OR)}_2 \tag{10.30}$$

In hydrolysis and chemical reactions, water is an essential parameter. Without the supply of water, the said two reactions can't proceed. Even in a humid atmosphere, hydrolysis-cum-interactions take place meaningfully. It is conjectured that a chemical reaction between two components becomes feasible in the presence of a trace quantity of atmospheric moisture. It is because the small amount of water is regenerated after the chemical polymerization of the two hydroxides. Therefore, once it starts, it will perpetuate repeatedly till completion of the reaction. It is time dependent. According to Yoldas and Partlow (1988), alkoxide mixtures formed a homogeneous oxide network when the chemical encounter rates with water were reduced significantly. In this process, AlOBu and TEOS were taken in an open flask, diluted in a nonaqueous solvent (5 wt % equivalent oxide), and stirred for 3 months to allow hydrolysis at a very slow rate with humid air in a methodical manner.

In humid atmosphere

$$\text{Al(OC}_4\text{H}_9)_3\ \text{Si(OC}_2\text{H}_5)_4 + \text{dry ethyl alcohol} \rightarrow$$
$$= \text{Al–O–Si}\equiv + \text{C}_4\text{H}_9\text{OH}\ \text{C}_2\text{H}_5\text{OH} \tag{10.31}$$

Stirring

Due to retardation of the hydrolysis process, the resultant copolymeric gel particle probably does not proceed further to develop a long-chain, cyclic-chain, or 3D cross-linked condensation product gel. Ideally, it simply produces an amorphous mullite gel of high-homogeneity network bonding.

Reaction 10.31 shows the 5th model of the polymeric gelation reaction as per Yoldas and Partlow (1988).

Nature of the precursor phase: In a water-free approach, G-152(ii) was synthesized by Chakraborty (1996a) as per the procedure of the molecular gel marked C made by Yoldas and Partlow (1988). It crystallized sharply to mullite with the exhibition of a pronounced exotherm at 980°C. It conformed with the DTA finding of the latter authors. The molecular gel also exhibited a small exotherm at 1270°C. To ascertain the cause of this exotherm, XRD analysis of the 980°C heated gel was conducted, which showed the crystallization of mullite (as the major phase), but it still contained a spinel phase in a minor amount. Yoldas and Partlow (1988) did not present any XRD pattern of it. The exhibition of the 980°C exotherm was related to both forms of mullite, and the 1270°C exotherm was explained as due to the subsequent polymorphic transition of spinel to o-mullite. The conclusion derived from these experimentations is that spinel formation could not be avoided even with the 5th model of the gelation process. Schneider et al. (1993) also noted a sharp peak at ~980°C but no peak near 1250°C in their Type I precursor. At ~1000°C, mullite appeared either as the only crystalline phase or together with a minor amount of poorly crystallized Al_2O_3 spinel (Fig. 2a) of them. The author's own observation is that the identification of a small amount of spinel phase is rather difficult and one could obviously skip its presence in an X-ray diffractogram chart unless the attached diffractometer had been operated at its maximum sensitivity. Quite similar is the observation that the thermal analysis equipment and its chart drive have to be run at their highest sensitivity to note the occurrence of the small broad peak at the second exotherm.

Under these circumstances, one can predict the presence of the polymeric gel synthesized even using a water-free approach; the moisture available in the humid air is sufficient to produce some quantity of other polymeric gel besides producing aluminosilicate copolymeric gel.

This view has been verified by the most significant TEM study of primary particles of Type I precursor by Schmucker and Schneider (1999). They showed the fractured surface of the dried gel (150°C) by SEM, which revealed the 20–50 nm sized spherulites with a uniform size distribution. the TEM study confirmed the existence of previously determined particles. In a higher magnification, they detected addition particles in the 5–10 nm range, as shown by Schneider et al. (1999) in the comparative Figs. 3 and 4. This observation indicates the presence of bimodular polymeric gel phases formed in the original Type I gel.

Taylor and Holland (1993) reported a high Pm/Ps value for a particular couple of components under a specific stirring period during the gelation process. Below and after this time period, t-mullite abruptly decreases. We have to find out the reasons for these two events. The major parameters arrived at by discussions are the following: (i) water used for hydrolysis, (ii) catalyst, that is, pH, (iii) the period of hydrolysis/chemical reaction process that leads to gelation, and (iv) the temperature of the reaction bath. Other examples are follows:

Sample A: Orefice and Vasconcelos (1997) showed the effect of pH and temperature of hydrolysis on the gelation of aluminum isopropoxide (AIP or $Al(O^iPr)_3$), which is as follows:

At 90°C, the gel time was rapidly reduced by decreasing pH. XRD analysis showed the formation of pseudoboehmite and thus the sol was pseudoboehmite colloids that displayed a plate- or fiber-like shape.

Contrarily, at room temperature, the gelation time was increased with the introduction of an acid, which resulted in the formation of spherical colloids that were amorphous in nature.

Thus, monophasic and diphasic mullite gels were produced by changing the temperature of hydrolysis of a mixture of AIP and TEOS. Ultimately, these led to the formation of mullite in the monophasic gel and Al-Si spinel in the diphasic gel on heating at ~1100°C.

Sample B: In all the cases below, Al-Si spinel (minor) is developed in addition to t-mullite, for example, the polymeric gel by Yoldas (1992). It was prepared at a neutral pH but use of water was necessary for the hydrolysis of TEOS. Even in the prehydrolysis of TEOS, water was invariably added (Chapter 2).

Sample C (gelation reaction of a mixture of two alkoxides): Addition of water is avoided, and hydrolysis occurs because of atmospheric humidity, for example, in the case of sample C prepared by Yoldas and Partlow (1988). The same procedure was repeated by Schneider et al. (1993) for Type I precursor and by Haque (2000) for SHI. In the Type I gel (NH) of Schneider et al. (1993), water was available because the mixed alkoxides in isopropanol were kept in a glove box for ~14 days and were in contact with air. It is expected that the hydrolysis process was very slow because of restricted supply of moisture from the air when water is the only predominant variable and pH effect is not considered. The question is, what is the gelation reaction in a water-free approach? In the Type III gel, hydrolysis of the double alkoxides TEOS and AlOBu was carried out with isopropanol and with high water content at pH < 10 (HB10), that is, two variables water and pH were chosen

The Type I precursor likely follows the 5th model of the polymeric gelation reaction as per Yoldas and Partlow (1988). However, the Type III precursor may follow other than the 5th model as discussed, since it was synthesized by using full prehydrolysis of TEOS as per Eq. 10.32.

pH ≥ 10

$$Si(OR)_4 + H_2O \text{ (excess)} \rightarrow Si(OH)_4 \text{ (monomeric)} + H_2O \qquad (10.32)$$

After the hydrolysis of TEOS with excess water, AlOBu diluted with isopropanol was added and hydrolyzed in a less basic condition (pH = 10).

A high concentration of water during hydrolysis might cause the formation of many intermediates, as suggested by Schmidt et al. (1984). Secondly, there is every possibility that hydrolysis reactions and self-polymerization reactions will be more rapid instead of monomeric, as assumed when TEOS is used as the SiO_2 source by Sakka and Kamiya (1982). Different types of intermediate products may form, for example, $[(OR)_3Si–O–Si(OR)_3]$, $[(OR)_3Si–O–Al(NO_3)_2]$, hydroxides of –Si–O–Si–O–, hydroxides of –Si–O–Si–O–Al–O–, 2D or 3D chains, and 3D cross-linked polymeric clusters. The resulting Al_2O_3-SiO_2 gels would consist of some mixtures of polymeric gel particles of different chain lengths and clusters instead of single polymeric gel particles.

The 6th model of the mixed gelation reaction as per Yoldas and Partlow (1988).

Sample D: Mazdiyasni and Brown (1972) used $Al(OC_3H_7)_3$ and $Si(OC_3H_7)_4$. Prochazka and Klug (1983) used 204 g $Al(O^iPr)_3$ dissolved in cyclohexane and 67 g of ethyl silicate also in the same solvent. To this mixture, 55 g of water per mol of each alkoxide was added. Somiya (1985) mixed $Al(O^iPr)_3$ and TEOS. Suzuki et al. (1990) used TEOS by partially hydrolyzing it with acidic water and alcohol, refluxing it at 70°C for 50 h, and then mixing and reacting it with a refluxed solution of AIP with isobutyl alcohol at 95°C for 6 h. Sakurai et al. (1988) prepared mullite powders from TEOS and $Al(O^iPr)_3$. Hirata et al. (1985, 1989) took a mixture of 37–62 g $Al(OPr)_3$ and 17–39 g TEOS corresponding to $xAl_2O_3 \cdot 2SiO_2$ ($x = 1, 2, 3, 4$) dissolved in benzene and refluxed it at 81°C–83°C) under dried N_2 flow. It was subsequently hydrolyzed with an equivalent amount of water (16–18 g) to form silica and $Al(OH)_3$. Low and McPherson (1989) used silicon tetraethoxide and AIP. Schneider et al. (1992) took TEOS and aluminum butyrate for the synthesis of a mullite precursor marked SGM by the sol-gel method. They homogenized the starting mixture with isopropanol and then hydrolyzed with the drop-by-drop addition of water, which produced a homogeneous white gel. Taylor and Holland (1993) used TEOS and ($Al(O^iPr)_3$). Hamano et al. (1985) used $Al(O^iPr)_3$ and TMOS. Paulick et al. (1987) used silicon and aluminum alkoxides. Mitachi et al. (1990) used ($Al(Oi\text{-}C_3H_7)_3$) and TMOS ($Si(OCH_3)_4$), and Sales and Alarcon (1996) used TEOS and ALOBu by the prehydrolysis technique. In the above cases of hydrolysis through the double-alkoxide system, the formation of the Al-Si spinel phase was a must and it occurred together with mullite. Accordingly, the prevention of alumina-silica segregation is especially problematic in dual-alkoxide systems. Hirata et al. (1989) noted Si–O–Al peaks in their "as-prepared gel" by IR spectroscopy. This observation indicates that aluminum hydroxide and silicon hydroxides might have reacted during condensation processes.

In all these cases of using the double-alkoxide route, the addition of acidulated water used for hydrolysis would certainly lead to homogeneity. And such things will be aggravated in a basic medium of hydrolysis using ammonia.

10.4 Mullite Gel by Coprecipitating Gel Methods

10.4.1 Both Water-Soluble Components Precipitated by NH_4OH

$AlCl_3$ and $SiCl_4$ were used by Horte and Wiegman (1956), Demediuk and Cole (1958), Crofts and Marshall (1967), and McGee and Wirkus (1972), as discussed in Chapter 7.

10.4.2 Water-Soluble Al Salt Precipitated by NH_4OH in the Presence of Organic Silica

The pH for this reaction is below 8 (Chapter 7). Examples are the colloidal gel by Li and Thomson (1991), the sample marked as CP-6 by Haque (2000), and the RH gel by Okada and Otsuka (1986). Huang et al. (1997) used $AlCl_3 \cdot 6H_2O$ and prehydrolyzed TEOS. Hyatt and Bansal (1990) processed comixing of ANN and TEOS in absolute alcohol.

10.4.3 Coprecipitation of Organic Al and Si Compounds

The reactions were carried out at basic pH (Chapter 7).

- The salts are precipitated using NH_4OH. Type II (HB13) by Schneider et al. (1993) was a colloidal gel, as was that by Yoldas (1988).
- Both organic salts are precipitated by using urea. Yamada and Kimura (1962) used silicon ethylester and aluminum ethylester to develop a colloidal gel. Orefice and Vasconcelos (1997) used $Al(O^iPr)_3$ and TEOS. They chose a higher gelation temperature of ~80°C.

10.5 Mullite Gel by the Diphasic Method

10.5.1 General Synthesis of a Diphasic Gel

These are discussed in Chapter 8. Hamano et al. (1985) mixed boehmite and silica sols. Ismail et al. (1986, 1987) used aqueous silica and boehmite. Huling and Messing (1989, 1990) used silica sol and boehmite sol. Klaussen (1990) synthesized mullite gel using a pseudoboehmite colloidal sol. Hyatt and Bansal (1990) prepared a diphasic gel by using AlO(OH) and Ludox. Pach et al. (1995) synthesized a diphasic $3Al_2O_3{\cdot}2SiO_2$ mullite gel using boehmite (10 nm) and silica sol. Wu et al. (1993) used submicrometer alumina powder and Ludox, Sugita et al. (1998) prepared mullite powder by a process similar to that used by Mizuno et al. (1990) and used a mixture of fume silica (Aerosil 200) in an aqueous solution of aluminum sulfate. Padmaja et al. (1998) used TEOS and boehmite with and without dehydroxylation at 400°C. Shinohara et al. (1986) mixed a colloidal boehmite suspension and a partially hydrolyzed TEOS solution. Jin et al. (2002) used prehydrolyzed TEOS and $ACl_3{\cdot}6H_2O$. Sonuparlak (1988) used colloidal boehmite suspensions and TEOS. Wei and Halloran (1988) synthesized a translucent white type of diphasic gel from TEOS and colloidal pseudoboehmite sol. Shiga et al. (1991) used urea as the precipitant from the colloidal silica and aluminum sulfate solution. Lee et al. (2002, 2003, 2004) and Kim et al. (2003) used ANN and colloidal silica in three pH conditions.

10.5.2 In situ Synthesis of a Diphasic Gel

The reaction in the presence of organic silica is carried out at a pH higher than 8 (see Chapter 8).

10.5.2.1 NH_4OH added to a mixture of TEOS and ANN

Some examples are CP-9 by Haque (2000), TEOS and ANN at pH > 10, G49 by Chakraborty (1994a). Hsi et al. (1989) developed a diphasic gel from boehmite and made a cross-linked 3D bulk gel.

In this basic condition, the rates of hydrolysis of both components are very fast, as discussed in Chapter 2 by various authors.

Hydrolysis of TEOS may generate highly condensed polymers during the gelation process, as suggested by Brinker et al. (1982). It may be spherical at a high-pH-catalyzed solution as per Pope and Mackenzie (1986). Keefer (1984) suggested highly cross-linked polymers in the base-catalyzed hydrolysis process of TEOS.

In the case of the precipitation of the Al component by the use of ammonia, as in gel CP-9, it is presumed that freshly formed aluminum hydroxide soon crystallizes as pseudoboehmite, which adsorbs TEOS at its surface during gelation. During warming, TEOS then hydrolyzes rapidly in the presence of ammonia and the bulk gel gets coated on the AlO(OH) gel. It is obvious that the phase transformation of such mixed hydroxides should follow the independent crystallization behaviors of their components. However, only γ-alumina was observed and cristobalite did not appear. It developed quite an abundant quantity of mullite at >1250°C and, thereafter, it increased slowly up to 1400°C. This can be explained as due to the development of an intermediate aluminosilicate phase prior to mullite nucleation. Its formation generally takes place during two major steps: (i) during dehydration and (ii) during calcination of the diphasic gel. It may be assumed that during gelation, a part of the newly formed AlO(OH) gel particle interacts with $Si(OH)_4$ gel due to coulombic attraction of two oppositely charge gel particles and then coalesce. It may so happen that some silicic acid particles attract a few –O(OH) groups of boehmite and form aluminosilicate hydroxide linkages at the surface with groups rich in silicon hydroxide in the interior. Similarly, by the interaction of the boehmite particle with a few $Si(OH)_4$ groups, aluminosilicate hydroxide may form at the outer surface layer with groups rich in Al–O–(OH) at the center. During drying and subsequent heating, inhomogeneous dehydroxylation continues with the development of more and more Si–O–Al linkages and in due course, they form an appreciable quantity of amorphous aluminosilicate phase, as shown in Chapter 14, along with mixtures of silica-poor and silica-rich regions. Equation 10.31 shows the gelation reaction in alkaline conditions according to Chakraborty (1994a).

It is evident that the Al-Si spinel phase is an inevitable intermediary phase in the thermal transformation of gels processed under alkaline conditions. The role of OH^{-1} during gelation of a mixture silica and alumina components can be explained in the following way. On addition of NH_4OH, both components will be hydrated and precipitated in the colloidal stage. For example, Al $(OH)_3$ colloids increase in size with increasing time; there is a good possibility of its agglomeration into varying sizes as noted in TEM studies. The formation of a cluster and the nucleation of different crystalline modifications, for example, bayerite, are generally dependent on processing variables—pH and temperature as per Hsi et al. (1989). Side by side, $Si(OH)_4$ micelles also polymerize when coprecipitated. The colloidal interactions of the two oppositely charged $Al(OH)_3$ and $Si(OH)_4$ macromolecules of different sizes may also lead to heterogeneity. The possible reaction will be as follows:

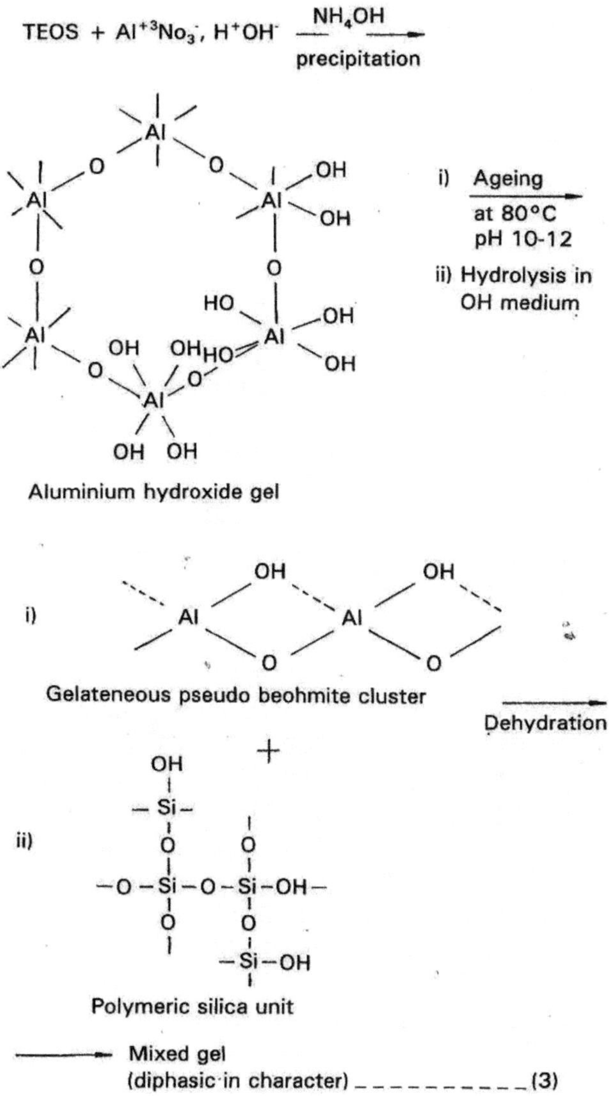

(10.33)

Equation 10.31 shows a schematic of the 7th model of the gelation reaction for G49 (Chakraborty, 1994a).

10.5.2.2 Urea added to a mixture of ANN and TEOS

In this case, a colloidal gel was produced. For example, the gelation reaction of ANN with TEOS was extensively studied by Jaymes and Douy (1995) by a solution NMR study for the system 0.6 M aqueous solution of ANN, which is acidic (pH = 2). At this pH, TEOS is hydrolyzed into silicic acid. According to Iller (1979), the kinetics of hydrolysis is maximum while that of the condensation of silicic acid is minimum. When TEOS was added to this solution and urea was not given, the mixed solution was clear and weak gelation occurred at 100°C. An energy-dispersive X-ray analysis of this gel obtained after filtering showed peaks only of Si, which indicates a lack of interaction between silicic acid and ANN. This result gives an answer as to why the gel marked 7030N synthesized by Chakraborty (1994a) behaved differently from other monophasic gels (Table 7.1).

To reveal the reaction chemistry of the synthesis of a mullite precursor from ANN and TEOS, Jaymes and Douy (1995) carried out an aqueous homogeneous precipitation technique. The whole process was monitored by ^{27}Al and ^{29}Si NMR to follow the local environment of Si and Al atoms in the sol and the gel state. In this process, TEOS was first hydrolyzed in an acidic aluminum nitrate solution. Ammonia was slowly generated in situ in the mixed solution by the thermal decomposition of added urea at ~80°C. The evolution of this process is a time-dependent one and consists of three steps:

1. Evolved ammonia raises the pH of the solution. As a result, silicic acid sol polymerizes and, simultaneously, Al^{+3} ions also polycondense. A Si NMR spectrum of the gel shows the characteristics of polymerized silica and three main resonances. The first one is at −110 ppm due to three dimensionally cross-linked tetrahedra. The second one is at −102 ppm due to chain- branching sites. On increasing the degree of hydrolysis (h) = OH/Al of Al, the bands at −110 and −102 reduce and are finally eliminated, while other resonances appear, at −90 and −78 ppm. A progressive depolymerization takes place. Jaymes and Douy (1995) are of the opinion that aluminum is incorporated into the silica network at the time

of depolymerization. Al NMR shows that aluminum atoms are first incorporated into the gel at octahedral (0 ppm) and tetrahedral sites.

2. With progress in time, the silica network depolymerizes.
3. The above-generated sol is again set to gel in ~1 week. Silicon and aluminum are not available in the filtrate, which indicates that gelation is complete and all the aluminum is incorporated in the silica gel and forms hydrated aluminosilicate. Thus, at least, the degree of hydrolysis is found to be a controllable parameter that can be monitored so that one can identify the stage of gelation.

10.5.2.3 Gelation course in acidic vs. basic condition

MI and MII gels were synthesized by Huang et al. (1997). The condition of synthesis of the mother gel from aluminum chloride hydrate and TEOS prior to the preparation of MI and MII is possibly equal to that SHIII of Haque (2000). Hydrolysis was carried out using a high concentration of water, acidic pH, hydrolysis temperature of ~70°C, and refluxing/hydrolysis time of ~8 h in the first step. It is largely possible that TEOS would hydrolyze and then condense to develop a 3D SiO_2 network as conceived. To this mix, ammonia was added until the pH became 9, resulting in a white precipitate. On pH adjustment to 1.5 and 11 in the second step by using HNO_3 and NH_4OH, the characters of the two gels developed are different. The controlling factor is the polymerized state of the $Si(OH)_4$ unit. Spinel forms in both cases. In MI, mullite formation took place earlier, at ~1200°C, along with some residual amount of spinel. In the case of MII, mullite formed after complete transformation of the spinel phase at ~1350°C following the diphasic way.

10.6 Mullite Precursor Synthesis by Other Processes

Precursors synthesized by spray drying, thermal decomposition, and spray pyrolysis are discussed in Chapter 9.

In the spray-drying process, TEOS and ANN were used in an alcoholic solution by Hamano et al. (1985), and in the spray-pyrolysis (SP) process, the same alcoholic solution mixture was used by Kanzaki and Tabata (1985). It is probable that components react with the formation of Si–O–Al linkages. So the resultant amorphous alumina and silica components left after pyrolysis may be a chemical compound of two component oxides instead of mixed oxides.

Sanz et al. (1991) heated an SP powder to below a temperature below the 980°C exotherm. The nature of the spray-pyrolyzed powder was analyzed by using ^{27}Al MAS-NMR spectroscopy. While an X-ray showed a precursor of an amorphous pattern heated to 400°C, the Al spectrum showed an extraordinary feature consisting of pronounced AlO_4 and AlO_5 peaks in comparison to a very less prominent peak of AlO_6. This clearly indicates that during SP operation itself, Al_2O_3 and SiO_2 react to develop major Al(IV)–O–Si and Al(V)–O–Si and minor Al(VI)–O–Si linkages. The spray-pyrolyzed precursor marked C of Jaymes et al. (1996) showed Al present as Al(VI) in the as-sprayed sample, which indicates that ANN did not decompose. As heating continued, Al(VI) decreased while the band corresponding to Al(IV) increased. Thus, it is easy to conceive that the pyrolyzed mass is an aluminosilicate other than free-oxide mixtures.

Aluminum sulfate, of course, does not decompose at a temperature below 80°C. As a result, the pyrolyzed residue of the reaction between TEOS and aluminum sulfate carried out by Suzuki et al. (1990) did not contain homogeneous oxide materials of SiO_2 and Al_2O_3. Rather aluminum sulfate would last up to 860°C and thereafter decompose into an oxide. Consequently, it did not form mullite at 1000°C but formed spinel and maybe γ-Al_2O_3 as an intermediate phase (Wheat et al., 1979). The gelation experiments also concluded that the hydrolysis of TEOS is the first step prior to the reaction with aluminum nitrate. To accomplish it, a hydrolysis agent, like alcohol, is essential in the presence of acidic salts. The time of hydrolysis is an important factor, as shown. The time/temperature of the gelation process is also a significant factor. Since the entire gelation process takes a few hours for completion, the intensity of the 980°C exotherm differs and the consequent phase development also varies. As a result, the mullite literature reports a great variation in phase development. A basic difference in those reactions perhaps is the variation in the reaction occurring during gelation. A similar type of bonding (Si–O–Al) may develop in the SP process as practiced by Kanzaki and Tabata (1985).

10.7 Summary

In major cases, TEOS is used for the synthesis of mullite gels and thus gel chemistry is largely dependent on its hydrolysis and later on its condensation behavior. A comprehensive literature on this subject on controlling the hydrolysis, condensation, size of the polymeric particles is presented. Thereafter, probable reactivity with other aluminum components in the solution-cum-gelation stage is discussed.

The gelation reactions in two major categories of mullite precursors, hydrolysis-gelation and hydrolysis-coprecipitation processes, have been enumerated to bring focus to the role of the gelation process versus the phase transformation processes.

In both gelation and coprecipitation methods, it is obvious that two components are reacting in the wet stage as the existence of Al(IV)–O groups and/or Al–O–Si bonds are noted by several researchers in their IR and ^{27}Al MAS-NMR spectroscopies.

ANN and TEOS are the two components chosen by a large section of researchers for the synthesis of mullite precursors. DTA studies indicate interaction between them in the solution stage at a temperature between 60°C and 80°C. This reaction is also found to be dependent on the alumina content. Al_2O_3 and SiO_2 in the ratio of 3:2 show the highest heat evolution, as noted in a preliminary experiment. The author has noticed that the temperature of occurrence of the reaction between the two components, ANN and TEOS, is exothermic. It is also shown that in an isothermal reaction condition, various mixtures of TEOS and ANN react vigorously at ~80°C in a water bath with the evolution of exothermic energy. The maximum heat evolution occurs in gel marked ISG72, which corresponds to the composition of 3:2 mullite.

In a dilute aqueous solution, the development of Si–O–Al(IV) is effective if the degree of polymerization of the $Si(OH)_4$ unit used from the silica component is less and preferably assumed to be monomeric in nature. The resultant gel transforms into mullite during the first exotherm itself.

The gelation mechanism during the synthesis of precursors from ANN and TEOS is noted to depend on the use of external water content and even on the presence of water of crystallization of the aluminum nitrate salt used. Accordingly, three models have been predicted for three cases of gel synthesis:

- When the water of crystallization ($9H_2O$) is roughly removed by repeated heating-solubilization technique, a specific model is described.
- In SH gel or monophasic gel synthesis, the water of crystallization is the only source of water causing hydrolysis of TEOS; a chain-like model of the aluminosilicate precursor is visualized.
- In the case of using excess external water, a 3D model of the aluminosilicate precursor is speculated.

In the coprecipitation process, the entire mechanism of gelation chemistry is different. It is more sensitive to the pH prevailing during the coprecipitation process. The alumina component is precipitated during this operation. It is amorphous at pH > 7. Obviously, Al_2O_3-SiO_2 gels prepared at pH 6–7, for example, the gel marked CP-6 or RH, show an amorphous character. At higher pH, the gel becomes diphasic in character since boehmite is precipitated out and is homogeneously mixed with silica gel. Therefore, the mixing state is changed only in two processes of precursor synthesis. This results in a great change in the phase transformation process. Accordingly, a new model of it is predicted. The gelation reaction in this case has been studied by ^{29}Si and ^{27}Al MAS-NMR studies by several researchers. The gelation reactions in acid and alkaline conditions are thus found to be absolutely different.

The mechanism of the SP process shows that the pyrolysis product of a dilute solution of ANN and TEOS is merely an aluminosilicate compound other than an intimate mixture of amorphous oxides. Both ^{27}Al and ^{29}Si MAS-NMR studies indicate the presence of predominant Al(IV)–O–Si bonds.

When AlOBu is taken as an alumina component, instead of the usual ANN, and along with TEOS is hydrolyzed using excess water, as in the colloidal gel method, at about neutral pH, the resultant gel does not show boehmite but the sizes of the two colloidal particles are assumed to be large enough, the result of which seems to be analogous to the thermal transformation behavior to a diphasic gel. A similar phase transformation behavior is noticed in the use of silica sol (synthesized from aqueous silica sol, Ludox, and silica fume–hydrolyzed TEOS) aged prior to the synthesis of mullite gels. Thus, the overall results conclude that the internal structure of the precursor or mixing state of the hydroxide components are the controlling factor in the overall high temperature reaction.

Problems

1. How far have the various problems of gel-mullite transformation (as stated in Chapter 12) been solved?
2. How far have the various controversies as pointed out in Chapter 12 been explained?
3. How far has the controversial spinel phase been characterized and difficulties removed?
4. Predictions by various scientific communities are explained by experimental findings and reasoning. Reader may observe new finding, if any out of the discussion presented in this chapter which have implications for phase transformation in the gel-mullite transformation reaction. Explain.
5. Problems arises when TEOS is used even with ANN. The relationship between the DTA exothermic peak height and the amount of mullite formed usually does not corroborate (Li & Thompson, 1991). Identify the problems.
6. There are problems even when TEOS is used with ALOBu at neutral pH. Spinel formation is found invariably during the heating process. Explain.
7. More problems occur when TEOS is used even with ALOBu at basic pH. Segregation of the aluminum component is a common phenomenon. How can this segregation be minimized?
8. There is no problem when silica sol (obtained from sodium silicate, or $SiCl_4$) is used. The DTA peak height corroborates mullite formation (Demediuk & Cole, 1958; Chakraborty and Ghosh, 1988). Explain.

References

1. H. Insley and R. H. Ewell, Thermal behavior of the kaolin minerals. *J. Res. Natl. Bur. Stand.*, **14**(5), 615–627 (1935).
2. J. Ossaka, Tetragonal mullite type phase from co precipitated gels. *Nature (London)*, **19**, 1000–1001 (1961).

3. A. K. Chakraborty and D. K. Ghosh, Synthesis and 980°C phase development of some mullite gels. *J. Am. Ceram. Soc.*, **71**(11), 978–987 (1988).
4. A. K. Chakraborty, Effect of pH on 980°C spinel phase-mullite formation of Al_2O_3-SiO_2 gels. *J. Mater. Sci.*, **29**, 1558–1568 (1994a).
5. A. K. Chakraborty, Role of hydrolysis water-alcohol mixture on mullitization of Al_2O_3-SiO_2 monophasic gels. *J. Mater. Sci.*, **29**, 6131–6138 (1994b).
6. Y. Hirata, H. Minamizono, and K. Shimada, Property of SiO_2-Al_2O_3 powders prepared from metal alkoxide. *Yogo Kyokai Shi*, **93**(1), 36–54 (1985).
7. S. Komarneni, R. Roy, C. A. Fyfe, and G. J. Kennedy, Preliminary characterization of gel precursors and their high-temperature products by 27 Al magic-angle spinning NMR. *J. Am. Ceram. Soc.*, **68**(9), C-243–C-245 (1985).
8. P. C. Carman, Constitution of colloidal silica. *Trans. Faraday Soc.*, **36**, 964–973 (1940).
9. H. R. Kruyt, ed. *Colloid Science*, Elsevier, New York, p. 26 (1952).
10. Y. Kubota and H. Takagi, Synthesis and properties of mullite, in *Advances of the 24th Symposium of Basic Science of Ceramics*. Paper No. 1808, Ceramic Society of Japan, Tokyo (1986).
11. M. Mizuno, M. Shiraishi, and H. Saito, Microstructure of bending strength of highly pure mullite ceramics, in *Ceramic Transactions*, Vol. 6, Mullite and Mullite Matrix Composites. eds. S. Somiya, R. F. Davis, and J. A. Pask, American Ceramic Society, Westervikke, OH, pp. 413–424 (1990).
12. K. Nishu, T. Yokuyama, T. Wataabe, and T. Tarutani, Characterization of amorphous aluminosilicate formed by absorption of silicic acid on aluminum hydroxide, in *Abstracts of the 24th Symposium of Basic Science of Ceramics*. Paper No. IB08. Ceramic Society of Japan, Tokyo (1989).
13. S. Komarneni and R. Roy, Solid-state ^{27}Al and ^{29}Si magic-angle spinning NMR of aluminosilicate gels. *J. Am. Ceram. Soc.*, **69**(3), C-42–C-44 (1986).
14. K. Okada, H. Otsuka, and S. Somiya, Review of mullite synthesis routes in Japan. *Ceram. Bull.*, **70**(10), 1613–1640 (1991).
15. K. Okada and N. Otsuka, Characterization of the spinel phase from SiO_2-Al_2O_3 xerogels and the formation process of mullite. *J. Am. Ceram. Soc.*, **69**(9), 652–656 (1986).

16. M. Yamane, S. Aso, S. Okami, and T. Sakaino, Low temperature synthesis of a monolithic silica glass by the pyrolysis of a silica gel. *J. Mater. Sci.*, **14**, 607–611 (1979).

17. M. Nogami and Y. Moriya, Glass formation through hydrolysis of $Si(OC_2H_5)_4$ with Nh_4OH and HC solutions. *J. Non-Cryst. Solids*, **37**, 191–201 (1980).

18. S. Sakka and K. Kamiya, The sol-gel transition in the hydrolysis of metal alkoxides in relation to the formation of glass fibres and films. *J. Non-Cryst. Solids*, **48**, 31–46 (1982).

19. E. Konrad, O. Bachle, and R. Singer, *Ann*, **143**, 118 (1867).

20. R. Aelion, A. Loebel, and F. Eirich, *J. Am. Chem. Soc.*, **72**, 5705 (1950).

21. E. Konard, O. Bachle, and R. Singer, *Ann*, **145**, 179 (1929).

22. R. G. Gossink, *Mater. Res. Bull.*, **10**, 35 (1975).

23. D. W. Schaefer and K. D. Keefer, in *Better Ceramics Through Chemistry*, eds. C. J. Brinker, D. E. Clark, and D. R. Ulrich, North-Holland, New York, p. 1 (1984).

24. H. Schmidt, H. Scholze, and A. Kaiser, Principles of hydrolysis, condensation and densification I. *J. Non-Cryst. Solids*, **63**, 1–11 (1984).

25. C. J. Brinker, K. D. Keefer, D. W. Schaefer, and C. S. Ashley, Sol-gel transition in simple silicates II. *J. Non-Cryst. Solids*, **48**, 47–64 (1982).

26. C. J. Brinker, K. D. Keefer, D. W. Schaefer, B. D. Kay, and C. S. Ashley, Sol-gel transition in simple silicates II. *J. Non-Cryst. Solids*, **63**, 45–59 (1984).

27. J. A. Pope and J. D. Mackenzie, Sol-gel processing of silica, II the role of the catalyst. *J. Non-Cryst Solids*, **87**, 185–198 (1986).

28. R. K. Iller, *The Chemistry of Silica*, Wiley, New York (1979).

29. B. E. Yoldas, in *Ultrastructural Processing of Advanced Ceramics*, eds. J. D. Mackenzie and D. R. Ulrich, Wiley, New York, p. 333 (1988).

30. D. R. Uhlmann, B. J. J. Zelinski, and G. E. Wenk, in *Better Ceramics Through Chemistry*, Vol. 32, eds. C. J. Brinker, D. E. Clark, and D. R. Ulrich, North Holland, New York, p. 59 (1984).

31. K. D. Keefer, in *Better Ceramics Through Chemistry*, eds. C. J. Brinker, D. E. Clark, and D. R. Ulrich, North Holland, New York, vol. 32, p. 15 (1984).

32. D. W. Schaefer and K. D. Keefer, in *Better Ceramics Through Chemistry*, eds. C. J. Brinker, D. E. Clark, and D. R. Ulrich, North Holland, New York, vol. 32, p. 1 (1984).

33. A. D. Irwin, J. S. Holmgren, J. Jonas, *Mater. Lett.*, **6**, 25 (1987).
34. A. D. Irwin, J. S. Holmgren, J. Jonas, *J. Mater. Sci.*, **23**, 2908 (1988).
35. B. E. Yoldas, Hydrolysis of aluminium alkoxides and bayerite conversion. *Appl. Chem. Biotechnol.*, **23**, 803–809 (1973).
36. B. E. Yoldas, A transparent porous alumina. *Ceram. Bull.*, **54**(3), 286–288 (1975).
37. B. E. Yoldas, Alumina sol preparation from alkoxides. *Ceram. Bull.*, **54**(3), 289–290 (1975).
38. A. K. Chakraborty, Reaction study of various mixtures of tetra ethyl orthosilicate and aluminum nitrate. *J. Mater. Sci.*, **43**, 5376–5384 (2008).
39. W. M. Jones and D. B. Fischbach, Novel processing of silica hydrolysis and gels. *J. Non-Cryst Solids*, **101**, 123–126 (1988).
40. A. Yasumori, M. Iwasaki, H. Kawazoe, M, Yamane, and Y. Nakamura, Nuclear magnetic resonance study of the structure of aluminosilicate gel and glass. *Phys. Chem. Glasses*, **31**(1), 1–9 (1990).
41. A. K. Bhattcharya, A. Hartridge, and K. K. Mallick, Inorganic aluminium precursors in the synthesis of mullite: an investigation. *J. Mater. Sci.*, **31**, 5551–5554 (1996).
42. M. Haque, Thesis, University of Kolkata (2000).
43. D. W. Hoffman, R. Roy, and S. Komarneni, Diphasic xerogels, a new class of materials: phases in the system Al_2O_3-SiO_2. *J. Am. Ceram. Soc.*, **67**, 468–471 (1984).
44. D. X. Li and W. J. Thomson, Effects of hydrolysis on the kinetics of high temperature transformations in aluminosilicate gels. *J. Am. Ceram. Soc.*, **74**, 574–578 (1991).
45. C. S. Hsi, F. S. Yen, and Y. H. Chang, Characterization of co-precipitated Al_2O_3–SiO_2 gels. *J. Mater. Sci.*, **24**, 2041–2046 (1989).
46. M. Fukuoka, Y. Onoda, S. Inoue, K. Wada, A. Nukui, and A. Makishima, The role of precursors in the structure of SiO_2-Al_2O_3 sols and gels by the sol-gel process. *J. Sol-Gel Sci. Technol.*, **1**, 47–56 (1993).
47. T. Tsuchiya, M. Fukuoka, T. Sci, and J. D. Mackenzie, *Seramikku ronnbunnshi*, **97**, 224 (1989).
48. D. X. Li and W. J. Thomson, Mullite formation kinetics of a single-phase gel. *J. Am. Ceram. Soc.*, **73**(4), 964–969 (1990).
49. H. Schneider, B. Saruhan, D. Voll, L. Merwin, and A. Sebald, Mullite precursor phases. *J. Euro. Ceram. Soc.*, **11**, 87–94 (1993).
50. I. Jaymes, A. Douy, D. Massiot, and J. P. Coutures, Characterization of mono- and diphasic mullite precursor powders prepared by aqueous

routes, ^{27}Al and ^{29}Si MAS-NMR spectroscopy. *J. Mater. Sci.*, **31**, 4581–4589 (1996).

51. B. E. Yoldas and D. P. Partlow, Formation of mullite and other alumina-based ceramics via hydrolytic poly-condensation of alkoxides and resultant ultra- and micro-structural effects. *J. Mater. Sci.*, **23**, 1895–1900 (1988).
52. J. A. Pask, X. W. Zhang, A. P. Tomsia, and B. E. Yoldas, Effect of sol-gel mixing on mullite microstructure and phase equilibria in the α-Al_2O_3-SiO_2 system. *J. Am. Ceram. Soc.*, **70**(10), 704–707 (1987).
53. A. J. Leonard, P. Ratnasamy, F. D. Declerck, and J. J. Fripiat, Structure and properties of amorphous silico-aluminous. Part 5. Nature and properties of silico-aluminous surfaces. *Disc. Faraday Soc.*, **52**, 98–108 (1971).
54. M. F. Bechtold, *J. Phys. Chem.*, **59**, 532 (1955).
55. K. Sinko and R. Mezei, Gelation of aluminosilicate systems under different chemical conditions. *J. Sol-Gel Sci. Technol.*, **21**, 147–156 (2001).
56. B. E. Yoldas, Effect of ultrastructure on crystallization of mullite. *J. Mater. Sci.*, **27**(24), 6667–6672 (1992).
57. B. E. Yoldas, Microstructure of monolithic materials formed by heat treatment of chemically polymerized precursors in the Al_2O_3-SiO_2 binary. *Ceram. Bull.*, **59**(4), 479–483 (1980).
58. A. K. Chakraborty, DTA characterization of three types of Al2O3-SiO2 gels made from TEOS-Al (OBU)3 mixture with variation of water. *Ceram. Int.*, **22**, 463–469 (1996a).
59. M. Schmucker and H. Schneider, Structural development of single phase (type I) mullite gels. *J. Sol-Gel Sci. Technol.*, **15**, 191–199 (1999).
60. A. Taylor and D. Holland, The chemical synthesis and characterization sequence of mullite. *J. Non-Cryst. Solids*, **152**, 1–17 (1993).
61. R.-L. Orefice and W.-L. Vasconcelos, Sol-gel transition and structural evolution on multicomponent gels derived from the alumina-silica system. *J. Sol-Gel Sci. Technol.*, **9**, 239–249 (1997).
62. I. Jaymes and A. Douy, Homogeneous mullite-forming powders from spray-drying aqueous solutions. *J. Am. Ceram. Soc.*, **75**(11), 3154–3156 (1992).
63. S. Kanzaki, H. Tabata, T. Kumazawa, and S. Ohta, Sintering and mechanical properties of stoichiometric mullite. *J. Am. Ceram. Soc.*, **68**(1), C-6–C-7 (1985).

64. K. S. Mazdiyasni and L. M. Brown, Synthesis and mechanical properties of stoichiometric aluminum silicate (mullite). *J. Am. Ceram. Soc.*, **55**(11), 548–552 (1972).

65. S. Prochazka and F. J. Klug, Infrared-transparent mullite. *J. Am. Ceram. Soc.*, **66**(12), 874–880 (1983).

66. S. Somiya, ed. *Mullite*, Uchida Rokakuho Publishing Co., Tokyo, Japan (1985).

67. H. Suzuki, H. Saito, Y. Tomokiyo, and Y. Suyama, Processing of ultrafine mullite through alkoxide route, in *Ceramic Transactions*, Vol. 6, Mullite and Mullite Matrix Composites. eds. S. Somiya, R. F. Davis, and J. A. Pask, American Ceramic Society, Westervikke, OH, p. 263 (1987).

68. O. Sakurai, N. Mizutani, and M. Kato, Preparation of mullite powders from metal alkoxides by ultrasonic spray pyrolysis. *J. Ceram. Soc. Jpn.*, **96**, 639–645 (1988).

69. Y. Hirata, K. Sakeda, Y. Matsushita, and K. Shimada, Preparation of fine SiO_2-Al_2O_3 powders by hydrolysis of mixed alkoxides. *Yogo Kyokai Shi*, **93**(9), 577 (1985).

70. Y. Hirata, K. Sakeda, Y. Matsushita, K. Shimada, and Y. Ishihara, Characterization and sintering behavior of alkoxide-derived aluminosilicate powders. *J. Am. Ceram. Soc.*, **72**(6), 995–1002 (1989).

71. M. Low and R. McPherson, The structure and composition of Al-Si spinel. *J. Mater. Sci.*, **7**, 1196–1198 (1989).

72. K. Hamano, Z. Nakagawa, G. Cun-Ji, and T. Sato, in *Mullite*, ed. S. Somiya, Uchida Rokakuho Publishing Co., Tokyo, Japan, p. 37 (1985).

73. K. Hamano, T. Sato, and Z. Nakagawa, Properties of mullite prepared by coprecipitation and microstructure of fired bodies. *Yogo Kyokai Shi*, **94**(8), 818–822 (1986).

74. H. Schneider, I. Merwin, and A. Sebald, Mullite formation from non-crystalline precursors. *J. Mater. Sci.*, **29**, 805–812 (1992).

75. L. A. Paulick, Y.-F. Yu, and T.-I. Mah, Ceramic powders from metal alkoxide precursors, in *Advances in Ceramics*, Vol. 21, Ceramic Powder Science, pp. 121–129 (1987).

76. S. Mitachi, M. Matsuzawa, K. Kaneko, S. Kanzaki, and Y. Tabata, Characterization of SiO_2-Al_2O_3 powder prepared from metal alkoxides. *Ceram. Trans.*, **6**, 275–286 (1990).

77. M. Sales and J. Alarcon, Synthesis and phase transformations of mullite obtained from SiO_2-Al_2O_3 gels. *J. Euro. Ceram. Soc.*, **16**, 781–789 (1996).

78. C. H. Horte and J. Wiegmann, Reaction between amorphous SiO_2 and Al_2O_3. *Naturwiss*, **43**, 9 (1956).

79. T. Demediuk and W. F. Cole, Exothermic Reaction of Metakaolin between 950° and 1,000°C. *Nature*, **181**, 1400–1401 (1958).

80. D. Croft and W. W. Marshall, A novel synthesis of alumino-silicates and similar materials. *Trans. Br. Ceram. Soc.*, **64**(3), 121–126 (1967).

81. T. D. McGee and C. D. WirKus, Mullitization of aluminum-silicate gels. *Am. Ceram. Soc. Bull.*, **51**, 577–581 (1972).

82. Y. X. Huang, A. M. R. Senos, J. Rocha, and J. L. Baptista, Gel formation in mullite precursors obtained via tetraethylorthosilicate (TEOS) prehydrolysis. *J. Mater. Sci.*, **32**, 105–110 (1997).

83. M. J. Hyatt and N. P. Bansal, Phase transformations in xerogels of mullite composition. *J. Mater. Sci.*, **25**, 2815–2821 (1990).

84. H. Yamada and S. Kimura, Studies on co-precipitates of alumina and silica gels and its transformations at higher temperatures. *Yogo Kyokai Shi*, **70**, 87–93 (1962).

85. M. G. M. U. Ismail, Z. Nakai, K. Minegishi, and S. Somiya, Synthesis of mullite powder and its characteristics. *Int. J. High Technol. Ceram.*, **2**, 123–134 (1986).

86. M. G. M. U. Ismail, Z. Nakai, and S. Somiya, Microstructure and mechanical properties of mullite prepared by the sol-gel method. *J. Am. Ceram. Soc.*, **70**(1), C-7–C-8 (1987).

87. J. C. Huling and G. L. Messing, Surface chemistry effects on homogeneity and crystallization of colloidal mullite sol-gels. *Ceram. Trans.*, **6**, 221–229 (1990).

88. J. C. Huling and G. L. Messing, Hybrid gels for homoepitactic nucleation of mullite. *J. Am. Ceram. Soc.*, **72**(9), 1725–1729 (1989).

89. G. Klaussen, Microstructural evolution of sol-gel mulite. *Ceram. Eng. Sci. Proc.*, **11**, 1087–1093 (1992).

90. L. Pach, A. Iratni, Z. Hrabe, S. Svetik, and S. Komarneni, Sintering and crystallization of mullite in diphasic gels. *J. Mater. Sci.*, **30**, 5490–5494 (1995).

91. J. Wu, M. Chen, F. R. Jons, and P. F. James, Mullite and alumina-silica matrices for composites by modified sol-gel processing. *J. Non-Cryst. Solids*, **162**, 197–200 (1993).

92. S. S. Sueyoshi and C. A. Contreras Soto, Fine pure mullite powder by homogeneous precipitation. *J. Euro. Ceram. Soc.*, **18**, 1145–1152 (1998).

93. M. Mizuno, M. Shiraishi, and H. Sato, Microstructure and bending strength of highly pure mullite ceramics. *Ceram. Trans.*, vol. 6, eds. S. Somiya, R. F. Davis, and J. A. Pask, p. 413 (1990).

94. P. Padmaja, G. M. Anilkumar, P. Krishna Pillai, A. D. Damodaran, and K. G. K. Warrier, Formation characteristics and densification behavior of diphasic mullite gels under various pH conditions. *Br. Ceram. Trans.*, **97**(5), 232–235 (1998).

95. N. Shinohara, D. M. Dabbs, and I. A. Aksay, Infrared transparent mullite through densification of monolithic gels at 1250°C. *Proc. SPIE*, **0683**, 19–24 (1986).

96. B. Sonuparlak, Sol-gel processing of infrared transparent mullite. *Adv. Ceram. Mater.*, **3**(3), 263–267 (1988).

97. X.-H. Jin, L. Gao, and J.-K. Guo, The structural change of diphasic mullite Ge studied by XRD and IR spectrum analysis. *J. Euro. Ceram. Soc.*, **22**, 1307–1311 (2002).

98. W.-C. Wei and J. W. Halloran, Phase transformation of diphasic aluminosilicate gels. *J. Am. Ceram. Soc.*, **71**(3), 166–172 (1988).

99. H. Shiga, M. G. M. U. Ismail, and K. Katayama, Sintering of ZrO_2 toughned mullite ceramics and its microstructure. *J. Ceram. Soc. Jpn.*, **99**, 798–802 (1991).

100. J.-E. Lee, J.-W. Kim, Y.-G. Jung, C.-Y. Jo, and U. Paik, Effect of precursor pH and sintering temperature on synthesizing and morphology of sol-gel processed mullite. *Ceram. Int.*, **28**, 935–940 (2002).

101. J.-E. Lee, J.-W. Kim, Y.-G. Jung, C.-Y. Jo, and U. Paik, Effect of precursor pH and composition on the grain morphology and size of mullite ceramics in aqueous system. *Mater. Lett.*, **17**, 3239–3244 (2003).

102. J.-E. Lee, J.-W. Kim, J.-E. Lee, Y.-G. Jung, and U. Paik, Mullite precursor synthesis in aqueous conditions: dependence of mullite crystallization and grain size and morphology on solution pH and precursor salt. *J. Mater. Res.*, **19**(4), 1133–1138 (2004).

103. J. W. Kim, J. E. Lee, Y. G. Jung, C. Y. Jo, J. H. Lee, and U. Paik, Synthesis behavior and grain morphology in mullite ceramics *with* precursor pH and sintering temperature. *J. Mater. Res.*, **18**(1), 81–87 (2003).

104. I. Jaymes and A. Douy, Homogeneous precipitation of mullite precursors. *J. Sol-Gel Sci. Technol.*, **4**, 7–13 (1995).

105. J. Sanz, I. Sobrados, A. L. Cavalieri, P. Pena, S. de. Aza, and J. S. Moya, Structural changes induced on mullite precursors by thermal treatment: a ^{27}Al MAS-NMR investigation. *J. Am. Ceram. Soc.*, **74**(10), 2398–2403 (1991).

106. T. A. Wheat, E. M. H. Sallam, and A. C. D. Chaklader, Synthesis of mullite by a freeze-dry process. *Ceramurgia Int.*, **5**(10), 42–44 (1979).

Chapter 11

Homogeneity of Al_2O_3-SiO_2 Gels/Precursors

11.1 Effect of Homogeneity

The question is, what should be the position of different mullite precursors synthesized by various researchers on the scale of chemical homogeneity of precursors?

11.1.1 Views on Homogeneous Al_2O_3-SiO_2 Gels vs. Mullitization and Atomic-Scale Mixing of Al and Si Atoms

Atomic-scale mixing: Generally, the precursors/xerogels that are directly crystallized to pseudotetragonal mullite at ~1000°C are predicted as the most homogeneously mixed alumina and silica components, the mixing being referred to as atomic-scale mixing of Al and Si atoms (Sundaresan & Aksay, 1991, and others). These precursors (as most homogeneous) showed the strongest exotherm at 980°C. Such types of precursors are usually obtained by the following procedures: (i) slow hydrolysis (SH), (ii) polymeric gel/Type 1, (iii) spray pyrolysis (SP), (iv) spray drying (SD), and (v) monophasic gel (obtained from cogelation of water-soluble components).

Mullite Formations: Analysis and Applications
Akshoy Kumar Chakraborty

ISBN 978-981-4877-05-3 (Hardcover), 978-1-003-03167-3 (eBook)
www.jennystanford.com

Nanometer-scale mixing: Besides most homogeneous precursors, various authors also noted segregation in the phase evolution process of some monophasic gels during heating, which resulted in spinel formation along with tetragonal mullite (t-mullite) formation. Some other precursor (a diphasic xerogel or a colloidal gel) prepared at pH > 10 did not exhibit 980°C exotherm but showed exotherm at ~1300°C, with the formation of mullite. Such precursors are predicted to produce less homogeneous gels. The mixing scale is shown by transmission electron microscopy measurement to be in the nanometer range.

A great number of combinations of silicon and aluminum components have been used for both aqueous and nonaqueous ways of synthesis of mullite precursors. These are mixtures of alkoxides, salts, and sols. Jaymes et al. (1994) indicated that the mullite phase formation depends on the scale of mixing of silicon and aluminum. Douy (2006) opined that mullite formation depends on the scale of homogeneity of mullite precursors. What does this mean?

- The scale of homogeneity at the molecular level: This molecularly mixed gel crystallized rapidly to mullite at ~980°C through a strong exotherm. This is obvious with the use of aluminum nitrate nonahydrate (ANN) or Al alkoxide as the component.
- The scale of homogeneity in the nanometer range (1–100 nm): This type of colloidal or diphasic gel slowly crystallized to mullite either at a temperature above 1200°C or at ~1270°C–1300°C, obviously through a less strong exotherm. Thus, in earlier days, two types of mixing (first by monophasic gelation way and second by colloidal sols) of components were envisaged.

11.1.2 Mixing Scale vs. Mullitization Temperature

Mullitization temperature is considered an important criterion to assess the mixing scale or the degree of aluminosilicate (–Si–O–Al–) bonds present in the precursor system (Sacks and Pask, 1982).

- They pointed out that temperatures in the range of 1600°C–1700°C are required to achieve complete mullitization when alumina and silica particles are mixed in the micrometer range.

- For nanosized mixing, the mullitization temperatures are generally reduced to 1300°C–1450°C.
- If the mixing scale is at the molecular level, mullitization temperatures of 1000°C–1100°C can be achieved. For instance, Huling and Messing (1991) attained complete mullitization of their molecular xerogel at around 1000°C. Such low mullitization temperatures indicate that all the aluminum atoms and silicon atoms are linked in the form of aluminosilicate bonds, eliminating the need for long-distance diffusion and bond rearrangement. However, when a cubic transition alumina, that is, γ-Al_2O_3/Al-Si spinel was allowed to form, epitactic nucleation of spinel occurred at about 1000°C and higher temperatures were needed for the formation of mullite. Here lies the problem of synthesizing mullite during the 980°C exotherm free of any sorts of spinel phase.

Structural transformation of the Al-Si spinel phase into mullite is in contrast to the concept of mullite formation through the scale of mixing of alumina-silica. The temperature gap between the spinel phase forms from the noncrystalline aluminosilicate precursor phase at ~980°C exotherm and then into mullite at ~1250°C exotherm, which cannot be explained by the scale-of-mixing concept.

11.1.3 Intensity of Exotherm of Mullite Formation Related to Al-Si Bonding

The 980°C exothermic peak may be assumed to be related to the alumina-silica gel structure, that is, the homogeneous development of Si–O–Al bonds in the gel network. The formation of this bond depends on the starting materials, for example, silicon and aluminum compounds, used during gelification/coprecipitation and/or pyrolysis and the techniques of synthesis followed. Yoldas (1992) imagined the following:

- The occurrence and the intensity of the 980°C peak appears to be closely related to the intimacy and homogeneity of the aluminum-silicon bonding in the network.
- The exothermic peak might be used for monitoring the nature of the precursor ultrastructure.

- The nature of the ultrastructure also determines what crystalline phase will emerge on heating such chemically polymerized gels. Yoldas presented three different ultrastructures schematically for three kinds of aluminosilicate derived from gels.

A definitive relation between sol-gel chemistry and mullite crystallization is essential for alumina-silica gel prepared from different components in solution (Huling & Messing, 1990). Yoldas (1992) also felt the need to directly quantify the alumina-silica homogeneity of the gel with the mechanism of its crystallization.

11.1.4 Interrelationship between 980°C Exotherm and Mullite Formation

A general observation is that the 980°C exotherm is related to the formation of mullite, as in a monophasic gel like the SH gel synthesized by Okada and Otsuka (1986), the Type I gel synthesized by Schneider et al. (1993), and the polymeric gel (PA) synthesized by Yoldas (1992). However, this interrelation does not always hold true. Several examples are cited in the paragraphs that follow.

Douy (1991) synthesized a mullite powder using various types of aluminum salts, for example, citrate, nitrate, chloride, sulfate, and different types of silicon alkoxides, followed by the organic gelation method with the use of polyacrylamide. It is noted that the 980°C exotherm is not correlated with mullite formation or spinel phase formation in a few cases.

Jaymes et al. (1996) made Precursor A with the organic gel-assisted method (Douy, 1991), wherein the reactivity of aluminum alkoxide was reduced by chelating it with citric acid. Then, the solution was gelled by the in situ formation of a polyacrylamide network. It exhibited the highest 980°C exotherm with the highest enthalpy release of 310 J/gm.

The question as to the extent to which citrate/urea or the organic route helps in the homogenization of gels remains unanswered. With these observations, Douy (1991) and Jaymes et al. (1996), however, indicated that there is no direct relation between the 970°C exotherm intensity in differential thermal analysis (DTA) and the powder's crystallization into mullite. According to them, the 970°C exotherm intensity and the nature of the crystalline phases obtained are thus strongly dependent on the Al salt and the Si alkoxide used for making the precursor solution. The thermochemical reactions taking place during the evolution of mullite gels with various amounts of alumina obtained from components from the same source were uncorrelated (Low & McPherson, 1989). Mullite gel with 63% alumina displayed a very broad exotherm, extending from 900°C to 1050°C, while it showed a sharp and distinct second exotherm at ~1280°C. Mullite gel of 72% alumina also exhibited a sharp exothermic peak at ~970°C, but it further showed a broad and diffused exotherm at ~1250°C. However, mullite gel with 80% alumina showed a relatively small and diluted exotherm at ~970°C and no second exotherm.

A polymeric gel reproduced by Cassidy et al. (1997) did not confirm the crystallization sequence as reported and synthesized earlier by Pask et al. (1987) or Yoldas (1992). When the latter authors noted a major amount of t-mullite and a minor quantity of spinel phase, Cassidy et al. (1997) showed an unexpected size of the exotherm at 980°C, with crystallization of broad X-ray peaks due to fine crystals of the spinel phase.

It was opined by Sundaresan and Aksay (1991) that the silica and alumina components of the precursors that showed direct mullitization at ~1000°C were mixed in a molecular state. However, a clear difference in the mullitization pathways of the two precursors obtained by aging them at 60°C and 100°C were observed by Okada et al. (1996). It is noted that the polymerization of silica increased with temperature but the octahedral structure of Al did not change with aging temperature. With this observation, Okada et al. (1996) believed that the precursors are not molecularly mixed in solution through direct mullitization, as observed in the sample aged at 60°C.

A change in the shape of the observed exothermic peaks at 980°C for the gel samples of L series was shown by Taylor and Holland (1993). As the homogeneity increased, the peak became broader. Moreover, with an increase in the homogeneity, a second exothermic peak occurred, at 960°C. However, the sample mark LH showed a very large 980°C exothermic peak, along with a shoulder at the usual 960°C peak, although it was highly homogeneous. This deviates from the normal trend.

11.2 Methods of Achieving Homogeneity in the Alkoxide Method of Mullitization

The primary attempt is to synthesize a chemically homogeneous high-purity mullite precursor by cogelation or coprecipitation techniques. To achieve this, the most important process is synthesis via the alkoxide route using both components as esters. Sol-gel synthesis with silicon and aluminum alkoxides in alcoholic solvents and also by the acid- or base-catalyzed method has been followed. There is a great limitation in the process in that the hydrolysis and polycondensation of silicon alkoxide are very low in comparison to those of aluminum alkoxide. So true copolymerization is unexpected or of secondary importance and there is every possibility of a rise in heterogeneity in the gel system. Rapid hydrolysis of Al alkoxides by the addition of excess water or in a basic condition results in heterogeneity. This is because self-condensation of alumina may be predominant and in consequence spinel formation may occur during heating processes. For example, Mazdiyasni and Brown (1972) used the above alkoxide components in a purely basic condition and noted a delayed mullitization. Major mullite formation occurred only at ~1200°C. A gel prepared using excess water showed spinel formation prior to mullitization at a temperature above 1150°C (Prochazka & Klug, 1983). On hydrolysis of a mixture of alkoxides at pH 8.2, spinel formation occurred at ~1100°C (Hirata et al., 1985). But on hydrolysis of a mixture of alkoxides by an aqueous ammonia solution, spinel phase formation was noted at the first stage, followed by mullite formation at ~1200°C exotherm (Low & McPherson, 1989; Mitachi et al., 1990). However, spinel (major) and mullite (minor) were noted to crystallize at ~988°C exotherm in the gel marked SGM by Schneider et al. (1992). Thus, in the majority of the cases, spinel is the abundant phase and mullite is the minor phase.

To circumvent this problem, many techniques have been followed by different researchers and these are cited one after another in this book. A theoretical solution was proposed by Sakka and Kamiya (1982). They discussed the difficulties in the copolymerization of two alkoxides during gelation processes and the problems faced in producing a homogeneous gel. For example, a more reactive alcoholate generally undergoes preferential hydrolysis than a less reactive alcoholate. In such a case, the metal alcoholate of less reactivity is partially hydrolyzed and thereafter the more reactive alcoholate is added for copolymerization as per the following scheme:

$$M2(OR')_m + M1(OR)_n + H_2O \rightarrow M1(OR)_{n-1} \rightarrow M1\text{–}O\text{–}M2 \quad (11.1)$$

Sakka and Kamiya (1982) suggested that another way would be gentle hydrolysis in which water is diffused in vapor form. For a copolymerization process, Klein and Garvey (1980) opined that it is preferable to hydrolyze one organometallic component, for example, tetraethyl orthosilicate (TEOS), in the presence of a suitable solvent and in the absence of water and allowing some polymerization to occur before adding a fast hydrolyzing component like AlOBu and water.

11.2.1 First Approach: Prehydrolysis Technique

This first approach was first adopted by Yoldas (1992), who chose the prehydrolysis technique of silicon alkoxide as discussed above and then added it to the aluminum alkoxide solution with a view to allow them to react readily and develop linkages. The question is, how does the partial hydrolysis method help in the homogenization of polymeric gels?

For example, the polymeric gel by Yoldas and Partlow (1988) and Pask et al. (1987) showed mullite (major) and spinel (minor). Even by exercising the partially hydrolyzed method, it is not possible to synthesize a purely homogeneous precursor since spinel formation cannot be avoided. Of course, mullite formation increases during exotherm, with a partially hydrolyzing technique. However, even with the use of prehydrolyzed TEOS, Suzuki et al. (1990) noted an Al-Si spinel phase at 900°C while heating their gel. There is a possibility of immediate condensation of the alkoxy silicon hydrate species formed after the hydrolysis of TEOS. So partially hydrolyzed TEOS would not help much in attaining homogeneity. Geradin et al. (1994) synthesized three types of mullite precursors to re-examine their crystallization paths. The MI precursor was synthesized by using prehydrolyzed TEOS in an aqueous acid solution at pH 1 and aluminum isopropoxide (AIP) while stirring at room temperature. It exhibited only one unique sharp exotherm, at 982°C, and showed the direct crystallization to mullite but with a small amount of spinel phase. As usual, this latter phase transformed to mullite after being annealed for 24 h. Although the M1 precursor behaves as a monophasic gel and is synthesized even on prehydrolysis of TEOS, spinel formation could not be suppressed. It is concluded that partial hydrolysis could not help in homogenization.

11.2.2 Second Approach: Reducing the Difference in the Hydrolysis Reaction Rates of Components

The second approach is to replace aluminum alkoxide by water-soluble salts, for example, ANN or aluminum chloride, and keeping TEOS as the usual source of the silicon component. Four parameters become important in controlling the hydrolysis of the two components and polycondensation of the silicon hydroxide monomers: (i) the amount of solvent, for example, alcohol or water, used, (ii) the acidic medium, (iii) the temperature of gelation, and (iv) the period of gelation. In the single-phase gel by Hoffman et al. (1984), spinel formation was noted in addition to mullite. The SH gel made by Okada and Otsuka (1986) using ANN and TEOS also showed t-mullite (major) with a spinel (minor) phase. Chakraborty and Ghosh (1988) and Chakraborty (1994a,b) noted that the quantity and nature of phase formed at 980°C from the gel prepared from TEOS and ANN are dependent on pH and on the amount of water/alcohol used during gel synthesis (Tables 7.1 and 7.7). A mixture of spinel and mullite phases at ~1000°C from the precursor collected at different periods of gelation condition was noted by Li and Thomson (1991). Even on replacing the organic Al source by ANN, as in the SH/condensation process, the resulting gel may achieving a high homogeneity but not to the full extent. The question still remains, what happens during slow gelation? A variation in the intensity of the exotherm with pH during gelation was noted by Huang et al. (1997). The gel prepared at pH 1.5 develops a spinel phase with a strong exotherm in DTA.

11.2.3 Third Approach: Matching the Hydrolysis Reaction Rates

In the third approach, it was felt that it is necessary either to reduce the difference in the precursor hydrolysis rates or to match the hydrolysis reaction rates such that preferential hydrolysis is prevented. For example, the hydrolytic activity of aluminum alkoxide was reduced due to its high rate of hydrolysis in the presence of water. A chelating agent, for example, β-diketone:acetylacetone, methlyacetone, or ethylacetoacetone, was used by Heinrich and Raether (1992). It acts to reduce the number of available condensation sites. To inhibit the hydrolysis and condensation of aluminum alkoxides, these three types of chelating agents are usually incorporated during the gelation process to retard hydrolysis and condensation of aluminum alkoxides. These chelating agents increase the number of blocked condensation sites. Moreover, these also control the size of primary particle. As a result, the reactivity of the aluminum alkoxides decreases, as well as the primary particle size. Consequently, a significant increase in t-mullite formation occurs (Table 6.2).

11.3 Methods of Achieving Homogeneity in Both Water-Soluble Salt Methods of Mullitization

11.3.1 pH

Precursors obtained from aqueous silica and alumina components in acidic pH showed direct mullitization at the single exotherm (Insley & Ewell, 1935; Ossaka, 1961; Chakraborty & Ghosh, 1988; Chakraborty, 1994; Schneider et al., 1992). The gel obtained from the same sources as above in a basic condition also produces complete mullitization (Horte & Wiegmann, 1956; Demediuk & Cole, 1958; Croft & Marshall, 1967).

Only in the following cases, the existence of a relation between the event of DTA exotherm—formation of mullite have been observed.

- Monophasic gel synthesized by aqueous method, Chakraborty and Ghosh (1988) establish the interrelation between the linearity of the 980°C.
- Demediuk and Cole (1958) noted the maximum heat of reaction in DTA analysis from the gel of 3:2 mullite composition.
- The exothermic peak is also found to be highest for the coprecipitated gel of 72 wt % Al_2O_3 composition (Chakraborty, 1979).
- Jaymes et al (1996) noted strongest exotherm at 980°C for their Precursor A in DSC traces.
- Later on Douy (2006) found values of maximum enthalpy of crystallization in the 60–70 mol % Al_2O_3 range of composition, where a maximum amount of mullite crystallizes.

The successful of this approach has led to the following conclusions:

- **Avoiding Aging of Component Mixture during Gelation or Coprecipitation**
  This fact is concurrent with the observation of Okada and Otsuka (1990). They prepared mullite thin films by the dip-coating technique using a solution prepared from TEOS and aluminum nitrate in ethanol, and precipitates were obtained by adding ammonia solution while vigorously stirring the solution. The mixture was left at room temperature for aging. Precipitates were collected after different aging periods, for example, <45 days, 45 days, and >100 days. The crystalline phase obtained by firing the mixture at 1000°C was mullite when the aging time of the solution (at room temperature) used for dipping was <45 days, whereas the spinel phase with the mullite formed appeared at a 45-day aging, and the mullite disappeared when the aging time of the solution exceeded 100 days, when only spinel formed. The reasons for obtaining spinel from the coated gel aged for more than 100 days may be the predominant condensation reaction of $Si(OR)_3(OH)$ rather than the interaction of Al^{+3} with the silica network (Fig. 7.11).

The interrelationship between the 980°C exothermic peak in DTA, its nature, and the amount of 980°C phase development of silica sols aged for increasing amounts of time was shown by Chakraborty (1994). The gel prepared with aluminum sulfate and silica sol with more aging time showed that (i) the intensity of the 980°C exotherm decreased rapidly from the highest value and then proceeded very slowly (Fig. 7.10a) and (ii) the amount of mullite decreased rapidly and thereafter the formation of Al-Si spinel commenced and then increased gradually (Fig. 7.10b).

- **Avoiding Silica Sol, Especially the Polymeric Form**

  The active reactant form of silica in a gelation reaction with its counterpart is $Si(OH)_4$. It is available in really different polymeric forms from silica sol and other sources, for example, from aqueous silicate or organic silicon compounds. There is every possibility of growth of a monomeric silica unit to form a cross-linked, branched, linear, cyclic, or 3D network from the hydrolysis of silicon compounds. These may reduce the reactivity. Accumulation of silicon hydroxide clusters ultimately leads to inhomogeneity or the partial formation of spinel other than the much-expected quantity of mullite.

  The mullitization processes of gels synthesized by two different methods in two different ways were reviewed by Okada et al. (1991). For example, Mizuno et al. (1990) used silica sol (obtained from fumed silica) and aluminum salts, such as chloride, nitrate, and sulfate, for mullite synthesis. Nishu et al. (1989) chose silica sol (obtained from silica gel on boiling with aqueous ammonia at pH 9) and Al(OH)3 sol (obtained by adding aqueous ammonia to aluminum chloride solution at pH 8). The mullitization processes in the two cases are different. Nishu et al. (1989) noted direct mullitization, whereas Mizuno et al. (1990) found mullite via spinel formation. The polymerization states of the Si ion in the two types of starting materials are largely different. The silica sol obtained from fume silica may contain a 3D framework of silica. In comparison, the silicic ion obtained from the depolymerization of the silica gel as measured by the absorption method and trimethylsilylation was found to contain monomeric silicon with a very little amount of dimer and trimer. The difference in the degrees of polymerization of Si ion may lead to a large difference in the mullitization process.

The polymerization of silica also depends upon the concentration of the silica present in the solution. Generally, a low concentration is suitable since the polymerization rate is slow in a dilute solution. The use of colloidal sols as taken by Yoldas (1992) showed a diphasic route of mullitization.

The development of the mullite precursor depends upon the pH range within which silica sol is generated by the hydrolysis of TEOS (Schneider et al., 1993). Schneider et al. (1993) adjusted the gelation condition in the pH range $10 < pH < 13$, and they opined that the formation of large silica-rich areas retards mullitization.

To reveal the local structure development, magic angle spinning–nuclear magnetic resonance spectroscopy (MAS-NMR) was carried out by Aoki et al. (1992) on the aging of the mixed solution. They noted that the polymerization of the silica unit proceeded with aging time with little change in aluminum coordination. They concluded that the polymerization of silica inhibited mullitization. Thus, any attempt to synthesize a homogeneous precursor using TEOS either by gelation or by coprecipitation methods seems troublesome. However, it was thought by various authors of mullite ceramics that the reaction conditions under which hydrolysis is promoted and polymerization is suppressed are one of the decisive factors for yielding chemically homogeneous precursors. It was also thought that the water:alkoxide ratio to control the hydrolysis process is a factor to obtain chemically homogeneous specimens. But in practice, polymerization starts soon after the hydrolysis of TEOS takes place. Accordingly, the author showed some examples, as stated above, in which water was avoided to avoid hydrolysis and achieve homogenization.

Jaymes et al. (1994) used silicon tetrachloride in place of TEOS and used AIP as before to synthesize a precursor that showed no Si resonance peak in NMR spectroscopy at ~110 ppm. The precursor showed a Si peak at –92 ppm only, which predicts the formation of a more homogeneous precursor. Moreover, the reproducibility of the phase transition process using TEOS as one of the components is to be noted and rechecked.

A comparison of the transformation of SP powder with that of polymeric and colloidal precursor powders was given by Sanz et al. (1991).

A polymeric gel prepared by Cassidy et al. (1997) as per the method of Yoldas and Partlow (1988) showed a different mode of phase transformation behavior.

Nieto et al. (1998) investigated the effect of heating on the phase and structural evolution of three kinds of mullite precursors obtained by varying the method of preparation: CA (a colloidal gel) and two polymeric gels PA and PB (made in acidic and basic conditions) synthesized by Yoldas (1992). The thermal changes of the precursors were compared.

Jaymes et al. (1996) compared the transformations of a monophasic gel, a diphasic gel, a gel made by the homogeneous precipitation technique using urea, and a spray-pyrolyzed precursor.

A polymeric gel synthesized by Rajendran et al. (1990) as per the procedure of Yoldas (1992) showed the formation of a spinel phase (major) and mullite (minor) at the 998°C exotherm, which is in contrast to the latter's observation.

- **Gelation of Alumina Component by Ammonia Affecting Phase Transformation Behavior**

  A large variety of alumina components have been used, and quite a high quantity of these are generally employed to achieve mullite stoichiometry. In an acidic medium, alumina is usually adsorbed on the surface of the silicic acid hydrate either as such or in a hydrolyzed state. There is every possibility of nonuniformity in the gelation processes. The situation is aggravated when a base is used as a catalyst. Segregation and/or aluminum hydroxide cluster formation is obvious. Precipitation of the alumina component becomes faster on increasing the pH. The result of inhomogeneity of this nature is impure mullitization.

  Therefore, rapid hydrolysis of AlOBu and Al(iPro) in the presence of a base as a catalyst and the subsequent condensation reaction have to be minimized to prevent segregation of the alumina component to develop an aluminum hydroxide gel or crystalline hydrates, like boehmite and bayerite.

11.3.2 Attempt to Produce Homogeneous Precipitation by Urea

Jaymes et al. (1995) considered that the neutralization of the mixed solution induces the hydrolysis of the aluminum nitrate and simultaneously condensation of the silicic acid also occurs to form a silica gel. They thought that chemical homogeneity may be achieved if the two phenomena are matched and copolymerization occurs. By slow and homogeneous precipitation in an aqueous solution induced by the in situ generation of ammonia by thermal hydrolysis of urea, they made a monophasic gel. It showed a strong and sharp exotherm at 975°C (109 KJ/mol) and formed pseudotetragonal mullite (Jaymes et al., 1996, Fig. 6.1). Densification also occurs, as in the case of the first gel. Homogeneity is perhaps achieved by this method. However, a sequential reaction of this precursor is found to be different from that of the precursor made by Yamada and Kimura (1962), although urea was used in both cases for coprecipitation of both components.

11.4 Achieving Homogeneity by the Soluble Salt–Organic Silicon Method of Mullitization

11.4.1 Varying Use of Solvent

The fourth approach is to control the hydrolysis of components by varying the use of water/alcohol as to avoid preferential hydrolysis of either the silica component or the alumina component. The phase development is sensitive to even the slightest change in processing variables, particularly the presence of water and alcohol, as discussed earlier. Both play a great role in hydrolysis and polycondensation of both component hydroxides. The effect of alcohol concentration on the crystallization exotherm at 980°C was shown by Hoffman et al. (1984). They showed that the single-phase gel prepared from "less ethanol," that is 2 mL/10.2 g of ANN, exhibited a sharp crystallization exotherm in comparison to the gel prepared from "more ethanol," that is, 30 mL/10.2 g of ANN, which exhibited a much less sharp exotherm. They assumed that an increase in alcohol may increase the gel volume and accordingly change the gel structure.

11.4.2 Amount of Water Used for Hydrolyzing Alkoxide

Water must be used cautiously. Gel 150 made by Chakraborty and Ghosh (1988), prepared from TEOS and ANN in excess of ethanol, showed spinel formation during the first exotherm. The gel marked GYW made by Taylor and Holland (1993) containing a ratio of $H_2O/-OR > 5$ showed a predominantly spinel phase. Heinrich and Raether (1992) showed that the gelation time is related to the water concentration. With increasing water content of the ANN-TEOS precursor solution, the amount of t-mullite decreased. Thus, their final conclusion was that the reactivity difference between the two reactants as well as the amount of water content are to be reduced to avoid clustering, as happens by the local enrichment of alumina, which may act as the nuclei for Al-Si spinel crystallization. Therefore, rapid hydrolysis of TEOS into large colloidal particles of silica by use of a large quantity of water is to be minimized.

11.4.3 Method or Rate of Addition of Water

This may affect homogeneity, for example, dropwise addition of water may create locally high-water-concentration areas and that may lead to the clustering of $Al(OH)_3$ gel out of the alumina component and also the formation of highly cross-linked $Si(OH)_4$ gel (Fonseca, 1997).

11.4.4 Time of Hydrolysis with or without Stirring Time

The hydrolysis period was varied by varying the amount of moisture available through different techniques (Yoldas & Partlow, 1988), for example, by using excess water with the aim to hydrolyze rapidly, by avoiding direct use of water but hydrolyzing components in humid air with a view to minimize accelerated hydrolysis and ensure slow gelation, and by stirring the component solution in an open flask for a long period to restrict hydrolysis. The first sample did not exhibit a 980°C exotherm, the second sample showed a moderate exotherm, and the third sample, marked C, showed strong 980°C exotherm. In these ways, Yoldas and Partlow (1988) gradually produced more and more homogeneous precursors (Fig. 12.3).

The Type I precursor made by Schneider et al. (1993) showed t-mullite formation together with a poorly crystallized spinel (minor) because in this case, direct use of water was avoided. Moreover, the velocity of the hydrolysis was reduced by keeping the mixed alkoxides diluted in isopropanol in contact with air for 14 days.

A systematic variation in water content during the gelation process results in great variations in the course of mullitization processes and likely their thermal events were noted (Chakraborty, 1996a). For example, with a decrease in the quantity of water, spinel formation decreased and mullite formation increased at the first exotherm from the sample marked G-152(i) to the sample marked G-152(ii) (Table 7.8).

The homogeneity of the sample is a function of stirring time (Taylor & Holland, 1993). The general trend is that after an initial level of inhomogeneity that is low in Pm/Ps, it increases suddenly, showing a high value of Pm/Ps (reaching a high value of homogeneity after a particular length of stirring time). Thereafter, it declined with further stirring. For example, the sample marked LB was stirred for 18.5 h and showed major t-mullite and minor spinel phase after heating to 980°C, while LD, which was stirred for 19.5 h, showed essentially spinel formation with just a trace of mullite. The reason is to be ascertained.

11.4.5 Rapid Solvent Removal during Spraying: An Important Factor?

The gel marked MP made by Fonseca (1997) is an example. The gelation time is in turn dependent upon (i) the concentration of the reactants present in the solution and (ii) the use of a large amount of solvents, which may be water or alcohol. Other examples include a series of gels—G-143, G-144, G-145, G-138, G-137, and G-142—made by Chakraborty and Ghosh (1988)

11.4.6 Evaporation of Solvents in a Water Bath or on a Hot Plate

The dependence of the 980°C exothermic peak intensities on gelation time was shown by Li and Thomson (1991). Their gels marked 2H and 1D at 980°C converted to almost identical amounts of mullite and spinel. But the DTA peak of gel 1D was sharper and much more intense than that of gel 2H. The reason is not known. Li and Thomson (1991) showed that the extent of mullite formation at the first stage, that is, at 980°C, is found to increase with an increase in the time of gelation. With these data, they conjectured that better mixing occurred with a longer gelation period. For example, at 980°C, the increase in the intensity of mullite proceeded in the following order: 2H < 1D < 2D < 2W < 2WC. However, at the second stage, that is at 1200°C, the increase in the intensity of mullite followed the reverse order: 2WC < 2W < 2D < 1D < 2H < CG. In other words, CG showed the highest mullite formation at the second exotherm.

If the 980°C exothermic peak is related to mullite formation, as stated above, it is obvious that the peak intensity should be dependent upon the quantity of mullite formation, which is also related to the gelation time. However, the 980°C peak intensity was found to be maximum in the case of the gel marked 1D and decreased in 2D and further decreased in 2W, that is, the reverse was true. In other words, with an increase in the gelation time, the peak intensity decreased. Where lies this discrepancy? The time and temperature of the gelation process may determine the interaction between the two hydrogels.

11.4.7 Role of the Alumina Component

Besides considering the role of the homogeneous network development of Si–O–Al bonds, gelation processes also depend on the aluminum component used as well as the nature of alkoxides taken. For example, Heinrich and Raether (1992) showed that a change in the nature of the alkoxides, for example, a change from TEOS to tetramethyl orthosilicate or tetrapropyl orthosilicate, affects t-mullite development.

Fukuoka et al. (1993) investigated the formation of Si–O–Al bonds vis-à-vis the microstructural homogeneity of the four binary sols and gels of SiO_2-Al_2O_3 of composition $80SiO_2 \cdot 20Al_2O_3$. They concluded that the Al(IV) ratio is related to the microstructural homogeneity of the gel (Fig. 13.6) and that the first exothermic peak of crystallization is also related to the gel structure, that is, the homogeneity of the Si–O–Al bonds in the gel network structure.

In the case of aluminum di(*sec*-butoxide) ethyl acetoacetic ester chelate (AC), the Si–O–Al bonds are formed in the sol state and the Al(IV) value in the dry state at 100°C is 66%, which gradually increases with an increase in the heat treatment temperature and becomes 92% at 800°C.

In the case of ANN, such bonds are not formed either in the sol or in the wet gel state. The bonds start to develop on drying. The Al(IV) value is 13% and rapidly rises to 90% at 800°C.

In the case of aluminum formoacetate (AF), Si–O–Al bonds are formed in the sol state. In the dry state, the Al(IV) value is 38% and finally becomes 87% at 800°C.

In the case of boehmite sol, such bonds are not formed in the sol solution. In a dry state, the Al(IV) value is 13% and reaches 63% at 800°C.

On the basis of these results, Fukuoka et al. (1993) ranked the microstructural homogeneity of the gel with aluminum compound used in the following order: AC > ANN > AF > boehmite sol.

In this approach, an attempt has been made to reduce the water content during gelation or use a water-free sol gel approach.

The gel marked G-31 synthesized by Chakraborty and Ghosh (1988) by eliminating the water of crystallizing of ANN prior to gelation with TEOS and the resultant gel showed complete mullitization during exotherm at ~1000°C. The gel marked GNW by Taylor and Holland (1993) prepared from TEOS and AIP in a nonaqueous solvent and without using external water showed t-mullite during exotherm. They classified gels as homogeneous or inhomogeneous depending on the relative amounts of mullite and spinel formed on heating the gels at 980°C, as measured by the X-ray technique. The ratio of the peaks of the (331) plane of mullite and the (440) plane of spinel was called the homogeneity coefficient. According to Yoldas and Partlow (1988), inhomogeneities in the oxide network could be brought down when chemical encounter rates with water of hydrolysis were reduced.

Mostly, in all processes of hydrolysis-cum-gelation or hydrolysis-cum-precipitation using an acid or a base as the catalyst, the resultant precursors developed varying quantities of spinel as an additional phase to mullite at 1000°C. The major reasons behind this may be the control of hydrolysis as per the different factors discussed above. In this connection, one point that has to be raised is the character of the spinel phase. The origin of the spinel phase has been found to be different in two sources. In the first case, it is produced from a monophasic gel during heating at and around the first exotherm, that is, 900°C–1000°C, and it exists up to ~1250°C and thereafter transforms to mullite. In the second case, it is formed from diphasic gel and colloidal gel at ~600°C and continues up to ~1300°C. Thereafter, it transforms to mullite. The formation of spinel affects the homogeneity of the two types of precursor systems.

11.5 Method of Achieving Homogeneity by Thermal Decomposition Process

In the sixth approach, instead of facing problems of cogelation, coprecipitation of two components (particularly concerning their hydrolysis), subsequent condensation, and their interaction during drying by various methods, some researchers thought about codecomposition or SD and SP techniques of two components to achieve homogeneous precursors.

11.5.1 Spray Drying

In the first method, Jaymes and Douy (1995) sprayed the mixed solution into a solution of ammonia in a nonaqueous solvent, for example, isopropanol, and a nearly monophasic mullite precursor was formed. It exhibited a very sharp and strong exotherm at 990°C along with an enthalpy of 80.5 KJ/mol, measured by differential scanning calorimetry. The precursor was amorphous until the occurrence of the exotherm and then crystallized into pseudo-t-mullite and a small amount of spinel phase. In a dilatometric study, it showed a single step of densification, which corresponds to the first exotherm in DTA. However, if the precipitate is made into an aqueous solution of ammonia or ammonium carbonate, the resultant precursor exhibits a small exotherm at 988°C (31.85 KJ/mol) and another exotherm at 1260°C (8.1 KJ/mol). A weakly crystalline spinel was noted after the first exotherm, whereas only orthorhombic mullite was noted after the second exotherm. It is not known why Jaymes and Douy (1995) designated this precursor as a diphasic one. Moreover, they did not present the X-ray pattern of the as-dried spray gel. This gel showed two steps of densification, which are related to two exotherms in the crystallization path of mullite.

Homogeneity was achieved by Bhattacharya et al. (1996) in a stepwise manner, as follows: by the colloidal sol-gel route, inorganic components were found to develop diphasic or inhomogeneous gel with particle sizes in the range of 50–100 nm. These precursors showed spinel formation and required high temperatures for mullitization. In the second method, they took aluminum chlorohydrate and by a solution NMR study, they reported the presence of three peaks: a six-coordinated Al at 0 ppm due to monomeric hexa-aquo Al^{+3}, a five-coordinated Al at 11 ppm due to a Keggin ion structure, and a four-coordinated Al at 63 ppm due to an $Al(OH)_4$ unit. They further indicated that the method of production influenced the proportions of the coordinated Al species. The monomeric and unsegregated species reacted and developed ~20 wt % of mullite at 800°C; the remaining aggregated species were unreactive and as such, the rest of the mullitization was delayed.

In the third method, the NMR study of a nitrate-based sol showed two distinct six-coordinated species, the major (60%) one at 0.6 ppm due to monomeric hexa-aquo Al^{+3} and the other (40%) at 4.9 ppm due to dimeric species and the absence of polymeric ions. The resultant gel showed equal concentration of mullite, spinel, and cristobalite.

In the fourth method, and to attain more homogeneity, Bhattacharya et al. (1996) spray-dried the sol to freeze the sol in situ to give 5–15 nm particles. The sol crystallized to single-phase mullite at the 920°C exotherm.

According to Jaymes and Douy (1992), the scale of homogeneity was at the atomic level when mullite formation was observed through a strong exotherm at 980°C. This is achieved when a freshly prepared mixture is spray-dried. Chemical homogeneity decreased as the exothermicity was reduced and spinel appeared at the cost of mullite.

11.5.2 Codecomposition

Instead of adopting precise control of the hydrolysis condition, a mixture of alkoxides, for example, alkaline phosphatase (ALP) and TEOS, with toluene as the solvent was thermally decomposed at ~300°C by Inoue et al. (1996) in an autoclave. It is claimed that the heterogeneity that probably arises in the hydrolysis process may not occur in the process of thermal decomposition of alkoxide mixtures. They made an interesting observation: On thermal decomposition of ALP alone in toluene, x-alumina was produced. On the other hand, when TEOS was alone in toluene, it did not decompose but remained as a liquid as usual. However, the combination of the two formed a powder, which indicates that ALP facilitated the decomposition of TEOS on one hand and inhibited the formation of x-alumina on the other. With these data, they predicted that ALP may interact with TEOS during thermal treatment in an autoclave. In support of this theory, they compared the infrared spectrum of this product with that of the precursor obtained by the hydrolysis of the above mixture and an absorption band at ~1000 cm^{-1} was considered as due to Si–O–Al bonds. This precursor exhibited a sharp exotherm in DTA, with the crystallization of pseudo-t-mullite. and a small amount of spinel phase at ~1000°C. There was another broad exotherm, beyond 1200°C, observed in the DTA trace, which was explained earlier as due to spinel-to-mullite formation. In fact, the mullite became more intense at ~1300°C and the spinel phase was found to have disappeared. This attempt was not fully successful in achieving a homogeneous powder since the spinel phase had crystallized.

11.5.3 Spray Pyrolysis

The SP method includes SP of a mixture of TEOS and ANN. This process was done by Kanzaki and Tabata (1985), Fukuoka et al. (1993), Jaymes et al. (1996), and Douy (2006).

11.6 Temperature Scale of Mullitization

11.6.1 Homogeneity of Mullite Precursors vs. Crystallization Path of Mullite Formation

On the basis of homogeneity of the mullite precursor, the author described as many as six mullitization paths of mullite formation (see Chapter 12 for details). The discussion further shows that the mullitization temperature could be brought down considerably. A stepwise reduction in its formation processes is shown in the following routes:

At ~1700°C

In the first oxide route: Corundum + Quartz → Mullite (11.2)

The reaction includes a solid-state reaction, long-range diffusion, generation of eutectic aluminosilicate glass, and nucleation and subsequent growth of mullite, with a residual glassy phase and unconverted reactant phases.

In the second route: The temperature lowers from as high as 1700°C to as low as 1300°C–1340°C in a diphasic gel system. In this method, two sols are mixed, for example, boehmite sol and silica sol, or boehmite is precipitated in situ in the presence of aqueous silica sol/organic silicon ester, etc. Particle sizes of the two hydroxide or sol components are on the order of nanometers instead of microns, as is usual in oxide mixtures. The diphasic route of mullitization is as per Eq. 16.12, Chapter 16.

In the third route: The temperature of mullitization comes further down from 1340°C to ~1250°C in some mullite precursors. In this method, generally a two-step process is followed for ensuring complete mullitization. Spinel formed in the first step at ~1000°C exotherm is transformed completely in the second step at 1250°C exotherm. The thermal transformation follows a route similar to that followed by the rapid hydrolysis gel as shown in Eqs. 16.6–16.9, Chapter 16.

In the fourth route: The temperature of mullitization further drops down from 1250°C to as low as 1000°C exotherm. In this method, generally one step is followed for mullite formation. This is probably a temperature scale of mullitization, with the size of the oxide particles-to-that of sol particles-to-that of the sizes of monomeric silicon hydroxide groups in aqueous or alcoholic solution. Thermal transformation follows as alike as monophasic gel, SH gel, polymeric gel, molecular gel, Type I gel and pyrolyzed precursor as shown in Eqs. 12.2–12.5, Chapter 12.

Thus, the temperature scale of mullitization apparently related to the size of the two components. The highest temperature of mullitization (~1700°C), drops from micron size of oxide particles to nano-size sol mixtures of diphasic gel at ~1340°C exotherm. It drops further to monophasic gel (second kind) at ~1250°C exotherm. The scale of temperature becomes lowest at ~980°C exotherm for monophasic gel (first kind) where the monomeric size of silicon hydroxide sol particles are predominant.

A literature survey points out some of the difficulties encountered in the first and second routes in achieving mullitization. By and large, it could be concluded from various ways of synthesis procedures as shown above that developing ultrahomogeneous mullite is exceedingly troublesome and a rather difficult task, particularly by the alkoxide route. Only in a few cases has direct mullitization been achieved.

Thus, the development of a homogeneous precursor seems to depend on the choice of the two component materials prior to gel synthesis.

- If both components are water soluble, silicon hydroxide would most probably be monomeric. Obviously, gelation could be done in a diluted solution of salts.
- If at least one component is organic, for example, TEOS and ANN, or both are organic, for example, TEOS and AlOBu, it is expected that hydrolysis would be restricted and no external water would be used and preferably the water of crystallization would be removed to a large extent prior to gelation in the atmosphere for a long duration of time.

11.6.2 As Per Homogeneity of Mullite Precursors in MAS-NMR Study

The best homogeneity in the angle of an MAS-NMR study is observed in the following cases: According to Lippmaa et al. (1980), a –108 ppm resonance is observed in heated mullite gels assigned to [Q4(0Al) sites. Its indication in a ^{29}Si spectrum of the gel heated at ~1000°C is due to the phase separation of the noncrystalline silica-rich alumina phase. Jaymes et al. (1994) synthesized a mullite gel by the hydrolysis of an organic aluminum compound, for example, AIP, in the presence of a water-soluble silica component, for example, $SiCl_4$. The ^{29}Si spectrum showed that –110 nm Si resonance is absent.

The spray-dried precursor C by Jaymes et al. (1996) did not show a –110 ppm resonance in their sample heated at 1000°C (Fig. 9.18).

Contrarily, Jaymes et al. (1995) synthesized a mullite precursor from ANN and tetraethoxysilane by the aqueous homogeneous precipitation technique using urea. However, it showed that a –110 nm Si resonance is present.

SP precursors of Kanzaki et al. (1985) were later studied by Sanz et al. (1991). It was observed that the noncrystalline aluminosilicate phase derived from the SP precursor crystallized into mullite directly at the first exothermic peak temperature. Therefore, this precursor is to be designated as the most homogeneous precursor.

On the contrary, in the literature, a large number of researchers noted the formation of t-mullite at 980°C in the DTA studies of their monophasic/polymeric/Type I mullite gels. However, XRD analysis of these heat-treated gels invariably showed a minor quantity of the Al-Si spinel phase along with t-mullite, for example, the SH gel by Okada et al. (1986), the polymeric gel by Yoldas (1992) and Pask et al. (1987), and the Type I gel by Schneider et al. (1993).

Therefore, the crystallization paths of amorphous mullite precursor powders (bulk composition close to 3:2 mullite) synthesized by various ways depend upon the scale of homogeneity of the primary amorphous phases.

- Most homogeneous: When aluminum and silicon polyhedral are most intimately mixed in such a way that the resultant precursor crystallizes into only mullite, with the exhibition of a sharp exotherm at 980°C.
- Homogeneous: Both mullite (as the major phase) and spinel crystallize together at the exotherm. The latter phase subsequently transforms into 1250°C.
- Less homogeneous: When only spinel likely crystallizes at 980°C and transforms further into mullite through a second exotherm, at 1250°C.

- Heterogeneous: No exotherm observed at 980°C. A broad exotherm observed at ~1300°C due to the formation of mullite from the transitional spinel phase developed earlier, for example, a diphasic gel.
- More heterogeneous: Microcomposites, for example, SiO_2(A) derived from TEOS or Ludox/γ-Al_2O_3-mixed gel.
- Microcomposite: Microcomposites made from silica sol/corundum.
- Oxide mixture, for example, corundum/quartz or SiO_2 mixture.

11.7 Summary

The homogeneity scales of different precursors synthesized by various researchers have been discussed, their transformation behaviors noted in the present investigation, and finally their mullitization scales indicated.

Depending on the exhibition of an exotherm in DTA and crystallization of mullite from various precursors, they have been grouped under the range from monophasic gels as homogeneous to diphasic gels as inhomogeneous in character. It is further assumed that the occurrence and intensity of the 980°C exotherm is related to homogeneity, that is, it is due to the formation of the Al–O–Si bond or the environment of aluminum in the precursor state. Finally, different models for different ranks of homogeneity are described.

The mullite formation temperature is a function of the mixing scale of two components. For a micrometer scale of mixing, it is >1600°C. For a nanometer scale of mixing, as in a diphasic gel, it is ~1300°C. For a molecular scale of mixing, as in a monophasic gel, it is ~1000°C. A group of researchers noted the 980°C exotherm, which is related to mullite formation, whereas others noted no direct relation between the intensity of the exotherm and mullite crystallization.

Different researchers have used several approaches to achieve homogeneity in their gels. These are state chronologically as follows:

1. In the alkoxide method of mullitization, the hydrolysis and condensation rates of TEOS and ALOBu are widely different and these result in the heterogeneity of the final precursor. Some techniques were introduced to minimize these

differences. Use of partially hydrolyzed TEOS was one such technique. Even then spinel formation is noted along with mullite. Chelating agents were used so as to retard hydrolysis and condensation of aluminum alkoxides with a view to match with the hydrolysis reaction rate of TEOS. Finally, an increase in mullite formation occurred.

2. Effect of water, pH, time/temperature, and aging as variables in processing of mullite gels/precursors
   (i) In the soluble salt–organic silicon method of mullitization, aluminum alkoxide was replaced by ANN with a view to reduce the problem of differences in the hydrolysis reaction rates of alkoxide. In this case too, phase development is very sensitive to two main processing variables—water and pH. These two variables control the hydrolysis behavior of TEOS to a great extent. Thus, the quantity of water used, its rate of addition, stirring time and time of hydrolysis of TEOS, the temperature of the hydrolysis medium, and the rate of solvent removal are controlling factors and are interrelated with phase development. pH also affects the final phase development to a large extent. Among the water-soluble components used, the microstructural homogeneity of the gel is of the following order: AC > ANN > AF > boehmite sol.
   (ii) Effect of aging: In the case of both water-soluble salts, direct mullitization at the single exotherm is noticed. Moreover, the linearity of the exotherm with the quantity of the mullite is noticed. This is claimed as the best approach to achieving homogenization of the mullite precursor system. The character of the silica sol used is very important. It would be monomeric as predicted.

       Instead of dealing with the various problems of cogelation/coprecipitation, codecomposition has been used to achieve homogeneous precursors. It shows mullite directly at the first exotherm. It is claimed to achieve the most homogeneous precursor.
   (iii) The mullitization temperature depends on the mixing scale of the two components or the development of aluminosilicate (–Si–O–Al–) bonds in the gelation and drying of the precursor system. SH techniques with

minimum water content (SHI gel) are considered to be the most important because most of the Al and Si atoms are linked in the form of aluminosilicate bonds. Only bond rearrangement takes place on heating the gel to the first exothermic peak temperature. Consequently, it develops early mullitization. In contrast, coprecipitation techniques at pH > 7 form Al-Si spinel as the intermediary phase, which exists till as high a temperature as ~1200°C and shows delayed mullitization.

3. Mixed monophasic and diphasic gels: There is every possibility of the formation of gels of two characters during the synthesis process of mullite gels. A variation in $Si(OH)_4$ colloidal particles occurs on polymerization or later on condensation. Similarly, a large variation in the size of $Al(OH)_3$ colloids will occur.

Finally, it is predicted that it is difficult to synthesize a perfectly homogeneous mullite gel by choosing any components or processing them in any conditions. Either, mullite will form predominantly, along with some quantity of Al-Si spinel; or a varying mixture of both will form; or only Al-Si spinel will form during the first exotherm.

Problems

1. To what extent does citrate/ urea or the organic route help in the homogenization of gels?
2. The mullite phase formation depends on the scale of mixing of silicon and aluminum. What does it mean?
3. Does water content have an impact on homogeneous mixing? If yes, to what extent?
4. Does the pH have an impact on homogeneity during gelation? If yes, to what extent?
5. Does the temperature have an impact on homogeneous mixing of components during gelation? If yes, to what extent?
6. Does the choice of components have an impact on the homogeneity? To what extent? Do combined processes really favor homogenization?
7. It may be concluded that none of the measures could help in the homogenization of gels since the mullite formed at 980°C in these cases is always mixed with additional spinel phase. Is it true?

References

1. S. Sundaresan and I. A. Aksay, Mullitization of diphasic aluminosilicate gels. *J. Am. Ceram. Soc.*, **74**(10), 2388–2392 (1991).
2. I. Jaymes, A. Douy, P. Florian, D. Massiot, and J. P. Coutures, New synthesis of mullite. Strucural evolution study by ^{17}O, ^{27}Al and ^{29}Si MAS NMR spectroscopy. *J. Sol-Gel Sci. Technol.*, **2**, 367–370 (1994).
3. A. Douy, Crystallization of amorphous spray-dried precursors in the Al_2O_3-SiO_2 system. *J. Euro. Ceram. Soc.*, **26**, 1447–1454 (2006).
4. M. D. Sacks and J. A. Pask, Sintering of mullite containing materials: I, Effect of composition. *J. Am. Ceram. Soc.*, **65**, 65–70 (1982).
5. J. C. Huling and G. L. Messing, Epitactic nucleation of spinel in aluminosilicate gelsand its effect on mullite crystallization. *J. Am. Ceram. Soc.*, **74**(10), 2374–2381 (1991).
6. B. E. Yoldas, Effect of ultrastructure on crystallization of mullite. *J. Mater. Sci.*, **27**(24), 6667–6672 (1992).
7. J. C. Huling and G. L. Messing, Surface chemistry effects on homoggeeity and crystalization of coloidal mullite sol-gels. *Ceram. Trans.*, **6**, 221–229 (1990).
8. K. Okada and N. Otsuka, Characterization of the spinel phase from SiO_2-Al_2O_3 xerogels and the formation process of mullite. *J. Am. Ceram. Soc.*, **69**(9), 652–656 (1986).
9. H. Schneider, B. Saruhan, D. Voll, L. Merwin, and A. Sebald, Mullite precursor phases. *J. Euro. Ceram. Soc.*, **11**, 87–94 (1993).
10. A. Douy, Organic gels preparation of silica-alumina powders, I. Mullite. *J. Euro. Ceram. Soc.*, **7**, 117–123 (1991).
11. I. Jaymes, A. Douy, D. Massiot, and J. P. Coutures, Characterization of mono- and diphasic mullite precursor powders prepared by aqueous routes, ^{27}Al and ^{29}Si MAS-NMR spectroscopy. *J. Mater. Sci.*, **31**, 4581–4589 (1996).
12. M. Low and R. McPherson, The origins of mullite formation. *J. Mater. Sci.*, **24**, 926 (1989)
13. D. J. Cassidy, J. L. Woolerey, J. R. Bartlett, and B. Ben-Nissan, The effect of precursor chemistry on the crystallization and densification of sol-gel derived mullite gels and powders. *J. Sol-Gel Sci. Technol.*, **10**, 19–30 (1997)
14. J. A. Pask, X. W. Zhang, A. P. Tomsia, and B. E. Yoldas, Effect of sol-gel mixing on mullite microstructure and phase equilibria in the α-Al_2O_3-SiO_2 system. *J. Am. Ceram. Soc.*, **70**(10), 704–707 (1987).

15. K. Okada, C. Aoki, T. Ban, S. Hayashi, and A. Yasumori, Effect of aging temperature on the structure of mullite precursor prepared from tetraethoxysilane and aluminum nitrate in ethanol. *J. Euro. Ceram. Soc.*, **16**, 149–153 (1996).

16. A. Taylor and D. Holland, The chemical synthesis and characterization sequence of mullite. *J. Non-Cryst. Solids*, **152**, 1–17 (1993).

17. K. S. Mazdiyasni and L. M. Brown, Synthesis and mechanical properties of stoiciometric aluminum silicate (mullite). *J. Am. Ceram. Soc.*, **55**(11), 548–552 (1972).

18. S. Prochazka and F. J. Klug, Infrared-transparent mullite. *J. Am. Ceram. Soc.*, **66**(12), 874–880 (1983).

19. Y. Hirata, H. Minamizono, and K. Shimada, Property of SiO_2-Al_2O_3 powders prepared from metal alkoxide. *Yogo Kyokai Shi*, **93**(1), 36–54 (1985).

20. S. Mitachi, M. Matsuzawa, K. Kaneko, S. Kanzaki, and Y. Tabata, Characterization of SiO_2-Al_2O_3 powdes prepared from metal alkoxides. *Ceram. Trans.*, **6**, 275–286 (1990).

21. H. Schneider, I. Merwin, and A. Sebald, Mullite formation from non-crystalline precursors. *J. Mater. Sci.*, **29**, 805–812 (1992).

22. S. Sakka and K. Kamiya, The sol-gel transition in the hydrolysis of metal akoxides in relation to the formation of glass fibres and films. *J. Non-Cryst. Solids*, **48**, 31–46 (1982).

23. L. C. Klein and G. J. Garvey, Kinetics of the sol/gel transition. *J. Non-Cryst. Solids*, **38–39**, 45–50 (1980).

24. B. E. Yoldas and D. P. Partlow, Formation of mullite and other alumina-based ceramics via hydrolytic polycondensation of alkoxides and resultant ultra- and micro-structural effects. *J. Mater. Sci.*, **23**, 1895–1900 (1988).

25. H. Suzuki, H. Saito, Y. Tomokiyo, and Y. Suyama, Processing of ultrafine mullite through alkoxide route, in *Ceramic Transactions*, Vol. 6, Mullite and Mullite Matrix Composities. eds. S. Somiya, R. F. Davis, and J. A. Pask, American Ceramic Society, Westervikke, OH, p. 263 (1990).

26. C. Gerardin, S. Sundaresan, J. Benziger, and A. Navrotsky, Structural investigation and energetics of mullite formation from sol-gel precursors. *Chem. Mater.*, **6**, 160–170 (1994).

27. D. W. Hoffman, R. Roy, and S. Komarneni, Diphasic xerogels, a new class of materials: phases in the system Al_2O_3-SiO_2. *J. Am. Ceram. Soc.*, **67**, 468–471 (1984).

28. A. K. Chakraborty and D. K. Ghosh, Synthesis and 980°C phase development of some mullite gels. *J. Am. Ceram. Soc.*, **71**(11), 978–987 (1988).

29. A. K. Chakraborty, Effect of pH on 980°C spinel phase-mullite formation of Al_2O_3-SiO_2 gels. *J. Mater. Sci.*, **29**, 1558–1568 (1994a).

30. A. K. Chakraborty, Role of hydrolysis water-alcohol mixture on mullitization of Al_2O_3-SiO_2 monophasic gels. *J. Mater. Sci.*, **29**, 6131–6138 (1994b).

31. D. X. Li and W. J. Thomson, Effects of hydrolysis on the kinetics of high temperatture transformations in aluminosilicate gels. *J. Am. Ceram. Soc.*, **74**, 574–578 (1991).

32. Y. X. Huang, A. M. R. Senos, J. Rocha, and J. L. Baptista, Gel formation in mullite precursor obtained via TEOS prehydrolysis. *J. Mater. Sci.*, **32**, 105–110 (1997).

33. T. Heinrich and F. Raether, Structural characterization and phase developmentof sol-gel-derived mullite and its precursors. *J. Non-Cryst. Solids*, **147–148**, 152–156 (1992).

34. H. Insley and R. H. Ewell, Thermal behavior of the kaolin minerals. *J. Res. Natl. Bur. Stand.*, **14**(5), 615–627 (1935).

35. J. Ossaka, Tetragonal mullite type phase from coprecipitated gels. *Nature (London)*, **19**, 1000–1001 (1961).

36. C. H. Horte and J. Wiegmann, *Naturwiss*, **43**, 9 (1956).

37. T. Demediuk and W. F. Cole, Exothermic reaction of metakaolin between 950° and 1,000°C. *Nature*, **181**, 1400–1401 (1958).

38. D. Croft and W. W. Marshall, A novel synthesis of alumino-silicates and similar materials. *Trans. Br. Ceram. Soc.*, **64**(3), 121–126 (1967).

39. A. K. Chakraborty, Formation of silicon aluminium spinel. *J. Am. Ceram. Soc.*, **62**(3–4), 120–124 (1979).

40. K. Okada and N. Otsuka, Formation process of mullite. *Ceram. Trans.*, 375–387 (1990).

41. K. Okada, N. Otsuka, and S. Somiya, Review of mullite synthesis routes in Japan. *Ceram. Bull.*, **70**(10), 1633–1640 (1991).

42. K. Nishu, T. Yokoyama, T. Watanabe, and T. Tarutani, Characterization of amorphous aluminosilicate formed by adsorption of silicic acid on aluminium hydroxide, in *Abstracts of the 27th Symposium on the Basic Science of Ceramics*. Paper No. IB08. Ceramic Society of Japan, Tokyo (1989).

43. M. Mizuno, M. Shiraishi, and H. Saito, Microstructure and bending strength of highly pure mullite ceramics. *Ceram. Trans.*, **6**, 413–424 (1990).

44. C. Aoki, T. Ban, S. Hayashi, and K. Okada, Analysis of mullitization process by NMR-analysis for sol-gel solution, in *Abstracts of the Annual Meeting of the Ceramic Society of Japan*, Paper No. 2E16. Ceramic Society of Japan, Tokyo (1992).

45. I. Jaymes, A. Douy, P. Florian, D. Massiot, and J. P. Coutures, Mew synthesis of mullite. Strucural evolution study by ^{17}O, ^{27}Al and ^{29}Si MAS NMR spectroscopy. *J. Sol-Gel Sci. Technol.*, **2**, 367–370 (1994).

46. J. Sanz, I. Sobrados, A. L. Cavalieri, P. Pena, S. de. Aza, and J. S. Moya, Structural changes induced on mullite precursors by thermal treatment: a ^{27}Al MAS-NMR investigation. *J. Am. Ceram. Soc.*, **74**(10), 2398–2403 (1991).

47. M. I. Nieto, G. Urretavizcaya, A. L. Cavalieri, and P. Rana, Structural changes in colloidal and polymeric aluminosilicate gels with mullite composition. *Br. Ceram. Trans.*, **97**(1), 17–23 (1998).

48. S. Rajendran, H. J. Rossell, and J. V. Sanders, Crystallization of a coprecipitated mullite precursor during heat treatment. *J. Mater. Sci.*, **25**, 4462–4471 (1990).

49. I. Jaymes, A. Douy, and D. Massiot, Synthesis of a mullite precursor from aluminum nitrate and tetraethoxysilane via aqueous homogeneous precipitation: an ^{27}Al and ^{29}Si liquid- and solid-state NMR spectroscopic study. *J. Am. Ceram. Soc.*, **78**(10), 2648–2654 (1995).

50. H. Yamada and S. Kimura, Studies on co-precipitates of alumina and silica gels and its transformations at higher temperatures. *Yogo Kyokai Shi*, **70**, 87–93 (1962).

51. A. M. L. M. Fonseca, J. M. F. Ferreira, I. M. M. Salvado, and J. L. Baptista, Mullite based compositions prepared by sol-gel techniques. *J. Sol-Gel Sci. Technol.*, **8**, 403–407 (1997).

52. A. K. Chakraborty, DTA characterization of three types of Al_2O_3-SiO_2 gels made from TEOS -Al (OBU)3 mixture with variation of water. *Ceram. Int.*, **22**, 463–469 (1996a).

53. M. Fukuoka, Y. Onoda, S. Inoue, K. Wada, A. Nukui, and A. Makishima, The role of precursors in the structure of SiO_2-Al_2O_3 sols and gels by the sol-gel process. *J. Sol-Gel Sci. Technol.*, **1**, 47–56 (1993).

54. I. Jaymes and A. Douy, Homogeneous precipitation of mullite precursors. *J. Sol-Gel Sci. Technol.*, **4**, 7–13 (1995).

55. A. K. Bhattcharya, A. Hartridge, and K. K. Mallick, Inorganic aluminium precursors in the synthesis of mullite – an investigation. *J. Mater. Sci.*, **31**, 5551–5554 (1996).
56. M. Inoue, H. Kominami, and T. Inui, Thermal decomposition of alkoxides in an inert organic solvent: novel method for the synthesis of homogeneous mullite precursor. *J. Am. Ceram. Soc.*, **79**(3), 793–795 (1996).
57. S. Kanzaki and H. Tabata, T. Kumazawa, and S. Ohta, Sintering and mechanical properties of stoichiometric mullite. *J. Am. Ceram. Soc.*, **68**(1), C-6–C-7 (1985).
58. S. Kanzaki, H. Tabata, and T. Kumazawa, Sintering and mechanical properties of mullite derived via spray pyrolysis. *Ceram. Trans.*, **6**, 339–351 (1990).
59. E. Lippmaa, M. Magi, A. Samoson, G. Engelhardt, and A. R. Grimmer, Structural studies of silicates by solid-state high resolution 29Si NMR. *J. Am. Chem. Soc.*, **102**(15), 4889–4893 (1980).

Additional Reference

E. Lippmaa, A. Samoson, and M. Magi, High-resolution aluminum-27 NMR of aluminosilicates. *J. Am. Chem. Soc.*, **108**, 1730–1735 (1986).

Chapter 12

Comparison of Thermal Transformation Processes of Six Mullite Precursors

In this chapter, both differential thermal analysis (DTA) curves and phase transformation studies by X-ray diffractometry (XRD) of various earlier activities have been thoroughly reviewed and it has been shown that the raw materials used and the choice of their synthesis generate three kinds of precursors: monophasic, monophasic of the other kind, and colloidal or diphasic. After critically analyzing those works, the kinds of precursors developed by various researchers are individually redesignated for the observation of the readers.

12.1 Experimental Techniques for Phase Evolution Studies of Mullite Precursors

Ceramists and mineralogists have shown a lot of interest in using the thermal analysis method to study the phase transformation of mullite gels. There have been numerous efforts in the investigation of the heating effects of different types of mullite gels. Various investigations and techniques used by various researchers over a period of more than 50 years are serially mentioned in Table 12.1.

Mullite Formations: Analysis and Applications
Akshoy Kumar Chakraborty

ISBN 978-981-4877-05-3 (Hardcover), 978-1-003-03167-3 (eBook)
www.jennystanford.com

Table 12.1 Physicochemical techniques used to study the transformation of mullite gels

No.	Experimental tool	Objective
1.	Crystallographic methods: Large angle X-ray scattering (LAXS) Radial electron distribution function (RDF)	Identify the crystallization temperatures of phases formed on the qualitative XRD technique and quantitative estimation of phases by QXRD
2.	Thermomechanical	Study the thermochemical reaction by DTA, DSC, DDTA, TMA, and DTMA
3.	Spectroscopic study	Use IR, Raman spectroscopy, and magic angle spinning–nuclear magnetic resonance to identify ^{27}Al and ^{29}Si MAS-NMR
4.	Chemical study	Use the alkali leaching technique to determine the presence of free silica of mullite gel calcined to ~1000°C
5.	Solid-state reaction studies	Determine the phase compositions of noncrystalline mullite precursor and Al-Si spinel phases
6.	Miscellaneous auxiliary studies	(i) Calculate the lattice parameter/ cell dimension of the Al-Si spinel and mullite (ii) Determine the of composition of mullite using Cameron's data

12.2 Variables Affecting Mullite Precursor Synthesis Methods and Phase Evolution

The studies mentioned in Table 12.1, made by various researchers, indicate that phase transformations of mullite gels are related mostly to the gelation and coprecipitation process. The different types of transformation paths as observed in the Al_2O_3-SiO_2 xerogel system are responsible for the following two major factors:

- The source of Al_2O_3 and SiO_2 components: The phase development of the above gels leading ultimately to mullite formation is shown to be a function of two major parameters:

(i) source and choice of silica and alumina components and (ii) processing of the precursor synthesis and variables. The following components are used for the synthesis of monophasic gels:

- Silica source: Sodium silicate, $SiCl_4$, aqueous silica sol, and Ludox.

  Organic silica source: Tetraethoxysilane, ethyl orthosilicate $Si(OC_2H_5)_4$, silicon ethyl ester, silicon tetrakis-isopropoxide, silicon ethoxide, and 3-(triethoxysilyl)-propyl-amine (TESPA)
- Alumina source: Aluminum salts like alumina citrate, aluminum sulfate, aluminum chloride, aluminum nitrate nonahydrate (ANN), and aluminum oxalate.

  Organic aluminum source: Aluminum isopropoxide [AIP, $Al(OPr)_3$], aluminum *sec*-butoxide [AlO(Bu)], and dibutoxy ethyl acetoacetate aluminum [$Al(OBu)_2$(AcAcEt)].
- Silica and alumina sources for synthesis of diphasic gels: Silica sol is generally synthesized by the method of dispersing of ultrafine particles, for example, silica fume, colloidal silica, Ludox, and Monsanto's Syton D30 colloidal silica (summer grade). Alumina sol can be prepared from a dispersion of γ-Al_2O_3, pseudoboehmite sol [AlO(OH)], aluminum di(*sec*-butoxide) ethyl acetoacetic ester chelate, aluminum formoxoacetate, Degussa "aluminum C" flame–hydrolyzed powder (99.6% Al_2O_3), Condea "disperal" pseudoboehmite powder (74% alumina), and Hoerscht aluminum chlorohydrate "Locron P" (47.5% alumina). The most popular combinations of components are ANN and tetraethyl orthosilicate (TEOS), $AlCl_3 \cdot 6H_2O$ and TEOS, aluminum sulfate and fume silica, and boehmite sol and fume silica.

- Processing techniques and their variables: The hydrolysis of silicon and aluminum compounds is very susceptible to the following variables: (i) the acidic or basic catalyst used, (ii) the temperature prevailing during hydrolysis, (iii) the time period of hydrolysis maintained, and (iv) the medium of hydrolysis chosen. These four factors greatly influence the rate of hydrolysis-polycondensation of organic sources of the two components. Consequently, the 980°C exothermic reactions and the mullitization processes of the Al_2O_3-SiO_2 gels vary. Even on choosing raw materials from the same source, a little variation in the synthesis technique will lead to mullite gels of different kinds, as noted by various authors, that transform in different ways. Out of these four parameters, the major processing parameters, namely, the medium of gelation or precipitation (i.e., the quantity of water or alcohol used) and the pH prevailing during the synthesis method are highlighted below. Details of the methods/techniques used vis-a-vis phase transformation of mullite precursors synthesized by (i) the monophasic gel route/slow-hydrolysis (SH) gel and polymeric gel/Type I, (ii) coprecipitated gel route, (iii) diphasic gel route, and (iv) spray pyrolysis (SP) or spray-drying route as shown in previous chapters (Chapters 2–7) are summarized. Precursors synthesized from various components and by different processing techniques are probably generated as six types of precursors. Their phase transformation is summarized at the end of Chapter 7.

12.3 Transformation Behaviors of Six Mullite Precursors Studied by DTA, XRD, IR, and MAS-NMR

The thermal behaviors of six representative mullite precursors chemically synthesized by various techniques and studied by different researchers by DTA, XRD, quantitative X-ray diffractometry (QXRD), infrared (IR), and magic angle spinning–nuclear magnetic resonance spectroscopy (MAS-NMR) are shown earlier, in Chapters 2–6. Of the two major variable parameters chosen by them, the importance of processing techniques is also narrated chronologically. It is noted below that the character of the precursor generated is a function of the processing parameters. Using components from the same source but varying the pH and/or the amount of water, some researchers synthesized two to three different kinds of mullite precursors. The author chose the same components but varied the processing methods and prepared as many as five different precursors. These follow five different ways of mullitization on heating. The type of precursor versus the mullitization paths are mentioned and chronologically compared and their DTA characteristics below.

Origin of the exothermic peaks of six categories of transformation: The exothermic peaks of six categories of precursors on the basis of studies are shown in previous chapters and summarized here.

- Some monophasic gels exhibited a significant 980°C exotherm, for example, the aqueous mullite gel by Chakraborty and Ghosh (1988), the slow hydrolysis (SH) gel made by Okada and Otsuka (1986), the polymeric gel of Yoldas (1992), and the Type 1 gel by Schneider et al. (1993).
- In the dried gel state, some monophasic gels did not show boehmite formation by XRD. These showed less significant 980°C exotherms but showed broad 1250°C exotherms, for example, the rapid hydrolysis (RH) gel made by Okada and Otsuka (1986), the CP gel made by Haque et al. (2000), the colloidal gel made by Li and Thomson (1991), precursor B made by Jaymes et al. (1996), and the RH gel synthesized by Colomban (1989).
- Some diphasic gels prepared from boehmite did not exhibit 980°C exotherms, for example, gels synthesized by Hoffman et al. (1984), M3 prepared by Geradin et al. (1994), and the colloidal gel made by Yoldas (1992). All these diphasic gels exhibited exotherms at 1270°C–1300°C.
- A mixture of single-phase and diphasic gels showed less significant 980°C exotherms and reduced exotherms at 1270°C than that by the diphasic gel made from pure boehmite.
- Microcomposite gels exhibited exotherms at as high as 1360°C, for example, AS-γ-Al_2O_3 by Bartsch et al. (1999) and the microcomposite gel made by Ivankovic et al. (2003).
- Spray-pyrolyzed or spray-dried precursors by Kanzaki et al. (1985), Douy (2006), and Chakraborty (2008) showed most significant 980°C exotherms, with the formation of the mullite phase, and showed the absence of any other exotherm at higher temperatures.
- The aqueous mullite gel by Chakraborty and Ghosh (1988) also exhibited an analogous 980°C exotherm as a spray-pyrolyzed precursor.

12.4 Transformation Processes of Precursors Synthesized under Varying Processing Conditions: A Few Examples

- Effect of the source of components: (i) Choosing a commercially available silica sol and four varieties of inorganic alumina sols—"Degussa" aluminum oxide C, Condea "dispersal" pseudoboehmite powder, Hoerscht aluminum chlorohydrate "Locron P," and an inorganic alumina sol prepared by extracting NO3–1 using tertiary alkyl amine from an aluminum nitrate solution—Bhattacharya et al. (1996) made different mullite gels and studied the effect of components on the exothermic reaction. (ii) From TEOS and each of four aluminum salts, Fukuoka (1993) synthesized mullite gels and these showed a linear relationship between the heat treatment temperature of these gels and varying coordination ratios of Al(IV)/Al(IV) + Al(VI) (Fig. 13.3).
- Effect of processing parameters: The synthesis of various precursors by different authors are chronologically presented below.

12.4.1 CM and SGM Gels

Schneider et al. (1992) carried out the mullitization of coprecipitated material (CM) and sol-gel material (SGM) gels.

- DTA data: The SGM sample was greatly different from the coprecipitated material (CM) sample synthesized by the aqueous sol-gel technique in terms of the DTA and thermogravimetric analysis (TGA) characteristics. When the SGM exhibited two exotherms, the CM showed only one exotherm.
- The aluminosilicate hydrate–aluminosilicate hydrate complexes originating in the CM and the polymeric SGM are probably not the same, although they are amorphous in XRD. Mullite formed from the aluminosilicate hydrate in the CM usually contained a lot of water, as noted from its final removal during TGA study, wherein a high-temperature water loss was

noted at 970°C at the endothermic D. This type of water loss was not observed in the case of the SGM sample.

- Major spinel formation occurred in the SGM sample at the first exotherm.
- Mullite formation: The result indicated that both CM and SGM, however, formed mullite and spinel phases at the first exotherm. Only the ratios of the two were different. A major amount of mullite formation took place at the first exotherm in the CM. Thus, the CM showed faster mullitization than the SGM. Also, as the associated spinel concentration was quite less in the CM and its subsequent transformation to mullite was also smaller in magnitude, as such a second exotherm is not found in the DTA trace. On the contrary, major spinel formation occurred in the SGM sample. A substantial amount of this spinel phase later transformed into mullite at the 1253°C exotherm. The DTA patterns in the high-temperature region for both SGM and kaolinite are similar. Thus, the precursor SGM behaves like other kinds of monophasic gels whereas a CM is a monophasic gel.

12.4.2 Precursors Synthesized from ANN and TEOS

Jaymes et al. (1996) compared the mullitization processes of five mullite precursors synthesized from ANN and TEOS in different media by different techniques. There were four variables in the processing conditions (see Chapters 3–6). According to differential scanning calorimetry (DSC) traces, they classified the precursors into two groups. The first group (A, C, and E) showed a strong exotherm at 980°C, and the enthalpy of this reaction varied between 240 and 310 J/g (Fig. 12.1), in comparison to the low data published by Hoffman et al. (1984) and Sen and Thiagarajan (1988).

The second group (B and D) showed two exotherms. Powder B was amorphous up to the first exotherm and crystallized into an aluminosilicate spinel–type phase. The enthalpy of crystallization was only on the order of 146 J/g. The second peak corresponded to the crystallization of the spinel-type phase into mullite at 1260°C, and the enthalpy of crystallization was only 64 J/g. However, on careful observation, one can recognize another peak below the second exotherm at 1260°C. According to Jaymes et al. (1996), powder D showed two exotherms, one at 924°C and one at 975°C, corresponding to the enthalpy of crystallization 170 J/g for the two combined peaks in place of one exothermic peak. At 1275°C, this precursor showed one weakly crystalline peak due to orthorhombic mullitization and the enthalpy of crystallization was on the order of 19 J/ g. They are of the opinion that the process of aging leads to heterogeneity of the two oxide systems. Probably, the reaction between the aluminate and silicate species occurs to a lesser extent. On the contrary, polycondensation of silicic acid may have taken place. On the basis of the corroborative data, the transformation sequences are given. It is obvious that powder D is a mixture of two precursors.

12.4.3 Mullite Gels from SiO_2 and Al_2O_3

Ivankovic et al. (2003) compared the thermal transformations of four mullite gels prepared from the SiO_2 component from the same source (TEOS) but the Al_2O_3 component from different sources by processing either with an excess of water or with a change of the pH by using ammonia. They studied the influence on the crystalline form, particle size, etc., during the heating process.

- For Gel A, they used ANN in excess of water for hydrolysis (nitrate/water molar ratio 1:32) and TEOS.
- For Gel B, to the alcoholic mixture of ANN and TEOS, they added 2 M of aqueous ammonia to raise the pH to 6.
- Gels C and D were prepared from γ-Al_2O_3 (Degussa) and γ-AlO(OH) (boehmite, "Disperal") as the alumina component mixed with TEOS as the Si precursor.

The DTA scans of these precursors are shown in Fig. 12.2.

Gel A was amorphous up to the first exotherm. It exhibited a sharp exotherm at ~977°C, with the formation of the Al-Si spinel phase, and thereafter it showed a second exotherm at ~1246°C, forming orthorhombic mullite (o-mullite).

The crystallization pathway of sample B was different. It first crystallized into a faint pattern of pseudoboehmite. It exhibited a faint exotherm at 991°C and then exhibited a second exotherm, as usual, at 1266°C, with the formation of mullite.

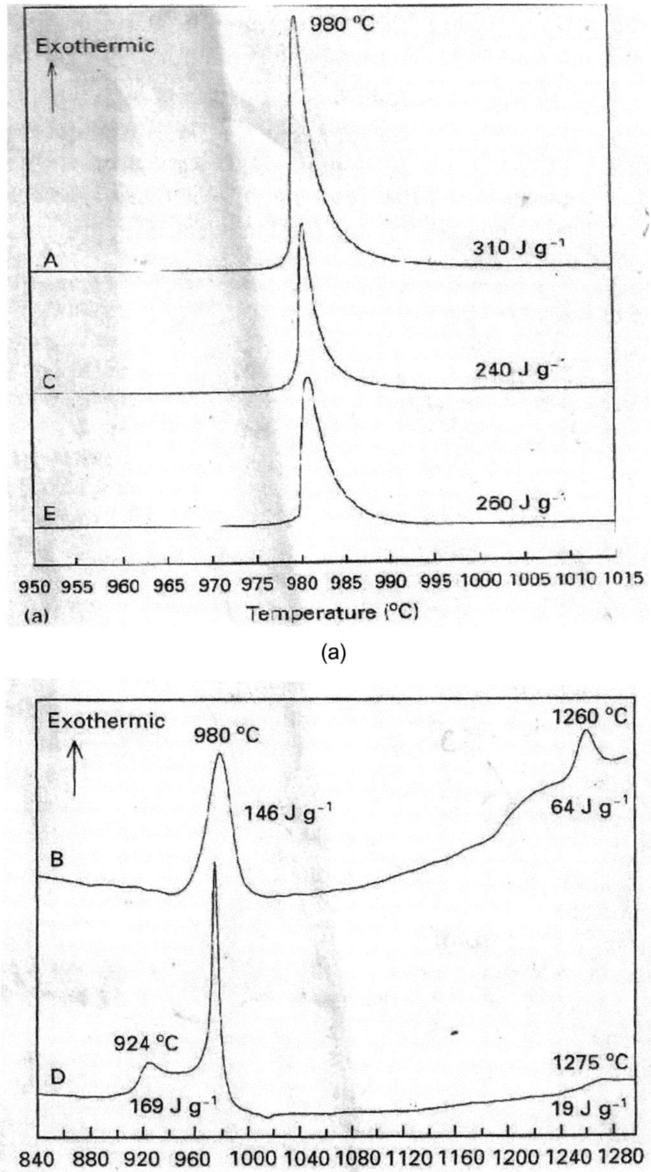

Figure 12.1 (a) DSC curves of precursors A, C, and E and (b) DSC curves of precursors B and D (Jaymes et al., 1996).

Samples C and D were diphasic in nature; the alumina component showed a different form. Sample C itself contained γ-Al_2O_3, whereas sample D developed γ-Al_2O_3 from the raw boehmite taken initially. The most interesting observation is that both these diphasic gels exhibited exotherms due to mullitization. It occurred in gel C at 1346°C and in gel D at 1361°C, which may be due to the choice of TEOS as the silica component and not Ludox, as used previously.

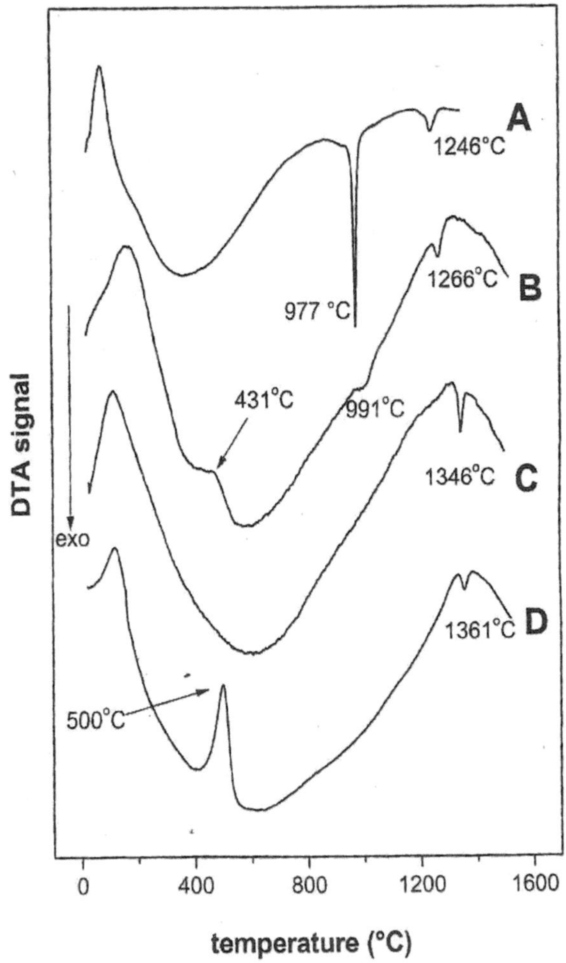

Figure 12.2 DTA scans of prepared precursors (Ivankovic et al., 2003).

Ivankovic et al. (2003) came to a few general agreements on the basis of the earlier studies, as follows:

- The mixing scale in the mullite sol-gel precursors might control the phase transformation sequence, temperature of mullite formation, and properties of the sintered mullite.
- The mixing scale also influences the density and microstructure development of the mullite ceramic. A monophasic gel is sintered at a temperature >1600°C. A colloidal gel/diphasic gel is sintered at a much lower temperature. A microcomposite consisting of alumina particles coated with SiO_2(A) achieves densification at a lower temperature, maybe due to viscous sintering.

12.4.4 Mullite Gels Synthesized by Three Methods

Yoldas (1992) compared three mullite gels—a polymeric gel and a colloidal gel (prepared by the hydrolysis and subsequent coagulation of organic Al and Si compounds, as per Chapter 4) and a nitrate gel (as per Chapter 3)—synthesized by three different methods in terms of their levels of homogeneity. He prepared the nitrate gel by using $Al(NO_3)_3{\cdot}9H_2O$ and $Si(OCH_3)_4$ in an alcoholic medium, as done by Hoffman et al. (1984), Okada and Otsuka (1986), and Chakraborty and Ghosh (1988). The phase transformation sequence of the nitrate gel is given in Table 12.2, and the DTA curves and XRD analysis of the three are given in Fig. 12.3.

Table 12.2 Phase transformation of the nitrate-derived gel (X-ray data is not given)

Temperature	Phase
At 1050°C	Crystallization of the spinel phase occurred along with a minor quantity of mullite.
At 1200°C	Mullite crystallization improved and diffuse peaks due to the spinel phase disappeared. The DTA curve exhibited two exothermic peaks, one at the usual 980°C and another at ~1230°C. Obviously, the disappearance of spinel is related to the occurrence of this peak to mullite.
At 1300°C	Fully developed mullite occurred.

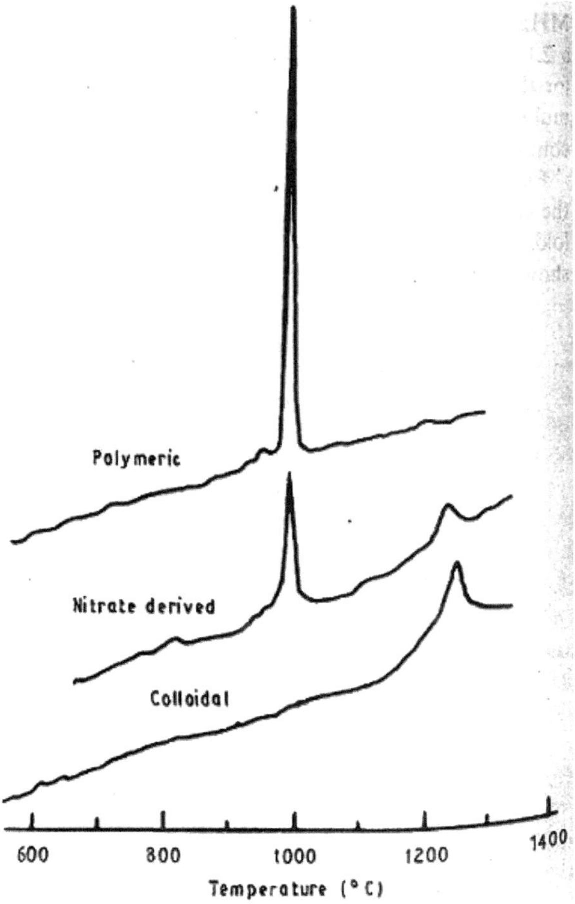

Figure 12.3 DTA curves of the aluminosilicate gels showing the exothermic peaks associated with mullite crystallization in these structures (Yoldas, 1992).

The relative ratio of mullite to spinel phase formed in the polymeric gel was found to be different from that in the nitrate-derived gel. Also, the ratio for the nitrate gel did not match that noted by Hoffman et al. (1984) and Okada and Otsuka (1986). However, the following are the conclusions:

- The occurrence and intensity of the 980°C peak appear to be closely related to the intimacy and homogeneity of the aluminum-silicon bonding in the network.

- The exothermic peak might be used for monitoring the nature of the precursor gel structure and finally conjecture that crystallization in these systems would be affected by the level of homogeneity, that is, whether alumina and silica bonding, as in the polymeric gel, is claimed as mostly homogeneous or less homogeneous and whether it is limited to the level of size of alumina colloids or whether there is molecular-level homogeneity without any bond formation, as in the nitrate-derived gel.

In conclusion, the following newer subject matters have evolved: In Chapter 7, the author depicted and described six different methods of synthesis of mullite gel/precursors from the past activities of various researchers. The synthesis of these is found to be largely dependent on (i) the choice of the two components, (ii) the processing parameters based on the component source, namely pH and water, and (iii) the drying condition of gels/precursors, etc.

12.4.5 Comparison of Monophasic Mullite Gels Made from TEOS and AIP by Different Techniques with Diphasic Gel

Geradin et al. (1994) compared the different behaviors of the thermal processes of two monophasic mullite gels made from components from the same source (TEOS and AIP) by different processing techniques with diphasic gel. The hydrolysis conditions were changed, particularly by adding water (Chapter 7). Their crystallization paths were also compared.

M1 precursor: Regarding the effect of hydrolysis conditions, the M1 precursor was synthesized by using prehydrolyzed TEOS in an aqueous acid solution at pH 1 and AIP while the solution was vigorously stirred at room temperature. It exhibited only one unique sharp exotherm, at 982°C, and showed direct crystallization to mullite and a small amount of spinel phase (as observed by 46° and 67° 2θ peaks in their XRD patterns). The latter phase transformed into mullite after the solution was annealed for 24 h (Fig. 12.5). Obviously, it is a polymeric gel.

M2 precursor: M2 was synthesized by mixing TEOS and AIP in 2-propanol at 60°C, which was hydrolyzed simultaneously by the addition of water. It exhibited two exotherms, one at 987°C and one at 1252°C. At 1000°C, poorly crystalline spinel was noted; it developed characteristic patterns of the δ-alumina phase. This precursor is likely to behave like other kinds of monophasic gels and might be analogous to the RH gel by Okada and Otsuka (1986).

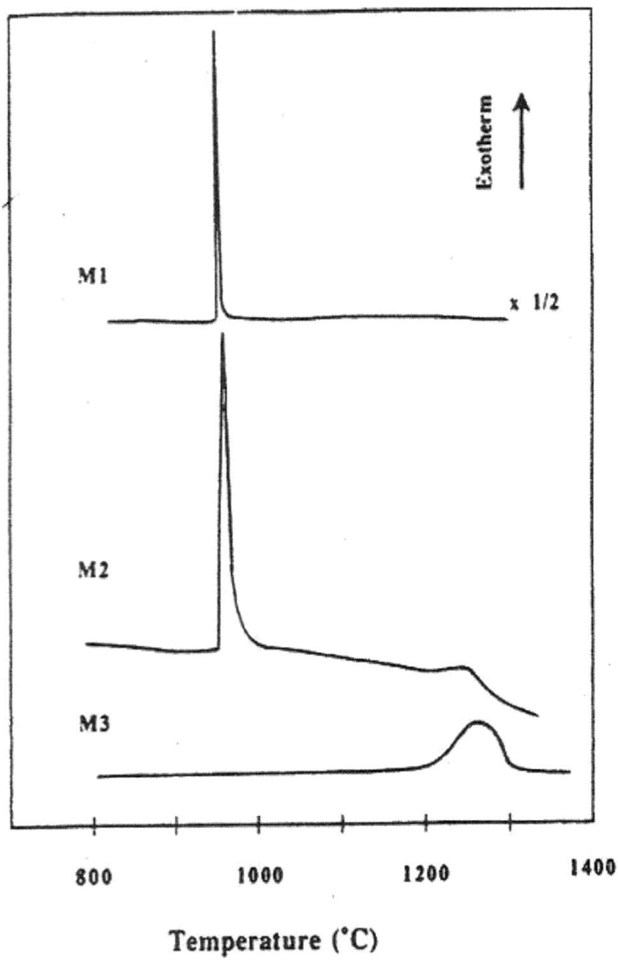

Figure 12.4 DTA heating curves of materials M1, M2, and M3 previously fired at 800°C for 6 h (Geradin et al., 1994).

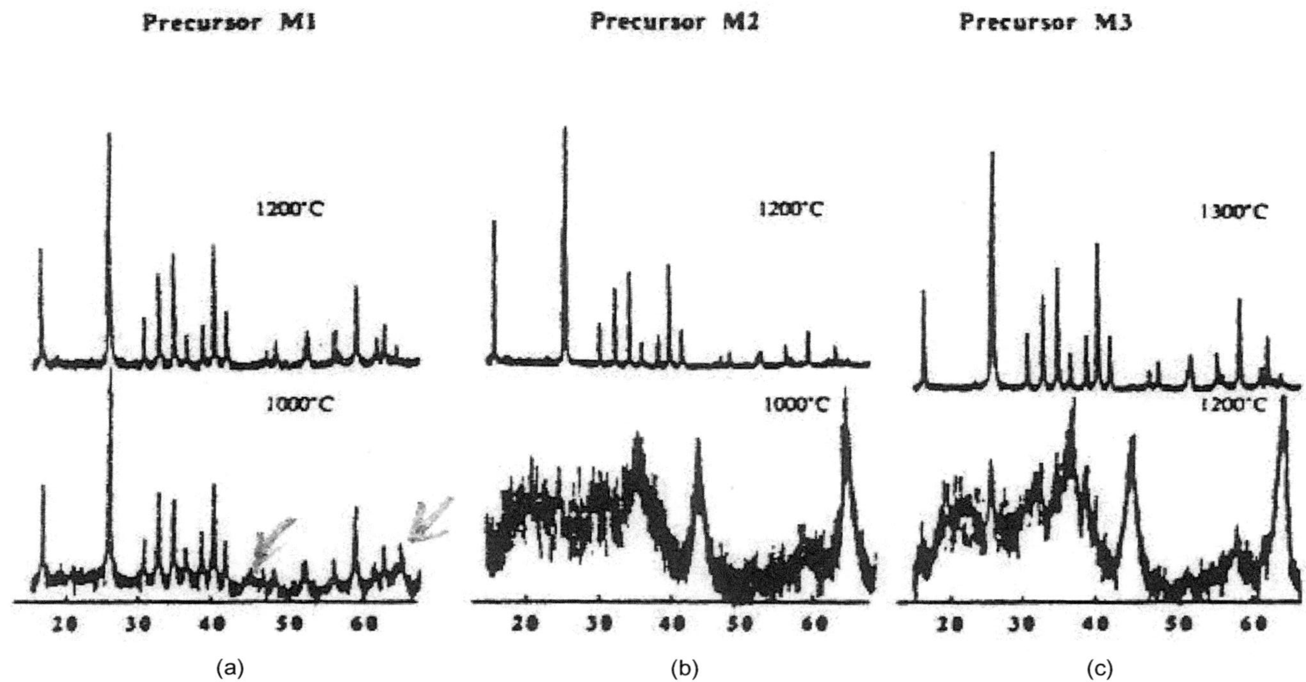

Figure 12.5 X-ray diffraction patterns of samples (a) M1, (b) M2, and (c) M3 after they were annealed to the given temperatures for 6 h. Below 900°C, the samples M1 and M2 were amorphous. The sample M3 had the same diffraction pattern for annealing temperatures between 800°C and 1200°C (Geradin et al., 1994).

M3 precursor: M3 was made by adding TEOS directly to an aqueous dispersion of boehmite and the dispersion stirred for 24 h. DTA scans of three precursors annealed at 800°C for 6 h are shown in Fig. 12.4. It exhibited only one high-temperature broad exotherm over the 1200°C–1295°C range. According to Geradin et al. (1994), it developed γ-alumina at ~800°C and existed above 1200°C. The reason was not, however, presented.

Geradin et al. (1994) opined that this predominantly formed δ-alumina as the intermediate alumina phase. This phase may be incorporated with silicon and is recharacterized by the author as a mixture of Al-Si spinel and may contain a small amount of silica incorporated θ-alumina (see Chapter 18). At 1300°C, Al-Si spinel decomposed and recrystallized into an abundant amount of the o-mullite phase. Thus, the three precursors on examination when they were being heated showed different crystallization sequences and all three paths finally reached the same crystal stage, that is, mullite. The two monophasic gels also showed different behaviors both in exhibition in DTA peaks and in mullite:spinel ratios at two exotherms. The diphasic which transformed to mullite with the exhibition of an unusually broad exotherm at 1295°C. The reason of this occurrence is to be ascertained..

12.4.6 Comparison of Crystallization Sequences of Four Mullite Gels

Cassidy et al. (1997) reproduced four mullite gels, polymeric, diphasic, colloidal (like RH gel), and monophasic (like SH gel), and compared them in terms of their crystallization sequences. They also studied the effect of the crystallization sequence on the densification of the precursor powder in each case.

- The polymeric gel prepared by Cassidy et al. (1997) showed a different behavior than that noted earlier by Yoldas (1992). Whereas Yoldas (1992) noted a major amount of tetragonal mullite (t-mullite) and a minor amount of spinel phase, these authors obtained broad peaks due to spinel in X-ray, showing a large 980°C DTA as usual, and the final transformation occurred at ~1200°C, when o-mullite formed but exotherm was not noted. Moreover, Cassidy et al. (1997) could not identify the reason for the spinel phase formation at 900°C in their polymeric gel. This gel was, however, synthesized by using a large quantity of deionized water (to the extent of 0.5 mol), in contrast to the synthesis procedure of Yoldas (1992). Densification occurred rapidly at the first crystallization step. Thereafter, it virtually ceased. This indicates that the crystallization of mullite opposes the densification of the material.

- The crystallization sequence of the colloidal gel was similar to the one noted for the polymeric gel. Since it promoted spinel formation at <900°C and induced crystallization of mullite during the exotherm at 1270°C, it densified slowly in the temperature range of 900°C–1200°C and it virtually ceased after 1200°C.
- The precursor synthesized by the nitrate method, wherein an ammonia solution was used for the precipitation of mixed gels, showed the usual 980°C exotherm with the formation of spinel and t-mullite. At ~1260°C, this t-mullite and spinel phase transformed into o-mullite without showing exotherm anymore. Like sequential crystallization, sintering was also shown to consist of two steps. First, rapid densification occurred at ~900°C and then it decreased dramatically. The second phase of densification occurred at 1100°C–1250°C. This gel largely resembles the gel prepared by Grofesik and Vago (1961).
- The gel prepared by the modified nitrate method showed t-mullite, with the exhibition of a large 980°C exotherm. The synthesis procedure may be like that of the SH gel made earlier by Okada and Otsuka (1986). Densification also occurred rapidly at the first stage. However, a very slow rate of densification was noted in three cases of gels at a temperature beyond 1200°C, which is in exception to the behavior of the polymeric gel. Two questions generally arise: (i) What should be the precursor structure that promotes the direct crystallization of mullite? (ii) Although both these authors studied gels of polymeric in nature, why phase transformation path of these vary?

Thus, the effect of precursor structure on the crystallization process in the four cases of gels varies from that shown by previous researchers. They agree with the general opinion that a homogeneous silica-alumina oxide network cannot be made from mixtures of two alkoxide components due to unmatched hydrolysis rates. In the modified nitrate method, both hydrolysis and condensation processes are allowed to be extremely slow by avoiding the use of extra water/alcohol and controlling the hydrolysis rate of TEOS by the presence of water of crystallization of ANN. The general observation is that crystallization shows an adverse effect on the densification of the four cases of gels.

12.4.7 Polymeric Gels Obtained by Changing Acidic Conditions to Basic Conditions

Nieto et al. (1998) compared the thermal changes two polymeric gels obtained by changing the acidic condition to a basic condition with those that a colloidal gel undergoes. Usually, the polymeric gel produces a mixture of t-mullite (major) and Al-Si spinel (minor) in contrast the colloidal gel, which doesn't develop t-mullite at ~1000°C. The phase transformation of the colloidal gel is analogous to that of the diphasic gel, as shown above. Nieto et al. (1998) investigated the effect of heating on the phase and structural evolution of two mullite precursors obtained by varying the method of preparations—polymeric gel PA and polymeric gel PB made under acidic and basic conditions (Yoldas, 1990; Yoldas and Partlow, 1988; described in Chapter 4)—and compared this with that of a colloidal gel. The thermal changes in the precursors were identified by the following procedures: (i) weight loss determination by TGA, (ii) thermal effects study by DTA, (iii) shrinkage behaviors by thermomechanical analysis, (iv) phase developments by XRD and transmission electron microscopy (TEM), (v) local structure by Fourier-transform infrared (FTIR) and NMR, (vi) change in the specific area by Brunauer–Emmett–Teller, (vii) skeletal density, etc. Dilatometric curves of the compact PA gel, the compact PB gel, and the colloidal gel showed major contraction at ~1250°C. At the second stage, there was a halt in the contraction profile, which is interpreted to be due to expansion during the mullitization process. At the third stage, rapid sintering started.

12.4.8 Mullite Powders Obtained by Four Methods at Two pH Conditions

Kumazawa et al. (1990, 1991) showed changes in the transformation paths of four different mullite powders of stoichiometric composition (71.8 wt % of Al_2O_3/212.2 wt % of SiO_2), obtained by the SP route, hydrolysis/gelation of a component mixture of ANN and ethyl silicate, and hydrolysis/gelation of $Al(OPr)_3$ and $Si(OMe)_4$, at two different pH conditions.

- The powder characteristics of the four precursors showed a difference in terms of exhibiting the 980°C exotherm, Al-Si spinel formation, and mullitization processes.
  - Powder S: In the SP method, mixed water-methanol solution of ANN and ethyl silicate was instantly decomposed to produce the powder. The spray-pyrolyzed powder exhibited a 980°C exotherm, attributed to the formation of mullite. Thus, the highest degree of homogeneity could be achieved by the SP method.
  - Powder C: In the hydrolysis/gelation method using components from the same source (ANN and ethyl silicate) coprecipitation was done by adding ammonia solution to the mixture.
  - Hydrolysis/gelation method: $Al(OPr)_3$ and $Si(OMe)_4$ underwent this method at two different pH conditions.

    Powder Ha: A mixture of $Al(OPr)_3$ and $Si(OMe)_4$ in a benzene solution was hydrolyzed at a neutral condition (pH = 7) by a solution of water/methanol (i.e., gelation by hydrolysis with H_2O) and marked as Ha. The Ha powder showed a single exothermic peak at 980°C of a lower magnitude than Powder S and formed Al-Si spinel in addition to predominant mullite.

    Powder Hb: A mixture of $Al(OPr)_3$ and $Si(OMe)_4$ underwent hydrolysis/gelation in different processing conditions (gelation by hydrolysis with H_2O and NH_4OH). When the mixture was hydrolyzed at pH 10 by the use of ammonia solution, the powder was marked as Hb.

All powders were amorphous to XRD in the dried state. On the other hand, two exotherms were observed, one of very less magnitude at ~980°C and one at ~1300°C for Hb and C. These peaks were assigned to be due to the formation of Al-Si spinel from an amorphous precursor and further transformation into mullite, respectively. The coprecipitated powder C behaves analogous to Hb. This precursor C, made by coprecipitation with H_2O and NH_4OH using ANN and $Si(OET)_4$, may be another kind of monophasic gel.

On the other hand, the precursor Hb was a mixture of monophasic and diphasic gels. The mullitization temperature of the as-synthesized amorphous powders increased in the order of S < Ha < Hb < C. The degree of homogeneity due to them decreased in the order of S < Ha < Hb < C.

- The powder characteristics of the four precursors obtained by different techniques of hydrolysis/coprecipitation/gelation were different in terms of the gel character. Kumazawa et al. (1991) synthesized mullite precursors from $Al(OPr)_3$ and $Si(OMe)_4$ by two methods: (i) gelation by hydrolysis with H_2O and (ii) gelation by hydrolysis with H_2O and NH_4OH.

  The powder characteristics were shown as follows: Two precursors influenced the mullitization course (Kumazawa et al., 1991). The gels obtained by gelation and/or coprecipitation method (samples marked A and B) showed two distinct exothermic peaks due to the crystallization of Al-Si spinel at 980°C and its further transformation into mullite at 1250°C. The intensities of the 980°C exothermic in the two cases were different, and the crystallization temperatures of the spinel phase in A and B, of course, varied. Thus variations in the nature of DTA curves and the corresponding changes in the thermal crystalline phases during heating of the two precursors are obvious due to a change in the synthesis technique. As soon as water was used for the hydrolysis of both alkoxides, $Al(OPr)_3$ and $Si(OMe)_4$, the resultant precursor formed a spinel phase instead of mullite. Further, as the hydrolysis condition changed from neutral water to ammoniated water, the intensity of the 980°C exotherm was much reduced and its course of transformation to mullite changed to the higher temperature side. This showed that ammoniated hydrolysis leads to the development of a partial diphasic character of the precursor gel. However, Kumazawa et al. (1991) showed that the powder processing routes for mullite synthesis did not affect their high-mechanical-strength properties when mullitization occurred fully prior to densification, for example, the precursor marked C synthesized by Kumazawa et al. (1991).

12.4.9 Transformation of Al_2O_3-SiO_2 Precursors Prepared by Different Techniques

Sanz et al. (1991) synthesized three alumina-silica (Al_2O_3-SiO_2) gels by SP, colloidal, and polymeric techniques at basic pH as per the process followed earlier by Kanzaki and Tabata (1985) and Yoldas (1992) and compared their transformation. They made polymeric gels in a basic condition only, unlike the acidic hydrolysis of TEOS adopted by Yoldas (1992) and marked those as SP, CA, and PB2. The DTA and TG curves were corroborated with the XRD studies of CA, PB2, and SP samples.

SP sample: It exhibited a sharp 980°C exotherm and no other exotherm in the DTA trace. In dilatometry, it showed one rapid expansive effect at the same temperature, and finally the XRD showed the sample to be amorphous on complete transformation to mullite.

PB2 sample: It showed an endotherm and a weight loss at ~130°C. Thereafter, it exhibited two exotherms of small magnitudes, one at 980°C and the other at 1230°C–1260°C. Accordingly, dilatometry showed two expansive effects at those temperatures. At ~1000°C, XRD showed four peaks, resembling the disputed spinel phase. Mullite formation was noticed at ~1200°C. along with the presence of a spinel phase of a reduced intensity. When the sample was further heated, to 1300°C, complete mullitization took place.

CA sample: It showed two endotherms and likely two times the weight loss, at 170°C and 490°C. DTA showed the absence both exotherms related to spinel/mullite formation at 980°C and spinel to mullite formation at ~1250°C. However, it showed a high-temperature exotherm at ~1285°C due to spinel-to-mullite crystallization. As such, it showed a more rapid contraction up to 1285°C and only after that an expansive effect was observed.

An ^{27}Al MAS-NMR study of the thermal evolutions of the three precursors SP, CA, and PB was done by Sanz et al. (1991), which showed considerable differences in the short range order. The NMR characterization of these three precursors showed different paths of mullitization.

Colloidal and polymeric samples heated between 300°C and 1000°C exhibited two main resonances, one at 60 ppm (tetrahedral) and one at 5 ppm (octahedral) coordination of Al. The pentahedral peak was small in the polymeric sample prepared in a basic condition, and it was absent in CA. Both CA and PB2 in the temperature range 1000°C–1300°C exhibited an increase in the tetrahedral component (58 ppm), with appearance of a new tetrahedral peak at 42 ppm, assigned to 3:2 mullite.

In contrast, three Al peaks, at 60, 35 (pentahedral), and 6 ppm, were noted in the SP samples heated between 400°C and 900°C. The intensities of the tetra- and pentahedral peaks were very high, and there was shifting of the peaks with a rise in temperature. The pentahedral peak disappeared completely with the appearance of the octahedral Al resonance and the overall Al spectra corresponded to o-mullite. These DTA, XRD, and MAS-NMR results suggest that in a basic condition, polymeric gel develops a diphasic character, resembling that of CA.

12.4.10 Comparison of Transformation of $3Al_2O_3$-SiO_2 Precursors with That of Oxide Mixtures

Hamano et al. (1986) first synthesized a coprecipitated powder from AIP and methyl silicate by using ammonia solution and examined the properties and microstructures of the fired bodies. Then they compared the transformation of the two $3Al_2O_3$-SiO_2 precursors (prepared by coprecipitation and colloidal method) with that of the route taken by the oxide mixtures. XRD of the coprecipitated material showed to be amorphous up to ~950°C. It crystallized at the 980°C exotherm to the spinel phase, and exclusively thereafter, this crystalline spinel phase was present up to < 1250°C and further transformed at the second exotherm, at ~1250°C, to mullite phase. When mullite was fired at 1780°C, it showed a high bulk density of 3.14 gm/cc. From 1500°C to 1690°C, the lattice parameters a_0 and b_0 decreased, but a_0 increased at 1780°C. At about 1780°C, the specimen obtained by coprecipitation showed the smallest amount of glass phase in comparison to the sol mixture and oxide mixtures.

12.4.11 Comparison of Transformations of Mullite Precursors Obtained from the Same Components under Five Processing Variables

Haque (2000) and then Chakraborty (2008) synthesized three types of SH gels and two types of coprecipitated gels (obtained from components from the same source) with five variables in the processing conditions. That is, they synthesized three kinds of monophasic mullite gels—by hydrolysis and subsequent gelation of the organic silicon component in the presence of a water-soluble Al component—at three different pH values (as per Chapter 3), and a coprecipitated mullite gel and an SD precursor at two different pH values. They then compared the transformations of the mullite precursors to identify the difference in the growth processes of the mullite.

The mullite formation processes of six different precursors were monitored by QXRD. These results are shown in Fig. 12.6.

SHI gel: Figure 12.6 shows the percentage of mullite formed from six precursors at different temperatures. At 1000°C, SHI formed a mullite phase only (~62.8%). The formation of a large quantity of mullite in this case at ~1000°C indicates that the precursor directly transformed into a mullite phase. The rest of the mass was still amorphous and called the silica-rich aluminous phase (A). The quantity of mullite phase rose to ~86.6 wt % at 1100°C and slowly further increased with an increase in temperature, to ~ 91.1% at 1600°C.

SHII gel: In this case, a major amount of mullite and a minor amount of Al-Si spinel were formed at the first exotherm. It developed mullite fully at the second exotherm. On comparison, the percentage of mullite formed at each temperature in this case was always less than in the case of SHI gel (Fig. 12.6).

SHIII gel: At 1000°C, the SHIII gel showed the presence of an Al-Si spinel phase and it formed traces of the mullite phase. At 1100°C, it developed mullite to the extent of 58.4%. The quantity of mullite increased with temperature and at 1600°C, the mullite content was 84.2%.

SHIV gel: In the temperature range of 1000°C–1100°C, the percentage of mullite formed was negligible and it contained only the Al-Si spinel phase. Mullitization occurred close to the second exothermic peak temperature. At 1200°C, rapid mullitization took place, amounting ~72.7%, and then it increased gradually from 1200°C to 1400°C. At 1600°C, the mullite content reached ~82%.

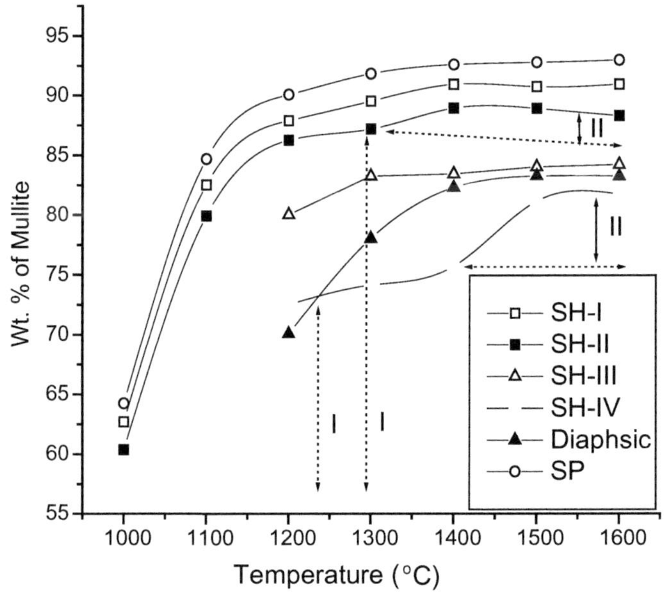

Figure 12.6 Growth process of mullite in different types of SH and coprecipitated gels.

Diphasic gel: Mullitization started abruptly at ~1200°C. At >1200°C, a slow increase in mullite formation occurred with an increase in the firing temperature up to 1600°C. In this case, the percentage of mullite formed at 1200°C was found to be less than that of SHIV, indicating that mullite formation in this case was delayed, and at 1600°C, it attained a slightly higher value of mullite than in SHIV.

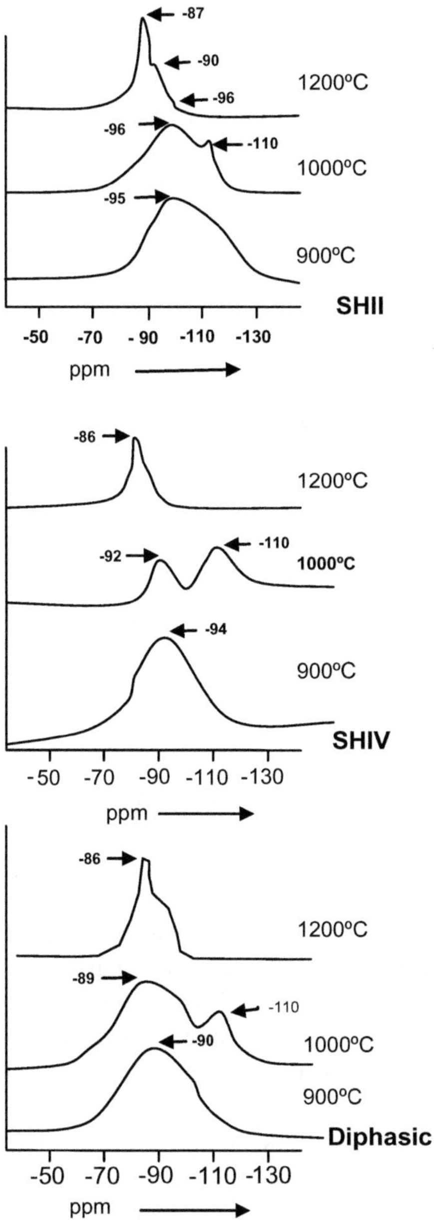

Figure 12.7 ^{29}Si MAS NMR data of monophasic and diphasic gels heat treated to 900°C, 1000°C, and 1300°C.

SP precursor: In this case, mullite formation took place rapidly in comparison to all other gels (Fig. 12.6). It showed a higher value of mullite at each temperature step in comparison to other precursors.

^{29}Si MAS-NMR data are shown in Fig. 12.7. As projected in the previous chapters and above, both parameters, component source and synthesis method, jointly lead to great variations in the 980°C phase development and the path of mullitization. Different components and various techniques applied are briefly shown above.

12.5 Various Gel-Mullite Transformation Processes: Problems and Questions

The concise results above show that both thermal evolution and phase transition are strongly influenced by various chemical processing methods. Component oxides from different sources produce mullite gels/precursors of different natures and types. The vast literature on the synthesis and phase development of mullite precursors indicates that the formation process of mullite is highly controversial. The results that follow, obtained by different authors, vary largely.

12.5.1 DTA and Densification Study

Yoldas and Partlow (1988) observed a high-temperature exotherm at 1230°C only due to crystallization of α-Al_2O_3. In the DTA studies of several monolithic gels and powders prepared from directly from AlOBu and synthesized in acetone, hexane with a few drops of acetic acid, pH 2 powder, and pH 9 powder, Colomban (1889) observed exotherm due to corundum crystallization only. However, any exotherm due to γ-Al_2O_3 was noted during the DTA study of the $Al(OH)_3$ gel.

- Variations by DTA results: It was noted that both temperatures of occurrence and variations in the intensities of the 980°C exotherms among precursors synthesized by gelation, coprecipitation, and codecomposition were largely different and mullitization processes also showed considerable variations among them (Hoffman et al., 1984; Horte &

Wiegmann, 1956; Demediuk & Cole, 1958; Kanzaki & Tabata, 1985).

- It is generalized that the DTA intensity at 980°C is related to crystallization into mullite. However, there are some inconsistencies too. In some cases, the 980°C exotherm was not shown to be related to mullite formation (Li & Thomson, 1991; Douy, 1991; Jaymes et al., 1996).
- Significant differences in the peak positions and heights of 980°C exotherms of two aluminosilicate gels prepared from TEOS and AlOBu and differing in their pore structures were noted by Fahrenholtz et al. (1991).
- There is no direct relation between the intensity of the 970°C exotherm and the powder crystallization into mullite (Douy, 1991). However, it was experimentally shown that the 980°C exothermic peak intensity has a distinct relationship with the formation of mullite (Chakraborty & Ghosh, 1988).
- Four types of mullite gels were reproduced by Cassidy et al. (1997), and the effect of crystallization sequence on the densification of the precursor powders was studied.
  (i) Polymeric gels prepared by Cassidy et al. (1997) showed a different behavior than that noted earlier by Yoldas (1992). Whereas Yoldas (1992) noted t-mullite (major) and a spinel phase (minor), these authors obtained broad peaks due to the spinel phase exhibiting a large 980°C DTA as usual and the final transformation occurred at ~1200°C, when o-mullite formed without exhibiting any exotherm.
  (ii) Moreover, it was not known why spinel phase formation was noticed at 900°C in this polymeric gel. This gel was, however, synthesized by using a large quantity of deionized water (to the extent of 0.5 mol), in contrast to the procedure followed by Yoldas (1992). Densification occurred rapidly at the first crystallization step. Thereafter, it virtually ceased. These results indicate that the crystallization of mullite opposes the densification of the material.
  (iii) The crystallization sequence of the colloidal gel was similar to the one noted the polymeric gel earlier. Since

it promoted spinel formation at < 900°C and induced crystallization of mullite during the exotherm at 1270°C, it densified slowly in the temperature range of 900°C–1200°C and it virtually ceased after 1200°C.

(iv) The precursor synthesized by the nitrate method, wherein an ammonia solution was used for the precipitation of mixed gels, showed the usual 980°C exotherm with the formation of spinel and t-mullite. At ~1260°C, this t-mullite and spinel phase transformed into o-mullite without showing exotherm anymore. Like sequential crystallization, sintering is also shown to consist of two steps. First, rapid densification occurred at ~900°C and then it decreased dramatically. The second phase of densification occurred at 1100°C–1250°C. This gel largely resembles the gel prepared by Grofesik and Vago (1961).

(v) The gel prepared by the modified nitrate method showed t-mullite, with the exhibition of a large 980°C exotherm. The synthesis procedure may be like that of the SH gel made earlier by Okada and Otsuka (1986). Densification also occurred rapidly at the first stage. However, a very slow rate of densification was noted in three cases of gels at a temperature beyond 1200°C, which is in exception to the behavior of the polymeric gel. Two questions generally arise: (i) What should be the precursor structure that promotes the direct crystallization of mullite? (ii) Although both these authors studied gels of polymeric in nature, why phase transformation path of these vary?

(vi) Thus, the effect of precursor structure on the crystallization process in the four cases of gels varies from that shown by previous researchers. They agree with the general opinion that a homogeneous silica-alumina oxide network cannot be made from mixtures of two alkoxide components due to unmatched hydrolysis rates. In the modified nitrate method, both hydrolysis and condensation processes are allowed to be extremely slow by avoiding the use of extra water/alcohol and controlling the hydrolysis rate of TEOS by the presence of water of crystallization of ANN. The general observation is that crystallization shows an adverse effect on the densification of the four cases of gels.

(vii) The intensity of the 980°C DTA exotherm (its nature and peak height) is not related to the composition of the mullite gel (Low & McPherson, 1989; Sales & Alarcon, 1996).

12.5.2 XRD Study

Crystallization processes of precursors synthesized by Pask et al. (1987) and Cassidy et al. (1997) showed differences although both were polymeric gels. Later, the authors noted the formation of only the spinel phase other than t-mullite in the temperature range of 900°C to 1200°C by XRD study (Cassidy et al., 1997, Table 1). But TEM results have confirmed the presence of nanocrystalline mullite. The Type I precursor synthesized by Schneider et al. (1993) from metal alkoxides by the SH method (by keeping the alkoxides in contact with air for 14 days) exhibited a 980°C exotherm with the formation of t-mullite and a small quantity of the spinel phase. However, in a different processing condition—by RH with excess of water using ammonia at $pH > 10$—the precursor named Type II formed boehmite and then formed spinel without exhibiting an exotherm.

Mullite gels (designated as colloidal gels as per experimental observations as shown in Chapters 2–7) prepared by Yoldas (1992), Pask et al. (1987), and Li and Thomson (1991); coprecipitated gel using urea prepared by Yamada and Kimura (1962); and coprecipitated gel using urea prepared by Campos et al. (2002) exhibit significant DTA peaks of different peak heights and variation in phase development behaviors (Table 12.2).

The exhibition of the third exotherm in lieu of the 980°C exotherm is characteristic of DTA traces of a diphasic gel as noted by several authors, including the present author (Figs. 12.6 and 12.7). However, the X-ray analysis of a raw diphasic gel must contain pseudoboehmite. Contrarily, this phase was not crystallized during the gelation process in the colloidal gel samples of Pask et al. (1987) and Campos et al. (2002).

Table 12.3 DTA and phase crystallization of some mullite gels synthesized by four different research groups

	Pask et al.	**Li and Thomson**	**Yamada and Kimura**	**Campos et al.**
	(1987)	**(1991)**	**(1962)**	**(2002)**
I. XRD of raw gel	Colloidal gel	Colloidal gel (CG)	Coprecipitated	Coprecipitated
	Amorphous(A)	(A)	(A)	–
2. DTA curve	1st exo – Nil	1st exo – present	1st exo – present	1st exo – Nil
	2nd exo –Nil	2nd exo –present	2nd exo – present	2nd exo – Nil
	3rd exo –	3rd exo – absent	3rd exo – absent	3rd exo occurred
	at 1270°C	–	–	in the range 1237°C–1350°C
3. Phase at	~1000°C- Spinel	At 1st exo – Spinel	At 1st exo – Spinel	Spinel up to 1100°C
	At 2nd exo	At 2nd exo – Mullite	to >1200°c-mullite	
	After 3rd.- Mullite exo.	–	–	After 3rd exo.- Mullite

12.5.3 IR Study

In the case of the SH gel and the Type I precursor, t-mullite formed at ~1000°C. With an increase in the firing temperature, the relative ratio of the 1130 cm^{-1} peak to the 1170 cm^{-1} peak changed (Fig. 8.1). This was explained as related to the change in the composition of the mullite from ~70 mol % of Al_2O_3 to 60 mol % of Al_2O_3. This prediction has to be checked.

In the case of the polymeric gel as prepared by Hirata et al. (1989) and in the case of the single-phase gel, abrupt crystallization occurred at ~1000°C with mullite (minor) and a spinel (major) phase. Concurrently, an IR study showed a sudden change in the frequency shift (Fig. 8.1). The reason is yet to be established.

In the case of the colloidal gel, a band was detected at 580 cm^{-1} at a temperature as low as 500°C in comparison to the polymeric gel, which showed no peak at the said temperature in FTIR spectra. This band was assigned to the formation of the spinel phase by Nieto et al. (1998).

12.5.4 NMR Study

The polymeric gel on heating at 500°C showed three Al sites, of which, the major Al sites were penta-coordinated (Yoldas, 1992, Fig. 5.10; Schmucker & Schneider, 1999). This indicated that the gel was homogeneous in character and developed an inorganic network prior to the first exotherm. As a result, a sudden change in the AlO_5 band was noted in the ^{27}Al MAS-NMR spectra when a major quantity of the t-mullite phase was developed. Why was there a sudden change?

Type I and the spray luminosi precursor also behaved as noncrystalline luminosilicate since these showed three Al NMR peaks (Sanz, 1991). At 500°C, of the three Al sites, the major Al sites were penta-coordinated, which easily and suddenly changed in their Al spectra with the development of a major quantity of mullite phase at the first exotherm (Yoldas, 1992, Fig. 9.3). The nature and intensities of the Al spectra corresponded to that of 3:2 mullite.

As in DTA and XRD results, a great variation was also observed in MAS-NMR studies. In the case of the monophasic gel, a change in the Al environment occurred on heating to the temperature of the first exotherm, where mullite first crystallized. The ratio of the tetrahedral Al to octahedral Al was noted to be 0.48 after deconvolution, which could not explain the formation of either 3:2 or 2:1 mullite (Komarneni & Roy, 1986).

Yoldas (1992, Figs. 9.3 and 9.11) also showed strikingly different Al spectra for heat-treated monophasic and diphasic mullite gels. The colloidal gel showed a difference in the chemical bonding in comparison to the polymeric gel (Schmucker & Schneider, 1999; Yoldas, 1992). Its Al spectrum at 500°C did not show pentahedral Al but showed a small peak of tetrahedral Al with a large octahedral Al (Yoldas, 1992, Fig. 9.11). Even heating at 1000°C, a little change in the spectrum was noted. It changed at 1300°C, when mullite developed.

In the case of Type II (may be diphasic in character) mullite formed at 1150°C–1250°C, and in the case of coprecipitated gel (Type III), mullitization started at temperatures greater than or equal to 1200°C. What significant changes occur when these two gels are heated further, up to 1650°C? Are they due to a change in the composition of the mullite or a change in the crystallinity of it or both in the said temperature range?

These precursors showed that the relative phase developments, for example, (i) the ratio of Al-Si spinel to mullite, (ii) the formation of mullite and its nature, (iii) the occurrence of different exotherms in DTA, and (iv) the different reaction paths leading to mullitization, depend on both the source materials of the silica and alumina components and the technique of gelation.

12.5.5 Water Used for Hydrolysis

Using water as a hydrolysis medium, Prochazka and Klug (1983), Suzuki et al. (1990), Sakurai et al. (1988), the SGM gel made by Schneider et al. (1992), and GYW by Taylor and Holland (1993) invariably showed spinel formation before mullitization. Therefore, the SGM gel, which formed below pH 7 and where water was used for the hydrolysis of TEOS and AlOBu, should be called a colloidal gel.

In the absence of water for hydrolysis, t-mullite was the predominating phase, for example, Yoldas's polymeric gel (1992), Type I by Schneider et al. (1993), and GNW by Taylor and Holland

(1993). Even in the presence of water at pH 7, the A series powder by Mitachi et al. (1990) showed t-mullite formation.

12.5.6 pH in the Gelation Process

Nieto et al. (1998) observed the difference in the phase evolution behaviors in their two polymeric gels synthesized in acidic and basic conditions.

Fonseca et al. (1997) synthesized a mullite gel from TEOS and ANN and correlated the exothermicity of the 980°C peak with use of an acid or basic catalyst. They showed that mullite formation was favored in an acidic condition with low water content.

12.6 Scientific Community: Controversies and Predictions

- Mullite precursor phase—its nature and structural characteristics: The thermal transformations of metakaolin and some types of mullite gels are analogous. Like metakaolinite, it is amorphous to X-ray and electron diffraction. The standard methods are inapplicable to the analysis of its structure. One is forced to rely upon indirect evidence for its elucidation.
- Endothermic dip (D): The D preceding the first exotherm plays an important role in the transitional stage to crystallization at the ~980°C exothermic reaction. This is generally ignored. General DTA and TGA curves do not provide convincing information on this matter.
- First exotherm: The 980°C exothermic event and identification of reaction products that result from the exothermic reaction are essential steps. Is the exotherm due to the spinel phase? Does a fraction of the mullite phase also contribute to this exotherm? Or is nucleation of the mullite solely responsible for the exotherm? The interpretation of this is still a matter of speculation.
- Spinel phase: Regarding the crystal structure and chemistry of the cubic spinel phase formed from the crystallization of mullite gels as well as metakaolinite, there is great

disagreement over its nature, the temperature at which transformation takes place, and the composition of this spinel phase.

- Primary mullite phase: What is the composition of the primary mullite formed at approximately the first exotherm, and does the ratio of Al and Si in the mullite change with increasing temperature? Determination of the lattice parameter of the primary mullite may be erroneous since all its intensity peaks are broad, requiring annealing prior to measurement (Okada et al., 1990). The formation of mullite of the tetragonal variety (t-mullite) at 1000°C–1050°C requires complementary proof.
- Secondary mullite: The evolution of mullite occurs via three possible routes, which are still not clear.
- The mechanisms of different paths of mullite formation are still not clear.
- The two more exotherms exhibited in the high-temperature region of DTA traces of some mullite gels synthesized by Chakraborty (1994) need explanation.
- Different characters of mullite precursors and their classifications are still under study.

These problems hinder arriving at a definite general conclusion in the interpretations of thermal transformations of different types of mullite gels. It is hoped that the pertinent problems in the gel-to-mullite reaction series are chronologically presented in the individual chapter of the book.

12.7 Predictions Made by the Scientific Community

There are a lot of predictions regarding the mullite gel-to-mullite transformation processes in the literature as a whole. These are not always a matter of speculation. Some of the findings are experimentally based. Yet, one theory does not explain other established results of the methods applied in the characterization process.

12.7.1 Characterization of the Spinel Phase

The main reasons for the difficulties in characterizing the phases are the following:

- It may be simple γ-Al_2O_3 by XRD, but an alkali leaching study predicts Al-Si spinel. The nature of the spinel phase, whether it is Al-Si spinel or Al-spinel-γ-Al_2O_3, is rather difficult to determine because it is poorly crystalline; the crystal size is small and shows only a few XRD peaks, those too very broad and diffuse; and even on heating for as long as 28 h at 950°C, it doesn't develop into a well-crystallized state. Now the reader will realize why it is so difficult to identify. Thus, the grain size of the spinel phase in the nanometer range makes the structural analysis most challenging. Not only this, the problem is critical considering the composition of Al- Si spinel, that is, the content of Si in the spinel phase.
- It is also predicted that the composition of spinel may be analogous to the composition of 3:2 o-mullite, and it is designated as cubic mullite (c-mullite), that is, it contains as high as 28 wt % of SiO_2. On the contrary, the spinel phase was ascertained to contain SiO_2 to the extent of 8–11 wt % only by analytical TEM–energy-dispersive X-ray spectroscopy (EDS) analysis. Therefore, the question is, how much silicon is present in the spinel phase?
- In most of the repetitive alkali leaching techniques, the end point of the SiO_2(A)-alkali reaction is not maintained properly, which may lead to an erroneous prediction in determining the silica content in the spinel phase by EDS and IR studies.
- MAS-NMR was applied to characterize disordered solids like Al_2O_3-SiO_2 precursor phases and the spinel phase to study the local atomic environment of Al and Si, and it is predicted that the 980°C exotherm is not associated with the formation of a particular phase.
- A large quantity of the noncrystalline aluminosilicate residual phase also remains as a coexisting phase in the 980°C decomposition product of the precursors. This clarifies how difficult it is to identify the spinel phase in terms of whether it is cubic Al_2O_3 or Si-bearing spinel, the reasons being its low

crystallinity and the fact that it is admixed with an average 20–30 wt % of the said noncrystalline aluminosilicate phase. Due to these reasons, a composition analysis of the spinel phase even by TEM, EDS, and MAS-NMR studies requires rechecking.

12.7.2 Mullitization Reaction

There are three distinct cases of mullite reactions: two types of monophasic gels and a diphasic gel. The mullitization process in other types of monophasic gels and in a diphasic gel may occur through a degree of segregation of silica and alumina by way of decomposition of the mullite precursor and then a more extensive diffusion process of Al and Si elements during the heating process to nucleate the mullite. The author, however, theoretically predicts a simple structural transformation of the orderly mixed Al-Si spinel crystal into the mullite phase. Thus, the contrast between the segregation-recombination path and the continuous structural transitional path is still a subject of study.

12.7.3 Analysis of Previous Studies and Probable Solutions

Both analysis and solutions of different problems as pointed out in this chapter will be continued in the remaining chapters of this book, with a conclusion provided at the end.

For comprehensive research and analysis, the complete knowledge of the following items is necessary:

- The basic theory on the structural order of the noncrystalline precursor phase and weakly crystalline intermediate spinel phases
- Mechanisms of mullite formation by all probable routes
- The effect of the processing variables versus the phase evolution of different routes of mullite production
- Characterization of mullite precursor phases
- Characterization of the intermediate 980°C heated spinel phase
- Characterization of the mullite phase

- Establishment of the different routes of mullite formation
- Classification of mullite precursors

12.8 Summary

High-temperature DTA scans of mullite precursors chemically synthesized by various techniques by different researchers have been compared with the phases crystallized during the heat treatment process by QXRD in this chapter.

Monophasic gel (obtained by the aqueous sol-gel method): A monophasic gel synthesized by the aqueous sol-gel method showed sudden crystallization to mullite after the exothermic peak temperature at 980°C (Insley & Ewell, 1935; Chakraborty & Ghosh, 1988; Ossaka, 1961). The quantity of mullite and the formation of any additional spinel phase seem to depend on the dilution of the silica sol used and the composition of the mullite gel. The role of silica sol is a criterion, and its dilution controls the size of the Si-OH colloid micelles.

Monophasic gel RH vs. monophasic gel SH: The monophasic gel/SH transformed into weakly crystalline mullite (major) and a spinel (minor) phase at ~1000°C. Contrarily, the RH gel crystallized into a weakly crystalline spinel phase (major) only. The criterion is the pH prevailing during the gel synthesis. The pH obviously influences the size of the colloids of both component hydroxides during polymerization and plays a role in the chemisorption process. Mullite formation ensued in the SH gel and increased rapidly at ~1000°C in the first step and then slowly in the second step, from 1100°C to 1400°C (Okada & Otsuka, 1986). However, in the RH gel, mullite formation ensued and increased rapidly at ~1100°C in the first step only.

Monophasic gel vs. diphasic gel: These two gels showed two radically different phase transformation sequences (Hoffman et al., 1984). The crystallinity and particle size of the alumina component play a major role. A single-phase mullite composition gel using tetraethoxysilane and ANN as precursors was repeatedly synthesized by Komarneni and Rustiser (1996). Similarly, a diphasic mullite gel was prepared from colloidal silica and boehmite using concentrated nitric acid to maintain the pH at 3. However, instead of drying in a water bath, these gels were dried in critical point drying conditions. The single-phase xerogel showed a 980°C exotherm while the counterpart did not.

Monophasic and colloidal gels: The effect of the hydrolysis condition, particularly the gelation time, on the mullite formation reaction from a number of single-phase gels and a colloidal gel made from components from the same source was studied by Li and Thomson (1991). They observed varying degrees of Al-Si-spinel-cum-t-mullite formation during the heating process at ~ 980°C. The following points are noteworthy:

- The intensity of the first exotherm in DTA is dependent on the time of gelation. It decreases with an increase in the gelation time from 1D to 2D to 2WC. The gelation time increases the size of hydroxide gel polymers.
- Mullite formation as shown in DXRD in a single-phase gel occurred in two stages, that is, at 980°C and at ~1200°C. Thus, mullite formation shows a direct relation with gelation time.
- The colloidal gel (amorphous) showed two insignificant exotherms, one at 980°C and one at 1250°C, and tentatively showed a one-step mullitization at ~1250°C.
- The diphasic gel did not exhibit a 980°C exotherm but transformed into mullite in one stage at the ~1300°C exotherm, which was higher than that in the colloidal gel. Mullite formation in the first stage appeared in the following order: 2WC > 2W > 2D > 1D > 2H.

Monophasic gels with variation in pH (acidic vs. basic): Huang et al. (1997) that the mullitization temperature is an important criterion to assess the mixing scale or the degree of aluminosilicate (–Si–O–Al–) bonds present in the precursor system. The pH prevailing during gelation also influences the mixing scale of the two silicon aluminum components. At pH 1.5, the gel structure was amorphous. With an increase in the pH of acidic gels, the T_m of the 980°C exotherm decreased. But at pH 11.5, the gel contained boehmite, bayerite, and gibbsite. The basic gels, on the other hand, did not exhibit any exotherm. The reason has to be ascertained.

Polymeric gel vs. colloidal gel vs. gels prepared by a water-free approach: The effects of precursor synthesis on the crystallization of the intermediate phases are different (Yoldas, 1980; Pask et al., 1987). The polymeric gel produced a mixture of t-mullite (major) and spinel (minor) in contrast the colloidal gel, which didn't develop t-mullite at ~1000°C. Secondly, the spinel phase transformed into mullite with the exhibition of a faint peak at 1235°C during the DTA run of the polymeric gel. But the spinel phase, which was noted at a lower temperature in the case of the colloidal gel, showed a broad exotherm at ~1267°C due to transformation into mullite by a separate mechanism.

The effect of hydrolysis on the intensity of the ~980°C DTA peak of the stoichiometric mixture of $Al(OC_4H_9)_3$ and $Si(OC_2H_5)_4$ in alcohol with (i) excess of water, (ii) short duration in a humid atmosphere condition, and (iii) long duration in humid atmosphere condition was studied by Yoldas and Partlow (1988). The result showed that as the water was reduced, the exothermic peak significantly increased. The inverse relation between the water used for hydrolysis of two alkoxides and the intensity of the 980°C exotherm predicts that the water used significantly affects the hydrolysis step of alkoxides, which has a direct relation with the gelation mechanism of the mixed gels.

Slow monolithic gel vs. rapid multiphasic powder synthesis: By the slow route, Colomban (1989) obtained mullite monoliths while a powder was obtained by the rapid route from the same component mixture. The monolith specimen exhibited a strong first exotherm and showed the formation of the spinel phase on heating at 1050°C. On continued heating to 1100°C, partial mullitization and an amorphous band at 0.4 nm were noted. Interestingly, the latter phase existed even on heating at higher temperature at 1100°C for 4 days. On heating at 1400°C for 1 day, purely crystalline mullite with complete elimination of the amorphous phase occurred. The monolith remained optically clear up to 1400°C. In comparison, in the rapid route, the coprecipitated powder exhibited a less intense 980°C exotherm. The reasons of the reaction processes starting from gel/powder synthesis to the exhibition of DTA events and finally to crystallization to mullite are to be established.

Coprecipitation powder diphasic gel vs. polymeric gel as per Yoldas (1992)

Rajendran et al. (1990) prepared two mullite gels, one by the coprecipitation technique using an ammonia solution at pH 9 and one by the polymeric precursor method. The formation of mullite from the coprecipitated powder required higher temperatures and was much slower than was the case for the polymeric precursor. The reason is that the former gel was diphasic in nature as it contained bayerite. It formed a weakly crystalline spinel phase with a minor quantity of θ-alumina at 1225°C on heating and an abundant quantity of mullite at ~1350°C, with the elimination of broad intensity peaks of the major spinel phase with θ-alumina as well. On the contrary, the polymeric gel (PB) decomposed to spinel at the first exotherm and then spinel crystallized to mullite at the second exotherm.

Polymeric gel and colloidal gel vs. nitrate gel

Besides polymeric and colloidal gels, Yoldas (1992) also synthesized a third kind of gel—a nitrate-derived gel—by using $Al(NO_3)_3{\cdot}9H_2O$ and $Si(OCH_3)_4$. The nitrate gel exhibited a major first exotherm due to spinel and a minor quantity of mullite. The disappearance of the former phase accounted for mullitization at the second exotherm. This result differs with that of Okada and Otsuka (1996). They proposed three different ultrastructure for three kinds of Al_2O_3-SiO_2 gels. Mostly, the homogeneous structure occurs in the polymeric gel. The colloidal gel is less homogeneous, and it is limited to the size of alumina colloids.

Polymeric gel (Type I) vs. diphasic gel (Type II) vs. coprecipitated gel (Type III): Type I (as per C by Yoldas & Partlow, 1988), Type II, and Type III were synthesized by Schneider et al, (1993) from two components from the same source but hydrolyzed by varying quantities of water at two different pH levels, which led to a change in the hydrolysis rates of the components. These showed a wide variation in the intensities of the 980°C/1250°C exotherms with a variation in the crystallization of mullite and spinel. For example, Type I (prepared with a very low water content and thus undergoing a very a hydrolysis) is probably monophasic in nature, exhibiting a sharp 980°C exotherm with the formation of a major quantity of t-mullite in addition to crystallization of the spinel phase (minor). The type III gel (prepared using a large quantity of water at pH < 10, where the hydrolysis was fast), probably a coprecipitated gel, exhibited the first exotherm of a lower intensity than Type I and crystallized to spinel only instead of major t-mullite, as noted in Type I. This intermediate spinel further transformed to mullite at the second exotherm. Type II (prepared using a large quantity of water at pH > 10, where the hydrolysis was very fast when boehmite was precipitated out), which is diphasic in nature, followed an absolutely different path to mullite compared to the other two cases of gels. Using component materials from the same source but differing in processing conditions, Ruscher et al. (1996) resynthesized Type I, Type II, and Type III, which showed varying precursor characteristics.

Polymeric gel (M1) vs. coprecipitated gel (M2) vs. diphasic gel (M3): Three types of mullite precursors were synthesized by Geradin et al. (1994) to examine their crystallization paths. Two mullite precursors were made from materials from the same sources (TEOS and AIP) but in different hydrolysis conditions, particularly the water quantity. Aqueous prehydrolyzed TEOS was used for the synthesis of M1 (as per Yoldas, 1992) at pH 1, whereas a mixture of TEOS and AIP in 2-propanol was hydrolyzed with water in the case of M2. These two gels showed different behaviors both in exhibiting DTA peaks and in the mullite:spinel ratio at the two exotherms. MIII was diphasic and transformed to mullite, exhibiting an exotherm at 1295°C via the δ-alumina phase as an intermediary phase. Thus, the three precursors on examination during heating showed different crystallization paths.

Phase transformation of mullite precursors synthesized at four variable processing conditions: Four types of mullite gels were resynthesized by Cassidy et al. (1997) to study the effect of the crystallization sequence on the densification of the precursor powders: polymeric, diphasic gel, colloidal (like RH gel), and monophasic (like SH):

- The polymeric gel prepared by them showed a difference in the phase transformation behavior and DTA exotherms than that noted by Yoldas (1992). The probable reason was the use of a larger quantity of water by Cassidy et al. (1997).
- The colloidal gel/diphasic showed a similar crystallization sequence and DTA exotherm to those shown in earlier cases.

- The nitrate gel prepared in the ammonia solution showed an analogous phase transformation to that prepared by Grofesik and Vago (1961).
- The precursor made by the modified nitrate method showed a thermal behavior and exhibitions by DTA analogous to the precursor prepared by Okada and Otsuka (1986). According to them, for the SH process, avoiding the use of additional water or alcohol might be beneficial for t-mullite formation at the first exotherm.

Monophasic gel in acidic vs. basic conditions/Rapidly dried acidic gels: Alumina-silica gels and powders for two compositions (in mol %), $60Al_2O_3{\cdot}40SiO_2$ and $64.3Al_2O_3{\cdot}35.7SiO_2$, from TEOS and ANN in varying conditions were reproduced by Fonseca et al. (1997). Besides considering the effect of acids, bases, and hydrolysis with water/alcohol, they emphasized the effect of drying of the hydrolyzed sol mixture in some ways. For example, (i) by evaporating the hydrolyzed sol mixture on a hot plate, (ii) by dropping it on a fixed hot plate, (iii) by dropping it on a rotating hot plate, and by reducing the size of the droplets by the spraying technique, they tried to correlate the exothermicity of the 980°C peak with mullite formation. On the basis of the results, they showed that thermal evolution processes are strongly dependent on the synthesis methods.

Colloidal gel (CA) vs. polymeric gels (PA and PB): The effects of heating on the phase and structural evolution of the colloidal gel (CA) and on two polymeric gels prepared in acidic and basic conditions (PA and PB) were compared by Nieto et al. (1998). Both latter gels showed spinel formation at the first exotherm and mullite formation at the second exotherm. Contrarily, spinel formation started earlier in the CA sample and mullite formation occurred at ~1200°C.

SP vs. hydrolysis/gelation vs. hydrolysis/coprecipitation: Four different mullite powders with stoichiometric composition (71.8 wt % of Al_2O_3/28.2 wt % of SiO_2) were prepared by different techniques—(S) SP, (Ha) hydrolysis with water, (Hb) hydrolysis with aqueous ammonia, and (C) hydrolysis/coprecipitation—by Kumazawa et al. (1990). On the basis of on the intensity of the 980°C exotherm, they correlated it with the degree of homogeneity of the precursors. It was reported to decrease in the order of Ha,

Hb, and C powders, while the spray-pyrolyzed powder was the most homogeneous and showed a sharp exotherm with the formation of t-mullite.

Mullite precursors in four different processing conditions: The nature of the DTA curve and thermal crystalline phase change of five precursors made from $Al(OPr)_3$ and $Si(OMe)_4$ were pursued by Kumazawa et al. (1991). They showed as many as three paths of mullitization for the five precursors: (i) precursors obtained by gelation/coprecipitation developed mullite with Al-Si spinel as an intermediary, (ii) mixed gels crystallized by the diphasic route, and (iii) mixed calcined powders formed mullite at ~1340°C, which was the highest temperature of mullitization. Thus, the DTA exothermic behaviors of these precursors are different.

Spray pyrolyzed, colloidal, and polymeric: Sanz et al. (1991) resynthesized alumina-silica gels by SP, which were colloidal as per the process followed earlier by Kanzaki and Tabata (1985) and Yoldas (1992). In addition, Sanz et al. (1991) synthesized a polymeric gel in a basic condition. The transformation process of SP was analogous to that in the earlier study of Kanzaki and Tabata (1985) and the transformation process of CA concurred with that of Yoldas (1992). However, the PB2 gel exhibited two exotherms and mullite formation occurred via the spinel phase.

Phase evolutions of five types of precursors: Jaymes et al. (1996) studied the phase evolutions of five different types of precursors.

Monophasic gel of three kinds vs. diphasic gel vs. SD precursor: Haque (2000) and later on Chakraborty (2008) resynthesized as many as six types of precursors—four monophasic gels, a diphasic gel, and an SD precursor—by choosing ANN and TEOS as the basic components and by modifying the processing parameters only. The results were as follows:

- There was a wide variation in the exhibition of the DTA exotherms, as shown in Table 20.1 and Fig. 7.20 of Chakraborty (2008).
- The phase transformation processes of these precursors were widely different. Probably, the mixing scales of the powder precursors were different, originating during the process of hydrolysis, which might control the evolution process.

- The comparative mullitization paths were also dissimilar as per Table 16.1 and Fig. 12.6 of Chakraborty (2008).

Utilizing components from the same sources (TEOS and ANN), Chakraborty (2008) was able to synthesize as many as six types of mullite precursors by changing processing methods. It is projected that both parameters, components source and synthesis methods, lead to great variations in the 980°C phase development and the path of mullitization. Besides the choice of components, the processing techniques applied also influence the occurrence of the first exotherm and spinel to mullite formation. Let us note the cases of variations in its development.

- Variation was observed in spinel to mullite development on varying the water/alcohol content, for example, differences were observed in the two kinds of mullite precursors synthesized by Taylor and Holland (1993) with two different water contents.
- Variation was observed in spinel to mullite development on varying the pH, for example, the monophasic gel prepared by Huang et al. at pH 1.5 and at pH 11 showed differences.
- Variation in spinel to mullite development by varying water/alcohol content was noted by Geradin et al. (1994).
- Variation in spinel to mullite development with organic-assisted methods was observed by Jaymes et al. (1986), for example, precursor B and precursor D.

Monophasic gel (with excess of water) vs. monophasic gel (in ammonia) or diphasic vs. composite gels: Both ANN and TEOS were chosen by Ivankovic et al. (2003) as the basic components. When hydrolysis was done with excess water, the resultant gel showed an Al-Si spinel phase at the sharp first exotherm and thence to mullite at the second exotherm. The crystallization path was different when hydrolysis was done in a basic condition. Pseudoboehmite was formed first, indicating the development of a gel of a basic character. They also studied the influence of the Al_2O_3 component as the source in the presence of TEOS on the mullitization pathway. For example, the gel made from γ-Al_2O_3 and TEOS showed a mullitization exothermic peak at 1346°C, which is much earlier than the peak temperature (1361°C) found in the gel prepared from boehmite and TEOS.

Final conclusions

When a water-free approach is used, that is, in the absence of water, t-mullite is the predominating phase, for example, Yoldas's (1992) polymeric gel, Type I by Schneider et al. (1993), GNW by Taylor and Holland (1993), and G31 by Chakraborty and Ghosh (1988).

Both 980°C exotherm and mullite formation seem the highest when the gel is made at pH 4–4.5 by the aqueous gel technique. Similar results were noted in the slowly dried mullite gel synthesized by the monophasic gel technique at pH 3–3.5 by Chakraborty and Ghosh (1988); the SH gel by Hoffman et al. (1984), Okada and Otsuka (1986), and Chakraborty and Ghosh (1988); and the polymeric gel G152 by Chakraborty (1996).

All aqueous mullite gels invariably show an exotherm of the highest intensity at 980°C (Insley & Ewell, 1935; Chakraborty & Ghosh, 1988; Schneider et al. 1992). As the quantity of water used during gelation increases, the resultant gel turns more and more diphasic (Yoldas & Partlow, 1988).

As the pH or acidity decreases or the pH increases to neutral, the intensity of the exhibited exothermic peak in DTA decreases (Okada & Otsuka, 1986; Kumazawa et al., 1990; Huang et al., 1997). As the pH crosses the neutral region and when the gel is prepared in the true basic range, an adverse change in the character of the gel occurs from polymeric/monophasic to diphasic.

References

1. A. K. Chakraborty and D. K. Ghosh, Synthesis and 980°C phase development of some mullite gels. *J. Am. Ceram. Soc.*, **71**(11), 978–987 (1988).
2. K. Okada and N. Otsuka, Characterization of the spinel phase from SiO_2-Al_2O_3 xerogels and the formation process of mullite. *J. Am. Ceram. Soc.*, **69**(9), 652–656 (1986).
3. B. E. Yoldas, Effect of ultrastructure on crystallization of mullite. *J. Mater. Sci.*, **27**(24), 6667–6672 (1992).
4. H. Schneider, I. Merwin, and A. Sebald, Mullite formation from non-crystalline precursors. *J. Mater. Sci.*, **29**, 805–812 (1992).

5. H. Schneider, B. Saruhan, D. Voll, L. Merwin, and A. Sebald, Mullite precursor phases. *J. Euro. Ceram. Soc.*, **11**, 87–94 (1993).
6. M. Haque, Thesis, University of Kolkata (2000).
7. D. X. Li and W. J. Thomson, Effects of hydrolysis on the kinetics of high temperature transformations in aluminosilicate gels. *J. Am. Ceram. Soc.*, **74**, 574–578 (1991).
8. I. Jaymes, A. Douy, D. Massiot, and J. P. Coutures, Characterization of mono- and diphasic mulite precursor powders prepared by aqueous routes, ^{27}Al and ^{29}Si MAS-NMR spectroscopy. *J. Mater. Sci.*, **31**, 4581–4589 (1996).
9. Ph. Colomban, Structure of oxide gels and glasses by infrared and Raman scattering part 2 mullite. *J. Mater. Sci.*, **24**, 3011–3020 (1989).
10. D. W. Hoffman, R. Roy, and S. Komarneni, Diphasic xerogels, a new class of materials: phases in the system Al_2O_3-SiO_2. *J. Am. Ceram. Soc.*, **67**, 468–471 (1984).
11. C. Geradin, S. Sundaresan, J. Benziger, and A. Navrotsky, Structural investigation and energetics of mullite formation from sol-gel precursors. *Chem. Mater.*, **6**, 160–170 (1994).
12. M. Bartsch, B. Saruhan, M. Schmucker, and H. Schneider, Novel low-temperature processing route of dense mullite ceramics by reaction sintering of amorphous SiO_2-coated γ-Al_2O_3 particle nanocomposites. *J. Am. Ceram. Soc.*, **82**(6), 1388–1392 (1999).
13. S. Sen and S. Thiagarajan, *Ceram. Int.*, **14**, 77–86 (1988).
14. H. Ivankovic, E. Tkalcec, R. Nass and H. Schmidt, Correlation of the precursor type with densification behavior and microstructure of sinteren mullite ceramics. *J. Euro. Ceram. Soc.*, **23**, 283–292 (2003).
15. S. Kanzaki and H. Tabata, Sintering and mechanical properties of stoichiometricmullite. *J. Am. Ceram. Soc.*, **68**(1), C-6–C-7 (1985).
16. A. Douy, Crystallization of amorphous spray-dried precursors in the Al_2O_3-SiO_2 system. *J. Euro. Ceram. Soc.*, **26**, 1447–1454 (2006).
17. D. J. Cassidy, J. L. Woolerey, J. R. Bartlett, and B. Ben-Nissan, The effect of precursor chemistry on the crystallization and densification of sol-gel derived mullite gels and powders. *J. Sol-Gel Sci. Technol.*, **10**, 19–30 (1997).
18. J. Grofesik and E. Vago, in *Mullite, Its Structure, Formation and Signeficance*, eds. J. Grofesik and F. Tamas, Publishing House of the Hungarian Academy of Sciences, Budapest, Hungary (1961).

19. M. I. Nieto, G. Urretavizcaya, A. L. Cavalieri, and P. Rana, Structural changes in colloidal and polymeric aluminosilicate gels with mullite composition. *Br. Ceram. Trans.*, **97**(1), 17–23 (1998).

20. B. E. Yoldas and D. P. Partlow, Formation of mullite and other alumina-based ceramics via hydrolytic polycondensation of alkoxides and resultant ultra- and micro-structural effects. *J. Mater. Sci.*, **23**, 1895–1900 (1988).

21. T. Kumazawa, S. Ohta, S. Kanzaki, and H. Tabata, Influence of powder characteristics on microstructural and mechanical properties of mullite ceramics. *Ceram. Trans.*, **6**, 401–411 (1990).

22. T. Kumazawa, S. Ohta, S. Kanzaki, and H. Tabata, Influence of powder characteristics on microstructural and mechanical properties of mullite ceramics (74 wt% Al_2O_3). *J. Jpn. Ceram. Soc.*, **99**, 1228–1233 (1991).

23. I. Sanz, Sobrados, A. L. Cavalieri, P. Pena, S. de. Aza, and J. S. Moya, Structural changes induced on mullite precursors by thermal treatment: a ^{27}Al MAS-NMR investigation. *J. Am. Ceram. Soc.*, **74**(10), 2398–2403 (1991).

24. K. Hamano, T. Sato, and Z. Nakagawa, Properties of mullite prepared by coprecipitation and microstructure of fired bodies. *Yogo Kyokai Shi*, **94**(8), 818–822 (1986).

25. C. H. Horte and J. Wiegmann, Reaction between amorphous SiO_2 and Al_2O_3. *Naturwiss*, **43**, 9 (1956).

26. T. Demediuk and W. F. Cole, Exothermic reaction of metakaolin between 950° and 1,000°C. *Nature*, **181**, 1400–1401 (1958).

27. M. Low and R. McPherson, The structure and composition of Al-Si spinel. *J. Mater. Sci.*, **7**, 1196–1198 (1989).

28. M. Sales and J. Alarcon, Synthesis and phase transformations of mullite obtained from SiO_2-Al_2O_3 gels. *J. Euro. Ceram. Soc.*, **16**, 781–789 (1996).

29. A. K. Chakraborty, An analysis of the phase evolution of six types of mullite gels. Unpublished data (2008).

30. J. A. Pask, X. W. Zhang, A. P. Tomsia, and B. E. Yoldas, Effect of sol-gel mixing on mullite microstructure and phase equilibria in the α-Al_2O_3-SiO_2 system. *J. Am. Ceram. Soc.*, **70**(10), 704–707 (1987).

31. H. Yamada and S. Kimura, Studies on co-precipitates of alumina and silica gels and its transformations at higher temperatures. *Yogo Kyokai Shi*, **70**, 87–93 (1962).

32. A. L. Campos, N. T. Silva, F. C. L. Melo, M. A. S. Oliveira, and G. P. Thim, Crystallization kinetics of orthombic mullite from diphasic gels. *J. Non-Cryst. Solids*, **304**, 19–24 (2002).

33. S. Komarneni and R. Roy, Solid-state ^{27}Al and ^{29}Si magic-angle spinning NMR of aluminosilicate gels. *J. Am. Ceram. Soc.*, **69**(3), C-42–C-44 (1986).

34. M. Schmucker and H. Schneider, Structural development of single phase (type I) mullite gels. *J. Sol-Gel Sci. Technol.*, **15**, 191–199 (1999).

35. S. Prochazka and F. J. Klug, Infrared-transparent mullite. *J. Am. Ceram. Soc.*, **66**(12), 874–880 (1983).

36. H. Suzuki, H. Saito, Y. Tomokiyo, and Y. Suyama, Processing of ultrafine mullite through alkoxide route, in *Ceramic Transactions*, Vol. 6, Mullite and Mullite Matrix Composities. eds. S. Somiya, R. F. Davis, and J. A. Pask, American Ceramic Society, Westervikke, OH, p. 263 (1990).

37. O. Sakurai, N. Mizutani, and M. Kato, Preparation of mullite powders from metal alkoxides by ultrasonic spray pyrolysis. *J. Ceram. Soc. Jpn.*, **96**, 639–645 (1988).

38. A. M. L. M. Fonseca, J. M. F. Ferreira, I. M. M. Salvado, and J. L. Baptista, Mullite based compositions prepared by sol-gel techniques. *J. Sol-Gel Sci. Technol.*, **8**, 403–407 (1997).

39. A. Taylor and D. Holland, The chemical synthesis and characterization sequence of mullite. *J. Non-Cryst. Solids*, **152**, 1–17 (1993).

40. S. Mitachi, M. Matsuzawa, K. Kaneko, S. Kanzaki, and Y. Tabata, Characterization of SiO_2-Al_2O_3 powdes prepared from metal alkoxides. *Ceram. Trans.*, **6**, 275–286 (1990).

41. D. X. Li and W. J. Thomson, Effects of hydrolysis on the kinetics of high temperatture transformations in aluminosilicate gels. *J. Am. Ceram. Soc.*, **74**, 574–578 (1991).

42. M. Fukuoka, Y. Onoda, S. Inoue, K. Wada, A. Nukui, and A. Makishima, The role of precursors in the structure of SiO_2-Al_2O_3 sols and gels by the sol-gel process. *J. Sol-Gel Sci. Technol.*, **1**, 47–56 (1993).

43. A. K. Chakraborty, DTA characterization of three types of Al_2O_3-SiO_2 gels made from TEOS-Al $(OBU)_3$ mixture with variation of water. *Ceram. Int.*, **22**, 463–469 (1996).

44. A. K. Bhattcharya, A. Hartridge, and K. K. Mallick, Inorganic aluminium precursors in the synthesis of mullite – an investigation. *J. Mater. Sci.*, **31**, 5551–5554 (1996).

45. B. E. Yoldas, Mullite formation from aluminum and silicon alkoxides, in *Ceram. Trans.*, Vol. 6, eds. S. Somiya, R. F. Davis, and J. A. Pask, American Ceramic Society, Westerville, OH, p. 255 (1990).

46. A. Douy, Organic gels preparation of silica-alumina powders. I. Mullite. *J. Euro. Ceram. Soc.*, **7**, 117–123 (1991).

47. W. G. Fahrenholtz, D. M. Smith, and J. Cesarano, Effect of precursor particle size on the densification and crystallization behaviour of mullite. *J. Am. Ceram. Soc.*, **76**(2), 433–437 (1993).

48. Y. Hirata, K. Sakeda, Y. Matsushita, K. Shimada, and Y. Ishihara, characterization and sintering behavior of alkoxide-derived aluminosilicate powders. *J. Am. Ceram. Soc.*, **72**(6), 995–1002 (1989).

49. K. Okada and N. Otsuka, Formation process of mullite. *Ceram. Trans.*, **6**, 375 (1990).

50. A. K. Chakraborty, Role of hydrolysis water-alcohol mixture on mullitization of Al_2O_3-SiO_2 monophasic gels. *J. Mater. Sci.*, **29**, 6131–6138 (1994).

51. J. Ossaka, Tetragonal mullite type phase from co precipitated gels. *Nature (London)*, **19**, 1000 –1001 (1961).

52. S. Komerneni and C. Ruscher, Single-phase and diphasic aerogels and xerogels of mullite: preparation and characterization. *J. Euro. Ceram. Soc.*, **16**, 143–147 (1996).

53. Y. X. Huang, A. M. R. Senos, J. Rocha, and J. L. Baptista, Gel formation in mullite precursors obtained via tetraethylorthosilicate (TEOS) pre-hydrolysis. *J. Mater. Sci.*, **32**, 105–110 (1997).

54. B. E.Yoldas, Microstructure of monolithic materials formed by heat treatment of chemically polymerized precursors in the Al_2O_3-SiO_2 binary. *Ceram. Bull.*, **59**(4), 479–483 (1980).

55. S. Rajendran, H. J. Rossell, and J. V. Sanders, Crystallization of a coprecipitated mullite precursor during heat treatment. *J. Mater. Sci.*, **25**, 4462–4471 (1990).

56. C. H. Ruscher, G. Schrader, and M. Gotte, Infra-red spectroscopic investigation in the mullite field of composition: $Al_2(Al_{2+2x}Si_{2-2x})O_{10-x}$ with $0.55>x>0.25$. *J. Euro. Ceram. Soc.*, **16**, 169–175 (1996).

57. H. Insley and R. H. Ewell, Thermal behavior of the kaolin minerals. *J. Res. Natl. Bur. Stand.*, **14**(5), 615–627 (1935).

Part II

Identification: Characterization of Four Phases of Gel-to-Mullite Transformation Series

Chapter 13

Nature and Its Characterization of the Noncrystalline Al_2O_3-SiO_2 Mullite Precursor Phase

13.1 Introduction

The first step in the transformation of mullite precursors is designated in the temperature range usually from room temperature to the onset of the first exothermic temperature. The following reactions take place: (i) dehydration of physically adsorbed moisture entrapped in the gel network at ~120°C, (ii) dehydroxylation of –OH bonds rapidly during first stage, with its continuance over a long range of temperature, (iii) decomposition of residual alkoxides at ~300°C–500°C, and (iv) decomposition and oxidation of unhydrolyzed ester groups and residual carbon. The precursor powders formed after drying monophasic gels and on heating diphasic gel at >400°C are amorphous in nature and exist up to the temperature of their crystallizations to mullite or Al-Si spinel. The characterization of the amorphous precursor powders is the subject of many discussions. The amorphous nature of the precursors creates great difficulty in their proper characterization. A few questions arise: What should be the nature of raw slow hydrolysis (SH) gel, dried SH gel, and heat-treated SH gel at a temperature below the first exotherm? Does any bond formation between SiO_2 and Al_2O_3 really

Mullite Formations: Analysis and Applications
Akshoy Kumar Chakraborty

ISBN 978-981-4877-05-3 (Hardcover), 978-1-003-03167-3 (eBook)
www.jennystanford.com

take place, and do these form aluminosilicate or do they remain as a mixture of silica and alumina oxides? If aluminosilicate formation takes place, what should be the approximate composition? Infrared (IR) and magic angle spinning–nuclear magnetic resonance (MAS-NMR) studies were performed to reveal the nature of dehydrated Al_2O_3-SiO_2 gels/precursors of various kinds of mullite gels prepared by different researchers. A large amount of information could be gathered by many studies—differential thermal analysis (DTA), thermogravimetric analysis (TGA), thermomechanical analysis (TMA), IR, radial electron distribution (RED), X-ray absorption fine structure (XAFS), large-angle X-ray spectroscopy (LAXS), ^{29}Si MAS-NMR, Fourier-transform infrared (FTIR) spectroscopy, and moisture evolution analysis (MEA)—to indicate (i) the presence of an Al–O–Si bond in the amorphous aluminosilicate phase and (ii) the presence of some bonded structural –OH groups in the later phase. Regarding the development of aluminosilicate bonds during dehydroxylation processes of a few kinds of mullite gels synthesized by various authors, some of the physicochemical techniques practiced by them are narrated next.

13.2 Chemical Techniques

To evaluate the amorphous nature of the gel structure and what happens during coprecipitation, coprecipitated gels of different SiO_2:Al_2O_3 ratios were heat-treated at the first stage and then characterized first by Insley and Ewell (1935) by comparing these with metakaolinite or vice versa. The following observations were noted:

- Both materials showed a sharp and intense heat effect at 925°C–985°C.
- Both were amorphous in the temperature range from ~700°C up to the start of the first exotherm.
- Heat treatment at a constant temperature just below 925°C reduced or completely eliminated the heat effects in both cases.
- The alumina component of the gel was completely soluble in HCl in both cases.

- Both formed kaolin to approximately the same extent when heated with water in an autoclave at ~300°C.

In view of these facts the authors concluded that coprecipitated gels of the composition $Al_2O_3 \cdot 2SiO_2$ and metakaolin are very similar in internal structure.

13.3 IR Technique

By the IR technique, a broad and weak band between 1000 and 1100 cm^{-1} related to Si–O bonds and an absorption band at about 1010 cm^{-1} corresponding to Al(IV)–O–Si bonds were usually noted. However, the latter band falls in the same region of the Si–O stretching vibration band. As such, detection of the Al–O–Si bond is rather difficult. The IR data of heat-treated coprecipitated gel (3:2 ratio) with heated component gels were compared by Yamada and Kimura (1962). Si–O stretching vibration for pure silica gel occurred at 1120 cm^{-1} and decreased to 1080 cm^{-1} in coprecipitated gel heated to 1000°C, which indicates a considerable effect possibly due to substitution. However, the extent of substitution was not predicted. The authors were of the opinion that during coprecipitation, Al^{+3} and Si^{+4} ions get bonded through O^{2-} and form alumina gel of silicic acid. Since a mechanical mixture of silica gel and alumina gel did not show any exotherm in exception to coprecipitated gels, it was concluded that heat energy that evolved during the process of crystallization of the substance is present in the amorphous state. Shifting of the absorption band of Si–O to a lower wave number was observed subsequently by Okada and Otsuka (1986), Chakraborty and Ghosh (1988), and several earlier researchers, namely Hirata et al. (1989) in their alkoxide-derived alumina-silicate powders heated to 900°C–1000°C, Fukuoka et al. (1993) in their heated gel obtained from tetraethyl orthosilicate (TEOS) and aluminum nitrate nonahydrate (ANN) at 400°C, Sales and Alarcon (1996) in their sample marked B heated to 1000°C, and Nieto et al. (1998). They explained this band to be due to tetrahedral Al–O–Si bond formation. Even rapid hydrolysis (RH) gels of compositions $SiO_2.3Al_2O_3$ and $SiO_2.2Al_2O_3$ made by Okada and Otsuka (1986) showed an absorption band at about 1030 cm^{-1}. They explained that the shift from a regular band at 1100 cm^{-1} for Si–O–Si bonds to 1030 cm^{-1} is due to substitution of alumina for silica in amorphous solids.

Three most significant observations were shown by IR, NMR, and elusion of aluminum of various Al_2O_3-SiO_2 gels synthesized using ANN as a component and heat-treated to different temperatures by Fukuoka et al. (1993):

- Two IR absorption bands, one due to a Si–O–Si bond at 1100 cm^{-1} and the other due to NO_3^{-1} groups at 1400 cm^{-1} in the absorption spectra of the gel over a range of temperature, are shown in Fig. 13.1. It shows the relation between heat treatment temperature and the ratio of the absorption intensity of NO_3^{-1} to Si–O–Si bonds.

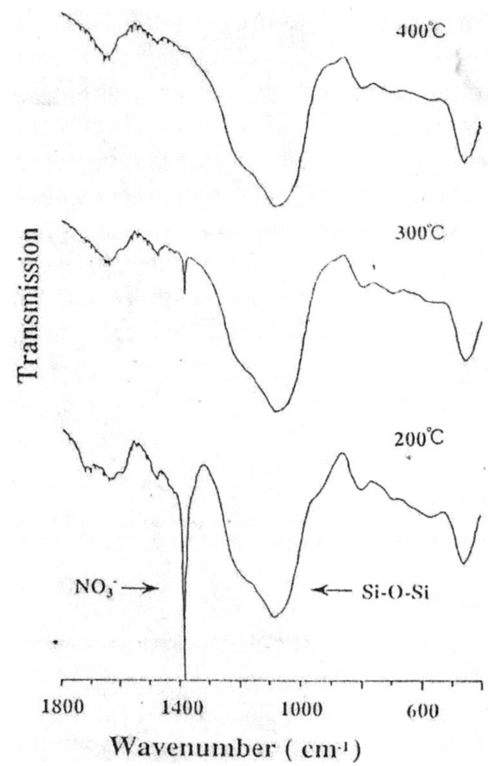

Figure 13.1 IR spectra of the gel using ANN heated at various temperatures (Fukuoka et al., 1993).

- Relations between heat treatment temperature and IR absorption intensity ratio of NO_3^{-1}/Si–O–Si and the ratio of Al(IV)/{Al(IV) + Al(VI)} are shown in Fig. 13.2 from data reported by Tsuchiya and Fukuoka (1989).

- Relations between heat treatment temperature and the ratio of Al(IV)/Al(IV) + Al(VI) obtained from NMR spectra of different gels synthesized from components from varying sources are shown in Fig. 13.3.

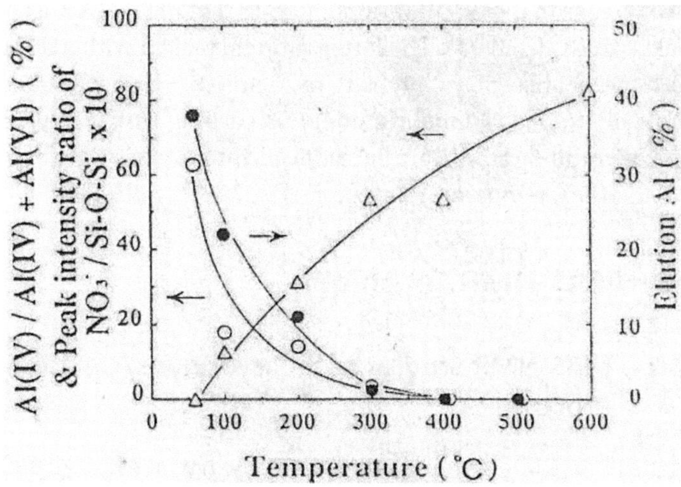

Figure 13.2 Relation between heat treatment temperature and IR absorption intensity ratio of NO_3^{-1}/Si–O–Si (•) and Al(IV)/{Al(IV) + Al(VI)}(Δ) (left) and elusion of aluminum (o) (right) (Tsuchiya et al., 1989).

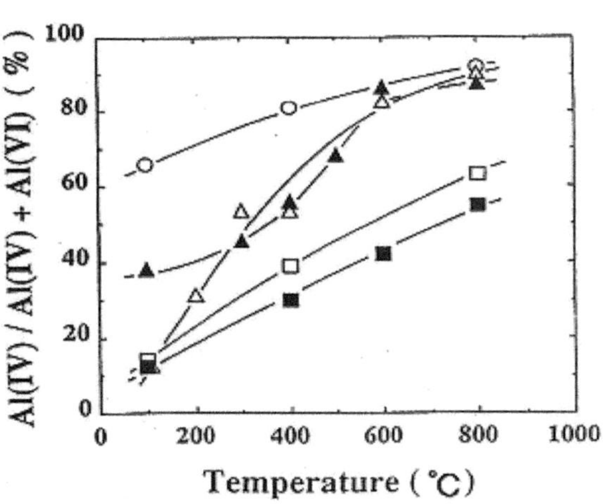

Figure 13.3 Relation between heat treatment temperature and Al(IV)/Al(IV) + Al(VI). o: AC, Δ: AN, □: BS, ▲: AF, and ■: Boehmite gel (Fukuoka et al., 1993).

Figure 13.2, as per Tsuchiya et al. (1989), also shows the amount of aluminum atoms exuded from the heated gel made from the same TEOS and ANN components into water. The change in the IR absorption intensity ratio of NO_3^{-1} to Si–O–Si showed a similar tendency as that in the amount of Al atoms exuded from the gel into water. Both ratios deceased with heating. This decrease was related to the increase in the Al(IV) ratio, similar to that in Fig. 13.3, as noted by Fukuoka et al. (1993). The latter ratio increased with a decrease in both the IR absorption intensity ratio of NO_3^{-1} to Si–O–Si and the amount of exuded aluminum atoms. According to Tsuchiya et al. (1989) aluminum atoms are introduced into a glass network in a 300°C to 400°C temperature range.

13.4 MAS-NMR Technique

13.4.1 MAS-NMR Studies of the Hydrolysis/Gelation Process

^{27}Al and ^{29}Si NMR spectra were used by Irwin et al. (1988) to examine the structure of aluminosilicate prepared by the sol-gel method from TEOS and $Al(OBu)_3$ in an isopropanol/water medium with no acid. Figure 13.4 shows a broader peak at 52 ppm, which is due to tetrahedral $[AlO_4]$ groups and a sharper resonance at 1 ppm from octahedrally coordinated aluminum. Thereafter the peak grew on heating at 450°C and up to 800°C, and then it became too broad to resolve.

In the case of pure silica gel, where Q2, Q3, and Q4 silicon sites were well resolved, the authors showed very large linewidths and contended that these spectra are indicative of a broad distribution of environments in the second coordination sphere of silicon. During heat treatment, the spectra showed increasing condensation. The shoulder at –92 ppm due to silicon atoms in Q2 sites disappeared, and the center of the main resonance shifted from –101 ppm (Q3 sites) to –107 ppm (Q4 silicon). Aluminosilicate prepared from the first case of gelation is shown in its ^{29}Si NMR spectra (Fig. 13.5). In addition to tetrahedral and octahedral aluminum peaks at 52 and 1 ppm, respectively, it showed an additional peak at 29 ppm, which was assigned as a 5-coordinate (AlO_5) group. The corresponding ^{29}Si resonance showed only one broad peak and no resolution of Q2, Q3, and Q4 sites. On heat treatment, the center of the peak shifted from –94 to 99.4 ppm.

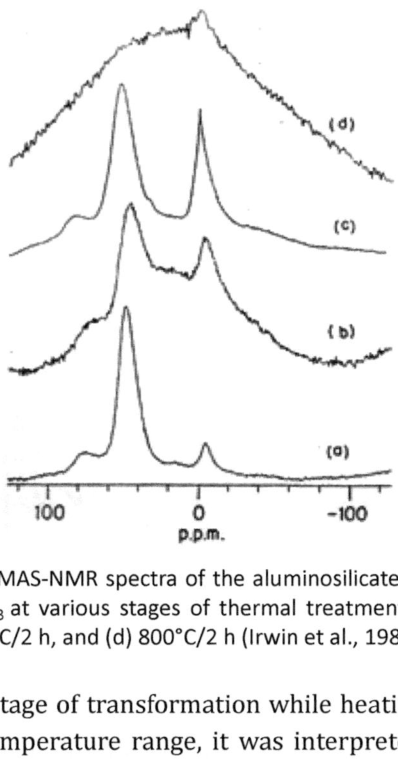

Figure 13.4 ^{27}Al MAS-NMR spectra of the aluminosilicate gel prepared from TEOS and $Al(OBu)_3$ at various stages of thermal treatment: (a) 40°C/6 h, (b) 150°C/6 h, (c) 450°C/2 h, and (d) 800°C/2 h (Irwin et al., 1988).

At the first stage of transformation while heating the gel in the 450°C–800°C temperature range, it was interpreted by ^{27}Al MAS-NMR spectra that in aluminosilicate gel, tetrahedral [AlO_4] existing in the oxide network are charge-balanced by [$Al(H_2O)$]$^{+3}$ groups based on a broader peak at 52 ppm, which is due to tetrahedral [AlO_4] groups and a sharper resonance at 1 ppm from octahedrally coordinated aluminum. The large linewidths observed in ^{29}Si NMR spectra were indicative of a broad distribution of environments in the second coordination sphere of silicon. The formation of Al–O–Si bonds at the time of gelation was predicted. On heat treatment, the center of the peak shifted from –94 to 99.4 ppm, which was in the range expected for the Si(2Al) environment.

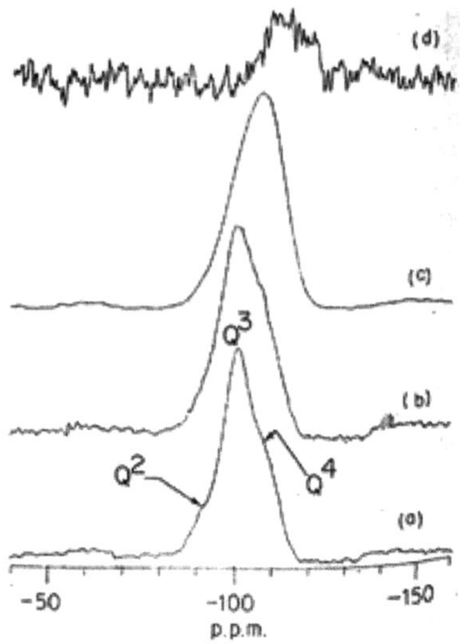

Figure 13.5 ^{29}Si MAS-NMR spectra of the aluminosilicate gel at various stages of heating: (a) 40°C/6 h, (b) 150°C/6 h, (c) 450°C/2 h, and (d) 800°C/2 h (after Irwin et al., 1988)

13.4.2 MAS-NMR Studies of Monophasic Gels (Hydrolysis/Gelation of Four Aluminum Components) vs. Diphasic Gel

Example 1: Si–O–Al bond development vs. component from the Al source

The evolution of the coordination number of aluminum and the formation of Si–O–Al bonds by ^{27}NMR in the sol-to-gel process was studied by Fukuoka et al. (1993) using TEOS and each of four aluminum compounds: aluminum di(*sec*-butoxide) ethyl acetoacetic ester chelate (AC), ANN, aluminum formoacetate (AF), and boehmite gel (BS) made from aluminum-i-propoxide, $Al(Oi\text{-}C_3H_7)_3$, as per Yoldas (1992).

- While mixing partially hydrolyzed TEOS with AC, small peaks appeared at 30 ppm and 50 ppm due to AlO_5 and AlO_4 sites, indicating the formation of Si–O–Al bonds at the first stage of hydrolysis. During the second stage, the intensity of the 50 ppm peak became stronger and this peak became even stronger when the Si-Al sol was gelled (Fig. 13.6).

- In the case of the ANN-TEOS mixture, no peak either due to AlO_4 or due to AlO_5 appeared in the sol state and even in the wet gel state left out for 40 days. Fukuoka et al. (1993) concluded that Al^{+3} could not be a part of the silica gel network and no Si–O–Al bond was generated.
- The AF-TEOS mixture showed three peaks, at about –15 ppm, 0 ppm, and 52 ppm, attributed to Al–O–Al bonds due to polymerization of AlO_6, usual AlO_6 groups, and Si–O–Al bonds (Fig. 13.6).
- In the case of BS, NMR spectra in the sol state and in the gel state showed only AlO_6 peaks. This indicated no bond formation in the dry gel state (100°C–800°C).

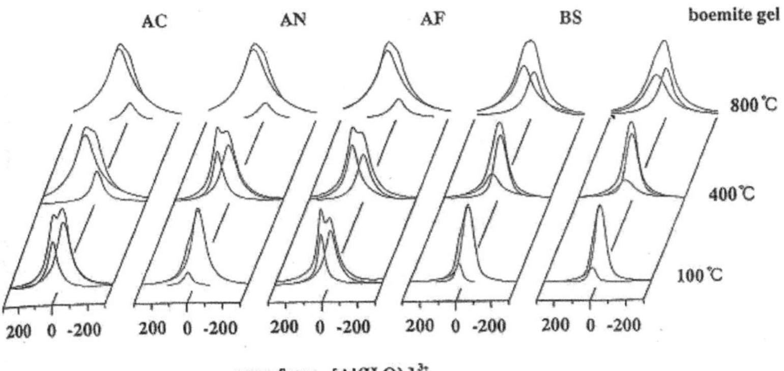

Figure 13.6 Change of ^{27}Al NMR spectra and their line shape analysis using various aluminum compounds heated at various temperature (Fukuoka et al., 1993).

When AC TEOS was heated, the Al(IV) ratio gradually increased with an increase in temperature. Its value increased from 66% at 100°C to 92% at 800°C.

When AF was used, its value increased from 38% at 100°C to 87% at 800°C.

When ANN was used, the Al(IV) ratio rapidly increased from 13% at 100°C to 90% at 800°C.

In the case of BS, the transformation of boehmite was predominant. The value of Al(IV) was on the order of 63% and there could be the formation of some Si–O–Al bonds (Fig. 13.3) during the heating process up to 800°C.

Bond formation depends on the source of the aluminum component used as the raw material. Fukuoka et al. (1993) correlated the ratio of Al(IV)/{Al(IV) + Al(VI)} bond development during the transformation of sol to gel and then to dried heat-treated gel to the microstructural homogeneity of the gel.

The evolution process of binary sols and gels of Al_2O_3-SiO_2 of the composition $80SiO_2 \cdot 20Al_2O_3$ prepared from different alumina and silica sources with help of NMR was studied by Fukuoka et al. (1993).

They showed that some amount of substitution of Al^{+3} for Si^{+4} may occur during the gel state. As a result the dry gel showed a 13% Al(IV) ratio.

When the dried gel was heated in the range 300°C–400°C, major bond development occurred because the Al(IV) ratio rapidly increased to ~90%. They conceived that the exhibition of the first exothermic peak was related to the homogeneity of the gel structure, which corresponded to the development of Si–O–Al bond formation in the gel network and the gel possessed good microstructural homogeneity.

The diphasic gel, probably prepared by using boehmite sol and TEOS, showed no Si–O–Al bonds in the sol state. The Al(IV) ratio increased gradually to ~60% with heat treatment. Consequently, the DTA curve showed no exotherm at 980°C and likely the microstructural homogeneity was far less than in previous cases.

However, the extent of formation of tetracovalent Al in 3:2 mullite gel ratio was not presented. Obviously, elaborate studies are required to throw more light on this aspect besides showing Si–O–Al linkages. Moreover, careful observation of Fig. 13.3 indicates that the change in Al(IV) was probably not similar to that in the case of transformation of pure boehmite component to γ-Al_2O_3. The curve of BS ran much above the curve of boehmite. Accordingly, it is expected that a small fraction of Si–O–Al linkages may form even in the case of BS gel during continued heat treatment process because the BS curve showed a relatively more quantity of Al(IV) than that present in γ-Al_2O_3.

Example 2: Formation of noncrystalline aluminosilicate precursor phase while heating mullite gels vs. the nature of the gel

Using MAS-NMR studies, Schneider et al. (1992) showed that in the temperature range of 600°C–800°C, the ^{29}Si NMR of the sol-gel material (SGM) showed a symmetrical ^{29}Si resonance at ~92 ppm and the coprecipitated material (CM) material showed the asymmetrical shape of the ^{29}Si resonance peak with a shoulder toward a more negative chemical shift at ~95 ppm. Moreover, NH (Type I) and HB10 (Type III) samples showed Si resonance at –85 ppm (Schneider et al., 1993). Both gels when heated to 900°C showed resonance at –95 ppm. The temperature-induced shifting to a more negative value was explained with the increase in the number of –Si–O–(Si,Al) bridges, corresponding to an increasing degree of condensation of the –Si–O–(Si,Al) network.

In the diphasic gel system, HB13 sample showed that the Si resonance consisted of two peaks, one at –80 ppm and one at –110 ppm. This gives a clear indication of the presence of two types of silicon environments in the diphasic gel system. The latter peak was described earlier as mostly due to Si–O–Si linkages. With further heating the peak at –80 ppm increased at the cost of reduction of the second peak, at –110 ppm. So the peak at –80 to –85 ppm may be conjectured as due to the formation of an aluminosilicate phase (Figs. 9.1 and 9.2). It can be interpreted that aluminosilicate formation takes place by gradual heat treatment of the diphasic gel system and this verifies the present solid-state reaction study of the author (2004) that a heated diphasic precursor is also aluminosilicate, as in an SH precursor.

Example 3: It was interpreted by Jaymes et al. (1996) that when aluminosilicate formation takes place on the gradual heat treatment of a diphasic gel system, then obviously noncrystalline aluminosilicate phase will develop on heating monophasic gels. ^{29}Si peak maxima were noted at –90 ppm for Si–O–Al bond formation in their heated monophasic gels/precursors marked A, C, and E (Figs. 9.14, 9.17, and 9.18). On the basis of the Si resonance spectral analysis of the observed Si peak maxima at approximately –90 ppm for their gels marked B and E, Jaymes et al. (1996) interpreted the formation of aluminosilicate during the heating of those gels.

Example 4: In ^{27}Al MAS-NMR spectroscopic studies, Yoldas (1992) showed different aluminum environments in polymeric and colloidal gels heated at ~ 500°C and suggested two different ultrastructures for the two gels.

Example 5: Two main resonances centered at –81 and –108 ppm with a shoulder at –100 ppm from two precursors marked MI and MII heated to 450°C were noted by Huang et al. (1997). The resonances were assigned to four aluminum atoms via oxygen {Q_o (4Al) sites} to {Q_4 (0Al) sites}. The shoulder at –100 ppm may be due to a Q_4(1Al) environment, as discussed earlier by Irwin et al. (1988) and Yasumori et al. (1990). They suggested that the ^{29}Si resonance at approximately –80 ppm might be due to the resonance of silicon in an Al_2O_3-rich noncrystalline phase. These reported results conclude that mullite precursor A is an alumina silicate that concurs with qualitative data of previous NMR results.

Example 6: ^{27}Al MAS-NMR for the characterization of single-phase and diphasic aluminosilicate gels was first applied by Komarneni et al. (1985). They showed that the tetrahedral coordination of Al increased with a decreasing Al:Si ratio in a monophasic gel but not in a diphasic gel. The ^{29}Si spectra of 3:1 monophasic gel showed a broad resonance at –103.3 ppm due to a disordered Si(Al) environment. The diphasic (3:1) gel on the other hand displayed two resonances, one at –111.8 and one at –100 ppm, due to a silica gel spectrum. These results indicated the presence of Al in the next environment of silica in the monophasic gel. The diphasic gel, on the other hand, confirmed nanoheterogeneity. However, the upper limit of substitution in the monophasic gel was not ascertained.

In summary, the following results are noted. An X-ray diffractometry (XRD) study shows an amorphous state only over a long range of temperature from the dried gel state up to the temperature of crystallization, at ~1000°C. A ^{29}Si MAS-NMR study, of course, reveals the following ways of solid-state transformation between two hydrated components of mullite gels as observed:

A gradual shift of Si resonance is noted. It indicates the condensation of $Si(OH)_4$ units toward the development of Si–O–Si linkages in the form of chains, branches, etc., during the heating process of precursor B (Fig. 9.15) of Jaymes et al. (1996).

Besides the formation of Si-O-Si linkages, a second process, involving the formation of an aluminosilicate phase, also occurs side by side. See the Si NMR patterns of Schneider et al. (1993), Huang et al. (1997), Jaymes et al. (1996), and Geradin et al. (1994).

A portion of amorphous A_2O_3 formed at the early stage shows both Al(4) and Al(6) bonds in the ^{27}Al MAS-NMR spectrum. The ratio of Al(4)/Al(6) changes on heating due to gradual development of aluminosilicate at the cost of amorphous or γ-Al_2O_3 with a rise in temperature.

Thus, the solid-state reaction continues by the diffusion process of Si and/or Al atoms on heating, for example, varying condensed silicate units and condensed alumina units, which form connected units.

A partial Si–O–Al bond formation, of course, takes place during the gelation process where substitution has already started (Fukuoka et al., 1993). Such bond development also occurs during the pyrolysis process (Sanz et al., 1991).

13.5 X-Ray Absorption Fine Structure Spectroscopy

Kamijo et al. (1994) tried to determine the local structure of mullite gels by using the XAFS technique. The changes in the local coordination characteristics of Al and Si atoms in the amorphous stage during the heating process of mullite gels between 60°C and 1200°C were determined. Finally, they compared the extended XAFS and X-ray absorption near-edge structure (XANES) of Al-K and Si-K edges of amorphous heat-treated mullite gels with reference samples. Generally, XANES spectra may be used at best as fingerprints for various treated samples. Figure 13.7a shows the XANES spectra of the Al K-edge for treated mullite gels compared with the same (b) for α-Al_2O_3, γ-Al_2O_3, $Al(OH)_3$, and amorphous Al_2O_3 as reference samples. Figure 13.8a shows the XANES spectra of the Si K-edge for treated mullite gels compared with the same (b) for nepheline, quartz, and SiO_2 glass as reference samples.

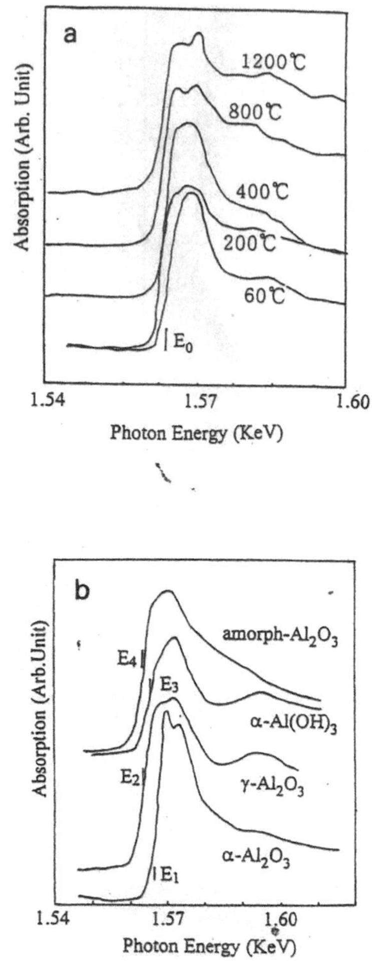

Figure 13.7 (a) Al K-XANES of mullite gels dried at 60°C and those heated at 200°C, 400°C, 800°C, and 1200°C. E0 indicates the edge energy of the spectra. (b) Al K-XANES of the reference samples (Kamijo et al., 1994).

Summarized results show that the two and even more types of bonding between Al and neighboring oxygen atoms exist in gels dried at 60°C when heat-treated between 200°C and 800°C and when heat-treated gel at 1200°C. Bond distances and the coordination number of Al show variations.

- In the gel dried at 60°C, the coordination number of Al is 6 and Al–O distance is ~1.81 Å.

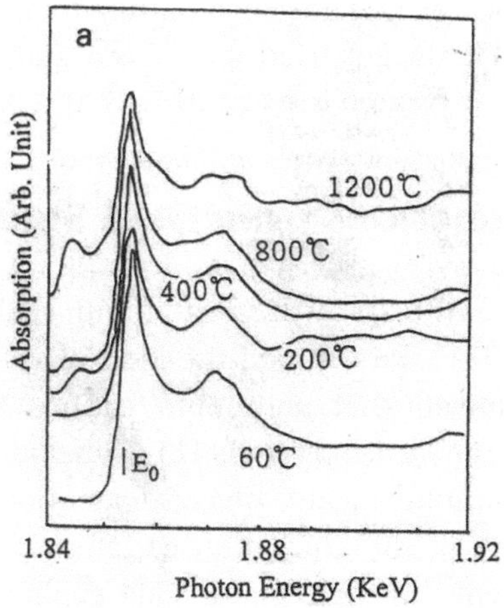

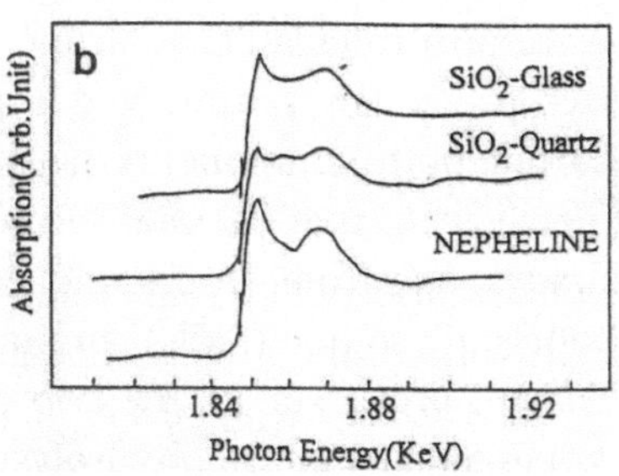

Figure 13.8 (a) Si K-XANES of mullite gels dried at 60°C and those heated at 200°C, 400°C, 500°C, and 1200°C. E_0 indicates the edge energy of the spectra. (b) Si K-XANES of the reference samples (Kamijo et al., 1994).

- In gels heat-treated up to 800°C, the coordination number of Al decreases to 4–5 and the Al–O distance is large and varies between 2.1 Å and 2.5 Å.
- In gels heated at 1200°C, the coordination number of Al increases to 6 again and the Al–O distance decreases to 1.86 Å.

According to Kamijo et al. (1994), there exist more than one type of bonding between Si and neighboring oxygen atoms. Only bond distances vary, and the coordination number of Si remains 4. In gels dried at 60°C, heat-treated between 200°C and 800°C and heat-treated at 1200°C, the Si–O bond distances are 1.60–1.64 Å and

1.92–2.07 Å, respectively. The coordination environment of oxygen around aluminum in the gel dried at 60°C is 6, and it is reduced to ~4 when the gel is heat-treated in the 200°C–800°C range.

13.6 RED Functions from LAXS Study

Example 1: By the RED study of silica-alumina gel heated to 500°C, Leonard et al. (1971) determined the kind of structures and their defects. They showed that aluminum cations substitute silicon cations in the tetrahedral position. This substantiates the view of noncrystalline aluminosilicate phase formation. Secondly, they showed that some of the aluminum cations were presented in a perturbed tetrahedral arrangement. This explained the source of acidity of the amorphous silico aluminous phase.

Example 2: RED function curves of mullite gels heat-treated to 800°C and 900°C were also calculated by Schneider et al. (1994). They marked the two previously synthesized Type I and Type III mullite gels made from both components from organic sources (TEOS and AlOBu) newly as I_{oo} and III_{oo} and secondly prepared two more mullite precursors, one from an organic source (TEOS) and the other from an inorganic source ($SiCl_4$), and marked them as I_{oi} and III_{oi}. For synthesis of precursor I_{oi}, the two components were admixed and then homogenized for 3 h. Hydrolysis was carried out with the addition of water and isopropanol in the ratio of 3:1. White flocculants and white vapors emerged during hydrolysis. A white gel with an opaque appearance was produced after 1 h. Precursor III_{oi} was made by the hydrolysis of TEOS and $SiCl_4$ and it was carried out by the addition of concentrated ammonium hydroxide. A violent release of white fumes was noted. Flocculent gels were formed, washed, and stored. Type I_{oi} and Type III_{oi} were heat-treated at 800°C, and the previous two precursors, Type I_{oo} and Type III_{oo}, were heat-treated at 900°C for LAXS measurements. On the basis of it, the RED function curves were calculated. The curves were divided into two sections, one falling below and the other above 6 Å.

- The short-distance RED function pattern (R < 6 Å) of SiO_2 glass displayed four strong peaks at the first, second, and third coordination spheres at 1.7 Å, 3.1 Å, 4 Å, and 5 Å assigned to distances Si–O(1), Si(1)–Si(2), Si–O(2), and Si(1)–Si(3) in the glass network. The short-distance RED function pattern (R < 6 Å) of Types I_{oo} and III_{oo} was very similar to SiO_2 glass, whereas the peak intensities of heat-treated Types I_{oi} and III_{oi} appeared to be different from the glass spectrum (Fig. 13.9). It was assumed that the short-range structural arrangements in the (oo) mullite precursors are less ordered than in the (oi) precursors.

- (Si^{+4},Al^{+3})–O(1) and (Si^{+4},Al^{+3})–O(2) distances for heated mullite precursors displayed longer distances, from 1.7 Å to 1.8 Å. This increment indicated isomorphous substitution of Si^{+4} by Al^{+3}. Similar was the cause for the increment of $(Si^{+4},Al^{+3})(1)$–$(Si^{+4},Al^{+3})(3)$ from 5 Å to 5.2 Å in Type I precursors. However, $(Si^{+4},Al^{+3})(1)$–$(Si^{+4},Al^{+3})(3)$ showed a decrease in distances from 5 Å to 4.9 Å, along with a decrease in the peak intensities in Type III precursors. Shortening of T(1)–T(3) was explained by denser packing of the tetrahedral network, possibly due to phase separation in the Type III precursor. See Schneider et al. (1994, Table 3).
- For long distances (R > 6 Å) the RED function pattern provided information on higher coordination spheres, corresponding to the structural long-range order of the precursor. Heated precursors of both Type I and Type III exhibited RED function bands near 6.5 Å and 8.7 Å, respectively, which indicated –Al–O–Si– matrix networks (Fig. 13.9). The Type III precursor displayed a new peak at 7.3 Å besides the two peaks noted as in Type I sample. These two peaks frequently split in the Type III precursor. Schneider et al. (1994) assumed that the degree of structural order of the said matrices increases within the sample sequence I_{oo} – I_{oi} – III_{oo} – III_{oi}. Regarding the occurrence of the peak at 7.3 Å, they further assumed that precursor I_{oi} contains very small heterogeneous areas embedded in the –Al–O–Si– matrix network structure. This result indicates the development of two phases when Type III precursors are being heated. This may be the reason for the display of different crystallization processes of the two precursors at ~1000°C.

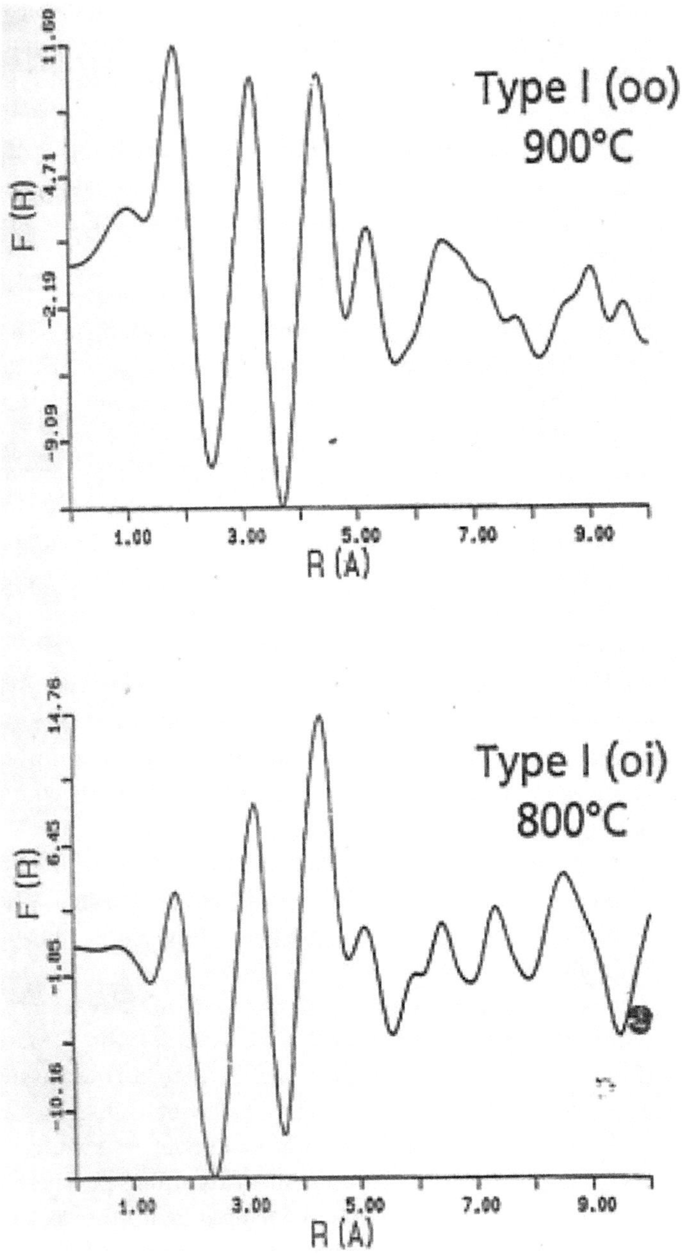
Type I (oo)
900°C
F (R)
11.60
4.71
-2.19
-9.09
1.00
3.00
5.00
7.00
9.00
R (A)
Type I (oi)
800°C
F (R)
14.76
6.45
-1.85
-10.16
1.00
3.00
5.00
7.00
9.00
R (A)

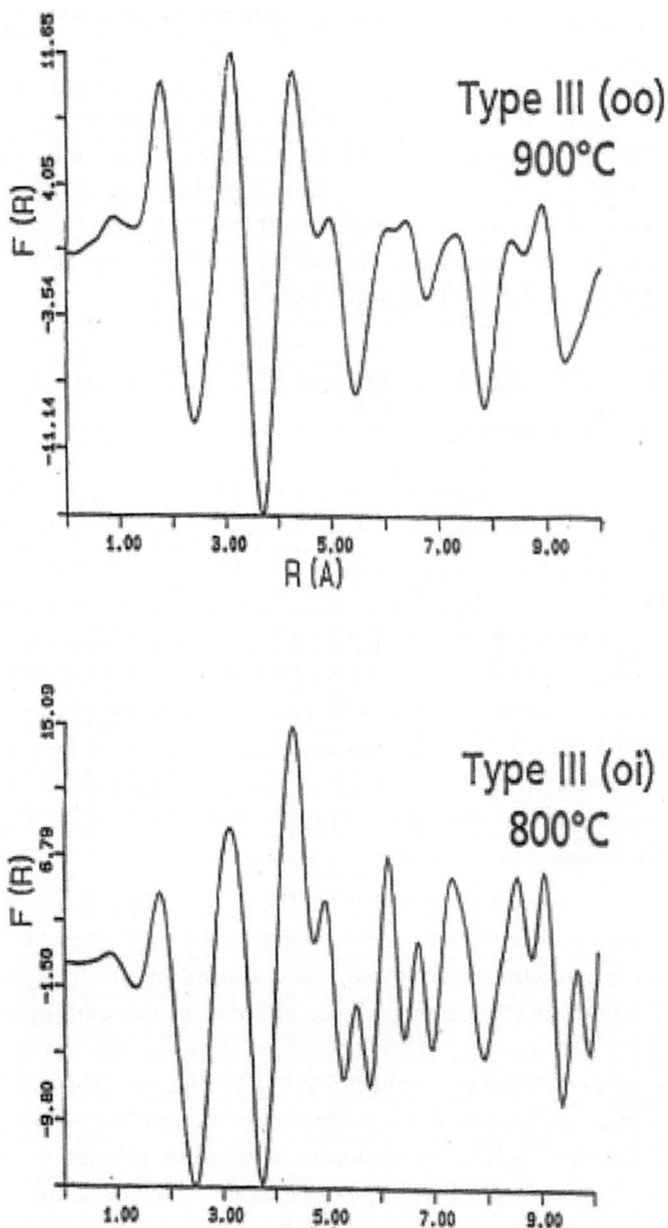

Figure 13.9 Radial electron distribution functions of Type I_{oo} and Type III_{oo} and Type I_{oi} and Type III_{oi} precursors, heat-treated just below the crystallization temperatures indicated (Schneider et al., 1994).

13.7 X-Ray Fluorescence Spectroscopy Study

A clear picture of changes in the coordination number of aluminum in amorphous silicoaluminas was presented by Leonard et al. (1964). According to Colomban (1989), both spinel and alumina phases were poor Raman scatterers for exhibiting silica spectra. He opined that there was no segregation during the transformation of the gel → orthorhombic mullite reaction series. A progressive ordering at the local scale might take place (Fig. 8.7).

13.8 Solid-State Reaction Study: A New Approach

The precursor powders formed after drying monophasic gels and on heating a diphasic gel at temperatures above 400°C are amorphous in nature and exist up to the temperature of their crystallization to mullite or Al-Si spinel. The characterization of the precursor A powders reviewed and placed above is a subject of many discussions. It would not be justified unless some approach other than spectroscopy could be taken to throw new light on the characterization of mullite precursor powder. It may be possible by performing a solid-state reaction study of both mullite gels and preheated precursor powders with $CaCO_3$ during heating followed by phase identification by XRD technique.

- An elaborate study regarding the nature of monophasic gels, for example, raw SH gel, dried SH gel, and heated SH gel at the temperature below the first exotherm and the question regarding the formation of bond between SiO_2 and Al_2O_3 with the development of noncrystalline aluminosilicate phase was carried out by Chakraborty (2004) by this new approach. The detailed results of solid-state reaction studies are placed in Chapter 15. Besides monophasic gel, the solid-state reactions of $CaCO_3$ heated with raw, dried, and calcined diphasic gels were also studied. X-ray patterns of raw, dried, and calcined (900°C) monophasic gels mixed with $CaCO_3$ and heated to 1200°C are shown in Tables 15.8 and 15.9 and Fig. 15.10a–c.

- Raw SH gels showed the formation of $C_{12}A_7$ (major) and C_3A (minor) phases during heating, which suggested that the aluminum component of the gel was present in a free state.
- Dried SH gel formed a major amount of calcium aluminosilicate phase (gehlenite) and a minor amount of $C_{12}A_7$, which indicated that the dried SH gel was noncrystalline aluminosilicate.
- Moreover, SH calcined between 400°C and 900°C also showed gehlenite in a major amount and traces of $C_{12}A_7$ phases. The formation of gehlenite rather than calcium aluminate indicated that the mother reactant (i.e., the calcined precursor phase in this case) should also be aluminosilicate phase.
- Raw diphasic gel behaved like raw monophasic SH gel and/or a mechanical mixture of amorphous silica and alumina. The author also showed the formation of gehlenite rather than calcium aluminate when the heat-treated diphasic precursor of mullite composition between 400°C and 900°C was further heat-treated with $CaCO_3$ at 1200°C (Table 15.10). It was concluded that the heated precursor was a noncrystalline aluminosilicate phase that substantiates that the diphasic gel also formed linkages at and above the dehydroxylation temperature, which continued until the crystallization stage was reached.
- The question is, what should be the approximate composition when aluminosilicate formation takes place in both monophasic and diphasic precursors? The reaction product may be either calcium aluminosilicate compound or simple calcium aluminates. The result, of course, showed that the product was a calcium aluminosilicate compound. To answer this question, the author used comparative tables (Tables 15.7–15.10) to show that all monophasic and all diphasic gels up to the composition of 3:2 mullite formed calcium aluminosilicate (gehlenite) rather than calcium aluminate. The formation of gehlenite rather than calcium aluminate by XRD indicated that both monophasic and diphasic precursors consisted of a noncrystalline aluminosilicate phase of composition nearly close to 3:2 mullite.

13.9 Alkali Leaching Study

Alkali leaching data and changes in amorphous band of heat-treated diphasic gels during XRD studies reveal the gradual development of Si–O–Al linkages with the formation of the noncrystalline aluminosilicate phase.

The alkali leaching data of three diphasic gels heat-treated to different temperatures are shown in Fig. 15.8. In the temperature range of 600°C to 900°C, the silica component of diphasic gels marked DG72 dissolved in a hot alkali solution to the extent of ~2.5 wt % only out of the 28 wt % of silica present originally in the gel state. This indicates that the silica component remained as a separate and individual component of the two-phase system, which after dehydration and decomposition of diphasic gels at ~400°C became unavailable in the hot alkali solution when it was heated to 900°C. Thus, in the said temperature range, a solid-state reaction between two components occurs, which leads to the development of a noncrystalline aluminosilicate phase during continued heating, as evidenced by alkali leaching study.

The nature of changes in the intensities of the X-ray amorphous bands when diphasic gels of three compositions are being heated is shown in Fig. 15.7 by Chakraborty (2008). In all cases of diphasic gels, the X-ray amorphous band first increased on heat treatment from 600°C up to ~900°C. It was very slow in the case of DG72, very fast in the case of DG40, and moderate in the case of DG66. The band intensity decreased sharply at ~1000°C due to the formation of the spinel phase, and thereafter it stayed up to 1200°C. The measurement of changes in the XRD intensity of the amorphous band as shown above indicated that a solid-state reaction between alumina (A) and silica (A) liberated after the dehydration and decomposition of the diphasic gel started while heating it at a temperature above 400°C.

13.10 Physicochemical Changes of Raw Mullite Gel up to the Final Dehydroxylation

Changes in refractive indices and estimated densities: The corresponding changes of these two physical parameters were shown by Okuno et al. (1997, Table 1). The refractive indices of mullite gel increase with the calcination temperature from 1.5052 at 150°C to 1.5437 at 800°C. Contrarily, the calculated density value (g/cc) decreases from 2.49 at 150°C to 2.68 at 800°C.

Changes in the chemical shift of ^{29}Si MAS-NMR: The characteristic changes of the Si spectrum of mullite gel from room temperature to the temperature of final dehydroxylation are shown by Mackenzie et al. (1996), Jaymes et al. (1996), and Okuno et al. (1997, in their Figs. 2 and 4).

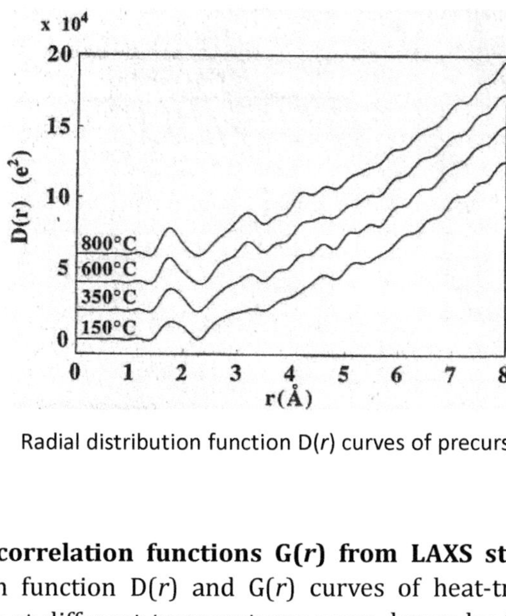

Figure 13.10 Radial distribution function D(*r*) curves of precursors (Okuno et al., 1997).

D(*r*) and correlation functions G(*r*) from LAXS study: Radial distribution function D(*r*) and G(*r*) curves of heat-treated Type I precursor at different temperatures were drawn by Okuno et al. (1997) and showed some prominent peaks (Figs. 13.10 and 13.11). In the as-prepared Type I mullite precursor (150°C) at the short-range level, that is, $R < 0.4$ Å, one prominent peak D(*r*) at $r = 0.18$ nm and three broad peaks of low intensity near 0.29, 0.32, and 0.42 nm were observed. The 0.18 nm peak was assigned to the T–O atomic pairs, whereas 0.29, 0.32, and 0.42 nm were associated with the O–O, T–T (i.e., T(1)–O(1)–T(2)), and T–O(2) (i.e., T(1)–O(1)–T(2)–O(2)) pairs. The T–O distance was 0.177 nm, which seems large. They attributed three general reasons for the Type I precursor heat-treated from 150°C up to 800°C.

- The T–O bond length, which is based on Si–O, Al–O, and O–O bonds and the ionic radii of the three ions
- The presence of Al(IV)–O and Al(VI)–O bonds, which depends upon the heat treatment temperature of the precursor
- The presence of -OH groups and organic radicles, which during decompositions of Si, Al-OH and Si, Al-R bonds in the precursors at different temperatures are responsible for increase of network combination as shown by RDF analysis

The question is, what will happen on heating the precursor to 800°C and beyond this temperature?

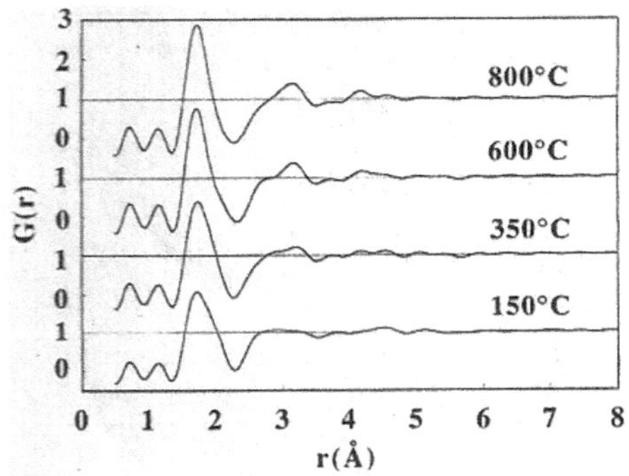

Fig. 5. G(r) curves of precursors.

Figure 13.11 G(*r*) curves of precursors (Okuno et al., 1997).

Okuno et al. (1997) showed that the amplitudes of both peaks increased and their widths decreased with increasing temperature. Both dehydroxylation of –OH groups and decomposition of organic groups occurred, leading to a decrease in the broadness of the peak and a decrease in the *r* value with an increase in the temperature of heating. The D(*r*) and G(*r*) curves of the precursor heated at 800°C were consistent with the RED function of the Type I precursor heated to 900°C. The T–O distance in alumina-rich glass (Al:Si ratio = 1.78) as found by Morikawa et al. (1982) was 0.179 nm. In comparison, the Type I precursor (Al:Si ratio = 3:1) showed a T–O distance of 0.174 nm at 800°C. For the T–T distance (middle range order), that is, r = 0.32 nm, the peak intensity increased with temperature. This was attributed to an increase in the network formation by the condensation of –OH groups. The peak shift indicated a successive decrease of bridging angles. For T–O(2), that is, the second neighboring oxygen, at 0.42 nm, the peak intensity increased with increasing temperature. This evidenced a gradual condensation of the precursor network. One more factor has to be considered—changes occurred in the coordination state of aluminum during the condensation process of the precursor. We can conclude that all these techniques demonstrate that monophasic Al_2O_3–SiO_2 gel at the first stage formed the amorphous aluminosilicate phase (marked as noncrystalline aluminosilicate phase) on being heating in the temperature range of 400°C (temperature of first dihydroxylation) and the temperature of the first exotherm, at 980°C. The results were reanalyzed by Schmucker and Schneider (1999) and also by Schneider et al. (1994). However, its composition is still unknown.

The inter-relation between heat treatment temperature and the ratio of Al(IV)/Al(IV) + Al(VI): Such a relationship is shown in Fig. 13.3 by Fukuoka (1993), which also concurred with the above result.

13.11 The Presence of –OH Groups in Noncrystalline Aluminosilicate Precursor Phase by FTIR and Its Role

The presence of hydroxyl groups in a partly dehydroxylated mullite precursor is an essential part of the noncrystalline aluminosilicate precursor structure. Their influence on the structural stability of the noncrystalline aluminosilicate phase vis-à-vis nucleation is not to be underestimated (Colomban, 1989). A few hydroxyls are chemically bonded with the partially dehydrated aluminosilicate structure. These bonded hydroxyls groups will be eliminated in two steps.

1. Gradually during the ongoing heating process
2. Sharply when the precursor is dehydroxylated to the fullest extent just before crystallization reaction at 980°C

The dehydration and dehydroxylation process of mullite precursors has been studied by Voll et al. (1998). FTIR powder spectra of the precursor preheated from 200°C to 900°C depict the following absorption bands:

- There are two weak absorption bands, one centered at 5160 cm^{-1} due to the H_2O combination mode and one at 4540 cm^{-1}

due to the (Si,Al)–OH combination mode in the region (5160–4540 cm^{-1}). See Voll et al. (1998, Fig. 2). The presence of these two bands proves that water is present as both H_2O molecules and OH groups.

- Among the two other absorption bands, one is a very strong bond due to the (H_2O, OH) stretching vibration centered at 3430 cm^{-1} and one is a distinct band due to the H_2O bending vibration at 1635 cm^{-1} in the other region (4000–1500 cm^{-1}). The differences in the intensities of the two bonds (one is weak in comparison to the other) also supports the presence of two peaks (see Voll et al., 1998, Fig. 3).

The existence of OH groups and their slow dehydroxylation versus heat treatment temperature were shown by Schneider et al. (1994). The temperature-dependent evolution of the peak intensities of the absorption bands corresponding to those of CO(H) (organic residuals), H_2O, and OH groups for Type I and Type III precursors was shown Schneider et al. (1994, Fig. 3). The intensity decrease of the OH absorption band in the Type I precursor showed the persistence of OH groups in the structure at or beyond 1000°C.

To investigate the relationship of protons to the Si atoms present on noncrystalline aluminosilicate and their possible retention during the thermal treatments, MacKenzie et al. (1996) studied ^{29}Si cross polarized (CP) spectra for heat treated Type I precursor. The semi-quantitative CP intensity was found to be constant to a temperature of 450°C. It decreased to 39% of its original value after being heated to 650°C. It decreased to ~31% after being reheated to 700°C. These results were explained by them as due to (i) the removal of silanol protons more or less uniformly during thermal treatment and (ii) retention of some amount of silanol protons in the structure up to the crystallization temperature. They further conjectured that these residual protons are not localized to any particular Si site.

13.12 Characteristics of Six Types of Precursor Phases

The nature and characteristics of noncrystalline precursor phases synthesized by six methods from different components using different processing techniques and pH and water contents may be different. Mullite precursors prepared by any the six processing techniques contained protonic moieties. The probable structures of dried gels of four models are tentatively described in Chapter 10 as follows:

- The 1st model equation of the gel made by the aqueous monophasic gel, which is an aluminosilicate hydrate complex phase (Eqs. 10.1 and 10.2).
- The 2nd model equation of the gel particle, which may consist of ethoxy silicate $-(OH)^{-1}$ aluminum$-(OH)^{-1}$ complex (Eq. 10.20).
- The 3rd model equation of gel A containing a unidirectional short polymeric chain.
- The 4th model equation of G-150 containing a 3D cross-linked chain.
- The 5th model of the polymeric gelation reaction as per Yoldas and Partlow (1988).
- The 6th model of the mixed gelation reaction as per Yoldas and Partlow (1988).
- A diphasic precursor made in the presence of boehmite—it is an aluminosilicate hydrate phase having 3D colloidal silica gel linked with aluminum hydroxide colloid/clusters of boehmite. The schematic model is shown in the 7th model gelation reaction for diphasic gel.
- A diphasic precursor made during aging—it is an aluminosilicate hydrate phase 3D colloidal silica unit linked with aluminum hydroxide colloids.
- A diphasic precursor made in the presence of large H_2O molecules—it is an aluminosilicate hydrate phase colloidal silica unit linked with giant polymeric aluminum hydroxide colloids particles.
- Pyrolysis of ANN and TEOS at 400°C, producing a noncrystalline aluminosilicate phase containing some water and OH^{-1} groups in the network structural phase. ^{27}Al MAS-NMR spectra show three peaks, at 65, 35, and 6 ppm, which correspond to tetra-, penta-, and octahedral Al configuration as per Sanz et al. (1991). These data strongly favor the development of chemical linkages between two oxides in this precursor and the formation of a noncrystalline aluminosilicate phase.

In these ways, the different natures and characteristics of the six precursor phases are tentatively described. It is, of course, certain that various precursors synthesized by different researchers will be of different characters too. Here are some examples:

- ^{27}Al MAS-NMR spectroscopy of two samples marked GNW and GYW were studied by Taylor and Holland (1993) and are shown in Figs. 9.9 and 9.10. The GYW xerogel shows varying amounts of three peaks, relating to Al(4), Al(5), and Al(6), of which octahedral Al is the preferred site. Contrarily, the GNW xerogel shows to be predominantly pentahedral.
- The aluminosilicate hydrate complexes originated in CM and SGM synthesized by Schneider et al. (1992) are not the same although they are amorphous in XRD.
- A gel made from silica sol and aluminum sulfate in an aqueous system (Chakraborty & Ghosh, 1988).

Such assumption of the varying nature of the noncrystalline aluminosilicate precursor phase is based on the analysis of varying physical properties noted during the heat treatment of mullite gels. From the crystallochemical point of view, it is worth referring to the large volume of publications on evolution processes of various kinds of mullite precursors as shown in the present book in Chapters 2–12 by the authors of different disciplines. The applications of many experimental techniques have been elaborated, and these studies focus on three major pieces of information. Of course, irrespective of the nature of the precursor phase, it must contain –OH groups for stabilization, as present in silica gel. Likely, the evolution of the amorphous phase in the temperature range of 100°C to 1000°C in the six types of precursor phases would be different.

The Nature of the Precursor Phase

A ^{29}Si MAS-NMR study generally shows the changes in the nature of noncrystalline alumina-silicate precursor structures of different kinds of mullite gels synthesized by various authors. The network structure obviously varies with heating of the raw dried gel to a temperature just below the 980°C exotherm. The entire process may consist of the following steps, which are also the main causes of its variation:

1. Removal of adhered water from the gel surface by dehydration: On drying, adhered moisture is removed from the clusters of the Si–O–Al network of the raw gel and it becomes a very soft powder. Loosely bound water dehydroxylates at two consecutive endotherms, as shown in DTA and differential thermo gravimetric analysis (DTGA) curves.
2. Removal of residual organics used as solvent for hydrolysis of esters and the decomposition of ethoxy groups, organic residuals, and other inorganic radicals NH_4^{+1}, NO_3^{-1}, Cl^{-1}, and CO_2: These reactions occur mostly at ~400°C. Combustion and oxidation of residual carbon may also take place at ~400°C. Sulfate radicals decompose where aluminum sulfate was used as one of the components. Evolved gas analysis was used for the combustion of organics in a microfurnace at a linear heating rate of 20°C per min. up to 900°C, and the generated effluent gas was analyzed using a mass spectrometer. The organic by-products of the gel decompose into cracking fragments such hexane, heptane, and cyclohexane. Beyond 350°C, the amounts of all organics rapidly decrease and they carbonize, with the samples becoming darker in color and finally decomposing into CO and water vapor. The decomposition of organics is dependent on the texture of the raw gel and its source. The dried powder sample changes its color to white at 150°C, to ochre at 250°C, to gray at 350°C–450°C, and finally to black at 550°C–650°C, indicating the presence of carbon. The gel piece sample preheated to 250°C–450°C becomes transparent.
3. Structurally bound water and OH^{-1} groups start undergoing dihydroxylation, as revealed by the continuous decrease of the band at 5160 cm^{-1} due to H_2O combination mode and simultaneously the decrease in the intensity of the 4540 cm^{-1} band due to (Si,Al)–OH groups combination mode. Consequently, the stretching vibration bands due to (H_2O, OH) at 3430 cm^{-1} and one H_2O bending vibration band at 1635 cm^{-1} also decrease with a continuous rise in temperature from 200°C to 600°C. Instead of more dehydroxylation, there is an increase in the band at 5160 cm^{-1} and the stretching vibration band at 3430 cm^{-1} becomes more prominent on heating the sample to 700°C. The increase in absorptivity of this sample may be due to strong absorption of water molecules at the

charge developed on the anhydrous surface of the precursor that has slowly contracted during the heating process. It is assumed that labile protons may connect to the H_2O cages and also make linkage between two oxide surfaces of the aluminosilicate precursor chain network. The prominence of these bands, however, decreases on heating the sample further, to 800°C–900°C, but they still persist.

4. As a result of dehydroxylation, partial condensation of the aluminosilicate precursor network continues, likely because of the shifting of the Si resonance peak from –84 to a more negative value of ~94 ppm. Schmucker and Schneider (1999, Fig. 6) have given a nice representative picture of the variation of the chemical shift of their Type I mullite gel from 150°C to 900°C. Due to the elimination of those hydroxyls and organic radicles as shown above, an increase in the number of pairs of bonds such as T–O, T–T, and T–O(2) slowly takes place with gradual heat treatment. As result, a slight change in density occurs analogous to heating of metakaolin. A question obviously arises as to why the bond distances of T–O do not decrease and why condensation does not take place to the fullest extent. What hinders these two effects? Is it due to the presence of some quantity of H^{+1} (protons) still structurally connected to Si^{+4} in the network chain?

13.13 Requirements for the Exhibition of a 980°C Exotherm

Question 1: What should be the structure of the noncrystalline precursor for exhibiting a sharp 980°C exotherm following an endothermic dip?

Question 2: What should be the criteria of attaining highest homogeneity for mullitization and absence of spinel at 980°C?

Examples

- ^{27}Al MAS-NMR spectroscopies of precursor A, precursor C, and precursor E are discussed here. Precursor A complies with the first condition, which shows that when the temperature of heat treatment approaches 980°C, Al(IV) is maximum and Al(VI) is at a low level. In the amorphous state, the relative intensities of the Al(IV) and Al(V) sites increase with temperature to the detriment of Al(VI). These are at their maximum before crystallization. Precursor A calcined at 750°C contains ~10% Al(VI), and this ratio falls to ~3% at 970°C, which is just before crystallization. The band of each site is broadened on its high field side. As shown the apparent intensity of (VI) includes the trailing contributions of Al(V) and Al(IV). At 1000°C, alumina-rich mullite crystallizes. Therefore, the structure of the noncrystalline precursor should be such that the total contents of distorted Al(IV) + Al(V) sites should be maximum and the content of Al(VI) would be very less (~3%). This would mean that after the final removal of H^+ it abruptly transforms to the mullite-type phase.

- A similar type of transformation happened in the case of spray-pyrolyzed powder as shown by Sanz et al. (1991). This is because spray-pyrolyzed powder synthesized at 400°C contains similar quantities of total Al(IV) and Al(V) content, with a very low quantity of Al(VI).
- One of the criteria of attaining the highest homogeneity for mullitization is to synthesize a precursor with chelating Al sourced from AlOBu by the sol-gel technique.
- Besides chelating and pyrolysis processes, in the following two sol-gel processes homogeneous precursors may develop. Precursor C shows that the bands corresponding to Al(IV) and Al(V) increase while Al(VI) decreases (Jaymes et al., 1996, Fig. 4). In Precursor E, each silicon is linked to three Al(VI) atoms by bridging oxygens and to one hydroxyl group. On being heated to a temperature above 300°C, the raw structure decomposes. Fig. 2(i, f) from Jaymes et al. (1996) shows that the precursor proceeds as per precursors A and C.

It is a matter of great curiosity how a noncrystalline aluminosilicate structure of high alumina content (Al_2O_3:SiO_2 = 3:2) may crystallochemically exist in a stable state from room temperature up to the temperature prior to crystallization.

The probable reasons based on coordination measurements of Al are as follows:

- The dual nature of the coordination states of aluminum is the first cause. An ^{27}MAS-NMR study of the mullite precursor showed that the total quantity of AlO_4 and AlO_5 bonds increases to ~97% on heating the raw state gel to the temperature of crystallization. See MAS-NMR data of precursors A, C, and E of the gel-derived material and spray-pyrolyzed oxide precursor.
- Both tetrahedral and pentahedral Al–O sites are strongly distorted since (Si,Al–O) bond distance measurements by RED by Schneider et al. (1994, Table 3) show the large atomic distance of T–O calculated from radial distribution function of Type I precursors previously heated to 900°C to the extent of 0.18 nm. Further LAXS studies by Okuno et al. (1997, Table 3) of the heat-treated Type I precursor heated over a range of temperature from 150°C to 800°C shows a large T–O distance value 0.177 nm, and it decreases slightly, to 0.174 nm, only at 800°C.
- On the basis of the two observations as stated above, it may be presumed that some Al–O–Si bond angles are large and some Al–O distances are elongated and may be truncated, as conceived by Freund (1967) during the evaluation of the structure of metakaolin. Tetrahedral Si and tetrahedral Al are edge-shared besides the usual corner-shared by three tetrahedras.
- Besides the three reasons pointed above, one of the most significant concepts is the presence of some quantity of H^{+1} (protons) for stabilization of this noncrystalline precursor structure for balancing the negative charge arising from the isomorph substitution of Si^{+4} by Al^{+3}.

Proposed Steps of Decomposition of Six Mullite Precursors

Overall the study of progressive heat-treated gels shows the following steps of transformation, which will be discussed in the forthcoming chapters:

1. Endothermic region at 500°C–600°C and the formation of noncrystalline aluminosilicate precursors by major dehydroxylation and /or organic removal.
2. Endothermic–exothermic region at 980°C, final minor dehydroxylation, and the decomposition of noncrystalline

aluminosilicate precursor structure with the evolution of the following phases:

- Crystallization of either the Al–Si spinel phase or the weakly crystalline mullite-type phase or both
- Noncrystallized mullite phase or spinel (A) as a residue
- Silica-rich noncrystalline aluminous phase

3. Two exothermic regions of mullite formation.
4. Disappearance of Al-Si spinel and formation of mullite and an increase in crystallinity of mullite from its poorly crystalline form developed at the first exotherm.
5. Nucleation and crystallization of the noncrystallized mullite phase after a solid-state reaction with the silica-rich noncrystalline aluminous phase into its well-developed form of mullite.

13.14 Summary

The IR data of heat-treated coprecipitated gel (3:2 ratio) were compared with that of pure SiO_2 by various authors. A shifting of Si–O stretching vibration for pure silica gel occurred at 1120–1080 cm^{-1} in coprecipitated gel when it was heat-treated. The phenomenon of shifting of the absorption band of Si–O to the lower wave number is explained as due to tetrahedral Al–O–Si bond formation. Even in RH gels of compositions $SiO_2{\cdot}3Al_2O_3$ and $SiO_2{\cdot}2Al_2O_3$, the shift from the regular band at 1100 cm^{-1} for Si–O–Si bonds to 1030 cm^{-1} was due to the substitution of alumina for silica in amorphous solids.

Besides IR studies, a large number of authors noted Si–O–Al bond formation by MAS-NMR spectroscopic studies. These studies also indicated symmetrical ^{29}Si resonance at ~92 ppm and the asymmetrical shape of the ^{29}Si resonance peak with a shoulder toward a more negative chemical shift at ~95 ppm, which accounts for Al–O–Si bond development in the noncrystalline aluminosilicate precursor phase. Even in the case of diphasic gels, gradual bond development occurred between two components.

A major breakthrough was noted in an MAS-NMR study that indicated that the tetrahedral coordination of Al increased with a decreasing Al:Si ratio in a monophasic gel but not in a diphasic gel (Komarneni et al., 1985). However, the extent of Al–O–Si bond detection by an isomorphous replacement process is rather difficult in both IR and MAS-NMR studies. And the presence of any free silica or Si–O–Si bond is difficult to predict in the presence of former bonds.

An XRF study showed a clear picture of changes in the coordination number of aluminum from 6 to 4 in amorphous silicoaluminas (Leonard et al., 1964).

An alkali leaching study by the author showed the absence of any free silica in the heated precursor, which conclusively predicted the formation of an aluminosilicate phase. Thus the chemical technique is more evident in characterizing an intermediate alumina-silica precursor structure.

To determine the local structure of mullite gels, the changes in the local coordination characteristics of Al and Si atoms in the amorphous stage during the heating process of mullite gels between 60°C and 1200°C have been determined by Kamijo et al. (1994) using the XAFS technique. According to them, there exist more than one type of bonding between Si and neighboring oxygens in gels dried at 60°C when heat-treated between 200°C and 800°C and when heat-treated at 1200°C. Si–O bond distances are of two types, 1.60–1.64 Å and 1.92–2.07 Å. The coordination environment of oxygen around aluminum in a gel dried at 60°C is 6, and it is reduced to ~4 when the gel is heat-treated in the range of 200° to 800°C.

Leonard et al. (1971) determined the kinds of structures and their defects and suggested that some of the aluminum cations were present in a perturbed tetrahedral arrangement to this disordered structure.

Two important observations were noted from RED function curves of mullite gels heat-treated at 800°C and at 900°C by Schneider et al. (1994).

- The phenomenon of substitution of Si^{+4} by Al^{+3} in heated Types I and III_{oi} precursors is confirmed by RED studies. This is evident from the observation of (i) an increase in the peak intensities at the four coordination spheres compared to silica glass and (ii) increment in bond distances from 1.7 Å to 1.8 Å.
- On higher coordination spheres corresponding to the structural long-range order of the precursor, the Type I precursor exhibited RED function bands at 6.5 Å and 8.7 Å,

which indicated –Al–O–Si– matrix networks. Thus, the structures of mullite precursors are different from that of pure silica glass.

Accordingly, it is conceived that the nature and characteristics of noncrystalline aluminosilicate precursor phases synthesized by the six methods from different components using different processing techniques and pH and water contents may be different. The probable structures of dried gels of four models are tentatively described in Chapter 10.

The radial distribution function D(*r*) and G(*r*) curves of heat-treated Type I precursor at different temperatures were drawn by Okuno et al. (1997). Prominent peaks assigned to the T–O distance at $r = 0.18$ nm and three broad peaks of low intensity near 0.29, 0.32, and 0.42 nm were observed. The T–O distance for Type I precursor (Al:Si ratio = 3:1) showed at 0.174 nm at 800°C. For the T–T distance (middle range order), that is, $r = 0.32$ nm, the peak intensity increased with temperature. This result was attributed to increasing network formation by the condensation of –OH groups. According to Okuno et al. (1997), a gradual condensation of the precursor network during heating of the precursor from room temperature to 800°C occurred probably with changes in the coordination state of aluminum. Both dehydroxylation of –OH groups and decomposition of organic groups occurred, leading to a decrease in the broadness of the peak and a decrease in the *r* value with an increase in the temperature of heating. This process results in an increase in the density and refractive index values with a gradual condensation process. However, its composition is still unknown and yet to be ascertained.

Significantly, by a solid-state reaction between raw, dried, and heat-treated precursors and CaO, the approximate composition of the noncrystalline aluminosilicate phase formed in both monophasic and diphasic gels has been ascertained by the author. The formation of gehlenite rather than calcium aluminate by XRD indicated that both monophasic and diphasic precursors consisted of a noncrystalline aluminosilicate phase of composition nearly close to 3:2 mullite.

In conclusion, all these techniques demonstrate that monophasic Al_2O_3-SiO_2 gel at the first stage formed an amorphous aluminosilicate phase (marked as a noncrystalline aluminosilicate phase) on heating in the temperature range of 400°C (the temperature of first dihydroxylation) and the temperature of first exotherm, at 980°C. The question is, how does such type of noncrystalline alumina-rich silicate structure still occur without crystallization even on heating at a high temperature?

Defects and disorder arise from the substitution effect, which explains the origin of negative lattice charge in the aluminosilicate structure (Tamele, 1950). The acidity of the amorphous silicoaluminous phase was also conformed as due to this effect (Leonard et al., 1971). This requires to be stabilized by some external means (see Chapter 2). For maintenance of electroneutrality, H^+ ions (protons) and/or aqueous $[Al(H_2O)]^{+3}$ groups are required to be connected with adjoining water and OH^{-1} groups present in the aluminosilicate structure. Thus, the Al_2O_3-SiO_2 precursor phase is a hydrated noncrystalline aluminosilicate.

References

1. H. Insley and R. H. Ewell, Thermal behavior of the kaolin minerals. *J. Res. Natl. Bur. Stand.*, **14**(5), 615–627 (1935).
2. H. Yamada and S. Kimura, Studies on co-precipitates of alumina and silica gels and its transformations at higher temperatures. *Yogo Kyokai Shi*, **70**, 87–93 (1962).
3. K. Okada and N. Otsuka, Characterization of the spinel phase from SiO_2-Al_2O_3 xerogels and the formation process of mullite. *J. Am. Ceram. Soc.*, **69**(9), 652–656 (1986).
4. A. K. Chakraborty and D. K. Ghosh, Synthesis and 980°C phase development of some mullite gels. *J. Am. Ceram. Soc.*, **71**(11), 978–987 (1988).
5. Y. Hirata, K. Sakeda, Y. Matsushita, K. Shimada, and Y. Ishihara, Characterization and sintering behavior of alkoxide-derived aluminosilicate powders. *J. Am. Ceram. Soc.*, **72**(6), 995–1002 (1989).
6. M. Fukuoka, Y. Onoda, S. Inoue, K. Wada, A. Nukui, and A. Makishima, The role of precursors in the structure of SiO_2-Al_2O_3 sols and gels by the sol-gel process. *J. Sol-Gel Sci. Technol.*, **1**, 47–56 (1993).
7. M. Sales and J. Alarcon, Synthesis and phase transformations of mullite obtained from SiO_2-Al_2O_3 gels. *J. Euro. Ceram. Soc.*, **16**, 781–789 (1996).

8. M. I. Nieto, G. Urretavizcaya, A. L. Cavalieri, and P. Rana, Structural changes in colloidal and polymeric aluminosilicate gels with mullite composition. *Br. Ceram. Trans.*, **97**(1), 17–23 (1998).

9. T. Tsuchiya, M. Fukuoka, T. Sci, and J. D. Mackenzie, *Seramikku ronnbunnshi*, **97**, 224 (1989).

10. A. D. Irwin, J. S. Holmgren, and J. Jonas, ^{27}Al and 28Si NMR study of sil-gel derived aluminosilicates and sodium aluminosilicates. *J. Mater. Sci.*, **23**, 2908–2912 (1988).

11. B. E. Yoldas, Effect of ultrastructure on crystallization of mullite. *J. Mater. Sci.*, **27**(24), 6667–6672 (1992).

12. H. Schneider, I. Merwin, and A. Sebald, Mullite formation from non-crystalline precursors. *J. Mater. Sci.*, **29**, 805–812 (1992).

13. H. Schneider, B. Saruhan, D. Voll, L. Merwin, and A. Sebald, Mullite precursor phases. *J. Euro. Ceram. Soc.*, **11**, 87–94 (1993).

14. I. Jaymes, A. Douy, D. Massiot, and J. P. Coutures, Characterization of mono- and diphasic mulite precursor powders prepared by aqueous routes, ^{27}Al and ^{29}Si MAS-NMR spectroscopy. *J. Mater. Sci.*, **31**, 4581–4589 (1996).

15. Y. X. Huang, A. M. R. Senos, J. Rocha, and J. L. Baptista, Gel formation in mullite precursor obtained via TEOS prehydrolysis. *J. Mater. Sci.*, **32**, 105–110 (1997).

16. A. Yasumori, M. Iwasaki, H. Kawazoe, M. Yamane, and Y. Nakamura, Nuclear magnetic resonance study of the structure of aluminosilicate gel and glass. *Phys. Chem. Glasses*, **31**(1), 1–9 (1990).

17. S. Komarneni, R. Roy, C. A. Fyfe, and G. J. Kennedy, Preliminary characterization of gel precursors and their high-temperature products by ^{27}Al magic-angle spinning NMR. *J. Am. Ceram. Soc.*, **68**(9), C-243–C-245 (1985).

18. C. Gerardin, S. Sundaresan, J. Benziger, and A. Navrotsky, Structural investigation and energetics of mullite formation from sol-gel precursors. *Chem. Mater.*, **6**, 160–170 (1994).

19. J. Sanz, I. Sobrados, A. L. Cavalieri, P. Pena, S. de. Aza, and J. S. Moya, Structural changes induced on mullite precursors by thermal treatment: a ^{27}Al MAS-NMR investigation. *J. Am. Ceram. Soc.*, **74**(10), 2398–2403 (1991).

20. N. Kamijo, N. Umesaki, K. Fukui, C. Guy, K. Tadanaga, M. Tatsumisago, and T. Minami, Soft X-ray XAFS: local structure of mullite gels prepared from modified aluminium alkoxides. *J. Non-Cryst. Solids*, **117**, 187–192 (1994).

21. A. J. Leonard, P. Ratnasamy, F. D. Declerck, and J. J. Fripiat, Structure and properties of amorphous silico-alumnous. Part 5. Nature and poperties of silico-auminous surfaces. *Disc. Faraday Soc.*, **52**, 98–108 (1971).

22. H. Schneider, D. Voll, B. Saruhan, J. Sanz, G. Schrader, C. Ruscher, and A. Mosset, Synthesis and structural characterization of non-crystalline mullite precursor. *J. Non-Cryst. Solids*, **78**, 262–271 (1994).

23. A. Leonard, S. Suzuki, J. J. Fripat, and C. De Kimpe, Structure and properties of amorphous silicoaluminus. I Structure from X-ray fluorescence spectroscopy and infrared spectroscopy. *J. Phys. Chem.*, **68**, 2008 (1964).

24. Ph. Colomban, Structure of oxide gels and glasses by infrared and Raman scattering part 2 mullite. *J. Mater. Sci.*, **24**, 3011–3020 (1989).

25. A. K. Chakraborty, Characterization of monophasic and diphasic mullite precursors by solid state reaction study. *Br. Ceram. Trans.*, **103**, 33–36 (2004).

26. A. K. Chakraborty, Si-incorporated alumina phases formed out of diphasic mullite gels. *J. Mater. Sci.*, **43**, 5313–5324 (2008a).

27. M. Okuno, Y. Shimada, M, Schmucker, H. Schneider, W. Hoffbauer, and M. Jansen, LAXS and Al-NMR studies on the temperature induced changes of non-crystalline single phase mullite precursors. *J. Non-Cryst. Solids*, **210**, 41–47 (1997).

28. H. Morikawa, S. Miwa, M. Miyake, F. Marumo, and T. Sata, Structural analysis of SiO_2-Al_2O_3 glasses. *J. Am. Ceram. Soc.*, **65**, 78–81 (1982).

29. M. Schmucker and H. Schneider, Structural development of single phase (type I) mullite gels. *J. Sol-Gel Sci. Technol.*, **15**, 191–199 (1999).

30. H. Schneider, D. Voll, B. Saruhan, M. Schmucker, T. Schaller, and A. Sebald, Constitution of the γ-alumina phase in chemically produced mullite precursors. *J. Euro. Ceram. Soc.*, **13**, 431–448 (1994).

31. D. Voll, A. Beran, and H. Schneider, Temperature-dependent dehydration of sol-gel-derived mullite precursors: an FTIR spectroscopic study. *J. Euro. Ceram. Soc.*, **18**, 1101 (1998).

32. K. J. D. Mackenzie, R. H. Meinhold, J. E. Patterson, H. Schneider, M. Schmucker, and D. Voll, Structural evolution in gel-derived mullite precursors. *J. Euro. Ceram. Soc.*, **16**, 1299 (1996).

33. F. Freund, Kaolinite–Metakaolinite, a model of a solid with extremely high lattice defect concentrations. *Ber. Deut. Keram. Ges.*, **44**(1), 5–13 (1967).

34. M. W. Tamele, Chemistry of the surface and the activity of alumina silica cracking catalysts. *Disc. Faraday Soc.*, **8**, 270 (1950).
35. B. E. Yoldas and D. P. Partlow, Formation of mullite and other alumina-based ceramics via hydrolytic polycondensation of alkoxides and resultant ultra- and micro-structural effects. *J. Mater. Sci.*, **23**, 1895–1900 (1988).
36. A. Taylor and D. Holland, The chemical synthesis and characterization sequence of mullite. *J. Non-Cryst. Solids*, **152**, 1–17 (1993).

Additional References

1. A. K. Chakraborty, Reaction study of various mixtures of tetra ethyl orthosilicate and aluminum nitrate. *J. Mater. Sci.*, **43**, 5376–5384 (2008).
2. E. Lippmaa, A. Samoson, and M. Magi, *J. Am. Chem. Soc.*, **108**, 1730 (1986).
3. E. Lippmaa, M. Magi, A. Samoson, G. Engelhardt, and A. R. Grimmer, Structural stidies of silicates by solid-state high resolution ^{29}Si NMR. *J. Am. Chem. Soc.*, **102**(15), 4889–4893 (1980).
4. J. Sanz, I. Sobrados, A. L. Cavalieri, P. Pena, S. de. Aza, and J. S. Moya, Structural changes induced on mullite precursors by thermal treatment: a ^{27}Al MAS-NMR investigation. *J. Am. Ceram. Soc.*, **74**(10), 2398–2403 (1991).

Chapter 14

Final Dehydroxylation of Noncrystalline Al_2O_3-SiO_2 Mullite Precursors

14.1 Intimacy of Mixing Al_2O_3/SiO_2 Powder, Homogeneity, and Agglomeration

It is recognized that a homogeneous chemical composition of the starting powder as well as high purity and fine-grained particles are the most important factors to obtain a dense sintered body of a precisely controlled microstructure. To achieve this, the ultimate processing technique would be chosen in such a way as to obtain homogeneous mixing at the molecular level. The molecular scale of mixing usually occurs after dissolution of two aluminum- and silicon-based salts in a suitable solvent and their homogeneous mixing. However, in the subsequent processing stage either in coagulation or in coprecipitation processes segregation starts in major cases. Finally, during the solvent removal process, as in drying and or in crystallization stages, more segregation takes place and it seems difficult to maintain the ideal mixing state in the subsequent course of heating.

Factors affecting homogeneity

The rate of condensation of silanol groups increases with an increase of pH (Iller, 1979; Keefer, 1984). The use of ammonia

Mullite Formations: Analysis and Applications
Akshoy Kumar Chakraborty

ISBN 978-981-4877-05-3 (Hardcover), 978-1-003-03167-3 (eBook)
www.jennystanford.com

enhances the growth of aluminum hydroxide, which ultimately leads to crystallization. tetraethyl orthosilicate (TEOS) is used for the synthesis of mullite gels, and thus the gel chemistry largely depends on its hydrolysis and later on its condensation behavior. Comprehensive literature on controlling hydrolysis and condensation and then developing polymeric particles of specific sizes is presented. The probable reactivity of silica sols with other aluminum components in the solution-cum-gelation stage is also discussed (see Chapters 10 and 11). Only a synopsis is presented here.

The gelation reactions in two major categories of mullite precursors, namely hydrolysis gelation and hydrolysis coprecipitation, have been enumerated to focus on some of their role in the future course of phase transformation processes. In both gelation and coprecipitation methods it was shown that two components react partially in the wet stage. The existence of Al(IV)–O groups and/or Al–O–Si bonds was noted by several researchers in their infrared (IR) studies and ^{27}Al magic angle spinning–nuclear magnetic resonance spectroscopy (MAS-NMR). Aluminum nitrate nonahydrate (ANN) and TEOS were the two components chosen by a large section of researchers for the synthesis of mullite precursor. A differential thermal analysis (DTA) study indicated interaction between them in the solution stage at a temperature between 60°C and 80°C. This reaction was also found to be dependent on the alumina content. Al_2O_3:SiO_2 in the ratio of 3:2 showed the highest heat evolution, as noted in the preliminary experiment of Chakraborty (2008b) (see Figs. 10.2 and Fig. 10.3, Chapter 10). In a dilute aqueous solution, the development of Si–O–Al(IV) is effective if the degree of polymerization of the $Si(OH)_4$ unit used from the silica component is less and preferably monomeric in nature (of course, experimental verification by molecular weight determination is needed). The resultant gel transforms to mullite early at the first exotherm itself. The gelation mechanism during the synthesis of precursors from ANN/AlOBu and TEOS and others was noted to depend on the use of external water content and even on the presence of water of crystallization of aluminum nitrate salt used, changes in pH value, etc. Accordingly, some models have been predicted for five cases of gel synthesis and have been shown by the author in Chapter 10, as follows:

- For the aqueous sol-gel method, a schematic 1st model equation is presented.
- When the water of crystallization ($9H_2O$) is roughly removed by a repeated heating-solubilization technique, as in the cases of G-31 and SHI gels, a complex Al–O–Si network is described.
- When the water of crystallization is the only source of water causing the hydrolysis of TEOS, as in monophasic gel synthesis of slow hydrolysis (SH)-type gels, a schematic 3rd model equation containing a unidirectional short chain of aluminosilicate precursor is visualized.
- When excess of external water is added, a schematic 4th model equation containing a 3D chain of aluminosilicate precursor is speculated.
- In the coprecipitation process, the entire mechanism of gelation chemistry is different. It is more sensitive to the pH prevailing during the coprecipitation process since the alumina component is precipitated during this operation. It was amorphous at $pH < 7$. Obviously, Al_2O_3-SiO_2 gels prepared at pH 6–7, for example, the gel marked CP6 or RH, showed an amorphous structure. At higher pH, the gel became diphasic in character since boehmite was precipitated out and was homogeneously mixed with silica gel. Therefore, the mixing state of the components in the two processes of precursor synthesis changes. This results in a great change in the phase transformation process. Accordingly, a new model of it is predicted. The gelation reaction in this case has been studied by ^{29}Si and ^{27}Al MAS-NMR studies by several researchers. The gelation reactions in acid and alkaline conditions are thus found to be absolutely different. A schematic 7th model of mixed gels diphasic in character is shown.
- In the spray pyrolysis process, the pyrolysis product of a dilute solution of ANN and TEOS is merely a noncrystalline aluminosilicate compound other than an intimate mixture of amorphous oxides. Both ^{27}Al and ^{29}Si MAS-NMR studies by Sanz et al. (1991) indicated the presence of predominant Al(IV)–O–Si bonds.
- When AlOBu or $Al(OPr)_3$ was taken as the alumina component instead of the much used ANN, along with TEOS, and hydrolyzed

using water, as in the colloidal gel method, at approximately neutral pH, the resultant gel did not show boehmite but the sizes of the two colloidal particles were assumed to be large, which resulted in a thermal transformation behavior analogous to that of the diphasic gel. Even when hydrolyzed with ammoniated water by using the former component from the same source, hydrolysis of the alumina component is very predominant and pseudoboehmite invariably forms in situ. Thus, from the overall results one can conclude that the internal structure of the precursor or the mixing state of the hydroxide components is the controlling factor in the overall high-temperature reaction. As the precursor structures differ, the transformation paths will automatically change. This is the origin of the different routes of phase transformation followed by mullite gels/precursors.

In all models of Al_2O_3-SiO_2 gels of noncrystalline aluminosilicate, the mullite precursor contains a large quantity of trapped water and surface water, pores, and OH^{-1} groups connected to Si or with Al cations. These are removed in steps, gradually, and are completely removed by ~1000°C during the heating process of the precursor. Before discussing the final dehydroxylation process at ~1000°C, the dehydration and dehydroxylation process that takes place over a long range of temperatures, from room temperature to ~1000°C, is first discussed on the basis of the following studies:

FTIR study

The said two processes of Type I mullite precursors have been studied by Voll et al. (1998) by three techniques: moisture evolution analysis (MEA), thermogravimetric (TG) studies, and spectroscopic studies.

The analytical H_2O^+ contents determined by MEA of precursors heated-treated to a temperature above 400°C show good agreement with the TG-determined weight losses of the precursor phase. Water content of 22.5 wt % was found in the precursor heated to 200°C for 15 h. It decreased to 14.2 wt % at 700°C. At this temperature the MEA curve showed a slight discontinuity and followed further decrease. A dehydration and dehydroxylation study of the static heat-treated precursor still showed the existence of 1.8 wt % of water content at 900°C after heating for 15 h.

Spectroscopic studies were done at two regions (5160–4540 cm^{-1} and 3430–1635 cm^{-1}). All bonds comprising H_2O and (Si,Al)-OH combination modes, (H_2O, OH) stretching modes, and H_2O bending modes gradually decreased on heat treatment of the precursor up to 600°C, abruptly increased at 700°C, and again started to decrease from 700°C to 900°C (experimentally chosen temperature).

Integrated absorbance of the band at 3440 cm^{-1} is correlated with the water content determined by MEA. The method also shows a slight discontinuity at 700°C, similar to that noticed in the H_2O combination and bending modes. The deconvolution of this band reveals four bands, attributed to two types of H_2O and two types of OH.

- The strong band II at 3440 cm^{-1} is attributed to the stretching vibration of one type of H_2O molecule.
- The weaker band I at 3585 cm^{-1} is attributed to the stretching vibration of one type of OH group.
- The relatively strong band III at 3226 cm^{-1} is attributed to the second type OH group.
- The weak band IV at 2961 cm^{-1} is attributed to the second type H_2O group.

According to Voll et al. the OH band position I at 3585 cm^{-1} and the H_2O band position II at 3440 cm^{-1} require no hydrogen bridging, whereas the OH band III at 3226 cm^{-1} and the H_2O band IV at 2961 cm^{-1} require strong hydrogen bonding. The spectra of the precursor preheated at a high temperature are mainly determined by band II at 3440 cm^{-1}, assigned to the stretching vibration of free H_2O. The results show that all four bands decrease in their intensities from the as-prepared precursor to precursors calcined to 700°C and then to 900°C. At a preheating temperature of 900°C, non-hydrogen-bonded molecular H_2O and minor amounts of non-hydrogen-bonded OH are the stable forms of water and although reduced are still present in significant amounts.

DTA study

DTA and TG curves of sol-gel and coprecipitated starting materials synthesized by Schneider et al. (1993, 1992) showed two small endotherms between 135°C and 229°C due to dehydration.

Differential scanning calorimetry (DSC) and TG analysis (TGA) of 50 mol % Al_2O_3 spray-dried powder and 90 mol % Al_2O_3 spray-dried powder synthesized by Douy (2006) showed two endotherms between ~200°C to 300°C due to dehydration of spray-dried precursors.

According to Colomban (1991) most of the OH^{-1} groups are lost between 600°C and 700°C. It was difficult to discriminate between the protonic species (OH^{-1}) and water at the external surface of the amorphous phase from inside the pores. Only a sufficient amount of OH^{-1} groups remains (~1 wt %), which prevents the above-mentioned models from collapsing and condensation of polymeric network and prevents precipitation.

Thus, the presence of a trace amount of –OH groups in the noncrystalline aluminosilicate precursor phase plays a role in stabilization the network structure against a collapse. This was noted by Colomban (1991) in his DTA and IR studies. Similarly the existence of non-hydrogen-bonded molecular H_2O and minor amounts of non-hydrogen-bonded OH was noted in the precursor preheated to 900°C by Voll et al. (1998). Thirdly, the retention of protons in ^{29}Si cross-polarized spectroscopy was observed by McKenzie et al. (1996). Accordingly, the final removal of residual protons and/or the final dehydroxylation of the precursor is the subject matter of this chapter.

14.2 Role of Protonic Species of the Noncrystalline Precursor Phase in Crystallization during a 980°C Exotherm

The exhibition of an exotherm by mullite gels/precursors of different classes depends upon the following factors:

- The development of a specific aluminosilicate structure containing the groups SiO_4^{-4} and OH^{-1} in the case of the gel obtained from Al sol and Si sol (Chakraborty, 1988; Schneider et al., 1992).
- Aluminosilicate hydrate obtained from ANN and TEOS; AlOBu and TEOS containing OH groups: The presence of OH groups acts to balance charge deficiency in the aluminosilicate

hydrate complex. Generally, these types of bonding are called hydrogen bonding. In fact, such bonding sustains in pure silica gel up to the point of rapid shrinkage, leading to crystallization preceded by dehydration, better known as deprotonization (see DTA of silica gel in Fig. 2.1). Analogous to rapid dehydroxylation versus shrinkage versus crystallization of silica gel, Al_2O_3-SiO_2 gel will also dehydrate rapidly as a result DTA, shown as endotherm marked D before the 980°C exotherm (called endo-exo assembly), as noted below by various authors.

Colomban (1991) correlated the glassy state to a protonic content level. The specific surface area, shrinkage, and weight loss followed homothetic laws. Departure of the last of the protonic species enhances catastrophic changes in the noncrystalline polymeric network. How the 980°C exothermic peak is exhibited and how the last of the protons are eliminated are discussed prior to discussions regarding the various routes of mullite formation.

14.2.1 Occurrence of an Endothermic Dip Just before the 980°C Exotherm

The exhibition of this usual phenomenon/event was rarely noted. Thermal analysis studies of the following authors are cited:

- During DTA and TGA studies, an endothermic dip D and a sharp weight loss of the precursor marked CM were noted by Schneider et al. (1992) and are presented in Fig. 14.1.
- In a mullite monolith synthesized from silicon butoxide and AlOBu by a slow gelation process, finally an endothermic event was observed just before an intense exotherm and the same was explained by Colomban (1989) as a "glass transition" anomaly.
- The exhibition of an endothermic dip (Figs. 7.6 and 7.8) was observed in DTA studies of kaolinite clay and in some monophasic gels marked G-31, Gel A, and SHI synthesized by Chakraborty (1994).
- An endothermic peak at D in DTA (Fig. 7.20) was observed in the SD precursor synthesized by Chakraborty (2008, unpublished).

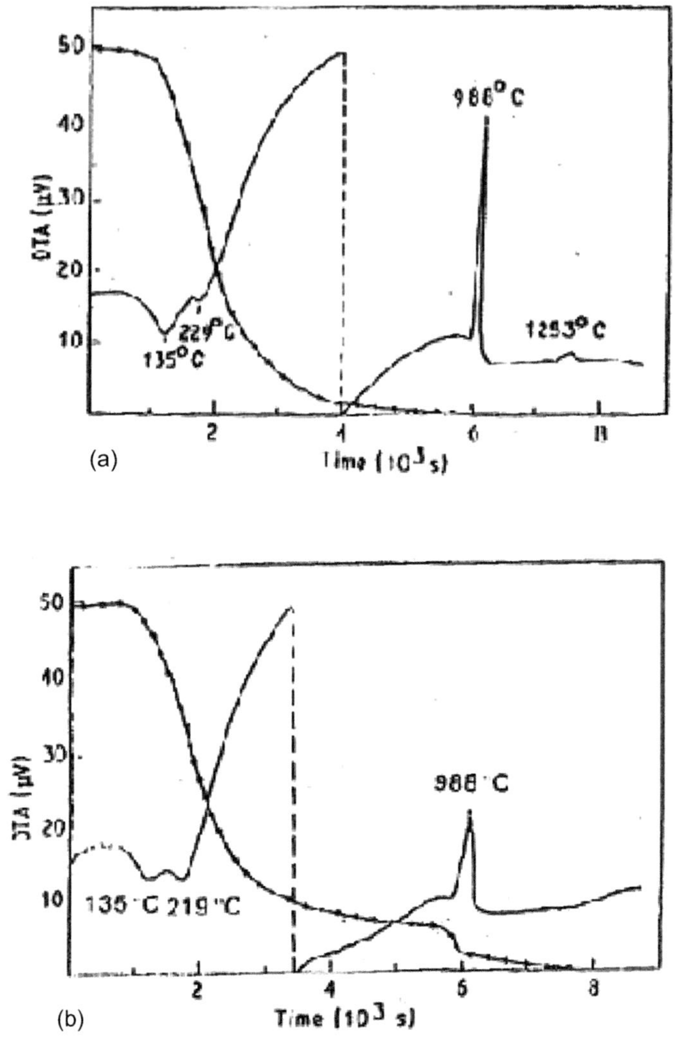

Figure 14.1 DTA and TG curves of (a) sol-gel and (b) coprecipitated starting materials (Schneider et al. 1992).

- Douy (2006) noted both an endothermic peak and a peak due to weight loss in DSC and differential thermo gravimetric analysis (DTGA) (Fig. 14.2) in his SD precursors. Both these peaks were prominently observed.
- Colomban and Mazerolles (1990) compared the DTA traces with dilatometric traces of alumina-silica gels of molar

ratios 9:1 and 3:7. The intense DTA peaks corresponding to crystallization were corroborated with 3% and 4% shrinkages in 8:2 and 7:3 Al_2O_3:SiO_2 gels, respectively. TGA traces of gels of formula $3Al_2O_3{\cdot}2SiO_2{\sim}1.23H_2O$ or $Al_6Si_2O_{10.77}$ $(OH)_{0.46}$ H_2O corresponded to the weight loss in some wt % due to the escape of large quantities of OH^{-1} (~1 wt %) as water in the same events.

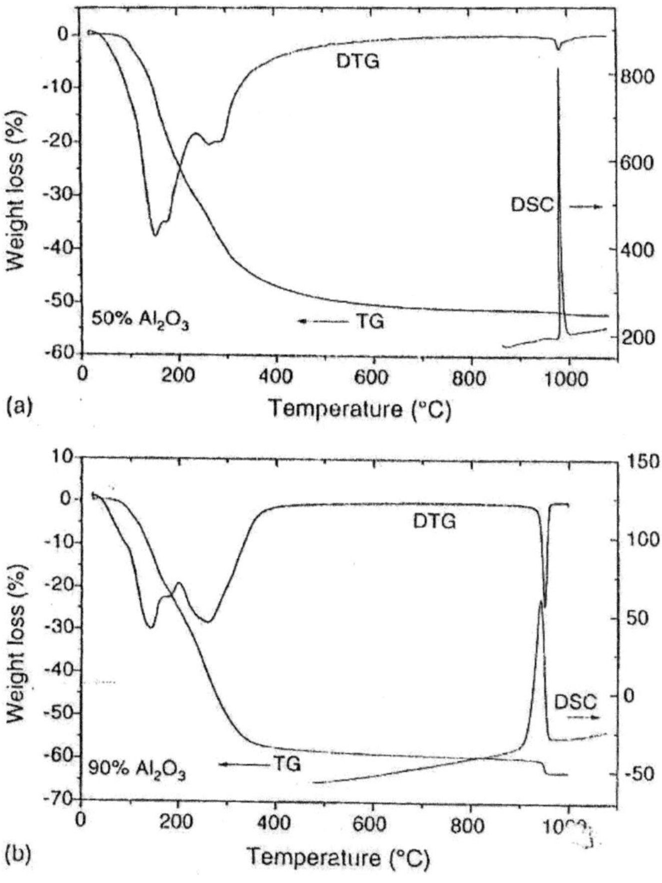

Figure 14.2 (a) DSC and TGA of 50 mol % Al_2O_3 spray-dried powder; heating rate = 5°C/min. (b) DSC and TGA of 90 mol % Al_2O_3 spray-dried powder; heating rate = 5°C/min. (Douy 2006).

The question is, why do the DTA, TGA, or DTG and dilatometric curves of different Al_2O_3-SiO_2 precursors exhibit the same phenomenon so sharply? It was shown that these precursors developed a noncrystalline aluminosilicate precursor phase (see Chapter 13) other than that present as a free oxide mixture. For example, a spray-pyrolyzed powder contains Al(IV) groups to a large extent (Sanz et al., 1991; Jaymes et al., 1996). It was also shown by Douy (2006) that the percentage of AlO_5 + AlO_4 groups was comparable with the precursor generally obtained by gelation techniques in Fig. 13.3 by Fukuoka et al. (1993) and also the presence of some stabilized protons H^+ in the aluminosilicate structure. This matter will be discussed gradually.

14.2.2 Final Dehydroxylation at the Endothermic Dip

The mechanism of removal of final (–OH) groups by dehydroxylation of the precursor ($3Al_2O_3 \cdot 2SiO_2 \times H_2O$) may be inhomogeneous and may follow a few steps.

Final dehydroxylation at the endo-exo assembly region

Final dehydroxylation-cum-crystallization processes

Accepter region	Donor region
$3Al_2O_3 \cdot 2SiO_2 \times H_2O$	$3Al_2O_3 \cdot 2SiO_2 \times H_2O$
(noncrystalline aluminosilicate partially hydrated precursor phase)	(noncrystalline aluminosilicate partially hydrated precursor phase)

Step 1: Proton migration begins

(at endothermic dip just before 980°C).

$\longrightarrow H_2O$

$\longleftarrow$

At endotherm (D)

Step 2: Al^{+3} cation diffusion starts

in the opposite direction.

(Step 3: Si^{+4} diffusion starts with

$\longrightarrow$

$\longleftarrow$

a counterdiffusion of Al^{+3} in the reverse direction.)

Step 4: Redistribution of cation ensues.

Phase separation occurs.

[Alumina-rich silica phase] (Noncrystalline, anhydrous, and unstable phase) (AS1)	Silica-rich aluminous phase (AS2) (^{29}Si MAS-NMR shows −110 ppm broad peak)

Repacking of oxygen starts immediately, resulting in a large contraction, as shown in TMA. The ratio of AlO_4:AO_6 changes as noted in ^{27}Al MAS-NMR. ^{29}Si MAS-NMR shows a −90 ppm broad peak.

Crystallization (sudden) ↓ at 980°C exotherm (sharp)

(i) Alumina-rich mullite and/or alumina-rich Al-Si spinel

(both are weakly crystalline)

\+

(ii) Residual noncrystalline alumina-rich mullite phase (AS3)

(as a major phase) (14.1)

The ratio of weakly crystalline mullite to poorly crystalline Al-Si spinel is dependent on the nature of the original aluminosilicate precursor phase, that is, a kind of monophasic gel, whether it is first type or the other type.

The dehydroxylation process starts at the ~950°C endothermic dip (D), where actually proton migration starts at the first step from the acceptor to the donor region and Al ions may migrate in the opposite direction for maintaining electroneutrality.

Silicon ion diffusion and counterdiffusion of Al ions ensue thereafter. This results in the generation of two separate phase boundary regions. The acceptor region becomes an anhydrous noncrystalline alumina-rich silica phase (named AS1). This AS1 phase is most unstable, and it immediately starts repacking of oxygen and a large contraction is noted in Fig. 14.3 (Yoldas & Partlow, 1988; Nieto et al., 1998). The ratio of the AlO_4 to AlO_6 groups changes, as shown in ^{29}Si MAS-NMR studies by Jaymes et al. (1996, Figs. 9.14–9.18). Its composition gradually approaches that of mullite slightly rich in alumina.

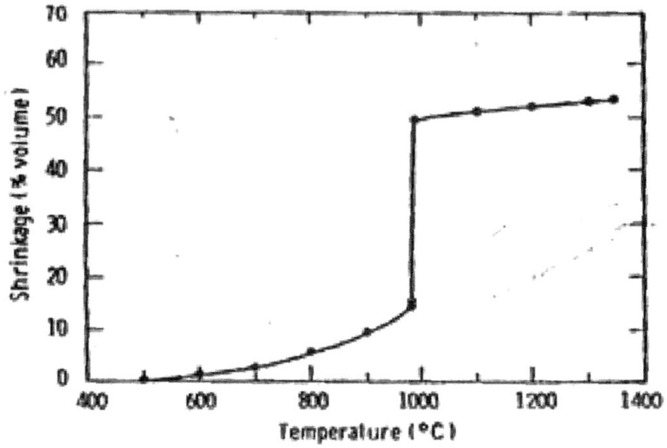

Figure 14.3 Monolithic amorphous $Al_6Si_2O_{11}$ gels produced by intimate polymerization of aluminum and silicon show a spontaneous shrinkage (~35%) at the mullite crystallization temperature.

Immediately crystallization ensues. Poorly crystalline mullite appears as a major phase, with Al-Si spinel as the minor phase, in the case of monophasic precursors, for example, SH gel/Type I precursor. On the contrary, the reverse of crystalline phases, for example, poorly crystalline Al-Si spinel as a major phase and weakly crystalline mullite as a minor phase, develop in the coprecipitated precursor, RH, and or Type III gels. The ratio of mullite and Al-Si spinel as crystalline phases formed depends upon the nature of the precursors used and their processing conditions based on two components chosen.

In both cases either noncrystalline mullite phase or noncrystalline spinel phase (AS3) as the major phase likely remains as a residue after completion of 980°C crystallization. This weakly crystalline mullite phase is to be designated as "nonstoichiometric mullite." The identification of this residual phase (AS3) is to be ascertained by ^{29}Si MAS-NMR.

The donor region contains a noncrystalline silica-rich aluminous phase (named AS2). ^{29}Si MAS-NMR indicates the liberation of this AS2 phase by −110 ppm resonance. Phase separation becomes complete.

14.2.3 Phase Separation Prior to Crystallization of Mullite and/or Al-Si Spinel

Segregation before the commencement of mullite

The aluminosilicate phase separates into alumina-rich and silica-rich phases as per the SiO_2-Al_2O_3 phase diagram. It shows the presence of a metastable immiscibility region. It is assumed that crystallization of mullite xerogel will follow on the basis of the phase diagram. Moreover, there is a lot of heterogeneity in the mullite gel system, which mostly depends upon the source of the two components and the processing techniques employed to synthesize xerogels.

Thus, ^{29}Si NMR generally shows two peaks, one at −89 ppm due to alumina-rich silicate (A) phase and the other at −110 ppm due to the alumina-poor silicate (A) phase. The alumina-rich silicate crystallized to mullite with a DTA exotherm at ~980°C. It is conjectured that the noncrystalline portion of the residual alumina-rich silicate phase may undergo a solid-state reaction with the noncrystalline alumina-poor silicate phase and crystallize further to mullite at a high temperature.

Segregation before the commencement of Al-Si Spinel formation

- The sol-gel material precursor of Schneider et al. (1992) showed a spinel phase besides some mullite at 1000°C. It showed two Si NMR peaks, one at −90 ppm and one at −110 ppm, indicating segregation, and it continued on heating.
- The MI sample of Huang et al. (1997) heated to 900°C showed that two Si NMR peaks are clumped together instead of splitting. It continued up to 1100°C.
- Precursors B and D of Jaymes et al. (1996) on heating at 1000°C crystallized to a spinel phase. The Si spectrum of B was broad and asymmetric. On decomposition into two lines, one

was centered at –95 ppm and the other at –108 ppm. Similar was the case for Precursor D. On simulation, two bands, one centered at –93 and the other at –108 ppm, were noted.

- Precursor M2 of Geradin et al. (1994) heated in the range 950°C–1100°C showed a broad signal, signals at –104 ppm and –85 ppm likely clumped together.
- Other examples are of SP precursors synthesized by Jaymes and Douy (1992) and Jaymes et al. (1995). Douy (2006) synthesized amorphous spray-dried precursors in the whole range of composition in the Al_2O_3-SiO_2 system.
- Precursor powder C is a spray-dried gel. Precursors synthesized by organic-assisted methods showed variation in spinel to mullite development (Jaymes et al., 1996).
- Precursor B showed large-scale segregation on heating from 750°C to 1200°C, as noted in ^{29}Si NMR (Fig. 9.15). ^{27}Al spectra at 1000°C and 1200°C are likely that of γ-Al_2O_3; Al(IV) was the major form, while Al(VI) occupied a higher level. This behavior is analogous to that observed by Hyatt and Bansal (1990) in their diphasic gel.
- From the choice of component sources and processing conditions, it might be assumed that Precursor D was a diphasic gel. Bigger molecules of colloidal silica gel were involved during aging prior to spray drying. At 500°C, Si spectra showed asymmetry (Fig. 9.16). This shape continued on heating, and at 1200°C, a large amount of spinel crystallized, as shown by a broad Si spectra being displayed at –90 ppm besides a free silica peak. At ~1275°C, spinel transformed and mullite crystallized. Transformations of these precursors are to be explained.

14.2.4 Absence of the Metastable Phase Separation Process during the Final Dehydroxylation Process

However, the configuration of the noncrystalline aluminosilicate precursor phase obtained during the heating process of mullite Powder C synthesized by Jaymes et al. (1996) would be nearly similar to that in the mullite lattice stage. This gel-derived noncrystalline precursor will not exhibit metastable phase separation but crystallize directly to the most stable state as 3:2 stoichiometric mullite. In the following cases, dehydroxylation does not likely proceed through phase separation:

- Monophasic gel sample marked S20 synthesized by Yasumori et al (1990).
- Monophasic gel synthesized by Chakraborty and Ghosh (1988) by the aqueous sol-gel method. Mullite gel synthesized from aqueous silica sol and ANN showed the following: (i) the gel of 3:2 mullite composition showed the highest 980°C exotherm; (ii) this gel did not form a spinel phase at the first exotherm; (iii) In this case, only 3:2 mullite crystallized and no 2:1 mullite; and (iv) both height of the first exotherm in DTA and XRD intensity of mullite are interrelated. It could be predicted that during the occurrence of the endo-exo region, the noncrystalline precursor phase directly crystallized to regular 3:2 mullite without any phase separation process phenomenon. Such segregation has been minimized or stopped in some cases of transformation of mullite precursors, for example, SP precursor. This precursor was synthesized by Jaymes et al. (1994) using $Al(OPr)_3$ and $SiCl_4$ in tetrahydrofuran and hydrolyzed by ^{17}O-enriched water.

14.3 Final Dehydroxylation of Noncrystalline Al_2O_3-SiO_2 Mullite Precursors

We will now look at the interrelation between the following physicochemical events and the final dehydroxylation of noncrystalline $2O_3$-SiO_2 mullite precursors (dehydroxylation processes from room temperature to final dehydroxylation at D).

DTA and DDTA

A strong DTA and differential DTA (DDTA) exothermic phenomenon are exhibited. Simultaneous thermal effects during the "glass to ceramic " crystallization phenomenon were shown by Chakraborty (1996b) in Fig. 15.1a,b in Chapter 15.

Dilatometric studies

Shrinkage or contraction behavior was noted in TMA and DTMA studies.

- Comparing sintering curves, a slight decrease in the lengths of the three compact CA, PB, and PA up to 800°C are noted, which may be associated with the weight loss as observed in the TGA curves (Nieto et al. 1998). In the polymeric sample (PA) a slight expansion was noted at 980°C.
- The crystallization of mullite from the amorphous Al_6 Si_2O_{13} polymer at 980°C exotherm was associated with ~35% linear shrinkage in the monolithic gel (Yoldas & Partlow, 1988), as shown in Fig. 14.3.
- A comparative shrinkage behavior (DTMA) was noted (Fig. 15.21) by Ivankovic et al. (2003).
- Corroboration of the 980°C exothermic effect with expansive effects in dilatometric curves of SP and PB2 samples was shown. The expansive effect of the spray-pyrolyzed sample was more rapid than that of polymeric sample PB2 at the temperature of exotherm (Sanz et al., 1991).
- A strong macroscopic shrinkage (≥8%) was noted by Colomban and Mazerolles (1990).

TGA and DTG

- They measure the surface, trapped water, and OH^{-1} content. Loss of protonic entities and condensation of the amorphous polymeric network (Colomban, 1990) were shown. A strong correlation between the departure of OH^{-1} defects from polymeric glassy phase with residual water content (0.5% in weight) was observed. A TGA jump, which may be related to the temperature increase occurring inside the material in relation with the exothermic reaction, was seen. A second explanation may be the loss of some quantity of "water" when crystallization or precipitation occurs. A sufficient quantity of OH^{-1} groups remained to the extent of ~1 wt % to ensure the stability of the amorphous structure of mullite (Colomban, 1990).
- TG and MEA studies by Voll et al. (1998) showed the removal of H_2O and hydroxyl groups with an increase in the temperature of heat treatment.

Density

Densification became drastic when the limit of the OH^{-1} level was reached below the stability range of the polymeric network. It is a temperature-dependent process and can be related to the energy needed to eliminate the last OH^{-1} defects, which stabilizes the disordered state (Colomban & Mazerolles, 1990). A drastic increase in the bulk density of the xerogel occurred and complete sintering took place when the precursor was heated at 1000°C (Yasumori et al., 1990).

Surface area

The rate of change of contraction was observed in dilatometry, which agrees with the drastic reduction in the specific surface area in between 900°C and 950°C. A relation between the content of the protonic species with a specific surface area of the polymeric gel network (Colomban & Vendange, 1992) was noted. A constant particle size during mullite formation was observed by X-ray, (Brunauer–Emmett–Teller) BET, and (transmission electron microscopy) TEM measurements of single-phase gel heated between 1213°K and 1473°K. The average value ranged from 17 nm to 16 nm in X-ray and from 13 nm to 32 nm in BET and was ~20 nm in TEM (Li & Thomson, 1990).

Pore and its size

The glassy phase was porous below 900°C (Colomban & Vendange, 1992).

Crystallization

The nucleation of mullite by electron transmission coupled with diffraction analysis showed that above 1000°C, the cubic-shaped phase precipitated out of the optically clear mullite sample. The mean size of the precipitate was 20 nm (Colomban, 1990). The crystallization of mullite or Al- Si spinel of varying ratios of both likely depends upon the character of the noncrystalline aluminosilicate precursor.

Microscopy

The micrograph remained characteristic of a "glassy phase" even after transition with particle size less than 5 nm.

IR and FTIR

A sudden IR spectral change during exothermic heat effect was noted (Hirata et al., 1989). Fourier-transform infrared spectroscopy study by Voll et al. (1998) and mass spectrometry studies by Mackenzie et al. (1996) showed the integral absorbances of the combination H_2O band centered at 5160 cm^{-1} and the integral band of (Si,Al)-OH^{-1} at 4540 cm^{-1} versus temperature (see Schmucker & Schneider, 1999, Fig. 5).

MAS-NMR

- Al resonance: A drastic change in Al coordination occurred. Penta-coordinated Al sites disappeared (Yoldas, 1992; Sanz et al., 1991) during the formation of the mullite type phase.
- Si resonance: A drastically sharp decrease in the Si resonance peak was noted. At 1000°C, it became −111 ppm, which may be attributed to Q_4(1Al) or Q_4(0Al) (Yasumori, 1990).

14.4 Conclusion

The following reactions have been speculated by various researchers in the past:

$$\begin{aligned}\text{Monophasic gel} &\xrightarrow[900^\circ\text{C}]{\text{at }800^\circ\text{C}} \text{Noncrystalline aluminosilicate} \\ &\qquad\text{(an anhydrous mullite precursor phase)} \\ &\xrightarrow{\text{at }900^\circ\text{C exotherm}} \text{t-mullite phase} + \text{SiO}_2\text{(A)} \end{aligned} \tag{14.2}$$

The conjecture that the reaction consisting of an intermediate anhydrous mullite precursor phase to t-mullite as shown above probably did not happen in practice is based on the following experimental observations and arguments:

- Generally, this dehydroxylation may be inhomogeneous in nature.

- The presence of some small quantity of –OH groups in the noncrystalline aluminosilicate precursor phase accumulated just before the temperature of first exotherm and their removal by dehydroxylation in the ongoing heating process must be considered prior to making any prediction on the gel $\rightarrow$ mullite reaction sequences.
- In such a dehydroxylation process, two zones, namely accepter and donor regions, are also assumed for consideration of proton movement vis-à-vis cation migration.
- After the final removal of OH^{-1} groups, an anhydrous unstable noncrystalline alumina-rich siliceous intermediate phase (AS1) and a noncrystalline silica-rich aluminous phase (AS2) were generated in the two regions.

The question is, what will happen on further heating? The precursor will, of course, crystallize. But that will depend on thermodynamic consideration and probably not be based on the kinetic factor. The crystalline phase is the most stable phase in this system, and that has to be primary objective. The reaction is given below.

Of course, mullite will crystallize, but in which form? It may be of any composition, from 3:2 to 2:1 to any up to point Q, as shown in Cameron's diagram (Fig. 16.6). As 3:2 is the most stable state in this system, it is assumed that stoichiometric mullite will crystallize out of the anhydrous alumina-rich mullite phase (AS1). But the circumstance is different. Since it is high in alumina content because of the inhomogeneous dehydroxylation process that occurred during the final elimination of some hydroxyl groups, the possible mullite phase may be nonstoichiometric alumina-rich mullite instead of stoichiometric (3:2) phase and the reaction at 980°C will be as follows:

[Alumina-rich silicate phase (AS1)]	$\xrightarrow[\text{exotherm}]{\text{at } 980^\circ\text{C}}$	(1) Mullite phase (alumina rich), (nonstoichiometric) + (2) Residual SiO_2 rich aluminous phase (AS2) + (3) Residual non-crystalline alumina-rich mullite phase (AS3)

(14.3)

- It is to be conceived that whatever be the source of components used and whatever be the process of precursor synthesis followed, the resultant noncrystalline aluminosilicate mullite precursor phase should pass through a final dehydroxylation stage. Thereafter, 3:2 mullite would obviously crystallize in due course of the heating process.
- It may be assumed that stable mullite with higher alumina content could not be synthesized by general sol-gel methodology at a 980°C exotherm.
- Several researchers thought that probably the homogeneity of the Al_2O_3:SiO_2 ratio from the solution level up to crystallization of the final mullite will be achieved in comparison to the formation of the same in Al_2O_3-SiO_2 oxide mixtures and predicted that such a sol-gel method of synthesizing mullite might be an excellent process. In practice, different results are observed during the dehydroxylation process and in the final elimination of hydroxyls from the noncrystalline mullite precursor phase as vividly analyzed in this chapter. Contrarily, chemical homogeneity would not probably be maintained from solution up to mullite grain since a phase separation process occurs at the intermediate stage.

References

1. R. K. Iller, *The Chemistry of Silica*, Wiley, New York (1979).
2. K. D. Keefer, in *Better Ceramics Through Chemistry*, eds. C. J. Brinker, D. E. Clark, and D. R. Ulrich, Elsevier, Amsterdam, New York, p. 15 (1984).
3. A. K. Chakraborty, Reaction study of various mixtures of tetra ethyl orthosilicate and aluminum nitrate. *J. Mater. Sci.*, **43**, 5376–5384 (2008b).
4. J. Sanz, I. Sobrados, A. L. Cavalieri, P. Pena, S. de. Aza, and J. S. Moya, Structural changes induced on mullite precursors by thermal treatment: a ^{27}Al MAS-NMR investigation. *J. Am. Ceram. Soc.*, **74**(10), 2398–2403 (1991).
5. Ph. Colomban, Protonic defects and crystallization of sol-gel (Si, Ge) mullites and alumina, in *Ceramics Today-Tomorrow's Ceramics*, ed. P. Vincenzini, Elsevior Science Publishers B.V., p. 599 (1991).

6. A. K. Chakraborty, and D. K. Ghosh, Synthesis and 980°C phase development of some mullite gels. *J. Am. Ceram. Soc.*, **71**(11), 978–987 (1988).
7. H. Schneider, I. Merwin, and A. Sebald, Mullite formation from non-crystalline precursors. *J. Mater. Sci.*, **29**, 805–812 (1992).
8. A. K. Chakraborty, Role of hydrolysis water-alcohol mixture on mullitization of Al_2O_3-SiO_2 monophasic gels. *J. Mater. Sci.*, **29**, 6131–6138 (1994b).
9. A. K. Chakraborty, An analysis of the phase evolution of six types of mullite gels. Unpublished data (2008).
10. A. Douy, Crystallization of amorphous spray-dried precursors in the Al_2O_3-SiO_2 system. *J. Euro. Ceram. Soc.*, **26**, 1447–1454 (2006).
11. Ph. Colomban and L. Mazerolles, SiO_2-Al_2O_3 phase diagram and mullite non-stoichiometry of sol-gel prepared monoliths: influence on mechanical properties. *J. Mater. Sci. Lett.*, **9**, 1077–1079 (1990).
12. I. Jaymes, A. Douy, D. Massiot, and J. P. Coutures, Characterization of mono- and diphasic mullite precursor powders prepared by aqueous routes, ^{27}Al and ^{29}Si MAS-NMR spectroscopy. *J. Mater. Sci.*, **31**, 4581–4589 (1996).
13. M. Fukuoka, Y. Onoda, S. Inoue, K. Wada, A. Nukui, and A. Makishima, The role of precursors in the structure of SiO_2-Al_2O_3 sols and gels by the sol-gel process. *J. Sol-Gel Sci. Technol.*, **1**, 47–56 (1993).
14. B. E. Yoldas and D. P. Partlow, Formation of mullite and other alumina-based ceramics via hydrolytic polycondensation of alkoxides and resultant ultra- and micro-structural effects. *J. Mater. Sci.*, **23**, 1895–1900 (1988).
15. M. I. Nieto, G. Urretavizcaya, A. L. Cavalieri, and P. Rana, Structural changes in colloidal and polymeric aluminosilicate gels with mullite composition. *Br. Ceram. Trans.*, **97**(1), 17–23 (1998).
16. Y. X. Huang, A. M. R. Senos, J. Rocha, and J. L. Baptista, Gel formation in mullite precursor obtained via TEOS prehydrolysis. *J. Mater. Sci.*, **32**, 105–110 (1997).
17. C. Gerardin, S. Sundaresan, J. Benziger, and A. Navrotsky, Structural investigation and energetics of mullite formation from sol-gel precursors. *Chem. Mater.*, **6**, 160–170 (1994).
18. I. Jaymes and A. Douy, Homogeneous mullite-forming powders from spray-drying aqueous solutions. *J. Am. Ceram. Soc.*, **75**(11), 3154–3156 (1992).

19. I. Jaymes, A. Douy, and D. Massiot, Synthesis of a mullite precursor from aluminum nitrate and tetraethoxysilane via aqueous homogeneous precipitation: an ^{27}Al and ^{29}Si liquid- and solid-state NMR spectroscopic study. *J. Am. Ceram. Soc.*, **78**(10), 2648–2654 (1995).
20. A. Yasumori, M. Iwasaki, H. Kawazoe, M. Yamane, and Y. Nakamura, Nuclear magnetic resonance study of the structure of aluminosilicate gel and glass. *Phys. Chem. Glasses*, **31**(1), 1–9 (1990).
21. A. K. Chakraborty, New data on thermal analysis of diphasic mullite gels. *J. Therm. Anal.*, **46**, 1413–1419 (1996b).
22. H. Ivankovic, E. Tkalcec, R. Nass, and H. Schmidt, Correlation of the precursor type with densification behavior and microstructure of sinteren mullite ceramics. *J. Euro. Ceram. Soc.*, **23**, 283–292 (2003).
23. D. Voll, A. Beran, and H. Schneider, Temperature-dependent dehydration of sol-gel-derived mullite precursors: an FTIR spectroscopic study. *J. Euro. Ceram. Soc.*, **18**, 1101 (1998).
24. Ph. Colomban and V. Vendange, Sintering of alumina and mullites prepared by slow hydrolysis of alkoxides: the role of the protonic species and pore topology. *J. Non-Cryst. Solids*, **147–148**, 245–250 (1992).
25. D. X. Li and W. J. Thomson, Mullite formation in kinetics of a single-phase gel. *J. Am. Ceram. Soc.*, **73**(4), 964–969 (1990).
26. Y. Hirata, K. Sakeda, Y. Matsushita, K. Shimada, and Y. Ishihara, Characterization and sintering behavior of alkoxide-derived aluminosilicate powders. *J. Am. Ceram. Soc.*, **72**(6), 995–1002 (1989).
27. K. J. D. Mackenzie, R. H. Meinhold, J. E. Patterson, H. Schneider, M. Schmucker, and D. Voll, Structural evolution in gel-derived mullite precursors. *J. Euro. Ceram. Soc.*, **16**, 1299 (1996).
28. M. Schmucker and H. Schneider, Structural development of single phase (type I) mullite gels. *J. Sol-Gel Sci. Technol.*, **15**, 191–199 (1999).
29. B. E. Yoldas, Effect of ultrastructure on crystallization of mullite. *J. Mater. Sci.*, **27**(24), 6667–6672 (1992).
30. H. Schneider, B. Saruhan, D. Voll, L. Merwin, and A. Sebald, Mullite precursor phases. *J. Euro. Ceram. Soc.*, **11**, 87–94 (1993).
31. Ph. Colomban, Structure of oxide gels and glasses by infrared and raman scattering part 2 mullite. *J. Mater. Sci.*, **24**, 3011–3020 (1989).
32. M. J. Hyatt and N. P. Bansal, Phase transformations in xerogels of mullite composition. *J. Mater. Sci.*, **25**, 2815–2821 (1990).
33. I. Jaymes, A. Douy, P. Florian, D. Massiot, and J. P. Coutures, Mew synthesis of mullite. strucural evolution study by ^{17}O, ^{27}Al and ^{29}Si MAS NMR spectroscopy. *J. Sol-Gel Sci. Technol.*, **2**, 367–370 (1994).

Chapter 15

Spinel Phase: A Concise Review

Evolution of the intermediate Al-Si spinel phase formed while heating of a variety of mullite precursors has been investigated for identification and quantitative estimation. In the mullite-gel-to-mullite reaction series, major future problems related to the mechanism of mullitization and its different paths of formation are due to mullite's improper identification and characterization. Some of the methods and techniques (QXRD, TEM/EDS, lattice constant, and alkali leaching studies) as used and reported by earlier authors for proper mullite identification and characterization and the resulting problems they faced are discussed in this chapter.

15.1 Formation of the Spinel Phase in Various Gels Observed by Earlier Authors

Spinel formation during heating at/around the first exothermic peak temperature was observed in different cases of mullite precursors. Sometimes, the precursor was found to crystallize as a major weakly crystalline phase, and in other cases it occurred as a minor phase along with weakly crystalline mullite phase in a major quantity. Variations in its development in different cases of mullite gels are outlined next.

Mullite Formations: Analysis and Applications
Akshoy Kumar Chakraborty

ISBN 978-981-4877-05-3 (Hardcover), 978-1-003-03167-3 (eBook)
www.jennystanford.com

Monophasic gels

In one type of monophasic gel: Spinel was formed in a minor quantity in slow hydrolysis (SH) gel synthesized by Okada and Otsuka (1986b, 1987). This gel resembles their model D, which directly crystallized to mullite (as a major phase) on being fired at 950°C. A similar nature of transformation was noted in Type I precursor synthesized by Schneider et al. (1993).

In the other type of monophasic gel: This gel was called rapid hydrolysis (RH) gel, where ammonia was used as a coprecipitating agent during synthesis. It formed spinel in a major quantity during heating at ~980°C exotherm in differential thermal analysis (DTA) (Okada and Otsuka, 1986b). They assumed it contained 8 wt % of SiO_2. Accordingly, the following reaction scheme for spinel formation at ~1000°C was proposed:

$$\underset{\text{Starting material}}{x\mathrm{SiO_2\cdot(I\text{-}x)\ Al_2O_3}} \xrightarrow{\sim 1000^\circ\mathrm{C}} \underset{\text{Spinel phase}}{m(\mathrm{SiO_2\cdot 6Al_2O_3})} + \underset{\text{Amorphous phase}}{n(\mathrm{6SiO_2\cdot Al_2O_3})} \quad (15.1)$$

Similar to RH gel, spinel as a major phase was also developed in a sol-gel material (SGM) by Schneider et al. (1992); Type II gel was synthesized by Schneider et al., (1993); and a precursor marked MI was developed by Huang et al. (1997). Even a polymeric gel synthesized by the hydrolysis of two organic Al and Si compounds (AlOBu/Al isopropoxide and tetraethyl orthosilicate [TEOS]) in a neutral to slightly acidic medium with subsequent cogelation showed a minor quantity of the spinel phase in addition to crystallization to mullite as a major phase (Pask et al., 1987; Yoldas, 1992). Subsequently, the formation of a spinel phase was noted by various authors from polymeric gels, for example, by Sanz et al. (1991) in their PB2 gel, by Geradin et al. (1994) in their M2 gel, by Cassidy et al. (1997) in their polymeric gel, and by Nieto et al. (1998) in their PA gel. The spinel phase was also developed in MP, MPP, and MAPIF gels synthesized by Fonseca et al. (1997). Chakraborty (2008) synthesized six different types of mullite precursors and showed the formation of different quantities of Al-Si spinel phase, the variation of which was dependent on the pH and the water used during the processing steps.

Coprecipitated gels

In the synthesis of coprecipitated gels utilizing both organic Al and Si compounds as source materials, Yamada and Kimura (1962), Hirata et al. (1985a), and Rajendran et al. (1990) noted spinel formation absolutely during the 980°C exotherm. Other researchers, namely Prochazka and Klug (1983), Suzuki et al. (1990), Low and McPherson (1989), Hirata et al. (1985b), and Hamano et al. (1985, 1986) noted spinel formation in their coprecipitated gel. Mitachi et al. (1990) showed spinel formation in their gel synthesized at pH 7 and at pH 10. Shiga et al. (1991) noted only spinel formation. Partial formation of the spinel phase was noted by Hirata et al. (1985a, 1989). Spinel phase formation in a coprecipitated gel prepared at pH 6–7 was noted by Chakraborty (1996b, 1997). In the sample marked CP6, the composition of spinel was conjectured by the author as analogous to that of 3:2 mullite.

Optically clear gels

It was shown by Colomban (1989) that an intense exothermic peak between 1020°C and 1030°C in the DTA trace occurred in the three mullite monoliths synthesized from aluminum *sec*-butoxide in hexane and in propanol using silicon methoxide, silicon propoxide, and silicon butoxide. The X-ray powder pattern of these monoliths showed amorphous material up to 900°C and then progressive transformation into a "γ-alumina-like" pattern when these monoliths were heat-treated to ~1000°C.

Colloidal gels

The colloidal gel synthesized by Yoldas (1992) was probably amorphous and did not crystallizes into pseudoboehmite in the dried gel state. It developed a spinel phase on being heated at 1000°C and above. The formation of the latter phase was also shown by Li and Thomson (1991) at the 980°C exotherm in their colloidal gel.

Diphasic gels

Diphasic gels synthesized by several researchers, as given in Chapter 4, showed spinel phase formation. Its formation was predominant in the diphasic gels and colloidal gel by Yoldas (1992), in the Type III (HB10 precursor) sample of Schneider et al. (1993), in the diphasic gel prepared by Hamano et al. (1985), in the M3 gels by Geradin et al. (1994), and in the gel marked Gel B by Ivankovic et al. (2003). Simply from the qualitative X-ray diffractometry (XRD) pattern of the spinel phase in the temperature range of 900°C–1000°C, a group of researchers conjectured this spinel phase as γ-Al_2O_3. These researchers further conjectured that the hypothetical solid-state reaction between γ-Al_2O_3 with silica (A) was responsible for mullite development at a high temperature. Experimental evidences in favor of γ-alumina are insignificant.

Spinel as the major phase was also developed in an "in situ" diphasic gel wherein the gel was synthesized at pH > 7, for example, the diphasic gel marked G 49 (Chakraborty, 1996b). Its composition changes on heating. It is conceived that it gradually takes up silica and finally reaches a 3:2 mullite-like composition. In this chapter, these two views will be discussed and the formation of Al-Si spinel will be highlighted with experimental results. It was conjectured by Chakraborty (1979, 1993, 2003a, and [unpublished], 2008) that the composition of spinel was analogous to the composition of 3:2 mullite. Under such circumstances, neither silicon incorporation nor silica evolution from the spinel structure is required to nucleate 3:2 mullite. Such type of mechanism may, of course, require discussion and experimental evidence. Accordingly, it is envisaged that the route of mullitization will be different among diphasic gels.

Various thoughts on spinel phase formation in a diphasic system

- Crystallization of both β-cristobalite between 1250°C and 1600°C and α-Al_2O_3 between 1300°C and 1700°C was noted by Hamano et al. (1985), which confirmed the true diphasic route in its crystallization behavior. According to these authors, the mullitization reaction was due to a solid-state reaction between α-Al_2O_3 and β-cristobalite at 1600°C other than the γ-Al_2O_3/amorphous SiO_2 reaction.

  Contrary to the above, the crystallizations of all transitional alumina phases occurred from boehmite (Al-O OH) and β-cristobalite from aqueous silica sol (Ludox) used during the synthesis of the diphasic gel even in acidic pH by Hyatt and Bansal (1990). On the other hand, mullite formation was noted by Li and Thomson (1991) at 1300°C instead of the very high temperature of 1600°C even when the gel prepared at pH 2 and both α-Al_2O_3 and β-cristobalite were crystallized at the intermediate stage of firing. Klaussen (1992) also prepared the gel at pH 1–2. Both Li and Thomson (1991) and Klaussen (1992) did not observe crystallization of the silica component. Later researchers noted exothermic peaks at 1188°C and 1254°C and claimed that these were due to the formation of δ-Al_2O_3 and θ-Al_2O_3 transition phases prior to mullite formation at the 1328°C exotherm.

- A diphasic gel was also synthesized by Wei and Halloran (1988) in an acidic condition using boehmite sol and TEOS. They noted that during heating, the alumina component crystallized to γ-Al_2O_3 or δ-Al_2O_3 and some amount of θ-Al_2O_3 but the silica component did not crystallize to β-cristobalite. With these observations, they believed that mullite formed by a solid-state reaction of δ-Al_2O_3 and amorphous silica during heating. Ismail et al. (1986) synthesized a diphasic gel at pH 1.8 using boehmite sol and silica sol.

  A different phase transformation behavior was reported by Sonuparlak (1988) when preparing a diphasic gel in an acidic or in a basic condition. The diphasic gel prepared in both conditions densified at ~1250°C with the development of the γ-Al_2O_3 spinel phase. When the gel was further heated to 1300°C for 1 h, only mullite was observed out of the gel synthesized in the acidic condition by using components from a similar source.

- Contrarily, intermediate Al-Si spinel phase formation was noted by Chakraborty (2003a) in a diphasic Al_2O_3-SiO_2 gel prepared in a basic condition before mullite formation. Thus, earlier studies show variations both in the phase transformation behavior and in the mullitization reaction path. The cause of the variations have not yet been explained.

15.2 Phase Transformation and Growth of Spinel Phase in "in situ" Diphasic Gels Observed by the Author

15.2.1 Synthesis and Transformation Behaviors of Diphasic Gels of Varying SiO_2 and Al_2O_3 Contents

15.2.1.1 Qualitative XRD study

Al_2O_3-SiO_2 diphasic mullite gels of varying Al:Si ratios were synthesized by Chakraborty (1993) and their phase transformation behaviors studied by qualitative and quantitative X-ray diffraction techniques. The results showed evidence of the formation of a noncrystalline aluminosilicate phase precursor together with slow crystallization of an Al-Si spinel phase of mullite-like composition up to 1250°C that existed up to ~1300°C. Table 15.1 shows the appearance and disappearance of phases on heat treatment of diphasic gels of various compositions at different temperatures.

- Silica gel (G-157) prepared from Ludox produces a strong 0.404 nm Bragg diffraction peak due to the formation of β-cristobalite. It starts crystallizing from 900°C onward.
- When a small quantity of aluminum nitrate nonahydrate (ANN) is used during gelation, the resultant gel G-168 does not develop boehmite but a drastic diminution in the intensity of cristobalite occurs. Thus, completely different pictures are observed when diphasic gel is heated.
- As the wt % of alumina in the gel increases to 7.5% in G-169, a trace quantity of boehmite results. On heat treatment, it is found that in G-169, cristobalite formation vanishes. XRD patterns of the gel heated to various temperatures are completely amorphous, that is, the alumina/boehmite portion of the gel does not show any crystallization. This result indicates that a noncrystalline aluminosilicate phase was generated when G-169 and other gels were being heated, by the solid-state reaction between dehydrated oxide components. It is conceived that the reaction is due to high free energy possessed by dehydrated boehmite, which an

Table 15.1 Sequential phase developments in some diphasic gels compared with pure gels

	Composition (wt%)		Phases present	Phases formed on heating to various temperatures with soaking						
Gel	SiO_2	Al_2O_3	at room temp.	400°C	600°C	750°C	900°C	1000°C	1100°C	1250°C
G-157	100	0	Amorphous (band) A	A	A	A	Cristobalite (A)	Cristobalite (vs)	Cristobalite (vs)	Cristobalite (vs)
G-168	97.5	2.5	Boehmite + amorphous band	A	A	A	Cristobalite (w)	Cristobalite (w)	Cristobalite (vw)	
G-169	92.5	7.5	Boehmite + amorphous band	A	A	A	A	A	A	A
G-170	85	15	Boehmite + amorphous band	A	A	Si–Al spinel (vw)	Si–Al spinel (vw)	Si–Al spinel (vw)	Si–Al spinel (vvw) + mullite (vw)	Mullite + cristobalite (w)
G-171	75	25	Boehmite + amorphous band	A	A	Si–Al spinel (vw)	Si–Al spinel (w)	Si–Al spinel (w)	Si–Al spinel (w) + mullite (vw)	Mullite (w) + cristobalite (s)
G-172	55	45	Boehmite + amorphous band	A	A	Si–Al spinel (vw)	Si–Al spinel	Si–Al spinel	Si–Al spinel + mullite (w)	Mullite (s) + cristobalite (ms)

(*Continued*)

Table 15.1 (*Continued*)

	Composition (wt%)		Phases present	Phases formed on heating to various temperatures with soaking						
Gel	**SiO_2**	**Al_2O_3**	**at room temp.**	**400°C**	**600°C**	**750°C**	**900°C**	**1000°C**	**1100°C**	**1250°C**
G-67	40	64	Boehmite + amorphous band	A	A	Si spinel	Si–Al spinel	Si–Al spinel	Si–Al spinel + mullite	Mullite (vs) + cristobalite (vvw)
G-173	31	69	Boehmite + amorphous band	A	A	Si spinel	Si–Al spinel	Si–Al spinel	Si–Al spinel + mullite	Mullite (vs)
G-64	20	80	Boehmite + amorphous band	A	A	Si spinel	Si–Al spinel	Si–Al spinel	Si–Al spinel + γ–Al_2O_3	Mullite (s) + Corundum (w)
G-174	10	90	Boehmite + amorphous band	A	γ–Al_2O_3	Si–Al spinel phase + γ–Al_2O_3	Si–Al spinel γ–Al_2O_3	Si–Al spinel + γ–Al_2O_3	Si–Al spinel + Corundum	Mullite (ms) + Corundum (s)
G-175	0	100	Boehmite	A	γ–Al_2O_3	γ–Al_2O_3 Al_2O_3 (vs)	Al_2O_3	$\gamma + \theta$ Al_2O_3	Corundum	Corundum (vs)

vs, very strong; ms, medium strong; s, strong; w, weak; vw, very weak; vvw, very very weak; A, amorphous contains enormous concentration of pores (Wilson, 1979) as well as high reactivity of a very low particle size (≤20 nm) silica phase. This view confirms the theoretical expectation of the formation of a glassy phase by the author as per the G-T diagram shown by Hoffman et al. (1984).

Chakraborty (1997) synthesized a diphasic mullite gel of composition Al:Si = 3:1 "in situ" and then hydrolyzed an aqueous solution of silica sol and aluminum salt with the help of NH_4OH solution. He did its DTA as well as qualitative phase identification and finally qualitative X-ray diffractometry (QXRD) study of the spinel phase.

15.2.1.2 DTA/DDTA studies of "in situ" diphasic gels

The DTA and derivative differential thermal analysis (DDTA) studies of the above gel were performed by Chakraborty (1996b). The resulting gel containing boehmite and silica gel first decomposes into amorphous phases, followed by an endotherm in DTA and loss in weight in thermogravimetric analysis (TGA). With a gradual increase in temperature, a solid-state reaction starts between newly liberated amorphous oxides, which leads to the gel's crystallization into an intermediate Al-Si spinel phase with the exhibition of a broad exotherm between 800°C and 1300°C (area marked ABC) with its T_m at 1180°C (Fig. 15.1a,b). On further heating, this newly formed spinel phase transforms into orthorhombic mullite with the exhibition of an exotherm at ~1320°C (area marked CDE). The thermal effects have been complemented by the derivative DTA peak run as shown in Fig. 15.1a and b. The exhibition of a new broad exotherm between 800°C and 1300°C in DTA characteristically shows two other broad peaks, AB'B (exo) and BB"C (endo), in the DDTA curve. The last exotherm, at ~1320°C in DTA, is in concurrence with the two peaks named CD'D and DD"E in the differential DTA curve. To study the sequence of crystallization of diphasic gels, a portion of the X-ray diffractogram for the intensity of the 0.139 nm peak of the Al-Si spinel phase for heated gels is shown in Fig. 15.2. The intensity increases with an increase in temperature to a value above ~1250°C. Thereafter, mullite formation ensues, as noted by the appearance of several peaks of it, namely 0.146 nm, 0.144 nm, 0.142 nm, and 0.140 nm, with a simultaneous decrease in the 0.139 nm peak of the Al-Si spinel phase. At 1300°C, Al-Si spinel and mullite are found to coexist.

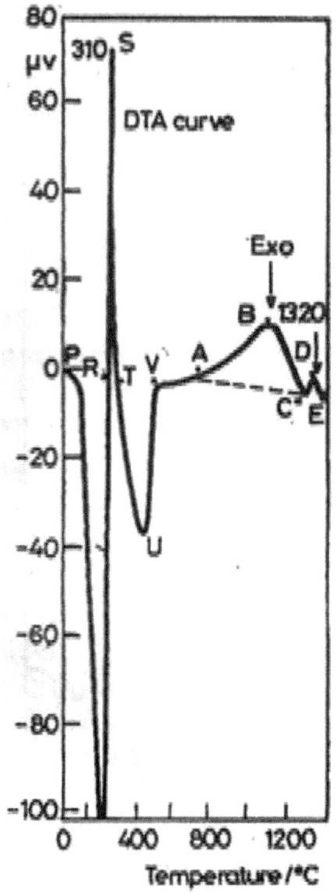

diphasic Al_2O_3–SiO_2 mullite

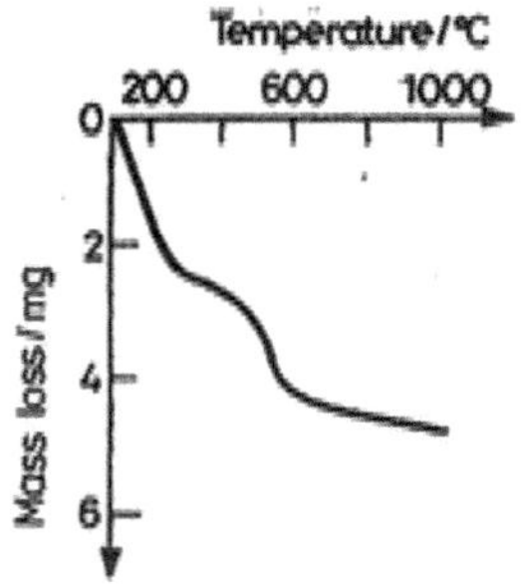

TG curve of G-49

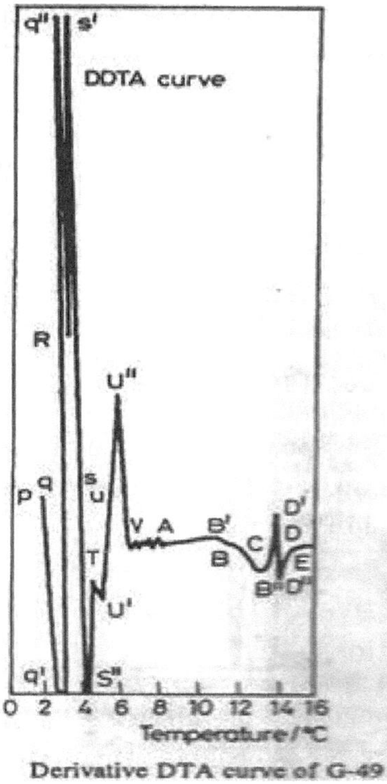

Figure 15.1 (a) DTA and TGA curves of diphasic Al_2O_3-SiO_2 mullite gel (G-49) and (b) derivative DTA curve of diphasic Al_2O_3-SiO_2 mullite gel (G-49) (Chakraborty, 1996b).

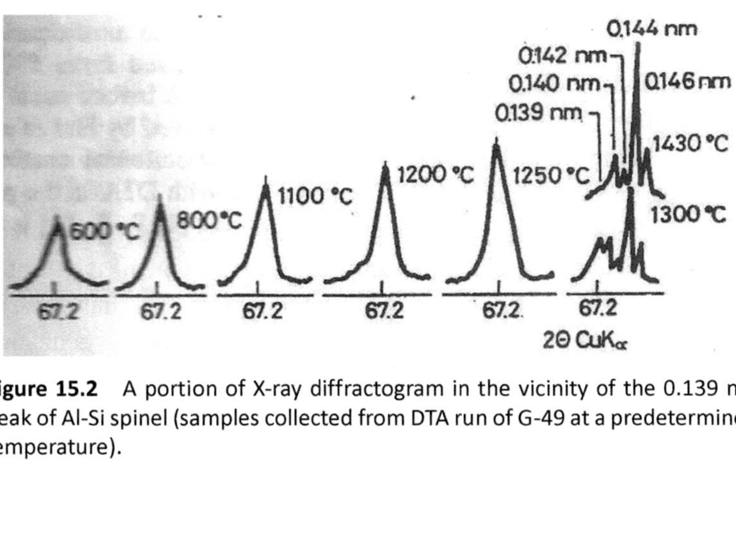

Figure 15.2 A portion of X-ray diffractogram in the vicinity of the 0.139 nm peak of Al-Si spinel (samples collected from DTA run of G-49 at a predetermined temperature).

Thereafter, mullite is the only crystalline phase to occur. This result substantiates the fact that after the dehydroxylation of boehmite and $Si(OH)_4$ gels, the two oxide phases SiO_2(A) and Al_2O_3(A) interact gradually and first develop the noncrystalline aluminosilicate phase, which slowly crystallizes with an increase in the temperature of heat treatment over a range of temperatures into Al-Si spinel instead of crystallizing into individual phases. The course of the crystallization process in between 800°C and 1300°C is with the concurrence of the exotherm in the same temperature range in DTA. Therefore, it is concluded that this peak is likely due to crystallization of the Al-Si spinel phase. Secondly, the transformation of this spinel phase into mullite accounts for the last exotherm, at ~1320°C (Fig. 15.2), which is in good agreement with both XRD, DTA, and DDTA results.

15.2.2 Qualitative XRD Study of Al-Si Spinel Phase Formation

The silica (A) component of the diphasic gel does not show any sign of crystallization to β-cristobalite during heat treatment of G-49. However, crystallization of the latter phase generally occurs on heating silica gel alone, with the exhibition of an exotherm at ~1060°C (Fig. 2.1). The alumina (A) component of G-49 does not show any sign of crystallization of α-Al_2O_3 in the long span of temperature up to 1350°C. Thus, hindrance of β-cristobalite and α-Al_2O_3 formation indicates that independent transformation processes of the two components become obscure, which demonstrates that silica diffuses into an alumina structure and vice versa and develops a noncrystalline aluminosilicate phase. In the matrix of this phase, Al-Si spinel starts crystallization, as noted by three broad XRD peaks (Fig. 15.3). More and more crystallization occurs on continued increase in temperature. A significant development of it is noted between 1000°C and 1200°C, as is evident from the sharp rise in the growth curve, which shows the comparative crystallization behaviors of pure boehmite gel and diphasic gel (Fig. 15.4a,b). The crystallization of such a large quantity of it as noted by three broad

XRD peaks is found to be in good agreement with the occurrence of a broad and large exothermic peak between 1000°C and 1300°C in DTA (Figs. 15.1 and 15.4). It was also shown further by Chakraborty (1994c) that Al-Si spinel formation is a function of temperature and concluded the spinel phase as Al-Si spinel phase formation other than γ-alumina. In comparison, the alumina (A) generated from the decomposition of boehmite crystallizes into γ-alumina, as noted by the rapid rise in the XRD peak intensity at the earlier temperature of ~400°C, as shown in its growth curves and Al-Si spinel phase (Fig. 15.4).

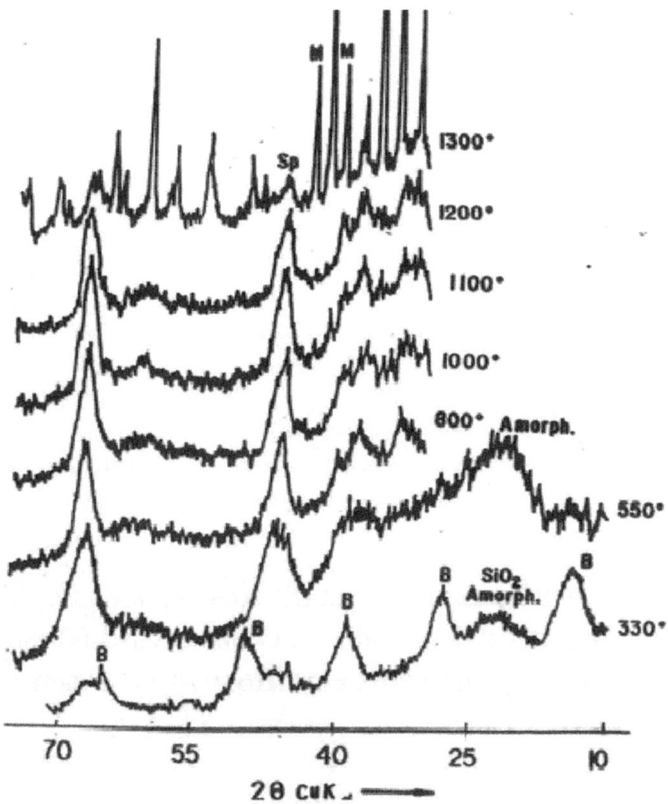

Figure 15.3 Portion of XRD recordings of diphasic gel (G-49) heat-treated to different temperatures. B = boehmite, Sp = spinel phase, Amorph = noncrystalline aluminosilicate phase, and M = mullite phase.

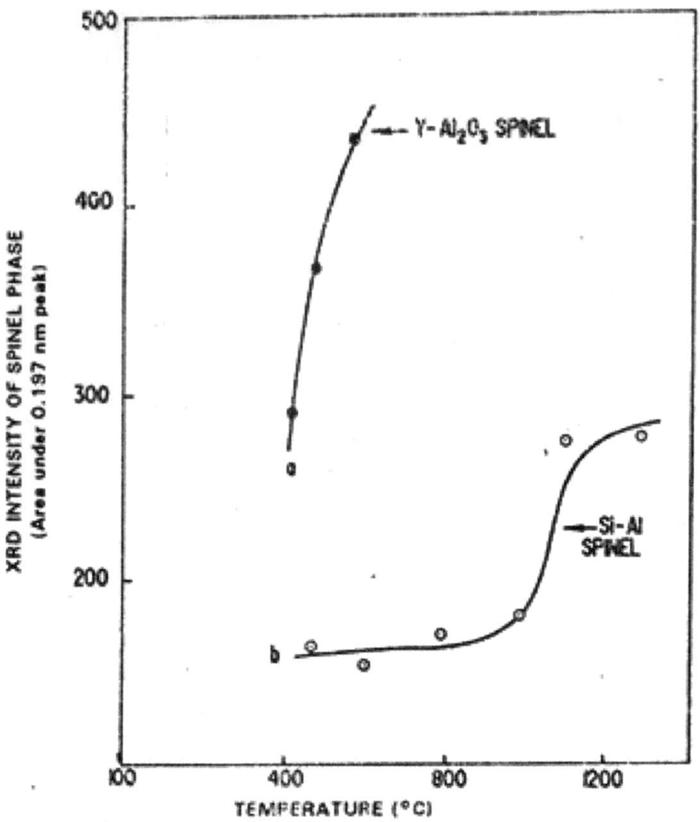

Figure 15.4 XRD intensity of spinel phases formed vs. firing temperatures. (a) γ-Al_2O_3 spinel from boehmite gel, and (b) Al-Si spinel from diphasic gel (G-49).

15.3 Variations in the Evolution Processes of Diphasic Gels, Mixed Oxides, and Mixed Gels of Different Compositions: Examples

15.3.1 Comparison of Phase Transformation of Diphasic Gels with That of a Chemical Mixture of γ-Al_2O_3 and Silica (A)

The spinel phase derived from three different compositions was first reinvestigated by Chakraborty (2005a). Phase evolution in heated diphasic gels was monitored by X-ray diffraction and is shown in Fig. 15.5.

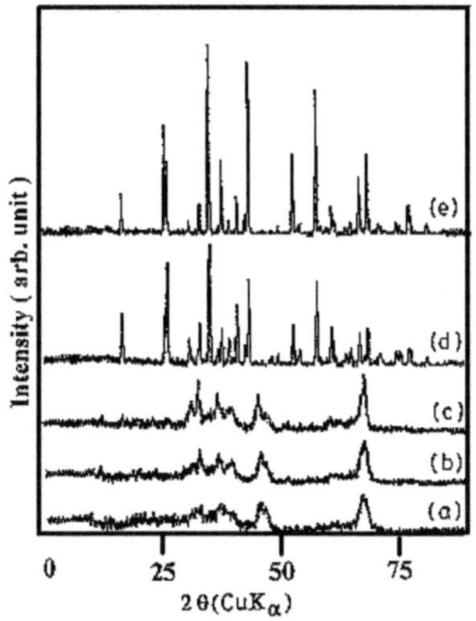

Figure 15.5 XRD pattern of diphasic gel DG72 fired at (a) 1100°C, (b) 1200°C, (c) 1400°C, and (d) 1600°C.

15.3.1.1 Phase evolution of three diphasic gels

An X-ray diffractogram of raw DG72 contains pseudoboehmite as the crystalline constituent in association with amorphous silica, as shown by an amorphous band at 22° 2θ. The other two gels also comprise a two-phase system. Gels synthesized by this process develop pseudocrystalline boehmite during the gelation period and as such these are called as "in situ" diphasic gels. In comparison, the earlier author [Hoffman et al. (1984), Jin et al. (2002), Wei and Halloran (1988), and Pach et al. (1997)] used crystalline boehmite powder as a component during synthesis of diphasic gels. Phase evolution in static heat-treated diphasic gels was monitored by X-ray diffraction and is shown in Fig. 15.5 and Table 15.2. At the first stage of transformation, stoichiometric diphasic mullite gel DG72 shows XRD peaks due to Al-Si spinel, and thereafter it forms purely mullite at a temperature above 1200°C. Like DG72, DG40 also forms Al-Si spinel and then forms mullite during heating at ~1200°C. On further rise in temperature, in addition to the presence of mullite, it forms cristobalite. At ~1600°C, cristobalite disappears and only mullite exists. DG85 shows XRD peaks due to γ-Al_2O_3 formation at ~1000°C. It starts forming θ-Al_2O_3 at ~1100°C. At ~1200°C, the XRD intensities of θ-Al_2O_3 increase and mullite formation starts. At ~1300°C, θ-Al_2O_3 decomposes and forms a substantial amount of mullite and a minor quantity of corundum. The intensities of mullite gradually increase with calcination temperature from 1300°C to 1500°C and thereafter the intensity remains unaltered. Corundum increases steadily with a rise in temperature up to 1600°C.

Table 15.2 Phase development in diphasic gels of three different compositions at various temperatures

Diphasic gel	Wt %	Wt %	Phases	Formed	On heating	(°C)
	SiO_2	Al_2O_3	1000	1200	1400	1600
DG40	59.85	40.08	Al-Si spinel	Mullite + cristobalite	Mullite	Mullite
DG72	27.79	72.1	Al-Si spinel	Mullite	Mullite	Mullite
DG85	14.46	85.27	Θ-Al_2O_3	Θ-Al_2O_3 + mullite	α-Al_2O_3 + mullite	Mullite + α-Al_2O_3

15.3.1.2 Phase evolution of mixed oxides

XRD analyses of mixed oxides (MOs) exposed to increasing temperatures are shown in Fig. 15.6. When they are heated at ~1050°C, the γ-Al_2O_3 component of the mixture transforms to θ-Al_2O_3 with partial formation of α-Al_2O_3. At ~1100°C, the ratio of the intensities of the two polymorphs only varies, for example, θ-Al_2O_3 decreases and α-Al_2O_3 increases. At ~1150°C, α-Al_2O_3 becomes the major phase, with a trace amount of unconverted θ-Al_2O_3. Thus, the two phases coexist when the MOs are heated over a long range of temperature. The SiO_2 component neither influences the crystallization behavior of alumina nor crystallizes during heating. The three diphasic gels of this study show three distinct cases of phase transformation, as shown in Table 15.2 and Fig. 15.5. In DG72, which has stoichiometric mullite composition, mullite is the only crystalline phase at a temperature above 1200°C. DG40, which is richer in silica, forms both mullite and cristobalite, and DG85, which is richer in alumna, develops mullite and corundum. The question is, how does mullitization occur in a diphasic gel? The answer lies in the characterization of the intermediate spinel phase. This is explained in the next two paragraphs. The present results address the basic question regarding the identification of the spinel phase and the course of mullite formation in three categories of diphasic gels. The concept of Al-Si spinel formation in the phase transformation of stoichiometric diphasic gel system is substantiated.

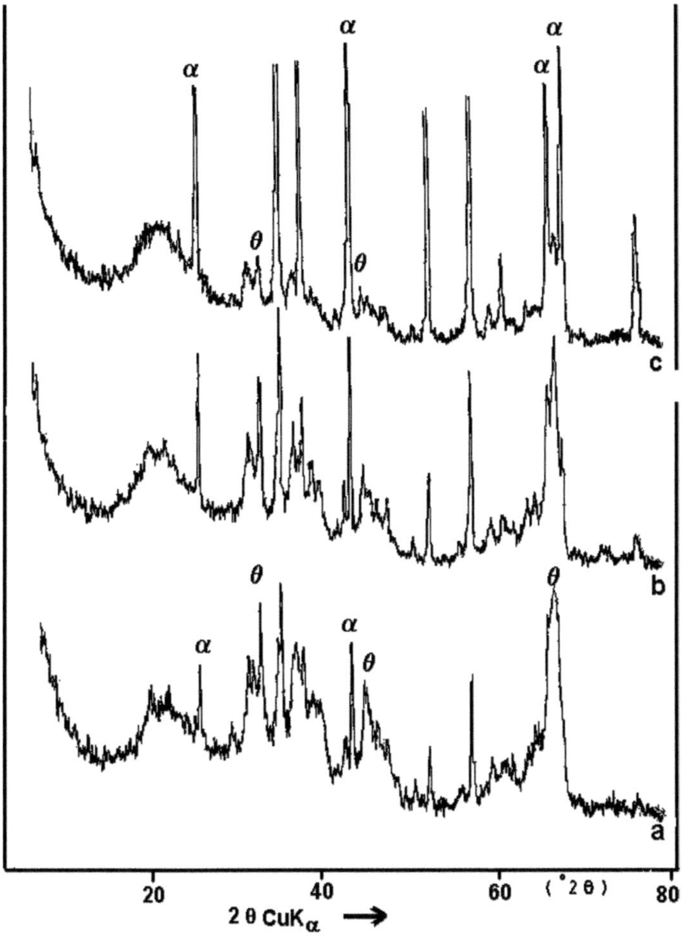

Figure 15.6 XRD patterns of mechanically mixed amorphous oxides marked MO show the formation of both θ-Al_2O_3 and α-Al_2O_3 when the oxides are heated to (a) 1050°C and (b) 1100°C and α-Al_2O_3 when the same are heated to (c) 1150°C.

15.3.2 Mode of Crystallization Behavior of Mixed Gels

Mixed gels MG24, MG40, MG66, MG72, and MG80 were synthesized by Chakraborty (2006). The phase evolution behaviors of these on heating are monitored along with the X-ray diffraction technique and summarized in Table 15.3.

Table 15.3 XRD results of Al_2O_3-SiO_2 mixed gels at different temperatures

Temp.	Alumina contents of mixed gels namely MG66, MG72, MG76, MG80, MG84, MG88, MG92, and MG95 heat treated in the range 800°C–1450°C (wt %)							
(°C)	66	72	76	80	84	88	92	95
1450	Mu + Cr(t)	Mu	Mu	Mu + Cor	Mu + Cor	Mu(vs) + Cor(s)	Mu(vs) + Cor(s)	Mu(vs) + Cor(s)
1400	Mu + Cr(m)	Mu	Mu, Cor(t)	Mu + Cor	Mu + Cor	θ(m)	θ(w, b)	θ(w, b)
1300	Mu + Cr(s)	Mu	Mu, Cor(t)	Mu + Cor	Mu + Cor	θ(m)	θ(m)	θ(m)
1250	Mu + Cr(s)	Mu	Mu, Cor	Mu(s) + θ(m)	Mu + Cor	θ(s)	θ(m)	θ(m)
1200	Mu	Mu	Mu	Mu(b) + θ(m)	Mu(b) + θ(m)	θ(s)	θ(s)	θ(s0)
1100	Sp	Sp	Sp, θ(t)	θ(w)	θ(w)	θ(m)	θ	θ(s0)
1000	Sp	Sp	Sp, γ(w)	γ(s)	γ(s)	γ(s)	γ(s), θ	γ(s)
900	Sp	Sp	Sp, γ(w)	γ(s)	γ(s)	γ(s)	γ(s)	γ(s)
800	A	A			γ(s)	γ(s)	γ(s)	γ(s)

s = strong, vs = very strong, m = medium, w = weak, t = trace, b = broad, A = amorphous, Mu = mullite, θ = θ-Al_2O_3, γ = γ-Al_2O_3, Sp = Al-Si spinel, Cr = cristobalite

At the first stage of transformation, MG40 and MG55 form a spinel phase at ~1000°C. This phase remains up to 1100°C. Thereafter, mullite develops during heating at ~1250°C. These gels form cristobalite in addition to the formation of mullite. At ~1500°C, cristobalite disappears and only mullite exists in MG40. MG66 also forms an intermediate spinel phase and then forms a large quantity of mullite. MG40 develops mullite at 1250°C, with no residual spinel phase, whereas MG66 forms mullite at 1200°C, with a trace amount of spinel as a residue. The pattern of this spinel is symmetrical, and it finally transforms to mullite at 1250°C. At 1500°C it develops traces of cristobalite. A mixed gel of stoichiometric mullite composition, MG72, first shows XRD peaks owing to spinel at 1000°C and then shows mullite at ~1200°C. MG80 shows XRD peaks owing to γ-Al_2O_3/spinel phase formation at ~1000°C. It starts to form θ-Al_2O_3 at ~1100°C. At ~1200°C, XRD intensities of θ-Al_2O_3 increase and mullite formation starts. At ~1300°C, θ-Al_2O_3 decomposes and forms a substantial amount of mullite and a minor quantity of corundum. The intensities of mullite gradually increase with calcination temperature from 1300°C to 1500°C and then the intensity remains unaltered. Corundum increases steadily with a rise in temperature up to 1500°C. The mixed gel mark MG84 evolves phases during heating, as in the case of MG80. Besides these results, trace amounts of cristobalite and corundum are detected when the mixed gels MG70 and MG76 are heated to 1400°C. The XRD results are summarized as follows:

- All mixed gels under the present study form a spinel-type phase during heating in the range 900°C to 1000°C. The phase remains stable up to as high as 1200°C in the cases of MG40 and MG55. But it exists only up to 1100°C in the case of mixed gels in the range between 70 and 76 wt % of alumina.
- The spinel phase formed while heating a mixed gel of a composition ranging from 80% to 95% of alumina thereafter transforms to the θ-Al_2O_3 phase at 1100°C–1200°C. The latter phase persists up to 1250°C in the case of mixed gels in the range from 80 wt % to 84 wt%. However, it remains stable even up to 1400°C in gels ranging from 88% to 95% of alumina.
- Silica-rich mixed gels MG40 and MG55 form cristobalite on being heated at a lower temperature of ~1250°C. Silica-poor mixed gels MG66 and MG70 develop cristobalite at 1500°C and 1400°C, respectively. However, a mixed gel with >72% of Al_2O_3 does not develop cristobalite on heating.
- Table 15.4 shows the changes in the breadth of the 0.137 nm peak of the spinel phase with an increase in the heat treatment temperature. The breadth of the 0.137 nm peak of the Al-Si spinel phase changes on heat treatment from 1000°C to 1250°C. An abrupt change occurs between 1200°C and 1250°C when mullite formation partially takes place. The remaining portion of the 0.137 nm peak after mullite formation

at 1250°C is nearly sharp and narrow, which may be due to Si-incorporated θ-Al_2O_3. Therefore, the initial broad peak at 0.137 nm may be considered as a mixture of Al-Si spinel and Si-incorporated θ-Al_2O_3. On further heat treatment, the latter phase transforms to corundum and mullite at ~1300°C. The intermediate spinel phases are characterized as follows:

- Spinel formed from gels with >28 wt % of SiO_2: X-ray diffractograms of raw mixed gels MG40, MG55, MG66, and MG70 show the development of only the Al-Si spinel phase during heating at ~1000°C and thereafter on further heating, the patterns show the formation of both mullite and cristobalite.

Table 15.4 Change in the breadth of the 0.137 nm peak of the spinel phase as a function of temperature

Sample	Heat treatment temperature (°C)	FWHM of 0.137 nm peak
MG80	1000	1.5818
	1200	1.3512
	1250	0.8478
MG84	1000	1.6168
	1100	1.3080
	1200	1.2795

- Spinel formed from gels with <28 wt % of SiO_2: However, it is observed that mixed gels with <28 wt % of SiO_2 (MG76, MG80, and MG84) do not crystallize to cristobalite on heating. Moreover, the θ-Al_2O_3 phase formed at the earlier temperature remains stable at temperatures as high as 1250°C, which is far above the normal crystallization temperature of θ-Al_2O_3 to α-Al_2O_3. Noting the high temperature stability, Yoldas (1980) believed earlier that the so-called θ-Al_2O_3 is substituted by silica. Therefore, it is assumed that the incorporation of silica into the alumina phase may start during the heating process of those gels at as low a temperature as 1000°C. It finally transforms on heating at 1100°C into silica-incorporated θ-Al_2O_3 in the case of MG80 and MG84 and exists up to

1250°C. However, silica-incorporated θ-Al_2O_3 formed in MG88 to MG95 remains stable even at a high temperature (up to 1400°C). The formation of a partial amount of mullite in addition to silica-incorporated θ-Al_2O_3 while heating MG80 and MG84 at 1250°C or above indicates that the structural transformation of the Al-Si spinel phase is likely to occur. Thus, the Al-Si spinel phase may coexist with silica-incorporated θ-Al_2O_3 in those heated gels. During further heating, the appearance of the Bragg diffraction peaks of the so-called spinel phase changes from broad to narrow. Simultaneously, the peak width of (440) reflection of the spinel phase decreases on heating the gel from 1000°C to 1200°C (Table 15.4). It is believed that the 0.137 nm peak may consist of two peaks. The first one is due to Al-Si spinel. It is broad and transforms to mullite on further heating. Part of the superimposed peak that still remains after mullite formation is the newly formed silica-incorporated θ-Al_2O_3. It decomposes at a high temperature to corundum and mullite, as noted in the case of MG80 and in other mixed gels of alumina-rich composition.

15.3.3 Crystallization of Al-Si Phase from Noncrystalline Aluminosilicate Precursor Phase during Heating of a Diphasic Gel

Three diphasic gels, marked DG40, D66, and DG72, were resynthesized and heat-treated in a static condition by Chakraborty (2008a). The XRD intensities of the 0.139 nm peak at 67° 2θ of Al-Si spinel and 0.164 nm peak at 55.8° 2θ of CaF_2 were scanned for calculation of the area of Al-Si spinel to CaF_2. Secondly, XRD intensities of the 0.40 nm peak at 22 °2θ of the amorphous band and the 0.164 nm peak of CaF_2 were scanned for calculation of the area of the amorphous band to CaF_2. The area of the amorphous hump and the integrated area under the peak of the spinel phase formed while heating diphasic gels were finally measured with the help of X'Pert graphic software and are shown in Table 15. 5.

Table 15.5 Changes in XRD intensities of the amorphous band and the Al-Si spinel phase derived from the heat treatment of four kinds of diphasic gels in the presence of CaF_2 as the celebrant

Gel mark	Heat treat temp. (°C)	XRD counts 22° 2θ band	XRD counts 55.8° 2θ CaF_2	Ratio of I_{band}/I_{CaF_2}	XRD counts 66° 2θ spinel	Ratio of I_{spinel}/I_{CaF_2}	Ratio of I_{spinel}/I_{band}
DG40	600	627.5	60	10.46	–	–	–
	700	952	68	14.0	92	1.36	0.096
	800	950.5	62	15.33	105	1.7	0.110
	900	909	57	15.95	190	3.33	0.209
	1000	882	56	15.75	193.7	3.46	0.219
	1100	977	70	13.96	273	3.91	0.279
	1200	998	72	13.86	383	5.33	0.383
DG66	600	159.6	56	2.85	57	1.01	0.357
	700	223.6	63	3.55	105	1.68	0.469
	800	230.6	61	3.78	195	3.2	0.845
	900	381.5	68	5.61	365	5.38	0.956
	1000	360.5	67	5.38	377	5.63	1.047
	1100	276.5	62	4.46	351	5.66	1.269
	1200	283	57	4.98	449	7.88	1.586
DG72	600	54.5	54	1.01	82	1.53	1.50
	700	71.3	57	1.25	199	3.5	2.791
	800	80	60	1.33	276	4.6	3.45
	900	88.6	62	1.43	340	5.51	3.837
	1000	131.6	59	2.23	339	5.75	2.575
	1100	172.84	58	2.98	387	6.68	2.339
	1200	170.8	56	3.05	495.6	8.85	2.901

15.3.3.1 Amorphous bands of the noncrystalline precursor phase and the noncrystalline aluminosilicate residual phase

The changes in the intensities of the X-ray amorphous bands developed during the heat treatment of diphasic gels of three compositions is shown in Fig. 15.7. It shows that in the first step, the intensity increases with heat treatment in the temperature range of 600°C–900°C for gels marked DG40, DG56, and DG72. Depending upon the composition of the gel, the plot shows that it increases slowly in the case of DG72, it is moderate in the case of DG66, and it is very fast obviously in the case of DG40. In the second step, it decreases sharply in the case of DG40 and DG66 in the temperature range of 900°C–1100°C. At the third step, it persists in the temperature range of 1100°C–1200°C. It slightly increases in the case of DG66. Contrarily, the band intensity gradually increases in the case of DG72 from 1000°C to 1200°C. The XRD intensity peak due to the spinel phase increases with heat treatment in the temperature range of 1000°C to 1200°C for gels marked DG40, DG66, and DG72, as given in Fig. 15.7.

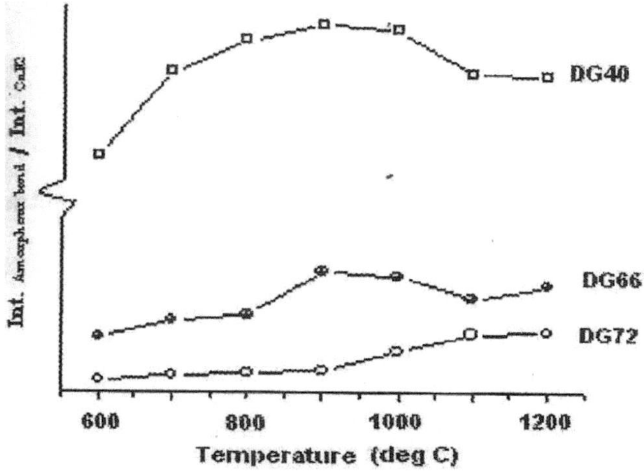

Figure 15.7 X-ray amorphous band of diphasic gels heat-treated to different temperatures (Chakraborty, 2008).

15.3.3.2 Al-Si spinel phase

In addition to this figure, Fig. 15.8 depicts a clear picture of changes in XRD intensity of the 0.139 nm peak of the Al-Si spinel phase over a long range of temperatures in a semiquantitative manner. Apparently, this figure consists of two portions. The first part of spinel formation occurs in the temperature range between 600°C and 900°C; thereafter, it halts; the second part of it starts from 1000°C and continues up to 1200°C.

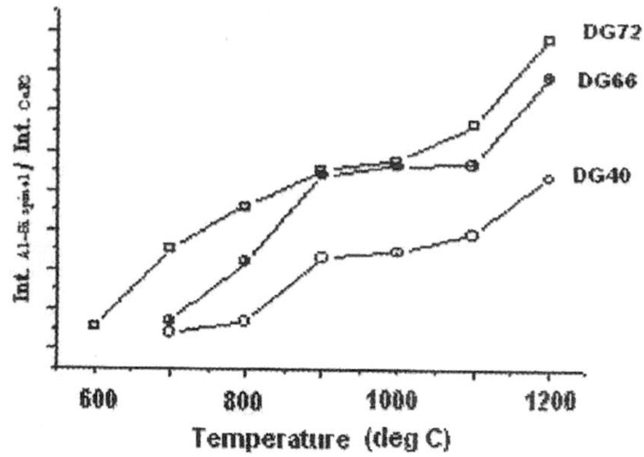

Figure 15.8 Ratio of $I_{Al\text{-}Si\ spinel}/I_{CaF2}$ of the diphasic gel versus heat treatment temperature (Chakraborty, 2008).

During the formation of the spinel phase, it is observed that the ratio of the XRD peak of the spinel phase to that of the amorphous band increases in the temperature range of 1000°C–1200°C. The area of the amorphous hump at 22° 2θ and the area under the peak of the spinel phase at 67° 2θ formed while heating diphasic gels were calculated with the help of X'Pert graphic software supplied by Philips and are shown in Table 15.5. It shows a comprehensive picture of the changes in the intensities of the amorphous band and also the spinel phase formed on heat treatment of three different diphasic gels on successive increase of temperature. It is noted that the intensity of the peak due to silica decreases with heat treatment in the case of the gels marked DG40 and DG66. It, however, increases in the case of DG72. Secondly, the intensity of the peak due to the spinel phase increases with heat treatment in all three cases of diphasic gels. Finally, the ratio of the intensity of the peak of the spinel phase to that of amorphous silica in gels of different compositions increases with heat treatment temperature from 1000°C to 1200°C.

The decrease in the intensity of amorphous silica with an increase in the peak of the spinel phase further indicates the incorporation of silica in the spinel structure to form an Al-Si spinel phase or a silica-incorporated δ- or θ-Al_2O_3 structure. Thus, this method supports the phase analysis of the spinel phase made above. The sudden decrease in the amorphous band is interrelated with the formation of the Al-Si spinel phase at 900°C–1000°C. Beyond 1000°C, a further increase of Al-Si spinel formation (Fig. 15.8) occurs. On heating to a temperature above 1200°C, it rapidly transforms to mullite.

Therefore, the decrease in the amorphous band due to the noncrystalline precursor phase may be related to the crystallization of the spinel phase, which is interpreted as the Si-substituted alumina phase (Al-Si spinel). These three practical studies further extend the proof of Al-Si spinel formation.

15.4 Characterization of the Spinel Phase by a Solid-State Reaction Study with CaO

The purpose of the study of the solid-state reaction of raw, dried, and calcined monophasic and diphasic gels with CaO on heating was to assess the probable composition of the mother reactant, that is, the intermediate spinel phase.

15.4.1 Characterization of Diphasic Gels and Precursors

Precalcined diphasic gels that contained a spinel phase were further calcined by Chakraborty (2003b) after mixing with $CaCO_3$ and then analyzed by XRD. Figure 15.9 and Table 15.6 show the phase development on heating such spinel-generated diphasic gels of two compositions mixed with $CaCO_3$. It was shown that all diphasic gels up the composition of 3:2 mullite form calcium aluminosilicate (gehlenite) rather than calcium aluminate. These results more definitely indicate that the spinel phase formed in a precalcined diphasic gel is Al-Si spinel other than γ-Al_2O_3.

Secondly, gehlenite formed in Si-rich gels up to approximately 3:2 mullite composition. Calcium aluminate started forming from spinel-derived diphasic gels of compositions beyond 3:2 composition of mullite. This observation further indicated that the probable composition of Al-Si spinel is analogous to the composition of 3:2 mullite.

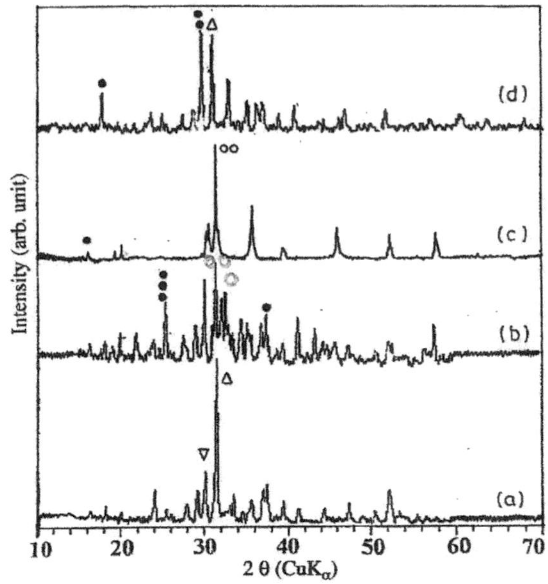

Figure 15.9 XRD pattern of diphasic gel DG72 heated to 1000°C, treated with CaO, and then reheated to 1200°C (Chakraborty, 2003b).

Table 15.6 Summary of phases developed on heating diphasic gels first to 1000°C and then treated with CaO and finally reheated to 1200°C

Sample characterisitc	Phases formed
DG40 heated to 1000°C/2 h, mixed with CaO in the wt. ratio of 1:1 and reheated to 1200°/2 h	Gehlenite (C_2AS)
DG44 heated to 1000°C/2 h, mixed with CaO in the wt. ratio of 1:1 and reheated to 1200°/2 h	Gehlenite (C_2AS)
DG67 heated to 1000°C/2 h, mixed with CaO in the wt. ratio of 1:1 and reheated to 1200°/2 h	Gehlenite (C_2AS)
DG72 heated to 1000°C/2 h, mixed with CaO in the wt. ratio of 1:1 and reheated to 1200°/2 h	Gehlenite (C_2AS)

Sample characterisitc	Phases formed
DG76 heated to 1000°C/2 h, mixed with CaO in the wt. ratio of 1:1 and reheated to 1200°/2 h	Gehlenite (C_2AS) + anorthite (CAS_2) + calcium aluminate (CA) + 12-calcium-7-aluminate ($C_{12}A_7$)
DG80 heated to 1000°C/2 h, mixed with CaO in the wt. ratio of 1:1 and reheated to 1200°/2 h	Gehlenite (C_2AS) + anorthite (CAS_2) + calcium aluminate (CA) + 12-calcium-7-aluminate ($C_{12}A_7$)
DG85 heated to 1000°C/2 h, mixed with CaO in the wt. ratio of 1:1 and reheated to 1200°/2 h	Gehlenite (C_2AS) + anorthite (CAS_2) + calcium aluminate (CA) + 12-calcium-7-aluminate ($C_{12}A_7$)
Mechanical mixture of (silica gel, Al_2O_3 (amorph) in the wt. ratio of 1200°C/2 h	Dicalcium silicate (β-C_2S) + α-Al_2O_3 + tricalcium aluminate (C_3A)
Raw DG72 mixed with CaO in the wt. ratio of 1:1 and heated to 1200°/2 h	Tricalcium aluminate (C_3A) + 12-calcium-7-aluminate $C_{12}A_7$)

15.4.2 Characterization of Monophasic and Diphasic Gels and Precursors

The solid-state reactions of $CaCO_3$ heated with raw, dried, and calcined monophasic and diphasic gels were studied. Two reaction products, gehlenite and calcium aluminates, were predominantly formed. On the basis of qualitative X-ray identification of phase development, the nature of the calcined precursor, the dried precursor, and raw monophasic and diphasic gels were assessed by Chakraborty (2004). Raw SH gels showed the formation of $C_{12}A_7$ (major) and C_3A (minor) phases during heating, suggesting that the aluminum component of the gel was present in a free state. Dried SH gels formed a calcium aluminosilicate phase (gehlenite) in a major amount and a minor amount of $C_{12}A_7$, which indicated that the dried SH gel was aluminosilicate. Moreover, SH calcined between 400°C and 900°C also showed gehlenite in a major amount and traces of $C_{12}A_7$ phases. X-ray patterns of raw, dried, and calcined (900°C) monophasic gels mixed with $CaCO_3$ and heated to 1200°C are shown in Tables 15.7, 15.8, and 15.9 and Figs. 15.10a, 15.10b, and 15.10c, respectively.

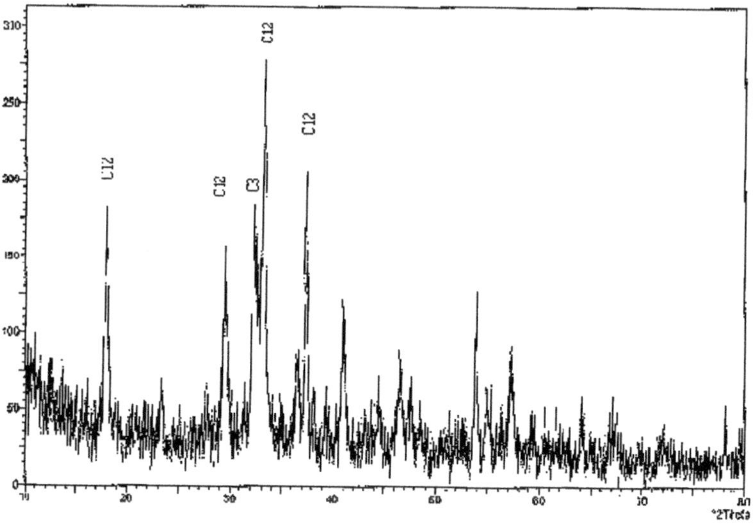

Figure 15.10a XRD patterns of preheated $CaCO_3$-mixed raw SH72 gel reheated to 1200°C (Chakraborty, 2004). $C_{12}A_7$ marked as C_{12}; C_3A marked as C_3.

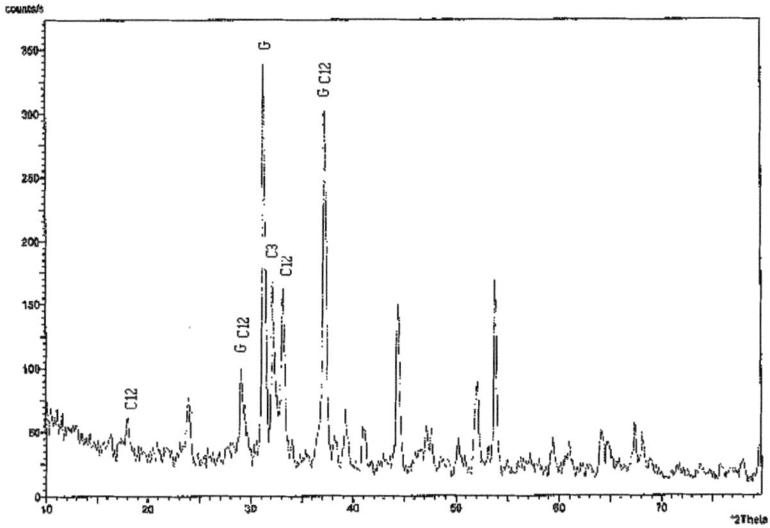

Figure 15.10b XRD patterns of preheated $CaCO_3$-mixed dried SH72 gel reheated to 1200°C.

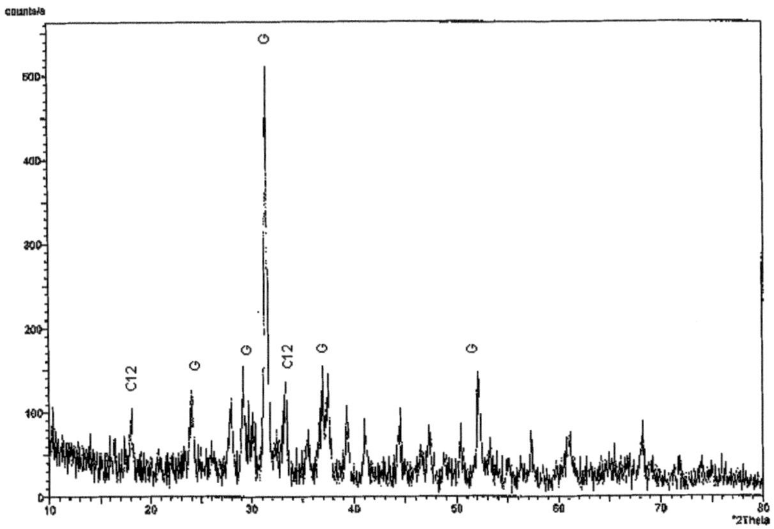

Figure 15.10c XRD patterns of preheated $CaCO_3$-mixed calcined SH72 gel reheated to 900°C (Chakraborty, 2004). Gehlenite marked as G; $C_{12}A_7$ marked as C_{12}.

The raw diphasic gel behaved as a raw SH gel and/or a mechanical mixture of amorphous silica and alumina. The author also showed the formation of gehlenite rather than calcium aluminate when the heat-treated diphasic precursor of mullite gel between 400°C and 900°C was further heat-treated with $CaCO_3$ at 1200°C (Table 15.9, Figs. 15.11a and 15.11b). It was concluded that the precursor was a noncrystalline aluminosilicate phase that substantiates that the diphasic gel also formed linkages at and above the dehydroxylation temperature, which continued until the crystallization stage was reached. Besides the solid-state reaction between the two oxide components, after the decomposition of boehmite at ~400°C a portion of alumina (A) crystallized to some quantity of γ-Al_2O_3 in some cases due to inhomogeneity. However, the formation of the noncrystalline aluminosilicate phase as an intermediate phase during heating those gels prior to mullite formation is generally ignored. The formation of gehlenite rather than calcium aluminate by XRD indicated that both monophasic and diphasic precursors consisted of a noncrystalline aluminosilicate phase of a composition close to 3:2 mullite. Therefore, it was concluded that the monophasic gel transformed to poorly crystalline mullite via an intermediate noncrystalline aluminosilicate phase during the 980°C exotherm. In fact, the SH, polymeric, and Type I gels do not directly transform to t-mullite during said exotherms, as stated by various researchers.

Table 15.7 Phases formed on heating to 1200°C (a) $CaCO_3$-pretreated raw SH gels and (b) $CaCO_3$-pretreated dried SH gels

Precursor characteristics	Amount of TEOS/ ANN/alcohol	Al_2O_3/SiO_2 ratio	Phases formed at 1200°C
	Raw SH gel		
SH62	2 mL/ 6.5 gm/ 10 mL	62.6/37.4	$C_{12}A_7$ (major), C_3A (minor)
SH68	2 mL/ 8.5 gm/ 10 mL	68.6/31.4	$C_{12}A_7$ (major)
SH72	2 mL/10.2 gm/ 10 mL	72/28	$C_{12}A_7$ (major), C_3A (minor)
SH74	2 mL/11.5 gm/ 10 mL	74.8/25.2	$C_{12}A_7$ (major), C_3A (minor)
SH76	2 mL/ 12.5 gm/ 10 mL	76.3/23.7	$C_{12}A_7$ (major), C_3A (minor), gehlenite (minor)
SH79	2 mL/ 15 gm/ 10 mL	79.4/20.6	C_3A (major), $C_{12}A_7$ (minor)
SH83	2 mL/20 gm/ 10 mL	83.7/16.3	C_3A (major), $C_{12}A_7$ (minor)
	Dried SH gel		
SH62	2 mL/6.5 gm/ 10 mL	62.6/37.4	Gehlenite (major), calcium silicate phase $C_{12}A_7$ (minor)
SH68	2 mL/8.5 gm/ 10 mL	68.6/31.4	Gehlenite (major), calcium silicate phase, $C_{12}A_7$ (minor) phase
SH72	2 mL/10.2 gm/ 10 mL	72/28	Gehlenite (major), C_3A and $C_{12}A_7$ (minor) phases
SH74	2 mL/11.5 gm/ 10 mL	74.8/25.2	Gehlenite (major), other minor phases

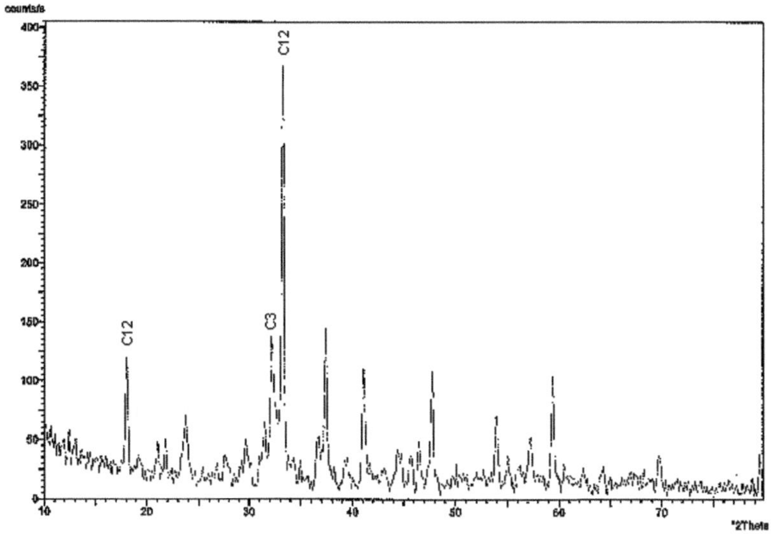

Figure 15.11a XRD patterns of $CaCO_3$-pretreated raw diphasic DG72 reheated to 1200°C.

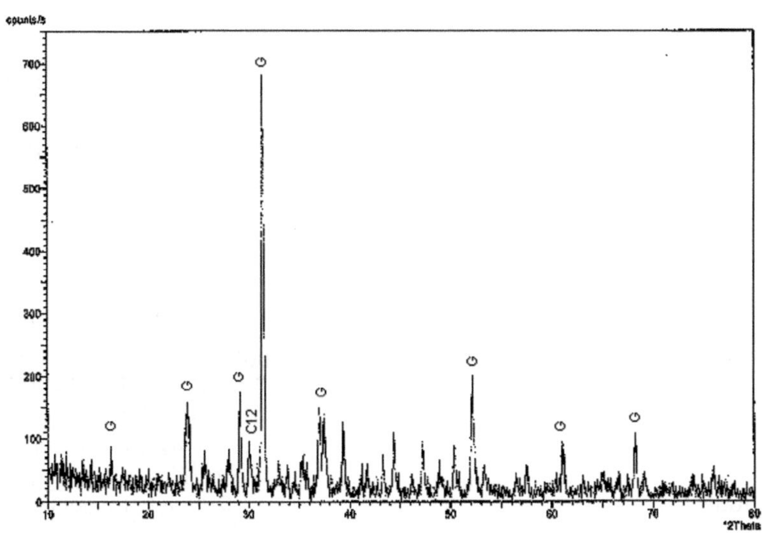

Figure 15.11b XRD analysis of $CaCO_3$-pretreated calcined diphasic DG72 gel reheated to 1200°C.

Table 15.8 Phases formed on heating $CaCO_3$-pretreated calcined SH gels to 1200°C

Precursor characteristics	Phases formed on heating at 1200°C
SH gel heated to 400°–900°C	
SH62	Gehlenite (C_2AS), calcium silicates
SH68	Gehlenite (C_2AS), calcium silicates
SH72	Gehlenite (C_2AS), $C_{12}A_7$ (trace)
SH74	Gehlenite (major), C_3A (trace)
SH76	Gehlenite (C_2AS), $C_{12}A_7$ (minor)
SH79	Gehlenite (C_2AS), $C_{12}A_7$ (minor)
SH83	Gehlenite (C_2AS) $C_{12}A_7$(medium)

Table 15.9 Phases formed on heating $CaCO_3$-pretreated raw and heated diphasic gels to 1200°C

Precursor characteristics	Amount of TEOS/ANN/ NH_4OH solution at pH 9 (wt %)	Al_2O_3/SiO_2	Phases formed at 1200°C
	Raw (DG) gel		
DG72	4 mL/20.4 gm	72/28	$C_{12}A_7$ (minor), C_3A, gehlenite (trace)
DG88	4 mL/60 gm	88.5/11.5	$C_{12}A_7$ (minor), C_3A, calcium silicates (major)
DG92	4 mL/80 gm	91.2/8.8	$C_{12}A_7$ (minor), C_3A, calcium silicate (major)
	DG gel heated to 400°C–900°C		
DG72			Gehlenite (major), $C_{12}A_7$ (minor), CAS (minor)
DG88			Gehlenite (major), $C_{12}A_7$ (minor), CAS (minor)
DG92			$C_{12}A_7$ (major), gehlenite (minor), CAS (minor)

Results also showed that both dried monophasic gel and dehydroxylated diphasic gel started to form calcium aluminate when the alumina content in the gels increased beyond that corresponding to the 3:2 mullite composition. This suggests that the composition of noncrystalline aluminosilicate in both types of gel may be close to that of 3:2 mullite.

15.5 Characterization of the Spinel Phase by the Alkali Leaching Method of Heat-Treated "in situ" Diphasic Mullite Gels

There are a lot of difficulties in the characterization of the cubic spinel phase by XRD methods, and there are some limitations too. It was newly conceived that Al_2O_3-SiO_2 gels of different compositions should obviously release SiO_2 (A) during their transformation to the cubic spinel phase. Accordingly, the objectives of the characterization of the cubic spinel phase were to (i) indirectly detect and estimate SiO_2 (A) liberated by the alkali extraction process from heat-treated gels of different Al_2O_3/SiO_2 ratios, (ii) isolate the cubic phase derived from these heat-treated gels to determine its composition, and (iii) characterize the isolated cubic phase. It was intended to determine the bulk composition of the spinel phase by alkali extraction studies.

An attempt to isolate Al-Si spinel: Coprecipitated gels of different Al_2O_3/SiO^2 ratios were heat-treated by Chakraborty (1979) at ~1000°C to convert into a weakly crystalline Al-Si spinel phase. This phase was subsequently characterized by an alkali leaching study. It was shown that the maximum amount of silica entering into the spinel phase is ~28 wt %. Figure 15.12 shows that different amounts of SiO_2 are detected since these can leach out during the alkali extraction process. Silica extraction curves of three heat-treated gel samples (A, B, and C) have two distinct parts. In the first stage, that is, after 30–40 min. of leaching, the SiO_2 extraction was comparatively rapid. In the next stage, that is, after ≥40 min., the SiO_2 dissolution was very slow and reached a more or less constant value although the leaching time increased. It was also noted that not the entire quantity of SiO_2 originally present in the 980°C heated mass was extractable when the leaching condition was enhanced by using a 10% NaOH solution. The alkali solution removed Al_2O_3

as well as SiO_2, and the extraction curves were quite unlike those of SiO_2. In all cases the percentage of Al_2O_3 in the extract increased very slowly with increased leaching time. Comparing the removal of SiO_2 and Al_2O_3 by the alkali extraction process of heated gels of A, B, and C, the major portion of SiO_2 that was solubilized selectively in the first stage of extraction could be designated as free amorphous SiO_2. A portion that seemed to be unextractable during continued leaching of 980°C heated gels could be termed as chemically bonded SiO_2. In sample C, for example, the intensity of the cubic pattern increased at 30–40 min. of leaching time because the cubic phase was concentrated by the removal of amorphous SiO_2. However, its intensity diminished gradually and a zeolite-type phase appeared when leaching of the heated sample was continued for up to 90 min. with 10% NaOH (Table 15.11).

Table 15.10 Result of alkali extraction experiments on aluminosilicate gels

			Nature of the residue			
			Chemical analysis (in anhydrous basis, wt %)			
Sample	SiO_2	Al_2O_3	Chemical composition liberated (wt %)	XRD	SiO_2	Al_2O_3
A	58.7	41.3	42.9	Cubic	27.6	72.4
B	49.5	50.5	29	Cubic	28.5	71.5
C	39.5	60.4	16.0–17.0	Cubic	27.9	72.1
D	28.7	71.3	0			
E	18.6	81.4	0			

Sample heated at 980°C for 4 h and leached with 5% NaOH in a boiling water bath for 30 min.

Table 15.11 X-ray data for residue from a gel leached for 90 min. showing a zeolite phase

dÅ	I/Io	dÅ	I/Io
10	1	2.79	2 (very broad)
8.25	Very weak	2.625	4
6.25	8 (broad)	2.575	4
4.2	5	2.54	6
3.6	10 (broad)	2.075	5 (broad)
3.18	Weak broad		

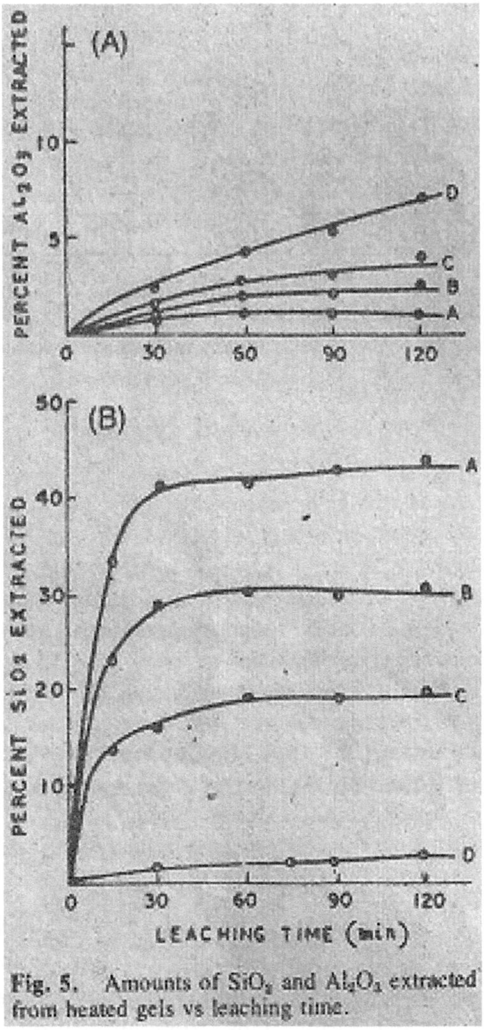

Figure 15.12 Amounts of SiO_2 and Al_2O_3 extracted from heated gels versus leaching time (Chakraborty, 1979).

It was shown that whatever be the initial composition of the gel, the leached residue became analogous to the composition of 3:2 mullite analytically and XRD showed that it consisted of a cubic spinel phase. In sample D, both SiO_2 and Al_2O_3 dissolved slowly and there was no rapid dissolution of SiO_2, that is, there was no free amorphous silica and all the SiO_2 was present in combination with

Al_2O_3 (Table 15.12). The course of formation of the zeolite phase from the 980°C heated gel followed a heterogeneous reaction mechanism, and it signified that the cubic phase was an aluminosilicate material, that is, Al-Si spinel.

Table 15.12 Phases developed on heating coprecipitated SiO_2-Al_2O_3 gels

	Crystalline phases formed on heating (°C)				
Sample	**980**	**1100**	**1200**	**1300–1400**	**1400**
A	Cubic phase	Cubic phase	M + C (major)	M + C (well developed	M
B	Cubic phase	Cubic phase	M + C	M + C	M
C	Cubic phase	Cubic phase	M + C	M + C	M
D	Cubic phase	Cubic phase	M	M	M
E	Cubic phase	Cubic phase + Cor	M + Cor	M + Cor	M + Cor

All samples are amorphous before the beginning of the 980°C peak.
Sample A (58.7% SiO_2 and 41.3% Al_2O_3); B (49.5% SiO_2 and 50.5% Al_2O_3); C (39.57% SiO_2 and 60.4% Al_2O_3); D (28.7% SiO_2 and 71.3% Al_2O_3) and E (18.6% SiO_2 and 81.4% Al_2O_3). These compositions are shown in Table 15.10.
M = mullite, C = cristobalite, Cor = corundum

Possible Enrichment of the Al-Si spinel phase by an alkali leaching study: Limitations of the characterization of the cubic phase by diffraction studies were discussed earlier by Chakraborty (1979). The problem was approached by studying the phase transformation of coprecipitated gel Al_2O_3-SiO_2 gels indirectly after extracting the amount of amorphous SiO_2 liberated during its formation at ~980°C and isolating the newly liberated cubic spinel phase by an alkali extraction technique. Coprecipitated gels of different Al_2O_3-SiO_2 ratios below the composition of 3:2 mullite showed a 980°C exotherm followed by crystallization of the spinel phase and liberation of silica. The alkali leaching process showed that whatever the initial Al_2O_3-SiO_2 ratio may be, it always forms a spinel phase of 3:2 mullite. Secondly, a DTA study showed that an Al_2O_3-SiO_2 gel of mullite composition exhibited the maximum exothermic effect and also showed the formation of the highest quantity of mullite on further heating. A composition other than 3:2 forms either cristobalite or corundum besides mullite as the gel contains silica or alumina in excess to that of 3:2 mullite. With these observations, the author conjectured that the composition of spinel will be analogous to that of 3:2 mullite. Thus the leached residue obtained after the first stage of leaching consists of both a Al-Si spinel phase (~80 wt %) and a noncrystalline aluminosilicate phase (~20 wt %). Low and McPherson (1989) repeated the alkali leaching experiments on the mullite gel heated to ~1000°C and the residue heat-treated to 1300°C for 2 h. They observed a considerable amount of corundum and 2:1 mullite, which indicated that the leaching period was extended beyond the schedule.

15.5.1 Leached Residue Characterization by Different Researchers Using Different Techniques: Supplementary Case Studies

15.5.1.1 TEM/EDS and a few drawbacks

The chemical composition of the residual phase was analyzed by Okada and Otsuka (1986b) using analytical transmission electron microscopy (TEM) with energy-dispersive X-ray (EDX)-type analyzer from a gel specimen fired at 1000°C and leached with 7 wt % of NaOH solution by boiling for 40 min. A value of ~8 wt % of SiO_2 was obtained by them. Accordingly, it was concluded that the spinel phase is not simple γ-Al_2O_3. However, the composition of the Al-Si spinel phase was first conjectured by them to contain 8 wt % of SiO_2 only. There may be two drawbacks in this study:

- Leaching in a boiling condition is against the procedure standardized by Chakraborty (1979). Increasing the silica extraction temperature from a boiling water bath condition to direct boiling temperature must attack the aluminosilicate phases. The reason for the low content of silica (~8 wt %) in the Al-Si spinel phase as noted by Okada and Otsuka (1986b) may be not extracting free silica up to the end point of the alkali extraction process, that is, they may have performed incomplete extraction.
- The spinel phase produced on heating 1S3A and 1S6A to 1000°C and above (Okada and Otsuka, 1986b) invariably contained γ-alumina besides the Al-Si spinel phase. Since in these samples, the alumina content is much above the composition of 3:2 mullite, the heat-treated samples of 1S3A

and 1S6A must have contained a mixture of γ-alumina and Al-Si spinel phases. It was unnecessary to perform the analysis of such type of alkali leached mass obtained after direct boiling condition with the help of analytical TEM with EDX-type analyzer.

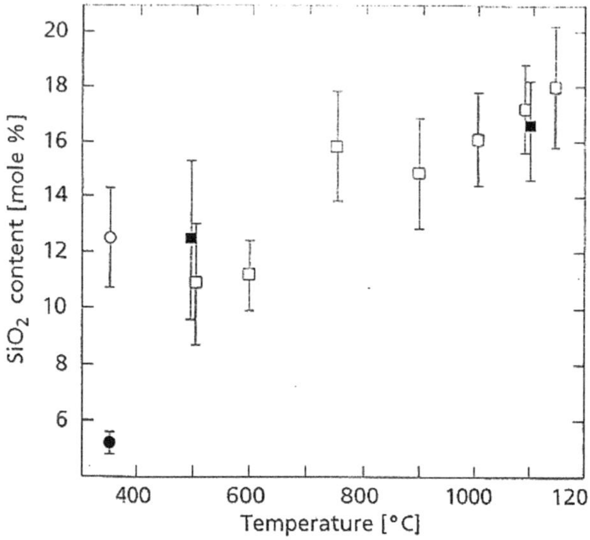

Figure 15.13 SiO_2 content of pseudoboehmite (350°C) and of γ-alumina spinel agglomerates (500°C–1150°C). Hollow symbols refer to not-leached samples, and filled symbols refer to NaOH-leached samples. The silica content has been determined by EDX (Schneider et al., 1994).

Using TEM, Schneider et al. (1994) showed that Type II gel consisted of rod-like pseudoboehmite crystals (20 nm) and spherical SiO_2 particles (≤0.5 nm), that is, it is purely diphasic in character. On heat treatment from 600°C to 900°C to 1100°C, the microstructure of the agglomerates and their diffractogram [selected area diffraction (SAD)] changed. A gradual increase in temperature led to a rise in crystallinity, growth of the spinel phase, and a change in shape from rod to isometric. The silica contents of the spinel phase developed on heating the gel at increasing temperatures was determined by EDX analysis for both unleached and leached samples and are shown in Fig. 15.13. The measured value of SiO_2 incorporated into γ-Al_2O_3 from the sample heat-treated at 1150°C was noted by them to be 18 mol % (11.5 wt %). The result indicated that even in the case of the diphasic gel system, Al-Si spinel formed gradually with increase in the firing temperature. These experimental data invalidate the concept of Hoffman et al. (1984) that the two phases remain independent during heating of the diphasic gel prior to crystallization. Secondly, its SiO_2 content increases and likely this spinel phase too would be Al-Si spinel. The content of the silica present in the diphasic gel heat-treated to 1100°C is more than 8 wt %, as shown by Okada and Otsuka (1986b) in the monophasic gel system.

Table 15.13 SiO_2 content of Type II gel after heat treatment in the temperature range of 500°C–1150°C and thereafter by alkali treatment shown by EDS study as per Fig. 15.13

	Pseudoboehmite Type II gel characteristics			
Heated at	**350°C**	**500°C**	**1100°C**	**1150°C**
Silica content in unleached gel (mol %)	12	~10.5	~16	18
Silica content in leached gel (mol %)	<6	12	~15.5	N D

Source: (Schneider et al., 1994)

TEM studies by Schneider et al. (1994) are cited next:

- It is observed that even in the diphasic gel system, incorporation of silicon in the γ-Al_2O_3 phase definitely occurs. Initially, the silica content present in the leached gel at 350°C was low (<6 wt % SiO_2), which indicates that before dehydroxylation of pseudoboehmite, interaction between the two hydroxide gels had already started and a low value was noted (Table 15.13). After complete dehydroxylation of Type II gel heat-treated beyond the decomposition of pseudoboehmite, a solid-state reaction started between the two newly formed noncrystalline oxide phases (both of which already contained some amount of silica/alumina) at ~500°C. As a result the amount of silica increased considerably (up to 12 mol %). This solid-state reaction increased more and more on further heating, up to 1100°C, when a value 15.5 mol % was noted. For heat treatment at temperatures beyond 1100°C, alkali leaching data and energy-dispersive X-ray spectroscopy

(EDS) analysis are necessary. It is conjectured that at these temperatures, more silica would be incorporated during the thermal evolution process on and before the onset of mullite crystallization at ~1300°C.

- As with the diffusion of Si into the γ-Al_2O_3 phase, counterdiffusion of Al into the silica phase might have occurred. It is necessary to do a SAD study in the adjoining amorphous phase.
- Transformation temperatures of Si-incorporated spinel to mullite need to be shown. Transformation generally occurs during the exotherm at ~1300°C. Therefore, an EDX analysis has to be done for determining the SiO_2 content in the sample that was heat-treated just before the onset of the above exotherm. Under any circumstances, 1150°C should not be chosen as the final temperature of Al-Si spinel crystallization to the fullest extent. Accordingly, the value of 18 mol %, which was taken as the final amount or the maximum Si incorporation, is a low value in contrast to the ~40 mol % as shown by Chakraborty (1979). Thus, proper collection of heat-treated diphasic gel samples and standard procedure of alkali leaching are necessary prior to EDX analysis for estimating the silica content of the Al-Si spinel phase. Over and above these studies, that by Schneider et al. (1994) is the most promising and confirms the incorporation of silica in alumina's structure.

15.5.1.2 IR technique and a few drawbacks

The IR spectral analysis of the Type II gel heat-treated to 900°C was done by Schneider et al. (1994). The spectra exhibited strong bands at ~1100 cm^{-1} and 470 cm^{-1} and a broad absorption area between ~820 cm^{-1} and 580 cm^{-1}. After the alkali leaching process, the strong bands at ~1100 cm^{-1} and 470 cm^{-1}, attributed to Si–O–Si stretching and O–Si–O bending vibrations, disappeared completely. On the other hand, the absorption region between ~820 c^{-1} and 580 cm^{-1}, maybe due to the spinel phase, did not change significantly. Simultaneously, a broad and weak absorption band appeared between 1080 cm^{-1} and 980 cm^{-1} (Fig. 15.14), which was interpreted by them to show the incorporation of silicon in the γ-Al_2O_3 phase. As shown in

Table 15.13, the presence of bonded silica in the diphasic gel heated to 900°C is >10 mol %. The absence of the IR absorption band at 1100 cm^{-1} and 470 cm^{-1} during leaching out of the Type II gel heat-treated at 900°C indicates that the leaching process adopted was more drastic and the alkali solution removed the bonded silica out of the spinel phase. Therefore, the leaching parameters need to be as per the standard procedure.

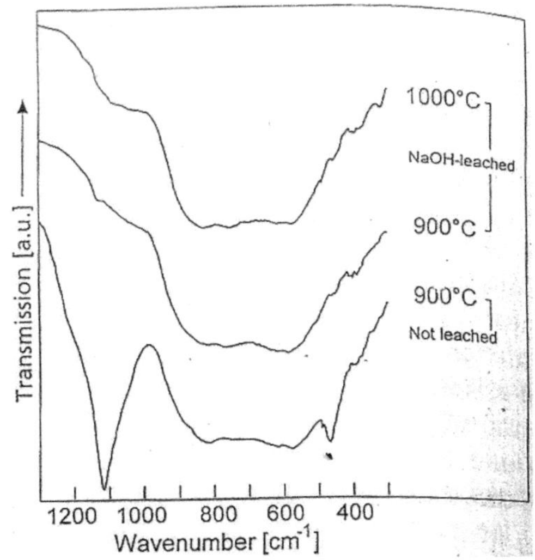

Figure 15.14 Infrared (IR) absorption spectra of precursors heat-treated at 900°C and 1000°C, not leached and NaOH-leached (Schneider et al., 1994).

Moreover, leaching had to be done on gel heat-treated at a temperature as high as ~1300°C, where Si incorporation into the γ-Al_2O_3 phase would be the maximum so Al-Si spinel formation is complete and mullitization sets in. Thus, EDX studies of heat-treated gel samples just up to 1150°C have led to inconclusive results regarding the composition of the Al-Si spinel phase.

Jin et al. (2002) also undertook an alkali leaching process but not did follow the scheduled process. Moreover, hydrolysis-coprecipitated powder (HCP) is a diphasic gel. Thirdly, they heat-treated the gel only up to 920°C and 1050°C and not up to 1250°C to 1300°C, where Al-Si spinel structurally transforms and mullitization occurs. In the temperature range between 920°C and 1050°C, silica incorporation into the γ-Al_2O_3 cubic phase structure is obviously less, as pointed out in Fig. 15.15. Therefore, IR studies of pre- and postleaching experimentations are to be repeatedly checked.

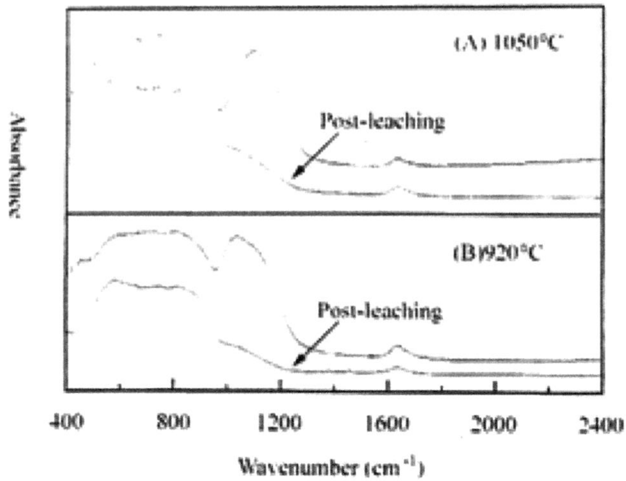

Figure 15.15 Infrared spectra of pre- and postleaching HCP calcined at 1050°C (Jin et al., 2002).

15.5.1.3 MAS-NMR Study and a few drawbacks

MAS-NMR study in the case of Type II gel synthesized by Schneider et al. (1994) showed the following changes: This precursor heat-treated to 350°C displayed weak ^{29}Si NMR signals at approximately –80 ppm, which according to them was due to the aluminum-rich compound containing silicon, and stronger signals at approximately –110 ppm, which was due to the silica-rich phase. With an increase in annealing temperatures from 900°C to 1000°C to 1100°C, the –80 ppm signal intensities increased significantly whereas the intensities of the –110 ppm peaks decreased (Fig. 15.16).

When the gel underwent the alkali leaching process, the band at –110 ppm disappeared completely but the peak at –80 ppm remained unaffected. ^{27}Al NMR spectra of the heated gel exhibited an intense peak at ~5 ppm and a weaker peak at ~60 ppm, which are attributable to the octahedral bond and the tetrahedral bond of aluminum, respectively. On the basis of these ^{27}Al NMR spectral results, Schneider et al. (1994) opined that these spectra are apparently similar to the pure γ-Al_2O_3 phase. A similar drawback was noted regarding the selection of heat treatment of the Type II gel as done in EDX studies by Schneider et al. (1994). It would be near 1300°C instead of 1100°C, as chosen both in the case of IR and MAS-NMR studies. The reasons are already shown. The presence of unaffected Si resonance at ~80 ppm after alkali leaching of the diphasic gel heated even up to 1100°C indicates the existence of an Al-Si spinel peak. The presence of an unaffected –80 ppm peak in the ^{29}Si MAS-NMR spectra and the appearance of an intense octahedral peak in comparison to a tetrahedral Al peak in ^{27}Al NMR leads us to the conclusion of the slow crystallization of the Al-Si spinel phase with the gradual heating process from 350°C to 1150°C of the Type II gel, which is diphasic in nature.

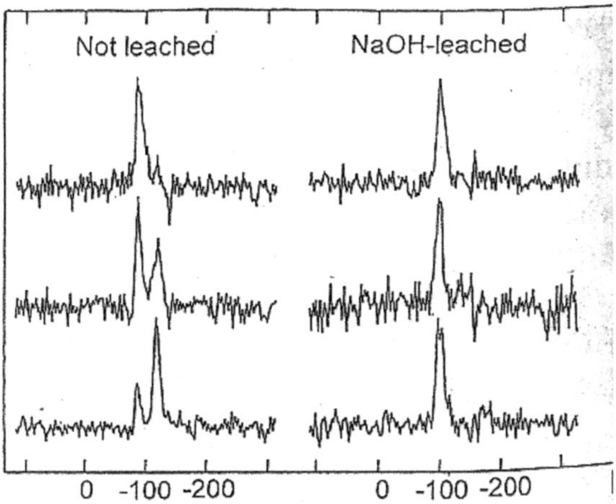

Figure 15.16 ^{29}Si MAS-NMR spectra of precursors heat-treated at 900°C, 1000°C, and 1100°C, not leached and NaOH-leached (Schneider et al., 1994).

15.6 Characterization of the Spinel Phase by QXRD Studies by Various Authors

15.6.1 Semiquantitative Amount of the Al-Si Spinel Phase by Earlier Studies

The ratios of the integrated intensity of the spinel phase and CaF_2 versus temperature for gels of different Al_2O_3-SiO_2 ratios were plotted by Okada and Otsuka (1986b). Growth of the spinel phase in SH and RH gels was a function of temperature. For two types of gels, the temperature of the start of spinel crystallization was found to lower as the composition became richer in alumina. SH gels of composition 1S2A and 1S3A started forming at ~900°C, whereas those of composition 1S5.4A, 1S9A, and 1S20A formed spinel between 700°C and 800°C. Similarly, RH gels of compositions 1S2A, 2S3A, 1S1A, 2S1A, 3S2A, 2.85S1A, and 3.5S1A all formed spinel on being heated for 24 h at ~900°C. On the other hand, the gels marked 1S3A, 1S6A, 1S9A, and 1S20A started to form spinel in the temperature range of 700°C and 800°C. Thus, both SH and RH gels of different Al_2O_3:SiO_2 ratios crystallized in two distinct temperature regions. Moreover, the rate of formation of spinel in the case of SH gel is faster than in the case of RH gel. These two experimental results shown by Okada and Otsuka (1986b) suggest that spinel forming at a lower temperature may be of a different nature than spinel forming at a higher temperature (at or above 900°C).

The intensity ratio of (440) reflection of the spinel phase and (400) reflection of CaF_2 against the bulk composition was plotted by Okada and Otsuka (1986b). According to them, the amount of spinel phase would be expected to be maximum where the composition of the spinel phase coincides with that of the starting materials. The maximum intensity was found near the composition $SiO_2 \cdot 6Al_2O_3$ in both SH and RH gels. With these observations, the chemical composition of the spinel phase was concluded as ~8 wt % of SiO_2 and the composition of the coexisting amorphous phase as $6SiO_2 \cdot Al_2O_3$, where the amount of spinel phase became zero.

It was further conceived by Wang and Thomson (1995) that the extent of spinel formation would be maximum when the composition of spinel coincides with the composition of the starting materials. Accordingly, it was shown that maximum amount of spinel was observed in the gel with an Al/Si ratio of 12:1, which corresponds to the spinel composition $6Al_2O_3 \cdot SiO_2$, as suggested by Okada and Otsuka (1986b). Moreover, the extent of spinel formation in diphasic gels 12:1 and 14:1 was very high compared to that in gels of lower Al:Si ratios, as noted by them.

The formation of the Al-Si spinel phase by TEM studies was indicated by Suzuki et al. (1990). They represented the unit cell of the spinel phase as (Al × [ ] 5.33) $(Si,Al)_8$ $Al_{10.67}$ O_{32}). By EDS analysis, the composition of the Al-Si spinel phase was shown to be stoichiometric mullite composition and its size was smaller than 10 nm.

15.6.2 Semiquantitative Amount of the Al-Si Spinel Phase by the Author

The author approached different techniques to estimate the Al-Si spinel phase using different standards. All these are presented in a stepwise manner.

15.6.2.1 Measurement of intensity of the Al-Si spinel phase

Manual: To derive the nature of the intermediate spinel phase present in gels heated to 1000°C, Chakraborty and Ghosh (1988) synthesized a series of coprecipitated Al_2O_3-SiO_2 gels heat-treated just to 1000°C for 1 h and areas of 0.139 nm peak of spinel were recorded by XRD technique. Next, the said intensities of spinel phases (measured manually using an XRD chart) formed were compared to the bulk composition of the gels. It was observed that the area of the said peak or, indirectly, the percentage of spinel formed from each gel is greater than that of the corresponding percent of theoretically expected for γ-Al_2O_3 crystallization. This result definitely leads to the conclusion that Si has entered in the tetrahedral part of the spinel phase by removing Al. The remaining Al formed γ-Al_2O_3 (Al spinel) in the usual way. As a result of the simultaneous crystallization of the Al-Si spinel and γ-Al_2O_3, joint XRD intensities of the 0.139 nm peak increase. These results show that the so-called spinel as observed by Okada and Otsuka (1986b) in both SH and RH gels of varying compositions contained a mixture of Al-Si spinel and γ-Al_2O_3 spinel.

Using CaF_2 as a standard: Instead of measuring the X-ray peak intensity manually, Chakraborty and Das (2003b) reinvestigated the composition of the spinel phase by comparing its intensity to the intensity of a standard for confirmation of the above XRD findings. It is shown that the ratio of the X-ray intensity (440) reflection of the spinel phase to that of CaF_2 versus the wt % of Al_2O_3 is linear (Fig. 15.17). A similar nature of curve is obtained when (400) reflection is taken. If the alumina component of the diphasic gel crystallizes theoretically to γ-alumina, then the two curves should superimpose. The calculated curve, however, lies above the theoretical curve. The observed higher values in all cases indicate that alumina (A) liberated just after the dehydration of boehmite at ~400°C reacted with SiO_2 (A) in the subsequent heating process and thereafter crystallizes into Si bearing a spinel phase. As a result of the entrance of Si into the spinel lattice, QXRD data should show a value higher than the amount of alumina taken during gel preparation.

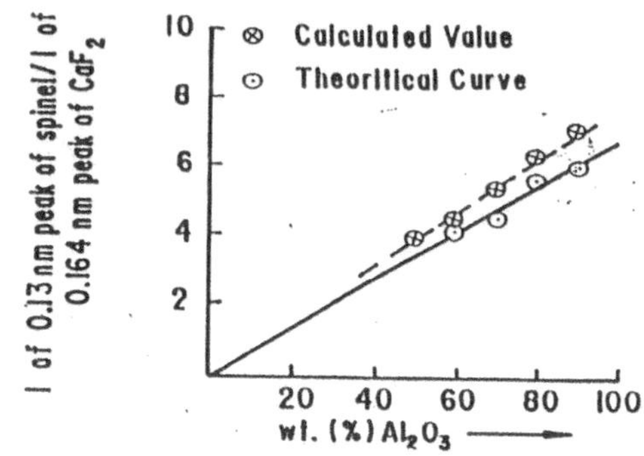

Figure 15.17 QXRD data of spinel of different Al_2O_3:SiO_2 ratios (Chakraborty and Das, 2003).

Using γ-Al_2O_3 as a standard: To substantiate the above result, diphasic gels of varying compositions of Al_2O_3-SiO_2 were synthesized by Chakraborty (2000, unpublished) by using ANN and ethyl orthosilicate and then heat-treated. The intermediate spinel phase was estimated by XRD technique using γ-Al_2O_3 as a standard and on considering the (440) reflection of the spinel phase. The quantity of spinel formed in each case of gel was calculated by assuming that γ-Al_2O_3 and Al-Si spinel may have equal structure factor since the difference in scattering factors for Si^{+4} and Al^{+3} is close (Fig. 15.18). The amount of spinel formed is shown in Table 15.14.

The estimated values are found to be higher than the percentage of alumina contents taken during gelation. Besides incorporation of Si in the alumina structure, noncrystalline aluminosilicate phase also exists in varying quantities, which depend on the initial composition of the gel, confirming the previous finding of development of the Al-Si spinel phase.

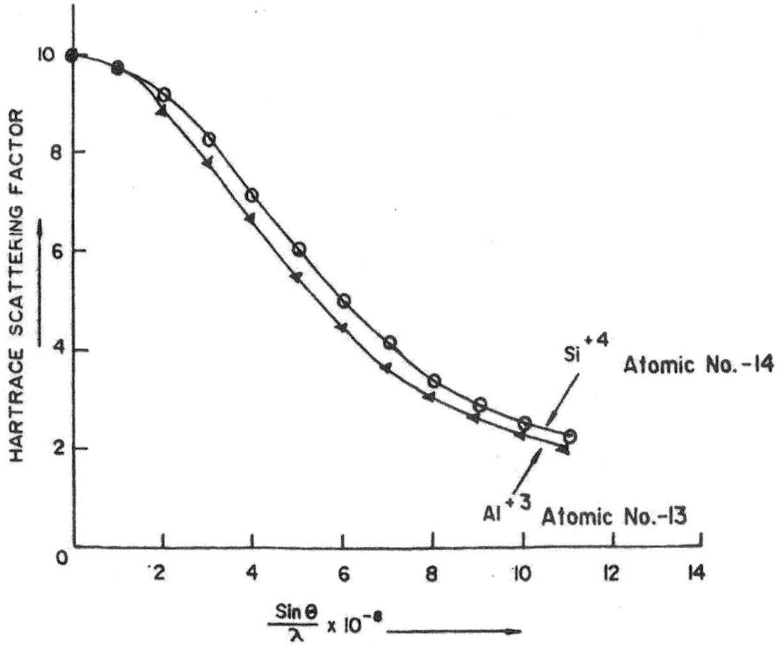

Figure 15.18 Scattering factor curves of Si^{+4} and Al^{+3} (Chakraborty, 2000, unpublished).

Table 15.14 Chemical composition of diphasic gels and QXRD data of these gels heated to 1000°C

	Composition of gel (approx)		Amount of precursor used			
Gel mark	Al_2O_3 wt %	SiO_2 wt %	TEOS ml	$Al(NO_3)_3$ $9H_2O$ gms	Quantity of Si–Al spinel estimated by QXRD technique	Residual amount of glassy phases associated with Si–Al spinel by difference
233	50	50	5	10	55.13	44.87
234	60	40	3.5	10	72.26	27.74
232	72	28	2	10	81.39	18.61
235	80	20	1.25	10	85.58	14.42
236	90	10	0.56	10	93.07	6.93

15.6.2.2 QXRD study of the spinel phase

By changing the source of the components, Al_2O_3-SiO_2 diphasic gel (Si:Al = 1:3) was synthesized by using ANN and Ludox instead of TEOS, as above, in a basic condition. It was heated dynamically at 10°C/min. The sample marked G-173 was collected at predetermined temperatures without allowing any soaking time. The spinel phase formed on heating (Al_2O_3 = 69%) was quantitatively estimated by the usual X-ray internal standard method using CaF_2 and γ-Al_2O_3 as a standard and by drawing a calibration curve. The amount of spinel is estimated to be 79–80 wt %, which is greater than the initial alumina content of the gel. These data exemplify two aspects. Firstly, they indicate that the spinel phase formed during the transformation of boehmite in the presence of silica is not γ-Al_2O_3. In other words, the entrance of some quantity of silica/Si^{+4} in the latter structure is responsible for the higher value due to the formation of ~79 wt % of the Al-Si spinel phase of mullite-like composition. Secondly, the remaining portion of silica remains in association with alumina and develops ~21 wt % of a noncrystalline aluminosilicate phase.

The quantitative data of the Al-Si spinel phase and the noncrystalline aluminosilicate phase shown in this method are comparable with the estimated data shown above by changing the source of the components, for example, Ludox for TEOS, during the synthesis of the diphasic gel.

15.6.3 Discussions of QXRD Results of the Al-Si Spinel Phase

Composition of the Al-Si spinel phase and effect of the Al/Si ratio on the formation of the spinel phase

Okada and Otsuka (1986b) presented that (i) the DTA exothermic peak was the highest in the SH xerogel for the composition $2SiO_2 \cdot 3Al_2O_3$ and (ii) the specimen fired just above the temperature of the exothermic peak at a heating rate of 10°C/min. showed the formation of mullite.

The author also observed the following:

- With a gradual increase in the amount of Al_2O_3 added to the coprecipitated gel, the Al-Si spinel phase crystallizes at first at ~1000°C, increases constantly from the beginning, and then reaches a maximum when the gel is of the composition Al:Si = 3:1. γ-Al_2O_3 will start crystallizing beyond this composition of the gel. The ratio of the Al-Si spinel phase to γ-Al_2O_3 spinel formation decreases with continued addition of Al_2O_3.

- Similarly, on the alumina side of the Al_2O_3-SiO_2 phase diagram, with a gradual increase in the percent composition of SiO_2 in the gel, Al-Si spinel will crystallize first due to the substitution effect, along with the formation of γ-Al_2O_3 when the gel is heated at ~1000°C only. The amount of Al-Si spinel formed would gradually increase, and γ-Al_2O_3 would likely be less as more and more SiO_2 is added. The relative proportion of the two spinel phases would, of course, depend on the composition of the gel. At just 28 wt % of SiO_2, the gel would form only Al-Si spinel in the largest amount. The formation of γ-Al_2O_3 ceases at that composition. Beyond this composition, only Al-Si spinel persists with continued addition.
- As the ratio of Al:Si increases, γ-Al_2O_3 crystallization increases (Al spinel [γ-Al_2O_3] starts forming at a lower temperatures, e.g., 800°C–900°C). Okada and Otsuka (1986b) and Wang and Thomson (1995) both made this observation earlier.
- γ-Al_2O_3 formed at ~1000°C transformed to polymorphs like Si-substituted θ-alumina with gradual heating of alumina-rich gels to ~1450°C. On careful observation of their figure, it is noted that spinel formation is more marked when the gel is heated in the range between 1000°C and 1300°C in the case of 12:1 gel and between 900°C and 1450°C in the case of 14:1 gel synthesized by Wang and Thomson (1995). As the higher alumina-rich gels crystallize first to γ-Al_2O_3 spinel and next to Si-incorporated θ-Al_2O_3 during the heating process, the formation of the maximum amount of the latter phase expected was at and around 6 wt % of silica, on the basis of the findings of Yoldas (1980). The reason for the high intensity of these two gels formed at a higher temperature as shown above is Si-incorporated θ-Al_2O_3 formation and not γ-Al_2O_3 or Al-Si spinel formation in the temperature range of 1300°C–1400°C. Some of the XRD peaks of these two phases are nearly common, and these superimpose at the same 2θ value, which makes identification difficult.

Therefore, the expectation that the extent of spinel formation would be maximum when the composition of the spinel coincides with the composition of starting materials seems unreasonable, as proposed by Wang and Thomson (1995) and by Okada and Otsuka (1886b). Secondly, the maximum extent of spinel formation noted by Okada and Otsuka (1986b) at approximately 8% silica concentration, as shown, is due to crystallization of both γ-Al_2O_3 (major quantity) and Al-Si spinel phases (minor quantities). And, thirdly, the maximum extent of spinel formation noted by Wang and Thomson (1995) at approximately 8% silica concentration, as shown in their figure, is due to the crystallization of Si-incorporated θ-Al_2O_3 and not due to spinel formation.

- A similar straight-line relationship curve between the intensity ratio of spinel to CaF_2 versus the wt % of Al_2O_3 was noted first by Okada and Otsuka (1986b). The X-ray intensity relationship with a bulk chemical composition of different Al_2O_3-SiO_2 gels was put forward. But they did not compare the intensity ratio of γ-Al_2O_3 spinel to CaF_2 versus the theoretical values of the wt % of Al_2O_3 for γ-Al_2O_3 gel. On the contrary, incorporation of Si into the alumina phase occurs.

15.6.4 Discussions on Al-Si Spinel Phase Composition First Suggested by Okada and Otsuka (1986b)

There are two problems regarding the solid solution in an Al_2O_3-SiO_2 system, which we will discuss next. One is Al_2O_3-SiO_2 solid solution, and the other is γ-Al_2O_3-SiO_2 solid solution.

Yoldas (1976), Okada and Otsuka (1986b), Wang and Thomson (1995), and McHale et al. (1997) all studied the Al_2O_3-SiO_2 solid solution. On the other hand, Chakraborty (1979), Chakraborty and Ghosh (1987) studied the γ-Al_2O_3-SiO_2 solid solution.

- In the case of the SH gel, the 980°C exothermic peak was shown by Okada and Otsuka (1986b) to be the sharpest and highest for the composition $3Al_2O_3 \cdot 2SiO_2$ (3A2S). They further noted that this peak intensity became diffuse and weakened as the composition deviated from it. Mullite formation was noted by XRD analysis of the specimen fired just above the temperature of the exothermic peak. A minor amount of spinel was also crystallized at the composition of $3Al_2O_3.2SiO_2$ and its reflection increased with an increase in the alumina content than 3:2 composition of mullite.

- On the other hand, a major quantity of the spinel phase and a minor quantity of mullite were crystallized in the RH gel of the composition $2.8.SiO_2{\cdot}Al_2O_3$, which showed the sharpest and highest exothermic peak. Okada and Otsuka (1986b) chose SH and RH gels of different compositions quite rich in alumina compared to 3A2S for recording the XRD intensity of the spinel phase in their study. Thereafter, they fired all alumina-rich compositions at increasing temperatures of heat treatment between 700°C and 1150°C.
- There are two questions to be raised: (i) When the objective is to characterize the composition of this minor quantity of spinel phase formed in the SH gel and/or a major quantity of the same phase in the RH gel, then why was the firing temperature of both gels above 980°C? (ii) Why were alumina-rich compositions beyond 3:2 stoichiometry selected for SH gels and compositions beyond 2.8SA taken for RH gels for quantitative X ray analysis?
- Okada and Otsuka (1986b) showed that the SH gel beyond 3:2 stoichiometry starts developing γ-Al_2O_3 (Al spinel) in addition to an Al-Si spinel phase. Therefore, the measured integrated XRD intensity ratio (as shown in Okada and Otsuka, 1986b, Fig. 2) of the spinel phase formed on heating both gels indicates that it must contain a mixture of Al spinel and Al-Si spinel phases.

  But the objective would be to determine the composition of the gel at which the maximum quantity of Al-Si spinel phase crystallizes, that is, the intensity of the XRD peak of Al-Si spinel phase should be maximum and thus heating is to be restricted to the exothermic peak temperature, that is, ~1000°C and not beyond this temperature.

  But a practical difficulty is that gels on heating at the exothermic peak temperature after 3:2 Al_2O_3:SiO_2 ratio on the aluminous side form more and more quantity of γ-Al_2O_3 in addition to the Al-Si spinel phase. Secondly, both show common XRD peaks. Accordingly, determination of XRD intensity exclusively for Al-Si spinel only in a mixture of the two phase poses a great problem. Similar also is the problem of selecting the end point of the γ-Al_2O_3-SiO_2 solid solution. The question is, how does one determine the difference in the XRD pattern intensities of the two spinel phases, namely γ-Al_2O_3 and Al-Si spinel?

- $SiO_2 \cdot 6Al_2O_3$(1S.6A) in both SH and RH gels when calcined to a high temperature deliberately crystallized to γ-Al_2O_3 (Al spinel). The ratio of Al-Si spinel to γ-Al_2O_3 changes with a change in the composition of starting materials and with heating temperature. So the chemical composition of the Al-Si spinel phase cannot be ascertained in this procedure, as done by Okada and Otsuka (1986b), and it is does not help establish the relationship between the maximum intensity ratio of ($I_{\text{spinel}}/I_{\text{CaF2}}$) and the bulk composition of the starting material, as shown in Okada and Otsuka (1986b, Fig. 3). However, the choice of this relationship is obvious for establishing an Al_2O_3-SiO_2 solid solution.

15.6.5 Discussions on XRD Analysis of the Spinel Phase by Wang and Thomson (1995)

Wang and Thomson (1995, Fig. 3) observed the spinel phase in gels of Al:Si ratios from 4:1 to 14:1. A big question is, what is the nature of this spinel phase, which crystallizes on heating the gels at a temperature between 850°C and 1400°C? According to them, the maximum extent of spinel phase was observed in the gel with the Al:Si ratio of 12:1. The question is, is this Al-Si spinel and does it form sharply at the 980°C exotherm? It is not Al-Si spinel but Si-substituted Θ-Al_2O_3 that occurs at a temperature as high as 1300°C–1400°C, as shown by Chakraborty (2008a).

On the basis of author data, spinel formation as shown in their Fig. 3 at ~1000°C is a mixture of γ-Al_2O_3 (Al spinel) and Al-Si spinel phases. This mixture transforms into mullite (from Al-Si spinel) and corundum (from γ-Al_2O_3), and this conversion is both temperature and composition dependent.

These authors did not observe the spinel phase from a single-phase gel with Al:Si ratios < 4:1 but observed it just at 4:1. This observation indicates two points:

- Al:Si in the ratio 3:1 forms only mullite at 980°C exotherm, and this composition is the end point of γ-Al_2O_3-SiO_2 solid solution series.
- Crystallization of some quantity of spinel phase at ~1000°C in the case of 4:1 ratio gel was characterized as θ-Al_2O_3 and γ-Al_2O_3. Thus, it is predicted that this gel does not form 2:1 mullite, as proposed by Wang and Thomson (1995), although Fig. 2 of them apparently shows the largest extent of mullite.

15.6.6 Discussions on the Quantitative Data of the Spinel Phase by Wei and Halloran (1988)

Moreover, the higher QXRD data of the spinel phase in the case of the gel marked CP-9 (whose alumina content in the gel was ~72 wt %) does not resemble the estimated data shown by Wei and Halloran (1988). They noted that the diphasic gel (Si:Al = 1:3) formed δ-Al_2O_3 in the intermediate stage and stated that it did not change and remained at a constant value of ~72 wt % Al_2O_3 prior to mullitization. This conjecture is against the incorporation data of silica in the alumina phase during the heating process of the diphasic gel as given experimentally by Wei and Halloran (1988). However, in practice, the spinel formation curve would actually show a higher quantity of spinel if Fig. 7 of Wei and Halloran (1988) is redrawn by joining the midpoints of the estimated quantities of spinel formed from the diphasic gel heated to different time/temperature schedules. They simply drew a dotted straight line through 72 wt % Al_2O_3 and concluded that the amount of spinel did not change on heat treatment of the diphasic gel. As Al-Si spinel forms in all cases, its exact composition could be ascertained only if the composition of the associated noncrystalline aluminosilicate is available. The composition of the former phase seems too difficult to ascertain since there is every possibility that the composition of the glassy phase will change due to the ongoing solid state reaction of SiO_2 (A) and Al_2O_3 (A) during heating. Table 15.14 further shows that the amount of glassy phase changes with the wt % of Al_2O_3 in the gels.

15.7 Characterization of the Spinel Phase by Lattice Parameter Measurement

The lattice parameters of the Al-Si spinel phase derived on heating Al_2O_3-SiO_2 gels of varying ratios was first determined by Yamada and Kimura (1962). It was shown that the lattice constant (LC) of the spinel-type crystal increases with a decrease in the value of the SiO_2/Al_2O_3 specimen. They suggested that the so-called spinel phase is not that of γ-alumina but a phase with various amounts of Si^{+4} substituted by Al^{+3} in the position of four coordination. Yamada and Kimura (1962) and Okada and Otsuka (1986b) faced difficulty in lattice parameter measurement of the spinel phase in terms of measuring the exact 2θ values. LC measurements show that the composition of spinel is a function of temperature and the composition of spinel varies with the Al_2O_3:SiO_2 ratio as per the study of Yamada and Kimura (1962).

The lattice parameter of the spinel phase was measured by the XRD technique by Okada and Otsuka (1986b). They showed the LC of the spinel phase as a function of composition, preparation technique used (whether it is SH or RH process), and firing temperature.

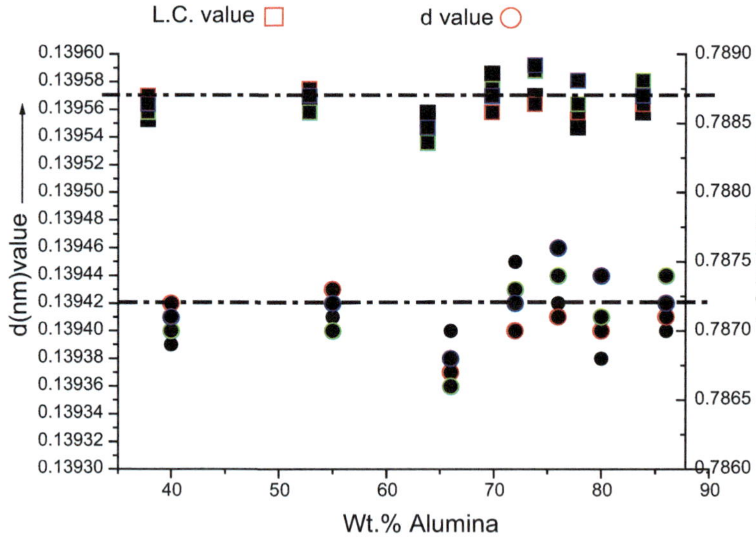

Figure 15.19 Changes in *d* (nm) values and/or the lattice constant value of the spinel phase versus the composition of the mixed gel.

The LC of θ-Al_2O_3 formed from aluminosilicate gel (Al_2O_3:SiO_2 = 3:2) at ~1300°C was measured by Low and McPherson (1989). They conjectured that θ-alumina indicates that the composition of the cubic phase is aluminosilicate. The unit cell size of Al-Si spinel (a = 0.7880 nm) is slightly smaller than that of γ-Al_2O_3 (a = 0.7900 nm), confirming the existence of a solid solution of silica in the lattice of γ-alumina to form an Al-Si spinel.

The lattice parameter of the spinel phase was calculated by Yamada and Kimura (1962), Hirata et al. (1985b), Low and McPherson (1989), and Hirata et al., (1989). They found that the value of the spinel phase produced by heating gels at 1100°C for 1 h was 7.894 ± 0.003 Å, compared with the value of 7.886 Å given by Brindley and Nakahira (1959). The LC values of the Al-Si spinel phase were also calculated by heating mixed gels of various compositions by Chakraborty (2006). The center of full width at half maximum of the (440) reflection was taken by XRD analysis with the help of graphics software of Philips. The LC values of spinel phase versus temperature curves of MG40, MG55, MG66, MG72, MG76, MG80, and MG86 were plotted (Fig. 15.21). The calculated LC value (0.78865) of spinel phase is lower than the LC value of pure γ-Al_2O_3 (0.7906). The reduction in the LC value of the spinel phase in comparison to that of γ-Al_2O_3 indicates the possible substitution of Al^{+3} by Si^{+4} at the tetrahedral site of the cubic spinel crystal, as proposed earlier by Brindley and Nakahira (1959). They obtained a phase with a spinel-type structure with a marked orientation by the decomposition of kaolinite at 950°C. Thereafter, they calculated the LC value of the spinel phase (0.7886) from both powder photograph and diffractometer records. This phase was considered as aluminum silicon spinel with vacant cation sites. A similar LC value is observed when heating mixed gels, and it is more likely to be Al-Si spinel, as designated earlier by Chakraborty (1979). Due to the prior occupation of Si^{+4} in the γ-Al_2O_3 structure at a tetrahedral position, subsequent transformation of the spinel phase to α-Al_2O_3 is hindered. This mechanism was considered by Stone and Tilly in the phase transformation of the pure Al_2O_3 system as a diffusion path blocked by an impurity cation (likely Si^{+4}) favoring a tetrahedral coordination. This value decreases with the composition of the mixed gels. What is the relationship between lattice constant change and spinel phase formation? It may be conjectured that as the diffusion of Si rises with increase in the heating temperature, γ-Al_2O_3 becomes richer in SiO_2, with a resulting decrease in the LC value (Fig. 15.19). So the highest concentration of SiO_2 lies in the narrow range of 70–76 wt % of Al_2O_3. Its average value is the composition of 3:2 mullite.

15.8 Interrelationship between Al-Si Spinel Phase and Mullite Formation: Case Studies by Different Researchers

The formation of Al-Si spinel and the formation of mullite are interrelated. It has been assumed that these two are polymorphs. As such, the formation of the former phase and its structural transformation into mullite are first described in this chapter for easy assessment of mullite development from different kinds of mullite precursors, described in the next chapter.

Example 1: The relationship between a coexisting Al-Si spinel phase and mullite formed during the first exothermic peak was shown by Yamada and Kimura (1962). Two remarkable observations are to be made. One is the commencement of Al-Si spinel and mullite phases. The other is the disappearance of the latter phase during the time-temperature schedule of heating. Changes in the X-ray intensity of the Al-Si spinel phase with variation in heat treatment temperature and rate of formation of mullite were shown. It is noted that Al-Si spinel formation is predominant at ~1050°C but mullite formation is lowest at the same temperature. Secondly, the rate of decay of the Al-Si spinel phase becomes faster at ~1180°C, where the rate of mullite formation starts predominating. With these observations, Yamada and Kimura (1962) suggested that mullite is formed by the thermal decomposition of spinel-shaped crystals.

Example 2: The relative changes in the X-ray diffraction intensity of the phases produced from alkoxide-derived SiO_2-Al_2O_3 powders on heating were shown by Hirata et al. (1985a). It indicated the interdependence of phase changes and also showed conclusively the occurrence of pronounced mullitization with the disappearance of the spinel phase at ~1300°C.

Example 3: Interrelated recordings and observations noted in the mullitization process of colloidal and diphasic gels and disappearance of Al-Si spinel and crystallization of the mullite DXRD curve were shown by Li and Thomson (1991).

Example 4: A sharp decrease in the spinel phase coincides with rapidity of crystallization of mullite, as shown by Sugita et al. (1998, Fig. 15.20).

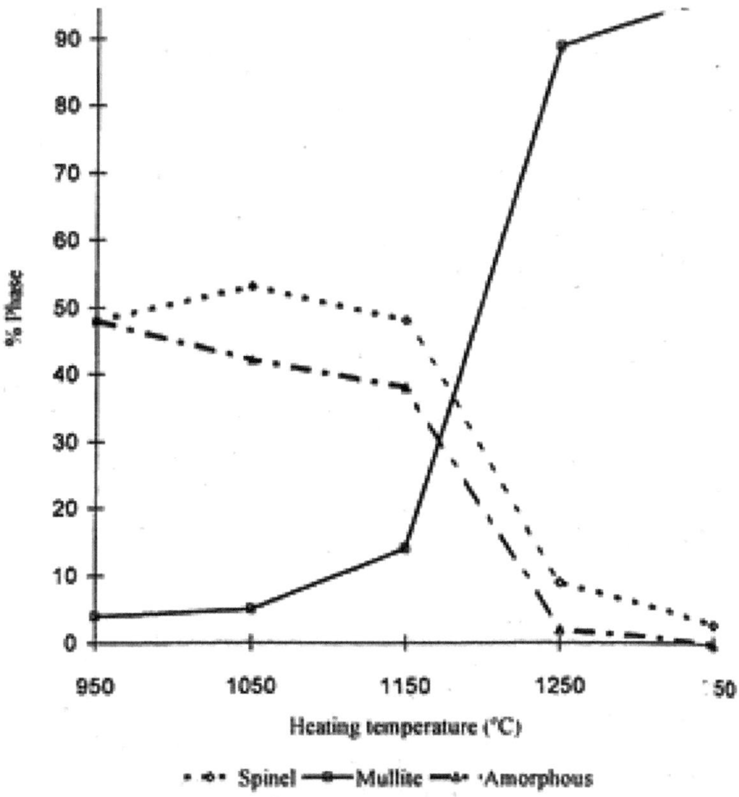

Figure 15.20 Change of phase composition versus heating temperature (Sugita et al., 1998).

Example 5: DTA and DDTA peaks of the transformation of Al-Si spinel to mullite (Fig. 15.1a,b) were shown by the author. The DTA result of the author was verified later, in subsequent studies by Huling and Messing (1991) and by Wang and Sacks (1996) in their respective diphasic gels. The latter authors also presented the formation of a broad exotherm between 1150°C and 1300°C, as shown by the author, which is explained as due to the crystallization of the Al-Si spinel phase prior to its transformation to mullitization at the 1332°C exotherm. Unfortunately, this phenomenon was not mentioned.

Example 6: TMA and DTMA peaks are interrelated with DTA and DDTA peaks of thermal events of mullite gels, as shown by Ivankovic et al. (2003) vide Fig. 15.21.

15.9 Controlling Factor in Spinel Formation

Huling and Messing (1991) showed that the intermediate spinel phase can forestall mullite formation as well as decrease the intensity of exotherm of the hybrid gel at ~1000°C. What controls the development of this spinel phase during heating at 1000°C in solution-derived gel or in colloidal gel? It is attributed solely to the heterogeneous alumina-silica mixing of gel components or maintenance of alumina-silica segregation during the heating process. According to them, both nucleation frequency of the spinel phase and its growth rate are influenced by gel processing. It is noted that even a few wt % of segregated alumina is sufficient for spinel to dominate crystallization at ~1000°C exotherm. They predicted that both direct mullite formation at exotherm and an intense DTA peak require complete molecular-scale mixing of two components. In view of this stringent requirement, they conjectured that most of the earlier sol-gel processes in the literature did not either yield a substantial amount of mullite or exhibit a sharp exotherm due to segregation of alumina and silica in alkoxide-derived gels and/or clustering in the homogeneity of gels. The spinel phase developed favorably due to the following processing parameters: (i) use of water, (ii) use of aged silica sol, (iii) use of colloidal alumina sol, and (iv) use of alumina phase as seed to selectively nucleate the spinel phase.

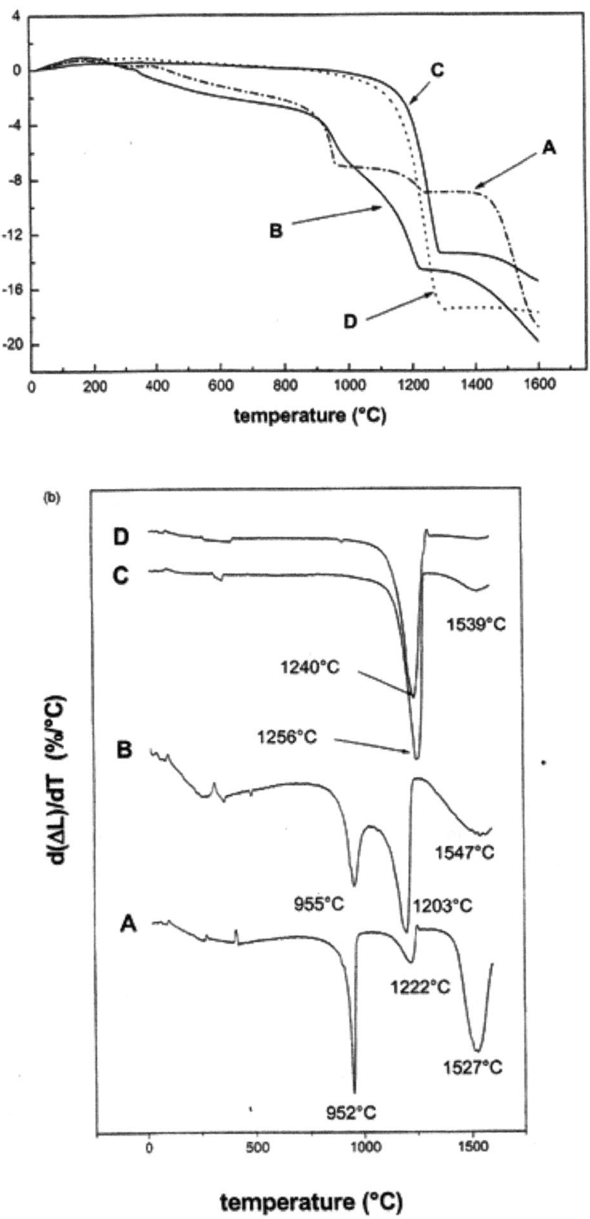

Figure 15.21 (a) Dilatometric curves of gels calcined at 700°C for 9 h. Nonisostatic pressure = 100–200 MPa. (A) Diphasic gel (Type III), (B) diphasic gel (combination of Type III and Type II), (C) diphasic gel (Type II, γ-Al_2O_3), and (D) diphasic (Type II, boehmite). (b) Derivatives of dilatometric curves shown in (a). (Ivankovic et al. (2003).

15.9.1 Role of SiO_2 in Phase Evolution Processes of Mullite Precursors

Besides alumina component as conjectured by Huling and Messing (1991), silica component also plays a major role in the phase transformation behavior of mullite precursors, especially in three aspects:

- Behavior of spinel synthesized by the aqueous gel method during DTA at the 980°C exotherm
- The mullite formation process at ~1250°C exotherm
- Exhibition of an endothermic dip just before the 980°C exotherm

A few examples are cited and these are listed below.

15.9.2 Role of Aqueous Silica Sol in Spinel Formation

- Aqueous silica sol prepared from sodium silicate was mixed with aluminum sulfate for the synthesis of Al_2O_3:SiO_2 gels of ratios of 1:4, 1:2, 1:1, 2:1, and 4:1 by Insley and Ewell (1935). $Al_2O_3{\cdot}2SiO_2$ gel similar to the composition of metakaolinite showed only mullite. As the Al_2O_3 content of the gel increased, γ-Al_2O_3 spinel formation started in addition to mullite.
- Aluminosilicate precursors coprecipitated from mixed aluminum and silicon chloride solutions by using ammoniated water were first prepared by Horte and Wiegmann (1956). Using a similar process, Demediuk and Cole (1958) prepared gels by mixing $SiCl_4$ and $AlCl_3{\cdot}6H_2O$ using a dilute NH_4OH solution. DTA of these gels showed orthorhombic mullite during 980°C exotherms. But peak heights varied considerably. During the coprecipitation process using aluminum chloride and silicon tetrachloride solution with ammonium hydroxide solution for the synthesis of mullite gel, McGee and Wirkus (1972) observed that silicon chloride is unstable and control of the chemical composition of the precipitates was difficult. Conversion of precipitates to high-purity mullite, however, occurs at ~1150°C. Similarly, segregation might have occurred to some extent during precipitation in the "flash hydrolysis" process used by Croft and Marshall (1967).

- Instead of ammonia, Ossaka (1961) chose excess of hexamethylenetetramine to synthesize coprecipitated gels from a mixture of sodium silicate and potassium aluminum sulfate in dilute aqueous sulfuric acid. Pseudotetragonal mullite formed during the sharp 980°C exothermic peak.
- Instead of using sodium silicate directly as a source component, Chakraborty and Ghosh (1988) first prepared silica sol from a sodium silicate solution mixed with a dilute aluminum sulfate solution. All mullite gels marked as G-104, G-96, and G-107 prepared in an acidic condition invariably showed a sharp 980°C exotherm and formed mullite directly.
- Using sodium aluminates and silica sol and by controlling pH at 7 with the addition of dilute H_2SO_4 at ~50°C, Schneider et al. (1992) synthesized a coprecipitated gel (CM). It shows mullite as the major phase at a temperature above 1100°C.
- Monosilicic acid solution and $Al(OH)_3$ sol were allowed to react at pH 9, and the resultant gel showed direct mullitization on being fired at 980°C by Nishu et al. (1989). This preparation method resembles that used by Chakraborty and Ghosh (1988). Polymerization and the subsequent growth of monomeric silicic acid seem negligible.

15.9.3 Role of Organic Silica Source in Spinel Formation

Monophasic gel prepared by Hoffman et al. (1984), SH gel made by Okada and Otsuka (1986b), polymeric gel synthesized by Yoldas (1992), and Type I and SGM gels prepared by Schneider et al. (1993) involved the use of TEOS as the organic silica component. In an attempt to prepare chemically homogeneous mullite precursors, the following measures were taken in these cases. Just sufficient amount of water/alcohol was added to completely hydrolyze alkoxides, for example, in sample C of Yoldas and Partlow (1988), Type 1 precursor of Schneider et al. (1993), GNW sample of Taylor and Holland (1993), and Sample C of Jaymes et al. (1996).

Water was added to alkoxide sufficiently slowly to avoid local polymerization of hydroxides, and care was taken such that the hydrolyzed solution is not milky white in appearance. An acidic medium was chosen over basic to hydrolyze components, and the pH was controlled. Even with these techniques, self-polymerization and condensation of hydrolyzed species could not be avoided or stopped, and these proceeded throughout the sol volume until gelification. Even during drying and in dehydroxylation stages, self-condensation continues. Consequently, in all these cases, a spinel phase invariably develops along with t-mullite at the 980°C exotherm and the ratio of the two phases, of course, varies. Not only the formation of associated spinel, a large quantity of the noncrystalline aluminosilicate phase also remains as a residue due to incomplete crystallization during the 980°C exothermic reaction. The reason is that the synthesis processes do not satisfy the basic structural feature of mullite, which is primarily made of chains of SiO_4 tetrahedral, AlO_4 tetrahedral, and AlO_6 octahedral units. In chemical processing, besides the formation of predominant Si–O–Al linkages, which is absolutely necessary for direct mullite nucleation, some other bonds, such as Si–O–Si and Al–O–Al, in addition to Si–O–Al bonds, are also generated as side-reactions.

Researchers and technologists must focus on minimizing these side-reactions as much as possible. As far as the analysis made in this chapter, the following measures are to be adopted: (i) spinel formation is to be avoided and (ii) the 980°C reaction should be completed such that the residual noncrystalline alumina-silicate phase does not remain.

- When silica sol/aqueous silica was taken as a water-soluble silica component, spinel crystallization was eliminated completely in samples synthesized by Chakraborty and Ghosh (1988).
- Spinel formation could also be eliminated with the use of TEOS if the mixed solution of TEOS and ANN was pyrolyzed. Thus the spray pyrolysis (SP) process of synthesizing mullite precursor or aqueous gelation process claims to be the most elegant among processes. Kanzaki et al. (1990) succeeded in achieving this by the SP method.

15.10 Summary

Formation of the spinel phase in different gels

RH gel by Okada and Otsuka (1986b), Type II gel synthesized by Schneider et al. (1993), and SGM gel of Schneider et al. (1992)

showed Al-Si spinel formation. A monophasic gel was synthesized by McPherson (1989) by a dynamic heat treatment procedure and by DXRD study, Li and Thomson (1991) noted an Al-Si spinel phase after the first exotherm.

PB2 gel by Sanz et al, 1991), M2 gel of Geradin et al. (1994), polymeric gel of Cassidy et al. (1997), PA gel of Nieto et al. (1998), and MI and MII gels of Huang et al. (1997) developed a spinel phase. The spinel phase also developed in MP, MPP, and MAPIF gels synthesized by Fonseca et al. (1997).

Coprecipitated gels made by Yamada and Kimura (1962), Hirata et al. (1985a), and Rajendran et al. (1990) showed spinel formation only. Chakraborty (1996, 1997) noted spinel formation in coprecipitated gel prepared at pH 6–7. The composition of spinel was conjectured by Chakraborty as analogous to the composition of 3:2 mullite. Other researchers, namely Prochazka and Klug (1983), Suzuki et al. (1990), Hirata et al. (1985a), Hamano et al. (1985, 1986), and Shiga et al. (1991), noted spinel formation in their coprecipitated gel. Mitachi et al. (1990) showed spinel formation in their gel synthesized at pH 7 and 10.

Fast hydrolyzed monoliths were X-ray amorphous up to 900°C and then underwent a progressive transformation into a γ-alumina-like pattern when heated to ~1000°C (Colomban,1989).

Colloidal gel of Li and Thomson (1991) showed Al-Si spinel exclusively at 980°C exotherm and mullite at 1200°C exotherm by DXRD study.

Colloidal gel of Yoldas (1992), Type III precursor by Schneider et al. (1993) and by Hamano et al. (1995), M3 gels by Geradin et al. (1994), and the gel marked Gel B by Ivankovic et al. (2003) showed spinel formation.

In in situ diphasic gel, Chakraborty (1996) showed that the composition of the spinel phase changes with heating and finally becomes analogous to 3:2 mullite.

Various thoughts on spinel phase formation in diphasic systems and limitations regarding its identification

There are three major thoughts on Al-Si spinel phase formation in a monophasic gel system versus γ-alumina formation in a diphasic system. The first group believes in intermediate Al-Si spinel phase formation. The second group conjectures γ-Al_2O_3 formation at

~1000°C and shows δ-Al_2O_3 formation, which exists in temperatures as high as 1200°C. The third group believes in densification prior to mullitization at ~1250°C, as mentioned above. Moreover, there are many difficulties in characterizing the cubic spinel phase by earlier methods and there are some limitations too. Chakraborty (1979) approached this problem from several different angles and approaches. The results are shown next.

First approach: Phase transformation of "in situ" diphasic gel

In this approach, the phase transformation behaviors of Al_2O_3-SiO_2 diphasic mullite gels of varying Al:Si ratios by qualitative X-ray diffraction techniques were studied by Chakraborty (1993). The following experimentations have been made:

- Comparison of transformation behaviors of diphasic gels of varying SiO_2 and Al_2O_3 contents with those of component gels: The effect of heat on the transformation behavior of silica gel with gradual addition and mixing of boehmite gel shows that initially cristobalite is the highest. It decreases slowly with increasing amount of the boehmite component and finally vanishes, and the gel develops a noncrystalline aluminosilicate phase. The appearance and disappearance of phases on heat treatment of diphasic gels of various compositions at different temperatures were noted. The evidence of the formation of a noncrystalline aluminosilicate phase precursor was shown.
- Comparative phase transformation study of boehmite gel vs. in situ diphasic gel: In this study the crystallization behaviors of γ-Al_2O_3 and the Al-Si spinel phase are shown in the growth curve. Crystallization of γ-Al_2O_3 occurs at ~500°C, and it is faster than the Al-Si spinel phase. A sharp rise in the growth of the spinel phase occurs in between 1000°C and 1250°C, identified by XRD study. The crystallization of such a large quantity of it is found to be in good agreement with the occurrence of a broad and large exothermic peak between 1000°C and 1300°C in DTA. Also, the delay in the rate of crystallization of the latter phase and that it coincidences with exothermic peak temperature confirms the introduction of silica in the γ-Al_2O_3 phase.

- DTA/DDTA studies: Diphasic gels and their components will change their heat content during the crystallization process on heating. With this view, thermal analysis (e.g., DTA and DDTA studies) was done by Chakraborty (1996). It was shown that the course of crystallization of an "in situ" diphasic gel to an intermediate Al-Si spinel phase, shown by XRD, concurs with the exhibition of a broad exotherm between 800°C and 1300°C in DTA. The rapid transformation of this newly formed Al-Si spinel phase to a mullite-type phase observed by XRD is in good agreement with exhibition of further exotherm at the same temperature, ~1320°C, in DTA and DDTA.

Second approach: Variations in the phase evolution process of diphasic gels, MOs, and mixed gels of different compositions

- Comparison of phase transformation of a mechanical mixture of γ-Al_2O_3 and silica (A) with that of diphasic gels: The phase evolutions of three diphasic gels were compared with XRD analyses of mixed oxides (MO) of alumina and silica (AS) heated at increasing temperatures. Polymorphic phases are gradually formed on heating the latter sample. Both θ-Al_2O_3 and α-Al_2O_3 phases are present during heating. Contrarily, such types of transformations are absent during the transformation of the diphasic gel DG72.
- Mode of crystallization of Al-Si spinel and the nature if its characteristic XRD peaks observed during heat treatment of mixed gel: As in the case of the in situ diphasic gel, the two component oxides of the mixed gel DG72 do not crystallize individually to either cristobalite or corundum. Also, the mixed gels on the silica-rich side of stoichiometric mullite composition develop cristobalite during heating. The mixed gels on the alumina-rich side of the solid solution of mullite crystallize to corundum. The breadth of the 0.137 nm peak of the spinel phase formed on heating mixed gels changes on heat treatment from 1000°C to 1250°C. Apart from the change in breadth, formation of θ-Al_2O_3 along with an Al-Si spinel phase has been identified. The former phase is found to be stable up to temperatures as high as 1250°C and it has been characterized as silica-incorporated θ-Al_2O_3 formed in a silica-poor mixed gel, for example, MG88 to MG95. It remains

stable even at a high temperature of up to 1400°C. Thus, it is concluded that the Al-Si spinel phase may coexist with silica-incorporated θ-Al_2O_3 in the heated gels. The XRD peak at 0.137 nm is superimposed by two such phases.

- Crystallization of the Al-Si phase from the intermediary noncrystalline precursor phase while heating a diphasic gel: The XRD intensity of the 0.139 nm peak at 67° 2θ of the Al-Si spinel phase and the same for the 0.40 nm peak at 22 °2θ of the noncrystalline/amorphous band were scanned and the area of the amorphous hump and the integrated area under the peak of the spinel phase formed while heating three representative diphasic gels were measured and finally plotted. It was found that the intensity of the amorphous band decreases in the temperature range of 900°C–1000°C. Contrarily, the XRD intensity of the peak due to the spinel phase increases with heat treatment in the same temperature range. The decrease in the amorphous band due to the noncrystalline precursor phase is thus interrelated to the crystallization of the spinel phase. This study gives an answer to the final question, how does Al-Si spinel crystallize? A crystallization study by XRD has been performed of the noncrystalline precursor phase (amorphous band) to the Al-Si phase formed during the heating process of the diphasic gel.

Third approach: Solid-state reaction of heated diphasic gels and spinel phase with CaO

Characterization of both monophasic and diphasic mullite precursors by a solid-state reaction study while heating a raw gel, a dried gel, and a gel preheated at 900°C and at 1000°C with $CaCO_3$ to 1200°C was followed by an XRD study. Chakraborty (2003b, 2004) identified separately the final phase or phases formed during the heating process.

- A monophasic gel dried at 170°C and one calcined at 900°C for 1 h react with CaO on heating and form gehlenite. Thus, the dried monophasic gel develops linkages and forms an aluminosilicate (A) phase.
- A raw diphasic gel reacts with CaO on heating and forms calcium aluminate. On the other hand, the diphasic gel heated to 400°C after completion of dehydroxylation of boehmite

and silica hydrogel also reacts with CaO and forms gehlenite. Thus, the diphasic gel likely develops linkages and forms an aluminosilicate (A) phase after dehydroxylation at ~400°C.

- Both dried monophasic gel and dehydroxylated diphasic gel start to form calcium aluminate when the alumina content in the gels increases beyond that corresponding to the 3:2 mullite composition, suggesting that both precursors consist of an amorphous aluminosilicate phase of a composition close to that of 3:2 mullite.
- The formation of gehlenite rather than calcium aluminate, shown by XRD, indicates that both monophasic and diphasic precursors consist of an aluminosilicate (A) phase of a composition close to 3:2 mullite. Therefore, it is concluded that the monophasic gel transforms to mullite via an intermediate aluminosilicate (A) phase during the 980°C exotherm. It is also concluded that the diphasic gel transforms to Al-Si spinel via an intermediate aluminosilicate (A) phase during continued heating prior to the exhibition of mullite at the ~1300°C exotherm.

Fourth approach: Characterization of the spinel phase of heat-treated "in situ" mullite gels by the alkali leaching method

To ascertain the composition of the spinel phase of heat-treated "in situ" mullite gels, Chakraborty (1979) synthesized both monophasic and diphasic Al_2O_3-SiO_2 gels of different compositions, heat-treated and finally alkali-leached. In this approach, free SiO_2(A) liberated on the transformation of gels of various compositions on being heated at ~1000°C was leached out by a standard alkali extraction technique. Different amounts of SiO_2(A) were detected during the alkali extraction process of these heated gels. The silica extraction curves generally show two distinct parts. At the first stage, that is, after 30–40 min. of leaching, the SiO_2 extraction is comparatively rapid, whereas in the next stage, that is, after ≥40 min. of leaching, SiO_2 dissolution is very slow and reaches a more or less constant value although the leaching time increases. However, the percentage of Al_2O_3 in the extract increases very slowly with increased leaching time. Comparing the removal of SiO_2 and Al_2O_3 by the alkali extraction process of heated gels A, B, and C, the major portion of SiO_2 solubilized selectively at the first stage of extraction can be designated as free amorphous SiO_2. Another portion, which seems to be unextractable during continued leaching of 980°C-heated gels, can be termed as chemically bonded SiO_2. A zeolite-type phase appears when leaching of the heated sample is continued for up to 90 min. with 10% NaOH. The alkali leaching process shows that whatever be the initial Al_2O_3-SiO_2 ratio, it always forms a spinel phase of the same composition as 3:2 mullite. The unleached/unextractable portion of chemically bonded silica is present in the Al-Si spinel phase whose composition is equivalent to 3:2 mullite. On the other hand, Schneider et al. (1994) noted in their EDS studies that the maximum SiO_2 incorporation into γ-Al_2O_3 with the sample heat-treated at 1150°C was 18 mol % (11.5 wt %). Jin et al. (2002) also carried out alkali leaching treatment of HCP powder at 1050°C and silica content of this post leaching samples were measured by X-ray fluorescence semi-quantitative analysis (XRF). It shows values contain 9–11 wt % SiO_2 which is due to nonscheduled extraction process practiced by them. Supplementary studies of the leached residue by TEM/EDS, IR, and MAS-NMR and their drawbacks are pointed out.

Prior to analysis of the chemical composition of the spinel phase by analytical TEM with EDX-type analyzer, Okada and Otsuka (1986b) leached a specimen fired at 1000°C with 7 wt % of NaOH solution by boiling it for 40 min. and obtained a value of ~8 wt % of SiO_2. The author pointed out that leaching in a boiling condition is against the standardized procedure and indicated that the increasing of silica extraction temperature from boiling water bath condition to direct boiling temperature must attack aluminosilicate. The reason for the low content of silica (~8 wt %) in the Al-Si spinel phase as noted by them was due to the practice of an incomplete extraction process.

Experimental procedures regarding the selection of the heat treatment temperature of the Type II gel and alkali leaching methodology chosen by Schneider et al. (1994) was low prior to supplementary studies on the characterization of the leached residue by both IR and MAS-NMR.

Schneider et al. (1994) also pointed out that the choice of the heat treatment temperature for the diphasic gel, at ~1150°C, was a reason for the crystallization of Al-Si spinel to its the fullest extent. The maximum crystallization generally occurs just before the occurrence of exotherm at ~1300°C due to mullitization. Therefore, the transformation temperature of Si-incorporated spinel to mullite must be shown by Schneider et al. (1994) in their diphasic gel sample.

Thereafter, EDX analysis of the SiO_2 content before that temperature has to be done. Under any circumstances, 1150°C should not be chosen as the final temperature. Accordingly, a value of 18 mol % should be taken as the final amount Si incorporation is a low value in contrast to 40 mol %, as shown by Chakraborty (1979).

In conclusion, the composition of Al-Si spinel equivalent to stoichiometric composition of 3:2 mullite was shown by Chakraborty (1979) by a standardized procedure of alkali leaching, and later Suzuki et al. (1990) in their TEM/EDS analysis confirmed it. And it is in contrast to the value of ~8 wt % obtained by Okada and Otsuka (1986b) in their TEM attached with EDX analyzer and also with the value (11.5 wt %) obtained by EDX analysis of unleached and leached samples by Schneider et al. (1994) in their IR and MAS-NMR studies.

Fifth approach: QXRD study of the spinel phase

XRD data of the Al-Si spinel phase regarding its appearance and disappearance are also limited (Okada and Otsuka, 1986b; Wang and Thomson, 1995; Chakraborty and Das, 2003). Both SH and RH gels of different compositions of Al_2O_3-SiO_2 were synthesized and heat-treated at different temperatures by Okada and Otsuka (1986b) to study the growth of the spinel phase. It was shown from the growth pattern that the rate of formation of spinel in SH gels is faster than in RH gels. From the plot showing the intensity of the spinel phase versus its composition, the maximum intensity was found near the composition $SiO_2{\cdot}6Al_2O_3$ in both SH and RH gels, which corresponds to ~8 wt % of SiO_2. The extent of spinel formation was also studied by Wang and Thomson (1995). They showed that the maximum amount of spinel was observed in the gel corresponding to $6Al_2O_3{\cdot}SiO_2$, as suggested by Okada and Otsuka (1986b). On the other hand, using TEM and EDS analysis, Suzuki et al. (1990) showed that the composition of the Al-Si spinel phase is similar to the stoichiometric composition of mullite. Contrarily, Figs. 15.2 and 15.5 of Chakraborty (1997) show how the intensity of the Al-Si spinel phase increases with heating temperature and at ~1300°C, its intensity suddenly changes to mullite. Figure 15.9 (Chakraborty, 2008) also depicts how the intensity of Al-Si spinel changes with heat treatment. Besides estimating the quantity of Al-Si spinel formed in each case of diphasic gels of different compositions,

Chakraborty (2000, unpublished) indicated the quantity of residual aluminosilicate phase (A), which is dependent on the initial composition of the gel (Table 15.14).

Besides estimating the composition of the noncrystalline aluminosilicate formed along with the Al-Si spinel phase, semiquantitative estimation of the latter phase by the usual X-ray internal standard method using CaF_2 and γ-Al_2O_3 as standard was done by Chakraborty (1993). The value of the noncrystalline aluminosilicate phase was noted to be ~20 wt %, and the remaining observed value for the Al-Si spinel content corresponded to 79%–81% in mullite gel G-173 of composition Al_2O_3 69%. The higher estimated data observed by QXRD analysis of the Al-Si spinel phase than the initial alumina content of the gel indicates that the spinel phase formed during the transformation of boehmite in the presence of silica is not γ-Al_2O_3. To corroborate this result, Chakraborty (2000, unpublished) re-estimated the contents of the spinel formed in mullite gels of varying compositions by XRD technique using γ-Al_2O_3 as standard and considering the (440) reflection of the spinel phase. In these cases, the estimated values of the Al-Si spinel phase are found to be higher than the percentages of alumina contents of all mullite gels (Table 15.14). This result definitely suggests the incorporation of Si in the alumina structure and thus confirms the previous finding of the development of the Al-Si spinel phase. A noncrystalline aluminosilicate phase also develops in varying quantities, depending on the initial composition of the gel. Chakraborty and Das (2003) recalculated and plotted the X-ray intensity of the Al-Si spinel phase versus γ-Al_2O_3. The calculated curve should not be found to superimpose; instead it lies above a theoretical curve (Fig. 15.19). The growth pattern of the Al-Si spinel phase is shown.

Sixth approach: Lattice parameter measurement study of the spinel phase

LC of Al-Si spinel phases formed on heating various Al_2O_3-SiO_2 gels by Yamada and Kimura (1962) showed an increase with a decrease in the SiO_2:Al_2O_3 ratio of the heated gel of specimen A, suggesting the substitution of Si^{+4} by Al^{+3} in the spinel lattice. The LC values of the spinel phase formed from SH and RH gels at increasing temperatures were measured by Okada and Otsuka (1986b). The unit cell size of Al-Si spinel was measured by Low and McPherson (1989), who showed that its value (a = 0.7880 nm) is slightly less than that of γ-Al_2O_3 (a = 0.7906 nm), which confirms the solid solution theory of silica in γ-Al_2O_3 structure. In comparison, the value of the spinel phase formed on heating the gel at 1100°C for 1 h was given by Yamada and Kimura (1962) and later by Hirata et al. (1985b) as 0.7894 ± 0.0003 nm, as against the value of 0.7886 nm given by Brindley and Nakahira (1959). Chakraborty (2006) measured the LC values of the Al-Si spinel phase derived from heating mixed gels of various compositions. The reduction in the LC value (0.78865 nm) of the gel-derived spinel phase compared to that for γ-Al_2O_3 (a = 0.7906 nm) indicates a possible substitution mechanism. Further, the lattice parameter value of the Al-Si spinel (a = 0.7900 nm) phase was found to be more or less constant, whatever the initial composition of the gel synthesized.

Seventh approach: Interrelationship of the Al-Si spinel phase and mullite formation

The interdependence of phase changes of Al-Si spinel and mullite was shown conclusively by the following observations: First, Al-Si spinel formation is predominant at ~1050°C but mullite formation is the lowest at the same temperature, as noted by Yamada and Kimura (1962). Second, the rate of decay of the Al-Si spinel phase becomes faster at ~1180°C, where the rate of mullite formation starts showing a predominance. It was suggested that mullite is formed by the thermal decomposition of spinel-shaped crystals. The occurrence of pronounced mullitization with the disappearance of the spinel phase at ~1300°C was noted by Hirata et al. (1985b). A similar behavior was noted by Sugita et al. (1998), who showed of a decrease in the spinel phase with mullitization.

Eighth approach: Effect of silica source on the development of the Al-Si spinel phase

Heterogeneous alumina-silica mixing of gel components during hydrolysis; segregation occurring during processing, especially at pH in the basic range; and continuance of alumina-silica segregated particles during drying and in the subsequent heating process are the controlling factors in spinel phase development. Both sources of aqueous silica sol and organic silica are highlighted in the spinel formation and their role in final mullitization are discussed.

Problems

1. The formation of two phases, namely $SiO_2{\cdot}6Al_2O_3$ and ($6SiO_2{\cdot}Al_2O_3$, is assumed by Okada and Otsuka (1986b, 1987). The following questions may be raised:
   - Why did corundum and cristobalite not crystallize on heating the diphasic gel?
   - Why did a combined mixture of θ-Al_2O_3 and α-Al_2O_3 not crystallize?

     The concept of mullite formation by a solid-state reaction of spinel ($SiO_2{\cdot}6Al_2O_3$) and silica-rich ($6SiO_2{\cdot}Al_2O_3$) phases could not explain why mullite does not crystallize at a temperature less than 1200°C.
   - Why is mullite formation so sudden and exhibits a broad exotherm at ~1250° C in DTA studies?
2. Okada and Otsuka (1986b) showed that for two types of gels, the temperature of the start of spinel crystallization was found to reduce as the composition becomes richer in alumina. Why? Identify the nature and then differentiate the spinel phases that crystallize on heating SH and RH gels below 900°C and at and above 900°C.

   What should be the composition of the residual amorphous Al_2O_3-SiO_2 phase left after crystallization of the spinel phase?
3. Why was the extent of spinel formation in diphasic gels 12:1 and 14:1 very high compared to other gels synthesized by Wang and Thomson (1995)?

References

1. K. Okada and N. Otsuka, Characterization of the spinel phase from SiO_2-Al_2O_3 xerogels and the formation process of mullite. *J. Am. Ceram. Soc.*, **69**(9), 652–656 (1986b).
2. K. Okada and N. Otsuka, Change in chemical composition of mullite formed from $2SiO_2\,3Al_2O_3$ xerogel during the formation process. *J. Am. Ceram. Soc.*, **70**, C-245–C-247 (1987).
3. H. Schneider, B. Saruhan, D. Voll, L. Merwin, and A. Sebald, Mullite precursor phases. *J. Euro. Ceram. Soc.*, **11**, 87–94 (1993).

4. H. Schneider, I. Merwin, and A. Sebald, Mullite formation from non-crystalline precursors. *J. Mater. Sci.*, **29**, 805–812 (1992).

5. Y. X. Huang, A. M. R. Senos, J. Rocha, and J. L. Baptista, Gel formation in mullite precursors obtained via tetraethylorthosilicate (TEOS) prehydrolysis. *J. Mater. Sci.*, **32**, 105–110 (1997).

6. J. A. Pask, X. W. Zhang, A. P. Tomsia, and B. E. Yoldas, Effect of sol-gel mixing on mullite microstructure and phase equilibria in the α-Al_2O_3-SiO_2 system. *J. Am. Ceram. Soc.*, **70**(10), 704–707 (1987).

7. B. E. Yoldas, Effect of ultrastructure on crystallization of mullite. *J. Mater. Sci.*, **27**(24), 6667–6672 (1992).

8. J. Sanz, I. Sobrados, A. L. Cavalieri, P. Pena, S. de. Aza, and J. S. Moya, Structural changes induced on mullite precursors by thermal treatment: a ^{27}Al MAS-NMR investigation. *J. Am. Ceram. Soc.*, **74**(10), 2398–2403 (1991).

9. C. Gerardin, S. Sundaresan, J. Benziger, and A. Navrotsky, Structural investigation and energetics of mullite formation from sol-gel precursors. *Chem. Mater.*, **6**, 160–170 (1994).

10. D. J. Cassidy, J. L. Woolerey, J. R. Bartlett, and B. Ben-Nissan, The effect of precursor chemistry on the crystallization and densification of sol-gel derived mullite gels and powders. *J. Sol-Gel Sci. Technol.*, **10**, 19–30 (1997).

11. A. M. L. M. Fonseca, J. M. F. Ferreira, I. M. M. Salvado, and J. L. Baptista, Mullite based compositions prepared by sol-gel techniques. *J. Sol-Gel Sci. Technol.*, **8**, 403–407 (1997).

12. M. I. Nieto, G. Urretavizcaya, A. L. Cavalieri, and P. Rana, Structural changes in colloidal and polymeric aluminosilicate gels with mullite composition. *Br. Ceram. Trans.*, **97**(1), 17–23 (1998).

13. A. K. Chakraborty, An analysis of the phase evolution of six types of mullite gels. (2008) unpublished.

14. H. Yamada and S. Kimura, Studies on co-precipitates of alumina and silica gels and its transformations at higher temperatures. *Yogo Kyokai Shi*, **70**, 87–93 (1962).

15. Y. Hirata, H. Minamizono, and K. Shimada, Property of SiO_2-Al_2O_3 powders prepared from metal alkoxide. *Yogo Kyokai Shi*, **93**(1), 36–54 (1985a).

16. S. Rajendran, H. J. Rossell, and J. V. Sanders, Crystallization of a coprecipitated mullite precursor during heat treatment. *J. Mater. Sci.*, **25**, 4462–4471 (1990).

17. A. K. Chakraborty, New data on thermal analysis of diphasic mullite gels. *J. Therm. Anal.*, **46**, 1413–1419 (1996b).

18. S. Prochazka and F. J. Klug, Infrared-transparent mullite. *J. Am. Ceram. Soc.*, **66**(12), 874–880 (1983).

19. H. Suzuki, H. Saito, Y. Tomokiyo, and Y. Suyama, Processing of ultrafine mullite through alkoxide route, in *Ceramic Transactions*, Vol. 6, Mullite and Mullite Matrix Composities. eds. S. Somiya, R. F. Davis, and J. A. Pask, American Ceramic Society, Westervikke, OH, p. 263 (1990).

20. M. Low and R. McPherson, The origins of mullite formation. *J. Mater. Sci.*, **24**, 926 (1989).

21. K. Hamano, Z. Nakagawa, G. Cun-Ji, and T. Sato, in *Mullite*, ed. S. Somiya, Uchida Rokakuho Publishing Co., Tokyo, Japan, p. 37 (1985).

22. K. Hamano, T. Sato, and Z. Nakagawa, Properties of mullite prepared by coprecipitation and microstructure of fired bodies, *Yogo Kyokai Shi*, **94**(8), 818–822 (1986).

23. S. Mitachi, M. Matsuzawa, K. Kaneko, S. Kanzaki, and Y. Tabata, Characterization of SiO_2-Al_2O_3 powdes prepared from metal alkoxides. *Ceram. Trans.*, **6**, 275–286 (1990).

24. H. Shiga, M. G. M. U. Ismail, and K. Katayama, Sintering of ZrO_2 toughned mullite ceramics and its microstructure. *J. Ceram. Soc. Jpn.*, **99**, 798–802 (1991).

25. Y. Hirata, K. Sakeda, Y. Matsushita, K. Shimada, and Y. Ishihara, Characterization and sintering behavior of alkoxide-derived aluminosilicate powders. *J. Am. Ceram. Soc.*, **72**(6), 995–1002 (1989).

26. Y. Hirata, K. Sakeda, Y. Matsushita, and K. Shimada, Preparation of fine SiO_2-Al_2O_3 powders by hydrolysis of mixed alkoxides. *Yogo Kyokai Shi*, **93**(9), 577 (1985b).

27. A. K. Chakraborty, Further studies on thermal transformation of diphasic Al_2O_3-SiO_2 gel. *Trans. Indian Ceram. Soc.*, **56**(1), 9–15 (1997).

28. Ph. Colomban, Structure of oxide gels and glasses by infrared and Raman scattering part 2 mullite. *J. Mater. Sci.*, **24**, 3011–3020 (1989).

29. D. X. Li and W. J. Thomson, Effects of hydrolysis on the kinetics of high temperatture transformations in aluminosilicate gels. *J. Am. Ceram. Soc.*, **74**, 574–578 (1991).

30. H. Ivankovic, E. Tkalcec, R. Nass, and H. Schmidt, Correlation of the precursor type with densification behavior and microstructure of sinteren mullite ceramics. *J. Euro. Ceram. Soc.*, **23**, 283–292 (2003).

31. A. K. Chakraborty, Formation of silicon aluminium spinel. *J. Am. Ceram. Soc.*, **62**(3–4), 120–124 (1979).

32. A. K. Chakraborty, Intermediate Si–Al spinel phase formation in phase transformation of diphasic mullite gel. *J. Mater. Sci.*, **28**, 3839–3844 (1993).

33. A. K. Chakraborty, Al-Si spinel phase formation in Al_2O_3-SiO_2 precursor-mullitization processes, paper presented at mullite workshop 2000 was held at Oban Town Hall, Oban, England (2000) unpublished.

34. A. K. Chakraborty, Si-incorporated alumina phases formed out of diphasic mullite gels. *J. Mater. Sci.*, **43**, 5313–5324 (2008a).

35. M. J. Hyatt and N. P. Bansal, Phase transformations in xerogels of mullite composition. *J. Mater. Sci.*, **25**, 2815–2821 (1990).

36. G. Klaussen, Microstructural evolution of so-gel mulite. *Ceram. Eng. Sci. Proc.*, **11**, 1087–1093 (1992).

37. W.-C. Wei and J. W. Halloran, Phase transformation of diphasic aluminosilicate gels. *J. Am. Ceram. Soc.*, **71**(3), 166–172 (1988).

38. M. G. M. U. Ismail, Z. Nakai, K. Minegishi, and S. Somiya, Synthesis of mullite powder and its characteristics. *Int. J. High Technol. Ceram.*, **2**, 123–134 (1986).

39. B. Sonuparlak, Sol-gel processing of infrared transperant mullite. *Adv. Ceram. Mater.*, **3**(3), 263–267 (1988).

40. A. K. Chakraborty, Preliminary study on the effect of pH on thermal transformation of some diphasic Al_2O_3-SiO_2 gels. *J. Sol-Gel Sci. Technol.*, **28**, 87–95 (2003a).

41. A. K. Chakraborty, Reinvestigation of Al-Si spinel phase in diphasic Al_2O_3-SiO_2 gel. *J. Am. Ceram. Soc.*, **88**(1), 134–140 (2005a).

42. S. J. Wilson, *Miner. Mag.*, **43**, 301 (1979).

43. D. W. Hoffman, R. Roy, and S. Komarneni, Diphasic xerogels, a new class of materials: phases in the system Al_2O_3-SiO_2. *J. Am. Ceram. Soc.*, **67**, 468–471 (1984).

44. B. E. Yoldas, Microstructure of monolithic materials formed by heat treatment of chemically polymerized precursors in the Al_2O_3-SiO_2 binary. *Ceram. Bull.*, **59**(4), 479–483 (1980).

45. A. K. Chakrabirty and S. Das, Al-Si spinel phase formation in diphasic mullite gels. *Ceram. Int.*, **29**, 27–33 (2003b).

46. A. K. Chakraborty, Characterization of monophasic and diphasic mullite precursors by solid state reaction study. *Br. Ceram. Trans.*, **103**, 33–36 (2004).

47. H. Schneider, D. Voll, B. Saruhan, M. Schmucker, T. Schaller, and A. Sebald, Constitution of the γ-alumina phase in chemically produced mullite precursors. *J. Euro. Ceram. Soc.*, **13**, 431–448 (1994).

48. X. H. Jin, L. Gao, and J. K. Guo, The structural change of diphasic mullite gel studied by XRD and IR spectrum analysis. *J. Euro. Ceram. Soc.*, **22**, 1307–1311 (2002).

49. Y. Wang and W. J. Thomson, Mullite formation from nonstoichiometric slow hydrolyzed single phase gels. *J. Mater. Res.*, **10**(4), 912–917 (1995).

50. A. K. Chakraborty and D. K. Ghosh, Crystallization behavior of Al_2O_3 in the presence of SiO_2. *J. Am. Ceram. Soc.*, **79**(3), C-46–C-48 (1987).

51. B. E. Yoldas, Thermal stabilization of an active alumina and effect of dopants on the surface area. *J. Mater. Sci.*, **11**, 465–470 (1976).

52. A. K. Chakraborty, Incorporation of silica in cubic γ-Al_2O_3 lattice, paper presented at 2nd Int. Natl. Workshop Mullite '94 was held in Irsee, Germany (1994c) unpublished.

53. J. M. McHale, K. Yurekli, D. M. Dabbs, A. Navrotsky, S. Sundaresan, and I. A. Aksay, Metastability of spinel-type solid solutions in the SiO_2-Al_2O_3 system. *Chem. Mater.*, **9**, 3096–3100 (1997).

54. G. W. Brindley and M. Nakahira, The kaolinite-mullite reaction series: III, The high-temperature phases. *J. Am. Ceram. Soc.*, **42**, 319–324 (1959).

55. S. S. Sueyoshi and C. A. Contreras Soto, Fine pure mullite powder by homogeneous precipitation. *J. Euro. Ceram. Soc.*, **18**, 1145–1152 (1998).

56. A. K. Chakraborty, Range of solid solutions of silica in spinel type phase. *Adv. Appl. Ceram.*, **105**(6), 297–303 (2006).

57. J. C. Huling and G. L. Messing, Epitactic nucleation of spinel in aluminosilicate gels and its effect on mullite crystallization. *J. Am. Ceram. Soc.*, **74**(10), 2374–2381 (1991).

58. K. Wang and M. D. Sacks, Mullite formation by endothermic reaction of α-Al_2O_3/SiO_2 microcomposite particles. *J. Am. Ceram. Soc.*, **79**(1), 12–16 (1996).

59. H. Insley and R. H. Ewell, Thermal behavior of the kaolin minerals. *J. Res. Natl. Bur. Stand.*, **14**(5), 615–627 (1935).

60. T. Demediuk and W. Cole, Exothermic reaction of metakaolin between 950° and 1,000°C. *Nature*, **181**, 1400–1401 (1958).

61. C. H. Horte and J. Wiegmann, Reaction between amorphous SiO_2 and Al_2O_3. *Naturwiss*, **43**, 9 (1956).

62. J. Ossaka, Tetragonal mullite type phase from co precipitated gels. *Nature (London)*, **19**, 1000–1001 (1961).

63. D. Croft and W. W. Marshall, A novel synthesis of alumino-silicates and similar materials. *Trans. Br. Ceram. Soc.*, **64**(3), 121–126 (1967).

64. Stone and Tilly.

65. A. K. Chakraborty, and D. K. Ghosh, Synthesis and 980°C phase development of some mullite gels. *J. Am. Ceram. Soc.*, **71**(11), 978–987 (1988).

66. K. Nishu, T. Yokoyama, T. Watanabe, and T. Tarutani, Characterization of amorphous aluminosilicate formed by adsorption of silicic acid on aluminium hydroxide, in *Abstracts of the 27th Symposium on the Basic Science of Ceramics*. Paper No. IB08. Ceramic Society of Japan, Tokyo (1989).

67. B. E. Yoldas and D. P. Partlow, Formation of mullite and other alumina-based ceramics via hydrolytic polycondensation of alkoxides and resultant ultra- and micro-structural effects. *J. Mater. Sci.*, **23**, 1895–1900 (1988).

68. A. Taylor and D. Holland, The chemical synthesis and characterization sequence of mullite. *J. Non-Cryst. Solids*, **152**, 1–17 (1993).

69. I. Jaymes, A. Douy, D. Massiot, and J. P. Coutures, Characterization of mono- and diphasic mulite precursor powders prepared by aqueous routes, ^{27}Al and ^{29}Si MAS-NMR spectroscopy. *J. Mater. Sci.*, **31**, 4581–4589 (1996).

70. K. Kanzaki, M. Ohashi, H. Tabata, T. Kurihara, S. I. Iwai, and S. I. Wakabayashi, in *Ceramic Transactions*, Vol. 6, Mullite and Mullite Matrix Composities. eds. S. Somiya, R. F. Davis, and J. A. Pask, American Ceramic Society, Westervikke, OH, p. 389 (1990).

71. T. D. McGee and C. D. Wirkus, Mullitization of aluminum-silicate gels. *Am. Ceram. Soc. Bull.*, **51**, 577–581 (1972).

Chapter 16

Mullite Phase: A Concise Review

Some of the methods and techniques (QXRD, and lattice constant studies and critical analysis of mullite phase) as used and reported by earlier authors for proper mullite identification and characterization and the resulting problems they faced are discussed in this chapter.

16.1 Introduction on Mullite Formation in Different Cases of Gels

The synthesis of following kinds of mullite precursors/gels namely (i) monophasic gels (prepared by two methods), (ii) coprecipitated gel, (iii) colloidal gel, (iv) spray pyrolysis/spray dried precursors, and (v) diphasic gel are concisely shown in Chapters 2–6 and 12. The first kind of monophasic gel forms mullite type phase at the first exotherm at ~980°C. The other kind of monophasic gel forms Al-Si spinel phase at the first exotherm and thereafter forms mullite type phase at the second exotherm at ~1250°C. Coprecipitated gel shows a varying ratio of spinel to mullite formation. Colloidal gel develops spinel as a major phase. Spray pyrolyzed powder crystallized to only mullite type phase at first exotherm. Diphasic gel forms a mullite type phase at ~1320°C exotherm via intermediate crystallization of Al-Si spinel phase occurred by a broad exotherm over 1000–1300°C in differential thermal analysis (DTA). The method of classification

Mullite Formations: Analysis and Applications
Akshoy Kumar Chakraborty

ISBN 978-981-4877-05-3 (Hardcover), 978-1-003-03167-3 (eBook)
www.jennystanford.com

of these gel/precursors is based on phase developments determined from associated DTA exotherms. Mullitization processes in all five to six cases are quite different. Probable variation in mullite formation processes of these six kinds of gels/powders studied by various authors are presented in this chapter.

16.1.1 Monophasic Gels

(i) **Synthesized by aqueous sol-gel method:** Monophasic gels prepared out of water soluble silica components in presence of Al salt were studied by (Insley and Ewell, 1935; Chakraborty and Ghosh, 1988). DTA exothermic phenomena of these gels showed highest rapidity with complete mullite formation. Such pronounced DTA effect and pre dominant mullitization of course did not happen in any other method of synthesis relative to any choice of two components and any processing technique. The mullite formed at first exotherm is to be called a mullite type phase. This view is verified with ^{27}Al MAS NMR studies of CM material synthesized during aqueous route by Schneider et al. (1992). It showed mullitization at 1100°C. However, they also showed that the measured ratio of AlO_4 to AlO_6 did not correspond to standard mullite. The ratio of AlO_4 to AlO_6 started changing during the heating process where. AlO_4 peaks tended to increase. At 1100°C–1650°C range, the ratio of AlO_4 to AlO_6 changed remarkably (Fig. 9.1). For example, AlO_4 increased at the cost of decrease of AlO_6. Moreover, splitting of AlO_4 occurred which corresponds to formation of well crystallized o-mullite. Ossaka (1961) noted only mullite formation at 980°C exotherm following this method which was also weakly crystalline in nature may be termed as mullite type phase as designated in other cases.

(ii) **Synthesized by monophasic gel method**: Single phase mullite gel marked less ethanol made by Hoffman et al. (1984) reported to content mullite on heating to 1070°C. It was interpreted that $Al_2O_3 . SiO_2$ gel prepared out of TEOS and ANN in acidic medium produce only mullite at the 980°C exotherm. Besides, the presence of later phase, Al-Si spinel was also identified in their given XRD pattern by Chakraborty (1986). Other than spray dried/spray pyrolyzed precursor or aqueous sol-gel precursor, any other sol-gel precursors hardly formed mullite directly at first exotherm. SH gel made out of organic silica source (TEOS) and ANN also exhibited strong 980°C exotherm with pre dominant alumina rich mullite but with traces of Al-Si spinel phase (Okada and Otsuka, 1986). Molecular gel (100 M) synthesized with use of TEOS and ANN by a procedure modified from Okada and Otsuka (1986) reported the crystallization of mullite only by Huling and Messing (1991). The formation of spinel was observed on 90 M and 75 M hybrid gels when heat treated post –1000°C exotherms as revealed in their XRD patterns.

(iii) Synthesized by polymeric gel method: In this method Yoldas (1992) showed spontaneous or extensive crystallization of mullite as revealed by XRD at the temperature pinpointed with a strong exotherm. Concurrently, a radical change in the transformation of Al from five to that for mullite was observed by MAS NMR studies (Fig. 9.3). However, the role of development of a small quantity of associated spinel phase by the later study remains to explain. The residual spinel phase left after 980°C transformed to mullite with further exhibition a broad exotherm at ~1250°C (Pask et al., 1987). Type I gel synthesized in an almost water free approach out of both organic components as source e.g., TEOS/AlOBu developed alumina rich mullite in addition to minor quantity of poorly crystalline spinel phase during 980°C exothermic phenomenon and called the precursor as most homogeneous (Schneider et al., 1993). But the second exotherm was noted neither in Type I precursor nor in SH gel of Okada and Otsuka (1986) although some quantity of Al-Si spinel phase was formed and noted by XRD in both the two heated precursors.

A sharp 980°C exotherm consequently to weakly crystalline nonstoichiometric alumina rich mullite (major) and alumina rich Al-Si spinel phase (minor) were noted by Haque (2000) during the phase transformation of monophasic gels marked SHI and SHII. An additional minor quantity of Al-Si spinel phase duly transformed to mullite type phase during the occurrence of second exotherm at 1250°C. This result concurs with that of Okada and Otsuka (1986) as noted in case of transformation of SH gel. The sudden crystallization of mullite as shown by the former author in case of SHI gel well corresponds to some characteristic IR changes occurred during structural transformation of monophasic gel at ~1000°C by Hirata et al. (1989). They showed that the IR absorption band too shifted abruptly to higher values from 1020 to 1134 cm^{-1} due to mullite formation with a minor amount of Al-Si spinel.

16.1.2 Coprecipitated (CP) Gel

(i) **In other type of monophasic gel method**: RH gel made out of same organic silica source (TEOS) and ANN exhibited moderate 980°C exotherm with predominant Al-Si spinel phase and with traces of weakly crystalline mullite phase. It thereafter completed a mullite transformation at the second exotherm (Okada and Otsuka, 1986). The mullite gel marked G-133 processed using ammonia as coagulant, showed both Al-Si spinel and mullite formation at ~1000°C (Chakraborty and Ghosh, 1988). Moreover, coprecipitated gel (CP) prepared at pH = 6–7 also mullitized rapidly at ~1200°C (Dafadar et al., 1998).

(ii) **Monophasic gel processed with ammonia:** It was made out of both organic components as source and generally processed with ammonia solution e.g., Type II gel of Schneider et al. (1993). Other researchers namely Prochazka and Klug (1983), Suzuki et al. (1990), Low and McPherson (1989), Hirata et al. (1985a), Hamano et al. (1985, 1986) and Mitachi et al. (1990) noted spinel formation first in their coprecipitated gels synthesized at pH = 7 and then mullitization at >1200°C. Hirata et al. (1985), Rajendran et al. (1990) also observed intermediate Al-Si spinel formation prior to mullitization in coprecipitated gels. Thus, mullite formation in this case of coprecipitated (CP) gels do not occur directly and rapidly during the first exothermic peak. The later gels crystallize first into an intermediate Al-Si spinel phase which transforms at second exotherm peak to mullite type phase. Partial formation of spinel phase which was noted by Hirata et al. (1985a), Hirata et al. (1989) at the first stage transformed to mullite on later stage of heating and added to earlier quantities of mullite formed.

(iii) **Slow monolith synthesis:** This optically clear monolith was obtained out of Al iso.Bu and silicon alkoxide and hydrolysed with propanol at atmospheric moisture for months by Colomban (1989). It exhibited 980°C exothermic phenomenon first with formation of Al-Si spinel type phase. Thereafter, this later phase transformed to a mullite phase during the occurrence of 1250°C exotherm. In coprecipitated gels made by Yamada and Kimura (1962), Al-Si spinel phase was first crystallized at 980°C exotherm and mullite formation occurred at second exotherm by the transition of later phase.

(iv) **Coprecipitated gel processed with ammonia at pH ≤ 7:** Out of this CP gel, Chakraborty (1996, 1997) first noted spinel phase formation during heating noncrystalline aluminosilicate precursor phase and next showed the growth of mullite which is analogous to that occurred on thermal transformation of kaolinite and course of mullitization took place by two simultaneous reaction paths.

16.1.3 Colloidal Gel

Colloidal gel (CG) synthesized by Li and Thomson (1991) exhibited a 980°C exothermic peak but without formation of mullite phase. DXRD study indicated sharp crystallization of Al-Si spinel crystallites which coincides with the 980°C exotherm. Spinel phase formation was also noted in colloidal gel of Yoldas (1992); in PB2 gel of Sanz et al., 1991); in M2 gel of Geradin et al. (1994); in polymeric gel of Cassidy et al. (1997); and in PA gel of Nieto et al. (1998). Spinel phase formation was also noted during heating MP, MPP and MAPIF gels synthesized by Fonseca et al. (1997) and in precursors marked MI & MII by Huang et al. (1997). Such spinel phase is an intermediary product of noncrystalline aluminosilicate precursor phase and final mullite.

16.1.4 Diphasic Gels

Progress of mullitization in different cases of diphasic gels synthesized in varying techniques are elaborately shown in Chapter 4. Details of phase transformation paths of different diphasic gels will be shown below.

16.1.5 Spray Pyrolysis/Spray Dried Precursors

Direct formation of mullite from amorphous SP precursor was noted first by Kanzaki et al. (1985). Mullitization took place followed by a very sharp exothermic peak at 970°C in DTA analysis.

16.1.6 Chelated Monophasic Gel

Monophasic precursor A prepared by Jaymes et al. (1994) however showed mullite formation directly and rapidly during heating noncrystalline aluminosilicate phase at first exothermic peak in a single step.

16.1.7 Different Views of Mullite Formation

To explain the causes behind (i) early mullitization and (ii) final mullitization of the monophasic gel e.g., SH gel, polymeric gel, Type I gels and diphasic gels, following proposals were put forwarded by different researchers.

(1) In monophasic gel

(i) Okada, Hoshi, and Otsuka (1986a) first showed a most rapid mullitization at 900°C–1000°C (to be called Path I). Thereafter, noted a gap of mullite formation between 1000–1100°C temperature range and a continued slow mullitization on further heating from 1150°C to 1300°C (to be called Path II). Based on qualitative and QXRD studies, they explained two steps of mullite formation in monophasic SH gel (Fig. 16.1).

(ii) Schneider et al. (1993) showed 980°C exotherm in DTA and silica rich amorphous phase formation in ^{29}Si MAS NMR spectroscopy. Schneider et al. (1992) previously reported that both CM material as like Type I precursor showed splitting of silicon resonance peak to –90 ppm and –110 ppm, respectively, when these gels were heated to 1000°C (Fig. 9.2). According to them, the later peak was due to silica rich phase and the former peak was due to alumina rich phase This observation suggested that phase separation prior to crystallization during exotherm is a regular phenomenon in case of Al_2O_3-SiO_2

precursors. They conceived regular mullitization proceeded by inter diffusion among two later phases on further heating.

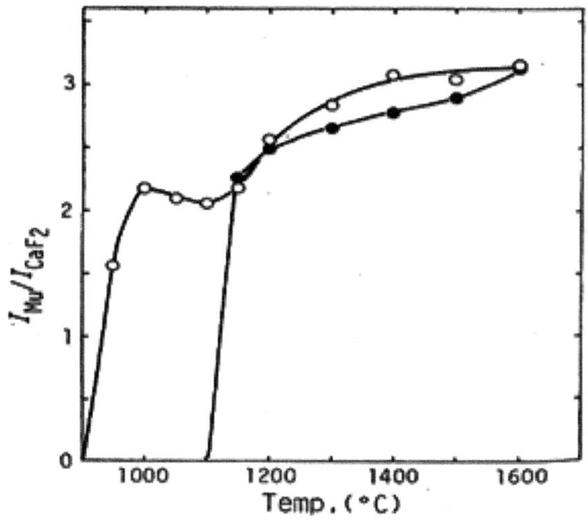

Figure 16.1 Formation of mullite from (o) SH and (•) RH xerogels as a function of firing temperature (Okada and Otsuka, 1986).

(2) In RH gel

(i) Regarding the formation of mullite, Okada and Otsuka (1986) first conjectured the solid state reaction between the intermediary Al-Si spinel phase formed at first exotherm and with SiO_2 (A) liberated during heating Al_2O_3-SiO_2 monophasic RH gel at second exotherm in DTA at ~1250°C by Path IV. This solid state reaction was noted to be predominant during static heating at ~1100°C (Fig. 16.1) and accordingly a rapid growth of mullitization was noted. Besides this step of the growth process, further mullite formation was continued on heating beyond 1100°C to 1600°C and mullitization in this region is to be called Path III.

(ii) In cases of Type II precursor and SGM gel synthesized by Schneider et al. (1993) and by (Schneider et al., 1992), two phases namely alumina rich spinel as major phase and noncrystalline silica at ~1000°C were observed. They thought that as the temperature increased beyond 1000°C, the amount of coexisting SiO_2 phase apparently decreased, while the amount of Si present in the Al_2O_3-rich compound did increase. These two segregated phases reacted at ≥1200°C and crystallized to mullite by Path IV. According to them, mullite nucleation and crystal growth occurred by short range migration of diffusing species. On long heat treatment, long range diffusion continued for crystal growth to occur and to produce well developed mullite.

(3) Diphasic gels

(i) When boehmite was used as a component, Hoffman et al. (1984) conjectured solid state reaction between γ-alumina and silica (A) is responsible for mullitization. Partial crystallization of component oxides e.g., cristobalite was noted by Hyatt et al. (1990) and corundum was seen by Hamano et al. (1985) prior to solid state reaction leading to mullitization in their diphasic gels.

(ii) During phase transformation study of "in situ" diphasic gel the present author (1996) speculates that (i) Si may be incorporated in alumina rich spinel phase and accordingly the stability of this spinel is increased from 1000°C to as high as at ~1300°C. Thereafter, structural transformation of this incorporated spinel is the sole cause of its sudden mullitization at the exotherm at ~1300°C in DTA.

(iii) Contrary, mullite formation might happen by dissolution of γ-/δ-alumina in vitreous matrix in gels (Wei and Halloran, 1988). These authors predicted that direct reaction between γ-alumina and SiO_2 (A) causes mullite formation by their EDS and TEM studies.

(iv) Growth of mullite in diphasic gel of three different compositions were compared with that for mixed γ-Al_2O_3 and silica powder marked (AS) and secondly with a mixture of Al_2O_3 (A) and SiO_2 (A) powder marked (MO) by Chakraborty (2005a). Estimated mullite formation profiles of three compositions of diphasic gels were compared with MO and AS samples and are shown in plots (Fig. 16.2).

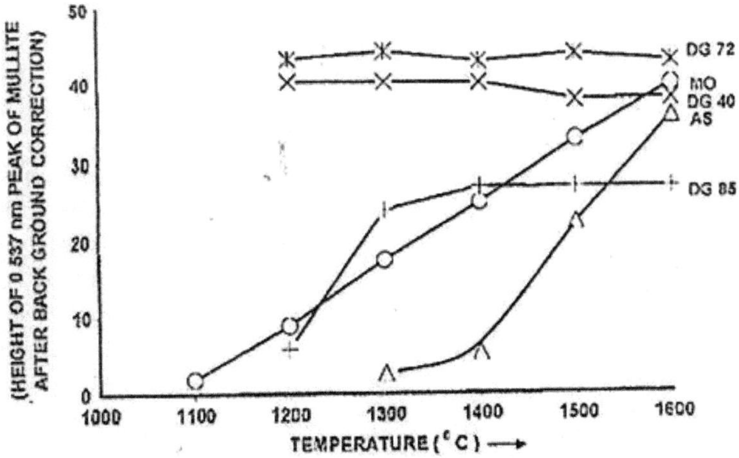

Figure 16.2 XRD intensity of mullite formed in mechanically mixed oxides AS, MO and three cases of diphasic gels with increasing temperature (Chakraborty, 2005a).

Growth curves of three diphasic gels show sudden mullitization during static heating at ~1200°C. In contrast to these, growth curves of these two ground mixtures (AS) and (MO) were parabolic in nature. These two curves were slow at ~1300°C, increased more with rise of temperature of heat treatment 1600°C.

Al-Si spinel formed out of diphasic gels of compositions silica rich (DG40) and stoichiometric gels (DG72) transformed rapidly to mullite and the amount formed remained unaltered even on continued heating. However, mullite formation in nonstoichiometric alumina rich diphasic gel (DG85) occurred by the disappearance of intermediate Al-Si spinel and Si—incorporated θ-phase at >1300°C. On the other hand, path of mullite formation in oxides mixtures proceeds in different ways due to differences in homogeneities of two components. The order is as follows: diphasic gel > MO > AS. Mullitization process in gels of stoichiometric, silica rich and alumina rich compositions of diphasic gels were shown schematically below (Fig. 16.5) by Chakraborty (2005a). Over all views of mullite formation in diphasic gels are summarized below.

First view:

A radically different behavior to that of single phase xerogels was noted on heating Al_2O_3–SiO_2 diphasic xerogel by Hoffman et al. (1984). For example, single-phase xerogel and kaolinite clay exhibited 980°C exotherm whereas diphasic gel did not do so. Secondly, monophasic gel formed mullite (as major) phase at first exotherm. On the other hand, the mullitization process occurred in diphasic gel in a different path than that occurred in single phase gel and kaolinite. It was conjectured by the later authors, and Komarneni and Roy (1986) that two discrete phases present originally in the later gel itself, e.g., boehmite and silica (A), crystallized independently up to 1000°C and developed spinel phase resembling γ-Al_2O_3 and silica (A) by XRD. Solid state reaction took place between those two phases of heating and developed mullite at higher temperature >1300°C in comparison to the rapid crystallization of mullite occurred in case of monophasic gel at ~980°C.

Second view:

According to Wei and Halloran (1988), Li and Thomson (1991), Sundaresan and Aksay (1991), mullite formation in diphasic aluminosilicate precursor occurred by the following solid state reaction at 1200°–1300°C instead of γ-Al_2O_3 and silica (A) as conceived by (Komarneni and Roy, 1986).

$$3(\delta\text{-}Al_2O_3) + 2\,SiO_2\,(A) \longrightarrow 3Al_2O_3{\cdot}2SiO_2 \qquad (16.1)$$

Wei and Halloran first believed that the two reactants i.e. δ-Al_2O_3 and SiO_2 (A) are in higher free energy state and more reactive but remain independently at a high temperature of ~1200°C without any substantial interaction. However, the said temperature (~1200°C) of existence of δ-Al_2O_3 is much above the temperature (~1000°C) as assumed by Hoffman et al. (1984). Secondly, these authors also believed in direct mullitization occurring out of the glassy phases generated by dissolving alumina phase in silica (A) and thirdly they emphasized that mullite formation occurs by a nucleation and growth mechanism in the silica rich matrix. The concept of intermediate glassy phase formation was also the view of Ismail et al. (1986, 1987). They noted a considerable quantity of mullite during the heating process. It was conjectured that alumina dissolved into a siliceous glassy phase and favors nucleation of mullite. According to Huling and Messing (1990) the diphasic gel marked gel A also transformed as per the sequence shown by Wei and Halloran. In this transformation process, crystallization of silica and alumina components to cristobalite and or corundum won't occur. What is the nature of the silica-rich matrix and how it is generated in the presence of alumina components, and the presence of any silica in δ-Al_2O_3 structure remains unanswered? Moreover, these authors however did not present any experimental evidence regarding the presence of free silica (A) associated with the δ-Al_2O_3 phase. By alkali leaching technique Chakraborty (1979) showed that free silica was not remained in free state in diphasic gel of stoichiometric composition of 3:2 mullite heated to ~1000°C. Is the processing parameter plays a role in developing siliceous glassy phase, at a high temperature ~1200°C prior to mullitization instead of independent crystallization of two components that is to be ascertained.

Third view:

According to Klaussen et al. (1992) the phase transformation behavior of the diphasic gel is noted to be different from researchers to researchers which show different ways of mullitization behavior. Hyatt et al shows that the mode of phase transformation of the above diphasic gel is completely different from those described by Wei and Halloran (1988), and Ismail et al. (1986, 1987). Unlike these authors, former authors noted the crystallization of both the two components into cristobalite and corundum prior to mullitization.

Fourth view:

On the contrary, Chakraborty and Ghosh (1993, 1979) made the following view. Secondly, the mullite formation process as per above Eq. 16.1 as predicted by those authors is different in context of transformation of in situ diphasic gel.

(i) During the heating process some diphasic gels showed no exothermic events at 980°C and 1250°C in DTA, respectively.

(ii) That mullitization behavior, particularly its growth pattern in the case of diphasic gel, was not analogous to that of oxide mixture of the two components because it is derived out of two discrete hydroxide particles present in the diphasic gel (Fig. 16.2). As in monophasic xerogel, diphasic gel also developed exclusively Al-Si spinel phase at intermediate stage (Chakraborty, 1997). Obviously, it was argued that the spinel phase forming in the diphasic system should be silicon substituted spinel and not simple γ-Al_2O_3. Accordingly, the possibility of the solid state interaction of γ-Al_2O_3 and SiO_2 (A) responsible for the formation of mullite in this diphasic gel may likely be questionable.

(iii) By solid state reaction studies as like as in monophasic gel with CaO, Chakraborty (2003a, 2004) experimentally reported the absence any part of free amorphous silica in dehydrated, and heat treated diphasic mullite gel from 500 to 1250°C (Fig. 15.9) and such silica was chemically bonded with alumina and present as a noncrystalline aluminosilicate phase. Al-Si spinel was slowly crystallized over a long range of temperature. Finally, this study also reaffirmed that intermediate Al-Si spinel is near to the composition of 3:2 mullite. It was designated as a cubic form of mullite (c-mullite). Likely, the formation and or nucleation process of mullite of different kinds of diphasic gels are different. In conclusion, there were distinctly four views of mullite formation reaction in diphasic gel.

(1) First one is solid state reaction between γ-, δ-, and α-alumina with SiO_2 (A) due to previous studies of (Hoffman et al., 1984; Schneider et al., 1993; Ivankovic et al., 2003).

(2) Alternatively by solid state reaction between crystalline components, e.g., corundum and cristobalite at high temperature.

(3) Third is due to views of (Wei and Halloran, 1988; Sundaresan and Aksay, 1991). During analysis of mullite formation kinetics, these authors believed the dissolution of γ- and or δ-Al_2O_3 phases in vitreous silicate phase is the cause of nucleation of mullite in diphasic gel.

(4) Fourth is due to polymorphic transformation of Al-Si spinel to mullite formation in diphasic gel at the third exotherm by the present author.

16.2 Mullite Formation Studies of Monophasic Gels

16.2.1 Quantitative XRD Study (QXRD) of Mullite Formation

(1) Monophasic gel:

The formation processes of SH and RH gel during heating at 800 to 1600°C with a soaking of 24 h was monitored by X-ray quantitative analysis (Okada, Hoshi, and Otsuka, 1986a). In SH gel, mullite formation occurred rapidly at 900°C to 1000°C. Mullitization ceased thereafter and showed a plateau between 1000°C up to 1100°C. Mullitization again increased slowly with rise of temperature (Fig. 16.1). They considered two stages for mullite formation of SH gel. First stage corresponds to the first step portion of the mullite formation curve between 900°C to 1000°C. According to them the mullite at this stage contained more than 60 mol % Al_2O_3. Second stage of mullite formation was started at 1100°C and increased slowly and gradually up to 1300°C. At this stage, the composition of mullite likely decreased to ~60 mol % Al_2O_3. A halt of mullitization is noticed at the temperature range of ~100°C between the first and second step of mullite formation in SH. Besides, Al-Si spinel phase (minor amount) was reported to crystallize along with mullite (major) during heating SH mullite gel at 1000°C by them. Mullite formation out of this minor phase is also to be considered.

SH mullite gels made by Li and Thomson (1991) out of organic silica source (TEOS) and ANN synthesized at different processing conditions exhibited strong 980°C and reported two steps of mullitization in samples (processed at different conditions, see Chapter 4) 1D, 2D, 2W and 2WC gels in both DTA and DXRD studies. Formation process of mullite for three kinds of gels (a) 2WC monophasic gel (Al/Si = 3/1), (e) for CG colloidal gel (Al/Si = 3/1), and (c) for DG3 diphasic gel (Al/Si = 3/1) at the exothermic temperatures are shown in DXRD studies. Out of these, precursors 2WC and 2W showed large intensity of mullite in their DXRD pattern. They tried to show the semi quantitative data on mullitization of different types of precursors. Gels of composition (Al/Si > 3/1) crystallized into mullite and corundum. Likely, other gels of composition (Al/Si < 3/1) produced mullite and cristobalite.

The composition of t-mullite was reported as a function of the Al/Si ratio by DXRD (Wang and Thomson, 1995). However, the composition of t-mullite and spinel are of much discussion by various authors. According to them a slowly hydrolyzed single phase gel with an alumina content to the composition of proposed t-mullite structure ($2Al_2O_3 \cdot SiO_2$) should form t-mullite of higher content. In diphasic gels having ratios 1/1, 6/1, 8/1 and 14/1, the extent of mullite formations were noted to swiftly increase nearly to the fullest extent. But the reason was unexplained.

(2) RH gel:

It developed mullite abruptly at 1100°C to 1150°C. and it also increased gradually with rise of temperature. The comparative mullite formation curves of SH and RH xerogels during heating processes as noted in (Fig. 16.1) predict mullitization in different behavior.

(3) Coprecipitated gel:

Mullite formation took place when it was heated above the temperature of the first exotherm where spinel crystallizes (Yamada and Kimura, 1962). The rate of its formation of mullite was slow at ~1000°C and it showed a steep increase when the temperature was increased to more than that of the second exotherm. In the temperature range between first and second the formation of mullite showed two dimensional grain growth of crystal nucleus. The formation of mullite in this temperature range as above is also extremely slow. They suggested that mullite is formed as a result of thermal decomposition of the spinel phase accompanied by autocatalysis in accordance with the equation $da/dt = ka^n(1\text{-}a)^m$. Major mullite formation starts at the second exotherm. At this temperature and above the rate of mullite is very fast and mullite shows a remarkably three dimensional grain growth of the crystal nucleus.

16.2.2 Mullite Formation of Six Precursors Based on Qualitative and Quantitative Analysis: A Case Study

Mullitization processes of so called three monophasic gels, one coprecipitated gel, one diphasic gel, and one spray dried (SD) precursor in all six precursors were monitored both by qualitative phase analysis and by using QXRD techniques (Chakraborty, 2008). The comparative results are shown earlier in Fig. 12.6 and in Table 16.1. Nature of the growth curves of mullite in six different cases of precursors, e.g., SHI to SD suggest three distinct cases as per qualitative phase analysis.

Table 16.1 XRD-quantitative percentages of mullite

Gel no.	Heat treatment temperature (°C)						
	1000°C	1100°C	1200°C	1300°C	1400°C	1500°C	1600°C
SHI	62.8 (approx.)	86.6	88.0	89.1	90.6	90.8	91.1
SHII	60.3 (approx.)	81.1	83.3	85.05	87.23	88.2	88.5
SHIII	0	58.4	79.3	83.2	83.34	84.07	84.2
CP6	0	0	72.7	74..2	75.6	81.30	82.4
CP9	0	0	70	78.1	82.5	83.3	84.5

(i) Major rapid mullite formation at ~1000°C is noticed in case of SHI by Path I. Subsequent slow rise in mullite formations occurred from 1200°C–1400°C by Path II. Subsequent more slow mullitization took place at >1400°C by Path III.

(ii) Direct Al-Si spinel formation at the first step at the exotherm ~1000°C, and polymorphic transformation of this Al-Si spinel to mullite formation at the second step at exotherm at ~1250°C are noticed in the case of RH gel, SHII and SHIV (CP6) by Path IV. The subsequent slow rise in mullite formation from 1300°C to on words 1400°C by Path III.

Growth process of SHI, SHIV shows two paths of mullite formations: The two paths of mullite formation consists of two portions. One is very rapid (Path I), second is rather slow and gradual (Path II). This later path as noted in case of SHI is more marked than that happens in mullite gel SHIV. Further, during the formation of mullite, an additional exotherm was exhibited by the later gel in DTA trace (Fig. 7.20).

(iii) Gradual development of Al-Si spinel phase and its subsequent polymorphic transformation to mullite at ~1300°C is observed in case of diphasic gel (CP9) by Path IV. Subsequent mullite formation beyond 1400°C by Path III.

(iv) SD precursor completes its mullitization at first exothermic temperature itself by Path VI.

16.2.2.1 First case of a monophasic gel: Formation of mullite phase at first exotherm

SHI gel: During heating it first dehydrates and then decomposes to a noncrystalline aluminosilicate precursor phase which remains amorphous up to ~900°C. Thereafter, it rapidly crystallizes to mullite type phase (62.8 wt % approx) at ~1000°C by Path I. In addition it develops some amount of noncrystalline silica rich aluminous phase (AS2) during phase separation stage. The later phase interacts with the mullite type phase on heating beyond 1100°C resulting in ~86.6 wt % o-mullite by Path II. This value increases slowly with rise of temperature to ~91.1% at 1400°C and that the formation of mullite on heating of SHI gels follow by two consecutive steps which conforms the earlier pioneering studies of Okada and Otsuka (1986). Beyond 1400°C a very slow rise in mullite is observed by Path III. Probably, the reaction is as follows:

SHI mullite gel

(i) Mullite gel (SHI) $\xrightarrow{\text{Temp. below last Endo.}}$ [Aluminosilicate precursor phase (noncrystalline)]

Aluminosilicate precursor phase $\xrightarrow[\text{At "D"}]{\text{Phase separation}}$ [Alumina rich silica phase]
Noncrystalline, anhydrous & unstable phase (AS1)
+
Silica rich aluminous phase (AS2)
(^{29}Si MAS-NMR shows –110 ppm broad peak)

At Path I (980°C Exo.),
[Alumina rich silica phase] $\longrightarrow$ (i) Alumina rich mullite (nonstoichiometric, weakly crystalline, major phase)
(ii) + Alumina rich mullite (noncrystalline residual major phase, AS3)
(iii) + Alumina rich Al-Si spinel (weakly crystalline minor phase)
+ Silica rich aluminous phase (AS2) (16.2)

At Path II (1100°C–1300°C) two following solid state reactions occur at parallel.
(ii) Alumina rich mullite phase (nonstoichiometric) + Silica rich aluminous phase (AS2) (partly)
$\longrightarrow$ Mullite type phase
At (1100°C–1300°C)
(iii) Alumina rich mullite (noncrystallineresidual phase, AS3) + Silica rich aluminous phase (AS2) (partly)
$\longrightarrow$ Mullite type phase (16.3)

At Path IV (1100°C–1300°C) by diffusion of Si
(iv) Alumina rich Al-Si spinel as minor phase + Silica rich aluminous phase (AS2) (partly)
$\longrightarrow$ [Al-Si spinel phase (3:2)] $\longrightarrow$ Mullite type phase (16.4)

At Path III (1300°C–1600°C)
Mullite type phase (collectively) $\longrightarrow$ o-Mullite (well crystalline) (16.5)

16.2.2.2 Second case of a monophasic gel: Formation of mullite phase by two paths

SHII gel: Table 16.1 and Fig. 12.6 shows the percentage of mullite formed on heating SHII gel at different temperatures. Besides the formation of mullite type phase (~60 wt %) in this case by Path I, Al-Si spinel phase (minor amount) forms at ~1000°C. It in turn transforms to additional mullite beyond the second exotherm by Path II and further slows mullitization from 1200°C to 1400°C by Path III and the quantity of mullite becomes ~88 wt %. In comparison, wt % mullite formed at each temperature in this case is always less than that of SHI gel.

16.2.2.3 Third case of a monophasic gel: Intense crystallization of Al-Si spinel phase

SHIII gel: Table 16.1 and Fig. 12.6 shows that a more intense crystallization of Al-Si spinel phase occurs during heating it at 1000°C along with traces of mullitization. At 1100°C, mullite formation starts. It rapidly transforms to mullite type phase (~79 wt %) at the second exotherm (Fig. 7.20) analogous to Path IV and then increases slowly to 84.2% at 1600°C by Path III.

These two precursors (SHII and SHIII) form both Al-Si spinel and mullite at ~1000°C. The ratio of the two are only different. SHII forms mullite (major) quantity and Al-Si spinel phase (minor) amount, whereas the reverse is true for SHIII precursors. Formation of mullite particularly in precursors SHII takes place rapidly at first exotherm by Path I. Another slow increment of mullite is noticed in SHII beyond 1200°C out of a small amount of spinel phase. Precursor III on the other hand crystallizes to mullite via spinel as a (major) phase formed at second exotherm by same Path IV. Besides this kind of mullite formation, DTA exhibits an additional exotherm to be designated as the third exotherm. Reason is given below.

16.2.2.4 Fourth case of coprecipitated gel: Formation of mullite occurs at second exotherms

SHIV/CP6 gel: Table 16.1 and Fig. 12.6 show the percentage of mullite formed from CP6 gel heated at different temperatures. In the temperature range 1000°C–1100°C, it shows the formation Al-Si spinel phase only and thus the percentage of mullite is found to be almost negligible. Mullitization occurs rapidly close to second exothermic peak temperature (Fig. 7.20). At ~1200°C, the percentage of mullite formed is ~72.7 wt % by Path IV and then increased gradually from 1200°C to 1400°C in the second stage of mullitization up to 82.0 wt % analogous to Path III. The small additional exotherm (called as third exotherm) was noted prior to the second exotherm in case of SHIII gel which is analogous to the similar exotherm as also appeared in DTA studies in kaolinite. In this step Al^{+3} cation diffusion restarts followed by counter diffusion of Si^{+4} starts in the opposite direction among silica rich aluminous phase (AS2) and noncrystalline alumina rich spinel phase as residue (AS5). It also accounts for mullitization at high temperature by Path Vas shown below.

RH mullite gel

(i) Rapid hydrolysis gel (RH gel) or Mullite gel (SHIV) or CP6	Temp. below last Endo. $\longrightarrow$	[Aluminosilicate precursor phase (noncrystalline)]
Aluminosilicate precursor phase	Phase separation $\longrightarrow$ At "D"	[Alumina rich silica phase] Noncrystalline, anhydrous & unstable phase (AS1) + Silica rich aluminous phase (AS2) (^{29}Si MAS-NMR shows −110 ppm broad larger peak than in SHI) see Fig. 9.20

Path I (at 1st Exo.) $\longrightarrow$ (i) Alumina rich Al-Si spinel phase (weakly crystalline as major phase)
Alumina rich silica phase (ii) Alumina rich mullite (weakly crystalline as minor phase)
(iii) Alumina rich spinel phase (noncrystalline second major residual phase, AS5) + Silica rich aluminous phase (AS2) (16.6)

Path IV (at 2nd Exo.)

(ii) [Alumina rich Al-Si spinel phase + Silica rich aluminous phase (AS2) (partly)
$\longrightarrow$ [Al-Si spinel phase] $\longrightarrow$ Mullite type phase
(Full Si^{+4} saturation occurs by diffusion and $Al_2O_3 : SiO_2$ ratio becomes = 3:2)

Path V (an additional exotherm before 2nd Exo.) see Fig. 7.20
(iii) Alumina rich spinel phase (AS5) + Silica rich aluminous phase (AS2) (partly)
$\longrightarrow$[Al-Si spinel phase (3:2)] $\longrightarrow$ Mullite type phase (16.7)

Further mullitization also occurs by solid state reaction as follows.
At Path II (1100°C–1300°C)
(iv) Alumina rich mullite + Silica rich aluminous phase (AS2) (partly)
$\longrightarrow$ Mullite type phase

At Path II
(v) Alumina rich spinel phase as residue (AS5) $\longrightarrow$ Mullite type phase
+ Silica rich aluminous phase (AS2) (partly) (16.8)

At Path III (1300°C–1600°C)
(vi) Mullite type phases (collectively) $\longrightarrow$ o-Mullite (well crystalline) (16.9)

16.2.2.5 Fifth case of diphasic gel: Formation of mullite in diphasic gels

CP9 gel: At static heating condition, mullitization of CP9 starts at ~1200°C. In this case, percentage of mullite formed is found to be less than that of CP6 at 1200°C indicating that mullite formation is delayed. However at 1600°C, the quantitative value of mullite attains close to CP6 (Table 16.1 and Fig. 12.6). In case of CP9, rapid mullitization takes place at exothermic peak temperature (~1300°C) by Path IV. Above this temperature, a slow increase in mullite formation occurs with an increase in firing temperature up to 1600°C by Path III. Sudden mullite formation occurs in two different classes of precursors (CP6 and CP9). The mother phases in these cases are of course different. In case of CP6 precursor, mullite formation is noticed at the temperature of second exotherm out of existing spinel phase formed earlier at ~1000°C. Whereas in case of CP9, mullite formation is noticed from intermediate aluminosilicate Al-Si spinel phase which is newly crystallized by solid state reaction of amorphous SiO_2 and Al_2O_3 components with exhibition of a broad exotherm over in the long temperature range between 850°C to ~1300°C. Thus the start temperature of mullitization in the two cases are different.

In case of SHI, mullite formed at the first stage at 980°C is nonstoichiometric and called as primary mullite by a group of researchers by noting its broad XRD pattern. On further heating beyond 1200°C it transforms into the secondary mullite which is also mullite type phase. In case of CP9, the first formed mullite which has formed itself at high temperature at 1300°C is also to be called as mullite type phase due to poorly crystalline nature. Mullite formation in this case is found to occur to a comparatively lesser extent than in case of CP6. Since, quantitative mullite content data (growth curve) of CP9 always lies below the growth curve of CP6. It is highly probable that mullite formation path/mechanism in both two coprecipitated gels are dissimilar in nature although mullite forms suddenly in two cases of precursors. In case of CP6 and in diphasic gel (CP9) mullitization occurs both by Path IV. In case of CP9, XRD intensitiy of mullite increases predominantly with rise of heating temperature at ~1300°C and the quantitative value comes to (~70 wt %). Beyond 1300°C, a slow increase in mullite formation occurs with increase in firing temperature up to ~83% mullite at 1600°C by Path III where mullite type phase transform to orthorhombic form

16.2.2.6 Sixth case in SD precursor: Formation of mullite in SD gels

In this case mullite formation takes place most rapidly and shows higher value of mullite at each temperature in comparison to other precursors (Fig. 12.6). A steep increment of mullite is noted for the formation of mullite directly out of noncrystalline aluminosilicate precursor phase up to ~1100°C and this route is to be called Path VI. In summary, the growth curves of mullite for six precursors (Fig. 12.6) show three different aspects of the mullitization process e.g., initiation temperature of its formation, it's paths and obviously mullitization mechanisms.

(i) Rapidity of mullitization of three precursors at the first exothermic peak at ~1000°C occurs in the following order. Spray dried precursor (SD) > SHI > SHII.

(ii) In addition to formation of mullite (major phase) as in SHI, SHII, Al-Si spinel phase (minor amount) forms at ~1000°C (Fig. 18.2) which transforms to mullite at the second exothermic peak by Path IV. Thus, mullitization which occurs in SHII by two consecutive steps agrees with the earlier findings of Okada and Otsuka (1986).

(iii) The ratios of Al-Si spinel to mullite formation during heating SHII and SHIII at ~1000°C vary greatly. For example, SHII forms mullite (major) amount and Al-Si spinel (minor amount) whereas the reverse is true for SHIII precursor.

(iv) In cases of SHIII and SHIV gels, mullite forms at the second exothermic peak temperature at ~1250°C out of Al-Si spinel phase rapidly forms earlier at ~1000°C. Reason of third exotherm occurs just before second exotherm at ~1150°C is explained as due to partial formation of mullite out of Al-Si spinel phase accumulated out of solid state reaction between residual noncrystalline alumina rich spinel phase (AS5) with silica rich aluminous phase (AS2) prior to occurrence of the said exotherm.

(v) In case of diphasic gel, mullitization takes place by the transformation of Al-Si spinel >1200°C which was accumulated gradually during the exotherm in the temperature range 850°C–1300°C by Path IV.

(vi) Intermediate Al-Si spinel phase formation is predominant in SHIII, SHIV and diphasic gel. Mullite formations in these cases found to occur to a comparatively lesser extent than in case of SHII. Since growth curves of mullite in those cases always lie below the growth curve of SHII.

(vii) Although sudden mullite formation occurs in three cases of precursors namely SHI, SHIV and diphasic gel (Fig. 12.6), starting phases are however different. In case of SHI, mullite formation takes place from the intermediate noncrystalline aluminosilicate precursor phase that existed prior to the 980°C exotherm. Whereas in case of SHIV (or CP6), mullite formation is noticed out of two phases namely (i). From existing intermediate Al-Si spinel phase at second exotherm (by Path IV), and (ii). From a large quantity of accumulated silica rich aluminous (A) phase on gradual heating (by Path III). The later step occurs negligible in case of SHII. In case of diphasic gel, mullitization starts from the accumulated Al-Si spinel phase at ~1320°C exotherm by Path IV and at higher temperature by Path III. Due to the difference in the start temperatures of mullitization, the nature of mullite formed varies. In the case of SHI, it may be nonstoichiometric in nature. On further heating beyond 1200°C it becomes orthorhombic, to be called as secondary mullite or o-mullite. Both in cases of SHIV and diphasic gel, mullite which is formed at temperature ~ 1300°C may be considered first as mullite type phase which transforms to orthorhombic on heating beyond it. Thus, the nature of mullite which are formed in these two precursors and the courses of its formations over a long temperature region are nearly the same (Fig. 12.6). But differences are observed in the formation temperature, transformation temperature of spinel phase and finally in their courses of transformation in two cases.

16.2.3 Mullitization Behavior of Oxide Mixture in Comparison to Diphasic Gel

Mullitization reactions e.g., from oxide mixture (Hamano et al., 1986); alumina-silica compact (Dekeyser, 1965) and (Staley and

Brindley, 1969) are shown to be controlled by diffusion. However, the mullitization reaction in oxide mixture and in diphasic is different. The reason may either due to small sizes of constituent oxide particles which are different among sources and that lead to different reactivity in solid state diffusional process and or due to incorporation of some silica in γ-Al_2O_3 phase forming Al-Si Spinel prior to form mullite in diphasic system of gel. It was noted that growth of mullite in diphasic gel systems was sigmoidal in nature (Wei and Halloran, 1988). On the other hand, growth curve was reported to be parabolic for oxide mixture (Aksay and Pask, 1975). The course of mullitization in two cases are dissimilar as shown in growth curves (Fig. 16.3) given by Chakraborty and Ghosh (1986).

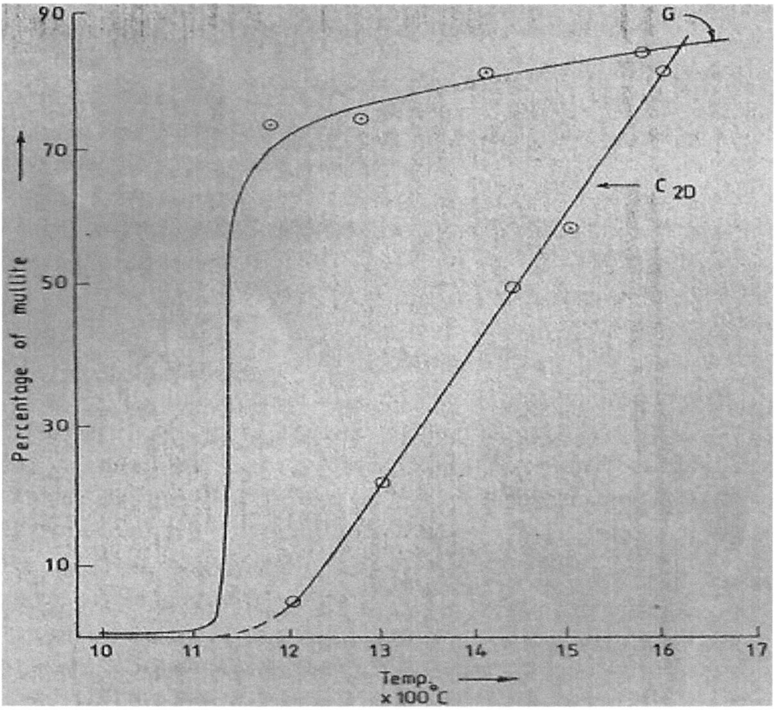

Figure 16.3 Comparison of mullite formation behaviors of synthetic Al-Si spinel ($3Al_2O_3{\cdot}2SiO_2$) to that of binary mixture of amorphous Al_2O_3 and SiO_2 in 3:2 ratio: curve G, Al_2O_3/SiO_2 gel (Al/Si = 3/1): curve C_{2D}, artificial mixture of amorphous Al_2O_3 and SiO_2 (Chakraborty and Ghosh, 1986).

16.3 Mullite Formation of Diphasic Gels (Previous Studies)

Synthesis of different kinds of diphasic gels have been made by various researchers and their DTA/mullitization processes as shown previously in Chapter 4 are listed below vide Table 16.2. It is shown how DTA events and phase evolution of different gels vary with both the choice of components and processing techniques of gelation made during synthesis.

Table 16.2 Diphasic gel synthesized by using boehmite sol as a component

1. Hoffman et al. (1984) first made diphasic gel out of silica sol, Ludox/boehmite sol	It did not show 980°C exotherm. A broad exotherm is noted at high temperature.
2. Hamano et al. (1985) prepared diphasic gel—by mixing boehmite and silica sols with HCl as catalyst	Noted a weakly crystalline Al-Si spinel phase in the range 1000°C–1250°C. Formation of mullite began at ~1250°C.
3. Ismail et al. (1986, 1987) mixed aqueous silica aluminum mono hydroxide sol at pH ~1.8. Gelation occurred by evaporation.	DTA showed exotherm at 1296°C due to mullite formation.
4. Shinohara et al. (1986) used colloidal boehmite suspension and partially hydrolyzed TEOS solution at pH = 2.5.	Noted mullite formation during heating at 1250°C for 2 h.
5. Sonuparlak (1988) used colloidal boehmite suspensions and TEOS to prepare mullite by the sol-gel process at two pH conditions. First one at acidic conditions (pH < 2) and other at pH ~6–7 condition using NH_4OH.	Showed varying degrees of densification.
6. Huling and Messing (1989, 1990) synthesized two diphasic gels from boehmite and silica sols Gel A and Gel B.	DTA reveals a small exotherm at ~1270°C due to mullite. Mullite formation occurs at higher temperature (~1290°C) in Gel A.
7. Klaussen et al. (1990) used pseudo boehmite and colloidal silica gelled by pH adjustment.	DTA exhibits one exothermic peak due to mullite at 1354°C.

(Continued)

Table 16.2 (*Continued*)

8. Wei and Halloran (1988) synthesized translucent white type diphasic gel out of TEOS and colloidal pseudo boehmite sol in acidic condition.	Mullite formation occurs at ≥1300°C.
9. Hyatt et al. (1990) prepared diphasic gel by using AlOOH and Ludox in HNO_3 medium.	Showed one exotherm at ~1300–1350°C due to mullite formation.
10. Su Yen et al. (1991) synthesized four xerogels by coprecipitation of ANN and TEOS at pH = 8.3.	Resultant gels were mixtures of monophasic and diphasic in characters.
11. Pach et al. (1995) synthesized diphasic gel using boehmite and silica sol at pH = 2.	Nucleation of mullite occurs at the interface of $SiO_2/\gamma\text{-}Al_2O_3$ particles at ~1240°C.
12. Wu et al. (1993) synthesized it from Ludox and sub micrometer alumina powder at pH = 5–6.	Noted densification at 1300°C prior to mullitization.
13. Anilkumar et al. (1997) synthesized three sets of conditions. TEOS and freshly peptized Boehmite sol.	Showed mullite development on pH dependence.
14. Sugita et al. (1998) prepared mullite powder out of fume silica (Aerosil 200) in aqueous solution of aluminum sulphate.	A very broad exothermic peak occurred at ~1150°C but mullite produced at a temperature >1250°C.
15. Ivankovic et al. (2003) synthesized two diphasic gels (C and D) out of $\gamma\text{-}Al_2O_3$ and γ-AlOOH as alumina components with TEOS.	Gel C exhibited exotherms due to mullitization at 1346°C and in gel D at 1361°C.

Besides, the synthesis of different kinds of diphasic gels as described above mullite formation out of "in situ" diphasic are given below.

16.3.1 Diphasic Gel Made Using NH_4OH at pH = 9 e.g., "in situ Gel" and It's Mullite Formation

(A) Mullite formation in stoichiometric diphasic gel

It was conjectured by Wei and Halloran (1988) that mullite formed in diphasic system by the direct reaction of γ- or $\delta\text{-}Al_2O_3$ and SiO_2 without the formation of intermediate aluminosilicate (A) precursor phase. It was shown that the growth curve of mullite in their diphasic gel is of sigmoidal in nature. However, it is conceived by Chakraborty (2008) in case of "in situ" diphasic gel that Si out of either amorphous silica or form AS diffuses into γ-Al_2O_3 structure and formed Al-Si spinel during heating diphasic gel in the temperature range ~400°C (where boehmite just decomposed and formed γ-Al_2O_3) to as high as ~1300°C (where crystallization ensues). Thereafter, mullitization occurred rapidly at temperature of exotherm at ~1300°C. The formation reaction was compared with mullitization behavior of kaolinite because it also formed Al-Si spinel but at lower temperature at ~980°C than that occurred in the temperature range between 1200°C to 1300°C for diphasic gel. The behavior of growth of mullite is analogous to that of kaolinite (Fig. 16.4). In case of phase transformation of kaolinite, a large quantity of noncrystalline aluminosilicate phase was noted prior to mullitization. A similar type of phase was generated during transformation of "in situ" diphasic mullite gel. Accordingly, formation of Al-Si spinel phase and existence of noncrystalline aluminosilicate residual phase during heat treatment process of diphasic gel is first of all discussed in detail thereafter their subsequent role to mullitization will be shown.

(i) **Formation of noncrystalline aluminosilicate phase**: The presence of some quantity of noncrystalline aluminosilicate phase was first pointed out by Okada and Otsuka (1986) in their heat treated SH and RH gels. Formation of it in case of diphasic gel heat treated in the range 600°C–1000°C was shown by XRD recording and the area of the X-ray amorphous hump was first measured manually by Ivankovak et al. (2003) and its variations with heat treatment temperature were shown. Using a software program the changes in areas of the X-ray amorphous hump vs. temperature were recalculated by Chakraborty (2008a). The author showed (i) how the quantity of noncrystalline aluminosilicate phase formation depend upon processing conditions using some of components of silica and alumina vide Table 15.1 and Table 15.5, and (ii) it's role in mullite formation.

(ii) **Formation of Al-Si spinel phase:** The XRD analysis of heat treated diphasic gel (Figs. 15.2, 15.3) of Chakraborty (1997) in the mullitization region showed that the two intensities of Al-Si spinel phase peaks gradually increased at first and then the intensities drastically decreased with consequent increase in the intensities of mullite at ~1300°C. At this temperature, X-ray peaks due to both Al-Si spinel and mullite were still coexisting. Thereafter, Al-Si spinel completely disappeared with predominant crystallization to mullite type phase. The growth curve (Fig. 16.4) of it showed that mullite rapidly formed at ~1300°C and thereafter it's growth still continued with further rise of temperature of heat treatment. The profile of growth of mullite is analogous to that of kaolinite. In the case of kaolinite, the growth curve of mullite consisted of two segments and accordingly it was explained earlier that mullitization took place by two simultaneous reaction paths out of two phases namely Al-Si spinel crystallized during exothermic peak temperature at ~980°C and from residual noncrystalline aluminosilicate phase (Chakraborty and Ghosh, 1991). Similar to kaolinite, the mullitization curve of diphasic gel also predicts two paths of mullitization.

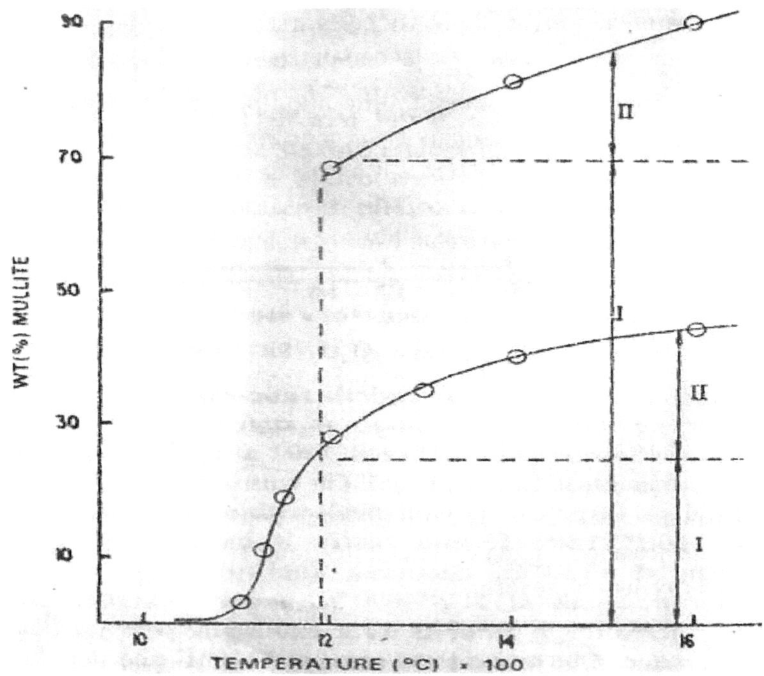

Figure 16.4 Percentage of mullite formation vs. heat treatment (a) upper curve shows mullitization of diphasic gel, G-49, (b) Lower curve shows mullitization of Bhandak Kaolinite (Chakraborty, 1997).

First step of mullitization: The rapid increase in the XRD intensity (during static heating) of mullite occurring at 1200°C–1250°C may be due to polymorphic transformation of accumulated Al-Si spinel (c-mullite) tentatively by Path IV.

$$\underset{\text{c-Mullite}}{3Al_2O_3{\cdot}2SiO_2} \xrightarrow[\sim 1320°C \text{ Path IV}]{\text{Exotherm}} \underset{\text{Mullite type phase}}{3Al_2O_3{\cdot}2SiO_2} \quad (16.10)$$

Second step of mullitization: The slow and gradual mullite formation followed by step rise in the high temperature region (1300°C to 1600°C) is accounted for mullitization by the nucleation from residual noncrystalline aluminosilicate phase accumulated during heating of diphasic gel between 900°C to 1300°C by Path III.

$$\underset{\text{phases (as residue} \sim 20 \text{ wt \%)}}{\text{Noncrystalline aluminosilicate}} \xrightarrow[\text{Nucleation and crystallization Path III}]{1300°C\text{–}1600°C} \underset{\text{o-mullite}}{3Al_2O_3{\cdot}2SiO_2} \quad (16.11)$$

Mullitization behavior in the category of in situ diphasic gel is thus unlike to the reported mullite formation reaction of γ-/δ-Al_2O_3 with SiO_2 (A) as proposed earlier by Wei and Halloran (1988) and others view (Table 16.2).

16.3.2 Comparative Mullite Formation Process of Diphasic Gels in Stoichiometric vs. Nonstoichiometric Compositions

(A) Stoichiometric diphasic mullite gel

In the latest study of diphasic gel system, Chakraborty (2005a) noted formation of mullite (determined from the height of the 0.537 nm X-ray peak) in diphasic gels and mixed oxides as a function of heat treatment temperature, is shown in Fig. 16.2. Figure 16.2 indicates that in static heating conditions, mullite formation in DG40 and DG72 diphasic gels started rapidly and attained a maximum quantity at ~1200°C and changed slightly on further heating. On the other hand, in DG85 mullite formation started just at ~1200°C. A moderate amount of mullite formed at ~1300°C and it increased gradually on heating up to 1500°C and then attained almost constant. Mullite formation in the three cases of fired diphasic gels is shown schematically in Fig. 16.5.

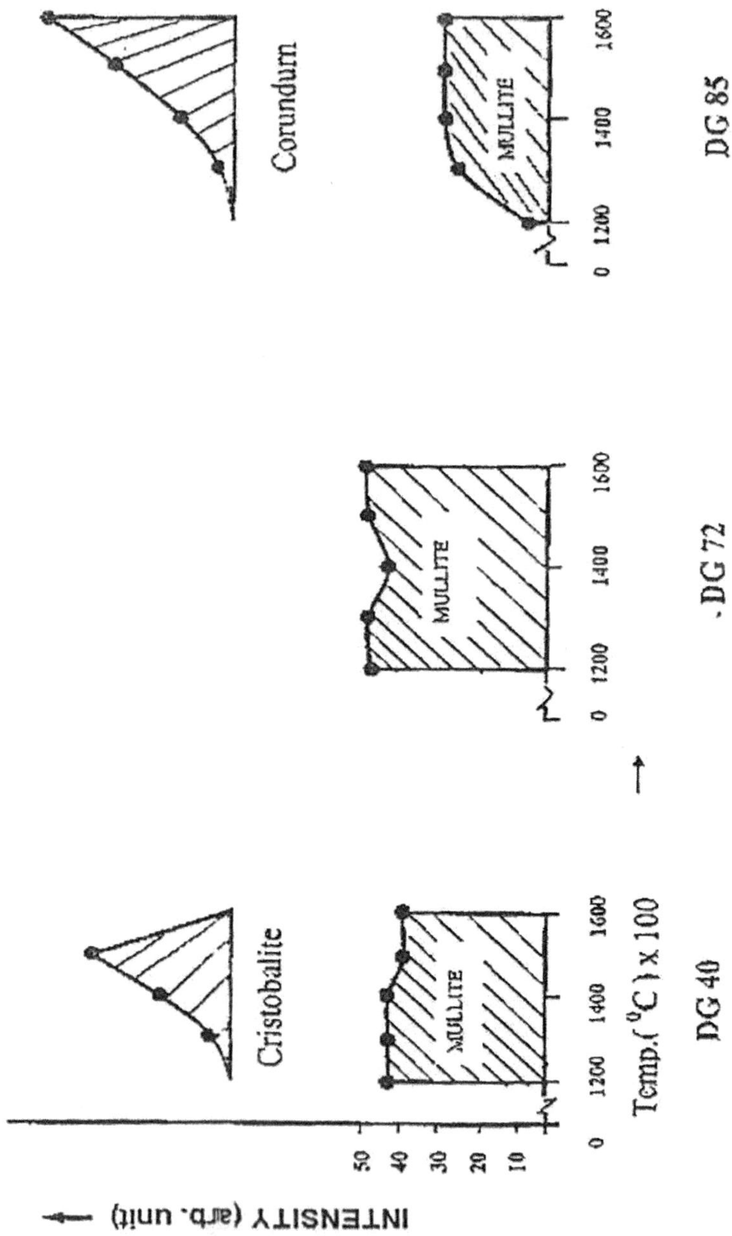

Figure 16.5 Schematic diagram showing the phase development of three cases of diphasic gels.

Besides mullite, other two crystalline phases formed during heating are also shown. The amount of phase formed corresponds to the height of the XRD peak at the respective temperature. Comparison of mullite formed in three cases of diphasic gels is shown. Also shown are the amounts of β-cristobalite formed in DG40 and corundum crystallized in DG85.

Mullite formation

It was shown that at static heating condition DG72 develops mullite abundantly at ~1200°C at the cost of complete elimination of Al-Si spinel phase (Fig. 16.1). This sudden mullite formation out of Al-Si spinel may indicate polymorphic nature of transformation which might be a feasible proposition. This theory is in contrast to the mullite formation by solid state reaction between α-Al_2O_3 and silica (A)/cristobalite which perhaps occurs through a long range diffusional process (Sundaresan and Aksay, 1991).

The unit cell formulae of Al-Si spinel of mullite like composition is as follows:

$[Si_{4.9}\,Al_{3.10}]$	$[Al_{11.7\,\square\,4.3}]$	O_{32}
Tetrahedral site	Octahedral site	Oxygen
IV	VI	
8A	16B	O_{32}

Thus, the early mullitization in DG72 may be explained as due to the preoccupation of Si atoms in tetrahedral sites of Al-Si spinel structure as shown in unit cell formulae given above. In ^{27}Al MAS NMR spectrum study, it was shown that during this transformation, only the ratio of AlO_4 to AlO_6 changed (Fig. 9.4a–c and Table 9.1). It is speculated that neither breaking of any bonds nor the formation of any new bond is essential for polymorphic transformation of Al-Si spinel of mullite like composition to weakly crystalline mullite type phase at ~1200°C. Low activation energy in the form of heat treatment just up to ~1200°C may be sufficient to rotate Al-O bonds to induce octahedral to tetrahedral configuration leading to mullite nucleation. In the process of crystallization, cristobalite and corundum did not appear at any stage. Cristobalite appeared when silica was present in excess to that of 3:2 mullite as in case of DG40. Similarly corundum appeared in case of DG85 where alumina was present in excess to that of 3:2 mullite.

The probable sequence of transformation of the diphasic gel (Al/Si = 3/1) may be described below. The phase evolution of this diphasic system proceeds as per following steps.

First step is the dehydration of hydrates of alumina at ~400°C in DTA to form γ-Al_2O_3 and the start of formation of noncrystalline aluminosilicate phase (traces of θ-alumina in some cases at higher temperature).

Second step in the temperature range 400°C to ~1000°C quantity of formation of noncrystalline aluminosilicate phase advances. At ~1000°C, phase separation slowly occurs as happens in case of monophasic gel, two phases are usually generated. First one is alumina rich spinel phase and the other is noncrystalline silica rich aluminous phase as evident from XRD and ^{29}Si resonance spectrum studies of diphasic gel made by different researchers. For example, Sample B, D of Jaymes et al. (1996).

Third step is the formation of a broad exotherm in DTA and a differential exotherm in DDTA noted by the author (Fig. 15.1a,b) over 1000 to 1300°C due to slow crystallization of silica enriched Al-Si spinel phase by the solid state reaction between alumina rich spinel phase with noncrystalline silica rich aluminous phase.

Fourth step is the exhibition of a small exotherm at 1335°C in DTA with crystallization of mullite out of structural transformation of Al-Si spinel phase. Details are given below.

(1) XRD & DTA study: On comparing XRD results with DTA observation in case of diphasic gel named CP9, Dafader (2000) observed above that crystalline boehmite phase of the diphasic gel first decomposed resulting to amorphous mixture of alumina and silica. The alumina component crystallized at ~600°C and gradually increased in its intensity on continued firing. At ~1270°C, DTA exhibited a broad exotherm of small magnitude. In this temperature region mullitization started forming. At ~1400°C, a large quantity of mullite is seen. But neither corundum nor cristobalite is developed at any stage of heating.

(2) MAS NMR study: This study too conforms this observation. Since both ^{27}Al resonance peaks as well as ^{29}Si resonance peaks correspond very well with classic mullite spectrum (Fig. 9.3). Now the question is: what way NMR study would help in resolving the two major questions generally raised by various researchers. What is happening during heating it between endothermic peak temperature ~400°C to onset of crystallization of spinel phase at 1000°C? And composition of spinel phase, whether it contains 28 wt % SiO_2? These two questions are first dealt with prior to mullitization of this gel.

(i) ^{29}Si spectrum study

Raw diphasic gel, it was noted that ^{29}Si peak which was found earlier at ~111.9 ppm in raw gel synthesized by Komarneni et al. (1986) found to shift on heat treatment, which indicates that Si–O environment due to free Si–$(OSi)_4$ and HO–Si $(OSi)_3$ changes.

At 900°C, ^{29}Si spectrum on deconvolution shows two broad peaks, one at –88.5 ppm and other at –98.5 which are in the region of Si(4Al)/Si(3Al) environment. Thus, shifting interprets the development of Si–O–Al (IV) linkages. Noncrystalline aluminosilicate phases of two varying SiO_2/Al_2O_3 ratios are being formed by solid state reaction between SiO_2 (A) and Al_2O_3 (A) components derived out of diphasic gel on heat treatment at ~400°C. Previous alkali leaching study of heated diphasic gel by Chakraborty (1979) showed gradual decrease in dissolution behavior of free silica component by NaOH solution with increase of heating temperature of diphasic gel. This result and the present MAS NMR data definitely predict the formation of noncrystalline aluminosilicate phase prior to spinel crystallization. The major solid state reaction would be the inter diffusion of Al in siliceous phase and counter diffusion of Si in aluminous phase and that lead to slow accumulation of aluminosilicate phases.

At ~1000°C, ^{29}Si spectrum becomes asymmetrical. The centroid at –89 ppm becomes more broad than in the previous case where spinel formation occurs (Chakraborty and McKenzie, 2006). The shoulder at –109 ppm on the right hand side of the asymmetry indicates liberation of noncrystalline silica rich aluminous phase at final dehydroxylation cum phase separation processes. It subsequently reacts with spinel phase where enrichment of silica occurs in spinel structure and thereafter transformed to mullite on further heating (Fig. 9.20).

At 1300°C, the broad ^{29}Si resonance at –89 ppm slightly shifted to a most intense narrow peak at –86.6 ppm. The other peak at –94.7 ppm also corresponds well with a minor peak reported at –94.1 to –94.3 ppm, the shoulder which is fitted to a peak at –90.3 ppm is also reported at –90 ppm. Lastly, the low intensity peak at –91.9 ppm is also due to well crystallized mullite. Thus, the continuance of silicon environment at –89 ppm resonance peak throughout the transformation process (from 900°C to 1000° to 1300°C) indicates that the structural transformation in the entire process is coherent. Similar, argument was already put forward by Colomban (1989) by his microscopic studies.

(ii) ^{27}Al spectrum study

The diphasic gel is originally a homogeneous mixture of pseudo boehmite gel and silica gel. On heat treatment e.g., at 900°C it shows AlO_4 groups to the extent of 25% along with usual AlO_6 and with some AlO_5 groups (Table 9.3). This is another supportive evidence of formation of noncrystalline aluminosilicate phase. Thus ^{27}Al MAS NMR spectrums are also very significant. The triplex Al spectrum conforms to aluminosilicate phase formation. But it is difficult to predict the amount of Al incorporation in the silica matrix. There may be a possibility that after attaining a definite temperature and after saturation of some concentration of alumina in noncrystalline phase, a portion of Al may exist as free alumina. On further heating to 1000°C, the AlO_5 group present in the aluminosilicate phase disappears. Consequently, the quantum of AlO_4 and AlO_6 increases to 31% and 69%, respectively, during spinel formation. On subsequent spinel to mullite transformation, AlO_4 further increases to 50%. Thus, in the entire transformation process, only the ratio of AlO_4 to AlO_6 smoothly changes without any discontinuity which is quite similar to the persistence of –89 ppm resonance peak throughout the transformation sequence. Only the nature of peak (its broadness to sharpness) changes on heat treatment. Therefore, the possibility of decomposition of newly developed aluminosilicate phase to free Al_2O_3 and free SiO_2 is ruled out. Further, ^{29}Si resonance peak at –89 ppm more close to resonance peak (–87 ppm) for mullite than for –80 ppm (theoretical value) for Al-Si spinel. However, mullite is not formed at 1000°C contrary XRD shows pronounced peaks due to spinel formation. Evidently, the broad –89 ppm peak and its size suggest that a major proportion of Si is combined with Al in the spinel lattice. But a question arises: why does the experimentally measured value not agree with theoretically calculated data? It is not clear now whether it is due to ill crystallized nature of spinel formed by substitution of a large extent of SiO_2 of the order of ~28% or so which causes enough strain in lattice as a result crystallite size did not increase but formed mullite even on static heating at the formation temperature.

If γ-Al_2O_3 formed other than Al-Si spinel, then there would be extensive segregation at 1000°C at the first stage and that might lead to liberation of greater extent of silica. Likely, -109 ppm peak intensity would increase at the cost of reduction of –89 ppm peak assigned to aluminosilicate. In the second stage of recombination reaction at ~1300°C between γ- Al_2O_3 and SiO_2 (A) would form mullite.

^{29}Si MAS NMR peak at –110 ppm peak was shown by different researchers. This result indicates that the glassy phase was invariably formed in diphasic gel. This peak was even observed beyond 1300°C by Jaymes et al. (1996). In "in situ" diphasic gel, this peak may be due to the aluminosilicate phase being richer in SiO_2 concentration. Since some amount of θ-Al_2O_3 is still noted in diphasic gel heated to 1300°C. Thus there may be two noncrystalline aluminosilicate phases one is silica rich and the other is alumina rich at 1000°C. The mullitization reaction behavior of CP9 gel is similar to fourth and or fifth cases as given above. However, the diphasic gel marked as Type II (fifth case of transformation) made by Schneider et al. (1993) did not correspond to analogous reaction course. It showed one small peak at –80 ppm indicates that aluminosilicate formation is started when heated at 350°C, the other peak at –110 ppm is due to pure amorphous silica liberated after dehydration of silicic acid gel. With continued heating the solid state reaction between silica (A) and Al_2O_3 (A) formed after decomposition of boehmite continued and as a result intensity of –80 ppm peak increased at the cost of decrease of the peak at –110 ppm as noted when it was heated at 900°C where γ-alumina developed and then continued to 1000°C. Thus, at 1000°C, the phase mixture may consist of an appreciable quantity of noncrystalline aluminosilicate phase, residual silica phase (A) and γ-Al_2O_3. The differences in the reaction course of the two cases, might be the variation in the processing conditions chosen by the various authors (Table 16.2). It is summarily given below.

(1) Decomposition of boehmite component at ~400°C.
(2) Solid state reaction between newly formed SiO_2 (A) and Al_2O_3 (A) and or interaction of hydroxyl groups among two components during dehydroxylation with generation of noncrystalline aluminosilicate phases 400°C to 1000°C.
(3) Phase separation into alumina rich spinel phase and formation of noncrystalline silica rich aluminous phase at ~1000°C.
(4) Continued diffusion of Si out of noncrystalline aluminosilicate (AS) into alumina rich spinel cubic lattice, and into γ-Al_2O_3 (if present) occur 1000°C to ~1300°C to crystallize

    Al-Si spinel phase following a broad exotherm till saturation with silica and its composition comes equivalent to 3:2 mullite. And presence of two residual noncrystalline aluminosilicate phases remain as residue.

(5) Polymorphic change of saturated Al-Si spinel to mullite through an exotherm at ~1300°C.

(6) Solid state reaction of two noncrystallization aluminosilicate phases to nucleate and then crystallized to mullite at 1300°C–1600°C. This result is derived by considering the result of phase separation into alumina rich spinel phase and formation of noncrystalline silica rich aluminous phase at ~1000°C from Si spectrum study and slightly modifying Eq. 16.11.

Diphasic mullite gel

The two tentative steps of mullitization in case of "in situ" diphasic gel vide Eqs. (16.10) and (16.11) as shown above are enumerated vividly and are shown below. To accomplish this, four steps of it's transformation as shown in the text are separated and increased up to as many as seven stages for clarity to readers.

1st stage

Boehmite gel [Al(OH)$_3$ gel + Si(OH)$_4$ gel] Diphasic gel	$\xrightarrow[\text{At} \sim 400^\circ\text{C}]{\text{Decomposition}}$	(i) Silica rich alumina phase (A) (ii) SiO_2 (A), Al_2O_3 (A) (iii) γ-Al_2O_3

2nd stage

(i) Silica rich alumina phase (A) (ii) SiO_2 (A), Al_2O_3 (A) (iii) γ-Al_2O_3	$\xrightarrow[\sim 900^\circ\text{C}]{\text{Solid state reaction}}$	(i) Aluminosilicate phase (noncrystalline) (ii) Si incorporated γ-Al_2O_3

3rd stage

(i) Aluminosilicate phase	$\xrightarrow[\text{At "D"} (\sim 1000^\circ\text{C})]{\text{Phase separation}}$	(i) Alumina rich spinel phase (weakly crystalline) (ii) Silica rich aluminous phase (noncrystalline) (iii) Si incorporated γ-Al_2O_3

4th stage

(i) Alumina rich spinel phase (ii) Silica rich aluminous phase (iii) Si incorporated γ-Al_2O_3	Continuous diffusion of Si ⟶ 1000°C–1300°C (A broad exotherm)	(i) Al-Si spinel (3:2) phase (weakly crystalline) (ii) Alumina rich silica phase (A) (A broad ^{29}Si resonance peak at –89 ppm) (iii) Silica rich aluminous phase (A) (A broad ^{29}Si resonance peak at –110 ppm) see Fig. 9.20 (v) Si incorporated θ-Al_2O_3 (minor) (iv) Silica incorporated γ-Al_2O_3 (minor)

5th stage

(i) Al-Si spinel (3:2) phase	Polymorphic transformation ⟶ Exotherm At ~1300°C–1340°C	Mullite type phase

6th stage

(ii) Alumina rich silica phase (A) (iii) Silica rich aluminous phase (A) (iv) Si incorporated θ-Al_2O_3	Solid state reaction ⟶ At >1300°C	Mullite type phase

7th stage

Mullite type phase (collectively)	Change of crystalline order ⟶ At 1300°C–1600°C	o-Mullite (well crystalline) (16.12)

Fifth stage

This temperature of exotherm shows higher than that occurs in case of RH gel. Thus, mullite formation temperature increases from RH category diphasic gel to gel containing amorphous $Al(OH)_3$ colloids to diphasic gel containing crystalline pseudo boehmite as phase.

(B) Nonstoichiometric diphasic mullite gel

It is necessary to discuss the effect of Al/Si ratio on the development process of mullite. Besides DG72, nonstoichiometric gels e.g., DG40 (a silica rich), and DG85 (an alumina rich) gels formed β-cristobalite or corundum besides mullite (Fig. 16.2). Growth characteristic of mullite in later two cases is different from the former.

Silica rich diphasic gel

Both DG40 and DG72 rapidly formed mullite during heating at >1200°C and then attained more or less constant (Fig. 16.2) with increase of temperature. So the mullitization process in two cases of diphasic gels is analogous. It indicates that Al-Si spinel is formed in both cases of gels. The excess portion of silica remained after the formation of Al-Si spinel phase first crystallized and subsequently disappeared. The precipitous drop in cristobalite in case of DG40 at ~1600°C may be due to formation of siliceous glassy phase. Similar observation was noted during the phase transition process of kaolinite.

Alumina rich diphasic gel

In the mullitization process of DG85, two observations are noted: (i) θ-Al_2O_3 existed at temperature as high as 1200°C. Contrary, this phase started transforming at ~1050°C and disappeared almost completely at ~1150°C to corundum during heating AS mixture. (ii) Mullite formation did not take place rapidly at 1200°C as happened in case of DG40 and DG72. It starts slowly at ~1200°C and it increases gradually from 1300°C with rise of temperature. The increase in stability of θ-Al_2O_3 indicates that Si may have incorporated in its structure and this results to hindrance to the normal phase transformation process of the boehmite component of the diphasic gel. The incorporation Si in θ-Al_2O_3 also obstructs further crystallization of θ-Al_2O_3 and as such a combined formation of both θ-Al_2O_3 and α-Al_2O_3 phases was not observed. Thus, Si incorporation in the alumina phase has two consequences. Silica is not left out in free state to crystallize to cristobalite during heating. Secondly, it changes the crystallization path of pure alumina form in the diphasic gel system. The Si incorporated θ-Al_2O_3 started disappearing at ~1300°C with formation of corundum and mullite. The mullite phase gradually increased from 1300°C up to ~1500°C and finally reached constant at ~1600°C.

The nature of mullite formation in three cases is also different. For example, splitting of 120 and 210 of mullite in case of DG40 and DG72 started at ~1300°C whereas the same occurred in case of DG85 at ~1600°C. These observations primarily indicate that composition of the initial gel may influence the formation temperature of mullite, its growth characteristics and finally the structure of it. Li and Thomson (1991) synthesized nonstoichiometric diphasic gels out of boehmite powder and TEOS and showed that mullite formation took place at ~1300°C for all samples and indicated that difference in composition had a little effect on the initial formation temperature or structure of mullite. With increase of temperature the 120 and 210 double peaks of mullite moved closer together in the gel with 6/1 ratio, while no change in the position of those peaks for 3/1 gel. This indicates that the lattice parameter of mullite changes with heat treatment of gel of varying Al/Si ratio. Yoldas (1992), Pask et al. (1987) noted a great difference in the crystallization behavior of polymeric and colloidal gels of similar Al/Si ratio when heated. The different behavior in mullitization paths, microstructure developments in two cases were explained by them and are due to the difference in ultra-structures of polymeric to colloidal gels. The basic difference occurred in gel structure was due to choice of chemical preparative conditions of the two gels out of same source components such as aluminum butoxide and TEOS. They noted that during the gel preparation stage direct chemical bonds are formed between silicon and aluminum, i.e., Al-O-Si, and size of domains becomes extremely consequential in crystallization. In the present case, diphasic gels were made out of the same raw materials such as ANN and TEOS and prepared out of the same preparative condition but with a difference in Al/

Si ratio only. It is speculated that the changes in Al/Si ratio may lead to differences in concentration of silica gel (A) to pseudo boehmite in the mixing state of three "in situ" diphasic gels that may lead to variations in the growth process of mullite as shown in Fig. 16.2. It may also affect the structure of mullite formed between 1200°C to 1600°C. It is shown that splitting of said two peaks occurs in DG85 at 1600°C which in contrast to the absence of the same in diphasic gel of 6/1 ratio by Li and Thomson (1991). This difference in observation may be due to the variation in heating schedules used in two cases. Mullite formation in two cases is also different. It is also shown that Al-Si spinel formed out of silica rich and stoichiometric diphasic gel transformed suddenly to mullite at ~1200°C and it attained a constant value even on continued heating. Accordingly, a new phase transformation sequence of diphasic Al_2O_3-SiO_2 gel is predicted. Mullite formation in case of nonstoichiometric alumina rich diphasic gel occurred by the decomposition of intermediate Si incorporated θ-Al_2O_3 phase at >1300°C as major quantity.

Hsi et al. (1989), Fischer et al. (1996), Huang et al. (1997) noted the formation of intermediate θ-Al_2O_3 in their phase transformation studies. All of them conjectured that θ-Al_2O_3 reacts with SiO_2 (A) to form mullite on further heating. On re-examination of their XRD pattern it is noted that the characteristic Bragg diffraction peaks resembling θ-Al_2O_3 were weak. So insignificant amounts of θ-Al_2O_3 might have formed and the residual pattern was due to a major quantity of spinel phase. If intermediate phase identified by them was at all θ-Al_2O_3 then a partial amount of α-Al_2O_3 should crystallize and formation of a joint mixture of both θ-Al_2O_3 and α-Al_2O_3 would be noted as like as happened in case of transformation of MO (Fig. 15.6). However, formation of a mixture of θ-Al_2O_3 and α-Al_2O_3 were not reported. Therefore, it is explained that a minor amount of θ-Al_2O_3 which was called by them may contain Si atoms as conceived by Yoldas (1976). On the other hand, abundant formation of mullite as noticed by Hsi et al. (1989), Fischer et al. (1996), Huang et al. (1997) in their X-ray studies indicated that the so-called spinel phase would be Al-Si spinel. Formation of θ-Al_2O_3 is obviously found when separate gels out of two components were mixed and heat treated at higher temperature. Phase changes of some mixed gels are shown below.

16.3.3 Transformation of Intermediate Al-Si Spinel Phase along with Θ-Al_2O_3 Formed in Mixed Gel to Mullite Phase

During monitoring the phase evolution studies of mixed gels MG24, MG40, MG66 MG72 and MG80 with X-ray diffraction, Chakraborty (2006) pointed out the characterization problem of the Al-Si spinel and Θ-Al_2O_3 phases. It was shown earlier that mixed gel in the heat treatment process form a mixture of phases e.g., (i) silica rich alumina phase (A), (ii) SiO_2 (A), (iii) γ-Al_2O_3, (iv) Al-Si spinel, (v) silica incorporated γ-Al_2O_3, θ-Al_2O_3, and (vi) alumina rich silica phase (A). Mullite formation process is explained on the basis of ^{29}Si MAS NMR results (Figs. 9.5 and 9.6). The possible mechanism of mullite formation of mixed gels seem to be different for two ranges of composition.

(A) Mixed gels of Al_2O_3 content up to 3:2 mullite

The transformation sequence of mixed gels of varying alumina content might take place mere tentatively in the following stages.

First stage: Solid state reaction took place between silica and alumina gels during heating and formed a noncrystalline aluminosilicate phase based on earlier observation of the present author. It was shown that mixed gel heated beyond 400°C reacted with $CaCO_3$ on further heat treatment at ~1200°C and formed gehlenite. This result indicated that two components of mixed gel reacted in the solid state and developed amorphous aluminosilicate phase. The composition of aluminosilicate would vary and it would also depend on the varying Al_2O_3 to SiO_2 ratio of the components chosen during synthesis of mixed gels. It might be a mixture of silica rich alumina phase (A) and alumina rich silica phase (A) phases.

Mixed mullite gel

Reactants	Conditions	Products	
1st stage [$Al(OH)_3$ gel + $Si(OH)_4$ gel (Mixed gel)]	$\xrightarrow[400°C–900°C]{\text{Decomposition}}$	(i) Silica rich alumina phase (A) (A broad ^{29}Si resonance peak at –89 ppm) (ii) SiO_2 (A) (iii) γ-Al_2O_3 (iv) Al-Si spinel alumina rich (v) Silica incorporated γ-Al_2O_3 (vi) Alumina rich silica phase (A) (A broad ^{29}Si resonance peak at –89 ppm) (vii) Silica incorporated θ-Al_2O_3	(16.13a)
2nd stage (i) Silica rich alumina phase (A) (ii) SiO_2 (A) (iii) γ-Al_2O_3 (iv) Al-Si spinel alumina rich (v) Silica incorporated γ-Al_2O_3 (vi) Alumina rich silica phase (A) (vii) Silica incorporated θ-Al_2O_3	$\xrightarrow[1200°C–1300°C]{\text{Solid state reaction \& crystallization}}$	(i) Al-Si spinel phase (weakly crystalline) (^{29}Si resonance peak at –89 ppm) Other residual phases e.g., [(ii) Silica richer alumina phase (A broad ^{29}Si resonance peak at –110 ppm) (iii) Alumina rich silica phase (A)] (iv) Si incorporated θ-Al_2O_3	(16.13b)

Second stage: Noncrystalline aluminosilicate phase crystallized to Al-Si spinel phase and left a residue which might be assumed as silica rich alumina phase. Composition of this phase is depending upon initial composition of two components chosen and thus SiO_2 content of it will vary. Formation of silica rich alumina phase during heating mixed gels was detected by the appearance of a resonance peak at –110 ppm in ^{29}Si MAS NMR spectrum. This phase is forming during heating mixed gels irrespective of composition.

3rd stage

(i) Al-Si spinel phase Other residual phases e.g., (ii) Silica rich alumina phase (iii) Alumina rich silica phase (A) (iv) Si incorporated θ-Al_2O_3	$\xrightarrow[\text{1300°C exotherm}]{\text{Phase transformation at}}$	(i) Mullite type phase (ii) Silica rich alumina phase (A) (iii) Alumina rich silica phase (A) (iv) Si incorporated θ-Al_2O_3 (16.14)

Third stage: Polymorphic transformation Al-Si spinel phase at ~1300°C to mullite.

Fourth stage: A solid state reaction between silica rich alumina phase and alumina rich silica occurred during heating at >1300°C. This reaction may be possible out of gels of composition e.g. 60 wt % silica up to composition like that of 3:2 mullite.

4th stage

(i) Silica rich silica phase (A) (ii) Alumina rich alumina phase (A) Other phases e.g., (iii) Si incorporated θ-Al_2O_3 (iv) β-Cristobalite (if formed)	$\xrightarrow[\text{Solid state reaction \& crystallization during heating at 1300°C–1600°C}]{}$	(i) Mullite (well crystallized) (^{29}Si resonance peak at –80, –87, –90 and –94 ppm) (16.15)

As discussed above the maximum content of silica ideally in Al-Si spinel will be 28 wt % when it is fully saturated. The composition of it would be the composition of mullite out of gels of alumina composition up to 3:2 mullite. It structurally transforms directly to mullite on heating. In practice, a small amount of silica rich alumina phase is also formed during heating MG72 at ~1000°C as noted by study of ^{29}Si MAS NMR spectrum (Fig. 9.6). Alumina rich silica (A) phase likely forms mullite by the solid state reaction of it with silica rich alumina phase. Even in other cases, when silica is more than 28%, mullite formation occurs as per Eq. 16.14. The residual highly silica rich alumina phase left after mullite formation transforms to cristobalite during further heating. The extent of cristobalite formation varies on further heating. This will decrease from gels of high silica content e.g., MG40 up to mixed gel MG72 (the composition of 3:2 mullite) where no cristobalite is seen (Table 15.3).

(B) Mixed gels of Al_2O_3 content higher than 3:2 mullite

Alumina contents of spinel phase formed out of gels for example, MG80, MG84 above the composition of MG72 are deficient in silica. In this case, some amount of silica incorporated θ-Al_2O_3 formed along with Al-Si spinel, silica rich alumina phase and alumina rich silica phases during the heating process. Al-Si spinel transformed to mullite (Path IV). Silica rich alumina phase and alumina rich silica phases on heating transformed to mullite (Path III) and lastly silica incorporated θ-Al_2O_3 decomposed on heating at 1250°C to develop some mullite (Path VII) and corundum (Table 15.3). The last stage of the transformation process is different and is shown in the fifth stage.

This result indicates that two phases namely silica incorporated γ-Al_2O_3 spinel (alumina rich) and silica rich alumina phase develop on heating MG80 gel. The formation of the later phase is shown by the appearance of –110 ppm peak in Si spectrum. The limit of SiO_2 content may be from 0 to 28%. Thus, a spinel of Al_2O_3 content above the composition of mullite (72% Al_2O_3) transforms as per above Eqs. 16.14–16.15. Only the fourth stage proceeds in the following way.

5th stage

(i) Alumina rich silica phase (A) $\xrightarrow[\text{reaction} >1300°C]{\text{Solid state}}$ (i) Mullite (^{29}Si resonance peak at –80, –87, –90 and –94 ppm)

\+

(ii) Silica rich alumina phase (A)

\+

(iii) Silica incorporated θ-Al_2O_3

Heating ↓ >1350°C

Corundum + Mullite (16.16)

Therefore, the range of SiO_2-γ-Al_2O_3 solid solution is 0–28 wt % although high surface area was noticed by the addition of 2–10 wt % SiO_2 to Al_2O_3 as shown earlier by Trimm (1983).

16.4 Lattice Constant Measurement Data of Mullite Phase, Relation of Unit Cell Parameter to Composition of Mullite/Solid Solution and Stoichiometry

Cell dimension of mullite

Many attempts have been made to obtain cell dimensions of mullite by the investigators of powder X-ray diffraction studies. Among them the works of Rooksby and Partridge (1939), Dekeyser (1965), Agrell and Smith (1960), Aramaki and Roy (1962), Majumder and Welch (1963), Cameron (1977) are most worthy for choosing both natural and synthetic mullite during their LC measurement. For synthetic mullite out of sol gel method, several researchers namely Yamada and Kimura (1962), Okada and Otsuka, Hirata et al., Low and McPherson, Sales and Alarcon are also significant. Mullite was discovered by Bowen and Greig (1924) and reported its chemical composition as ($Al_2O_3/SiO_2 = 3/2$).

Different varieties of mullite

The conversion process of sillimanite to mullite was suggested by Taylor (1932) by the replacement of (SiO_4) group for the similar (AlO_4) group based on X-ray examination of some natural and synthetic mullites, and porcelain. Bradley and Roussin (1932) distinguished the existence of different varieties of mullites based on the X-ray investigation of mullite samples synthesized over a range of compositions containing 72 to 80% by weight of alumina.

Composition of mullites

Rooksby and Partridge (1939) analyzed the high diffraction angle regions of X-ray patterns of mullite which comprises a series of α-doublets. These reflections arise from atomic planes of comparatively high (*hkl*) values which are very sensitive to lattice changes and small changes in axial ratios. The movements of two sets of doublets at 0.8457 and 0.8429 Å; one doublet at 0.8169 Å; and three doublets at 0.7951, 0.7934 and 0.7918 Å are particularly remarkable. A mullite sample of composition approximately 78% containing a maximum proportion of alumina may present as a solid solution in mullite. They concluded that mullite exists as a two-component solid solution, i.e., $Al_2O_3/SiO_2 = 3/2 \sim 2/1$. The chemical formulae of $3Al_2O_3{\cdot}2SiO_2$ corresponds to 28.2% SiO_2 and 71.8% by weight of Al_2O_3. As Al_2O_3 content increases, a progressive change in the lattice occurs from 72 to 78% which corresponds to formulae $2Al_2O_3{\cdot}SiO_2$. This view is supported by Tromel et al. (1957). Change in LC values of mullite with addition of alumina are also discussed by several authors.

Tromel et al. (1958) and Gelsdrof et al. (1958) showed a linear relationship between the magnitude of the *a* lattice parameter of the unit cell to alumina content.

Agrell and Smith (1960) classified mullite and showed the existence of a correlation between both the *a* and *c* cell dimensions with composition.

The latter authors first calculated the lattice parameter of natural and synthetic mullite samples by X-ray diffraction procedure using Si as celebrant. They estimated *a*, *b*, and *c* values by considering following (*hkl*) peaks e.g. (041), (401), (331), (002), (250), (520) and using the relation between lattice parameters with inter planar spacing (d) as per orthorhombic system ($1/d^2 = h^2/k^2 + k^2/b^2 + l^2/c^2$) and plotted the cell volume against the cell edge *c*. The values for mullite show considerable variation and show two extreme trends e.g., $3Al_2O_3{\cdot}2SiO_2$ on one hand and $2Al_2O_3{\cdot}SiO_2$ on the other hand and all values fall between the two curves. Small individual variations of *a*, *b*, and *c* are obviously there. Atomic substitution of aluminum increases *a* and *c* but does not effect *b*. Substitution of Fe and Ti in natural mullite shows a slight initial decrease in *a* and *b* followed by a subsequent increase; *c* increases uniformly for all concentrations. All the data lie in the solid solution range between $3Al_2O_3{\cdot}2SiO_2$ to $2Al_2O_3{\cdot}SiO_2$ and synthetic ferrian mullite. They noted two extreme compositions: 3:2 to 2:1 composition (ratios of Al_2O_3:SiO_2) and finally, axial ratios were plotted by them. Well crystallized mullite specimens give a consistent pattern of variation from D mullite ($3Al_2O_3{\cdot}2SiO_2$) to S mullite ($2Al_2O_3{\cdot}SiO_2$).

Crystallinity of mullite

The poorly crystallized samples however show erratic variations. It is considered by Agrell and Smith (1960) that it is due the existence of correlation between the state of structural order (as expressed by sharpness of reflections) and the axial ratio a/b. The positions of the axial ratios of the mullite specimens HS76, HS77 and HS78 (supplied by Hill and Roy) was explained by a low state of order. The X-ray studies of mullite and the unit cells of it from several sets of reflections were calculated by Aramaki and Roy (1962). They stated a considerable variation in unit cell measurements among different earlier authors which are unaccountable. Mullite synthesized at one composition shows different spacing during heating at different temperatures. That the changes in cell dimension of mullite caused by variation of the Al_2O_3/SiO_2 ratio from 3/2 to 2/1 might be equal to the changes caused by heat treatment of mullite of a fixed composition (Aramaki and Roy, 1962). This phenomenon is explained as likely related to Al-Si order to disorder transformation.

Composition vs. cell parameter relationship

Besides showing the existence of relation between its composition with the cell parameter of mullite as were measured by Tromel et al. (1958) and then Gelsdrof et al. (1958), Majumder and Welch (1963) also noted *a*-parameter of mullite is related to chemical composition of it. They showed a linear relationship with a dimension of the unit cell in mullite synthesized out of melt other than that obtained from solid state reaction. Numerous investigations have been done on the constitution of mullite and it's melting behavior. Much disagreements or many conclusions have been observed regarding the composition of mullite crystals by various researchers. Composition of it may be variable Al/Si ratio. They synthesized mullite of composition 3:2 alumina–silica mix by oxy-acetylene flame fusion process and second by melting followed by growing slowly in iridium high temperature microscopic stage. Thirdly, by melting ternary mixes of alumina, silica and calcium oxide followed by acid purification process. The cell dimensions of mullite of all samples were calculated from d values corresponding to the ×1 component of diffraction maxima of some reflections as per procedure of Agrell and Smith (1960). The variation in the unit cell dimensions of mullite shows as a function of their Al/Si ratio. It further shows that Al substitution for Si results in an increase of both *a* and *c* dimensions of the mullite unit cell. It also indicates that the Al/Si ratio of melt grown mullite bears a linear relationship with a dimension of the unit cell. In addition, they also studied the effect of heat treatment on mullite prepared from melts. Mullite may undergo a change on long heat treatment and may precipitate or exsolve either cristobalite or corundum. The lattice parameters of heat treated samples may increase with increased temperature, but the extent of *a* dimension change is smaller than in the previous case. The partial reason may be Al-Si order-disorder as pointed out by Aramaki and Roy (1962).

Various mullite samples were collected by Cameron (1977) under widely differing experimental conditions from melt and, some from natural sources followed by grinding and HF treatment. Lattice parameters measurement were made from a least-squares computer program. The *a* (cell edge) vs. *V* (cell volume) for mullite of different sources were plotted and from the linearity of the *a*–*V* plot, he suggested that there is some relation between cell dimensions and either chemical composition of Al-Si order or both. The *a*–*X* plot also shows linearity. With this observation he suggested a method of determining alumina content from the cell edge value. They expressed the stoichiometric mullite phase as under $[Al_2]^{vi}(Si_{2-2x}Al_{2+2x})^{iv}{}_{10-x}$. It corresponds to a continuous solid solution of mullite of variable composition of $x = 0.17$ to 0.59. They also measured lattice parameters of mullite from different sources. He thereafter outlined a method for determining the alumina content of mullite by establishing a linear relationship between *a* cell edge with Al_2O_3 composition in the range 57 to 72 mol %. Based on the findings that change in *a* axis value of mullite is very much sensitive to variation of (i) mole ratio of Al_2O_3 to SiO_2 and (ii) to calcination conditions, mechanism of mullitization in monophasic and diphasic gels.

16.4.1 Lattice Parameter Measurement of Mullite Obtained by Solid State Reaction

Several synthetic mullite samples synthesized under various conditions are also used for lattice parameter measurement studies (Cameron, 1977).

(i) Mullite samples crystallized from liquids can have compositions ranging from 57 to 71 mol % Al_2O_3 depending both on temperature of crystallization and cooling rate. For a given cooling rate, the alumina content increases with temperature. Rapid quenching at the high temperature end produces more aluminous mullite.

(ii) Mullite formed by reaction in the solid state shows temperature dependent compositions over a range of 60–66 mol % Al_2O_3. To determine the alumina-rich end he attempted to synthesize two mullite run samples limiting compositions of them are compared and following observations were noted.

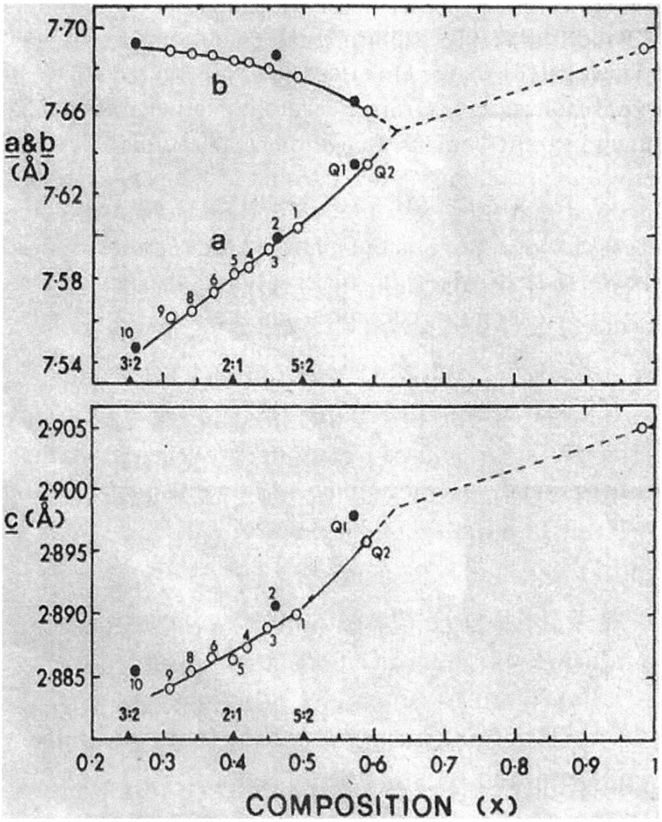

Figure 16.6 Variation of lattice parameters with mullite composition, with speculative extrapolations to the pure alumina end member (Cameron, 1977).

They reported that cell dimensions for mullitization in the composition range 60–71 mol % Al_2O_3 vary linearly with composition (Fig. 16.6). When best fit **a**- and **b**-lattice parameter curves inclusive of Q1 and Q2 data points are extrapolated on continuation, these two curves cross near $x = 0.63$ which is about 79 mol % Al_2O_3 composition of mullite. However, there is no direct evidence by way of synthesis regarding orthorhombic ↔ tetragonal transformation at 79 mol % Al_2O_3 composition.

The stoichiometric mullite phase as written below predicts to a crystalline phase with $[Al_2]^{vi}(Si_{2-2x}Al_{2+2x})^{iv}O_{10-x}$ variable composition/solid solution (Angel and Prewitt, 1986). Cameron showed above the possibility of obtaining mullite containing 79 mol % Al_2O_3 corresponding to $x = 0.65$ from the diagram Fig. 16.6. However, mullite containing 77.3 mol % Al_2O_3 is not presently available. We will see whether it could be synthesized by recent sol-gel technique.

16.4.2 Lattice Parameter Measurement of Mullite Obtained by Monophasic Gel/Coprecipitated Gel Method

(1) Metastable tetragonal mullite-like phase (t-mullite) with $a = b$ was first designated by Ossaka (1961). He prepared it as low as 1000°C, it showed considerable broad peaks and then presented the data of metastable tetragonal mullite like phase. The composition of it varies between 60–66 mol % Al_2O_3.

(2) That the alumina content of mullite and the amount of glassy phase increased with increasing temperature in case of gel of theoretical mullite composition (60 mol % Al_2O_3) were noted by Prochazka and Klug (1983) during their phase diagram studies. Based on the lattice parameter measurements of mullite, the composition of mullite was predicted by Okada and Otsuka (1986), Klug et al. (1987, 1990), Hirata et al. (1985a), Hirata et al., 1989) and these are chronologically shown below.

(3) Lattice constant data (LC) of mullite formed out of calcining SH and RH gels at different temperatures were calculated by least-squares method using RSLC-3 program first by

Okada and Otsuka (1986) and then compared these results with Cameron's data (1977). The lattice constants together with peak width of the 121 reflection data are also shown in Fig. 16.7.

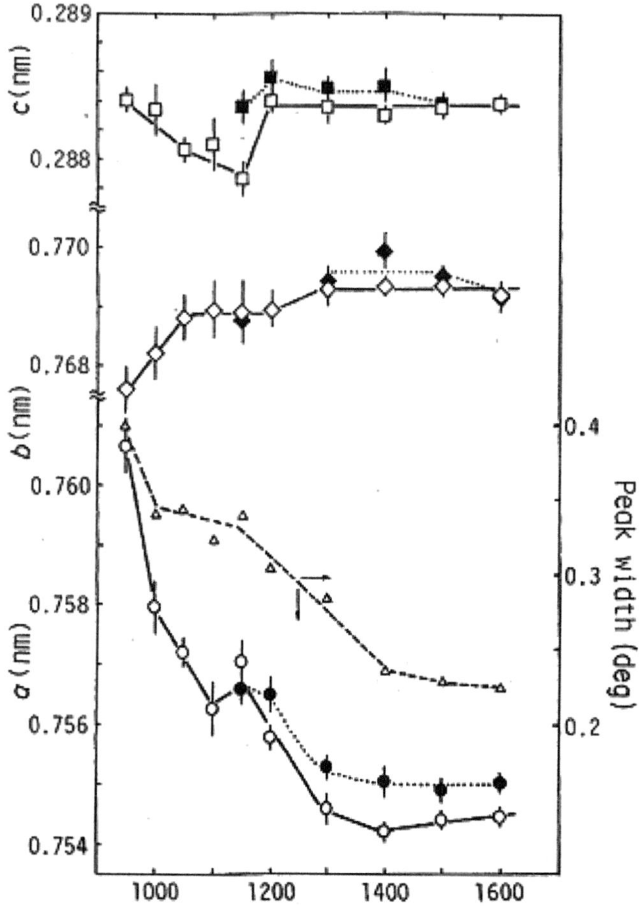

Figure 16.7 Lattice constants and peak width for 121 reflection of mullite as a function of firing temperature: Open symbols for SH and closed symbols for RH xerogels (Okada and Otsuka, 1986). (Δ, □, ◊, o) SH gel and filled in symbol, (●) RH gel.

During the LC measurement process, the length of *a* shortened rapidly in SH gel in two steps, one at 950 to 1100°C and the other at ~1150 to 1300°C. The lattice parameter changed slowly beyond this temperature. The length of *b* showed converse behavior to *a* while *c* shortened gradually from 1000° to 1150°C, but suddenly increased at 1200°C to the initial value, remaining constant thereafter. Variations of the lattice constants of RH gel with temperature changed similarly to those of SH gel. The estimated chemical composition of mullite was found to change from ~71 to 60 mol % in SH gel and 64 to 61% in RH gel as the firing temperature increased. They indicated that the tetragonal alumina rich mullite formed upon heating mullite gels at 1000°C gradually transformed to 3:2 mullite at ~1400°C.

Secondly, they also noted a correlation between variation in *a*, the IR spectra, and chemical composition conforming Cameron (1977).

(4) The compositions of mullite were obtained from measured lattice constant *a* value and on the basis of Cameron's data by Hirata et al. (1985a) out of SiO_2-Al_2O_3 powders of different ratios on heating at different temperatures. The variation in the lattice constants of mullite produced on firing powders at 1400°C and 1500°C with composition of the powders were further shown by Hirata et al. (1989). With these data they're calculated the composition of mullite formed at those two temperatures. They firmly observed that lattice constant value *a* showed the greatest composition dependence.

(5) On considering (401), (041) and (002) reflections of mullite using high purity silicon powder as an internal standard Suzuki et al. (1990) determined the lattice parameters. They showed that LC values of *b* and *c* are almost constant and coincided with the value reported in JCPDS 15-776 below 1500°C. The *a* axis value was slightly higher than that of stoichiometric mullite. On prolonged heating for 10 h at 1400°C to 1500°C, the *a* axis value decreased and coincided with the stoichiometric value.

(6) Lattice constants of gels marked SGM and CM heated to different temperatures were calculated by Schneider et al. (1992) and it was also shown that change in lattice parameters values were function of annealing temperature; *a*

value decreased while *b* increased slightly, and *c* value did not change with heating. They designated the mullite formed at the first exothermic temperature in both cases of SGM and CM are Al_2O_3 rich phase and likely pseudo tetragonal in nature. They emphasized that the fact of gradual transformation of the unstable, low temperature Al_2O_3 rich mullite to stable high temperature orthorhombic form was essentially controlled by annealing temperature.

Mullite type phase (~74 wt % Al_2O_3 and 26 wt % SiO_2) was synthesized by Schneider and Rymon-Lipinski (1988) out of TEOS and AlOBu by dissolving in isopropyl alcohol in absence of water and then heating between 950°C to 1000°C. They observed good coincidence between X-ray powder diffractograms of normal mullite and mullite type phase with respect to position and intensity of X-ray reflection as well as some deviations. For example, lack of 120/210, 240/420, 041/401, and 250/520 reflection pair splitting on the X-ray diffractograms indicated that the length of *a* and *b* constants have become equal. Symmetry is tetragonal with respect to normal orthorhombic mullite. Secondly, t-mullite was slightly broadened due to structural disorder. Thirdly, lattice constant calculation showed a_{TM} spacing of t-mullite occur in the range 7.612 to 7.659 Å, while c_{TM} spacing comes to the range 2.886–2.896 Å. Thus, c'_{TM} value of tetragonal mullite is very close to that of orthorhombic (OM) 3/2 mullite (c_{OM} = 2.883). The length of a'_{TM} lies between a'_{OM} and b'_{OM} values of orthorhombic mullite (a_{OM} = 7.543 Å and b_{OM} = 7.702 Å). The mean cell volume (V_{TM}) of tetragonal mullite (~167.4 A^3) is similar to that of orthorhombic mullite (V_{OM} = 167.4 A^3). Thus, primary mullite which is crystallized during monophasic gel at 1000°C shows no splitting of peaks with Miller indices of 120, 240, 041, and 250. Two possible ways either with increase of soaking time or increase in calcination temperature may give rise to a clear angular separation of diffraction lines. The metastable mullite becomes highly crystalline and stable.

The plot of *a* lattice constant vs. temperature by Schmucker and Schneider (1999) showed that mullite formed on heating Type I precursor at ~900°C is reached in Al_2O_3 (>70 mol %) as per Cameron data. The *a* axis value approached stoichiometric mullite composition after heat treatment 1500°C which verifies the result of Okada and Otsuka (1988) and Geradin et al. (1994). Schneider, Fischer, and Voll (1993), Fischer, Schneider, and Voll (1996) of course showed a metastable mullite phase with $a > b$.

(7) Other researchers noted similar observations. For example, Baranwal et al. (2001) who noted the splitting of the (120) (210) peaks at 26.15° 2θ in t-mullite into 25–97° and 26.26° 2θ in o-mullite during the heating process. Formation of low crystalline mullite patterns was also noted by Huang et al. (1997). The (120) and (210) crystal planes of mullite are indistinguishable.

16.4.3 Changes in LC of Mullite Formed Out of Monophasic Gels of Different Compositions

(1) Monophasic gels of three different compositions were synthesized by Yamada and Kimura (1962) and it was observed the following. Lattice constants of mullite formed from coprecipitated gel of the same composition differ depending upon heating temperature and time. Coprecipitated gel of different compositions under the same conditions even differ in LC.

(i) Higher the temperature of heating of sample, lower are the LC of a and b values, and higher is the constant c of the mullite formed. Lattice parameter values of mullite measured by them showed fluctuation in the following range. a = 7.54–7.56 Å, b = 7.69–7.70 Å, c = 2.88–2.89 Å. Mullite is said to attain a standard stable state corresponding to that heating temperature. With increase of heat treatment temperature, the value of a and c of mullite decrease, whereas the value of b showed an increase.

(ii) When a coprecipitated gel was heated at ~1300°C or 1400°C for a long time, the LC values converged to a constant value. They are of the opinion that with passage of time the crystallinity of mullite formed went to increase gradually and finally it reached *a* lattice stage for that particular temperature. They plot the converging value of the lattice constant a/b vs. c/b in Smith's chart. Three samples A, B, and C show the convergence in different regions. That for B and C nearly conforms with the position that corresponding component used by Agrell and Smith (1960). According to them, these results indicate that the chemical composition of mullite shows the existence of a solid solution rich in Al_2O_3 rather than $Al_2O_3/SiO_2 = 3/2$. However, the convergence point of A falls out of Smith's diagram.

(iii) With increase in Al_2O/SiO_2 ratio, the value of a is increased and b is decreased. It was concluded by them that the time variations of a and c might be mainly due to change in chemical composition and time variations of b might be due to the process of ordering of lattice. From the plot of the converging value of the lattice constant a/b vs. c/b in Smith's chart, they indicated the existence of a solid solution rich in Al_2O_3. The possibility of mullite containing more silica than 3:2 is also indicated. They reviewed that LC of mullite shows the variation. Summarily, these variations may be due to various causes.

Composition of mullite itself. It forms solid solutions containing alumina.

Even in case of mullite of the same composition, LC of mullite varies due to increase in degree of crystallization of mullite.

During the formation process of mullite out of coprecipitated gel of a fixed composition on heating, LC differs depending on heating temperature and time.

During the formation process of mullite out of coprecipitated gels differing in compositions, LC differs under the same condition on heating.

(2) The lattice constants of mullite obtained by heat treatments at different temperatures of three monophasic aluminosilicate gels in the range 63 to 80 wt % Al_2O_3 were measured by Low and McPherson (1989). They showed that a straight line relationship between cell parameter, a, and cell volume, C_v fitted Cameron's data. This result further established the existence of solid solutions between (3:2) and (2:1) mullites.

They showed that some of mullite peaks at (120) and (210) lines as also peaks at (250) and (520) lines are not distinctly split rather than broad and diffuse on heating monophasic gels <1200°C. Splitting occurs on heating ≥1300°C. Characteristically, it occurred with the disappearance of Al-Si spinel phase when abundant mullite appeared.

Thus, various earlier researchers correlated this change with variation in the composition of mullite obtained in case of monophasic gel following Cameron (1977). According to them, mullite crystallized at ~1000°C termed as primary mullite, is pseudo–tetragonal in nature since they did not observe spitting of 121 and other peaks. Clear angular separation was only noted on further increase of temperature. The tetragonal mullite (66 mol % Al_2O_3) changes to orthorhombic (60 mol % Al_2O_3) on further heating at ~1600°C. In the coprecipitation process, mullite was found to appear at 1200°C i.e., much above 1000°C at which RH gel generally started mullitization. Furthermore, splitting of 121 peaks was not also observed at that temperature. With gradual increase of firing temperature more and more clearer angular separation of 120/210 diffraction lines became evident. Two questions may be raised : whether mullite formed at ~1200°C is pseudo–tetragonal in nature and the observed change in *a* lattice parameter during further heating is related to any change in composition as assumed earlier. Cameron plotted *a* cell edge vs. *V* (cell volume) and secondly *a* cell edge vs. X (composition of mullite) chosen from different sources such as melt grown and of naturally occurring. He noted that both curves are linear. With these relationships, he suggested that changes in *a* cell edge were related to dual causes. It may either be chemical composition of mullite or to Al–Si ordering or both. However, he had taken mullite samples which are mostly fused or sintered at high temperature instead of just ensuring mullite phase (unstable in nature) crystallized out of monophasic and or coprecipitated gels.

(3) That the primary crystallized mullite formed on heating two precursors marked A and B at 1000°C are poorly crystalline (Sales and Alarcon, 1996). As like above researchers, they

called this as pseudo-tetragonal mullite because of the absence of splitting of peaks of Miller indices of 120, 240, 041, and 250. They also noted two ways for increasing crystallinity of mullite. With these observations they were of the opinion that at ~1000°C mullite seems to be tetragonal and that it transformed into a truly orthorhombic one upon increasing either holding time or heating temperature which also give rise to a clear angular separation of diffraction lines. The reason for nonsplitting of peaks is due to poor crystallization of mullite and consequently showing low resolution. Secondly, variation of lattice parameters of mullite with increasing temperature for samples marked as A, B, C synthesized by them were also noted. The change of the parameter *a* was shown to be a function of temperature.

16.4.4 Lattice Parameter Measurement of Mullite Obtained by Diphasic Gels Method

Diphasic mullite gels of different compositions were synthesized by (Klug et al., 1987) using AlO(OH) and TEOS as starting materials in ammoniacal medium prior to their phase diagram study. They measured the lattice parameter of mullite formed at high temperature by a computer programmer using some nonoverlapping peaks of mullite with those of alumina and standard silicon. Plot of *a* spacing varied as a function of composition between 72–77 wt % Al_2O_3 which agreed with Cameron's curve.

The lattice parameter of mullite, prepared out of fume silica and aluminum silicate during calcination in the range of temperature of 1250°C to 1700°C was carried out by Mizuno and Saito (1989). The parameter *b* of poorly crystallized mullite formed at 1250°C was 7.682 Å which is lower than *b* value (7.690 Å) of mullite formed at 1300°C. The parameter *a* (7.559 Å) for mullite formed at 1250°C decreased to 7.541 Å for mullite obtained on calcinations at 1400°C and 1500°C.

Changes in lattice parameter values of mullite in case of heating diphasic gel was shown by Chakraborty (2005b) in Fig. 16.20 as shown below.

16.4.5 Composition of Alumina Rich Mullite

A spectacular observation was made on the formation of newly formed mullite type phases at different steps of the phase evolution process of mullite precursors (Schneider et al., 1993) and later by Fisher et al. (1996).

At the first step at 700°C to 900°C—a mullite type phase was noted at an annealing temperature of 700°C which exhibits broad X-ray peaks of low intensity. X-ray line intensities increased out of coexisting large quantities of noncrystalline phase on heating. The broad line widths became sharper. The degree of structural order increased at 900°C. LC measurement showed $a > b$ in lattice constant. The chemical composition of this low temperature mullite was determined from the usual linear relationship between a lattice constant value and molar content of Al_2O_3 and that corresponds to very high in Al_2O_3 content (~88 mol %).

At the second step at 1000°C—a new type of mullite phase developed and Al_2O_3 content of it increased to 92 mol % during heating at 1000°C.

At the third step at 1250°C—mullite exhibits $a < b$ lattice constant during measurement and that corresponds to decrease in Al_2O_3 content from 73 mol % to (std. mullite) during heating.

16.5 Changes in LC of Mullite Formed Out of Five Different Compositions of Diphasic Gels: A Case Study

The changes in nature and character of mullite formed during transformation in case of diphasic gels were reexamined by Chakraborty (2005b) following the determination of lattice parameters, strain values of mullite etc.

1. Lattice constant data

Variation of lattice constants data of mullite with increasing temperatures for diphasic gels of three different compositions are shown in Table 16.3.

(a) Silica-rich gel (DG40), analogous to the composition of kaolinite: The length of *a* decreases markedly with increase of firing temperature and then remains more or less constant on further heating. The length of *b* also decreases but to a lesser extent than *a*. The length of *c* shortens gradually on heat treatment. The nature of changes in lattice parameters agrees with those reported by Okada and Otsuka (1986). Irrespective of the composition and nature of the gel, whether monophasic or diphasic, in the secondary mullite formation stage *a* axis length decreased with increase of firing temperature and then remained constant.

(b) Gel of sillimanite composition (DG56): The axial value of *a* decreases gradually while those of *b* and *c* remain almost constant throughout the entire range of temperature.

(c) Gel of 3:2 mullite composition (DG72): The value of *a* remains almost constant throughout the range of temperature, *b* and *c* values increase slightly on heating. In earlier study, Suzuki et al. (1990) noted constant *b* and *c* values during heating their aluminosilicate gel obtained by partial hydrolysis method.

(d) Alumina-rich gel (DG85): The value of *a* increases slightly but subsequently decreases on heating.

Gel DG40 shows higher *a* and *c* values during heating than other gels. On continued heating these values approach those of conventional mullite. Lattice constant of mullite formed from gels having mullite composition and from alumina rich compositions show larger values of orthorhombic mullite (OM) shown in JCPDS 15-776 (a_{OM} = 0.75456 nm, b_{OM} = 0.76898 nm and c_{OM} = 0.28842 nm). Small axial differences may be due to differences in chemical composition among the diphasic gels. Specifically, the percentage of alumina in mullite controls *a* value. Variation of lattice constants of mullite heated at 1600°C is a function of the chemical composition of diphasic gels (Fig. 16.8) and the data resemble those of Hirata et al. (1989).

This result indicates that values of mullites produced in DG72 and DG85 at 1600°C are higher than that noted in case of DG40 and DG56. It is concluded that mullite formed in former gels may be rich in alumina and may be differ from the starting compositions.

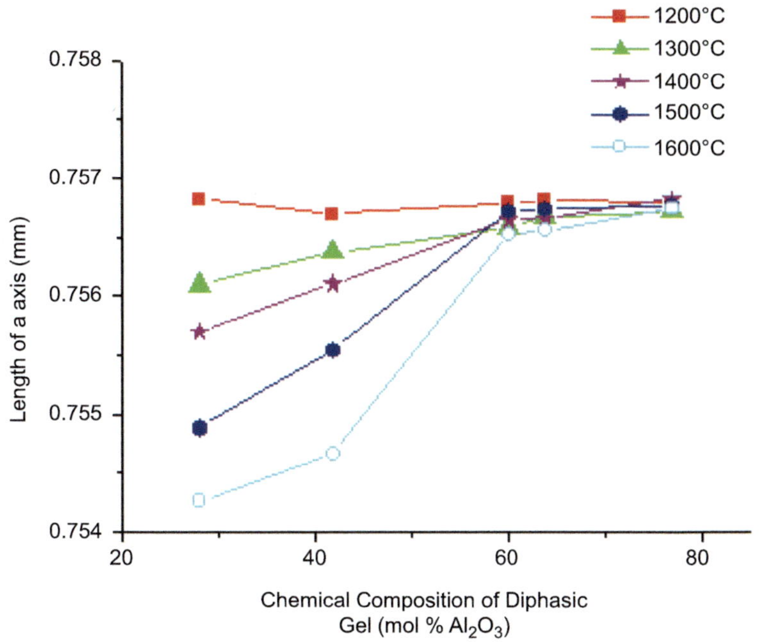

Figure 16.8 Changes in *a* axial length of mullite formed out of diphasic gels a function of temperature of heat treatment.

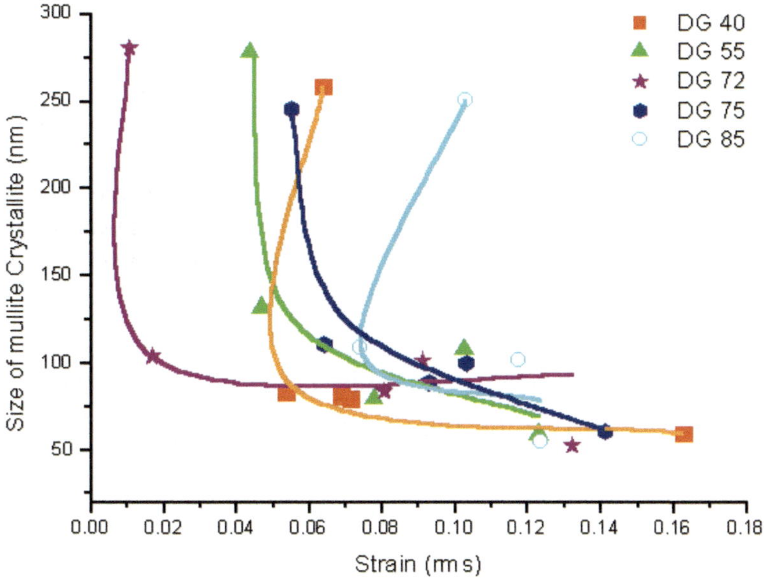

Figure 16.9A Interrelationship between crystallites size vs. strain of mullite formed during heating DG40 at different temperatures.

2. Size and Strain values of mullite

The size of mullite crystallites and its strain value are both functions of heat treatment temperature and are shown in Table 16.3. The inverse relationship among size and strain is apparently noted upon heating DG40 at different temperatures are shown in Fig. 16.9.

Average particle sizes of mullite were measured by Li and Thomson for a diphasic gel (Al/Si = 3:1) heated to 1250°C/1 h and found to be ~50 μm. In comparison, the size of mullite in DG40 heated at the same temperature is ~63 μm by Quasar analysis.

3. Peak width

Changes in peak width for the 121 reflection of mullite derived from DG40 diphasic gel are also a function of firing temperature (Fig. 16.10).

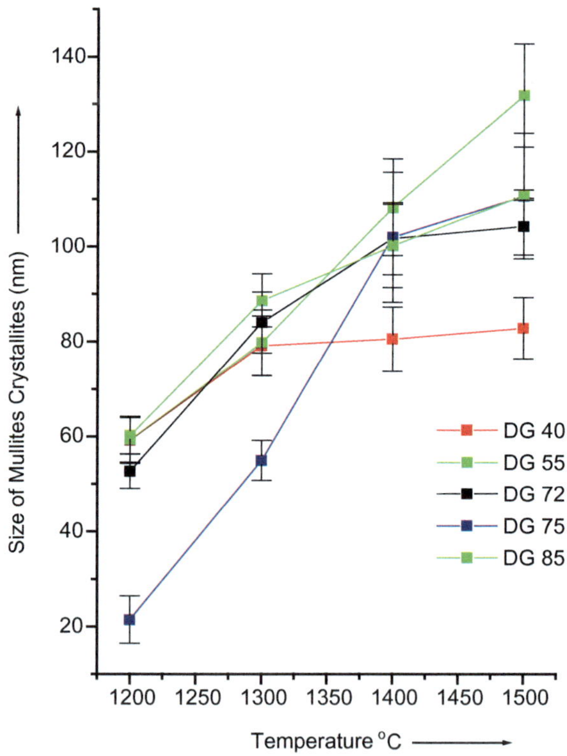

Figure 16.9B Size of mullite crystallites developed in mullite gels of different compositions vs. temperature.

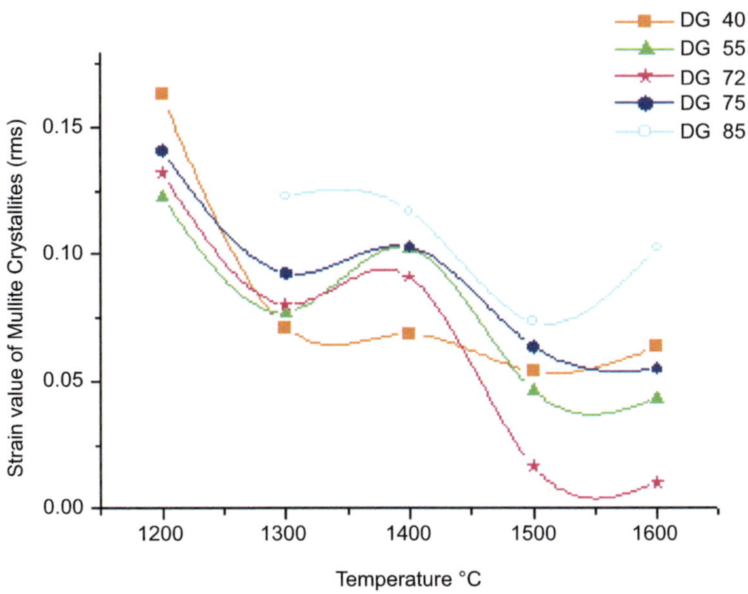

Figure 16.9C Variation in strain values of mullite formed out of diphasic gels at different temperatures.

The decreasing peak width with increasing calcination temperature verifies earlier observations of Okada and Otsuka (1986). This is explained to be due to increasing crystallinity of mullite at higher temperatures.

In the case of DG40, *a* lattice parameter value of mullite decreases with increase of temperature from 1200°C to higher temperatures (Table 16.3). This result verifies the earlier finding that the change in *a* axis parameter of mullite is a function of heat treatment of aluminosilicate gels. XRD studies show that mullite formation in DG40 starts at ~1200°C, and remains almost constant up to 1600°C, the final temperature of the heating schedule. However, lattice parameters, crystallite size and strain values of crystalline mullite change in the temperature range 1200°C to 1600°C. Figure 16.10 shows that size of mullite when it starts to crystallize is very small and this newly formed mullite is highly strained. These two parameters change in the reverse manner with increasing temperature i.e., strain decreases while size increases. This behavior definitely indicates that mullite changes in structural order. This may be the reason for the change in *a* lattice constant value. With the rise of temperature mullite crystals become well crystallized.

Table 16.3 Lattice parameters of four diphasic gels fired at different temperatures

Gel mark	Heating temperature (°C)	Lattice constants of mullite (Å)		
		a (error)	*b* (error)	*c* (error)
DG40				
	1200	7.5683 (0.00223)	7.6988 (0.00219)	2.8875 (0.00068)
	1300	7.5559 (0.00272)	7.7004 (0.00215)	2.8865 (0.00061)
	1400	7.5570 (0.00261)	7.6949 (0.00210)	2.8822 (0.00049)
	1500	7.5489 (0.00173)	7.6948 (0.00162)	2.8850 (0.00050)
	1600	7.5426 (0.00183)	7.6924 (0.00180)	2.8879 (0.00056)
DG 56				
	1200	7.5625 (0.00235)	7.6968 (0.00228)	2.8879 (0.00071)
	1300	75638 (0.00222)	7.6959 (0.00227)	2.8871 (0.00061)
	1400	7.5611 (0.00163)	7.6969 (0.00161)	2.8868 (0.00049)
	1500	7.5555 (0.00126)	7.6944 (0.00124)	2.8861 (0.00039)
	1600	7.5467 (0.00089)	7.6956 (0.00087)	2.8856 (0.00027)
DG72				
	1200	7.56613 (0.00219)	7.69824 (0.00211)	2.88658 (0.00066)
	1300	7.56720 (0.00190)	7.69296 (0.00197)	2.88676 (0.00057)
	1400	7.56649 (0.00150)	7.69539 (0.00145)	2.88695 (0.00045)
	1500	7.56719 (0.00112)	7.69716 (0.00113)	2.88844 (0.00036)
	1600	7.56532 (0.00102)	7.69791 (0.00102)	2.88783 (0.00032)

Gel mark	Heating temperature (°C)	Lattice constants of mullite (Å)		
		a (error)	*b* (error)	*c* (error)
DG85				
	1200	7.56400 (0.02285)	7.69379 (0.01891)	2.88762 (0.00525)
	1300	7.56720 (0.00243)	7.71932 (0.00250)	2.89325 (0.00060)
	1400	7.57157 (0.00217)	7.69135 (0.00211)	2.88665 (0.00063)
	1500	7.56003 (0.00183)	7.69627 (0.00182)	2.88671 (0.00055)
	1600	7.56751 (0.00164)	7.69516 (0.00179)	2.88562 (0.00054)

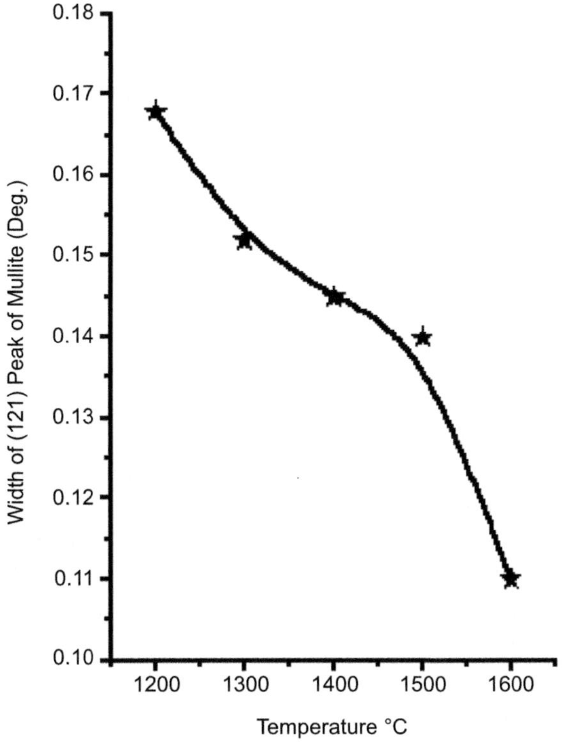

Figure 16.10 Changes in the peak width of mullite during heating DG40.

Lattice parameter measurements of mullite samples have been done during the start of its formation from 1200°C to higher temperatures. Therefore, the nature of the mullite samples used by Cameron are certainly different from those of the present investigation. He used quite well developed mullite crystals differing only in composition within or near the solid solution range. In the title study mullite develops from mixed alumina - silica precursors from their incipient stage to highly crystalline form. It is obvious that the reason for the changes in mullite lattice parameters could differ between the two studies. In the course of heating process of diphasic gels, particle sizes of mullite when started to form at ~1200°C are quite small, of the order of 50 to 70 μm but increase gradually and then rapidly at 1500°C to 1600°C. The reverse occurs for strain values i.e., initially formed mullite is highly strained and strain decreases on increase of calcination temperature. This is more marked at and above 1500°C. So it would not be necessary to compare the lattice constant data of strained and small size mullite which is just developing at lower temperature at <1200°C with that of highly crystalline mullite taken by Cameron. It is better to assume that changes in lattice parameters of mullite observed in the title study are mostly due to changes in structural order rather than to changes in composition. In support of this view, the present study shows that the width of (120) of mullite decreases with temperature when diphasic gel marked DG72 is heated gradually (Fig. 16.11). The splitting of 120/210 reflection pairs increase when fired at increasing temperature. Therefore, mullite which develops at ~1200°C in four cases of diphasic gels, namely a silica rich composition, sillimanite like composition, ideal 3:2 composition and an alumina rich composition would be a mullite type phase. A high temperature is required to transform from this mullite type form to orthorhombic mullite. In all three cases lattice parameter measurement of mullite shows $a < b$ values. The probable change in lattice constant with heating is due mainly to increase in crystallinity of mullite phase by gradual relief of strain present during its incipient formation stage than to changes in chemical composition. For example, ^{27}Al resonance spectra of a diphasic precursor D heat treated to different temperatures (after Jaymes et al., 1996). It first crystallizes into a spinel phase and a small amount of mullite. Resonance pattern shows a strong band of AlO_6 compared to AlO_4.

At 1275°C when it completely crystallized to mullite, AlO_4 increased in contrast to AlO_6. Surprisingly the ratio increased further at 1600°C even after mullitization occurred at an earlier temperature.

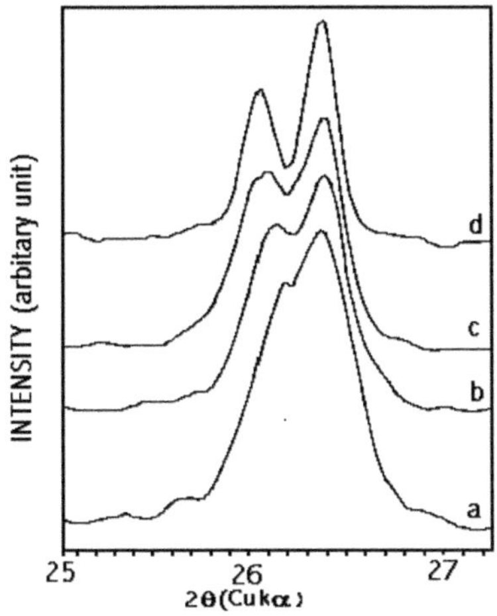

Figure 16.11 Resolution of 120/210 peak of mullite with increase of heat treatment temperature of DG40: (a) 1200°C; (b) 1300°C; (c) 1400°C; (d) 1600°C.

16.6 Heat of Reaction/980°C Crystallization Enthalpy of Various Mullite Gels and Variation of Activation Energy for 980°C Crystallization

Heat of reaction data of mullite precursors synthesized by different processing conditions by various authors vary to a great extent.

(i) Coprecipitated gel:

Geometry of 980°C exotherm of coprecipitated gel varies as SiO_2/Al_2O_3 ratios vary. An equimolar mixture of the two gel shows the highest heat of reaction, whereas that of the mullite ratio was ½ lower (Horte and Wiegman, 1956). A maximum heat of reaction at the composition of mullite (0.67 SiO_2:Al_2O_3) was noted by Demediuk and Cole (1958). However, a maximum heat of reaction was noted at 0.5 SiO_2/Al_2O_3 by West and Gray (1958). Maximum of exothermic peak height in gel of composition of 3:2 mullite was shown by Chakraborty (1979) in (Fig. 16.12). The highest exotherm indicates maximum crystallization of Al-Si spinel phase. The decreased peak height in gels of other than mullite composition can be explained as due to increased SiO_2 or Al_2O_3.

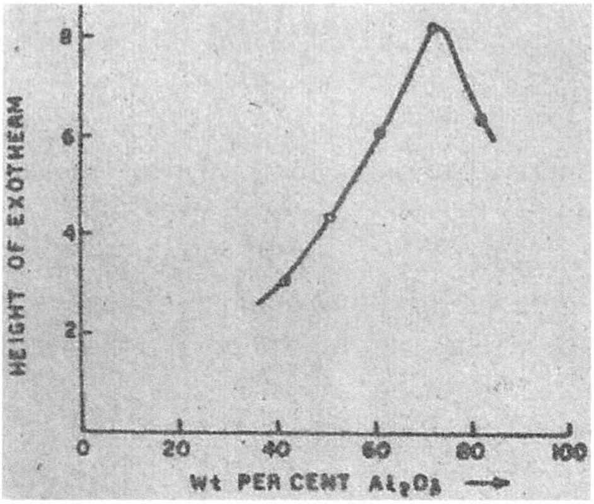

Figure 16.12 Exothermic peak height vs. amount of Al_2O_3 in coprecipitated gels.

(ii) Monophasic gel precursor:

The formation of mullite (major amount) and spinel phase (quantity) was noted by Hoffman et al. (1984) who obtained enthalpy value of the exotherm to be −85 J/gm. Contrary, a sharp exothermic peak at ~975°C in the DTA curve for monophasic gel was attributed to major formation of Al-Si spinel phase by XRD analysis by Hyatt et al. (1990). The enthalpy change, $-\Delta H$, corresponding to the above exotherm was determined from the peak area for a sample which had been pre calcined at 490°C and was calculated as −166 J/gm. The enthalpy of crystallization of sol-gel precursor to mullite was also measured by Geradin et al. (1994) using a twin calorimeter of the Tian Calvet type. It measures the heat flow through a sensitive thermopile which surrounds the sample chamber and it is separated from an alloy block maintained at 703°C. The thermopile records a change in voltage corresponding to small temperature variation occurring due to heat effect of the crystallization of the precursor sample. The enthalpies of formation of the precursors M1, M2, and M3 from γ-alumina and fused silica glass as functions of the annealing temperature are compared. The enthalpies of formation of the mullites from a mixture of γ-alumina and fused silica (quartz) were endothermic. In case of M1, the enthalpy value obtained by drop solution calorimetric technique was −112 KJ/mol. The abrupt enthalpy decrease was correlated with crystallization of mullite with complete disappearance of penta coordinate aluminum to stable four and six environments.

(iii) Spray pyrolyzed precursors:

1. DTA analysis of three samples of compositions namely 68A (67.8% Al_2O_3), STD (72.1% Al_2O_3) and 78A (77.8% Al_2O_3) were carried out by Kanzaki et al. (1990), Kumazawa et al. (1990, 1991). Two significant observations were noted by them.
   (i) Apparently, the height of sharp exotherm was found highest in STD sample and it fell on other sides of STD sample.
   (ii) Mullite formations only above 980°C were noted in silica rich and STD samples. However, the alumina rich sample showed corundum in addition to mullite.
2. Jaymes and Douy (1992) noted enthalpy changes of three following precursors.
   Powder A = Enthalpy of 102 kJ/mol (240 J/g) for first exotherm (exo).
   Powder C = Enthalpy for first exo [58 J/g (first peak) and 111 J/g (second peak)]. Second exo = 40 J/g
   Powder B = Enthalpy for first exo [46 J/g (first peak) and 137 J/g (second peak)]. do- (not observed) no
3. Jaymes and Douy (1995) also noted enthalpy changes of three following precursors.
   Mull A = Enthalpy of 80.5 kJ/mol for first exo, not observed - second exotherm.
   Mull B = Enthalpy of 31.85 kJ/mol for first exo, enthalpy of 8.1 kJ/mol for second exo
   Mull C = Enthalpy of 109 kJ/mol for first exo

4. Jaymes et al. (1996) too noted enthalpy changes of three following precursors.
   Enthalpy at first exotherm for samples
   A = 310 J/g; E = 260 J/g; and C = 240 J/g
   B = 146 J/g for first exo; and 64 J/g for second exo
   D = 169 J/g for first exo; and 19 J/g for second exo

A list enthalpy values by different authors are also tabulated below.

Table 16.4 Enthalpy value of the 980°C exotherm of various precursors

	Researchers	Enthalpy value (−ΔH)
1.	Hoffman et al. (1984)	85 J/g
2.	Hyatt et al. (1990)	166 J/gm
3.	Kanzaki et al. (1985)	240 J/g (102 kJ/mol
4.	Sen and Thiagarajan (1988)	250 J/gm
5.	Gerardin et al. (1994)	112 kJ/mol
6.	Tkalcec et al. (1999)	105 kJ/mol
7.	Douy (2006)(for 60 mol % Al_2O_3)	290J /g or −124 kJ/mol

(a) Geometry of 980°C exotherm

(i) Highest DTA peak height was noted by Demediuk and Cole (1958), Chakraborty (1979) of their coprecipitated gels.
(ii) Highest 980°C exotherm was noted by Kanzaki et al. (1990), Kumazawa et al. (1986) in their STD (72.1% Al_2O_3) of SP precursor samples.
(iii) Highest 980°C exotherm enthalpy was obtained based on DSC data (Douy, 2006).

(b) Temperature of crystallization of mullite ~1280/1300°C (Douy, 2006)

(c) Activation energy

Activation energy of crystallization of mullite formation data which was calculated out of exothermic peak noted in DSC experiments and using Kissinger equation Douy (2006).

16.7 Summary

Mullitization in all six cases of mullite gel synthesized by various researchers in cases of (i) monophasic gels, (ii) coprecipitated gel, (iii) colloidal gel, (iv) spray pyrolysis/spray dried precursors, and (v) diphasic gel are quite different.

In monophasic gel system studied by earlier authors

Based on qualitative and QXRD studies, Okada and Otsuka (1986) explained two steps of mullite formation in monophasic SH gel (Fig. 16.1). Schneider et al. (1993) first pointed out the formation of alumina rich mullite by XRD during heating Type I precursor at 980°C exotherm and silica rich amorphous phase by MAS NMR. Regular mullitization proceeded by inter diffusion among two later phases on further heating. SH mullite gels made by Li and Thomson (1991) exhibited strong 980°C and two steps of mullitization in samples 1D, 2D, 2W, and 2WC gels in both DTA and DXRD studies. They compared the formation process of mullite for three kinds of gels (a) 2WC monophasic gel (Al/Si = 3/1), (e) for CG colloidal gel (Al/Si = 3/1), and (c) for DG3 diphasic gel (Al/Si = 3/1) at the exothermic temperatures as shown in DXRD studies.

In RH gel studied by earlier authors

The solid state reaction between the intermediary Al-Si spinel phase formed and with SiO_2 (A) liberated during heating Al_2O_3-SiO_2 monophasic RH gel at first exotherm is responsible for predominant mullitization during heating at ~1100°C (Fig. 16.1) and accordingly DTA analysis exhibited a second exotherm at ~1250°C as shown by Okada and Otsuka (1986). Besides this rapid mullitization, the growth of mullite is continued on heating to 1600°C.

In case of in case of SGM gel and Type II precursor synthesized by (Schneider et al., 1992), and Schneider et al. (1993), two segregated phases (alumina rich spinel as major phase and noncrystalline silica) reacted at ≥1200°C, nucleate and formed mullite by short range migration of diffusing species. On long heat treatment, long range diffusion continued for crystal growth to occur and to produce well developed mullite.

In six different types of precursors synthesized by author

Nature of the growth curves of mullite e.g., SHI to SD suggest three distinct cases by Chakraborty (2008). Growth curves (Fig. 12.6) show that these are different in the following aspects e.g., initiation temperature of its formation, it's paths and obviously mullitization mechanisms.

(i) Most rapid mullite formation out of noncrystalline precursor phase is noticed at ~1000°C exotherm in case of SHI, SHII and SD by Path I. Rapidity of mullitization of these precursors occurs in the following order. Spray dried precursor (SD) > SHI > SHII.

(ii) In addition to formation of mullite (major phase) as in SHI, SHII, Al-Si spinel phase (minor) forms at ~1000°C (Fig. 18.2) which transforms at the second exothermic peak by Path IV and the amount of mullite increases gradually to slight extent with further rise of temperature. Besides these mullitization processes mullite formation also occurs on heating at >1200°C by usual Path III.

(iii) Direct Al-Si spinel formation at the first step at ~1000°C, and transformation of silica enriched Al-Si spinel to mullite formation at the second step at ~1200°C are noticed in case of SHIII and SHIV/CP6 by Path IV.

(iv) The ratios of Al-Si spinel to mullite formation during heating SHII and SHIII at ~1000°C vary largely. For example, SHII forms mullite (major amount) and Al-Si spinel (minor amount) whereas the reverse is true for SHIII precursor. Mullite forms at the second exothermic peak temperature at ~1250°C by Path IV out of Al-Si spinel phase rapidly forms earlier at ~1000°C. Prior to second exotherm, an additional amount of mullite formation occurs at the occurrence of third exotherm by Path V.

(v) In case of diphasic gel, mullitization takes place by the transformation of Al-Si spinel by Path IV at >1300°C which was accumulated gradually during the exotherm in the temperature range 1000°C–1320°C. Details are given below.

(vi) Intermediate Al-Si spinel phase formation is predominant in SHIII, SHIV and diphasic gel and mullite formations found to occur to a comparatively lesser extent than in case of SHII. Since growth curves of mullite in those cases always lie below the growth curve of SHII.

(vii) Although sudden mullite formation occurs in three cases of precursors namely SHI, SHIV and diphasic gel (Fig. 12.6), starting phases are however different. In case of SHI, mullite formation takes place from the intermediate noncrystalline aluminosilicate precursor phase that existed prior to the 980°C exotherm. Whereas in cases of SHIII and SHIV, mullite formation is noticed out of two phases namely (i). From intermediate Al-Si spinel phase at second exotherm (by Path IV), and (ii). By solid state reaction between large quantities alumina rich phase and silica rich aluminous (A) phase accumulated on gradual heating (by Path V following the third exotherm). The later step occurs negligible in case of SHII. In case of diphasic gel, mullitization starts from the accumulated Al-Si spinel phase at ~1320°C exotherm by Path IV and at higher temperature by Path III. Due to the difference in the start temperatures of mullitization, the nature of mullite formed varies. In the case of SHI, it may be nonstoichiometric in nature. On further heating beyond 1400°C it becomes orthorhombic, called as secondary mullite or o-mullite. Both in cases of SHIV and diphasic gel, mullite which is formed at temperature >1200°C may be considered as mullite type phase which changes to orthorhombic in nature on further heating. Thus, nature of mullite formed in these precursors at a long temperature region is of two different forms and the courses of its formations are also varying (Fig. 12.6).

Various concept of mullitization in three types of diphasic gels by different authors

(1) First one is solid state reaction between γ-, δ-, and α-alumina with SiO_2 (A) due to previous studies of (Hoffman et al., 1984; Schneider et al., 1993; Ivankovic et al., 2003).

(2) Second was conjectured by Wei and Halloran (1988) that mullite formed in diphasic system by the direct reaction of γ- or δ-Al_2O_3 and SiO_2 without the formation of intermediate aluminosilicate (A) precursor phase. It was noted that growth

of mullite in diphasic gel system was sigmoidal in nature. On the other hand, it was reported to be parabolic growth curve for oxide mixture (Aksay and Pask, 1975). During analysis of mullite formation kinetics, it is believed that the dissolution of γ- and or δ-Al_2O_3 phases in vitreous silicate phase is the cause of nucleation of mullite in diphasic gel (Sundaresan and Aksay, 1991).

(3) Third one is conceived by Chakraborty (2008) in case of "in situ" diphasic gel and Type II precursor.

Mullitization of in situ diphasic gel studied by DTA and QXRD. It is assumed that Si out of noncrystalline aluminisilicate (AS) diffuses into γ-Al_2O_3 or alumina rich spinel structure and formed Al-Si spinel during heating diphasic gel with exhibition of a broad exotherm in the temperature range ~1000°C to as high as ~1300°C (just before mullite crystallization). Thereafter, mullite formation occurred rapidly at temperature of exotherm at ~1300°C. This sudden mullite formation out of Al-Si spinel may indicate polymorphic nature of transformation which might be a feasible proposition.

The growth curve (Fig. 16.4) of it showed that mullite rapidly formed at ~1300°C and thereafter it's growth still continued with further rise of temperature of heat treatment. The profile of growth of mullite is analogous to that of kaolinite. In the case of kaolinite, the growth curve of mullite consisted of two segments and accordingly it was explained earlier that mullitization took place by two simultaneous reaction paths out of two phases namely Al-Si spinel crystallized during exothermic peak temperature at ~980°C and from residual noncrystalline aluminosilicate phase (Chakraborty and Ghosh, 1991). Similar to kaolinite, the mullitization curve of diphasic gel also predicts two paths of mullitization.

The author showed (i) how the quantity of noncrystalline aluminosilicate phase formation takes place in diphasic gel system and how it is depending upon processing conditions using same of components of silica and alumina vide Table 7.9 and Table 11.14, (ii) it's role in mullite formation.

Mullitization of the diphasic gel studied by MAS NMR studies. For example, at 900°C, ^{29}Si spectrum on deconvolution shows two broad peaks, one at −88.5 ppm and other at −98.5 which are in the region of Si(4Al)/Si(3Al) environment. Thus, shifting interprets the development of Si–O–Al(IV) linkages i.e., noncrystalline aluminosilicate phases of two varying SiO_2/Al_2O_3 ratios are being formed by solid state reaction between SiO_2 (A) and Al_2O_3 (A) components derived out of diphasic gel on heat treatment at ~400°C. At 900°C the ^{27}Al MAS NMR spectrum shows AlO_4 groups to the extent of 25% along with usual AlO_6 and with some AlO_5 groups (Table 9.1). This is another supportive evidence of formation of noncrystalline aluminosilicate phase. At ~ 1000°C, the ^{29}Si spectrum becomes asymmetrical. The centroid at –89 ppm becomes more broad than in the previous case. The shoulder at –109 ppm on the right hand side of the asymmetry indicates phase separation occurring during crystallization of alumina rich spinel with liberation of noncrystalline silica rich aluminosilicate phase. The later phase subsequently reacted with an alumina rich spinel phase formed earlier over a long range of temperature following a broad exotherm into Al-Si spinel phase of composition may be equal to 3:2 of standard mullite. At 1000°C, spectral study shows that the AlO_5 group disappears. Consequently, the quantum of AlO_4 and AlO_6 increases to 31% and 69%, respectively, during spinel formation. The changes in the ratio of Al(IV) to Al(VI) is explained as due to crystallization of Al-Si spinel of composition (3:2) by the introduction of silica out of noncrystalline aluminosilicate phase into alumina rich spinel phase during heating by long range diffusion process.

At 1300°C, the broad ^{29}Si resonance at –89 ppm slightly shifted to a most intense narrow peak at –86.6 ppm. The other peak at –94.7 ppm also corresponds well with a minor peak reported at –94.1 to –94.3 ppm, the shoulder which is fitted to a peak at –90.3 ppm is also reported at –90 ppm. Lastly, the low intensity peak at –91.9 ppm is also due to well crystallized mullite. Thus, the continuance of silicon environment at –89 ppm resonance peak throughout the transformation indicates that the structural transformation in the entire process is coherent. Obviously by the subsequent Al-Si spinel to mullite transformation process, AlO_4 further increases to 50% at the cost of decrease of AlO_6. Thus, in the entire transformation process, only the ratio of AlO_4 to AlO_6 smoothly changes without any discontinuity which is quite similar to the persistence of –89 ppm resonance peak throughout the transformation sequence. Only the nature of peak (its broadness to sharpness) changes on heat treatment. Further, ^{29}Si resonance peak at –89 ppm more close to resonance peak (–87 ppm) for mullite with change of –80 ppm (theoretical value) for Al-Si spinel phase and thus verifies polymerization concept.

This theory is in contrast to the mullite formation by solid state reaction between α-Al_2O_3 and silica (A)/cristobalite which perhaps occurs through a long range diffusional process (Sundaresan and Aksay, 1991). The course of mullitization reaction in oxide mixture and in diphasic is thus different (Chakraborty and Ghosh, 1986).

Formation of intermediate θ-Al_2O_3: Hsi et al. (1989), Fischer et al. (1996), Huang et al. (1997) in their phase transformation studies conjectured that θ-Al_2O_3 reacts with SiO_2 (A) to form mullite on further heating. On re-examination of their XRD pattern it is noted that the characteristic Bragg diffraction peaks resembling θ-Al_2O_3 were weak. So insignificant amounts of θ-Al_2O_3 might have formed, the residual pattern was due to a major quantity of spinel phase and a fraction of θ-Al_2O_3 is incorporated with silica. It started disappearing at >1300°C with formation of corundum and mullite. It was shown by Chakraborty (2006) that mixed gel in the heat treatment process develops a mixture of phases e.g., (i) silica rich alumina phase (A), (ii) SiO_2 (A), (iii) γ-Al_2O_3, (iv) Al-Si spinel, (v) silica incorporated γ-Al_2O_3, θ-Al_2O_3, and (vi) alumina rich silica phase (A). The mullite formation process is explained on the basis of ^{29}Si MAS NMR results (Figs. 9.5 and 9.6). The possible mechanism of mullite formation of mixed gels is found to be different for two ranges of compositions viz. mixed gels of Al_2O_3 content up to 3:2 mullite and mixed gels of Al_2O_3 content higher than 3:2 mullite.

(1) **Lattice parameter measurement of mullite obtained from melt and natural sources:** Cell dimensions of mullite by powder X-ray diffraction studies have been investigated by various earlier authors. It was concluded that mullite exists as a two-component solid solution, i.e., $Al_2O_3/SiO_2 = 3/2 \sim 2/1$. With increase of Al_2O_3 content, a progressive change in the lattice occurs from its chemical formulae of $3Al_2O_3 \cdot 2SiO_2$ to formulae $2Al_2O_3 \cdot SiO_2$. (i) In monumental studies, mullite was further classified and showed the existence of a correlation between both the *a* and *c* cell dimensions with composition. Out of large number of mullite samples obtained from melt and, some from natural sources, Cameron (1977) measured

lattice parameters of mullite comprehensively and plotted (i) *a* (cell edge) vs. *V* (cell volume) and (ii) *a*–*X* for mullite of different sources. From the linearity of the *a*–*V* plot, and *a*–*X* plot, he suggested that there is some relation between cell dimensions and either chemical composition of Al-Si order or both. The stoichiometric mullite phase is expressed as $[Al_2]^{vi}(Si_{2-2x}Al_{2+2x})^{iv}O_{10-x}$. It corresponds to a continuous solid solution of mullite of variable composition of $x = 0.17$ to 0.59.

(2) **Lattice parameter measurement of mullite obtained by solid state reaction:** Several synthetic mullite synthesized under various conditions are also used for lattice parameter measurement studies (Cameron, 1977). It is reported that cell dimensions for mullitization in the composition range 60–71 mol % Al_2O_3 vary linearly with composition (Fig. 16.6).

(3) **Lattice parameter measurement of mullite obtained by monophasic gel/coprecipitated gel method:** Metastable tetragonal mullite-like phase (t-mullite) was first designated by Ossaka (1961) after preparing mullite as low as 1000°C, which showed considerable broad peaks with $a = b$.

Lattice constants of mullite formed from coprecipitated gel of three different compositions were measured by Yamada and Kimura (1962) and showed that the values differ depending upon heating temperature and time. They are of the opinion that with passage of time the crystallinity of mullite formed went to increase gradually and finally it reached *a* lattice stage for that particular temperature. They plotted the converging value of the lattice constant a/b vs. c/b in Smith's chart and indicated the existence of a solid solution rich in Al_2O_3. That the decrease of lattice constant value with increase of temperature was also shown by Hirata et al. (1985b) and others during the heat treatment process of SiO_2-Al_2O_3 powders.

Based on the measurements lattice parameter of mullite, the composition of mullite was predicted by several researchers e.g., Okada and Otsuka (1986), Klug et al. (1987), Hirata et al. (1985a), Hirata et al., 1989). LC of mullite formed out of calcining SH and RH gels at different temperatures were calculated by least-squares method using RSLC-3 program by Okada and Otsuka (1986) and then compared their results

(Fig. 16.7) with Cameron's data (1977). During mullitization, the length of *a* shortened in SH gel in two steps, one at 1000°C to 1100°C and the other at 1300°C to 1400°C. The lattice parameter remained nearly constant beyond this temperature. The length of *b* showed converse behavior to *a* while *c* shortened gradually from 1000°C to 1150°C, but suddenly increased at 1200°C to the initial value, remaining constant thereafter. Variations of the lattice constants of RH gel with temperature changed similarly to those of SH gel. The estimated chemical composition of mullite was tentatively given and was found to change from ~71 to 60 mol % in SH gel and 64 to 61% in RH gel as the firing temperature increased. They indicated that the tetragonal alumina rich mullite formed upon heating mullite gels at 1000°C gradually transformed to 3:2 mullite at ~1400°C.

The compositions of mullite were obtained from measured lattice constant *a* value and on the basis of Cameron's data by Hirata et al. (1985a). They firmly observed that lattice constant value showed the greatest composition dependence. Suzuki et al. (1990) showed that *a* axis value was slightly higher than that of stoichiometric mullite. On prolonged heating for 10 h at 1400 to 1500°C, the *a* axis value decreased and coincided with stoichiometric value.

Schneider and Rymon-Lipinski (1988) observed good coincidence between X-ray powder diffractograms of normal mullite and mullite type phase obtained via gelation process with respect to position and intensity of X-ray reflection as well as some deviations. For example, lack of 120/210, 240/420, 041/401, and 250/520 reflection pair splitting on the X-ray diffractograms indicated that the length of *a* and *b* constants have become equal. Symmetry is tetragonal with respect to normal orthorhombic mullite. They emphasized that the fact of gradual transformation of the unstable, low temperature Al_2O_3 rich mullite to stable high temperature orthorhombic form was essentially controlled by annealing temperature which verifies result of Okada and Otsuka (1988) and Geradin et al. (1994).

The lattice constants of mullite obtained by heat treatments at different temperatures of three monophasic aluminosilicate gels in the range 63 to 80 wt % Al_2O_3 were measured by Low and McPherson (1989) and showed that a straight line relationship between cell parameter, *a*, and cell volume, C_v fitted Cameron's data. That some of mullite peaks at (120) and (210) lines as also peaks at (250) and (520) lines are not distinctly split rather than broad and diffuse on heating monophasic gels <1200°C. Splitting occurs on heating ≥1300°C. Characteristically, it occurred with the disappearance of Al-Si spinel phase when abundant mullite appeared.

(4) **Lattice parameter measurement of mullite obtained by diphasic gels method:** The lattice parameter of mullite formed out of diphasic mullite gels of different compositions were measured by (Klug et al., 1987) at high temperature by a computer programmer using some non-overlapping peaks of mullite with those of alumina and standard silicon. Plot of *a* spacing varied as a function of composition between 72–77 wt % Al_2O_3 which agreed with Cameron's curve. Mizuno and Saito (1989) showed parameter *b* of poorly crystallized mullite formed at 1250°C was 7.682 Å which is lower than *b* value (7.690 Å) of mullite formed at 1300°C. The parameter *a* (7.559 Å) for mullite formed at 1250°C decreased to 7.541 Å for mullite obtained on calcinations at 1400°C and 1500°C.

Variation of lattice constants data of mullite with increasing temperatures from 1200°C to on words for diphasic gels of three different compositions are shown by Chakraborty (2005b). (i) The length of *a* axis value of mullite formed out of silica-rich gel (DG40), analogous to the composition of kaolinite decreases markedly with increase of firing temperature and then remains more or less constant on further heating. Variation of lattice constants of mullite heated up to 1600°C is a function of temperature (Fig. 16.8) and the data are resembling to those of Hirata et al. (1989). (ii) The value of *a* of mullite formed from gel of 3:2 mullite composition (DG72) remains almost constant throughout the range of temperature and (iii) The LC value of *a* formed out of alumina-rich gel (DG85) increases slightly but subsequently decreases on heating.

The size of mullite crystallites and its strain value are both functions of heat treatment temperature and are shown in Table 16.3. The inverse relationship among size and strain is apparently noted upon heating DG40 at different temperatures are shown in Fig. 16.9. Variation in strain values of mullite formed out of diphasic gels at different temperatures is shown in Fig. 16.9C.

Changes in the peak width of mullite during heating DG40 is shown in Fig. 16.10. The decreasing peak width with increasing calcination temperature verifies earlier observations of Okada and Otsuka (1986). It shows that the size of mullite when it starts to crystallize is very small and this newly formed mullite is highly strained. These two parameters change in the reverse manner with increasing temperature i.e. strain decreases while size increases. This may be the reason for the change in *a* lattice constant value. With the rise of temperature mullite crystals become well crystallized.

The nature of the mullite samples used by Cameron (1977) are certainly different from mullite formed out of diphasic gel at increasing temperatures. He used quite well developed and stable mullite crystals differing only in composition within or near the solid solution range. In contrary, mullite has been developed from mixed alumina-silica precursors from their incipient stage to highly crystalline form. Roles of crystallite size, strain and crystallinity are comparatively less in those mullites than that of unstable incipient mullites. Accordingly, the changes in lattice parameters of mullite values differ between the two cases. In the course of heating process of diphasic gels, particle sizes of mullite when started to form at ~1200°C are quite small, of the order of 50 to 70 μm but increase gradually and then rapidly at 1500°C to 1600°C. The reverse occurs for strain values i.e., initially formed mullite is highly strained and strain decreases on increase of calcination temperature. This is more marked at and above 1500°C. So it would be cautious to compare the lattice constant data of strained and small size mullite which is just developing at lower temperature at <1200°C with that of highly crystalline mullite taken by Cameron. It is better to assume that changes in lattice parameters of mullite observed in case of diphasic gel are mostly due to changes in structural order rather than to changes in composition. In support of this view, the present study shows that the width of (120) of mullite decreases with temperature when diphasic gel marked DG72 is heated gradually (Fig. 16.11). The splitting of 120/210 reflection pairs increase when fired at increasing temperature. Therefore, mullite which develops at ~1300°C in four cases of diphasic gels, namely a silica rich composition, sillimanite type composition, ideal 3:2 composition and an alumina rich composition would be a mullite type phase. In these cases lattice parameter measurement of mullite shows $a < b$ values. The probable change in lattice constant with heating at higher temperature > than 1600°C is due mainly to increase in crystallinity of mullite phase by gradual relief of strain present during its incipient formation stage than to changes in chemical composition and with changes in crystalline form to orthorhombic mullite

Heat of reaction/980°C crystallization enthalpy of various mullite gels: Heat of reaction data of different mullite precursors are dependent upon two following parameters.

(i) Processing conditions of the gel/precursor synthesis whether it is synthesized by (i) gelation process or (ii) coprecipitation process or (iii) spray dried (SD) or spray pyrolysis process.

(ii) Method of measurement made by various authors. Whether it is determined from peak height as reported by Demediuk and Cole (1958), Chakraborty (1979), Kanzaki et al. (1990), Kumazawa et al. (1986), and Douy (2006) or from peak area of 980°C exotherm (Hyatt et al., 1990) or directly measured by using a twin calorimeter of the Tian Calvet type (Geradin et al., 1994). As a result a great variation to a large extent was noted (Table 16.4).

References

1. H. Insley and R. H. Ewell, Thermal behavior of the kaolin minerals. *J. Res. Natl. Bur. Stand.*, **14**(5), 615–627 (1935).

2. A. K. Chakraborty and D. K. Ghosh, Synthesis and 980°C phase development of some mullite gels. *J. Am. Ceram. Soc.*, **71**(11), 978–987 (1988).

3. H. Schneider, I. Merwin, and A. Sebald, Mullite formation from non-crystalline precursors. *J. Mater. Sci.*, **29**, 805–812 (1992).

4. J. Ossaka, Tetragonal mullite type phase from co precipitated gels. *Nature (London)*, **19**, 1000 –1001 (1961).

5. D. W. Hoffman, R. Roy, and S. Komarneni, Diphasic xerogels, a new class of materials: phases in the system Al_2O_3-SiO_2. *J. Am. Ceram. Soc.*, **67**, 468–471 (1984).

6. A. K. Chakraborty and D. K. Ghosh, Comment on "diphasic xerogels, a new class of materials phases in the system Al_2O_3-SiO_2. *J. Am. Ceram. Soc.*, **69**(8), C-202–C-203 (1986). Reply by S. Komarneni and Rustom Roy, *ibid*, **69**(8), C-204 (1986).
7. K. Okada and N. Otsuka, Characterization of the spinel phase from SiO_2-Al_2O_3 xerogels and the formation process of mullite. *J. Am. Ceram. Soc.*, **69**(9), 652–656 (1986).
8. J. C. Huling and G. L. Messing, Epitactic nucleation of spinel in aluminosilicate gels and its effect on mullite crystallization. *J. Am. Ceram. Soc.*, **74**(10), 2374 –2381 (1991).
9. B. E. Yoldas, Effect of ultrastructure on crystallization of mullite. *J. Mater. Sci.*, **27**(24), 6667–6672 (1992).
10. J. A. Pask, X. W. Zhang, A. P. Tomsia, and B. E. Yoldas, Effect of sol-gel mixing on mullite microstructure and phase equilibria in the α-Al_2O_3-SiO_2 System. *J. Am. Ceram. Soc.*, **70**(10), 704–707 (1987).
11. M. H. Dafadar, Thesis, Calcutta University (2000).
12. M. H. Dafadar, S. Das, and A. K. Chakraborty, Effect of heat on Al_2O_3-SiO_2 gel co-precipitated at different pH. *Trans. Ind. Ceram. Soc.*, **57**(4), 100–105 (1998).
13. Y. Hirata, K. Sakeda, Y. Matsushita, K. Shimada, and Y. Ishihara, Characterization and sintering behavior of alkoxide-derived aluminosilicate powders. *J. Am. Ceram. Soc.*, **72**(6), 995–1002 (1989).
14. H. Schneider, B. Saruhan, D. Voll, L. Merwin, and A. Sebald, Mullite precursor phases. *J. Euro. Ceram. Soc.*, **11**, 87–94 (1993).
15. S. Prochazka and F. J. Klug, Infrared-transparent mullite. *J. Am. Ceram. Soc.*, **66**(12), 874–880 (1983).
16. H. Suzuki, H. Saito, Y. Tomokiyo, and Y. Suyama, Processing of ultrafine mullite through alkoxide route, in *Ceramic Transactions*, Vol. 6, Mullite and Mullite Matrix Composites. eds. S. Somiya, R. F. Davis, and J. A. Pask, American Ceramic Society, Westervikke, OH, p. 263 (1990).
17. M. Low and R. McPherson, The structure and composition of Al-Si spinel. *J. Mater. Sci.*, **7**, 1196–1198 (1989).
18. Y. Hirata, H. Minamizono, and K. Shimada, Property of SiO_2-Al_2O_3 powders prepared from metal alkoxide. *Yogo Kyokai Shi*, **93**(1), 36–54 (1985a).
19. K. Hamano, Z. Nakagawa, G. Cun-Ji, and T. Sato, in *Mullite*, ed. S. Somiya, Uchida Rokakuho Publishing Co., Tokyo, Japan, p. 37 (1985).

20. K. Hamano, T. Sato, and Z. Nakagawa, Properties of mullite prepared by co-precipitation and microstructure of fired bodies. *Yogo Kyokai Shi*, **94**(8), 818–822 (1986).

21. S. Mitachi, M. Matsuzawa, K. Kaneko, S. Kanzaki, and Y. Tabata, Characterization of SiO_2-Al_2O_3 powders prepared from metal alkoxides. *Ceram. Trans.*, **6**, 275–286 (1990).

22. S. Rajendran, H. J. Rossell, and J. V. Sanders, Crystallization of a co-precipitated mullite precursor during heat treatment. *J. Mater. Sci.*, **25**, 4462–4471 (1990).

23. Y. Hirata, K. Sakeda, Y. Matsushita, and K. Shimada, Preparation of fine SiO_2-Al_2O_3 powders by hydrolysis of mixed alkoxides. *Yogo Kyokai Shi*, **93**(9), 577 (1985b).

24. Ph. Colomban, Structure of oxide gels and glasses by infrared and Raman scattering part 2 mullite. *J. Mater. Sci.*, **24**, 3011–3020 (1989).

25. H. Yamada and S. Kimura, Studies on coprecipitates of alumina and silica gels and its transformations at higher temperatures. *Yogo Kyokai Shi*, **70**, 87–93 (1962).

26. A. K. Chakraborty, New data on thermal analysis of diphasic mullite gels. *J. Therm. Analysis*, **46**, 1413–1419 (1996).

27. A. K. Chakraborty, Further studies on thermal transformation of diphasic Al_2O_3-SiO_2 gel. *Trans. Indian Ceram. Soc.*, **56**(1), 9–15 (1997).

28. D. X. Li and W. J. Thomson, Effects of hydrolysis on the kinetics of high temperature transformations in aluminosilicate gels. *J. Am. Ceram. Soc.*, **74**, 574–578 (1991).

29. J. Sanz, I. Sobrados, A. L. Cavalieri, P. Pena, S. de Aza, and J. S. Moya, Structural changes induced on mullite precursors by thermal treatment: a ^{27}Al MAS-NMR investigation. *J. Am. Ceram. Soc.*, **74**(10), 2398–2403 (1991).

30. C. Gerardin, S. Sundaresan, J. Benziger, and A. Navrotsky, Structural investigation and energetics of mullite formation from sol-gel precursors. *Chem. Mater.*, **6**, 160–170 (1994).

31. D. J. Cassidy, J. L. Woolerey, J. R. Bartlett, and B. Ben-Nissan, The effect of precursor chemistry on the crystallization and densification of sol-gel derived mullite gels and powders. *J. Sol-Gel Sci. Technol.*, **10**, 19–30 (1997).

32. M. I. Nieto, G. Urretavizcaya, A. L. Cavalieri, and P. Rana, Structural changes in colloidal and polymeric aluminosilicate gels with mullite composition. *Brit. Ceram. Trans.*, **97**(1), 17–23 (1998).

33. A. M. L. M. Fonseca, J. M. F. Ferreira, I. M. M. Salvado, and J. L. Baptista, Mullite based compositions prepared by sol-gel techniques. *J. Sol-Gel Sci. Technol.*, **8**, 403–407 (1997).

34. Y. X. Huang, A. M. R. Senos, J. Rocha, and J. L. Baptista, Gel formation in mullite precursors obtained via tetraethylorthosilicate (TEOS) pre-hydrolysis. *J. Mater. Sci.*, **32**, 105–110 (1997).

35. S. Kanzaki and H. Tabata, Sintering and mechanical properties of stoichiometric mullite. *J. Am. Ceram. Soc.*, **68**(1), C-6–C-7 (1985).

36. I. Jaymes, A. Douy, P. Florian, D. Massiot, and J. P. Coutures, Mew synthesis of mullite. Structural evolution study by ^{17}O, ^{27}Al and ^{29}Si MAS NMR spectroscopy. *J. Sol-Gel Sci. Technol.*, **2**, 367–370 (1994).

37. M. J. Hyatt and N. P. Bansal, Phase transformations in xerogels of mullite composition. *J. Mater. Sci.*, **25**, 2815–2821 (1990).

38. W.-C. Wei and J. W. Halloran, Phase transformation of diphasic aluminosilicate gels. *J. Am. Ceram. Soc.*, **71**(3), 166–172 (1988).

39. A. K. Chakraborty, Reinvestigation of Al-Si spinel phase in diphasic Al_2O_3-SiO_2 gel. *J. Am. Ceram. Soc.*, **88**(1), 134–140 (2005a).

40. S. Komarneni and R. Roy, Solid-state ^{27}Al and ^{29}Si magic-angle spinning NMR of aluminosilicate gels. *J. Am. Ceram. Soc.*, **69**(3), C-42–C-44 (1986).

41. S. Sundaresan and I. A. Aksay, Mullitization of diphasic aluminosilicate gels. *J. Am. Ceram. Soc.*, **74**, 2388–2392 (1991).

42. J. C. Huling and G. L. Messing, Surface chemistry effects on homogeneity and crystallization of colloidal mullite sol-gel, *Ceramic Transactions*, Vol. 6, Mullite and Mullite Matrix Composites. eds. S. Somiya, R. F. Davis, and J. A. Pask, American Ceramic Society, Westervikke, OH, p. 221 (1990).

43. G. Klaussen, Microstructural evolution of sol-gel mullite. *Ceram. Eng. Sci. Proc.*, **11**, 1087–1093 (1992).

44. M. G. M. U. Ismail, Z. Nakai, K. Minegishi, and S. Somiya, Synthesis of mullite powder and its characteristics. *Int. J. High Technol. Ceram.*, **2**, 123–134 (1986).

45. M. G. M. U. Ismail, Z. Nakai, and S. Somiya, Microstructure and mechanical properties of mullite prepared by the sol-gel method. *J. Am. Ceram. Soc.*, **70**(1), C-7–C-8 (1987).

46. A. K. Chakraborty, Intermediate Si-Al spinel formation in phase transformation of diphasic mullite gel. *J. Mater. Sci.*, **28**, 3839–3844 (1993).

47. A. K. Chakraborty, Formation of silicon-aluminium spinel. *J. Am. Ceram. Soc.*, **62**, 120 (1979).

48. A. K. Chakraborty and S. Das, Al-Si spinel phase formation in diphasic mullite gels. *Ceram. Int.*, **29**, 27–33 (2003a).

49. A. K. Chakraborty, Characterization of monophasic and diphasic mullite precursors by solid state reaction study. *Br. Ceram. Trans.*, **103**(1), 33–36 (2004).

50. H. Ivankovic, E. Tkalcec, R. Nass, and H. Schmidt, Correlation of the precursor type with densification behavior and microstructure of sintering mullite ceramics. *J. Euro. Ceram. Soc.*, **23**, 283–292 (2003).

51. Y. Wang and W. J. Thomson, Mullite formation from nonstoichiometric slow hydrolyzed single phase gels. *J. Mater. Res.*, **10**(4), 912–917 (1995).

52. A. K. Chakraborty, An analysis of the phase evolution of six types of mullite gels. Unpublished (2008).

53. W. L. De Keyser, Reaction at the point of contact between SiO_2 & Al_2O_3. *Sci. Ceram.*, **2**, 243 (1965).

54. W. G. Staley, Jr. and G. W. Brindley, Development of non-crystalline material in sub solidus reactions between silica and alumina. *J. Am. Ceram. Soc.*, (11), 616–619 (1969).

55. I. A. Aksay and J. A. Pask, Stable and metastable equilibria in the system SiO_2-Al_2O_3. *J. Am. Ceram. Soc.*, **58**(11–12), 507–512 (1975).

56. N. Shinohara, D. M. Dabbs, and I. A. Aksay, Infrared transparent mullite through densification of monolithic gels at 1250°C. *Infrared Opt. Trans. Mater. Proc. SPIE*, **683**, 19–24 (1986).

57. B. Sonuparlak, Sol-gel processing of infrared transparent mullite. *Adv. Ceram. Mater.*, **3**(3), 263–267 (1988).

58. J. C. Huling and G. L. Messing, Hybrid gels for homoepitactic nucleation of mullite. *J. Am. Ceram. Soc.*, **72**(9), 1725–1729 (1989).

59. Fu-Su Yen, C. S. Hsi, Y. H. Chang, and H. Y. Lu, Mullite formation from xerogels of (0.84–2.2) $Al_2O_{3.1}SiO_2$. *J. Mater. Sci.*, **26**, 2150–2156 (1991).

60. L. Pach, A. Iratni, Z. Hrabe, S. Svetik, and S. Komarneni, Sintering and crystallization of mullite in diphasic gels. *J. Mater. Sci.*, **30**, 5490–5494 (1995).

61. J. Wu, M. Chen, F. R. Jons, and P. F. James, Mullite and alumina-silica matrices for composites by modified sol-gel processing. *J. Non-Cryst. Solids*, **162**, 197–200 (1993).

62. G. M. Anilkumar, P. Mukundan, A. D. Damodaran, and K. G. K. Warrier, Effect of precursor pH on the formation characteristics of sol-gel mullite. *Mater. Lett.*, **33**, 117–122 (1997).

63. S. Sugita, Sueyoshi, and C. A. Contreras Soto, Fine pure mullite powder by homogeneous precipitation. *J. Euro. Ceram. Soc.*, **18**, 1145–1152 (1998).

64. A. K. Chakraborty, Si incorporated alumina phases formed out of diphasic mullite gels. *J. Mater. Sci.*, **43**, 5313–5324 (2008a).

65. A. K. Chakraborty and D. K. Ghosh, Kaolinite-mullite reaction series. The development and significance of a binary aluminosilicate phase. *J. Am. Ceram. Soc.*, **74**(6), 1401–1406 (1991).

66. I. Jaymes, A. Douy, D. Massiot, and J. P. Coutures, Characterization of mono- and diphasic mullite precursor powders prepared by aqueous routes, ^{27}Al and ^{29}Si MAS-NMR spectroscopy. *J. Mater. Sci.*, **31**, 4581–4589 (1996).

67. A. K. Chakraborty and K. J. D. McKenzie, Unpublished (2006).

68. Ph. Colomban, Structure of oxide gels and glasses by infrared and Raman scattering part 2 mullite. *J. Mater. Sci.*, **24**, 3011–3020 (1989).

69. C. S. Hsi, F. S. Yen, and Y. H. Chang, Characterization of co-precipitated Al_2O_3-SiO_2 gels. *J. Mater. Sci.*, **24**, 2041–2046 (1989).

70. R. K. Fischer, H. Schneider, and D. Voll, Formation of aluminium rich 9:1 mullite and its transformations to low alumina mullite upon heating. *J. Euro. Ceram. Soc.*, **16**, 109–111 (1996).

71. B. E. Yoldas, Thermal stabilization of an active alumina and effect of dopants on the surface area. *J. Mater. Sci.*, **11**, 465–470 (1976).

72. A. K. Chakraborty, Range of solid solutions of silica in spinel type phase. *Adv. Appl. Ceram.*, **105**(6), 1–7 (2006).

73. D. L. Trimm, *Appl. Catal.*, **7**, 249 (1983).

74. H. P. Rooksby and J. H. Partridge, X-ray study of natural & artificial mullite. *J. Soc. Glass Technol.*, **23**(100), 338–346 (1939).

75. S. O. Agrell and J. V. Smith, Cell dimensions, solid solution, polymorphism, and identification of mullite and sillimanite. *J. Am. Ceram. Soc.*, **43**(2), 69–76 (1960).

76. S. Aramaki and R. Roy, Revised phase diagram for the system Al_2O_3-SiO_2. *J. Am. Ceram. Soc.* **45**(5), 229–241 (1962).

77. A. J. Majumder and J. H. Walch, New data on synthetic mullite. *Trans. Br. Ceram. Soc.*, **62**(80), 603–613 (1963).

78. W. E. Cameron, Composition and cell dimensions in mullite. *Am. Ceram. Soc. Bull.*, **56**, 1003–1011 (1977).

79. N. Bowen and J. W. Greig, The system Al_2O_3-SiO_2. *J. Am. Ceram. Soc.,* **7**, 238–254 (1924).

80. W. H. Taylor, The structure of sillimaite and mullite. *Z. Krist,* **68**, 503–521 (1928).

81. G. Tromel, H. K. Obst, K. Konopicky, H. Bauer, and I. Patzk, Untersuchungehin system SiO_2-Al_2O_3 (system silica-alumina). *Ber. Deut. Keram. Ges.,* **34**(12), 397–402 (1957).

82. G. Tromel, H. K. Obst, K. Konopicky, H. Bauer, and I. Patzk, Investigation in the system silica alumina. *Ber. Deut. Keram. Ges.,* **35**, 108 (1958).

83. G. Gelsdorf, H. Muller-Hesse, and H. E. Scghwiete, einlagerungsversuche an synthetischem mullit und substitutions versuche mit galliumoxyd und germanium oxyd II (additive experiments on synthetic mullite and substitution experiments with gallium oxide and germanium oxide II), *Arch. Eisenhuttenw,* **29**, 513–519 (1958).

84. R. J Angel and C. T. Prewitt, Crystal structure of mullite: a re-examination of the average mullite structure. *Am. Mineral,* **71**, 1476–1482 (1986).

85. S. Prochazka and F. J. Klug, Infrared-transparent mullite. *J. Am. Ceram. Soc.,* **66**(12), 874–880 (1983).

86. F. J. Klug, S. Prochajka, R. H. Doremus, Aumiosilicate phase diagram in mullite region. *J. Am. Ceram. Soc.,* **70**, 750–759 (1987).

87. F. J. Klug, S. Prochazka, and R. H. Doremus, Alumina-silica phase diagram in the mullite region, in *Ceramic Transactions,* Vol. 6, Mullite and Mullite Matrix Composites. eds. S. Somiya, R. F. Davis, and J. A. Pask, American Ceramic Society, Westervikke, OH, p. 15 (1990).

88. H. Suzuki, H. Saito, Y. Tomokiyo, and Y. Suyama, Processing of ultrafine mullite through alkoxide route, in *Ceramic Transactions,* Vol. 6, Mullite and Mullite Matrix Composites. eds. S. Somiya, R. F. Davis, and J. A. Pask, American Ceramic Society, Westervikke, OH, p. 263 (1990).

89. H. Schneider and T. Rymon-Lipinski, Occurrence of Pseudo tetragonal mullite. *J. Am. Ceram. Soc.,* **71**, C-162–C-164 (1988).

90. M. Schmucker and H. Schneider, Structural development of single phase (type I) mullite gels. *J. Sol-Gel Sci. Technol.,* **15**, 191–199 (1999).

91. H. Schneider, R. K. Fischer, and D. Voll, Mullite with lattice constants a>b. *J. Am. Ceram. Soc.,* **76**, 1879–1881 (1993).

92. R. F. Fischer, H. Schneider, and D. Voll, Formation of aluminium rich 9:1 mullite and its transformation to low alumina mullite upon heating. *J. Eur. Ceram. Soc.,* **16**, 109–113 (1996).

93. R. Baranwal, M. P. Villar, R. Garcia, and R. M. Laine, Flame spray pyrolysis of precursors as a route to nano-mullite powder: powder

characterization ad sintering behavior. *J. Am. Ceram. Soc.*, **84**(5), 951–961 (2001).

94. M. Low and R. McPherson, The structure and composition of Al-Si spinel. *J. Mater. Sci.*, **7**, 1196–1198 (1989).
95. M. Sales and J. Alarcon, Synthesis and phase transformations of mullite obtained from SiO_2- Al_2O_3 gels. *J. Euro. Ceram. Soc.*, **16**, 781–789 (1996).
96. M. Mizuno and H. Saito, Preparation of highly pure fine mullite powder. *J. Am. Ceram. Soc.*, **72**, 377–382 (1989).
97. A. K. Chakraborty, Structural parameters of mullite formed during heating of diphasic mullite gels. *J. Am. Ceram. Soc.*, **88**(9), 2424–2428 (2005b).
98. C. H. Horte and J. Wiegmann, Reaction between amorphous SiO_2 and Al_2O_3. *Naturwiss*, **43**, 9 (1956).
99. T. Demediuk and W. F. Cole, Exothermic reaction of metakaolin between 950° and 1,000°C. *Nature*, **181**, 1400–1401 (1958).
100. R. R. West and T. J. Gray, Reactions in silica-alumina mixtures. *J. Am. Ceram. Soc.*, **41**(4), 132–136 (1958).
101. K. Kanzaki, M. Ohashi, H. Tabata, T. Kurihara, S. I. Iwai, and S. I. Wakabayashi, in *Ceramic Transactions*, Vol. 6, Mullite and Mullite Matrix Composites. eds. S. Somiya, R. F. Davis, and J. A. Pask, American Ceramic Society, Westervikke, OH, p. 389 (1990).
102. T. Kumazawa, S. Ohta, S. Kanzaki, and H. Tabata, Influence of powder characteristics on microstructural and mechanical properties of mullite ceramics (74wt.% Al_2O_3). *J. Jpn. Ceram. Soc.*, **99**, 1228–1233 (1991).
103. T. Kumazawa, S. Ohta, S. Kanzaki, and H. Tabata, Influence of powder characteristics on microstructural and mechanical properties of mullite ceramics. *Ceram. Trans.*, **6**, 401–411 (1990).
104. I. Jaymes and A. Douy, Homogeneous mullite-forming powders from spray-drying aqueous solutions. *J. Am. Ceram. Soc.*, **75**(11), 3154–3156 (1992).
105. I. Jaymes and A. Douy, Homogeneous precipitation of mullite precursors. *J. Sol-Gel Sci. Technol.*, **4**, 7–13 (1995).
106. S. Sen and S. Thiagarajan, *Ceram. Int.*, **14**, 77–86 (1988).
107. E. Tkalcec, D. Hoebbel, R. Nass, and H. Schmidt, Structural changes of mullite precursors in presence of polyethyleneimine. *J. Non-Cryst. Solids*, **243**, 233–243 (1999).

108. A. Douy, Crystallization of amorphous spray-dried precursors in the Al_2O_3-SiO_2 system. *J. Euro. Ceram. Soc.*, **26**, 1447–1454 (2006).

109. I. Jaymes, A. Douy, D. Massiot, and J. P. Coutures, Characterization of mono- and diphasic mullite precursor powders prepared by aqueous routes, ^{27}Al and ^{29}Si MAS-NMR spectroscopy. *J. Mater. Sci.*, **31**, 4581–4589 (1996).

Chapter 17

Nature of Residual Noncrystalline Aluminosilicate Phases Associated with Mullite Formation

17.1 Nature of Noncrystalline Al_2O_3-SiO_2 Mullite Precursor Phase Prior to 980°C Exotherm

(1) Earlier studies

Nature of this noncrystalline mullite precursor phase prior to exhibition of the 980°C exotherm has been studied by various physicochemical methods.

(i) Chemical techniques indicate the formation of aluminosilicate compounds in both monophasic and diphasic precursors heated to the said temperature range other than present as two discrete component phases. TGA shows a small endotherm prior to crystallization at 980°C exotherm.

(ii) Shifting of Si–O stretching vibration at 1120 cm^{-1} to the lower wavenumber 1080 cm^{-1} in IR study detects the presence of Al–O–Si bond formation due to formation of aluminosilicate during heating of precursors.

Mullite Formations: Analysis and Applications
Akshoy Kumar Chakraborty

ISBN 978-981-4877-05-3 (Hardcover), 978-1-003-03167-3 (eBook)
www.jennystanford.com

(iii) Si–O–Al bond formation was confirmed by MAS-NMR spectroscopic studies which shows symmetrical ^{29}Si resonance at ~92 ppm to asymmetrical shape of ^{29}Si resonance peak with shoulder towards more negative chemical shift at ~95 ppm. ^{27}Al MAS-NMR study indicated the increment in AlO_4 bond in heated Al_2O_3–SiO_2 gels with increase of silica content of it. Even in diphasic gel system a broad Si spectrum is noted at ~ –80 to –90 ppm and it is explained as due to development of a large concentration of amorphous aluminosilicate phase may be rich in alumina. Changes in coordination amount of aluminium from 6 to 4 in amorphous silica-aluminas was indicated in XRF studies

(iv) Structure of aluminosilicate precursor was studied by following techniques.

   (a) X-ray absorption fine structure (XABFS) technique shows the existence of more than one type of bonding between Si and neighbouring oxygens in gels dried at 60°C to heat treated between 200–800°C. The coordination environment of oxygen around aluminium in gel dried at 60°C is 6 and it is reduced to ~4 when gel was heat treated in the same temperature range.

   (b) RDF reported (i) increase in peak intensities at four coordination spheres compared to silica glass and (ii) increment in bond distances from 1.7 to 1.8 Å due to substitution phenomenon of Si^{+4} by Al^{+3} to precursors heated to 800°C–900°C.

   (c) The radial distribution function D(r) and G(r) curves of heat treated Type I precursor during heating of precursor from RT to 800°C indicated a gradual condensation of the precursor network probably with changes in coordination state of aluminium. During dehydroxylation of –OH groups and decompositions of organics occurred resulting in decrease in broadness of the peak and decrease of r value from higher side to words smaller value.

(v) Finally, due to substitution effect a negative lattice charge in aluminosilicate structure has been created. H^+ ions (proton) and or aqueous $[Al(H_2O)]^{+3}$ groups are connected from adjoining water and OH^{-1} groups present in aluminosilicate structure. Thus, Al_2O_3–SiO_2 precursor phase is an hydrated noncrystalline aluminosilicate (see Chapter 13 for details).

(2) By alkali extraction technique

In the present chapter detail of leaching technique is described to conform the nature of precursor. Three diphasic gels marked DG40, D66, and DG72 were synthesized as before, heat treated to different temperatures as static condition and leached with alkali as per standard procedure Chakraborty (1978, 1979). 0.5 g of ground sample was treated with 10 ml of 5 wt % alkali solution in boiling water bath condition for 40 min duration for each case. After treatment, it was centrifuged and washed. The centrifugation and washing were analyzed for silica and alumina for gravimetric analysis. Percent of leached silica and alumina of different diphasic gels vs. heat treatment temperature are plotted in Fig. 17.1. The plot shows the following observations,

(i) In the temperature range 600°C–900°C, the silica content of diphasic gel DG72 dissolves in hot alkali solution to the extent of only ~2.5% out of 28% present in bulk gel. This indicates that silica component which remained as separate and individual component of the two phase system after dehydration and decomposition of diphasic gels at ~400°C becomes insoluble in hot alkali solution when it is heated to 600°C. In contrast, the dried diphasic gel showing amorphous band and boehmite by XRD shows almost complete solubility of silica part of it. This result definitely concludes that the X-ray amorphous band of gel heat treated to >600°C is not free amorphous silica as assumed by various earlier researchers.

(ii) At ~1000°C, Fig. 17.1 (lower curve) shows that a portion of silica liberates as free amorphous silica during phase transformation of those diphasic gels during heating. Thereafter, it becomes more or less constant even with further heating. The approximately quantity of silica (A) dissolved at 1000°C in hot alkali solution in three cases of gels are different, and it obeys the following order DG40 > DG56 > DG72.

To determine the composition of evolved silica (A) phase, the alkali extract is also analyzed for alumina content. The Fig. 17.1 (upper curve) shows that there is no sudden increase in wt %

of alumina content in the extract. However, a certain amount of alumina dissolved in all cases of diphasic gels heat treated to 1250°C by alkali attack of Al-Si spinel phase. The extracted values lie in the following order DG73 > DG56 > DG40. The minor amount of alumina obtained in cases of leaching three diphasic gels heated up to 1250°C which conforms that the said exsolved silica is free amorphous silica.

(iii) At 1000°C–1200°C, both silica and alumina constituents are available in the leachate. It is interpreted that the amorphous phase which remains after the transformation of aluminosilicate (A) phase to Al-Si spinel phase is a noncrystalline aluminosilicate phase (A) phase and not pure siliceous (A) phase.

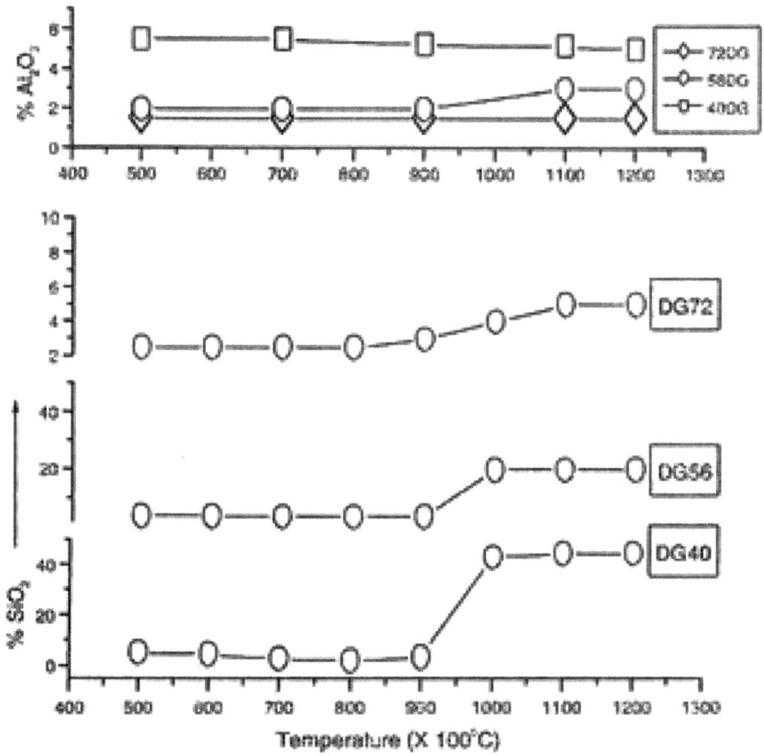

Figure 17.1 Existence of amorphous phase formed out of heating three kinds of diphasic gels.

(3) Identification of X-ray amorphous band

Phase transformation study of different raw mullite gels were performed by Chakraborty and Ghosh (1987), and Ivankovick et al. (2003). The presence of amorphous hump is observed (Figs. 15.3 and 18.2).

(4) Amorphous band-to-noncrystalline aluminosilicate precursor phase

Intensity of amorphous band at the first step increases with heat treatment in the temperature range 600°C–900°C for gel marked DG40, DG56, and DG72 (Fig. 15.7). Depending upon the composition of the gel, the plot shows that it increases slowly in case of DG72, it is moderate in case of DG66 and it increases very fast obviously in case of DG40. This result indicates the development of noncrystalline aluminosilicate precursor phase out of the solid state reactants of two oxide components.

17.2 Liberation of One Noncrystalline Aluminosilicate Phase & One Residual Phase during Final Dehydroxylation Process at 980°C Exotherm

Identification of amorphous band due to (i) noncrystalline silica rich aluminous phase during phase separation and (ii) noncrystalline alumina rich silicate residual phase after crystallization of Al-Si spinel or mullite phase.

At the earlier studies, noncrystalline aluminosilicate precursor phase accumulated during progressive heating of mono phasic gel/precursor transforms to either alumina rich mullite or Al-Si spinel phase at the first exothermic peak temperature. It also reported the existence of some quantity of noncrystalline alumina rich silicate phase (AS3) as residue after said crystallization. But this data is unavailable. Possibility of the presence of some quantity of amorphous silica rich aluminous phase during Al-Si spinel phase formation in case of RH gel was first shown by Okada and Otsuka (1986). It's chemical formulae was assumed as $6SiO_2 \cdot Al_2O_3$.

Quantity of this noncrystalline aluminosilicate phase (AS2) left as residue after 980°C crystallization is to be ascertained.

The area of the X-ray amorphous hump due formation of such noncrystalline aluminosilicate phases in the temperature range 600°C–1300°C for heated diphasic gel was first measured manually by Ivankovak et al. (2003) and its variations are shown in Fig. 18.2. Subsequently, using a software programme the changes in areas of the X-ray amorphous hump were recalculated by Chakraborty (2008a) as shown in (Fig. 15.7). In the case of monophasic gel SHIV or RH gel, the transformation likely proceeds as described earlier and is shortly reproduced as follows.

Noncrystalline aluminosilicate precursor phase (amorphous band)	980°C exotherm ⟶ 2nd step	Al-Si spinel phase + Residual noncrystalline	1100 to ~1200°C ⟶ 3rd step
Al-Si spinel phase (more intense) + (amorphous band decreases)	~1250°C exotherm ⟶ 4th step	Mullite type phase aluminosilicate phase (amorphous band)	

In the second step, Fig. 15.7 *s*hows that intensity of amorphous hump decreases sharply in case of DG40 and DG66 in the temperature range 900°C–1100°C due to crystallization of Al-Si spinel phase. After it's crystallization, an amorphous band is still observed as a residual phase. At the third step, XRD intensity peak due to spinel phase increases with heat treatment in the temperature range 1000°C–1200°C for all three gels marked DG40, DG66, and DG72. Moreover, a certain portion of the amorphous band still persists in the temperature range 1100°C–1200°C.

Identification of noncrystalline aluminosilicate phase in the light of MAS-NMR study (AS2)

(1) A mixture of mullite (major) and Al-Si spinel phase (minor) phase were crystallized during heating CM sample to 1000°C (Schneider et al., 1992, 1993). ^{29}Si MAS-NMR showed a large −110 ppm Si resonance peak which was designated by the author as due to liberation of a phase separated

phase resembling some quantity of noncrystalline silica rich aluminous phase.

(2) In the case of SGM sample, the –110 ppm Si resonance peak was observed still up to 1100°C (Schneider et al., 1993). The resonance data at higher temperature e.g., 1200°C or 1300°C are unavailable.

(3) Quite larger –110 ppm Si resonance peak was also noted by Gerardin et al. (1994) during heating sample M2 at ~1000°C. It shifted with increase of heating temperature from 1000°C and still remained at 1100°C. Similar observation is noted in case of sample MI. –110 ppm Si resonance peak appears just on heating at 1000°C which also persists even at 1100°C.

(4) Sample B of Jaymes et al. (1996) also showed –110 ppm Si resonance peak. Both the composition and quantity of noncrystalline phase related to Si resonance at –110 ppm may vary. Peak areas of the Si resonance between –90 to 110 ppm are different.

(5) Sample B of Jaymes et al. (1996) on heating at 980°C exotherm formed Al-Si spinel and shows large Si resonance at –110 ppm in 1000°C heated sample. It remained even on heating at 1200°C (Fig. 9.15). Similarly, Sample D/SD precursor made out of aged silica sol showed similar resonance at 980°C exotherm with crystallization of Al-Si spinel phase to 1275°C exotherm the intensity of this –110 ppm peak gradually increased on heat treatment from 750°C to 1200°C and it still exists as high as 1400°C (Fig. 5.16).

(6) Mullite phase crystallizes during 980°C exotherm in sample A (Sales and Alarcon, 1996). However, they observed a more exothermic peak in the temperature range 1200°C–1300°C. The formation of further mullite can be explained by the fact that noncrystalline aluminosilicate phase might have present earlier and it played a role which takes part in transformation into mullite.

(7) In Type III (HB13) precursor initially showed Si resonance due to noncrystalline aluminosilicate phase other than free silica as assumed by Schneider et al. (1993). It decreased on further heating prior to formation of spinel phase and showed

peak resembling residual noncrystalline aluminosilicate phase (Fig. 9.12).

(8) In case of MI precursor made by Huang et al. (1997), a broad Si spectrum is noted at ~ –80 to –90 ppm and it is explained as due to development of a large concentration of amorphous aluminosilicate phase may be rich in alumina. It is assumed that Al cation has more access to diffuse easily into coagulated silica phase during heating. It is expected as per the process of synthesis as laid down by the above authors. It was shown that Al^{+3} remained as a counter ion and was absorbed on the surface of the silica gel particles.

17.3 Composition of Residual Noncrystalline Aluminosilicate Phase in the Light of MAS-NMR Study

Joint formation of two types of noncrystalline phases AS2 and AS3

This chapter focuses on the nature of two noncrystalline intermediate aluminosilicate phases prior to mullitization.

(1) Anhydrous silica rich aluminous phase (AS2) generated during inhomogeneous dehydroxylation process of noncrystalline precursor phase in the donor region at the temperature at "D".

(2) Anhydrous alumina rich silica phase (AS1) generated in the acceptor region which crystallizes abruptly to weakly crystalline alumina rich mullite and or alumina rich Al-Si spinel at 980°C. A residual noncrystalline alumina rich mullite phase (AS3) will remain after completion of first exotherm (AS3) in the acceptor region (see Chapter 14).

Thus, two noncrystalline phases will jointly account for a total of aluminosilicate (A) phase. The amount of these associated amorphous phases has been estimated by QXRD which imparts a new light in the reaction course of precursor - mullite formation series.

Approximate value of noncrystalline aluminosilicate phases

Chakraborty and Ghosh (1978) assumed that ~7.5% alumina might have reacted with silica to form noncrystalline aluminosilicate phase. Chakraborty (1993) and Chakraborty (2003b) first indicated the formation of noncrystalline aluminosilicate phase during heat treatment of diphasic gels of various compositions to different temperatures (Table 15.6). The author further quantitatively estimated by the usual X-ray internal standard method the quantity of spinel phase formed on heating G-173 (Al_2O_3 = 69%) at >1000°C. The estimated value of spinel (79–80 wt %) is found to be greater than the initial alumina content of the gel. Likely, the approximate quantity of the noncrystalline aluminosilicate phase comes to ~21 wt %. The quantity of aluminosilicate residue present in different cases is dependent on the processing condition of the gel. The residual aluminosilicate phase (21 wt %) may nucleate to mullite at high temperature.

17.4 Formation of Noncrystalline Aluminosilicate Phases during Mullite Formation Process in Six Cases of Precursors

XRD study shows that substantial quantity of mullite type phase is formed on heating monophasic gels e.g., SHI and SHII at ~1000°C and major quantities of Al-Si spinel from SHIII and SHIV at 1100°C (Table 16.1) by Chakraborty 2008. By QXRD studies, Chakraborty (2008) showed the amount of mullite type of phase formed at ~1000°C is 60–63% approximately in SHI and SHII gels. Therefore, the residual phase is obviously aluminosilicate (A) phase. So the observation of amorphous band beyond mullite and or Al-Si spinel formation may be due to the presence of newly formed noncrystalline silica rich alumina phase (AS2) and residual noncrystalline aluminosilicate (AS3) phase formed during dehydroxylation process of gel at earlier temperature range.

It is conjectured that this phase is more likely silica rich aluminous (A) phase other than free silica (A) alone as assumed by various earlier authors. Alkali extraction study by Chakraborty (1979) and solid state reaction studies of heat treated gels at 1000°C with CaO by Chakraborty (2003a, 2004) indicate the absence of free silica (A) as a residual phase in gels heat treated to 980°C (see Chapter 15).

In this chapter, the author will further show (i) how the quantity of residual noncrystalline aluminosilicate phase formation depend upon processing conditions using same components of silica and alumina as sources, the amount of joint formation of AS2 and AS3 is assessed from mullitization growth curve (Fig. 12.6) and it is found to be in the following order.

$$SHI < SHII < SHIII < SHIV.$$

pH noted during the gelation process of those gels is as follows.

$$SHI = 4, SHII = 3 - 2.8, SHIII = 2.8 - 2, SHIV = 6.$$

These results show that the remainder part of silica rich aluminous (A) phase during crystallization of aluminosilicate precursor phase at the first exotherm of above four precursors is pH related. It is most predominant in the case of SHIV when gelation was done at pH = 6 and it is least in SHI when it was made at pH = 4. With further decrease of pH the formation of silica rich aluminous (A) phase increases.

In ^{29}Si MAS-NMR study, intense broad peak centered at –104 ppm spectrum of SHIV may correspond to this phase (Fig. 9.20). Thus, a considerable amount of this phase (AS2) develops and it remains as residue during heating of those gels. Theoretically, it is expected that full conversion of amorphous precursor phase to crystalline phases would not happen during heating at ~980°C. There is every possibility of presence of certain amount of residual noncrystalline alumina rich-silicate phase (AS3) in all cases of gels namely SHI, SHII, SHIII synthesized.

17.5 Role of the Large Quantity of Noncrystalline Aluminosilicate Phase in Mullitization

Noncrystalline aluminosilicate residual phase-to-mullite phase
The existence of these two residual phases out of four phases as constituents of the 980°C exothermic equation must be well considered. Their role in mullite formation are given partly in Chapters 16 and 19.

For example,

(1) Mullite crystallizes out of noncrystalline aluminosilicate phase as shown in DTA curve (Fig. 7.20) shown by Chakraborty (2008) for gel marked SHIII and SHIV (Table 16.1).

The growth curve shown by Chakraborty (2008) predicts two consecutive steps of mullitization occur in both cases of SHIII and SHIV (Fig. 12.6).

Besides the transformation of Al-Si spinel by Path IV, an additional mullite formation occurs slowly in the high temperature region by Path V. The second mullitization step is explained as owing to mullite nucleation in this residual silica rich aluminous (A) phase.

(2) DTA curve of (a) G-150 compared with (b) kaolinite clay Fig. 2.2 (Chakraborty, 1994). These two samples show similar thermal events. Moreover, the thermal transformation of G-150 is analogous to that of kaolinite (see Chapter 16).

These two phases plays a great role during overall mullitization processes particularly in the second step of mullitization where the solid state reaction processes of two phases (AS2 and AS3) start on further heating which lead into some clues to sequential phase transformation of different types of mullite precursors.

(i) Mullite formation reaction in the case of SHI, are shown in Chapter 16 vide Eqs. 16.2–16.5.

(ii) Mullite formation reactions in cases of RH gel/SHIV/CP6 are shown in Chapter 16 vide Eqs. 16.6–16.9.

(iii) Mullitization behavior in the event in the category of in situ diphasic gel, residual noncrystalline aluminosilicate phase takes part in the final growth process of mullite as shown in Eq. 16.11.

References

1. A. K. Chakraborty and D. K. Ghosh, Crystallization behavior of Al_2O_3 in the presence of SiO_2. *J. Am. Ceram. Soc.*, **70**(3), C-46–C-48 (1987).

2. H. Ivankovic, E. Tkalcec, R. Nass, and H. Schmidt, Correlation of the precursor type with densification behavior and microstructure of sinteren mullite ceramics. *J. Euro. Ceram. Soc.*, **23**, 283–292 (2003).

3. K. Okada and N. Otsuka, Characterization of the spinel phase from SiO_2–Al_2O_3 xerogels and the formation process of mullite. *J. Am. Ceram. Soc.*, **69**(9), 652–656 (1986).
4. A. K. Chakraborty, Si-incorporated alumina phases formed out of diphasic mullite gels. *J. Mater. Sci.*, **43**, 5313–5324 (2008a).
5. H. Schneider, I. Merwin, and A. Sebald, Mullite formation from non-crystalline precursors. *J. Mater. Sci.*, **29**, 805–812 (1992).
6. H. Schneider, B. Saruhan, D. Voll, L. Merwin, and A. Sebald, Mullite precursor phases. *J. Euro. Ceram. Soc.*, **11**, 87–94 (1993).
7. C. Gerardin, S. Sundaresan, J. Benziger, and A. Navrotsky, Structural investigation and energetics of mullite formation from sol–gel precursors. *Chem. Mater.*, **6**, 160–170 (1994).
8. I. Jaymes, A. Douy, D. Massiot, and J. P. Coutures, Characterization of mono- and diphasic mulite precursor powders prepared by aqueous routes, ^{27}Al and ^{29}Si MAS-NMR spectroscopy. *J. Mater. Sci.*, **31**, 4581–4589 (1996).
9. M. Sales and J. Alarcon, Synthesis and phase transformations of mullite obtained from SiO_2–Al_2O_3 gels. *J. Euro. Ceram. Soc.*, **16**, 781–789 (1996).
10. Y. X. Huang, A. M. R. Senos, J. Rocha, and J. L. Baptista, Gel formation in mullite precursors obtained via tetraethylorthosilicate (TEOS) prehydrolysis. *J. Mater. Sci.*, **32**, 105–110 (1997).
11. A. K. Chakraborty, Intermediate Si–Al spinel formation in phase transformation of diphasic mullite gel. *J. Mater. Sci.*, **28**, 3839–3844 (1993).
12. A. K. Chakrabirty and S. Das, Al–Si spinel phase formation in diphasic mullite gels. *Ceram. Int.*, **29**, 27–33 (2003b).
13. A. K. Chakraborty, Formation of silicon-aluminium spinel. *J. Am. Ceram. Soc.*, **62**, 120 (1979).
14. A. K. Chakraborty, An analysis of the phase evolution of six types of mullite gels. Unpublished (2008).
15. A. K. Chakraborty, Characterization of monophasic and diphasic mullite precursors by solid state reaction study. *Br. Ceram.Trans.*, **103**, 33–36 (2004).
16. A. K. Chakraborty, Role of hydrolysis water-alcohol mixture on mullitization of Al_2O_3-SiO_2 monophasic gels. *J. Mater. Sci.*, **29**, 6131–6138 (1994).

Part III

Critical Analysis

Chapter 18

Critical Analysis and Characterization of Spinel Phase

18.1 Introduction

A freshly prepared neutral silica sol reacts with alumina sol to form an aluminosilicic acid with decrease of pH (Tamele, 1950). The resultant hydrogen form of aluminosilicic acid gel contains many surface hydroxyl groups and trapped water. Free and adsorbed water molecules are eliminated at the final stage of heating at ~200°C. When further heat treated, the chemically bonded OH groups react with each other and are removed gradually as H_2O vapor with simultaneous shrinkage of the gel material. The question at this stage is: what will happen when this partially dehydrated aluminosilicic acid gel material loses all its hydroxyl groups on further heating? DTA analysis of this aqueous aluminosilicic acid hydrate gel exhibited a sharp exothermic peak at 980°C, which is a crystallization phenomenon. Quite a similar DTA observation is noticed when monophasic mullite gel synthesized out of aluminum nitrate nonahydrate (ANN) and TEOS and is analyzed. Based on the nature of crystalline phases formed either Al-Si spinel or mullite by XRD, two theories are put forward.

Mullite Formations: Analysis and Applications
Akshoy Kumar Chakraborty

ISBN 978-981-4877-05-3 (Hardcover), 978-1-003-03167-3 (eBook)
www.jennystanford.com

First theory

At the initial stage, this dried monophasic mullite gel showed reduced XRD peaks of aluminum nitrate and amorphous silica band by Li and Thomson (1991). On heat treatment of gels at 1000°C in some cases, X-ray diffraction analysis of it identified three broad peaks which resemble near to γ-Al_2O_3 phase, a cubic spinel structure. According to the first theory, sudden crystallization of it was predicted as the reason for the occurrence of a 980°C exothermic peak in DTA analysis. During synthesis of mullite gel, Chakraborty and Ghosh (1988) dried it at ~60°C either in water bath condition or in a thermostat controlled air oven for a long period of time. At this drying stage, Al^{+3} out of ANN solution likely entered into the amorphous structure of silica gel formed during hydrolysis of TEOS, developed linkages and formed aluminosilicate hydrate. During heating gels between 400°C–900°C, Fukuoka et al. (1993) confirmed the formation of Al-O-Si bonds with formation of noncrystalline aluminosilicate structure by their IR and MAS-NMR studies (Figs. 13.1 and 13.3). Due to earlier introduction of silica both at drying and firing stages into alumina structure, it was assumed that a part of silica is likely a constituent of γ-Al_2O_3 spinel structure formed during the decomposition of aluminosilicate precursor at 980°C. Thus, two possibilities may be considered regarding the nature of this 980°C heated cubic spinel phase.

(1) In the first possibility, amorphous Al_2O_3 and amorphous SiO_2 components of the mullite gel may crystallize to γ-Al_2O_3 at the first exotherm.

(2) In second possibility, there may be a complete breakdown of SiO_2-Al_2O_3 linkages of newly generated aluminosilicate phase followed by crystallization of amorphous alumina part into cubic γ-Al_2O_3 (Al spinel) structure at the first exotherm.

(3) In the third possibility, it is thought that there may not be complete segregation of aluminosilicate structure into γ-Al_2O_3 and SiO_2(A). As the former phase generates from noncrystalline aluminosilicate material, it should accommodate some Si^{+4} ions in substitution of Al^{+3} ions in tetrahedral positions of γ-Al_2O_3 spinel structure. Accordingly, it is predicted that silicon bearing γ-Al_2O_3 spinel (called Al-Si spinel) is the 980°C

crystalline phase. Thus, two conjectures have been developed simultaneously and exist in mullite literature.

Second theory

In some other cases of mullite precursor XRD analysis showed direct crystallization of poorly crystalline mullite (major quantity) in association with γ-Al_2O_3 or Al-Si spinel phase as minor quantity at 980°C exotherms (Hoffman et al., 1984); Okada and Otsuka, 1986). The nature of this mullite whether it is t-mullite or o-mullite is a subject of many investigations. Thus, there are two disputes regarding the phase evolution of mullite gels due to formation of either of the two crystalline phases. When spinel crystallizes, the dispute arises on the nature or composition of it i.e., whether it is γ-Al_2O_3 spinel or Al-Si spinel. When mullite rapidly forms, the dispute concerns too on its composition may be orthorhombic form (o-mullite) or tetragonal (t-mullite) or any other.

Besides the crystallization of weakly crystalline phases either t-mullite or spinel or a mixture of t-mullite and spinel phase as mentioned above during 980°C exotherm of the monophasic precursors, a residual quantity of noncrystalline aluminosilicate phase is also remaining in association with those crystalline phases. Diphasic gel too in the second step of transformation forms spinel phase over a long range of temperature leaving behind residual noncrystalline aluminosilicate phase. Thus major problems are the characterization of

(i) Detection of the noncrystalline aluminosilicate phase,
(ii) 980°C heat treated spinel phase, and
(iii) The composition of incipient mullite phase.

The spinel phase would be called Al-Si Spinel. Since the dispute lies only in the composition of the spinel phase other than the old concept of Al spinel (γ-Al_2O_3) of the following formulae.

$$[Al\ 8]IV\ [Al\ 13.33 <> 2.66]\ VI\ O_{32}$$

Substitution of silicon in the tetrahedral site of cubic γ-alumina structure is feasible and an acceptable fact from the point of view of crystal chemistry. It is suggested by Chakraborty (1979) the formula of the Si substituted spinel resembles the composition analogous to 3:2 mullite as follows.

$$[Si\ 4.9\ Al\ 3.1]IV\ [Al\ 11.7 <> 4.31]VI\ O_{32}$$

This spinel phase has both AlO_6 and AlO_4 groups in octahedral and tetrahedral sites in contrast to the hypothetical spinel of Brindley and Nakahira (1959) which contains only AlO_6 groups in octahedral site and no AlO_4 groups in tetrahedral part of the spinel structure.

$$[\mathrm{Si}\ 8\ ]\mathrm{IV}\ [\mathrm{Al}\ 10.66 <> 5.33]\ \mathrm{VI}\ \mathrm{O}_{32}$$

By the IR absorption studies Low and McPherson (1988, 1989) believes the formulae of spinel is of the composition analogous to that of 2:1 mullite having the following formulae

$$[\mathrm{Si}\ 4\ \mathrm{Al}\ 4]\mathrm{IV}\ [\mathrm{Al}\ 12 <> 4]\ \mathrm{VI}\ \mathrm{O}_{32}$$

Isolation of spinel phase out of mullite gel marked HW1 heat treated to 1100°C by HF (5 wt %) treatment to remove amorphous silica was done by Hwang et al. (2000). They reiterated the composition of intermediate spinel phase to the composition alike to that of 2:1 mullite which transformed to t-mullite at 1100°C and then to o-mullite at 1250°C. A schematic presentation of said spinel was shown in Fig. 8 of them.

Even in diphasic gel system of synthesis under different set of conditions e.g., (i) using different choice of source components of silica and alumina namely Ludox, fume silica, aqueous silica sol, TEOS, boehmite, peptized alumina sol, ANN; (ii) pH and (iii) water/alcohol mixture used, the nature of the spinel phase crystallized is highly controversial and still a subject of much studies. Simply, on the basis of DTA analysis and qualitative X-ray diffractometry a group of researchers believe in the solid state reaction occurring between γ- or δ-alumina with silica (A) to form mullite. In this chapter, the following experimental data as presented below show that spinel formed in both two cases of mullite gels is Al-Si spinel and its composition is analogous to composition of o-mullite which transforms directly to mullite.

18.2 Identification and Characterization of Spinel Phase: A Summary

18.2.1 Indirect Evidences Al-Si Spinel

The following indirect evidence indicates that spinel phase formed out of monophasic and in diphasic gel contains some quantity of silicon and it is to be designated as Al-Si spinel.

(i) Retardation of the crystallization of alumina polymorphs observed during the heating process of some Al_2O_3 gels (Iller, 1964; Yoldas, 1976).

(ii) Changes in the crystallization temperature of corundum in some cases of Al_2O_3-SiO_2 gels were noted by Yoldas (1976).

(iii) Inhibition of nucleation of θ-, κ-, and α-Al_2O_3 in presence of silica were observed by Wei and Halloran (1988) in the phase transformation process of their diphasic gel.

(iv) Suppression of formation of corundum and cristobalite: Independent crystallization sequence of the alumina component is modified in presence of silica and vice versa (Chakraborty, 1987). Similar observation is noted when kaolinite clay is heated. When heated, pure aluminum hydroxide transforms to corundum through some of its intermediate polymorphs. Question is: what is the cause of this suppression? What is the relevance of it to the studies of synthesis of mullite out of monophasic or diphasic route? The reason might be due to

(i) Formation of some quantity of noncrystalline aluminosilicate phase when a monophasic or diphasic Al_2O_3-SiO_2 mullite gel was heat treated to higher temperature.

(ii) Substantial incorporation of silica in γ-Al_2O_3 structure resulting in Al-Si spinel formation and suppression of crystallization of either corundum or cristobalite.

As such Chakraborty (2005c) conjectured that the so called γ-Al_2O_3 spinel phase exist as intermediary phase prior to mullite formation of those gels is a Si substituted spinel or to be called as Al-Si spinel phase other than simple γ-Al_2O_3.

(v) Formation of **silica incorporated θ-Al_2O_3:** As like as in Al_2O_3-SiO_2 monophasic gel, on heating mixed gel (Boehmite and TEOS mixture), and impregnated gel made out of (γ-Al_2O_3 and TEOS mixture), silica (A) probably enters into the parent alumina structure and develops silicon incorporated stable θ-Al_2O_3 phase and Al-Si spinel phases (Chakraborty, unpublished data). Consequently, a hindrance of usual polymorphic transitions is noted at intermediate temperatures. Heating to temperature (>1400°C) is required for subsequent transformation to either corundum or mixture of corundum and mullite depending upon the concentration of silica doped in those gels. The existence of θ-Al_2O_3 phase far above 1000°C as noted in both cases of gels definitely indicates that a part of γ-Al_2O_3 which transformed to θ-Al_2O_3 is stabilized as per the concept of Iller (1964) and Yoldas (1976). It is the role of some wt % of silica which is introduced into the θ-Al_2O_3 structure and delayed further crystallization to α-Al_2O_3. Possibly silica is also introduced in the γ-Al_2O_3 structure during the heating process in both kinds of gels. The phenomenon of silica incorporation is more marked when larger concentrations of silica are used as dopant in both cases. Besides formation of θ-Al_2O_3, Al-Si spinel phase development also claims for stabilization of active alumina. Even in case of oxide mixture derived out of (amorphous alumina and amorphous silica obtained from decomposition of ANN from TEOS), crystallization of both the components are considerably delayed. Reason may be due to diffusion of aluminum atoms into silica phase and vice–versa.

(vi) According to Low and McPherson (1989) the formation of the intermediate θ-Al_2O_3 phase in the transformation processes indicated the composition of the cubic phase as aluminosilicate. Since, it was stabilized by the presence of silica in the lattice structure of γ-Al_2O_3 (Yoldas, 1976).

(vii) Crystallization temperatures of Al-Si Spinel and γ-Al_2O_3 are widely different (Low and McPherson, 1989). γ-Al_2O_3 crystallizes at ~870°C with no apparent exotherm while Al-Si spinel formed at 970°C with a sharp exotherm.

(viii) At high temperature γ-Al_2O_3 polymorphically transformed to α-Al_2O_3. While Al-Si spinel polymorphically transformed to mullite type phase at 1250°C exotherm (Chakraborty, 1988).

(ix) **Noncrystallization of a joint mixture of θ-Al_2O_3 and α-Al_2O_3**: It was shown that alumina component of (AS) polymorphically transformed on heating. It developed a varying mixture of θ-Al_2O_3 and α-Al_2O_3 at the temperature range of 1050°C–1200°C (Fig. 15.6). In comparison, DG72 won't show the formation of a mixture of θ-Al_2O_3 and α-Al_2O_3 (Fig. 15.5 and Table 15.3). This result indicates that alumina component of the "in situ" diphasic gel reacted with silica component during heating process to a large extent and hinders independent crystallization to either cristobalite or a mixture of θ-Al_2O_3 and α-Al_2O_3 (Chakraborty, 2005a). It is expected that during the gelation process, enhanced mixing of boehmite gel and silica gel might have taken place. During dehydration and dehydroxylation processes between 100°C–400°C, two gels changed into a homogeneous mixture of amorphous oxides. During gradual heat treatment in between 400°C to 1000°C, solid state reaction among two reactive oxides might occur preferably and predominantly instead of transformation into independent crystal phases. This leads to crystallization of Si bearing spinel phase at the intermediate stage. Thus, Si incorporation modifies crystallization of pure alumina form in diphasic gel systems. It was concluded that (i) mechanical mixture of alumina and silica showed a combined phase of θ-Al_2O_3 and α-Al_2O_3 invariably before mullite formation. Whereas diphasic gel of stoichiometric mullite composition won't show such crystallization sequence. (ii) The result that the components of diphasic gel, did not crystallize jointly to a mixture of θ-Al_2O_3 and α-Al_2O_3 may be a confirmatory test for the absence of θ-Al_2O_3 and formation of Al-Si spinel at the intermediate stage of heating.

Crystallization pathways of diphasic gels of different compositions have been studied by X-ray analysis. The precise XRD chart shows the following observations. On direct comparison of the characteristic XRD peaks of spinel phase formed out of DG72 at ~1000°C to that from pure γ-Al_2O_3 and δ-Al_2O_3 are shown in Table 18.1. It is not probable by X-ray/electron diffraction data (2θ vs. intensity values) to distinguish γ-alumina from Al-Si spinel because of the proximity of atomic sizes of Si and Al and as a consequence both have nearly identical crystal structure and nearly close lattice parameters.

(i) XRD intensities of some peaks of pure γ- and δ-Al_2O_3 are largely different from spinel phase,

(ii) Some of the peaks of it are missing. Absence of some characteristic XRD reflections of spinel phase. All characteristic peaks of it do not correspond to either γ- or δ-Al_2O_3 (JCPDS file Nos. 46-1215, 04-0877, 47-1770, 46-1131) so far it's intensities and peak positions are concerned. Thus, the intermediate spinel phase derived out of DG72 could not be identified as γ-Al_2O_3 or as δ-Al_2O_3. Moreover, XRD data of heated DG72 indicates that the pattern does not show crystallization of corundum even on heating gel at 1200°C.

The XRD intensity 0.140 nm peak of spinel phase increases gradually with increase of heat treatment temperature.

The spinel phase exists as high as 1200°C. It may be possible that Si^{+4} substitutes tetra covalent aluminum of γ-Al_2O_3 phase during heating process of Al_2O_3-SiO_2 diphasic gel and forms a silica incorporated γ-Al_2O_3 (called as Al-Si spinel). This process may result in an increase of strain in spinel lattice which may be the reason for the loss of some characteristic XRD peaks of spinel phase and poisoning of further phase transformation of it to corundum.

XRD intensities of some peaks of spinel phase formed out of heating DG40, DG66 and DG72 are largely different from pure γ- and δ-Al_2O_3. Some of the peaks of it are missing. All characteristic peaks of it do not correspond to either γ- or δ-Al_2O_3 (JCPDS file Nos. 46-1215, 04-0877, 47-1770, 46-1131) so far its intensities and peak positions are concerned. Thus, the intermediate spinel phase derived out of DG72 could not be conclusively identified as γ-Al_2O_3 or as δ-Al_2O_3. Thus patterns of the intermediate spinel are different from those of pure γ- or δ- or θ-Al_2O_3 and it is identified as Al-Si spinel phase (slightly rich in alumina). Lastly at ~1200°C, Al-Si spinel phase rapidly transforms to mullite during heating >1250°C.

18.2.2 Direct Evidences of Al-Si Spinel

(i) Characterization of spinel phase by alkali leaching method of heat treated mullite gels (Chakraborty, 1979) (see Chapter 15).

(ii) Solid state reaction study of Spinel phase by CaO (Chakraborty, 2003a) and (Chakraborty, 2004) (see Chapter 15).

18.2.3 Characterization of Al-Si Spinel Phase Formed Out of Two Mullite Precursors in the Light of MAS-NMR Study: An Example

18.2.3.1 Al-Si spinel phase formation in case of monophasic gel

HB10 precursor: On heating it at 900°C and on subsequent ^{29}Si MAS-NMR spectral study, Schneider et al. (1993) showed a sharper peak at −95 ppm due to development of homogeneous noncrystalline aluminosilicate precursor phase (Fig. 9.12). The later phase exhibited 980°C exotherm in DTA analysis. At 1000°C, it showed crystallization of the spinel phase by XRD and silicon resonance spectra became broad with splitting. The peak maximum appeared at –80, –100 and –110 ppm, respectively. This complex resonance peak, according to them, is attributed to the beginning of phase segregation. The later peak at –110 ppm may be due to noncrystalline silica rich aluminous phase other than pure noncrystalline SiO_2 phase The reason behind it is discussed by the author in Chapter 16 and pointed out the difficulties in leaching of silica out of mullite gel and or from HB10 heat treated at 1000°C with alkali solution. Obviously, the remaining resonance peaks at –80 ppm with its maximum at –100 ppm may correspond to Al-Si spinel phase. Schneider et al. (1993) too suggested that –80 ppm resonance might be due to the presence of silicon in an alumina rich noncrystalline phase or to the incorporation of silicon in γ-Al_2O_3 structure.

18.2.3.2 Al-Si spinel phase formation in case of diphasic gel

Example 1

HB13 xerogel: This gel on heating from 350°C to onward temperatures showed the ^{29}Si MAS-NMR peak at –80 ppm which gradually increased in intensity with increase of temperature, whereas that of –110 ppm peak due to noncrystalline silica decreased in reverse way (Fig. 9.12) of Schneider et al. (1993). This indicates that partly dehydroxylated silicon hydroxide phase reacts with the alumina hydroxide component of the dehydrated diphasic gel during continued heating process and results to gradual development of noncrystalline aluminosilicate precursor phase which is likely related to increment of above –80 ppm resonance by inter diffusion at 900°C. At 1000°C crystallization of spinel phase occurred. The ^{29}Si spectrum pattern depicted more intense –80 ppm resonance with a residual peak at –110 ppm resonance due to formation of noncrystalline silica rich aluminous phase. Therefore, spinel phase which is crystallized on heating the HB13 xerogel at 1000°C is obviously related to the same broad –80 ppm resonance peak as that of noncrystalline aluminosilicate precursor phase. It is necessary to perform ^{29}Si MAS-NMR study in case of diphasic gel heat treated at more higher temperatures beyond 1000°C as was done by them due to following reasons.

(i) It was shown by Chakraborty (1996b) that on increasing the heat treatment temperatures from 1000°C to as high as 1280°C–1300°C more crystallization of noncrystalline aluminosilicate phase into spinel phase occurred as XRD peaks of both 400 and 440 peaks were increased with temperature of heat treatment (Figs. 15.2 and 15.3) of Chakraborty, 1997).

(ii) Secondly, it may be observed that the –110 ppm peak which was previously due to SiO_2(A) phase now becomes related to noncrystalline silica rich aluminous phase would remain partially.

Thus, the intensity of Si spectrum due to noncrystalline aluminosilicate precursor phase would decrease with gradual heating. At the same instance intensity of the Si spectrum of spinel phase which occurred at the same position (–80 ppm) should increase. The training of increase of this spectrum would remain observable up to 1300°C, i.e., up to the onset of DTA exotherm due to Al-Si spinel–mullite crystallization.

Example 2

MI and MII gels: Both these two gels on heating at 950°C and at 1100°C showed spinel as crystalline phase only. Si spectrum of MI at 950°C showed overlap of signals of Q_0(4Al) to Q_4 (0Al) sites (Huang et al 1997). Major signal centered in the range –80 to 95 ppm and another at –108 ppm. At 1100°C both samples displayed three signals at –91, –101 and –109 ppm. Whereas the same spectrum of MII at 950°C showed intense peak at –83 ppm. This broad signal likely related to formation of spinel phase and noncrystalline aluminosilicate phases.

Example 3

CM and SGM materials: These two materials heats treated at 900-to-1000°C showed ^{29}Si resonance near –95 ppm and another at –110 ppm (Fig. 9.2) by (Schneider et al., 1992, 1993). They indicated phase separation into SiO_2 poorer (~90 ppm) and SiO_2-richer domains (~110 ppm) out of highly distorted SiO_2- and Al_2O_3-rich units (–92 to –95 ppm). Sol-gel samples CM and SGM heated between 900°C to 1100°C showed the presence of both spinel and mullite phases by XRD. Out of these, spinel phase was predominant in SGM material while mullite was the major phase in CM material. Obviously, –90 ppm resonance may be attributed to spinel phase in SGM material. Secondly, this resonance likely corresponds to weakly crystalline mullite in the CM sample. It could be concluded that the mullite phase is inherited from the spinel phase, both are likely to be polymorph. On comparing Si spectrum of Type I gel heated to 1000°C (–90 ppm) for crystallization of mullite phase with same of SGM heated to 1100°C (–90 ppm for formation of spinel as major phase) a reasonably, similar conclusion could be drawn, i.e., spinel is polymorph of mullite.

The CM sample showed more mullite and very less quantity of spinel at 1000°C and as such –110 ppm and –90 ppm peaks were well separated instead of clumping. When the –110 ppm peak is designated as due to a silica phase containing some alumina as discussed earlier, then it is obvious that some amount of alumina rich noncrystalline aluminosilicate might have there in 1000°C heated samples. The corresponding Si resonance peak for this aluminosilicate (–94 ppm) may be present within the broad spectrum in the region of –90 ppm. Further mullitization on heating above 1000°C may be due to the following reasons. Two noncrystalline aluminosilicate phases, e.g., alumina rich and silica rich possibly diffuse into each other with rise of temperature and nucleate mullite. As a result –110 ppm peak is found to decrease. Similar nature of transformation was shown by Jaymes et al. (1996) for their gels marked A and F; Type I precursor of Schneider et al. (1993).

Example 4

Precursor B: ^{29}Si spectra of the precursor B synthesized at 750°C shows broad and asymmetric peak (Jaymes et al., 1996). On resolving it, two peaks, one centered at –89 ppm (major) and other at –105 ppm (minor), may be related to aluminosilicate network development (Fig. 9.15). At 1000°C, the spinel phase was only crystallized, and the Si spectrum on decomposition showed two peaks, centered on –95 ppm for spinel phase and –108 ppm for noncrystalline silica phase. ^{27}Al spectrum showed Al(VI) content was largely predominant which corresponded to spinel formation. They attempted to estimate the chemical composition of the spinel phase and found tentatively to be 71.4 mol % Al_2O_3 and 28.6 mol % SiO_2 with an atomic ratio Al:Si = 5:1 which indicates Al-Si spinel concept.

Example 5

Precursor M2 and M3: The Si environment of precursor M2 showed a broad distribution of silicon sites (Geradin et al., 1994). After annealing at ~900°C, the spectra were characteristic of amorphous aluminosilicate structures with a random distribution of silicon and aluminum units. At 1000°C, phase segregation occurred and spectrum became asymmetric. The predominant signal was centered at −104 ppm. Another broad signal centered around −85 ppm is most interesting in the nature of silicon environment which is noted during the course of annealing process of M2 precursor. It might be characteristic of $Si(OAl)_4$ sites present in the spinel phase observed by XRD.

M3 is a diphasic gel containing boehmite initially, did not exhibit 987°C and or 1252°C exotherms as in cases of M1 and M2 precursors, but formed mullite at ~1295°C exotherm. A wide Si spectrum of it was noticed in calcined samples at the temperature range 800°C to 1000°C. The predominant signal in the range −80 to −98 ppm approximately is the characteristic of noncrystalline aluminosilicate phase developed on heating M3 up to 1000°C. A narrowing of peak occurred which predicts that a large quantity of noncrystalline aluminosilicate phase was reacted γ-Al_2O_3 phase and resulted to more amount of weakly crystalline spinel. The position of maxima of the spectrum was still continued in samples annealed from 800°C to 1000°C to 1200°C. Therefore, Si should be the part of spinel structure or it may be concluded that silicon has entered in γ-Al_2O_3 phase. Thus, first part of transformation process of

Noncrystalline aluminosilicate phase (1) -to-spinel phase (2) -to-mullite (3) transformation process is coherent. The remaining second part of transformation of noncrystalline aluminosilicate phase to mullite will be the nucleation process as shown by the author vide Eq. 16.11 (see Chapter 16).

Arguably, a fraction of θ-Al_2O_3 which was noted during the heat treatment process of precursor M3 and the same observation was noted by other researchers namely Huling and Messing (1987). This θ-Al_2O_3 form may likely be silicon incorporated too as discussed earlier in Chapter 16. Therefore, 980°C spinel phase formed in monophasic gels and even spinel phase formed over a long temperature interval in diphasic gel is also Al-Si spinel.

Moreover, Okada et al. (1986), Schneider et al. (1994) disbelieved the γ-alumina hypothesis in transformation processes of both monophasic and diphasic gels and put forwarded tangible evidence by EDS TEM studies to prove that spinel phase was not purely Al-spinel (γ-Al_2O_3) and it contained 8–11 wt % SiO_2 in its structure. Thus, the view of crystallization of spinel as silicon incorporated spinel of Al-Si spinel is more and more evident as per the examples of former researchers shown above. However, the composition of the spinel phase is most probably analogous to composition of mullite other than the composition suggested by later authors (see comments of their studies in Chapter 15).

18.3 Evidence of Formation of Si Incorporated θ-Al_2O_3 as Coexisting Phase with Al-Si Spinel Phase during Transformation of Mixed Gels of Varying Compositions by Deconvolution of XRD Peaks and by Measurement of Peak Width

18.3.1 Precision Identification by Deconvolution of XRD Peak of Spinel Phase

Crystallization pathways of diphasic gels of different compositions have been studied by X-ray analysis. Figure 18.1a,b shows the XRD patterns after smoothening and deconvolution treatments of DG72 gel heated to different temperatures. XRD intensity peaks at 46° 2θ and 66° 2θ corresponding to spinel phase derived out of gels heat treated at 700°C–900°C appear single in nature. Each of those peaks was resolved on deconvolution as the heat treatment temperature increased. For example, two peaks are observed at 46° 2θ region during peak search operation using graphics software, first one at 45.71°, second one at 47.08° which resembles near to θ-Al_2O_3 (Fig. 18.1c). Similar observation was noted during profile fitting of the said XRD pattern (Fig. 18.1d) at 66° 2θ region using X'Pert Profit software. These observations apparently indicate that silica incorporated γ-Al_2O_3 likely transforms to silica incorporated θ-Al_2O_3 during gradual heat treatment from 1000°C–1200°C (Chakraborty, 2008a).

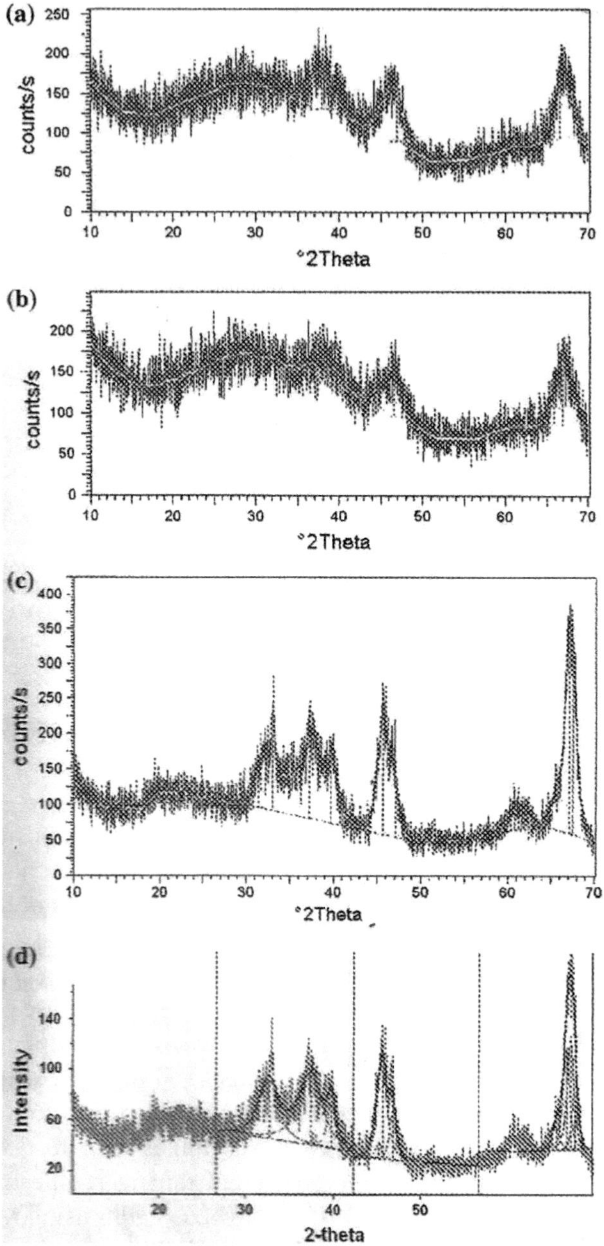

Figure 18.1 (a) DG72 gel heated to 700°C shows a single peak at 46° 2θ and 66° 2θ; (b) DG72 gel heated to 900°C shows a single peak at 46° 2θ ad 66° 2θ; (c) DG72 gel heated to 1200°C shows two single peaks at 46° 2θ ad 66° 2θ; (d) the profit data DG72 gel heated to 1200°C shows peaks around 46° 2θ and 66° 2θ (Chakraborty, 2008a).

18.3.2 Measurement of Peak Width

The mullite formation behavior of mixed gels namely MG40, MG55, MG66, MG70, MG72, MG76, MG80 and MG84 by X-ray diffractogram technique and measurement of the breadth of 0.139 nm peak of spinel phase were attempted by Chakraborty (2006). Besides the formation of Al-Si spinel phase, some quantity of θ-Al_2O_3 is also formed as an additional transitional phase and these two phases have common Bragg diffraction peaks (0.139 nm peak). On analyzing the nature of XRD peak of spinel phase and it was shown that the breadth of 0.139 nm peak of spinel phase changes with increase of heat treatment temperature (Table 15.5). The author concluded that the phase development of any mixed gel at ~1000°C will be a mixture of Al-Si spinel, silica incorporated γ-Al_2O_3, and some amount of silica rich alumina phase (A). The amounts of each phase formed are dependent on the initial composition of the gel and heat treatment temperature. The mixed gels of compositions lie in the silica side of mullite stoichiometry formed only by the Al-Si spinel phase on heating (Table 15.6). On the other hand a mixture of Al-Si spinel and silica incorporated θ-Al_2O_3 was developed on heating mixed gels of compositions falling on the alumina side of mullite stoichiometry.

The existence of this phase above 1000°C was also noted earlier by Iller (1963) and Yoldas (1976) in case of heating Al_2O_3-SiO_2 gels and explained the stabilizing effect as the role of some wt % of silica which was introduced into the θ-structure and delayed corundum formation. Formation of θ-alumina is the reason why identification of Al-Si spinel poses a problem. The mullite formation course also changes due to its existence.

Iller (1963), Wakao and Hibino (1962), and Saito et al. (1998) also showed the existence of θ-Al_2O_3 in gels made out of colloidal particles increased the temperature to the extent of 100°C–200°C more than pure gel components. This θ-Al_2O_3 phase might have formed out of Si-incorporated γ-Al_2O_3 and likely this θ-phase is Si incorporated. Thus, the phase development of any mixed gel or diphasic gel at ~1000°C will be Al-Si spinel, silica incorporated γ-Al_2O_3, and some amount of noncrystalline silica rich alumina phase. The amounts of each phase formed are dependent on the initial composition of the gel, it's homogeneity and heat treatment temperature. Therefore, the development of Al-Si spinel phase (major)—to a mixture of

Al-Si spinel and silica incorporated θ-Al_2O_3 during heating of mixed gels depend upon whether its composition lie in the silica side—to alumina side of mullite stoichiometry.

On comparing the XRD patterns of diphasic gel in the temperature range 900°C–1200°C, the incorporated θ-Al_2O_3 phase was also noted to crystallize with diminished XRD peaks of it (Hulling and Messing, 1990).

Table 18.1 Phase identification of diphasic mullite gel during heating

γ-Alumina			**δ-Alumina**			**θ-Alumina**		**Diphasic mullite gel heated to 1000°C**		**Diphasic mullite gel heated to 1200°C**	
d(Å)	I	hkl	d(Å)	I	hkl	d(Å)	I	d(Å)	I	d(Å)	I
			7.97	8	100						
			6.58	10	101						
			5.85	1	002						
						5.46	10				
			5.07	20	111						
			4.71	3	102						
			4.95	20	112						
4.55	10	111									
						4.53	30				
			3.75	2	201						
			3.56	7	210						
						3.53	3				
			3.40	10	211						
			3.28	15	202						
			3.21	10	113						
			3.03	10	212						
						2.84	80			2.84	18
			2.783	30	203						
2.782	15	220									
			2.737	30	104;221						
						2.72	100	2.73	19	2.71	44

γ-Alumina			**δ-Alumina**			**θ-Alumina**		**Diphasic mullite gel heated to 1000°C**		**Diphasic mullite gel heated to 1200°C**	
			2.593	70	114						
						2.564	20				
			2.457	70	311						
						2.439	90			2.41	44
2.398	90	440						2.39	42		
2.387	35	311									
						2.314	60				
			2.311	40	312						
2.283	20	222									
			2.277	30	223	2.254	50	2.27	31	2.26	33
			2.156	25	115						
						2.018	60				
1.977	100	400	1.989	70	400			1.98	64	1.98	57
			1.950	65	006	1.955	15	1.94	49		
						1.907	35			1.92	34
			1.833	2	412						
						1.801	15				
			1.796	7	225					1.79	2
						1.773	7				
			1.759	3	421						
						1.734	5				
			1.730	1	413						
			1.701	4	422						
						1.621	7				
			1.602	15	226;117						
			1.572	1	306						
						1.568	8				

(*Continued*)

Table 18.1 (*Continued*)

γ-Alumina			δ-Alumina			θ-Alumina		Diphasic mullite gel heated to 1000°C		Diphasic mullite gel heated to 1200°C	
			1.543	10	316						
						1.542	30				
1.521	10	511								1.52	10
						1.510	8	1.51	15		
			1.507	20	512						
						1.486	20				
			1.462	8	335;326						
			1.435	2	522						
						1.426	15				
			1.407	60	440			1.40	100		
						1.404	40				
1.398	90	440	1.392	100	406	1.388	90			1.39	100
						1.295	10				
						1.285	10				
						1.259	6				
			1.248	10	525						
			1.235	10	209						

18.4 Stepwise Process of Crystallization of Al-Si Spinel Phase

The author demonstrated the following experimentations regarding crystallization of the so-called spinel phase during heating of both monophasic gel and diphasic gel of different compositions (as given in Chapters 15 and 16).

(1) Synthesis of Al-Si spinel phase and it's partial isolation by alkali leaching technique (Chakraborty, 1979).

(2) Comparative phase transformation studies of Al-Si spinel vs. γ-Al_2O_3-SiO_2 mixture.

(3) Semi quantitative estimation of Al-Si spinel phase by QXRD technique.

It was shown that silica has been incorporated in to γ-Al_2O_3 structure and formed Al-Si spinel other than corundum, the composition of it is analogous to the composition of 3:2 mullite which is due to suppression of phase transformation of alumina system as reviewed earlier. It is still required to have good support from the studies of various researchers. Of course, the concept of γ-Al_2O_3 spinel formation is nowadays a foregone conclusion and the formation of Si bearing spinel phase is more and more advancing. Only a little dispute remains regarding the composition of intermediate spinel phase which will be either silicon substituted alumina rich spinel phase of probable composition (~$6Al_2O_3.SiO_2$) or Al-Si spinel of composition analogous to composition of 3:2 mullite.

(4) It is felt necessary to measure the XRD intensity of the amorphous band and to explain the role of this amorphous silica phase in the formation process of Al-Si spinel phase. The role of alumina components affecting the mullitization behavior of four different gels were noted by Ivankovic et al. (2003). It was pointed out by them that if Al-Si spinel instead of γ-Al_2O_3 was formed, the amount of silica (A), i.e., the intensity of the amorphous band at 22° 2θ should be decreased. With this idea in mind they first calculated the integrated area of spinel peak at 46° 2θ and same for silica peak for heated mullite gel samples. In this method they observed that hump of probable silica (A) is decreased at the same time the intensity of spinel phase is increased and no mullite is formed. Accordingly, they concluded that Gel B (diphasic, Type II & III) formed Al-Si spinel at the first exotherm and it was formed by the interaction of SiO_2 (A) with alumina phase. Recharacterization of the intermediate spinel phase formed on heating diphasic mullite gels was performed by Chakraborty (2008a) similarly by measuring the integrated area of the intensity peak of it and amorphous band present during heating diphasic gels by XRD technique as per the unique approach of former authors. Instead of manual selection of background prior to calculate XRD peak intensity (Fig. 18.2), it was done with the help of X'Pert Graphics and Profit soft wares. The author will discuss

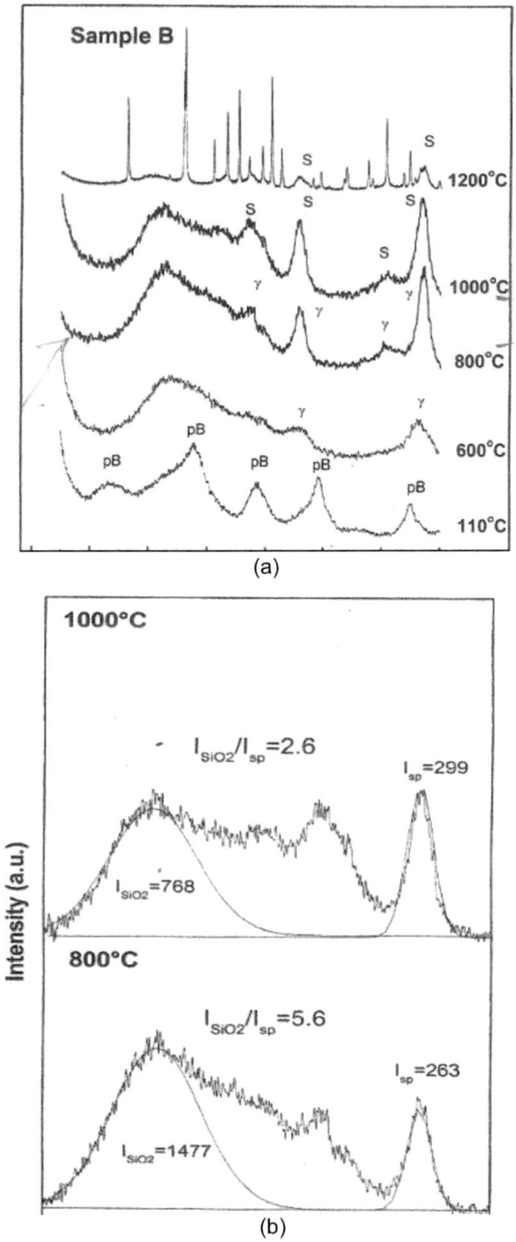

Figure 18.2 (a) XRD patterns of the sample B heated to DTA up to the temperature given in the picture. pH = pseudo boehmite, γ = γ-Al_2O_3 and S (both phases γ-Al_2O_3 and Al-Si spinel. (b) Intensity ratios of amorphous SiO_2 hump at 22° 2θ CuK_α (I_{SiO2} and spinel line at $2\theta \sim 45.6°$ (I_{SP}) at 800°C and 1000°C, respectively. Full lines are the best fit of the corresponding intensities (Ivankovic et al., 2003).

below the characterization of spinel phase based on the measurement of changes of X-ray intensities of amorphous band and spinel phase. Prior to this discussion the formation of aluminosilicate (A) phase out of decomposition of boehmite and silicon hydroxides during heating and its role to further transformation is first of all pointed out.

18.4.1 Formation of Noncrystalline Aluminosilicate Phase during Dehydration and Decomposition of Diphasic Gel (400°C–600°C)

At the first stage of transformation at room temperature to ~500°C:

Diphasic gel which initially contained hydrated silica gel mixed with pseudo crystalline boehmite dehydrates at ~400°C endotherm. In this temperature range, Chakraborty (1996b) observed a broad endothermic peak at DTA and a loss in TGA analysis (Fig. 15.1a). XRD analysis technique revealed that the boehmite component decomposed and developed an amorphous matrix with SiO_2 (A). This phase has been characterized as a noncrystalline aluminosilicate phase. The composition of it changed with heat treatment temperatures and was dependent upon the initial Al_2O_3:SiO_2 ratio. The nature of it was evaluated by two processes.

(i) By noting the changes in XRD intensity of the amorphous band and
(ii) By studying the dissolution behavior in hot alkali solution.

Corroborating two results viz the changes in increase in XRD intensity of amorphous band and nondissolution behavior of SiO_2 component of diphasic gel (Chakraborty, 1979), it is interpreted that during dehydration and decomposition of diphasic gel a gradual development of Si–O–Al linkages occur and which leads to formation of noncrystalline aluminosilicate phase by solid state reaction between oxides of two components on continued heating. Due to inhomogeneity arising during gel synthesis, a fraction of alumina (A) may crystallize to γ-alumina.

In the second stage of transformation (600°C–900°C) range:

Both noncrystalline aluminosilicate (as major) phase and γ-alumina (as minor) phase coexist. But it was experimentally shown by Chakraborty (2004) that two components of diphasic gel are not present homogeneously as amorphous mixtures, rather an amorphous chemical compound constituting two oxides was likely developed. Since calcium aluminate phase did not crystallize instead gehlenite formation occurred when heat treated diphasic precursor of mullite gel was further heat treated with $CaCO_3$ at 1200°C (Tables 15.7, 15.8, 15.9 and 15.10). With this observation, it was concluded that the precursor was a noncrystalline aluminosilicate phase which substantiates the present study. However, the formation of noncrystalline aluminosilicate phase as an intermediate phase during heating of those gels prior to mullite formation is generally ignored by researchers. The formation of gehlenite rather than calcium aluminate by XRD indicated that both monophasic and diphasic precursors consist of noncrystalline aluminosilicate phase of composition close to 3:2 mullite (Chakraborty, 2004). Therefore, it is concluded that monophasic gels, SH gel, Polymeric gel, Type I gel also transformed to mullite via an intermediate noncrystalline aluminosilicate precursor phase during 980°C exotherm.

18.4.2 Crystallization of Noncrystalline Aluminosilicate Phase to Al-Si Spinel Phase and Interrelationship between Area of the Amorphous Hump at 22° 2θ to that of Peak at 67° 2θ of Al-Si Spinel Phase

During the third stage of transformation between 900°C to 1300°C:

Al-Si spinel formation takes place. Corroborating the nature of the curves as shown in the Figs. 15.7 and 15.8, it is interpreted that noncrystalline aluminosilicate phase crystallizes into spinel-like phase in two ways as follows.

(i) In the temperature range between 600°C–900°C, spinel formation occurs slowly in all three cases of diphasic gels. In a parallel reaction, some amount of silica also incorporates into the γ-Al_2O_3 phase and develops Si incorporated γ- Al_2O_3 and Si incorporated θ-Al_2O_3 phases.

(ii) It was shown in Table 15.6 that the XRD intensity ratio of Al-Si spinel phase to amorphous band increased in the entire range of temperature from 600°C to 1200°C for DG40 and DG66 and up to 1100°C for DG72, respectively. In the temperature range between 900°C to 1000°C, two phenomena occur simultaneously. Intensity of X-ray amorphous band decreases (Fig. 15.7), crystallization of accumulated noncrystalline aluminosilicate phase leading to formation of Al-Si spinel phase starts (Fig. 15.8). As these two events are interrelated, it is interpreted that crystallization of Al-Si spinel occurs beyond 900°C from noncrystalline aluminosilicate phase itself.

(iii) The existence of XRD band at 22° 2θ during heat treatment of diphasic gels beyond 1000°C confirms the presence of remaining noncrystalline silica rich aluminous phase. Table 15.6 shows that ratio of intensity of Al-Si spinel to amorphous band decreases beyond 1100°C in case of DG72 which supports amorphous phase development. Thus, diphasic gel in general may be conceived as a micro composite which during heating in the temperature range 600°C–900°C develops noncrystalline aluminosilicate phase and some quantity of Al-Si spinel phase and silica incorporated θ-Al_2O_3 phase. At ~1200°C, the remaining noncrystalline aluminosilicate phase fully crystallizes to Al- Si spinel phase. A residual amorphous band due to noncrystalline silica rich aluminous phase remains in association with these phases.

18.4.3 Mechanism of Transformation of Alumina Phase Content of Al_2O_3-SiO_2 Gel

(1) First mechanism: During heating diphasic Al_2O_3-SiO_2 gel, it is assumed that interaction takes place among two reactants resulting in the formation of noncrystalline aluminosilicate phase. This is evident from the increase in XRD peak area of amorphous band at 22° 2θ with decrease in wt % of Al_2O_3 of diphasic gel of composition from DG72 to DG66 to DG40 (Fig. 15.7). This glassy phase seems to retard further transformation of the alumina component. This mechanism may be the cause of suppression of θ- to α-Al_2O_3 conversion.

(2) Second mechanism: In the transformation of pure θ- to α-Al_2O_3, two steps may occur in the nucleation process as suggested by Clark and White (1950), Bye and Simpkin (1974), and Xue and Chem (1992).

(i) Change in mobility of O^{-2} ion arrangement, i.e., close packed cubic system (CCP) to hexagonal close packed system (HCP);

(ii) Migration of aluminum cation from tetrahedral to octahedral interstices.

The matter of inhibition in nucleation of α-Al_2O_3 may be due to the possibility of formation of solid solution of silicon in γ-Al_2O_3 has already been discussed. Moreover, tetrahedral substitution of Si^{+4} by replacing Al^{+3} is a feasible proposition from a crystallographic point of view. Once Si enters into the lattice, migration of Al^{+3} within the alumina grain is not likely initiated as necessary for polymorphic phase change to α-Al_2O_3. Thus, both the two mechanisms are operative in the present case of evolution of diphasic gels. Al-Si spinel and or silicon incorporated δ- or θ-Al_2O_3 formed during the heating process of diphasic gel and subsequent structural transformations of these phases are the cause of mullite formation.

18.4.4 Critical Studies on Intermediate Noncrystalline Phase-to-Al-Si Spinel Phase-to-Mullite

Previous literature on (i) characterization of amorphous band, (ii) other ways of the crystallization process of it to Al-Si spinel phase, (iii) on the composition of it and its final transformation to o-mullite are discussed below.

(i) Characterization of amorphous band

It is shown that amorphous bands exist in the transformation process of diphasic gels are not free amorphous silica as assumed by various earlier researchers. Okada et al. (1986) earlier revealed that the chemical composition of the amorphous phase coexisting with the spinel phase was not merrily pure silica, but it contained alumina and it might be corresponded to $6SiO_2.Al_2O_3$ tentatively. By ^{29}Si MAS-NMR studies, Jaymes et al. (1996) and Schneider et al. (1993) noted the appearance of a new peak at ~ −110 ppm during the transformation of diphasic gel marked B and HB13 heat treated to 1000°C. They characterized this −110 ppm Si NMR peak as amorphous silica without presenting any experimental evidence in its favor. According to Engelhadt and Michel (1987), in aluminosilicate, the silica tetrahedra with four bridging oxygen atoms are also classified into five species based on the number of adjacent aluminum atoms bonded to oxygen (see Chapter 9).

As free silica was noted to be absent as per the experimental alkali leaching data as shown by the author, the −110 ppm peak is characterized as noncrystalline silica rich aluminous phase in view of ^{29}Si MAS-NMR data which is near to Q_4(1Al) region. The present study supports the earlier MAS-NMR study of Chakraborty (2006).

(ii) Characterization of spinel phase

By transmission electron microscopy, Schneider et al. (1994) noted the occurrence of large spherical particles (≤0.5 μm) of silica phase and fine-grained aggregates of pseudo-boehmite particles (~20 μm) in their Type II gel. They showed the decay of those spherical particles during heating and conjectured the theory of incorporation of silicon into the agglomerates of γ-alumina. They suggested that the diffusion process of Si at the surfaces of the γ-alumina crystallites was related to temperature of heating. But it was not known from which phase Si diffuses out. Was it from free silica or from aluminosilicate (A) phase? Present study shows that free silica is not available for diffusion reaction as assumed. It was suggested by Ivankovic et al. (2003) that the formation process of Al-Si spinel phase by the solid state reaction between amorphous silica with γ-alumina. In the contest of this assumption, the above reaction shown by them does not occur experimentally. Since amorphous silica phase is absent as one of the products of transformation. Therefore, the concept of Al-Si spinel formation suggested by them seems untenable. In the temperature range 600°C to 900°C, the band due to noncrystalline aluminosilicate phase formation increased. Apparent relation between the decrease in amorphous band with formation of Al-Si spinel shown by them needs explanation. They did not present any mechanism by which Al-Si spinel develops. It was shown that diphasic gel at the first stage decomposed into a noncrystalline phase mixture consisting of alumina and silica. XRD pattern at this stage showed an amorphous band. Thereafter, a solid state reaction among the two amorphous phases occurred that led to development of a more noncrystalline aluminosilicate phase. The reaction might have occurred during dehydroxylation of boehmite and dehydration of silicon hydroxide gel on heating (Chakraborty and Ghosh, 1987). In consequence, a slight increase in XRD intensity of amorphous band at 22° 2θ of diphasic gel in the temperature range >400°C occurred. In some instances, a weak intensity of γ-Al_2O_3 was also noted along with later phase at early stages of heating. At the next step, the former phase crystallizes to Al-Si spinel in the temperature range 900°C to 1000°C and as a result the band intensity decreases at first. Intensity of it again increased slightly with further heating to 1200°C owing to noncrystalline aluminosilicate formation as shown in Fig. 15.8. Thus, the amorphous band still existed in association with Al-Si spinel and with some unconverted γ-Al_2O_3 or θ-Al_2O_3 at high temperature. Even in case of monophasic gel, the author earlier (2004) showed the formation of noncrystalline aluminosilicate phase as an intermediary phase prior crystallization of either mullite or a mixture of mullite and Al-Si spinel phase or solely Al-Si spinel phase. Therefore, the decrease in intensity of amorphous band as observed in case of Ivankovic et al. (2003) is correlated to crystallization of noncrystalline aluminosilicate phase itself to Al-Si spinel phase transformation pathway as shown for their gel sample marked B and not due to solid state reaction between silica (A) with γ-alumina.

18.4.4.1 Discussion on spinel formation studies of Ivankovic et al. (2003)

Figure 15.7 indicates that free silica should not exist at temperature at ~800°C as conceived by Ivankovic et al. (2003). This is true in the entire temperature range between 600°C (the lowest temperature chosen in the present experiment) and 1200°C. Moreover, they did not succeed to detect changes in areas for other two heat treated diphasic gels marked C and D synthesized from γ-Al_2O_3, Degusa and (γ-AlOOH) boehmite of dispersal varieties respectively. These are due to manual selection of background prior to calculating peak intensities. It is thus required to re-examine the applicability of this unique concept for characterizing spinel phase formed out of different source materials under varying processing techniques.

On careful observation of their Figs. 5a and 5b, two apparent observation are given below.

(i) That both the intensity peak due to spinel phase 0.197 nm (46° 2θ Cu K_{α}) and 0.139 nm (67° 2θ Cu K_{α}) increased with heat treatment from 600°C to 1000°C to 1200°C and,

(ii) The intensity of the amorphous band at 22° 2θ Cu K_{α} decreased from 600°C to 1000°C to 1200°C for heat treated sample mark D.

On the contrary, they mentioned that the amount of amorphous silica was almost constant in this temperature range and intensity of γ-Al_2O_3 increased with temperature. These are obviously due to their unsystematic background selection and calculation of peak area there of (Fig. 18.2). However, the ratios of the intensity peak of spinel phase to that of amorphous band of gel samples of different compositions increased with heat treatment temperature from 900°C–1200°C (Fig. 15.8). Thus, decrease in intensity of amorphous band width increase of the peak of spinel phase further confirms the incorporation of Si out of aluminosilicate (A) phase in spinel structure to form Al-Si spinel phase.

Therefore, the concept of the formation of Al-Si spinel phase as analyzed above is supported, and it is apparent that amorphous band gradually decreased form 600°C–1200°C. It is obvious that decrease in amorphous band is to be correlated to more and more development of spinel phase which should be conjectured as Si substituted alumina phase instead of simple δ- or θ-Al_2O_3. So it is believed that diffusion of Si out of noncrystalline aluminosilicate phase into γ-Al_2O_3 structure resulting to the development of Al-Si spinel and silicon incorporated δ-Al_2O_3 in one hand and counter diffusion of Al out of γ-Al_2O_3 into noncrystalline aluminosilicate phase likely occurs on the other hand to form silica rich alumina phase.

(iii) Composition of nanocrystalline spinel phase formed at different temperatures. EDX analysis showed that silica content of the spinel phase was 12 mol % at 500°C and increased up to 18 mol % (11.5 wt %) at 1150°C (Schneider et al., 1993). This data apparently agree with 8–10 wt % silica as shown by Okada and Otsuka (1987) and later with 10 wt % silica by Sonuparlak et al. (1987), and ~7 wt % silica by Gerardin et al. (1994). However, neither of these values fit well with present author's data (1979) and that of Suzuki et al. (1990) nor with data shown by Low and McPherson (1989). It is concluded by the author (2006) that noncrystalline aluminosilicate phase developed during continued heating diphasic gel. The major quantity of spinel phase is formed out of this noncrystalline aluminosilicate phase by sudden decrease in XRD intensity of it and it remains stable up to as high as 1300°C during the phase evolution process. It subsequently transformed to mullite on heating just above 1300°C by exhibiting an exothermic peak in DTA. Some amount of noncrystalline silica rich alumina phase remains as residue.

18.4.4.2 Discussion on microstructural studies of Wei and Halloran (1988)

According to Wei and Halloran (1988) the diphasic gel did not form Al-Si spinel phase but generated δ-Al_2O_3. Based upon the direct observation of TEM and AEM of the microstructure of the heated gel they conjectured the following.

(i) That mullite formation takes place between a siliceous amorphous phase and δ-Al_2O_3 at ~1200°C.

(ii) They however showed that mullite grains appeared to be individually nucleated and the nucleation density near the surface was the same as that at the center of the specimen, i.e., a case of uniform nucleation and then spontaneous crystallization of mullite instead of engulfing alumina particles in siliceous matrix.

(iii) The Al/Si ratio of the mullite product was the same as in the bulk composition of the diphasic gel. i.e., mullite nucleation occurs within the so-called alumina grains.

Based on these results, the author however explained that in presence of glassy siliceous phase, silica very easily introduced into δ-Al_2O_3 structure and developed Al-Si spinel phase which instantaneously nucleates to mullite. This may be a case of polymorphic transformation of Al-Si spinel to mullite as shown in the present study. If there is at all a case of diffusion then the diffusion distance is equivalent to the size of the alumina particle, i.e., ~30 nm only.

18.4.4.3 Discussions on mullitization study of HB13 and sample B by Schneider et al. (1993) and Jaymes et al. (1996)

Mullite precursor marked HB13 synthesized by Schneider et al. (1993) contained pseudo-boehmite and as such it was equivalent to diphasic gel. This heating at 350°C showed two ^{29}Si NMR resonances at -80 and -110 ppm, respectively. The intensity of -80 ppm peak increased with heat treatment and formed a spinel phase. While -110 ppm peak decreased and still persisted at ~1000°C which may be due to residual silica rich aluminosilicate (A) phase (see Chapter 9) instead of free amorphous silica.

The depiction of -110 ppm peak in the ^{29}Si NMR spectra of their diphasic gel sample marked B (a diphasic gel) heated to 1000°C was also noticed by Jaymes et al. (1996). It could be explained that the band at ~110 ppm in MAS-NMR study is due to Si–O–Al bonds of residual noncrystalline silica rich alumina phase. This disappears on further heating during mullitization. This suggests us to believe that a fraction of silica rich alumina silicate phase reacts with alumina rich silica phase (i.e., Si incorporated δ- or θ-Al_2O_3 as shown in the present study) and nucleates mullite as an alternative procedure.

18.5 Variations in Quantity of Formations of Al-Si Spinel Phase from Six Cases of Precursor: A Case Study

The formation of Al-Si spinel phase from six cases of precursors was studied by XRD (Chakraborty, 2008) (see Chapter 15). The elaborative investigation showed variations in the intensity of Al-Si spinel phase formed out of precursors synthesized by either of the two methods.

In the gelation process, it is shown that the precursor marked SHI did not form a spinel phase at 980°C exotherm. On the other hand, SHII formed a minor amount of spinel with an abundant quantity of mullite when heated. Fullest extent of spinel occurred when precursors marked SHIV and SHIII heated to the temperature of 980°C (Table 18.2). Therefore, intensity of spinel phase formation out of monophasic precursors synthesized by gelation method proceeds in the following ways.

SHIV > SHIII > SHII and it is absent in SHI.

In the coprecipitated method, spinel formation is pH dependent. SHIV prepared at pH = 6 formed predominant spinel formation when heated at the 980°C exotherm. As the precipitation pH is increased from 6 to as high as 9 in "in situ" diphasic gel, sharp 980°C exotherm is not exhibited. Spinel phase started to crystallize at a temperature of 900°C and it gradually increased with continued heating as shown in XRD recordings and in (Fig. 15.3). After phase separation at ~1000°C, it forms alumina rich Al-Si spinel (minor) and remains noncrystalline aluminosilicate phase (major). DTA trace of it showed a broad exotherm over a long range of temperature (Fig. 15.1). Thus, the mode of spinel development in coprecipitated gel prepared at pH = 9 (i.e., in diphasic gel) is significantly different from that of slow hydrolysis gels (SHII and SHIII) and coprecipitated gel synthesized at pH = 6. Spinel phase developed in cases of SHII, SHIII and SHIV transformed on heating and exhibited a second exothermic peak at ~1250°C in DTA. Whereas a higher transformation temperature as high as 1320°C is noted for the exhibition of exotherm in case of diphasic precursor.

An extensive study has been made to the processing and characterization of sol-gel derived mullite. It is thought that mixing the scale of the two components actually controls the reaction course to mullite. In fact processing parameters followed for sol-gel precursors really guide the path of phase transformation and temperature of mullite formation. Based on different sets of conditions, there exist as many as 5 different transformation paths leading to mullite formation out of monophasic and diphasic gels (see Chapter 16).

Table 18.2 Phase composition of monophasic and diphasic gels during heating process at 980°C

Nature of the gel	980°C Phase developed after phase separation
Monophasic gel SHI	M (metastable nonstoichiometric mullite) (major) + AS (noncrystalline aluminosilicate phases, AS2 & AS3)
SHII	M (major) + alumina rich Al-Si spinel (SP) (minor) + AS (AS2 & AS3)
SHIII	M (minor) + SP (major) + AS (AS2 & AS3)
SHIV/CP6 coprecipitated gel made at pH ~6	SP (major) + AS (AS2 & AS3) (major)
"In situ" diphasic gel prepared at pH ~9/CP-9	At 900°C, γ-Al_2O_3, SP and aluminosilicate precursor phase. At ~1000°C forms SP (minor) + (noncrystalline aluminosilicate phase (major)

18.6 Composition of Spinel Phase by Different Researchers: Supplementary Case Studies

18.6.1 TEM Study

(1) The chemical composition of the phase was determined by Okada et al. (1986) followed by three steps of experimentations. In the first step, mullite gel was fired at 1000°C. In the second step, the so called associated amorphous siliceous was leached out with 7 wt % NaOH solution by boiling for 40 min. In the third step, the content of bonded silica in the spinel phase was analyzed by studying the leached mass in analytical TEM with EDX-type analyzer and obtained a value of ~8 wt % SiO_2.

(2) It was shown by Schinieder et al. (1994) that during heating diphasic gel, Si is gradually introduced into the γ-alumina lattice as evidenced by TEM study made. They estimated the silica content of the nano crystalline spinel phase at different temperatures on a heating precursor by EDX analysis. The silica content of the spinel phase at 500°C is ~12 mol %. It increases to 18 mol % at 1150°C (Fig. 15.13) of them.

With this result they suggested that the maximum silica incorporation into gamma alumina is 18 mol % (11.5 wt %) at 1150°C. The presence of ~8 wt % SiO_2 in the spinel phase. This value fetches the composition of the formulae of spinel as $6Al_2O_3SiO_2$ which was earlier suggested by Okada and Otsuka (1986). The views of TEM studies of both Okada and Otsuka (1986) and Schinieder et al. (1994) are commented by the author already shown in Chapter 15.

(3) However, by electron microscopic studies Suzuki et al. (1987) presented the electron diffraction pattern of Al-Si spinel phase. They noted the composition of the spinel phase obtained at calcination of the monophasic gel/precursor at 900°C/2 h by EDS in TEM. They represented the unit cell of the spinel phase as $[(Al_{x\,\square\,5.33\text{-}x})\ (Si, Al)_8\ Al_{10.67}O_{32}]$. By EDS analysis, the composition of Al-Si spinel phase was shown to be stoichiometric mullite composition. (and its size are smaller than 10 nm. This spinel phase converted to yield ultrafine mullite during heating at 1200°C with the same stoichiometric composition which conformed the composition of intermediate spinel analogous to the composition of 3:2 mullite as shown by the author.

(4) The formation of spinel-like phase during heat treatment of his slow hydrolyzed mullite monolith after the completion of DTA exothermic phenomenon was noted by Colomban (1990). No segregation of silica (A) phase in his Roman spectroscopic study was observed.

18.6.2 IR Study

Direct confirmatory proof in support of presence of Si in the spinel lattice:

IR spectrum of 1000°C heated gels are different for four classes of gels (Hirata et al., 1998). The IR absorption band due to Si–O stretching in case of pure silica shift from 1120 cm^{-1} to 1080 cm^{-1} in heated coprecipitated gel. Substitution of Al^{+3} for Si^{+4} considerably lowers the stretching frequency (Sales and Alarcon (1996), Yamada and Kimura, 1962). This substantiates Al-Si spinel formation theory. By similar IR spectroscopic study Okada et al. (1986) also conjectured that substitution of SiO_2 by Al_2O_3 in spinel structure is plausible. But how much quantity of silica substituted is not ascertained by IR spectroscopic studies.

It was noted by Low and McPherson (1989) from IR study that the spectrum of cubic phases obtained from mullite and alumina gel heat-treated at 1000°C and 900°C are dissimilar in relation to their bond strength and their composition. For example, absorption bands due to AlO_4 tetrahedra are observed to exist in both the cubic phases however with differences in composition. For that reason the two spectra are not identical. However, concrete conclusions regarding the identification of spinel phase and revealing its composition are necessary by further IR technique.

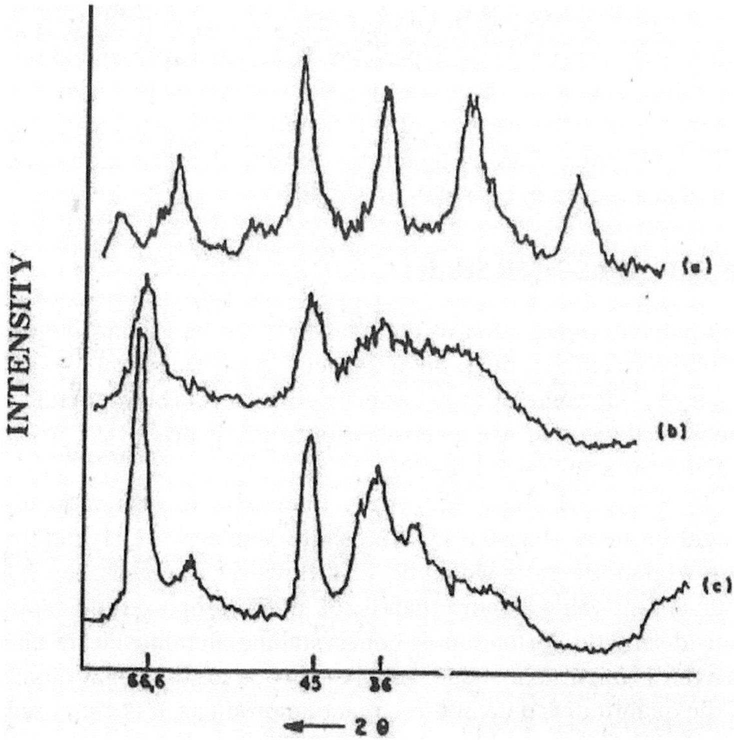

Figure 18.3 X-ray diffractograms of mullite xerogels prepared at 90°C (a) and heat treated at 500°C (b) and 1100°C (c) (Orefice and Vasconcelos, 1997).

It was shown by Schneider et al. (1993) that the strong bands at ~1100 cm^{-1}, 470 cm^{-1}, attributed to Si-O-Si stretching, O-Si-O bending vibrations disappear completely in the IR spectra of the Type II gel heat treated to 900°C and on subsequent alkali leaching process. The remaining absorption region between 820 cm^{-1} and 580 cm^{-1} may be due to the spinel phase which did not change significantly. Simultaneously, a broad weak absorption appeared between 1080 to 980 cm^{-1} (Fig. 15.14) which was interpreted by them as incorporation of silicon in γ-Al_2O_3 phase.

Phase transformation of mullite xerogel obtained out of pre hydrolyzed TEOS and Al-isopropoxide at hydrolysis temperature of 90°C showed diphasic in character by Orefice and Vasconcelos (1997). The XRD of this diphasic gel submitted to heat treatment at 1100°C is shown in Fig. 18.3.

The resultant spinel phase was identified by FTIR. The spectra showed the presence of a peak at 1070 cm^{-1} corresponding to crystallized spinel of this diphasic gel. They argued this peak is not associated with the presence of Si-O-Si bonds but due to the presence of Si-O-Al bonds. They assigned this bond to the formation of Al-Si spinel phase even in case of diphasic gel.

18.6.3 MAS-NMR Study

Computer deconvolution of the spectra given by Komarnine et al. (1985) showed that the ratio of tetrahedral to octahedral Al is 0.48. However, this value does not concur either with the ratio 1.34 for theoretical value of Al for tetrahedral to octahedral of 2:1 or with 1.25 for the same for 3:1 mullite.

If this spinel phase be γ-Al_2O_3 then tetra to octa ratio of Al would be more than 0.6 by MAS-NMR. Similarly, if 2:1 mullite is formed then the ratio would be 1.34. However, the measured value is 0.48. This value is more than 0.265 of the proposed Al-Si spinel. Considering the formation of noncrystalline aluminosilicate phase as a third constituent and for that reason and for the nonavailability of the quantitative data of three reaction products, it is not possible to compute what should be the theoretical AlO_4/AlO_6 ratio and then to compare with the measured value. In the corroborative XRD and ^{29}Si MAS-NMR techniques studies of phase transformation of Al_2O_3-SiO_2 mixed gels of different compositions by (Chakraborty, 2006) the following conclusions were made.

Table 18.3 Percentage average AlO_4 and AlO_6 content in Al-Si Spinel, γ-Al_2O_3 mullite (3:2) and mullite (2:1)

Spinel phase contents	**Al-Si spinel $2Al_2O_3.3SiO_2$**	**Al-Si spinel $3Al_2O_3.2SiO_2$**	**γ-Al_2O_3**		**Mullite (3:2) $2Al_2O_3.3SiO_2$**	**Mullite (2:1) $2Al_2O_3.SiO_2$**
AlO_4 (%)	0	21	25	37.5	55	58.3
AlO_6 (%)	100	79	75	62.5	4	41.7
Ratio (AlO_4/AlO_6)		0.265	0.33	0.6	1.24	1.34

(1) Incorporation of silica into γ-Al_2O_3 phase leads to the formation of Al-Si spinel phase. The maximum amount of silica incorporated into γ-Al_2O_3 structure to develop spinel phase is equivalent to that present in 3:2 mullite (Fig. 9.6). Beyond this amount, the added silica will remain as free noncrystalline silica rich aluminous phase and forms cristobalite thereafter.

(2) Thus, the composition of Al-Si spinel is similar to composition of 3:2 mullite. It remains constant irrespective of the composition of the mixed gels. Average LC value of it is 0.7887 nm.

18.6.4 Quantitative X-Ray Diffraction Analysis (QXRD)

Okada et al. (1986) measured the chemical composition of the spinel phase by QXRD technique and noted a value of ~8 wt % SiO_2 for monophasic gel, e.g., SH and for other monophasic gel, e.g., RH types. This result was vividly discussed in Chapter 15. Contrary, the author predicted the composition of Al-Si spinel is equivalent to that of 3:2 mullite.

Variation of the quantity of spinel phase in different precursors: Regarding the varying amount of spinel phase formation Chakraborty and Das (2003) noted the following observations.

(i) That the measured X-ray intensities of spinel phase formed in different cases of diphasic gels are shown to be greater than that expected for γ-Al_2O_3 formation even in case of diphasic Al_2O_3-SiO_2 gel system (Fig. 15.17). The exhibition of higher X-ray intensity value for each composition indicates the possibility of substitution of Al by Si in spinel structure.

(ii) Gels of high alumina composition show θ- and k-polymorphs of Al_2O_3 besides some γ-, δ- and silicon bearing spinel. As a consequence, there is every possibility of showing reduced X-ray intensity of 0.139 nm peak.

(iii) Some amount of amorphous phase always coexisted with the spinel phase as shown in Table 15.2 by Chakraborty, 2003. This might be another reason for showing reduced X-ray intensity due to formation of a really lower quantity of spinel phase.

As Al-Si spinel is forming in all cases, the exact composition of it could be ascertained only if the composition of associated noncrystalline aluminosilicate phase is available. The composition of the former phase seems too difficult to ascertain. since there is every possibility to change the composition of the glassy phase due to the ongoing solid state reaction of SiO_2 (A) and Al_2O_3 (A) during heating. Table 15.2 further showed that amount of glassy phase changes with the wt % of Al_2O_3 in the gels. All these results indicate the possibility of variation in the amount of Al-Si spinel phase formation in the diphasic system which seems as a most inhomogeneous system of Al_2O_3-SiO_2 gel.

Possibility of formation of Al-Si spinel of varying composition: It is conjectured by the author that composition of Al-Si spinel is ascertained in the following ways may change from 0 to 28% silica. It is interpreted that the formulae of spinel would start from (right hand side of Al_2O_3-SiO_2 phase diagram up to the end of inclined trend of the liquidus curve, i.e., peritectic point). The range of existence of Al-Si spinel may lie from 100 wt % Al_2O_3 up to incongruent melting point of mullite i.e., up to 72 wt % Al_2O_3. At 28% silica composition, Al-Si spinel transforms to 3:2 mullite as like polymorphic change. Some experimental observations in support of the above view are cited below.

(i) In corporation of Si in alumina phase (Schneider et al., 1993) in their MAS-NMR studies vide Fig. 15.13.
(ii) Increasing in intensity of both 400 and 440 peak on heating (Chakraborty and Das, 2003) vide Fig. 15.17.
(iii) Intensity of spinel phase rises with concentration of SiO_2 (Dowy, 2006) vide Fig. 18.11.
(iv) Decrease in **d** nm value/LC value of spinel with increasing the concentration of SiO_2 (Chakraborty, vide Fig. 15.20).

Accordingly, the reaction that siliceous phase (AS2) or other aluminosilicate rich in silica phase as noted in Si NMR, interacts/ Si of it diffuses into γ-Al_2O_3 spinel of (0.0 wt % in silica) during heating holds true. Therefore, Al-Si Spinel which is formed at 980°C exotherm in case of RH gel is to be called as unsaturated Al-Si spinel. Gradually, it becomes saturated by taking Si out of noncrystalline aluminosilicate phases. It is assumed that when it is saturated with silica (when Al_2O_3:SiO_2 becomes 3:2) it instantly transforms to mullite with exhibition of an exotherm.

18.6.5 Polymorphic Transformation of Al-Si Spinel to Mullite

As discussed above, diphasic gels during heating at ~1200°C consist of Al-Si spinel phase (major), silica incorporated θ-Al_2O_3 (minor), and noncrystalline silica rich alumina phase. The amount of phases vary with the composition of diphasic gel chosen, e.g., DG40, DG66 and DG72. At the third stage (> 1250°C–1300°C) mullitization commences. It is explained that Al-Si spinel formed out of these three diphasic gels polymorphically transforms to mullite type phase during heat treatment at ~1300°C. This mullite transforms to regular well crystallized 3:2 mullite on further heating > 1400°C. Schematic diagram of phase development of DG40 and DG72 are shown in Fig. 16.5 (Chakraborty, 2005a).

18.7 Solid Solution Range of Silica in γ-Alumina Phase to Form Al-Si Spinel Phase: Illustrative Examples

18.7.1 Introduction to Solid Solution Range of Silica in Al-Si Spinel Phase

Regarding solid solution range of silica in Al-Si spinel phase the main point of discussion is: how the end point of spinel type solid solution could be visualized. To detect solid solution range of silica-to-γ-Al_2O_3 phase, following two criteria are taken into consideration during the phase development process of mixed gel at different stages of heat treatment.

(i) The existence of intermediate forms namely γ-Al_2O_3, Al-Si spinel and θ-Al_2O_3 phases in the phase transformation process of various silica doped Al_2O_3 are first of all identified by XRD.

(ii) Secondly, it is to be noted the temperatures of crystallization of either cristobalite and/or corundum which just ensues out of various compositions of mixed gel.

Accordingly, the author investigated these two points by critically re-examining the results of Table 15.4. As shown above γ-Al_2O_3 or Al-Si spinel usually forms during heating mixed gel between 900°C–1000°C.

Formation of cristobalite in mixed gels:
Mixed gels namely MG40, MG66 and MG70 develop cristobalite in addition to mullite during heating beyond 1250°C. The intensity of cristobalite formation among those gel changes in the temperature range 1250°C–1500°C. At ~1400°C, intensity of it gradually decreases with a decrease in the concentration of silica of the mixed gel and minimum quantity of it is noted in case of MG70. After words, cristobalite formation ceases as noted during heating MG72. This indicates that besides formation of Al-Si spinel as the only crystalline phase, MG70 also forms a small amount of siliceous phase during heating it at 1000°C. The former phase thereafter forms mullite and excess portion of siliceous phase crystallizes to traces of cristobalite at 1400°C. On considering the composition of MG70, it is assumed that less than 30% silica is the maximum possible content of silica in the γ-Al_2O_3-SiO_2 solid solution limit. Let the amount of silica would be near to 28% approximately considering MG72 where no cristobalite is formed during heating.

Formation of corundum in mixed gels:
Mixed gels namely MG84, MG80 and MG76 form corundum in addition to mullite on heating at 1400°C to 1500°C. A minimum concentration of corundum formation is observed in case of MG76 at 1400°C. This indicates that besides formation of the Al-Si spinel phase at 1000°C to 1100°C, a portion of alumina is present separately as γ-Al_2O_3. This phase later crystallizes to silica incorporated θ-Al_2O_3 and finally to corundum and mullite during heating at higher temperature. The Al-Si spinel phase transforms to mullite as usual. On considering the composition of MG76, it is suggested that much below 76% alumina is the maximum possible content of alumina content in the γ-Al_2O_3-SiO_2 solid solution limit.

Thus, two important points emerge out during the critical analysis of the XRD results. Formation of cristobalite is just ensued during heating 70 wt % alumina gel (MG70) in comparison to other mixed gels to 1400°C. Second, crystallization of corundum is just detected when 76 wt % alumina gel is heated in comparison to other aluminous mixed gels. Based on the second finding it is suggested that the amount of SiO_2 in γ-Al_2O_3-SiO_2 solid solution would be below 30 wt % maximum. Let the amount of alumina possibly incorporated with SiO_2 and form Al-Si spinel phase would be much less than 76%. Thus the maximum wt % of silica accommodation in the γ-alumina phase has been determined and it will be much > 24 and slightly < 30 wt %.

Some of the supportive evidence is cited here from earlier studies. It was shown by Kanzaki et al. (1990) that mullite was observed as the only crystalline phase when heated at ~1600°C for calcined powder in the composition range from 60–74 wt %. Later on, it was also shown the formation of single phase mullite for precursor containing 74 wt % and mullite + corundum for 75–78 wt % Al_2O_3. Formation of corundum out of gel containing 76.6 wt % Al_2O_3 >1550°C was also shown by Hirata et al. (1989). The formation of corundum in gel containing 75.5 wt % or greater Al_2O_3 content at 1650°C was noted by Lee et al. (1992). On the basis of comparative XRD studies of crystallization behavior of two components, quantity of SiO_2 more than 24 wt % in solid solution with γ-Al_2O_3 as derived is feasible. During the process of solid solution, Al-Si spinel forms as a resultant phase. The value agrees well with the trend of low LC data of Al-Si spinel phase formed out of heating of single crystal kaolinite as shown by Brindley and Nakahira (1959). The lattice parameter of it was found to decrease with concentration of alumina up to ~72 wt % alumina of the mixed gel. The average value of it is nearly equal to the composition of 3:2 mullite.

Previous conjectures on solid-solution range:

Three earlier data are first of all presented thereafter two following new points are considered in arriving into a final conclusion regarding SiO_2-γ-Al_2O_3 solid solution range.

(1) Solid solubility of SiO_2 in the spinel phase derived out of both SH and RH gels was first pointed out by Okada and Otsuka (1986). According to them, the chemical composition of spinel phase was found to be around $SiO_2.6Al_2O_3$ based on X-ray quantitative analysis where the intensity of spinel phase was noted to be maximum. This is equivalent to 8 wt % SiO_2. They of course considered the formation of noncrystalline

aluminosilicate phase along with weakly crystallization of spinel phase but conceived the composition of it as silica rich ($6SiO_2.Al_2O_3$) phase which lacks any experimental support.

(2) Subsequently by EDX technique, Schneider et al. (1994) determined the SiO_2 content present in γ-Al_2O_3 and showed that the maximum amount of 11.5 wt % SiO_2 could be incorporated into γ-Al_2O_3 structure when boehmite-SiO_2 mixture was heated to 1150°C.

(3) Later on based on ^{29}Si MAS-NMR, Gerardin et al. (1994) showed that Al_2O_3-SiO_2 gel obtained by intimate mixing formed an alumina–rich spinel phase at ~980°C, which contained 7 wt % SiO_2.

Interpretations of previous studies on solid solution of silica in alumina phase to form Al-Si spinel phase in siliceous side vs. aluminous side of mullite stoichiometry is first of all discussed.

18.7.2 Siliceous Side of Mullite Stoichiometry

Range of existence of SiO_2-γ-Al_2O_3 solid solution in diphasic gel system has been discussed on the basis of crystallization behavior of alumina and silica gel components to corundum, cristobalite, and mullite as studied by Chakraborty (1993). The first observation was shown in Chapter 15 that the boehmite component of the gel hinders the crystallization sequence of the silica gel component of the diphasic gel (Al_2O_3 –7.5%, reminder SiO_2) during heating. It will not transform to its own high temperature modification and thus indicates the formation of a noncrystalline aluminosilicate phase in Al_2O_3-SiO_2 gel system. Secondly, it was also shown that on heating mullite gel containing 28% SiO_2, the boehmite component will not transform topotactically to θ-, κ-alumina modifications, as it does when it is present in the free state. Crystallization behavior of diphasic gel shows differently. But it will form Al-Si spinel as an intermediate phase and then transforms to mullite on further heating rather than forming corundum. Figure 18.4 shows the appearance and its disappearance of spinel phase to mullite transformation (Chakraborty, 1993 and 2003).

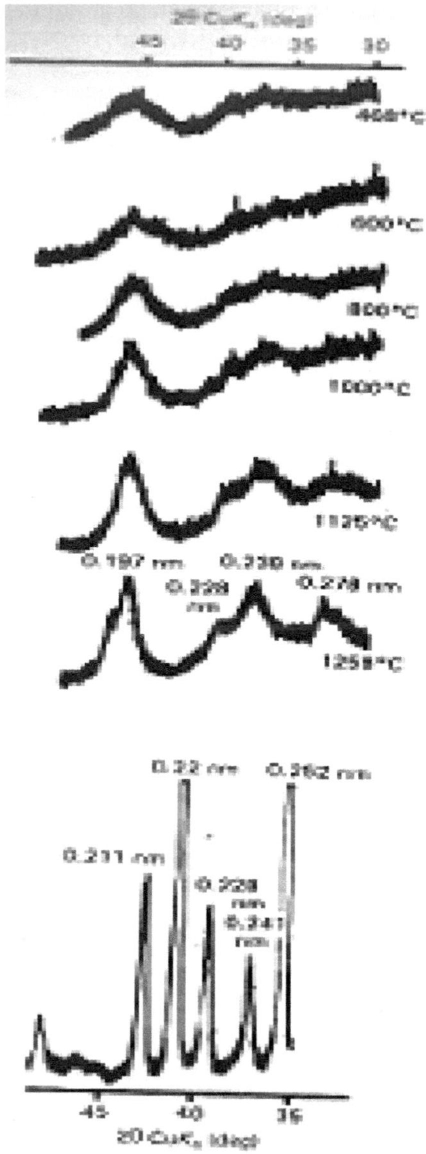

Figure 18.4 Portion of XRD patterns of G-173 heated to different temperatures at 10°C/min and 0 h soaking in all cases (Chakraborty, 1993).

The mode of development of Al-Si spinel phase formed on heating G-173 is shown in Table 15.1. It is qualitatively observed that the area of 0.197 nm Bragg diffraction peak (46° 2θ CuK_α) increases with increasing temperature up to 1258°C. Thereafter, mullite formation ensues as is shown by the appearance of 0.255, 0.32 nm diffraction peaks. At 1325°C, drastic reduction in the XRD intensity of 46° 2θ CuK_α peak of spinel phase takes place with consequent development of large amounts of mullite.

Formation of cristobalite and corundum in diphasic gels:
The intensity of cristobalite decreases from G-171 to G-172 to G-67 and cristobalite is not observed in G-173 when heating such gels to 1200°C. The intensity of corundum decreases from pure $Al(OH)_3$ gel G-175 to G-174 to G-64 and corundum is not obtained in heated gel marked G-173. Variations in the XRD intensities of cristobalite and corundum of different gels are shown in Fig. 18.5.

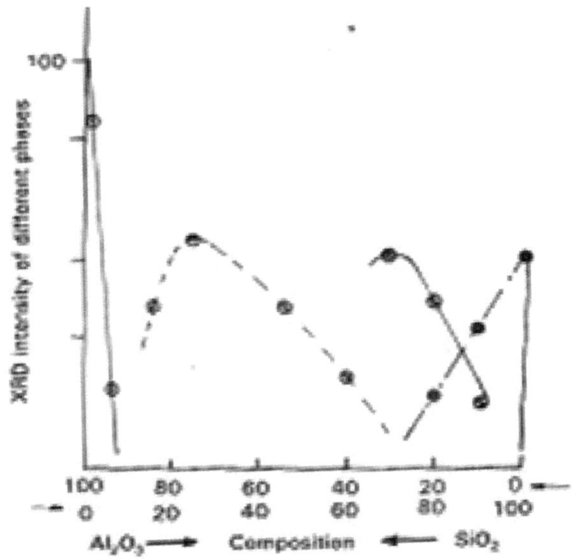

Figure 18.5 Variation of XRD intensity of corundum and *β*-cristobalite (arbitrary scale) vs. composition of various diphasic gels. (⦿) 0.404 nm peak of cristobalite; (x) 0.537 nm peak of mullite and (•) 0.208 nm peak of corundum (Chakraborty, 1993).

By identifying the mode of crystallization of cristobalite and corundum during heating of diphasic gels having varying ratios of silica and alumina contents, the concentration of Si^{+4} in Al-Si could be ascertained. Table 15.1 shows that the silica component of the gel will not transform to cristobalite on heating to 1250°C, from diphasic gels having 0%–28% silica content. It is crystallized only when silica is in excess of that the composition of 3:2 mullite (Fig. 18.5). After that, its crystallization increases with increasing silica content of the diphasic gel. Pure boehmite gel G-175 forms only γ-Al_2O_3 on heating. Its intensity decreases more with further addition of silica. At 3:2 mullite composition, α-Al_2O_3 is not found to form at 1250°C, but it develops only a maximum amount of mullite. When either of the percentages of SiO_2 or Al_2O_3 is higher than its percentage in 3:2 mullite, the resulting gels will crystallize to either β-cristobalite or α-Al_2O_3. In the studies of siliceous range of compositions, it is observed that the limit of Al-Si spinel is more likely to lie in the composition range from 0–28 wt % SiO_2 content. Therefore, the spinel phase form which mullite develops during the course of heating may be of the composition of 3:2 mullite.

18.7.2.1 Enthalpy data

Enthalpy data conform to the solid solution range data of the author. Since the highest enthalpy of crystallization of mullite out of SD precursor corresponds to ~60 mol % Al_2O_3 close to 3:2 mullite. The value is determined by: (i) 290 J/g = –124 kJ/mol (Douy, 2006); (ii) 112 kJ/mol (Geradin, 1994); (iii) 105 kJ/mol (Taklace, 1998); and by (iv) 240–310 J/g (Jaymes et al. (1996) (see Table 16.4).

18.7.3 Aluminous Side of Mullite Stoichiometry

(1) The suppression of phase transition of alumina due to silica additives is a well-known phenomenon and this fact has been reported by various authors. Generally, the γ-to-α transformation temperature is determined experimentally by the versatile method of DTA. Applying this technique, Yoldas (1976, 1980) used monolithic alumina gel and showed that crystallization temperature of intermediate θ-Al_2O_3 which transformed to corundum was affected at the beginning due to presence of certain percent of silica and then optimized the thermal stabilization effect of active alumina. He showed that on doping with silica, θ- to α-Al_2O_3 transformation

temperature of alumina gel is increased to ~80°C through structural incorporation of silica in θ-Al_2O_3. Maximum stabilization temperature of about 1380°C was noted around 6% SiO_2. The maximum solubility of silica in θ-Al_2O_3 is thus considered as ~6 wt % silica.

Beyond 6% silica addition, a decrease of DTA temperature as well as its relative intensity was observed and the silica doped alumina gel formed an aluminosilicate compound, i.e., mullite. Figure 18.6 shows the effect of silica doping on the crystallization temperature of alumina phase.

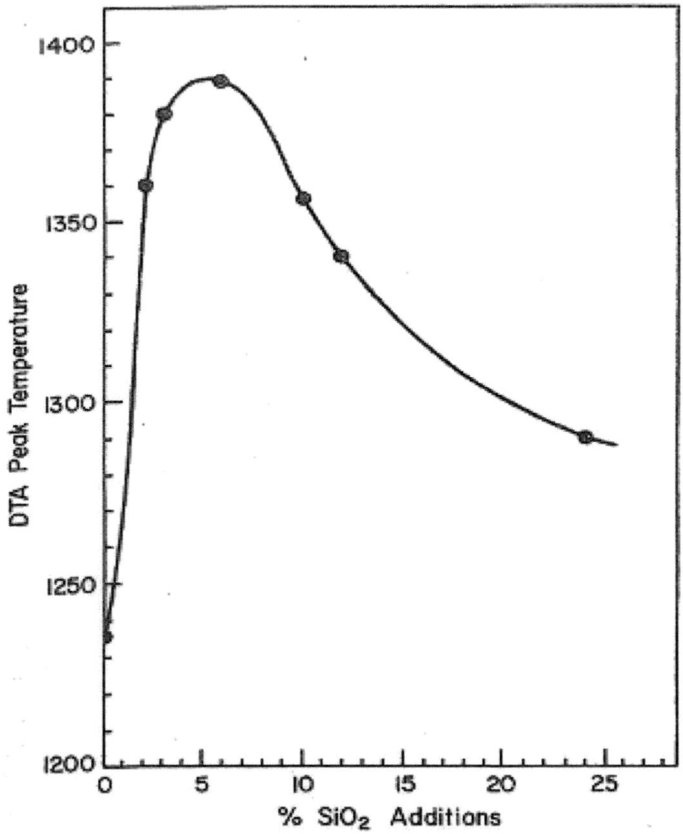

Figure 18.6 The effect of SiO_2 concentration on the DTA peak temperature (Yoldas, 1976).

(2) Addition of 5 wt % (8.2 mol %) SiO_2 suppressed phase transformation of alumina component. The persistence of θ-phase up to 1400°C/2 h was noted by Horiuchi et al. (1999).

(3) Suppression of crystallization of alumina was also observed by McHale et al. (1997). They synthesized alumina-silica precursor out of 95% ethanol solution of ANN and TEOS by usual spray drying technique and studied the followings.

(i) During static heating, they showed that pure Al_2O_3 precursor starts partial transformation of α-alumina at 1273°K (1000°C). In comparison, silica doped samples from (2–10 wt %), the minimum temperature necessary for crystallization of later α-alumina phase was greater than 1473°K (1200°C). This demonstrates the suppression of crystallization temperature of α-alumina of the order of 200°K by silica doping.

(ii) During dynamic heating study, they made DTA analysis of both pure Al_2O_3 precursor and doped spray dried precursors. While pure precursor exhibits DTA exotherm at ~1273°K due to crystallization of the γ-alumina spinel phase. Silica doped precursors showed solid solutions and started crystallization of spinel. The temperature of its formation determined by DTA is shown in (Fig. 18.7). The curve consists of two portions.

In the first part, the peak temperature increases linearly with increase of silica content of the precursor up to 10 wt %. This indicates that silica doping apparently stabilizes the γ-alumina spinel phase as a result the temperature of crystallization increases up to ~100°K (1150°K–1250°K) approx. They showed the suppression of crystallization may be a kinetic phenomenon other than thermodynamic. Apparently, the presence of silicon in solid solution poses a kinetic barrier for early crystallization.

In the second part, the temperature rises slowly up to 16 wt % SiO_2 (last amount used in the experiment). With this amount, the precursor crystallizes to mullite directly without spinel solid solution at ~1273°K as detected by XRD.

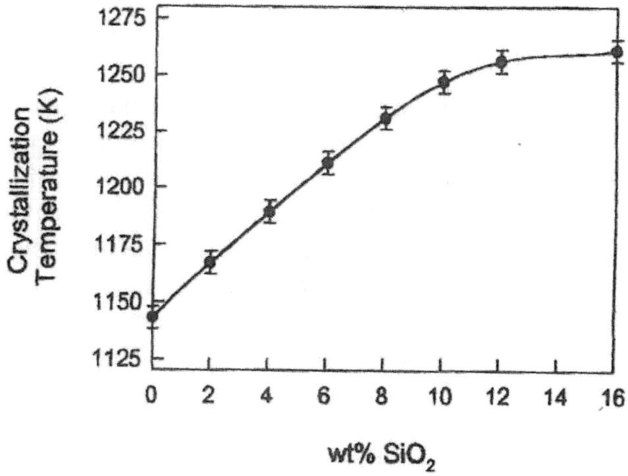

Figure 18.7 Temperature of crystallization of a spinel-type phase aluminosilicate solid solution from amorphous aluminosilicate precursors as a function of SiO_2 content. Crystallization temperatures determined via DTA. The temperature reported for the 16 wt % SiO_2 sample corresponds to mullite formation (McHale et al., 1997).

Basic system of alumina and alumina doped silica in the above two cases are different. McHale et al. studied amorphous alumina to amorphous alumina-silica spray dried precursors and their object of study is different from that of Yoldas who used aluminum monohydroxide sol to alumina-silica precursor derived out of sol-gel procedure. Former authors noted the following.

(i) The formation of γ-Al_2O_3 during calcination of 0% silica sample occurs at 1073°K while precursors containing 2–10 wt % SiO_2 at 1073°K were amorphous. Suppression occurs.

(ii) These precursors on calcinations at 1273°K develop the γ-Al_2O_3 spinel phase which corresponds to an exothermic event observed in DTA at 1173°K (an increase of 100°K).

(iii) 16 wt % SiO_2 precursor develops mullite at 1273°K.

(iv) The temperature of crystallization of spinel type phase vis-a-vis the temperature of occurrence of DTA exotherm is found to increase with increase of silica content up to ~10% and thereafter the curve becomes flattens.

(v) It is noted that silica doped amorphous alumina did not form corundum below 1473°K whereas 0% silica sample developed corundum partially at 1273°K.

They suggested that the range of spinel type solid solution would lie between 2–10 wt % SiO_2 from the linear portion of the curve of crystallization temperature as judged from DTA vs. wt % SiO_2 doped curve. Question is: at what basis the upper limit of solid solution range was chosen by them at 10 wt % SiO_2? Presentation of both DTA plots and XRD patterns of DTA analyzed samples are necessary. It is also necessary to run the crystallization temperature vs. wt % SiO_2 curve of S D precursors beyond ~28 wt % SiO_2 content.

The portion of curve between 2–10 wt % SiO_2 (Fig. 18.7) seems analogous to the portion of the first stage of the curve (Fig. 18.6) showing increase of temperature of crystallization of corundum from 1200°C to 1380°C. This first portion of curve was noted around 6% SiO_2 content due to formation of silica incorporated θ-Al_2O_3 in case of Yoldas (1976). It was also shown that the ability to retain surface area is greatest at 5–7 wt % SiO_2 and the surface stabilizing effect of SiO_2 parallels its effect on the α-Al_2O_3 transformation. Similarly, this first portion of curve was noted around 10% SiO_2 content due to formation of silica incorporated γ-Al_2O_3 in case of McHale et al. (1997). Question remains, how end point in the solid solubility limit could be ascertained and what methodology to be followed? How the upper limit of silica incorporation in γ-Al_2O_3 phase could be understood.

Before studying the solubility effect of silica in γ-Al_2O_3 structure, it is necessary to have prior knowledge about the phase evolution behavior of Al_2O_3-SiO_2 system both in case of mullite precursor obtained out of sol-gel process and from spray pyrolyzed process.

18.7.3.1 Starting temperature of mullite formation

(1) Sol-gel precursor

DTA study shows that an increase in silica concentration caused a corresponding increase in α-Al_2O_3 transformation temperature (Fig. 18.8) of a monolithic alumina material (Yoldas, 1976). Following results were noted with silica addition.

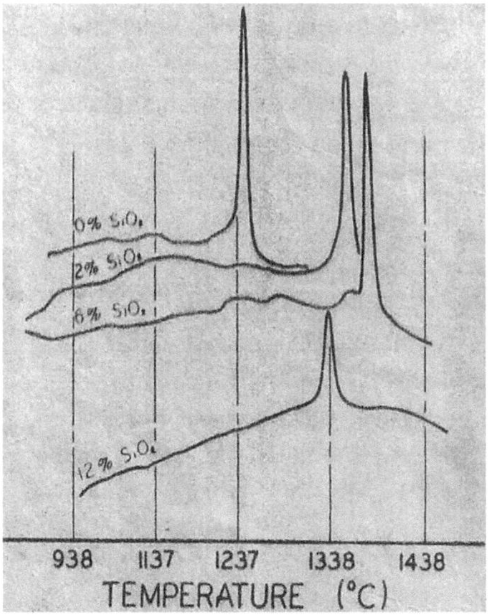

Figure 18.8 The DTA curves of the material doped with various amounts of silica (Yoldas, 1976).

(i) Pure alumina material

The sharp DTA exotherm observed in undoped material at 1200°C is due crystallization of α-Al_2O_3. XRD analysis showed that the crystalline phase of the material was δ-Al_2O_3 below the temperature of exotherm and α-Al_2O_3 above the exotherm (1260°C).

(ii) 2% doped material

It showed an exotherm at higher temperature of ~1360° C with absence of peak at 1200° C due to α-Al_2O_3.

(iii) 6% doped material

It showed double exothermic peaks between 1350°C–1400°C. First small exotherm occurred at ~1340°C and second sharp peak situated at ~1380°C. At this stage, crystalline structure consisting of δ-Al_2O_3 as well as original pore morphology and high surface area of the calcined material are retained.

(iv) 10 wt % silica doped material

By electron microscopic study Yoldas (1980), it showed that it resembled the same porosity and pore morphology at 1400°C as displayed in original state. XRD is consistent with the retention of the original δ-Al_2O_3 phase. Surface property likely showed a high surface of this material.

(v) 12–14 wt % silica doped material

Mullite formation likely starts at the occurrence of a small exotherm to ~1340°C and it shifts to lower temperature and with reduced intensity at ~1338° C with where much mullite formation occurred.

A drastic change in microstructure occurred when silica concentration was increased to ~20% due abundant mullite crystallization as criss-crossing plates.

This mullite formation temperature likely coincides with the exhibition of exotherm at ~1300°C in case of diphasic gel (Fig. 15.1a,b) as shown by Chakraborty, 1996b). Extrapolated value of exothermic peak of 28 wt % SiO_2 from curve (Fig. 18.6) of Yoldas is 1280°C.

(2) SD precursor

Mullite formation was observed by McHale et al. (1997) in their SD precursor containing 16 wt % SiO_2 at ~1000°C.

18.7.3.2 Crystallization temperature of mullite

Crystallization temperature of alumina components in spray dried precursors made by Douy (2006) noted the existence of θ-phase to a temperature of as high as ~1700°C/1 h for precursors with 70 and 75 mol % alumina.

The increase of the crystallization temperature of alumina to corundum and increase in stabilization domain of transition alumina polymorphs—one aspect of study. Whereas the determination of SiO_2-γ-Al_2O_3 solid solution—is another aspect of study/design. Douy (2006) observed the following.

Formation of cristobalite:

Pure silica sample crystallizes to only cristobalite at 1400°C. 80 mol % SiO_2 precursor sample does not form cristobalite but develops to mullite. This phenomenon is even more marked in 50 mol % silica samples when heated to 1700°C for 1 h. where mullite is detected. Samples in 40-to-30 mol % silica range, highest intensity of mullite is detected without any crystallization of either cristobalite or corundum.

Formation of corundum:
In a pure alumina sample, corundum is formed at 1100°C /1 h. With addition of 5 mol % SiO_2, transition alumina phase (θ-alumina) phase occurs after heat treatment at 1400°C/1 h. With increase of silica content, intermediate transition alumina phase found to crystallize in 85, 80, 75 and traces in 70 mol % samples at 1400°C/1 h. This phase remains still in appreciable quantity in 75 mol % sample and traces in 70 mol % sample at 1700°C/1 h. Samples in 70 mol % alumina (30 mol % silica) formed the least quantity of transitional alumina.

Solid solution range:
Both mullite and corundum crystallizes together at 1650°C in 85 mol % alumina. At 1700°C in 80 mol % alumina, minor quantity of corundum, transitional alumina are noted. Major quantity of mullite is seen at 75 mol % alumina precursors. Thus, corundum decreases with decrease of alumina up to 70 mol %. Only a trace of the alumina phase is still noted. In conclusion, end point of SiO_2-γ-Al_2O_3 solid solution should lie at a composition

(i) where crystallization of alumina polymorph is zero,
(ii) where crystallization of cristobalite is also zero and,
(iii) Al-Si spinel formation would be maximum.

Largest amount of Al-Si spinel phase is noted in diphasic gel of composition of mullite corresponding to ~80 wt % (see Chapter 15 amount determined by QXRD technique). Based on these three critical criteria, the following observations are made.

(i) 10 wt % data as maximum solid solution limit as predicted by McHale et al. in their Fig. 18.7 is underestimated. Since SD precursor containing 10 wt % silica data forms corundum on calcination at 1573°K (Table 1 of them).
(ii) Similar underestimated values are observed in determination of solid solubility of SiO_2 (8 wt %) in spinel phase derived out of both SH and RH gels suggested by Okada and Otsuka (1986) in alkali leaching study.
(iii) Invariably, maximum amount of 11.5 wt % SiO_2 data shown in case of diphasic gel heat treated to 1150°C and analyzed by

Schneider et al. (1994) in EDX study, and 7 wt % SiO_2 value as given by Gerardin et al. (1994) in their ^{29}Si MAS-NMR studies of Al_2O_3-SiO_2 gel are required to be recalculated.

In the aluminous side region (Fig. 18.9) shows that SD precursor starts crystallization on mullite below 1000°C (Douy, 2006). Both spinel and mullite coexist in precursor ~70 mol % Al_2O_3.

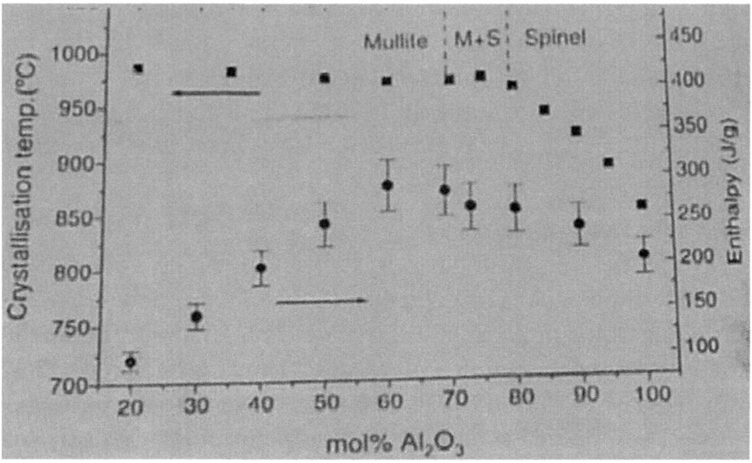

Figure 18.9 Temperature of crystallization at 5°C/min, domains of composition where mullite, mullite and spinel or spinel crystallize, and enthalpy of crystallization.

At the initial stage of silica doping, it likely formed a silica incorporated alumina phase (δ-Al_2O_3). Its intensity increases with continued substitution of silica for alumina. At 16 wt % (~70 mol % Al_2O_3) silica addition, McHale et al. (1997) noted direct mullite formation. Of course, Douy (2006) noted by XRD that mullite formation starts at ~80 mol % alumina along with spinel phase. Both forms coexist up to ~70 mol % alumina. Then spinel disappears entirely and a large quantity of mullite crystallizes at 60 mol % alumina composition (Fig. 7 of him). Following these observations regarding appearance, existence and subsequent disappearance of spinel to 3:2 mullite phase, the maximum limit of silica addition to γ-alumina spinel could be assigned as 40 mol % (~28 wt %) silica (Fig. 9 of him) and composition of spinel is assigned as analogous to stoichiometric mullite. The theory of Chakraborty that Al-Si spinel and 3:2 mullite are polymorphs is held true from the first experimentation of Douy (2006).

The enthalpy data:

The enthalpy curve of Douy (2006) shows that the enthalpy value increases with addition of SiO_2 as well as its crystallization temperature. This value (1394 kJ/mol) reaches a maximum at ~60 mol % Al_2O_3 composition.

Demediuk and Cole (1958) and Okuda and Otsuka (1886) noted the exothermic peak area was highest for gels of composition 3:2 mullite. Chakraborty (1979) showed maximum heats of reaction in composition of 3:2 mullite (Fig. 16.4). With these data in mind, it is further conceived that Al-Si spinel of varying composition would form in the mutual substitution process. In this connection, two data of Douy are most interesting. Figure 18.10 of his shows that heating rate likely favors the crystallization of either spinel or mullite due to difference in kinetics. First experiment, rapid heating of 75 mol % alumina samples favor spinel formation, while slow heating produces mullite. In both the cases, crystallization of either spinel or mullite takes place out of the same noncrystalline aluminosilicate intermediate precursor phase derived out of spray drying technique. Similar result was shown in thermal transformation of kaolinite.

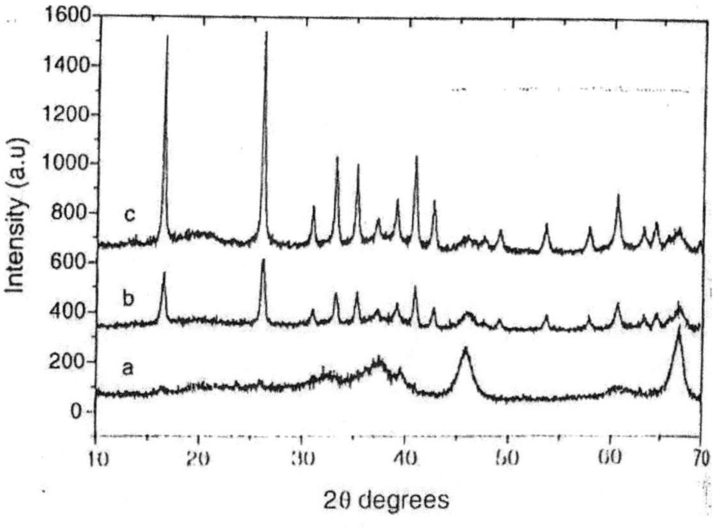

Figure 18.10 XRD patterns of the 75 mol % Al_2O_3 sample after crystallization at different heating rates: (a) putting directly at 1100°C; (b) 5°C/min; (c) 0.1°C/min to 950°C for 10 h (Douy, 2006).

In the second experiment this nature of crystallization does not hold for precursors of 80 mol % alumina samples. These two results perhaps conclude two mechanisms. Al-Si spinel obtained during heating of 80 mol % alumina is not fully or nearly saturated with silica and thus unable to transform into mullite even during slow heating process in his second experiment

Douy (2006) indicated the possibility of 25 mol % SiO_2 is likely incorporated in the γ-alumina phase. However, the presence of even some quantity of spinel in addition to formation of major quantity of mullite in 70 mol % alumina sample in slow heating cycle indicates that needle of silica saturation points more to words left hand side of the curve, i.e., to words less alumina composition than 70 mol %. So the upper limit of silica solubility would be approximately near to 60 mol % alumina. Moreover, SD precursor shows highest enthalpy at 60 mol % alumina composition (Fig. 18.9).

18.7.3.3 Nucleation mechanism

Addition of SiO_2 to alumina causes a hindrance to α-Al_2O_3 crystallization. Question is: what should be the mechanism behind it? Yoldas (1976) shows chemical polymerization and demonstrated Si-O-Al bond development during synthesis process of Al_2O_3-SiO_2 gel. The later gel leads to formation of noncrystalline aluminosilicate phase on heating (Leonard et al., 1971).

According to McHale et al. (1997), silica addition appears to provide mainly a kinetic hindrance to α-Al_2O_3 crystallization.

Saito et al. (1998) are of the opinion that retardation effect is due to formation of film or coating of SiO_2 (A) deposited on the surface of γ-alumina during impregnation suggested that these two crystalline phases may provide nucleation sites for early nucleation. On the other hand, acceleration both in early exhibition of DTA peak and crystallization of corundum are explained due to following reasons

(1) Silicon out of quartz or cristobalite does not diffuse into γ-alumina as happens easily in case of amorphous SiO_2 (obtained out of hydrolysis of TEOS or from fume silica).

(2) Further, aluminum of γ-Al_2O_3 penetrates into quartz or cristobalite particles and causes formation of eutectic liquid phase. This is more pronounced in cristobalite rather than in quartz due to comparatively openness of structure of cristobalite.

It is expected that the newly formed liquid phase showed DTA exotherm at lower temperature as happens in case of use of Na, K, Ti, and B oxides. Survey report of Yoldas (1976) showed that these oxides progressively decrease the exothermic peak temperature.

Polymorphism between spinel and mullite

Spinel formed at 1000°C between 70–80 mol % alumina containing material vide Fig. 18.9 are alumina rich and thus composition of it and mullite are not likely analogous. These are different in structure. Moreover, crystallization kinetics of mullite and spinel are different. The heating rate favors the formation of either of the two (Fig. 18.10 of Douy (2006). In a similar way, formation of some quantity spinel phase at early temperature at the first exotherm were also observed in coprecipitated gels, RH gel etc. It becomes saturated gradually on heating with surrounding noncrystalline aluminosilicate rich in silica phase by diffusion process. Thereafter, this silica saturated Al-Si spinel phase (c-mullite) converts to 3:2 o-mullite at ~1300°C in a polymorphic way.

18.8 Summary

Controversies:

Two theories have been put forward regarding the crystallization phenomenon of the sharp exothermic peak at 980°C of mullite gels which may be either Al-Si spinel or mullite by XRD. Further two possibilities are considered regarding the nature and composition of this 980°C heated cubic spinel phase. It may be cubic γ-Al_2O_3 (Al spinel containing ~8 wt % silica) or may be silicon bearing spinel analogous to the composition equivalent to 3:2 mullite (called Al-Si spinel). When mullite rapidly forms, the dispute concerns too on its composition may be orthorhombic form (o-mullite) or tetragonal (t-mullite) or any other.

Evidences:

Some of the direct evidences regarding the formation of Al-Si spinel, e.g., alkali leaching method of heat treated mullite gels and solid state reaction study of spinel phase by CaO indicated the composition of it as analogous to composition of 3:2 mullite by the author in the years (1979, 2003 and 2004).

Composition of spinel phase:

Al-Si spinel phase formation and regarding its composition in case of monophasic gel and in diphasic gels by different authors are pointed out.

(1) The chemical composition of the spinel phase was measured by QXRD technique and noted a value of ~8 wt % SiO_2 for two kinds of monophasic gels, e.g., SH and RH types by Okada et al. (1986).

(2) In case of heating HB10 precursor at 900°C and on subsequent ^{29}Si MAS-NMR spectral study, Schneider et al., (1993) suggested that −80 ppm resonance might be due to the presence of silicon in an alumina rich noncrystalline phase or to the incorporation of silicon in γ-Al_2O_3 structure.

(3) Both CM and SGM materials at 900-to-1000°C showed ^{29}Si resonance near −95 ppm and another at −110 ppm. The −90 ppm resonance peak may be attributed to spinel phase in SGM material (Schneider et al., 1992). This resonance also corresponds to weakly crystalline mullite in the CM sample. It could be concluded that the mullite phase is inherited from spinel phase, both are likely to be polymorph.

(4) ^{29}Si spectra of the precursor B synthesized by Jaymes et al. (1996) at 1000°C crystallized to spinel phase only. They attempted to estimate the chemical composition of it and found tentatively to be 71.4 mol % Al_2O_3 and 28.6 mol % SiO_2 with an atomic ratio Al:Si = 5:1 which indicates Al-Si spinel concept.

(5) Okada et al. (1986), Sonuparlak et al. (1987), Low and McPherson (1989), Schneider et al. (1994), and Geradin et al. (1994) disbelieved the γ-alumina hypothesis in transformation processes of both monophasic and diphasic gels and put forwarded tangible evidence by EDS TEM studies to prove that spinel phase was not purely Al-spinel (γ-Al_2O_3) and it contained 7–11 wt % SiO_2 in its structure. Thus, the view of crystallization of spinel as silicon incorporated spinel is more and more advancing. A few of the confirmatory evidences on Al-Si spinel of 3:2 mullite composition are shown below.

(6) X-ray intensities of spinel phase formed in different cases of diphasic gels were measured by Chakraborty and Das

(2003) and these intensities are shown to be greater than that expected for γ-Al_2O_3 formation even in case of diphasic Al_2O_3-SiO_2 gel system (Fig. 15.17). The exhibition of higher X-ray intensity value for each composition indicates the possibility of substitution of Al by Si in spinel structure.

(7) With the corroborative XRD and ^{29}Si MAS-NMR studies of phase transformation of Al_2O_3-SiO_2 mixed gels of different compositions by the author (2006), incorporation of silica into γ-Al_2O_3 phase leads to the formation of Al-Si spinel phase is conformed. It is shown that the maximum amount of silica incorporated into γ-Al_2O_3 structure to develop spinel phase is equivalent to that present in 3:2 mullite (Fig. 9.5). Beyond this amount, the added silica will remain as free noncrystalline silica rich aluminous phase and forms cristobalite thereafter.

(8) Suzuki et al. (1987) presented the electron diffraction pattern of Al-Si spinel phase. They noted the composition of the spinel phase obtained at calcination of the monophasic gel/precursor at 900°C/2 h by EDS in TEM with similar stoichiometric mullite composition. This spinel phase converted to yield ultrafine mullite during heating at 1200°C with the same stoichiometric composition which conformed the composition of intermediate spinel analogous to the composition of 3:2 mullite as shown by the author.

(9) By FTIR study, Orefice and Vasconcelos (1997) showed the presence of a peak at 1070 cm^{-1} corresponding to crystallized spinel of the diphasic gel heat treated at 1100°C. They argued this peak was not associated with the presence of Si-O-Si bonds but due to the presence of Si-O-Al bonds. They assigned this bond to the formation of Al-Si spinel phase even in case of diphasic gel and substantiated the concept of the author.

Evidences of θ-Al_2O_3:

Formation of θ-Al_2O_3 along with Al-Si spinel formation has been made by precision identification by deconvolution of XRD peak of spinel phase and measurement of peak width as follows.

(1) XRD intensity peaks at 46° 2θ and 66° 2θ corresponding to spinel phase derived out of heat treated gels were resolved on deconvolution. For example, two peaks are observed at 46° 2θ region during peak search operation using graphics software,

first one at 45.71°, second one at 47.08°, respectively. Similar observation was noted during profile fitting of the said XRD pattern using X'Pert Profit software which resembles near to θ-Al_2O_3.

(2) On analyzing the nature of XRD peak of spinel phase and it was shown that the breadth of 0.139 nm peak of spinel phase changes with increase of heat treatment temperature. The author concluded that the phase development of any mixed gel at ~1000°C will be a mixture of Al-Si spinel, θ-Al_2O_3.

(3) The existence of this θ-Al_2O_3 phase above 1000°C was noted by Iller (1963) and Yoldas (1976) and explained as a stabilizing effect. The role of some wt % of silica which was introduced into the θ-structure, delayed corundum formation and thus designated as silica incorporated θ-Al_2O_3.

(4) It was shown that silica has been incorporated during heating into γ-Al_2O_3 structure and formed Al-Si spinel other than corundum. Obviously, alumina also diffuses into noncrystalline silica and develops noncrystalline aluminosilicate phase. Ivankovic et al. (2003) measured the integrated area of spinel peak at 46° 2θ and amorphous band for silica peak at 22° 2θ for heated mullite gel samples manually and to explain the role of it in the formation process of Al-Si spinel phase. Instead of manual selection of background prior to calculate XRD peak intensity (Fig. 18.2), it was done with the help of X'Pert Graphics and Profit soft wares by the author and shown that ratios of the intensity peak of spinel phase to that of amorphous band of gel samples of different in compositions increased with heat treatment temperature from 900°C–1200°C (Fig. 15.9). Thus, decrease in intensity of amorphous band width increase of the peak of spinel phase further confirms the incorporation of Si out of aluminosilicate (A) phase in spinel structure to form Al-Si spinel phase.

Crystallization of Al-Si spinel phase:

The author demonstrated the stepwise process of crystallization of Al-Si spinel phase during heating of both monophasic gel and diphasic gel of different compositions.

(1) Formation of noncrystalline aluminosilicate phase during dehydration and decomposition of diphasic gel (400°C–600°C).

At the first stage of transformation at room temperature to ~500°C, diphasic gel which initially contained hydrated silica gel mixed with pseudo crystalline boehmite dehydrates followed by a broad endothermic peak at DTA and a loss in TGA analysis and started development of an amorphous matrix with SiO_2 (A) as revealed by XRD analysis. This phase has been characterized as a noncrystalline aluminosilicate phase.

(2) In the second stage of transformation (600°C–900°C) range, more quantity of noncrystalline aluminosilicate phase developed out of two amorphous oxide components and a portion of alumina oxide crystallized to γ-alumina and likely these two phases coexist.

(3) During the third stage of transformation between 900°C to 1300°C, crystallization of aluminosilicate (A) phase to Al-Si spinel phase which was started initially at 600°C increases on heat treatment (Fig. 15.2). At ~1000°C intensity of spinel phase attains constant (Fig. 15.8). In the above temperature range, two parallel reactions occur.

  (i) In the first parallel reaction, two phenomena occur simultaneously during heating. The intensity of X-ray amorphous band due to accumulated noncrystalline aluminosilicate phase decreases vis-à-vis crystallization of it to formation of Al-Si spinel phase continuously increases up to 1000°C. A noncrystalline silica rich aluminous phase appears by phase separation at this stage. Interrelationship between the area of the amorphous hump at 22° 2θ to that of the peak of Al-Si spinel phase formed at 67° 2θ is noted up to this temperature. During the second phenomenon crystallization of Al-Si spinel occurs beyond 1000°C increases slowly with heating beyond 1200°C (Fig. 15.8). DTA analysis observed broad exothermic events in the temperature 1000°C–1300°C and DDTA run showed a differential peak (Fig. 15.1a,b). The occurrence of this broad exotherm is explained as due to incorporation of silica from noncrystalline aluminosilicate phase itself into un saturated Al-Si spinel phase for saturation.

(ii) In the second parallel reaction, some amount of silica of noncrystalline aluminosilicate phase also incorporates into residual γ-Al_2O_3 phase and develops Si incorporated γ-Al_2O_3 and Si incorporated θ-Al_2O_3 phases. A portion of residual amorphous band due to silica rich aluminous noncrystalline phase remains in association with these phases.

(4) During the fourth stage of transformation above 1300°C, DTA analysis exhibits a strong exotherm due to transformation of accumulated Al-Si spinel to crystallization of mullite type phase. Two mechanisms of transformation of alumina phase content of Al_2O_3-SiO_2 gel are discussed. Al-Si spinel and or silicon incorporated δ- or θ-Al_2O_3 formed during the heating process of diphasic gel and subsequent structural transformations of these phases are the cause of mullite formation.

Intensity of Al-Si spinel phase vs. method of processing:
Intensity of Al-Si spinel phase formed out of precursors synthesized by gelation method or by coprecipitation method shows variations. Intensity of spinel phase formation out of monophasic precursors synthesized by gelation method proceeds in the following ways.

SHIV > SHIII > SHII and it is absent in SHI.

In the coprecipitated method, spinel formation is pH dependent. SHIV prepared at pH = 6 formed predominant spinel formation when heated at the 980°C exotherm. The "in situ" diphasic gel prepared at pH = 9 however exhibited a broad exotherm over a range of temperature. Thus, the mode of spinel development in coprecipitated gel prepared at pH = 9 is significantly different from that of slow hydrolysis gels (SHII and SHIII) and coprecipitated gel synthesized at pH = 6. Further, spinel phase developed in cases of SHII, SHIII and SHIV transformed on heating and exhibited a second exothermic peak at ~1200°C in DTA by path IV due to mullitization. Whereas a higher transformation temperature as high as 1320°C is noted for the exhibition of exotherm due to mullitization in case of diphasic precursor. There is a possibility Al-Si spinel of varying composition. The range of existence of Al-Si spinel may lie from 100 wt % Al_2O_3 up to incongruent melting point of mullite, i.e., up to 72 wt % Al_2O_3. Based on different sets of methods of synthesis of precursors, there exist as many as 5 different transformation paths leading to mullite formation out of monophasic and diphasic gels.

Solid-solution range:
Question remains, how end point in the solid solubility limit could be ascertained and what methodology to be followed? How the upper limit of silica incorporation in γ-Al_2O_3 phase could be understood.

The maximum solubility of silica in θ-Al_2O_3 is considered as ~6 wt % (Fig. 18.6). Beyond 6% silica addition, a decrease of DTA temperature as well as its relative intensity was observed (Yoldas, 1976, 1980). McHale et al. (1997) showed the range of spinel type solid solution would lie between 2–10 wt % SiO_2 from the linear portion of crystallization temperature as judged from DTA vs. wt % SiO_2 doped curve (Fig. 18.7).

However, the portion of curve between 2–10 wt % SiO_2 (Fig. 18.7) as shown by McHale et al., 1997 seems analogous to the portion of the first stage of the curve around 6 wt % (Fig. 18.6) as given by Yoldas (1976, 1980) showing increase of temperature of crystallization of corundum from 1200°C to 1380°C.

Two criteria are taken into consideration for detecting solid solution range of silica-to-γ-Al_2O_3 phase. The composition at which crystallization of either cristobalite and/or corundum which just ensues out of various compositions of mixed gel at different stages of heat treatment are noted. The end point of SiO_2-γ-Al_2O_3 solid solution should lie at a composition where crystallization of alumina polymorph is zero. With these observations regarding appearance, existence and subsequent disappearance of spinel to 3:2 mullite phase, the maximum limit of silica addition to γ-alumina spinel could be assigned as 40 mol % (~28 wt %) silica (Fig. 9 of Douy 2006) and composition of spinel is assigned as analogous to stoichiometric mullite.

Based on critical phase identification studies, it is suggested that the amount of SiO_2 in γ-Al_2O_3-SiO_2 solid solution would be nearly equal to the composition of 3:2 mullite. During the process of solid solution, Al-Si spinel forms as a resultant phase. The lattice parameter of it was found to decrease with concentration of alumina up to ~72 wt % alumina of the mixed gel.

In the studies of siliceous range of compositions, it is observed that the limit of Al-Si spinel is more likely to lie in the composition range from 0–28 wt % SiO_2 content (Chakraborty, 1993). In aluminous range, DTA study shows that an increase in silica concentration caused corresponding increase in α-Al_2O_3 transformation temperature (Fig. 18.8) of a monolithic alumina material (Yoldas, 1976). Maximum stabilization temperature of about 1380°C was noted around 6% SiO_2. Beyond 6% silica addition, a decrease of DTA temperature as well as its relative intensity was observed and the silica doped alumina gel formed an aluminosilicate compound, i.e., mullite.

Mullite formation likely starts at the occurrence of small exotherm to ~1340°C and it shifts to lower temperature and with reduced intensity at ~1338°C on further doping with 12% silica where much mullite formation occurred. A drastic change in microstructure occurred when silica concentration was increased to ~20% due to abundant mullite crystallization. Extrapolated value of exothermic peak of 28 wt % SiO_2 from curve (Fig. 18.6) of Yoldas is 1280°C. This mullite formation temperature likely coincides with the exhibition of exotherm at ~1300°C in case of diphasic gel (Fig. 15.1a,b) as shown by (Chakraborty, 1996b) where maximum quantity of mullite formation occurs. Thus data addresses the end point in the solid solubility limit.

In support of this conclusion enthalpy result are cited here. Highest enthalpy of crystallization of mullite out of SD precursor corresponds to ~60 mol % Al_2O_3 close to 3:2 mullite. This data conform the solid solution range data of the author (see Section 16.6).

Polymorphysim:

Both spinel and mullite are likely analogous in composition and likely polymorphs. But these are different in structure. Moreover, crystallization kinetics of mullite and spinel are different. Therefore, the heating rate would favour the formation of either of the two (Fig. 18.10).

References

1. M. W. Tamele, Chemistry of the surface and the activity of alumina silica cracking catalysts. *Disc. Faraday Soc.*, **8**, 270 (1950).

2. D. X. Li and W. J. Thomson, Effects of hydrolysis on the kinetics of high temperature transformations in aluminosilicate gels. *J. Am. Ceram. Soc.*, **74**, 574–578 (1991).
3. M. Fukuoka, Y. Onoda, S. Inoue, K. Wada, A. Nukui, and A. Makishima, The role of precursors in the structure of SiO_2-Al_2O_3 sols and gels by the sol-gel process. *J. Sol-Gel Sci. Technol.*, **1**, 47–56 (1993).
4. A. K. Chakraborty and D. K. Ghosh, Synthesis and 980°C phase development of some mullite gels. *J. Am. Ceram. Soc.*, **71**(11), 978–987 (1988).
5. D. W. Hoffman, R. Roy, and S. Komarneni, Diphasic xerogels, a new class of materials: phases in the system Al_2O_3–SiO_2. *J. Am. Ceram. Soc.*, **67**, 468 -471 (1984).
6. K. Okada and N. Otsuka, Characterization of the spinel phase from SiO_2–Al_2O_3 xerogels and the formation process of mullite. *J. Am. Ceram. Soc.*, **69**(9), 652–656 (1986).
7. A. K. Chakraborty, Formation of silicon–aluminium spinel. *J. Am. Ceram. Soc.*, **62**, 120 (1979).
8. G. W. Brindley and M. Nakahira, The kaolinite-mullite reaction series: Ill, the high-temperature phases. *J. Am. Ceram. Soc.*, **42**, 319–324 (1959).
9. M. Low and R. McPherson, The origins of mullite formation. *J. Mater. Sci.*, **24**, 926 (1989).
10. M. Low and R. McPherson, The structure and composition of Al-Si spinel. *J. Mater. Sci.*, **7**, 1196–1198 (1988).
11. U.-Y. Hwang, J.-W. Lee, J.-H. Choi, H.-S. Park, S.-J. Yoo, H.-S. Yoon, and Y.-R. Kim, Synthesis of spherical pre-mullite particles by sol-gel method and mullitization mechanism of pre-mullite III. Crystal structure and composition of spinel phase and mullitization mechanism. *Hwahak Konghak*, **38**(5), 669–675 (2000) (*Journal of the Korean Institute of Chemical Engineers*).
12. R. K. Iller, Effect of silica on transformations of fibrillar colloidal boehmite and gamma alumina. *J. Am. Ceram.*, **47**(7) 339–341 (1964).
13. B. E. Yoldas, Thermal stabilization of an active alumina and effect of dopants on the surface area. *J. Mater. Sci.*, **11**, 465–470 (1976).
14. W.-C. Wei and J. W. Halloran, Phase transformation of diphasic aluminosilicate gels. *J. Am. Ceram. Soc.*, **71**(3), 166–172 (1988).
15. A. K. Chakraborty and D. K. Ghosh, Crystallization behavior of Al_2O_3 in the presence of SiO_2. *J. Am. Ceram. Soc.*, **70**(3) C-46–C-48 (1987).

16. A. K. Chakraborty, Structural parameters of mullite formed during heating of diphasic mullite gels. *J. Am. Ceram. Soc.*, **88**(9) 2424–2428 (2005c).
17. A. K. Chakraborty, Unpublished data.
18. A. K. Chakraborty and S. Das, Al-Si spinel phase formation in diphasic mullite gels. *Ceram. Int.*, **29**, 27–33 (2003b).
19. A. K. Chakraborty, Characterization of monophasic and diphasic mullite precursors by solid state reaction study. *Brit. Ceram. Trans.*, **103**, 33–36 (2004).
20. H. Schneider, B. Saruhan, D. Voll, L. Merwin, and A. Sebald, Mullite precursor phases. *J. Euro. Ceram. Soc.*, **11**, 87–94 (1993).
21. A. K. Chakraborty, Reinvestigation of Al-Si spinel phase in diphasic Al_2O_3-SiO_2 gel. *J. Am. Ceram. Soc.*, **88**(1), 134–140 (2005a).
22. A. K. Chakraborty, DTA characterization of three types of Al_2O_3–SiO_2 gels made from TEOS–Al $(OBU)_3$ mixture with variation of water. *Ceram. Int.*, **22**, 463–469 (1996a).
23. A. K. Chakraborty, Further studies on mullitization of diphasic gel. *Trans. Ind. Ceram. Soc.*, **56**(1), 9–15 (1997).
24. Y. X. Huang, A. M. R. Senos, J. Rocha, and J. L. Baptista, Gel formation in mullite precursors obtained via tetraethylorthosilicate (TEOS) prehydrolysis. *J. Mater. Sci.*, **32**, 105–110 (1997).
25. H. Schneider, I. Merwin, and A. Sebald, Mullite formation from non-crystalline precursors. *J. Mater. Sci.*, **29**, 805–812 (1992).
26. I. Jaymes, A. Douy, D. Massiot, and J. P. Coutures, Characterization of mono- and diphasic mullite precursor powders prepared by aqueous routes, ^{27}Al and ^{29}Si MAS-NMR spectroscopy. *J. Mater. Sci.*, **31**, 4581–4589 (1996).
27. C. Gerardin, S. Sundaresan, J. Benziger, and A. Navrotsky, Structural investigation and energetics of mullite formation from sol–gel precursors. *Chem. Mater.*, **6**, 160–170 (1994).
28. J. C. Huling and G. L. Messing, Epitactic nucleation of spinel in aluminosilicate gels and its effect on mullite crystallization. *J. Am. Ceram. Soc.*, **74**(10), 2374 –2381 (1991).
29. A. K. Chakraborty, Si-incorporated alumina phases formed out of diphasic mullite gels. *J. Mater. Sci.*, **43**, 5313–5324 (2008a).
30. A. K. Chakraborty, Range of solid solutions of silica in spinel type phase. *Adv. Appl. Ceram.*, **105**(6), 297–303 (2006).
31. Y. Sato, T. Takei, S. Hayashi, A. Yasumori, and K. Okada, *J. Am. Ceram. Soc.*, **81**, 2197 (1998).

32. H. Ivankovic, E. Tkalcec, R. Nass, and H. Schmidt, Correlation of the precursor type with densification behavior and microstructure of sintered mullite ceramics. *J. Euro. Ceram. Soc.*, **23**, 283–292 (2003).

33. A. K. Chakraborty, New data on thermal analysis of diphasic mullite gels. *J. Therm. Anal.*, **46**, 1413–1419 (1996b).

34. G. Bye and G. T. Simpkin, *J. Am. Ceram. Soc.*, **57**, 367 (1974).

35. L. A. Xue and I. W. Chem, *J. Mater. Sci. Lett.*, **11**, 443 (1992).

36. F. W. Clark and J. White, *Trans. Br. Ceram. Soc.*, **49**, 305 (1950).

37. G. Engelhadt and D. Michel, *High Resolution Solid State NMR of Silicates and Zeolites*, John Wiley and Sons, New York, 1987, p. 54.

38. H. Schneider, D. Voll, B. Saruhan, M. Schmucker, T. Schaller, and A. Sebald, Constitution of the γ-alumina phase in chemically produced mullite precursors. *J. Euro. Ceram. Soc.*, **13**, 431–448 (1994).

39. B. Sonuparlak, Sol–gel processing of infrared transparent mullite. *Adv. Ceram. Mater.*, **3**(3), 263–267 (1988).

40. H. Suzuki, H. Saito, Y. Tomokiyo, and Y. Suyama, Processing of ultrafine mullite through alkoxide route. *Ceram. Trans.*, Vol. 6, Edited by S. Somiya, R. F. Davis, and J. A. Pask. *Am. Ceram. Soc.*, Westerville, OH (1990), p. 263.

41. A. K. Chakraborty, Unpublished work (2008).

42. A. K. Chakraborty, Further studies on thermal transformation of diphasic Al_2O_3-SiO_2 gel. *Trans. Indian Ceram. Soc.*, **56**(1), 9–15 (1997).

43. K. Okada and N. Otsuka, Change in chemical composition of mullite formed from $2SiO_2\,3Al_2O_3$ xerogel during the formation process. *J. Am. Ceram. Soc.*, **70**, C-245–C-247 (1987).

44. Ph. Colomban, Structure of oxide gels and glasses by infrared and Raman scattering part 2 mullite. *J. Mater. Sci.*, **24**, 3011–3020 (1989).

45. M. Sales and J. Alarcon, Synthesis and phase transformations of mullite obtained from SiO_2-Al_2O_3 gels. *J. Euro. Ceram. Soc.*, **16**, 781–789 (1996).

46. Y. Hirata, K. Sakeda, Y. Matsushita, K. Shimada, and Y. Ishihara, Characterization and sintering behavior of alkoxide-derived aluminosilicate powders. *J. Am. Ceram. Soc.*, **72**(6), 995–1002 (1989).

47. H. Yamada and S. Kimura, Studies on co-precipitates of alumina and silica gels and its transformations at higher temperatures. *Yogo Kyokai Shi*, **70**, 87–93 (1962).

48. R.-L. Orefice and W.-L. Vasconcelos, Sol-gel transition and structural evolution on multicomponent gels derived from the alumina-silica system. *J. Sol-Gel Sci. Technol.*, **9**, 239–249 (1997).

49. S. Komarneni, R. Roy, C. A. Fyfe, and G. J. Kennedy, Preliminary characterization of gel precursors and their high-temperature products by ^{27}Al magic-angle spinning NMR. *J. Am. Ceram. Soc.*, **68**(9), C-243–C-245 (1985).

50. A. Douy, Crystallization of amorphous spray-dried precursors in the Al_2O_3-SiO_2 system. *J. Euro. Ceram. Soc.*, **26**, 1447–1454 (2006).

51. S. Kanzaki, M. Ohashi, H. Tabata, T. Kurihara, S. I. Iwai, and S. I. Wakabayashi, in *Ceram. Trans.*, Vol. 6, p. 389, *Mullite and Mullite Matrix Composites*. Edited by S. Somiya, R. F. Davis, and J. A. Pask, *Am. Ceram. Soc.*, Westervikke, OH, 1990.

52. S. Kanzaki, H. Tabata, and T. Kumazawa, p. 339, ibid.

53. A. K. Chakraborty, Intermediate Si-Al spinel formation in phase transformation of diphasic mullite gel. *J. Mater. Sci.*, **28**, 3839–3844 (1993).

54. E. Tkalcec, D. Hoebbel, R. Nass, and H. Schmidt, Structural changes of mullite precursors in presence of poly-ethyleneimine. *J. Non-Cryst. Solids*, **243**, 233–243 (1999).

55. B. E. Yoldas, Thermal stabilization of an active alumina and effect of dopants on the surface area. *J. Mater. Sci.*, **11**, 465–470 (1976).

56. B. E. Yoldas, Microstructure of monolithic materials formed by heat treatment of chemically polymerized precursors in the Al_2O_3-SiO_2 binary. *Ceram. Bull.*, **59**(4), 479–483 (1980).

57. T. Horiuchi, L. Chem, T. Sugiyama, K. Suzuki, and T. Mori, A novel alumina catalyst support with high thermal stability derived from silica modified alumina gel. *Catal. Lett.*, **58**, 89–92 (1999).

58. J. M. McHale, K. Yurekli, D. M. Dabbs, and A. Navrotsky, **9**, 3096–3100 (1997).

59. T. Demediuk and W. F. Cole, *Nature*, **181**, 1400–1401 (1958).

60. A. J. Leonard, P. Ratnasamy, F. D. Declerck, and J. J. Fripiat, Structure and properties of amorphous silico- alumnous. Part 5. Nature and properties of silico-aluminous surfaces. *Disc. Faraday Soc.*, (52), 98–108 (1971).

61. Y. Wakao and T. Hibino, Effects of metallic oxides on alpha-transformation of alumina. *Nagoya Kogyo Gijutsu Shikensho Hokoku*, **11**(9), 588–595 (1962).

Chapter 19

Critical Analysis and Characterization of Mullite Phase

19.1 Introduction

After thorough review of the studies of a large section of earlier researchers and the author's own experimental results, five to six processes of mullitization have been presented elaborately in previous chapters. During phase transformation processes of those different kinds of mullite gels, following observations have been made.

(1) It has been shown that as the nature of phase developments at four exotherms e.g., 980°C, 1150°C, 1250°C and 1300°C exhibited in DTA varies, their four steps of mullitization processes result in great variations.

(2) Composition of mullite vis-à-vis nature of it formed at first, second, and lastly at fourth exothermic peak temperatures also show differences. There are likely two–three distinct types of mullite phases.

 (i) Alumina rich metastable nonstoichiometric-mullite crystallizes during the first exotherm in SH gel.

 (ii) Alumina rich mullite formed partially at first exotherm in RH and some SHIII/SHIV gels.

Mullite Formations: Analysis and Applications
Akshoy Kumar Chakraborty

ISBN 978-981-4877-05-3 (Hardcover), 978-1-003-03167-3 (eBook)
www.jennystanford.com

(iii) Mullite type phase crystallizes at intermediate stages at the third exotherm (~1150°C), and at the second exotherm (1250°C) in SHIV gel.

(iv) Lastly, mullite type phase also formed at fourth exothermic peak temperature (1300°C) in diphasic gel.

The composition of these forms of mullite require proper identification and characterization.

Possible nature of mullite

Author showed that all the XRD peaks of mullite formed as low as at ~1000°C in case of monophasic gel are broad, some are shifted, some are very weak and Bragg diffraction peaks (120)/(210) are not resolved. It is called a poorly crystalline metastable mullite. Researchers are very much concerned for its characterization and likely it's changes in composition at different stages of heating and finally try to establish its solid solution range. Due to openness of mullite structure an extensive solid solution is possible. Aluminium and many other ions can enter into the mullite structure as solid solution in different concentrations. Aluminium forms a series of solid solutions starting from Al_2O_3:SiO_2 ratio of 2:1 (mullite) to 1:0 (i-Al_2O_3) and all the compositions of mullite and mullite-Al_2O_3 solid solution was expressed in the following manner as $(Al_{2+2x}Si_{2-2x})$ $^{iv}Al_2{}^{vi}O_{10-x}$ (Angel and Prewitt, 1986).

To incorporate the vacancy existing the mullite structure the above formula was partially modified as $(Al_{2+2x}Si_{2-2x})^{iv}Al_2{}^{vi}\ O_{10-x/2}$ $V_{x/2}$, where iv and vi denote the tetrahedral and octahedral sites respectively in the mullite structure, V denotes the oxygen vacancy and 'x'; stands for the number of Si-replaced by the same manner of Al. This formula suggests that several mullites are possible as x varies from 0 to 1. Of these α-mullite 3:2 ($3Al_2O_3$:$2SiO_2$) with $x = 0.25$ and β-mullite 2:1 ($2Al_2O_3$:SiO_2) with $x = 0.4$ are most common. With $x = 0.98$, the mol % of alumina became 99%, this is a high alumina content compound known as i-Al_2O_3 (iota-alumina) which was supported by Foster (1959). It is a tetragonal polymorph of Al_2O_3 having X-ray profile similar to that of mullite and commonly known as SiO_2-free mullite.

19.2 Characterization of 980°C Mullite and Solid Solution: Different Views

Characterization

To characterize the nature of tetragonal mullite formed in case of phase transformation of first type of monophasic (SH) mullite gel, Okada et al. (1986) first monitored the formation process of it (i) by IR studies, (ii) by lattice parameters measurement and then determined the composition of mullite using the relationship between Al_2O_3 content of mullite and its lattice constants. The chemical analysis of it was also estimated with analytical electron microscopy (Okada et al., 1991). The refinement of the structure of it was studied with help of the Rietveld method (Okada and Otsuka, 1987; Ban and Okada, 1992; Ban et al., 1996) and finally reported t-mullite which corresponds to Al_2O_3 rich composition unlike to 3:2 composition of mullite based on the following three methods of characterization.

(1) **Characterization by IR study:** Correlation between variation in a, the IR spectra, and chemical composition was also shown by Cameron (1977) as a first method of deriving the composition of mullite. According to Cameron (1977) the IR absorption bands in the range 1000–1200 cm^{-1} are related to the chemical composition of mullite. A good correlation between variation in a axis and IR band with chemical composition was established by Okada et al. (1986) during the phase evolution process of monophasic gel (Fig. 8.1).

(2) **Characterization by a lattice constant measurement:** Relationship of unit cell parameter vs. composition of mullite/solid solution and stoichiometry were first shown by Cameron (1977). Accordingly, the composition of mullite by lattice parameter calculation was predicted on the basis of lattice constant a vs. composition plot of Cameron as a second method. Based on this interrelationship, composition of primary mullite and changes in its composition during heating processes are determined (see Figs. 16.6 and 16.7).

(3) **Characterization of mullite by AEM/EDS measurements:** The chemical composition of mullite was estimated by Okada

and Otsuka (1987) directly from analytical TEM and indirectly from data estimated from the lattice constant using Cameron's relation (Fig. 3) of them. At 950°C, the composition of mullite was reported to 66 mol %. This value gradually decreased to 59.6 mol % on heating mullite gel to 1400°C. Based on the basic data of Cameron (1977), some of the past researchers too tentatively made following assumptions regarding the composition of mullite at various stages of heating of their mullite gels.

(i) That the mullite formed during first exotherm might be in pseudo tetragonal form (t-mullite) with release of certain portion of silica (A) out of noncrystalline aluminosilicate precursor phase (Okada et al., 1986; Schneider et al., 1993, 1994; Fischer et al., 1993; Fischer et al., 1993; Sanz et al., 1991; Jaymes et al., 1996).

(ii) SH gel with an alumina content close to the composition of t-mullite structure should form t-mullite to higher alumina content (Wang and Thomson, 1995). It is reported that the maximum extent of mullite obtained in the gel with an Al/Si ratio of 4/1 was consistent with Okada's work.

(iii) The released portion of silica (A) reacted with t-mullite on further heating to develop o-mullite i.e., silica was re-introduced into t-mullite structure and developed o-mullite. Accordingly, they speculated the following 980°C reaction on a tentative basis.

$$\underset{\text{(Monophasic)}}{\text{Mullite gel}} \xrightarrow[\text{Exotherm}]{980°\text{C}} \underset{(65\text{ mol \% }Al_2O_3)}{\text{t-mullite}} + SiO_2\text{ (A)} \xrightarrow{1300°\text{C}} \underset{(60\text{ mol \% }Al_2O_3)}{\text{o-mullite}} \tag{19.1}$$

Besides Okada et al. (1986), progressive changes in lattice parameters particularly (a value) of mullite in the heating process of mullite gels was also noted by several researchers. Ossaka (1961), Schneider and Rymon-Lipinski (1988), Schneider et al. (1992), Sales and Alarcon (1996), showed gradual decrease in a value from 1000°C to ~1400°C. The changes in lattice constant values together with peak width of the 121 reflection were shown by Okada and Otsuka (1986). It was shown that the length of a axis of newly formed mullite phase formed in monophasic (SH) gel is shortened at two steps during heating. First one at 1000°C to 1100°C and second at 1300°C to 1400°C. The value remained constant after this temperature (Fig. 16.7). On comparing the gradual changes in *a* value with Cameron's data (1977), the chemical composition of mullite was found to decrease from ~71 to 60 mol % in SH gel and 64% to 61% in RH gel as the firing temperature increased. It was concluded that the alumina rich mullite formed on heating mullite gels at 1000°C gradually transformed to 3:2 mullite at ~1400°C.

That the *a* axis value is slightly higher than that of stoichiometric mullite was noted by various other researchers e.g., Hirata et al. (1985a), Klug et al. (1987), Hirata et al. (Sept., 1985), Hirata et al. (1989), Low and McPherson (1989), Mizuno and Saito (1989) Suzuki et al. (1990), Schneider et al. (1992). It of course decreased on prolonged heating for 10 h at 1400°C to 1500°C, and coincided with stoichiometric mullite. Thus, annealing temperature/time plays a major role in controlling LC values.

Solid solution

Solid solubility of mullite has been established and estimated by Rooksby and Partridge (1939) to be from the 3:2 to the 2:1 ratio. According to Aramaki and Roy (1962) the extent of equilibrium solid solution in mullite at solidus temperature was from approx. 60 mol % Al_2O_3 (3:2 ratio) to 63 mol % Al_2O_3. Metastable solid solution could be prepared up to 67 mol % Al_2O_3. Lattice parameter values of mullite measured by Yamada and Kimura (1962) showed fluctuations in the following range. a = 0.754–0.756 nm, b = 0.769–0.770 nm and c = 0.288–0.289 nm. The variation of these results are the function of (i) temperature of heat treatment; (ii) composition of gel and (iii) duration of heating. When a coprecipitated gel was heated at ~1300°C for a long time, the lattice constant values converged to a constant value. They were of the opinion that with passage of time the crystallinity of mullite formed increases gradually until reaching constant value characteristic of a particular temperature i.e., mullite attains standard stable state corresponding to this heating temperature. The converging values of LC of samples Al_2O_3/SiO_2 = 3/2 and Al_2O_3/SiO_2 = 2/1 are plotted in Smith's chart which conform it. This indicates that the chemical composition of mullite formed showed the existence of a solid solution rich in Al_2O_3 rather than Al_2O_3/SiO_2 = 3/2.

The unit cell edge, *a* against the composition of mullite (mol % Al_2O_3) was also plotted by Low and McPherson (1989). The composition of mullite formed from monophasic gels at various temperatures and from varying amounts of alumina seemed to lie on a straight line. Plot showed the variation in composition of mullite (mol %) as a function of temperature. The plot of cell parameter, *a* and cell volume, Cv was also linear and fitted with Cameron's data. With this observation they suggested that there was some very simple relation between cell dimensions and either chemical composition of Al-Si order. This result also established the existence of solid solutions between (3:2) and (2:1) mullite.

However, the above transformation processes (Eq. 19.1) seem doubtful and difficult to conclude due to the following two reasons. (1) These two mechanisms are mere conjectures and based on *a* lattice parameter measurement data and also on MAS-NMR, IR spectra analysis results of poorly crystallized mullite derived out of mullite gels/precursors at ~1000°C only. (2) Other constituents of heat treated precursors phases formed at 980°C are not properly characterized and moreover their coefficients are not taken into account during making above speculations and those are elaborately analysed first in this chapter prior to characterization of mullite phases.

19.3 Problems of Characterization of Weakly Crystalline Intermediate Mullite Phase vis-à-vis 980°C Reaction

Three major techniques e.g., (i) lattice constant determination by XRD, (ii) identification of –110 ppm in Si resonance peak by MAS-NMR, and (iii) recording of changes of AlO_4 to AlO_6 by IR studies were employed by various researchers for characterization and identification of nature of three kinds of mullite e.g., t-mullite, or cubic mullite (c-mullite) and o-mullite those are likely crystallized at three different exothermic regions of two types of monophasic precursors of mullite. Moreover, major researchers believe direct mullite formation out of mullite precursor during the first exothermic reaction in DTA. However, these researchers neither considered the presence of some quantity of Al-Si spinel phase in association of t-mullite nor characterized the presence of two associated noncrystalline aluminosilicate phases. One is silica rich aluminosilicate phase (AS2) and other is alumina rich silica phase (AS3) left as residue besides mullite formation. Prior to the characterization of any form of mullite, it is necessary to characterize first those two later phases and moreover it is required to identify the occasional formation of any θ-Al_2O_3 phase along with Al-Si spinel phase at any stage of transformation of mullite gels.

Readers may see how Al-Si spinel phase has been identified. The first part was already shown in Chapter 15. Second part consists of two noncrystalline aluminosilicate phases from which mullite generally develops on subsequent heating above 1200°C which is shown in Chapter 16. The remaining third part i.e., formation of θ-Al_2O_3 phase was also identified and its role to mullitization was stated in Chapter 16. These are cited in this chapter and will be discussed.

19.3.1 Presence of Noncrystalline Aluminosilicate Residual Phase

The existence of two residual phases as the phase constituents of 980°C exotherm and it must be well considered which led to a clue to sequential phase transformation of different kinds of mullite precursors. Readers see Chapter 15 for details. The highest conversion of t-mullite was achieved by Li and Thomson (1991) was 70% during heating 2WC gel @ of 10°C/min at 1000°C. Question is: why mullite formation from stoichiometric single phase gel never riches full conversion at 1000°C? Chakraborty (2008) also noted the rapid formation of mullite type phase on heating monophasic gel (SHI) as like Okada et al. (1986) to the extent of 62.8 wt % approx. at ~1000°C. It indirectly predicts that a large amount of noncrystalline aluminosilicate as a residual phase is in existence at said temperature.

19.3.2 Progressive Changes in Lattice Parameter of Mullite

Changes in LC curve of SH mullite gel (Fig. 16.7) consists of 3–4 steps as per data of Okada et al. (1986). It is required to compare it with the mullite formation curve and then proper explanation is to be given accordingly.

19.3.3 Progressive Changes in Crystallinity of Mullite

It is shown that all X-ray diffractogram peaks of primary mullite which was first crystallized are slightly broadened when monophasic gel was heat treated at 1000°C. The reason may be due to structural disorder (Fig. 16.9). Secondly, there are absence of splitting of peaks with Miller indices of 120, 240, 041 and 250 (Schneider and Rymon-Lipinski, 1988). This indicated that the length of *a* and *b* constants have become equal. Symmetry is tetragonal with respect to normal orthorhombic mullite. The mean cell volume (V_{TM}) of tetragonal mullite (~167.4 A^3) was similar to that of orthorhombic mullite (V_{OM} = 167.4 A^3). The reason to nonsplitting of peaks are due to poor crystallization of mullite and consequently showing low resolution as pointed out by Sales and Alarcon (1996). These authors suggested two ways for increasing crystallinity of mullite (i) with increase of soaking time led to splitting of the peak (ii) increase in calcination temperature might also give rise to a clear angular separation of diffraction lines and that the metastable mullite became highly crystalline and stable. Even in lattice parameter measurement of mullite obtained by coprecipitated gel method, Yamada and Kimura (1862) noted the variations in changes of lattice parameter of mullite. Variation of lattice parameters of mullite with increasing temperature was noted by Baranwal et al. (2001). Chakraborty (2005c). It was shown that resolution increased with heating (Fig. 16.11). *a* parameter vs. composition plot indicated that mullite formed in alumina rich gels may be rich in alumina (Fig. 16.8).

19.3.4 Progressive Changes of AlO_4 to AlO_6 Ratios of Mullite

19.3.4.1 Observed by IR study

A regular shift of Al-O-Si absorption band at ~1020 cm^{-1} in IR spectroscopy occurred during heating of monophasic gel. For example, the first shift occurred abruptly to higher values from 1020 to 1134 cm^{-1} due to mullite formation. It again shifted gradually to a higher energy field at 1165 cm^{-1} and with a shoulder at 1130 cm^{-1} associated with AlO_4 and SiO_4 tetrahedra vibrations (Hirata et al., 1989). A remarkable observation due to Okada et al. (1986) who noted that initially $A_{1130} > A_{1170}$, then $A_{1130} \cong A_{1170}$ and, finally $A_{1130} < A_{1170}$ for SH xerogels as shown in Fig. 8.1. This IR result was used to interpret the mullitization process of SH which showed major mullite crystallization during occurrence of first exotherm. According to them this mullite on comparing with previous data of mullite of Cameron contained more than 60 mol % alumina. On continued heating at higher temperature, the data might correspond to ~60 mol % alumina and that coincided with the bulk composition of the SH gel.

19.3.4.2 Observed by ^{27}Al NMR study

The Al spectra of two types of monophasic gels show distinctly different spectra during the heating process in DTA studies.

(i) **In case of formation of weakly crystalline mullite at first exotherm:** Mullite is the resultant crystalline phase formed during heating MI precursor at 980°C exotherm (Geradin et al., 1994). In ^{27}Al NMR spectroscopy study, preferentially Penta coordinated Al in of the noncrystalline aluminosilicate precursor phase changed into resonance consisting of three types of environments, characteristic of mullite Al(IV), Al(IV*), and Al(VI). However, both tetrahedral and octahedral peaks nearly resemble mullite. Since the relative proportions were different from the well crystallized state of mullite. Question is: why Al(IV) to Al(VI) ratio of poorly crystallized mullite does not fit to the same for well crystallized variety? At the second stage, the ratio of tetrahedra to octahedra was also increased from precursor calcined at 950°C to 1200°C to the value nearly corresponding to well crystallized mullite. The probable distinction between t- and o-forms of mullite could be differentiated by ^{27}Al MAS-NMR study by the measurement of the apparent ratio "Al(IV)/Al(VI)". According to Jaymes et al. (1984) the ratio increases on heating the gel from 1000°C to 1300°C.

(ii) **In case of formation of Al-Si spinel at first exotherm:** Significantly more octahedral coordinated aluminum was noted when Al-Si spinel phase crystallized out of noncrystalline aluminosilicate precursor phase at 980°C. In both the cases, the apparent ratio "Al(IV)/Al(VI)" in MAS-NMR increased on heating from 1000°C to 1300°C (Komerneni et al., 1985), Yasumori et al. (1990), Jaymes et al. (1996), Geradin et al. (1994).

19.3.4.3 Alignment of ATEM and Lattice Parameter Curves

The nature of the change of two profiles one by ATEM and other by XRD are nearly the same except the 950°C data. The reason may be sought for. There is every possibility of error shooting to composition evolution since mullite formed at ~1000°C is adhered with a small amount of Al-Si spinel and large amount of noncrystalline aluminosilicate phases (Chakraborty, 2006) as was shown in Chapter 15. The value of *a* axis parameter changes very swiftly with a change in temperature interval of ~25°C. The specific reason behind such a drastic shift in *a* axis value is to be ascertained.

19.3.4.4 Progressive shift of Si resonance peak in ^{29}Si NMR study

A regular shift of ^{29}Si NMR band during the thermal treatment of this gel was noted by Jaymes et al. (1995). They observed a replacement of a broad resonance at –90 ppm due to intermediate aluminosilicate precursor phase to –86, –90, –94 and –80 ppm and a broad signal at –110 ppm in the evolution process at ~1000°C.

19.3.5 Tentative Prediction of the Nature of 980°C Mullite Phase

Further predictions are based on other observations.

(1) The Al(4)/Al(6) ratio of mullite phase changed and raised on continued heating of mullite gel as shown by (Jaymes et al., 1996).

(2) Particularly, ^{46}Al resonance peak due to *T* (tetrahedral) increased with heating temperature (Sanz et al., 1991; Chakraborty, 2005); Yossumori, 1990).

(3) It was believed by the several researchers that 2:1 mullite formed at 980°C may reacted with siliceous phase (due to

presence of −110 ppm peak in Si NMR of single phase mullite gel heated to 1000°C) during heating >1300°C to form 3:2 mullite.

Based on the points as discussed above, it is concluded that the 2:1 composition of mullite crystallized at 980°C seems doubtful. The cause of disagreements are presented below.

(4) Jaymes et al. (1996) noted three Al sites in Al spectrum of precursor marked A heated at 1000°C, one octahedral site (~0 ppm), and two tetrahedral sites denoted by *T* (60 ppm) and *T** (44 ppm). At this temperature, the precursor completely crystallizes to mullite (Al:Si = 4:1). However, on heating at 1300°C when Al:Si = 3:2 a slight increase in ratio of Al(IV)/Al(VI) occurs which is contrary to the variation of the chemical composition within the mullite solid solution.

(5) Moreover, in case of Precursor *C* heated at 1000°C of Jaymes et al. (1996), Si resonance peak at −110 ppm was absent, accordingly siliceous (A) phase was not existing. There was no question of interaction between mullite (formed at 980°C) with siliceous phase on heating at between 1000°C–1300°C. It may be assumed that weakly crystalline mullite increases its crystallinity on progressive heating. As a result, *a* axis value changes, the increase of AlO_4 groups of *T* sites increases. Generally, in case of heating other monophasic gels beyond 1000°C, increase of siliceous phase in 2:1 mullite will theoretically decrease AlO_4 groups. But it is not observed during ^{27}Al MAS-NMR spectroscopic study of Precursor *C*. On the other hand, an increase of bands attributed to AlO_4 groups due to o-mullite occurs.

(6) Furthermore, Chakraborty (2003a, 2004) showed two experimental results.

 (i) Solid state reaction studies of monophasic mullite gel in its dehydrated form, and in heat treated samples from 500°C to 1100°C with calcium oxide (CaO) conclusively reported the complete absence of release free amorphous silica and further concluded that these three phases derived out of Al_2O_3-SiO_2 precursor during heating processes namely (i) noncrystalline aluminosilicate precursor phase, (ii) Al-Si spinel phase and (iii) residual noncrystalline aluminosilicate phase are all compounds (Fig. 15.10a–c) and Tables 15.8 and 15.9.

(ii) By alkali leaching study, it was also shown that there is no free amorphous silica in mullite gel heat treated to 1000°C (Fig. 15.12 and Table 15.12).

In view of the author's two experimental findings, the decomposition of noncrystalline aluminosilicate precursor phase into t-mullite and SiO_2 (A) at 980°C exotherm as predicted by some researchers vide Eq. 19.1 needs more verification.

During characterization of mullite gel heat treated to ~1000°C by MAS-NMR study, the observed –110 ppm of ^{29}Si MAS-NMR peak requires further examination. It is very broad instead of sharp in nature. Classification of this peak due to $Q_4(0Al)$ only was suggested by Engelhadt and Michel (1987). As this peak occurs in a long range of resonance, it is thought that instead of free silica (A), it may be characterized as a noncrystalline aluminosilicate phase rich in silica. In this connection the author refers to the study of Risbud et al. (1987) who presented ^{29}Si MAS-NMR spectra of silica glasses which shows a wide range of composition over a wide range of chemical shifts. Thus, this –110 ppm band is to be designated as a silica rich aluminous phase other than free amorphous silica as assumed by various researchers.

(iii) It could be argued that if the 980°C phase is 2:1 mullite, it should be stable and exist far above 1000°C during the heating process. Furthermore, to assess the nature of 980°C heated mullite formed out of monophasic gel, it is necessary to put more attention to the following phenomenon.

1. How poorly crystalline mullite is crystallized during the occurrence of endo-exotherm assembly region at 980°C? One must have to consider first the final dehydroxylation mechanism
2. The formation of unstable noncrystalline anhydrous aluminosilicate intermediate phase (AS1) formed at the point "D" by decomposition of mullite precursor noted in DTA and DDTA analysis just

prior to crystallization at 980°C exotherm. Thus the intermediate product should be looked into.

Answers of these questions are discussed in Chapters 14 and 15 (readers may go through those points).

19.4 Discussion on Earlier Results on Changes in Composition of Primary Mullite to Regular Mullite

Now, the next questions remain are (i) the entire course of mullite gel-to-mullite formation process during gradual heating heat treatment and (ii) it's mechanism of transformation. QXRD data of mullite formed on heating two types of monophasic and diphasic are also scare. Okada and Ousuka (1986) of course showed mullite formation over a range of temperature in their pioneering mullitization studies. The process in case of monophasic gel is analysed and it consists of two distinct steps. Initial step is sharp at first exotherm which is followed by a plateau up to a certain temperature at ~1100°C thereafter a slow and gradual mullite formation process with rise of temperature (Fig. 16.1).

19.4.1 Mullitization in First Type of Monophasic Gels e.g., SH Gel of Okada et al.; Type I of Schneider et al.; SHI of Haque and Chakraborty

At the final stage of inhomogeneous dehydroxylation process at the temperature of endothermic dip (D) ~950°C, proton migration starts from acceptor to the donor region followed by counter migration of Al^{+3} ion takes place in the opposite direction. Thereafter, silicon ion diffusion and counter diffusion of Al^{+3} ion occur which results to generation of two separate phase boundary regions consisting of

(1) Noncrystalline alumina rich silica as an unstable anhydrous phase (named as AS1) in the acceptor region and
(2) Noncrystalline silica rich aluminous phase (named as AS2) in the donor region (see Chapter 14).

The unstable former phase (AS1) crystallizes at 980°C exotherm to

(i) nonstoichiometric alumina rich mullite phase (weakly crystalline as major);
(ii) alumina rich Al-Si spinel (poorly crystalline as minor) phase; and
(iii) a noncrystalline alumina rich mullite (as the second major residual phase (named as AS3) remains after crystallization of two former phases.

Regular mullite formation reactions on heating after endo-exothermic region-to-1600°C are given earlier in Eqs. 16.2–16.5.

In the case of monophasic gel, the major crystalline phase formed at 980°C is mullite (weakly crystalline). Whereas in coprecipitated gel, RH gel or in some colloidal gel, the main crystalline phase is Al-Si spinel (Table 19.1). Now the question is: what should be the composition of two crystalline phases? Reasons are as follows. Regarding mullite is concerned, it may be t-mullite or o-mullite or other. Further, regarding the Al-Si spinel phase is concerned, composition may be analogous to composition of 3:2 mullite or alumina rich. As per thermodynamic data of liberation of heat measured by a large section of researchers as shown in (Table 16.4), the highest ($-\Delta H$) value goes to 3:2 ratio of mullite at the occurrence of first exotherm. Accordingly, it is theoretically expected that the 980°C exotherm is correlated to the crystallization of large quantities of mullite which may be orthorhombic form although a large section of researchers speculate it as tetragonal form of mullite. But the direction of crystallization should come from thermodynamic or kinetic reasoning.

Formation of spinel as invariable phase

It could be presumed that except aqueous —ANN gel method, in all other gel preparation methods using TEOS/ANN couple, and or double alkoxides or others gel processing methods of using different components, noncrystalline alumina rich silica phase (AS1) will likely generate in phase separation process during endo-exothermic region of DTA (see Chapter 14). Some of the experimental results of earlier authors are cited below. These three methods (monophasic, SH, and polymeric/Type I gel) of synthesis of mullite gels always show the formation of Al-Si spinel as (minor phase) along with crystallization of weakly crystalline mullite as major phase.

Table 19.1 Name of authors studying composition of mullite formed out of three types of mullite gels at three different exotherms

Author	Gel type	Mullite first exotherm (°C)	Formation second exotherm (°C)	Temperatures third exotherm (°C)
Okada and Otsuka (1986)	SH gel monophasic first type	Exhibited	Nil	Nil
Schneider et al. (1992)	Aqueous gel	Exhibited	Nil	
Sales and Alarcon (1996)		Sample A exhibited sharp exotherm	Nil	
Jaymes et al. (1996)	Monophasic gels	Precursor A exhibited	Nil	
Okada and Otsuka (1986)	RH gel monophasic, other type	Exhibited	Exhibited	
Yamada and Kimura (1960)	Coprecipitated gel	Exhibited	Exhibited	
Suzuki et al. (1990)				
Hirata et al. (1989)		Exhibited		
Low and McPherson (1989)	Aluminosilicate gels	-72 (wt % Al_2O_3) gel-highest exotherm -80 (wt % Al_2O_3) – medium exotherm -63 wt % Al_2O_3.broad insignificant exotherm	- A broad in significant exotherm - Nil - A medium exotherm	
Klug et al. (1990)	Diphasic gel			
Chakraborty (1988)	Diphasic gel			Exhibited

Accumulation noncrystalline aluminosilicate phases

As minor quantity of Al-Si spinel phase formed in those cases at first exotherm, it is almost certain that quite a substantial quantity of noncrystalline silica rich aluminous phase (AS2) must be present through phase separation step during final dehydroxylation of noncrystalline mullite precursor (see Chapter 14). ^{29}Si silicon spectral analysis supports the presence –100 ppm resonance peak due to presence of noncrystalline silica rich aluminous phase (AS2) other than free amorphous silica in MAS-NMR study of a variety of mullite gel heat treated at ~1000°C.

With these experimental findings, it is logical to conclude that mullite which crystallizes at first exotherm is alumina rich as speculated by various earlier researchers. However, the proposed weakly crystalline alumina rich mullite should not be due to 2:1 mullite. Since there is no reason to crystallize a metastable phase during the first exotherm when there is every possibility of crystallization of stable and regular mullite. Moreover, the former mullite does not exist on calcinations of mullite gel at temperature >1200°C. Question is: what is this phase and what is the composition of this weakly crystalline mullite phase formed at 980°C? It is assumed that either (i) the composition of weakly crystalline mullite phase formed at 980°C during heating mullite precursor varies as per the diagram (Fig. 16.6) of Cameron (1977) which predicted the probability of mullite over a long range of composition or (ii) quantity of residual noncrystalline alumina rich mullite phase (AS3) changes from precursor to precursors.

Possibility of mullite of varying composition

Monophasic gel synthesized by different researchers shows variation of intensity of first exotherm. It's enthalpy of crystallization shows a great variation in experimental value by different studies. Precursors of the same composition (A, C, E) of Jaymes et al. (2006) even show differences in $-\Delta H$ values (Figs. 12.1 and 12.2). Similar is the finding in case of SH gels. Interestingly, literature data reveal that 980°C DTA exotherms in case of polymeric gels also show a great variation in peak height.

Even quantity of noncrystalline silica rich alumina phase (AS2) and its composition to show changes as per MAS-NMR data of

Geradin (1994), Jaymes et al. (1995) and clumping of Si MAS spectra of Schneider et al. (1992), Huang (1997). Thus, it could be concluded that the newly crystallized disordered mullite like phase may have varying composition. There are two groups of researchers who rely on either t-mullite or o-mullite formation in heated mullite gels at 1000°C. Thus, there arises some possibilities.

(A) Tetragonal mullite: The formation of t-mullite was suggested by Ossaka (1961). According to Schneider et al. (1993), distinct t-mullite with ($a = b$) really exists. Formation of such mullite will require a complete rearrangement of the atomic positions so as to achieve tetragonal symmetry. Precisely, it is a "pseudo-tetragonal metric" since the symmetry of the crystal structure clearly stays orthorhombic as suggested by Schneider and Rymon-Lipinsky (1988).

Regarding the metastability of the pseudo-tetragonal form of mullite, Cameron (1977) showed the possible way of obtaining t-mullite when $x = 0.65$ which corresponds to 79 mol % Al_2O_3. Reported data is concerned, there is one way to synthesize it of having 77.3 mol % Al_2O_3 by rapid solidification from melt. However, mullite of such variety is not reported for its synthesis from sol-gel route (Sales and Alarcon, 1996). Formation of mullite is indirectly suggested by Okada and Oksuka (1986) as tetragonal which is based on (i) cell dimensions measurements of mullite, and (ii) by calculation of composition of mullite using TEM attached with EDX analyzer, and (iii) lack of splitting of Bragg diffraction peaks with Miller indices of (120) and (210), 240, 041 and 250 XRD doublet reflections of mullite poorly crystallized form at 1000°C. With these results, Li and Thomson (1991), Wang and Thomson (1995), Jaymes et al. (1996) also postulate the formation of t-mullite.

(B) Orthorhombic mullite: Structural changes of 3:2 mullite of weakly crystalline variety to well-ordered mullite is the cause of change of *a* lattice value (Aramaki and Roy, 1962). Transformation of the first kind of monophasic gel (SH) was first reported by Okada and Otsuka (1986). Details in progressive changes of it are given first.

(C) Progressive changes in quantitative value of mullite formed over a long range of temperature:

At first step (900°C–1000°C): In SH xerogel mullite formation occurred at ~900°C and rapidly increased at 1000°C by Path I. Both X-ray peak width and *a* lattice constant were very high at 900°C. The length of the *a* axis value of just formed mullite decreased on firing to 1000°C and then to 1100°C. Composition of Al_2O_3 was estimated from the measured length of *a* axis value and speculated to contain ~70 mol % Al_2O_3 and called the newly formed mullite as alumina rich mullite.

At second step (1000°C–1100°C): Intensity of mullite formed became constant between 1000°C to 1100°C i.e., mullitization did not occur at this stage. In the temperature range 1050°C to 1150°C. Both peak width (Δ) value and *a* axis value (O) are also remained almost constant [see Fig. 16.7 or Fig. 2 of Okada et al. (1986)].

At third step (1100°C–1300°C): Above 1100°C, further mullite formation ensued and continued up to 1300°C by Path III. Both values (width and *a* axis) were shortened again on heating between 1150°C–1300°C. Converse behaviour of length of *b* axis was noted at three above mentioned steps. The length of the *c* axis was shortened gradually from 1000°C to 1150°C and reached quickly at ~1200°C and then remained constant with increase of firing temperature. Question is: what is the cause of further rapid formation of mullite at the third step?

At fourth step (1300°C–1600°C): Both values of were shortened very slowly on heating between 1300°C–1600°C. As shown in discussion of final dehydroxylation of mullite gel and after rapid crystallization of newly formed mullite at and around 1000°C, two other residual phases are present e.g., (i) noncrystalline silica rich aluminous phase (AS2), (ii) noncrystalline alumina rich mullite phase (AS3) as a second major phase as shown already. These phases would play a role in the slow mullitization process at third and fourth steps.

Recharacterization of –90 ppm Si resonance peak

^{29}Si MAS-NMR spectra of the first type of a monophasic gel heated to 1000°C is reanalysed as given below. There are two broad Si resonance peaks. A comparatively large broad peak is centred at –90 ppm and another small broad peak situated in the range –100 to –120 ppm in the Si resonance spectrum. Various earlier researchers designated these two peaks as only due to poorly crystalline alumina rich mullite phase (assumed as metastable 2:1 mullite) and noncrystalline silica phase respectively (as per Eq. 19.1). However, the poorly crystalline alumina rich mullite phase shows broad Si resonance peak at –90 ppm (Jaymes et al., 1996, the latest data). It is inherited from it's previous of noncrystalline alumina silicate mullite precursor phase via the intermediate noncrystalline anhydrous unstable alumina rich silica phase (AS1) generated just after complete dehydroxylation (shown in Chapter 14). Considering the shape, it's nature and broadness of the Si resonance peak stationed at ~ –90 ppm, it could be interpreted that this peak is not due to poorly crystalline alumina rich mullite phase alone as assumed but this entire broad peak should contributed jointly by the following three phases namely (i) alumina rich mullite itself as weakly crystalline phase (major), (ii) alumina rich Al-Si spinel as weakly crystalline phase (minor), and (iii) residual noncrystalline alumina rich mullite phase (AS3) phase.

Possible solid state reactions

The other noncrystalline silica rich alumina phase (AS2) is related to peak at –100 to –120 ppm range. It is most logical to conclude the two following solid state reactions lead to mullitization during the course of heating.

(1) Diffusion of Si^{+4} out of residual noncrystalline silica rich alumina phase (AS2) into so-called poorly crystalline alumina rich mullite phase occurs on continued heating at the third step.

(2) Inter diffusions between the three phases namely (i) noncrystalline alumina rich mullite phase (AS3), (ii) alumina rich Al-Si spinel phases (minor) and (iii) noncrystalline silica rich aluminous phase (AS2) also take place to crystallize further mullite at later third and fourth steps.

19.4.2 Mullitization in Other Type of Monophasic Precursors e.g., RH Gel of Okada et al.; Type III of Schneider et al.; CP Gel and SHIII of Haque and Chakraborty

Regular mullite formation process on heating after endo-exothermic region-to-1600°C is given below.

In this case, Al-Si spinel is the major crystalline phase which interacts with two other associated major phases likely (i) noncrystalline silica rich aluminosilicate phase (AS2) and (ii) residual noncrystalline alumina rich spinel phase (AS5). As a result its composition changes gradually on continued heating and it likely riches the composition equivalent to 3:2 mullite. The enriched Al-Si spinel (called as cubic mullite or c-mullite) transformed to weakly crystalline mullite at second exotherm by Part IV and at third exotherm by Part V. The next courses of transformation during heating will follow accordingly as per first type of monophasic gel as per mullitization by Part III. The overall transformation process consists of five steps and are shown in Eqs. 16.6–16.9 (see Chapter 16).

19.4.3 Discussions on Mullitization in Some Other Type of Monophasic Gel of Previous Authors

Examples:

(1) In mullite gel of composition equals 72 wt %, Low and McPherson (1989) observed Al-Si spinel phase at 980°C exotherm and o-mullite at second exotherm at ~1250° C in DTA. The later temperature is much higher (~250°C) than 1000°C where weakly crystalline primary mullite formed in SH gel. Thus, this gel is designated as another type of monophasic gel. Lattice constant *a* value of mullite formed on heating of this gel at 1200°C to other increasing temperatures were measured, plotted and showed a straight line relationship between cell parameter, *a*, and cell volume, Cv as depicted in their (Fig. 3) and (Fig. 4). These two results follow between the lines of best fit obtained by Cameron (1977) and Schneider and Wohlenben (1981). Further, the composition of

mullite as a function of temperature was also shown in their (Fig. 5). Accordingly, they first interpreted that mullite formed initially at ~1200°C instead at 1000°C approaches to (2:1) composition which gradually approaches to 3:2 composition with increasing temperature at 1500°C as shown by Okada et al. (1986).

Interestingly, this mullite starts to form at 1200°C and this monophasic gel of other type also follows Cameron rule. Not only this mullite gel shows the straight line composition relationship, but also it follows the change of 1110 to 1165 cm^{-1} IR pattern upon increasing heat treatment temperature as per Cameron's second observation. These two data emphasized the earlier results shown by Okada et al. (1986) in their Fig. 6c.

In the third step when 72 wt % gel sample heat was treated from 1200°C to 1500°C, a straight line relationship between cell parameter *a* and cell volume, C_v fitted Cameron's data (Fig. 16.6) was also shown as a second observation by Low and McPherson (1989). It further established the existence of solid solutions between (3:2) and (2:1) mullites. These two results verify the works of Cameron obviously. However, there are no plots of *a* spacing vs. composition of gels of varying alumina between 67 mol % to 60 mol % (2:1 to 3:2) mullite composition at a fixed temperature ~1000°C. This is very much needed in view of drawing conclusive evidence of the characterization of primary mullite formed at 980°C which is one of the most controversial issues and not the mullite which is crystallized at second exotherm or above.

(2) The variation in the lattice constants of mullite *a* with the dependence of composition was reported by Hirata et al. (1989). Formation of mullite exclusively during heating of gel no 1 (1.03/2 see Table 1) at 1000°C indicates full homogeneity of the gel. Contrary, as the alumina content of the gels increases as in gel no 2 and in gel no 3 (nearly mullite composition) spinel formation starts as an additional phase to mullite during heating to 1000°C. This result indicates the arise of inhomogeneity in later two gels. What is the cause behind it? What is the cause of changes in relative ratio of

formation of spinel to mullite at 1000°C? It may be assumed that either alumina gel or silica hydrogel segregates during hydrolysis and polymerization process and produces bigger molecules and hinders homogeneity of resultant mixed Al_2O_3-SiO_2 gels.

(3) Refer to gel (Sample A) of high silica concentration of Sales and Alarcon (1996) vide Table 1. This gel at 1400° and at 1500°C, shows *a* value of the order of 7.542 and 7.543 (Å), respectively. These data suggest that mullite formed in sample A is not alumina rich in contrast to sample B and sample C of them. In latter two cases much alumina are in solid solution with end mullite. But they predicted as per Cameron that mullite formed in all three samples B and C inclusive of A at 1000°C was Al_2O_3-rich mullite and fully developed at 1400°C. The chemical composition of mullite changed as a function of calcinations temperature as per Cameron rule. Application of Cameron's rule may be a positive step in ascertaining the composition of weakly crystalline mullite. But the changes in the shape of reflection pair 120/210 and others on gradual heat treatment may be due to dual causes. (i) Decrease of *a* value in accordance with composition-temperature relationship and (ii) Poor-to-high order of crystallization of mullite. For example, in case of Sample A of Sales and Alarcon (1996), the contribution of order-disorder phenomenon is the reason of *a*, vs. temperature change plot. In mullitization processes of Sample A and gel no 1 (1.03/2 of Hirata et al. (1989), change in *a*, value with temperature application of Cameron's rule should not be applicable. Since spinel phase won't crystallize as a second phase to mullite at 1000°C. Moreover, *a* value of mullite at 1300°C to 1400°C is nearly equal to 3:2 mullite. In sample B and C, the global part in *a*, vs. temperature change plot consists from the contribution of order-disorder phenomenon as well as from the contribution from *a*, vs. composition of Al_2O_3 plot corresponding to Cameron relationship.

Questions are:

(i) Why monophasic gel, SH gel, Polymeric gel or Type I gel forms spinel phase invariably with weakly crystalline mullite (as a major amount) at first exotherm?

(ii) What are reasons for formation of varying ratio of spinel to mullite formation on heating mullite gels at 980°C?

(iii) How mullite formation occur in monophasic gel, SH gel, Polymeric gel or Type I gel at the second step, when weakly crystalline mullite (major) and spinel (minor) phase formed earlier at the first stage at 980°C exotherm?

Detail experimentation and reasoning are required. It may occur by solid state interaction between two residual amorphous phases namely AS3 and AS2 with AS1 on further heating. This reaction may be the reason for mullitization at the second step. There are two steps of mullite formation in monophasic gels (Fig. 16.1) and there are also two major steps of decrease of *a* value vs. temperature shifts in the overall graph of lattice parameter *a* value as a function of temperature (Fig. 16.7).

(1) After first step of rapid alumina rich metastable mullite formation

Just after completion of 980°C exotherm and formation of large quantity of primary mullite formation occurred on static heating at ~900°C. Further mullite formation ceases in the temperature range 1000°C to 1100°C. After this this first step of mullitization, LC constant *a* value however decreased sharply in between 950°C to 1100°C. The rapid shift of *a* value vs. temperature is explained as due to changes of very low crystallite size and high strain value of incipient poorly crystallized alumina rich mullite (Fig. 16.7) and not due to changes in composition of it. In this region the composition of mullite does not change as there is no solid state reaction between AS1 with either AS2 or AS3. Cameron's rule could not be applied in determination of changes in composition of primary mullite due to change in *a* LC value. Since rapid change of *a* value is not related to mullite formation. Contrary, this change of *a* axis is only a function of temperature and not a function of both time and temperature. Since *a* axis data of primary remained unchanged even on soaking for 1000 h (Okada et al., 1990) which showed the stability of primary mullite in transient state and the trend of *a* LC value vs. temperature for several samples (Fig. 3) of them. Thus, primary mullite should neither t-mullite (2:1 mullite variety) nor o-mullite. The designation of it as alumina rich metastable mullite remains valid.

(2) At the second step cessation of mullitization

After completion of the first step of rapid mullitization i.e., just after completion of 980°C exotherm, further mullite formation stops in the temperature range 1000°C to 1100°C. This refers to the plateau region of the mullitization curve from 1000°C–1100°C (Fig. 16.1). After a rapid drop of *a* L C value in the range 950 to 1100°C, change of *a* value vs. temperature plot from 1100°C–1150°C was more or less ceases (Fig. 16.7). Both plots of mullite formation and change of *a* lattice parameter remain almost constant within a short interval of temperature and likely superimpose. In this region again solid state reaction between primary mullite with silica rich aluminous phase was not even started and inter diffusion of Si^{+4} and Al^{+3} among two phases did not take place. This discontinuous change of *a* value vs. temperature plot of Okada et al. (1986) was verified by lattice parameter measurement of Type I precursor heated at different temperatures and reported in Fig. 2.23 (see Schneider's book, p. 36).

(3) At the third step of mullitization between alumina rich metastable mullite with noncrystalline aluminosilicates

It occurs above 1100°C by the start of solid state reaction between heat treated alumina rich primary mullite with AS2 and AS3 to mullite type phase itself as per Eqs. 16.2 and 16.5. On heating between 1150°C–1300°C, both values (width and *a* axis) were shortened again. This may be due to dual causes (i) partly to increase of crystallinity due to increase of heating temperature only from 1150°C to 1300°C i.e., disorder to order transformation and/or (ii) due to change of composition of primary mullite to secondary mullite i.e., from alumina rich mullite to mullite type phase by the said solid state reaction at temperature of third step.

(4) At the fourth step of structural change of mullite type phase

A comparatively small change of *a* parameter is due to further transformation of mullite type phase to 3:2 mullite with continued heating between 1300°C to 1600°C. This change in a LC is due (i) rapid increase of crystallinity due to increase of heating temperature only from 1300°C to 1600°C i.e., increase in order of mullite crystals and/or (ii) due to change of composition of mullite type phase to standard 3:2 mullite by the said solid state reaction at temperature of fourth step. In support of this views, results of both Al and Si MAS-NMR resonances significantly shown by Schneider et al. (1992) for their CM and SGM samples (Figs. 9.1 and 9.2) and that of Jaymes et al. (1996) for their sample Precursor *C* (Fig. 9.19) are cited and are referred below.

19.4.4 Effect of Heat on Changes in Mullite Type Phase at 1200°C–1600°C

(1) MAS-NMR data

(i) It is reported in those studies that after completion of the second stage of mullitization of monophasic gels, XRD reveals crystalline mullite as the only phase. But NMR resonance doesn't confirm well developed mullite. Much structural changes occur in the temperature range 1200°C to 1400°C and 1400°C to 1600°C. For example, a CM sample on calcination at 1100°C tends to complete mullitization at the second exothermic stage. The ratio of AlO_4 to AlO_6 does not correspond to standard mullite. The ratio of AlO_4 to AlO_6 starts changing. AlO_4 peak tends to increase. At 1100°C–1650°C range, the ratio of AlO_4 to AlO_6 changes remarkably.. For example AlO_4 increases at the cost of decrease of AlO_6. Moreover, splitting of AlO_4 occurs. Si resonance peak at −110 ppm already vanishes out on firing before 1400°C with crystallization of pure mullite only. The Si resonance peaks around −90 ppm (−87, −90 and −94) become more and more sharper and narrower. Both XRD and Si and Al spectrum concur with that of ideal 3:2 mullite.

(ii) Almost similar observation is noted in the case of SGM sample. When SGM exhibited two exotherms. Spinel phase formation was major in SGM sample, and due to that intensity of tetrahedral Al resonance was found less (as expected for spinel of Chakraborty's spinel model). The substantial amount of this spinel phase later on transformed to mullite at 1253°C exotherm analogous to RH gel. Mullite formation in SGM takes place via intermediary spinel phase in one step at second exotherm. It was observed that mullitization attains completion at 1100°C to 1250°C (as per ^{29}Si resonance data is concerned) (Fig. 9.1a). ^{26}Al resonance showed three broad peaks at –5, 45, and 62 ppm which are the characteristics of typical mullite resonances. But this mullite which is formed at the second exotherm is not attained to a stable state. It is to be called as mullite type phase instead of o-mullite as designated as in mullite obtained at first exotherm out of mullite gel synthesized out of aqueous gel route. It requires calcination to be done far above 1100°C. At 1100°C–1650°C range, profound changes occur in the ratio of AlO_4 to AlO_6. For example, AlO_4 increases at the cost of decrease of AlO_6. Moreover, splitting of AlO_4 occurs resembling o-mullite.

(iii) Strikingly, differences in octahedral to tetrahedral Al resonance peaks of precursor sample C heated to 1000°C and that of sample heated to 1400°C are observed. At 1000°C when mullite formation is over Si spectra showed one resonance band and a shoulder near –94 ppm due to mullite. At 1400°C, an increase in ratio of Al(IV)/Al(VI) due to well crystallized mullite development occurred. Si resonance bands resemble characteristic orthorhombic mullites. The spectrum consisted of one main resonance at ~ –80 analogous to sillimanite type SiO_4 tetrahedra and another peak at ~ –90 and finally a shoulder at ~ –94 ppm, respectively. At this stage ^{29}Si MAS-NMR pattern corresponds to data of mullite structure analysis of Ban and Okada (1993).

(2) IR data

(i) As like MAS-NMR, similar changes are observed in IR studies done by Hirata et al. (1985a), Sales and Alarcon (1996).

At ~1200°C, mullite formation proceeded rapidly with transformation of the intermediate Al-Si spinel phase by Hirata et al. (Jan., 1985) as noted in XRD analysis. At this stage the ^{27}Al spectrum shows broad absorption bands. Absorption bands due to mullite clearly observed when heat treated at 1400°C The absorption band at 1165 cm^{-1}

and with a shoulder at 1130cm^{-1} associated with AlO_4 and SiO_4 tetrahedral vibrations were distinctly observed. More clear band ratio of 1165 cm^{-1} to 1130 cm^{-1} is also observed. Besides, two octahedral peaks at 600 cm^{-1} and 550 cm^{-1} and one tetrahedral peak at 740 cm^{-1} become more distinct. At this stage IR result corroborates the formation of well crystallized mullite as observed in XRD patterns.

(ii) At the first stage of mullite formation at ~1000°C/15 h in sample B, Sales and Alarcon 1996) noted the IR absorption band at 1130 cm^{-1} stronger than 1165 cm^{-1} band due to development of mullite type phase. On further heating to 1300°C, the intensity of the band is reversed but the bands are still broad. The band is centred at 1165 cm^{-1} with a shoulder at 1130 cm^{-1} associated with SiO_4 and AlO_4 tetrahedral vibration modes respectively. In this case orthorhombic mullite is fully developed but peaks are broad which needs further heating. Contrary to calcination of sample B at 1000°C/15 h in which case mullite type phase was only observed. Thus transformation of mullite type phase to o-mullite requires further heating. Calcination at higher temperature is more useful to change it from disorder to order/stable crystallization of mullite and such change may not be related to change in composition of it. Therefore, change in *a* axis parameter with heat treatment is not due to change in composition rather due to structural perfection of mullite lattice.

19.4.5 Composition of Mullite Formed Out of Diphasic Gel

In diphasic gel system, Klug et al. (1987) obtained a lattice parameter of mullite and plotted *a* spacing with wt % alumina and showed the variation of *a* as a function of composition between 72–77 wt % Al_2O_3. The former authors heat treated precursor at high temperature as high as 1650°C. According to them in diphasic gel system, the plot of *a* spacing varied as a function of composition between 72–77 wt % Al_2O_3 which agreed with Cameron's curve.

Mullite crystallizes out of Al-Si spinel phase at ~1300°C in four cases of diphasic gels would be mullite type phase which transforms to orthorhombic form on heating >1300C vide Eq. 16.12

as suggested by Chakraborty (2005c). The probable change in lattice constant with heating beyond the said formation temperature is due mainly to increase in crystallinity of mullite phase by gradual relief of strain present during its incipient formation stage than to changes in chemical composition (see Chapter 16 regarding lattice constant measurement).

Even in this case of diphasic gel, newly formed mullites at ~1300°C show high *a* axis value. And change in *a* axis value shows linearly with temperature. Thus, it is observed that whenever mullite formed out of first type of SH gel at primary stage at 1000°C, or from other type of mullite gel at ~1250°C and even from diphasic gel at ~1300°C is of low crystalline variety and measured *a* lattice value is quite high. Is to be concluded that mullite formations occurred at each incipient stage at three exothermic peak temperature are of high alumina content? According to the present author, application of Cameron rule to determine the composition by change in *a* axis value on heating of newly formed needs more review and revision.

19.5 Summary

The followings are the three methods of characterization of primary mullite made by Okada and Otsuka (1987). (1) Characterization by *a* lattice constant measurement—where composition of mullite was estimated from the interrelationship between *a* lattice constant data to composition of mullite diagram. (2) Characterization by IR study where chemical composition of it was shown by the correlation between variation in *a* axis value with ratio of two IR bands. (3) By direct estimation of the chemical composition of mullite from analytical TEM and indirectly from data estimated from the lattice constant using Cameron's relation.

That the view of primary mullite formed during first exotherm might be in pseudo tetragonal form (t-mullite) with release of certain portion of silica (A) out of noncrystalline aluminosilicate precursor phase as characterized above by Okada et al. (1986) has been extended by several others namely Sanz et al., 1991; Schneider et al., 1993, 1994; Wang and Thomson, 1995; and Jaymes et al., 1996 and others.

Besides LC measurement, IR observations and AEM/EDS studies regarding characterization of mullite as t-mullite, these above authors also conjectured the detection of SiO_2 (A) as a second constituent phase of 980°C exotherm by observing −110 ppm peak in ^{29}Si MAS-NMR spectrum. Accordingly, they further conjectured that the released portion of silica (A) reacted with t-mullite on further heating to develop o-mullite i.e., silica was reintroduced into t-mullite structure and developed o-mullite as speculated by the following 980°C reaction (Eq. 19.1) on a tentative basis. Contrary, the above 980°C exotherm as well as mullite transformation processes seem doubtful. Since, other constituent phases and it's coefficient are not considered during making speculations. Interpretation of *a* lattice constant data to composition of primary mullite needs fresh review.

Problems of characterization of weakly crystalline intermediate 980°C mullite phase are due to following reasons.

(1) Presence of three associated phases namely (i) Al-Si spinel phase (minor quantity), (ii) noncrystalline alumina rich silica phase left as residue (AS3), and (iii) noncrystalline silica rich aluminous phase (AS2) in association with alumina rich primary mullite.

(2) Progressive changes in the quantitative value of mullite formed over a long range of temperature are not considered.

(3) Variation in the course of mullitization reaction among different types of mullite gels are to be seen.

(A) Course of mullitization in first type of monophasic gels e.g., SH, Type I, SHII

Mullitization stages in the first type of monophasic gels consist of the following steps.

Generation of noncrystalline alumina rich silica phase (AS1) will occur likely in the final dehydroxylation process during the endo-exothermic region of DTA. Thereafter, transformation proceeds as follows.

(i) Phase separation process leading to

  (a) generate noncrystalline silica rich aluminous phase (AS2) in the donor region as shown by the presence −100 ppm resonance peak on ^{29}Si silicon spectral analysis.

(b) Crystallization of noncrystalline alumina rich silica phase (AS1) at 980°C exotherm to alumina rich metastable mullite, weakly crystalline variety of Al-Si spinel phase (as first evidenced phase) by XRD.

(c) Noncrystalline alumina rich mullite residual phase (AS3) as second major phase.

(ii) Solid state reaction between alumina rich mullite with noncrystalline phases (AS2 and AS3) to crystallization of further quantity of mullite during progressive heat treatment from 1100°C up to 1300°C (as evidenced by QXRD of mullite and elimination of −110 ppm resonance band).

(iii) Development of crystalline order of mullite on further heating from 1300°C to 1600°C (as evidenced first by XRD peak separation/resolution, second by narrowing of both ^{27}Al and ^{29}Si resonance MAS-NMR peaks of o-mullite, third by rapid increase in size of mullite crystallites with increase of temperature (Fig. 16.9B), fourthly by decrease of high strain values of mullite crystallites in the range of 1400°C to 1600°C (Fig. 16.9C).

(B) Correlation between *a* axis data with change of quantity of mullite at increasing temperature

There are four steps of transformation of the first kind of monophasic gel (SH) to mullite as noted from pioneering studies of Okada and Otsuka (1986).

At first step: Mullite formation just started at 900°C and rapidly increased at 1000°C. Measured *a* lattice constant value were very high at 900°C. Instead of increase of mullite content, the length of the *a* axis of just formed mullite decreased on firing to 1000°C and continued up to 1100°C. An inverse relation between the length of the *a* axis with mullite formation is thus observed.

At second step: Intensity of mullite formed became constant between 1100°C to 1150 °C. In this temperature range *a* axis value also remained almost constant. Compare Fig. 16.1 and Fig. 16.7.

At third step: Mullite formation ensued further above 1100°C which continued up to 1300°C. The question regarding the cause of further rapid formation of mullite at third step is explained.

a axis value was further shortened rapidly on heating between 1100°C–1300°C.

At fourth step: Shortening of *a* axis value takes place slowly from 1300°C–1600°C. Beyond 1400°C crystallinity of mullite increases.

(C) Correlation between IR data with change of quantity of mullite at increasing temperature

At the occurrence of 980°C exotherm and with formation of mullite at 1000°C the wave numbers for $\nu_{Si\text{-}O}$ shift abruptly from 1012 to 1130 cm^{-1} with structural change (Okada et al., 1986). They noted the absorption ratio of $A_{1130} > A_{1170}$ of this weakly crystalline mullite phase (Figs. 8.1 and 16.1).

Between 1100°C–1150°C, the ratio of absorption did not change significantly. Ratio became more or less equal i.e., the absorption ratio of $A_{1130} = A_{1170}$.

At 1100°C–1300°C, the Al-O-Si band further shifts to higher energy field on gradual heating powders when absorption band at 1170 cm^{-1} and with a shoulder at 1130 cm^{-1} associated with AlO_4 and SiO_4 tetrahedral vibrations are distinctly observed and XRD showed well crystallized mullite formation. Absorption ratio was reversed i.e., $A_{1130} < A_{1170}$.

At 1400°C–1600°C, prominent absorption peaks 1130 cm^{-1} and 1170 cm^{-1} were observed and crystallinity of o-mullite increased significantly.

(D) Correlation between MAS-NMR data with change of quantity of mullite at increasing temperature

Generally at 980°C exotherm, weakly crystalline mullite is the resultant crystalline phase. In ^{27}Al NMR spectroscopy study, preferentially Penta coordinated Al in of the noncrystalline aluminosilicate precursor phase changed into resonance consisting of both tetrahedral and octahedral peaks. The ratio did not resemble mullite fully. The relative proportions characteristic of mullite Al(IV), Al(IV*), and Al(VI) were different from well the crystallized state of mullite. ^{29}Si NMR band during the thermal treatment of this gel at 1000°C showed two bands, large one at ~90 ppm and another small broad peak at –110 ppm.

At the second stage, the ratios of tetrahedra to octahedra was also increased from precursor calcined at 950°C to the value corresponding to well crystallized mullite formed at >1200°C. Broad ^{29}Si resonance at –90 ppm shift to –86, –90, –94 and –80 ppm characteristic of mullite (Jaymes et al., 1995). Another small broad peak at –110 ppm disappears.

(E) Effect of heat treatment of first type of monophasic gels at temperature range between 1400°C to 1600°C

In the above temperature range marked changes of nature of AlO_4 to AlO_6 peaks from broad to narrow by ^{27}Al NMR study and in Si resonance peaks of mullite in ^{29} Si NMR study were seen (Fig. 12.7).

Marked changes in crystallinity of mullite from its incipient stage at ~1000°C to 1400°C and from 1400°C to 1600°C was shown in Fig. 16.9B,C. Secondly changes in the peak width of mullite during heating as shown in Fig. 16.10 along with resolution of 120/210 peak of mullite with increasing heat treatment temperature: (a) 1200°C; (b) 1300°C; (c) 1400°C; (d) 1600°C were observed.

(F) Course of mullitization in other type of monophasic precursors e.g., RH gel, CP gel, Type III, SHIII

Mullitization stages in other types of monophasic gels consist of followings.

After phase separation as like as shown in case of first type of monophasic gel, crystallization of noncrystalline alumina rich silica phase (AS1) at 980°C exotherm to alumina rich Al-Si spinel phase as weakly crystalline variety is evident by XRD.

Solid state reaction between alumina rich Al-Si spinel phase with noncrystalline phases (AS2 and AS3) on continued heating enriches silica, composition of it becomes equivalent to 3:2 mullite at the first step. This spinel is called c-mullite which transforms polymorphically and rapidly to mullite type phase at the temperature of the second exotherm as observed in DTA.

At the second step, inter diffusion starts by Al^{+3} and Si^{+4} cations between (noncrystalline alumina rich mullite residual phase, AS3) and Silica rich aluminous phase (AS2) occurs and nucleates mullite. The growth of mullite continues in the range 1200°C–1400°C.

The next courses of transformation during heating will follow accordingly as per first type of monophasic gel. Development of crystalline order of mullite on further heating from 1400°C to 1600°C occurs.

(G) Course of mullitization in diphasic gels

Diphasic gel in the mullitization region showed that the two intensities of Al-Si spinel phase peaks gradually increased at first and then the intensities drastically decreased with consequent increase in the intensities of mullite at ~1300°C (Figs. 15.2, 15.3). At this temperature, X-ray peaks due to both Al-Si spinel and mullite were still coexisting. Thereafter, Al-Si spinel completely disappeared with predominant crystallization to mullite type phase. The growth curve (Fig. 16.4) of it showed that mullite rapidly formed at ~1300°C and thereafter its growth still continued with further rise of temperature of heat treatment (Chakraborty, 1997). The profile of growth of mullite is analogous to that of kaolinite.

In the case of kaolinite, the growth curve of mullite consisted of two segments and accordingly it was explained earlier that mullitization took place by two simultaneous reaction paths out of two phases namely Al-Si spinel crystallized during exothermic peak temperature at ~980°C and from residual noncrystalline aluminosilicate phase (Chakraborty and Ghosh, 1991). Similar to kaolinite, the mullitization curve of diphasic gel also predicts two paths of mullitization.

Thus growth process of first type, other type of monophasic gels and in diphasic gel show variation mullite formations paths. Growth process of SHI, SHII, and SHIV show two paths of mullite formations. The mullite growth process of mullite gel SHI made by the author consists of two portions, one is very rapid, second one is rather slow and gradual. These two events are more marked in mullite gel SHIV. Accordingly during the formation of mullite, an additional exotherm was exhibited in DTA trace (Fig. 7.20).

Problem of characterization of t-mullite and o-mullite during heating monophasic gels at first exotherm has been discussed elaborately

(1) The explanation of the two plateau regions of mullitization curve from 1000°C–1100°C (Fig. 16.1) and change of *a* value vs. temperature plot from 1100°C–1150°C (Fig. 16.7) Okada et al. (1986) are presented.

Both plots of mullite formation and change of *a* lattice parameter remain almost constant and likely superimpose. In this region solid state between primary mullite with silica rich aluminous phase was not even started and inter-diffusion of Si^{+4} and Al^{+3} among two phases did not take place. This is based on the discontinuous change of *a* value vs. temperature plot of Okada et al. (1986) and the same observation was that of Schneider et al. (1994). Regarding the first step of 980°C phase transformation, it has been concluded as follows.

(2) Cameron's rule could not be applied in determination of composition of primary mullite since rapid change of *a* value is not related to mullite formation; contrary this change of *a* axis is only a function of temperature.

(3) Primary mullite as designated as alumina rich nonstoichiometric metastable mullite remains valid.

In the second step of mullitization, the newly formed mullite phase which occurs above 1200°C by the solid state reaction between heat treated metastable alumina rich mullite with AS2 and AS3 phases is weakly crystalline mullite type phase itself. Change of *a* parameter of this incipient mullite with continued heating between 1200°C to 1400°C may be due to increase of crystallinity i.e., disorder to order transformation. Thus the reason for changes in lattice parameters of mullite at two major stages during the heating process are explained.

In the solid solution range, it is observed that LC data *a* of end mullite of several studies show larger value like 7.552 Å. This result interprets the formation of mullite SS from the earlier stage of crystallization of 72 wt % mullite gels which may be monophasic RH or gel of diphasic character. Gels of higher alumina content synthesized by Low and McPherson (1989). Sales and Alarcon (1996) of course reported high *a* value. The applicability of Cameron's rule is noted in the lattice parameter measurement study in the solid solution range of mullite by Klug et al. (1987) and Klug et al. (1990). It is hopeful that a varying composition of mullite will be made finally out of Al_2O_3-SiO_2 gels of different starting composition in solid solution range.

Therefore, weakly crystalline mullite formed out of SH at first exotherm of nonstoichiometric mullite symmetry and weakly crystalline mullite formed out of RH at second exotherm and from diphasic gel at 1300°C exotherm are likely of mullite type symmetry.

References

1. R. J. Angel and C. T. Prewitt, Crystal structure of mullite: a re-examination of the average mullite structure. *Am. Mineral.*, **71**, 1476–1482 (1986).
2. Jr. P. A. Foster, Nature of alumina in quenched cryolite-alumina melts. *J. Electrochem. Soc.*, **106**, 971–975 (1959).
3. K. Okada and N. Otsuka, Characterization of the spinel phase from SiO_2-Al_2O_3 xerogels and the formation process of mullite. *J. Am. Ceram. Soc.*, **69**(9), 652–656 (1986).
4. K. Okada, N. Otsuka, and S. Somiya, Review of mullite routes in Japan. *Ceram. Bull.*, **70**(10), 1633–1640 (1991).
5. K. Okada and N. Otsuka, Change in chemical composition of mullite formed from $2SiO_2.Al_2O_3$ xerogel during the formation process. *J. Am. Ceram. Soc.*, **70**(10), C245–C247 (1987).
6. T. Ban, S. Hayashi, A. Yasumori, and K. Okada, Calculation of metastable immiscibility in the Al_2O_3-SiO_2 system. *J. Mater. Res.*, **11**, 1421–1427 (1996).
7. T. Ban and K. Okada, Structure refinement of mullite by the Rietveld method and a new method for estimation of chemical composition. *J. Am. Ceram. Soc.*, **75**, 227–230 (1992).
8. W. E. Cameron, Composition and cell dimensions in mullite. *Am. Ceram. Soc. Bull.*, **56**, 1003–1011 (1977).
9. H. Schneider, B. Saruhan, D. Voll, L. Merwin, and A. Sebald, Mullite precursor phases. *J. Euro. Ceram. Soc.*, **11**, 87–94 (1993).
10. H. Schneider, D. Voll, B. Saruhan, M. Schmucker, T. Schaller, and A. Sebald, Constitution of the γ-alumina phase in chemically produced mullite precursors. *J. Euro. Ceram. Soc.*, **13**, 431–448 (1994).
11. R. K. Fischer, H. Schneider, and M. Schmucker, Crystal structure of Al-rich mullite. *Am. Mineral.*, **79**, 983–990 (1994).
12. R. K. Fischer, H. Schneider, and D. Voll, Formation of aluminium rich 9:1 mullite and its transformations to low alumina mullite upon heating. *J. Euro. Ceram. Soc.*, **16**, 109–111 (1996).
13. A. K. Chakraborty, Unpublished work XRD.

14. J. Sanz, I. Sobrados, A. L. Cavalieri, P. Pena, S. de Aza, and J. S. Moya, Structural changes induced on mullite precursors by thermal treatment: a ^{27}Al MAS-NMR investigation. *J. Am. Ceram. Soc.*, **74**(10), 2398–2403 (1991).
15. I. Jaymes, A. Douy, D. Massiot, and J. P. Coutures, Characterization of mono- and diphasic mullite precursor powders prepared by aqueous routes, ^{27}Al and ^{29}Si MAS-NMR spectroscopy. *J. Mater. Sci.*, **31**, 4581–4589 (1996).
16. Y. Wang and W. J. Thomson, Mullite formation from nonstoichiometric slow hydrolysed single phase gels. *J. Mater. Res.*, **10**(4), 912–917 (1995).
17. J. Ossaka, Tetragonal mullite type phase from coprecipitated gels. *Nature (London)*, **19**, 1000–1001 (1961).
18. H. Schneider and T. Rymon-Lipinski, Occurrence of Pseudo tetragonal mullite. *J. Am. Ceram. Soc.*, **71**, C-162–C-164 (1988).
19. H. Schneider, I. Merwin, and A. Sebald, Mullite formation from non-crystalline precursors. *J. Mater. Sci.*, **29**, 805–812 (1992).
20. M. Sales and J. Alarcon, Synthesis and phase transformations of mullite obtained from SiO_2- Al_2O_3 gels. *J. Euro. Ceram. Soc.*, **16**, 781–789 (1996).
21. Y. Hirata, H. Minamizono, and K. Shimada, Property of SiO_2-Al_2O_3 powders prepared from metal alkoxide. *Yogo Kyokai Shi*, **93**(1), 36–54 (1985).
22. F. J. Klug, S. Prochajka, and R. H. Doremus, Alumiosilicate phase diagram in mullite region. *J. Am. Ceram. Soc.*, **70**, 750–759 (1987).
23. Y. Hirata, K. Sakeda, Y. Matsushita, and K. Shimada, Preparation of fine SiO_2-Al_2O_3 powders by hydrolysis of mixed alkoxides. *Yogo Kyokai Shi*, **93**(9), 577 (1985).
24. Y. Hirata, K. Sakeda, Y. Matsushita, K. Shimada, and Y. Ishihara, Characterization and sintering behavior of alkoxide-derived aluminosilicate powders. *J. Am. Ceram. Soc.*, **72**(6), 995–1002 (1989).
25. M. Low and R. McPherson, The structure and composition of Al-Si spinel. *J. Mater. Sci.*, **7**, 1196–1198 (1989).
26. M. Low, and R. McPherson, The origins of mullite formation. *J. Mater. Sci.*, **24**, 926 (1989).
27. M. Mizuno and H. Saito, Preparation of highly pure fine mullite powder. *J. Am. Ceram. Soc.*, **72**, 377–382 (1989).

28. H. Suzuki, H. Saito, Y. Tomokiyo, and Y. Suyama, Processing of ultrafine mullite through alkoxide route. *Ceram. Trans.*, Vol. 6, Edited by S. Somiya, R. F. Davis, and J. A. Pask, *Am. Ceram. Soc.*, Westerville, OH (1990) p. 263.

29. H. P. Rooksby and J. H. Partridge, X-ray study of natural & artificial mullite. *J. Soc. Glass Tech.*, **23**(100), 338–346 (1939).

30. S. Aramaki and R. Roy, Revised phase diagram for the system Al_2O_3-SiO_2. *J. Am. Ceram. Soc.*, **45**(5), 229–241 (1962).

31. H. Yamada and S. Kimura, Studies on co-precipitates of alumina and silica gels and its transformations at higher temperatures. *Yogo Kyokai Shi*, **70**, 87–93 (1962).

32. D. X. Li and W. J. Thomson, Effects of hydrolysis on the kinetics of high temperature transformations in aluminosilicate gels. *J. Am. Ceram. Soc.*, **74**, 574–578 (1991).

33. A. K. Chakraborty, An analysis of the phase evolution of six types of mullite gels. Unpublished (2008).

34. R. Baranwal, M. P. Villar, R. Garcia, and R. M. Laine, Flame spray pyrolysis of precursors as a route to nano-mullite powder : powder characterization ad sintering behavior. *J. Am. Ceram. Soc.*, **84**(5), 951–961 (2001).

35. A. K. Chakraborty, Structural parameters of mullite formed during heating diphasic mullite gels. *J. Am. Ceram. Soc.*, **88**(9), 2424–2428 (2005c).

36. K. Okada, Y. Hoshi, and N. Otsuka, Formation reaction of mullite from SiO_2-Al_2O_3 xerogels. *J. Mater. Sci. Lett.*, **5**, 1315–1318 (1986).

37. C. Gerardin, S. Sundaresan, J. Benziger, and A. Navrotsky, Structural investigation and energetics of mullite formation from sol-gel precursors. *Chem. Mater.*, **6**, 160–170 (1994).

38. S. Komarneni, R. Roy, C. A. Fyfe, and G. J. Kennedy, Preliminary characterization of gel precursors and their high-temperature products by ^{27}Al magic-angle spinning NMR. *J. Am. Ceram. Soc.*, **68**(9), C-243–C-245 (1985).

39. A. Yasumori, M. Iwasaki, H. Kawazoe, M, Yamane, and Y. Nakamura, Nuclear magnetic resonance study of the structure of aluminosilicate gel and glass. *Phys. Chem. Glasses*, **31**(1), 1–9 (1990).

40. I. Jaymes, A. Douy, P. Florian, D. Massiot, and J. P. Coutures, Mew synthesis of mullite. Structural evolution study by ^{17}O, ^{27}Al and ^{29}Si MAS NMR spectroscopy. *J. Sol-Gel Sci. Technol.*, **2**, 367–370 (1994).

41. A. K. Chakraborty, Range of solid solutions of silica in spinel type phase. *Adv. Appl. Ceram.*, **105**(6), 1–7 (2006).
42. I. Jaymes, A. Douy, and D. Massiot, Synthesis of a mullite precursor from aluminum nitrate and tetraethoxysilane via aqueous homogeneous precipitation: an ^{27}Al and ^{29}Si liquid- and solid-state NMR spectroscopic study. *J. Am. Ceram. Soc.*, **78**(10), 2648–2654 (1995).
43. A. K. Chakraborty, Reinvestigation of Al-Si spinel phase in diphasic Al_2O_3-SiO_2 gel. *J. Am. Ceram. Soc.*, **88**(1), 134–140 (2005a).
44. I. Jaymes and A. Douy, Homogeneous precipitation of mullite precursors. *J. Sol-Gel Sci. Technol.*, **4**, 7–13 (1995).
45. A. K. Chakraborty and S. Das, Al-Si spinel phase formation in diphasic mullite gels. *Ceram. Int.*, **29**, 27–33 (2003a).
46. A. K. Chakraborty, Characterization of monophasic and diphasic mullite precursors by solid state reaction study. *Br. Ceram. Trans.*, **103**, 33–36 (2004).
47. G. Engelhadt and D. Michel, *High Resolution Solid State NMR of Silicates and Zeolites*. John Wiley and Sons, New York, 1987, p. 54.
48. S. H. Risbud, R. J. Kirpatrick, A. P. Taglialavore, and B. Montez, Solid state NMR evidence of 4-, 5-, 6-fold aluminium sites in roller-quenched SiO_2-Al_2O_3 glasses. *J. Am. Ceram. Soc.*, **70**, C10–C12 (1987).
49. Y. X. Huang, A. M. R. Senos, J. Rocha, and J. L. Baptista, Gel formation in mullite precursors obtained via tetraethylorthosilicate (TEOS) pre hydrolysis. *J. Mater. Sci.*, **32**, 105–110 (1997).
50. H. Schneider and K. Wohlleben, Microchemical composition and cell dimensions of mullite from refractory grade: South American bauxites. *Ceram. Int.*, **7**, 130–136 (1981).
51. K. Okada and N. Otsuka, Formation process of mullite. *Ceram. Trans.*, **6**, 375 (1990).

Chapter 20

Emergence of Only Three Routes of Phase Transformation Sequences of Mullite Precursors and Critical Analysis of Intermediates

20.1 Introduction

As mentioned in Chapters 2–7 and 12, different methodologies particularly variation of content of hydrolysis water, pH, aging time and temperature of gelation inclusive the choice of two components as source have been tried by various authors with a view to synthesized mullite precursors. Some researchers use TEOS and ANN to synthesize two types of precursors; some use the same components and prepare three types precursors; and some others follow different methodologies for precursor syntheses. Some other amorphous mullite precursors follow four different types of crystallization paths and these depend on the scale of homogeneity of the amorphous phase as per review of Douy (2006). In these ways, a huge volume of precursor developments have been synthesized in various routes by authors of the mullite community. Thermal sequences of these different mullite precursors are found to be different as revealed by the studies of DTA analysis, qualitative and quantitative XRD, IR and TEM/EDS and MAS-NMR etc.

Mullite Formations: Analysis and Applications
Akshoy Kumar Chakraborty

ISBN 978-981-4877-05-3 (Hardcover), 978-1-003-03167-3 (eBook)
www.jennystanford.com

On the basis of various methods of synthesis of precursors, and their phase transformation studies, maximum possible routes of mullitization processes are needed to ascertain. To accomplish it, the author has prepared as many as six different types of precursors out of the most popular source of raw materials (e.g., TEOS and ANN). A variation in paths of thermal evolution of these six mullite precursors are already shown in Chapters 7 and 16 and are compared with transformation processes of above mentioned precursors of earlier researchers. In this chapter the following are highlighted.

1. Views of mullite formations in different cases along with these six precursors are discussed and generalized in contest previous predictions.
2. The vast methodologies of synthesis of precursors made by different researchers are put and accommodated into maximum three possible routes namely, (i) acidic route, (ii) basic route and (iii) pyrolysis routes.
3. It is also demonstrated that whatever be the components used, and processing techniques followed, final structural evolution processes of mullite will lie in either any of the six evolution processes as shown by author.

20.2 Mullite Precursors Synthesized by Acidic Medium as First Route of Mullitization and Their Phase Evolution

In this chapter, phase transformations of all precursors are presented based on the critical analysis as shown in previous two chapters (Chapters 18 and 19) regarding final characterization of spinel phase as Al-Si spinel and formation of two kinds of mullites at first exotherm and at second exotherms regions as metastable nonstoichiometric mullite and mullite type respectively.

20.2.1 Phase Transformation of Aqueous Mullite Gels Showing Complete Mullitization Reaction in a Single Step at 980°C Exotherm

Monophasic gel prepared out of silica sol and aluminum sulphate formed poorly crystalline mullite type phase at the 980°C exotherm only (Chakraborty and Ghosh, 1988). It was further noted by XRD that the intensity of 980°C exotherm and the quantity of mullite formed are directly proportional. Both the DTA intensity and XRD intensity of mullite are dependent on pH of aqueous solution. This mullite type phase transformed to o-mullite on heating >1400°C.

20.2.2 Phase Transformation of Monophasic Gels Showing Mullitization Reaction Mostly in a Single Step at 980°C Exotherm

Monophasic gel of (Hoffmann et al., 1984); single phase gel (SH) of (Okada et al., 1986); polymeric gel of (Yoldas, 1992); molecular gel of (Huling and Messing, 1989, 1991); Type I gel of (Schneider et al., 1993) and SHI gel of (Haque, 2000) showed (i) the exhibition of most significant 980°C exotherm in DTA analysis, and (ii) direct crystallization of a large quantity of weakly crystalline metastable mullite as major phase and Al-Si spinel as minor phase in the XRD study. Accordingly, it is concluded that all these monophasic mullite precursors SH, Polymeric, Type I, and (Precursors A, C and E of Jaymes et al., 1996) show similar phase evolution behavior on heating and follow similar route of mullitization. Controversial primary mullite phase weather it is t-mullite or o-mullite as raised by different authors is already discussed elaborately and presented in Chapter 19 and its composition is also critically analyzed there. Henceforth, it will likely be designated as metastable nonstoichiometric mullite.

Using a similar source of two components (ANN and TEOS), six different kinds of precursors namely SHI, SHII, SHIII, SHIV or CP6, CP9 and spray dried (SD) with variation of process parameters were synthesized by Haque (2000) and later on by Chakraborty (2008). This single method resembles six separate processes tried by above mentioned authors. Six mullite precursors showed different characteristics when studied and correlated by DTA, XRD, IR and MAS-NMR. During studies in phase evolution processes, a considerable amount of noncrystalline aluminosilicate phase was noted which remained as residue in all those cases of gels during heating at first exotherm (Haque, 2000). This residual phase was associated with metastable mullite phase in case of heating SHI gel; metastable mullite phase and alumina rich Al-Si spinel phase in case of heating SHII gel and in SHIII gel (Table 20.1).

Table 20.1 Phase composition of monophasic and diphasic gels during heating process at four exothermic peak temperatures

	Phases developed			
Nature of gel	**Phase developed at 980°C exotherm (first exotherm)**	**Phase developed at 1150°C exotherm (third exotherm)/phase developed at 1250°C exotherm (usual second exotherm)**	**Phase developed at 1300°C exotherm (fourth exotherm)**	**Phase developed at 1600°C**
SHI	M (metastable nonstoichiometric alumina rich mullite), major + AS (noncrystalline aluminosilicate phases, AS2 & AS3) (major)	Mullite type phase (weakly crystalline major phase)		o-mullite
SHII	M, major + Alumina rich Al-Si spinel (SP), minor + AS (AS2 & AS3) major	Mullite type phase (weakly crystalline major phase)		o-mullite
SHIII	Alumina rich Al-Si spinel (major) + Partial formation of alumina rich mullite (trace) + Noncrystalline silica rich aluminous phase (AS2) + Noncrystalline alumina rich siliceous phase (AS3)	Two consecutive exothermic peaks at 1150°C, and at 1250°C respectively due to (i) Mullite type phase formation at intermediate stage at third exotherm at 1150°C. (ii) Mullite type phase formation at intermediate stage at second exotherm (1250°C).		o-mullite

SHIV/CP6 prepared at pH = 6	Alumina rich Al-Si spinel (major) + Noncrystalline silica rich aluminous phase (AS2) + Noncrystalline alumina rich siliceous phase (AS3)	Two consecutive exothermic peaks at 1150°C, and at 1250°C respectively due to (i) Mullite type phase formation at intermediate stage at third exotherm at 1150°C. (ii) Mullite type phase formation at intermediate stage at second exotherm (1250°C).		o-mullite
"In situ" diphasic gel prepared at pH = 9	Absence of 980°C exotherm. At 900°C Al-Si Spinel (minor), γ-Al_2O_3 (minor). Noncrystalline silica rich aluminous phase (major) formed between 400°C–1000°C. At 1000°C-Alumina rich Al-Si spinel (minor) + Noncrystalline silica rich aluminous as major phase.	Exhibition of a broad exotherm due to crystallization of Al-Si spinel (cubic mullite) as major phase between 1000°C–1300°C + Noncrystalline aluminosilicate (as residual phase)	Al-Si spinel (cubic mullite) -to- Mullite type phase (major) formation due to fourth exothermic peak temperature (1300°C) + Noncrystalline aluminosilicate (as major residual) phases	o-mullite
SD	Mullite type (major) phase			o-mullite

The tentative phase transformation of SHI where mullite formation is predominantly high and Al-Si spinel is a minor phase is given in Chapter 16 vide Eqs. 16.2–16.5. Whereas SHIII crystallizes to Al-Si Spinel (major) phase, mullite (minor) phase and with considerable amount of noncrystalline aluminosilicate phase. Mullite formation of it was given in Eqs. 16.7–16.8 in Chapter 16. Factors affecting those transformations are explained in this chapter as due to development of the former noncrystalline aluminosilicate phase as given at primary stage.

Role of residual noncrystalline aluminosilicate phase:
As like Okada et al. (1986), the mullite formation process during heating in the high temperature region in the above case of gel was indicated by Haque (2000). It was explained that the gradual rise of mullite at that region was due to further mullitization from residual noncrystalline phase (AS2) with minor quantity of Al-Si spinel phase. By alkali leaching study it was concluded that this amorphous phase is most likely noncrystalline aluminosilicate phase (Table 16.1 and Fig. 15.12) other than free silica (A) alone as assumed by various authors. Therefore, mullite formation in case of SHII should consist of two parallel paths. The first rapid step is undoubtedly due to formation out of noncrystalline aluminosilicate precursor phase at 980°C to metastable mullite phase. The second gradual step on heating >1100°C accounts for mullitization out of the residual noncrystalline aluminosilicate phases (AS2, AS3) and Al-Si spinel phase remained after the exotherm as shown in (Fig. 12.6) of Chakraborty (2008).Therefore, monophasic precursor phase made out of same TEOS and ANN as components during heating at 980°C exotherm did not directly transform to mullite in full extent, residual precursor still contents in association with a lot of noncrystalline alumina rich silica phase and noncrystalline silica rich aluminous phase, respectively. Huling and Messing (1992) showed the microstructure of alumina-silica gel heated to 900°C/2 h. It displayed an interconnecting two-phase morphology characteristic of a phase separated material. This view was also noted in rapidly quenched aluminosilicate glasses (MacDowell and Beall, 1969). On heating at 1000°C alumina rich amorphous phase transformed predominantly to metastable mullite phase. The role of the remaining silica rich phase and residual alumina rich mullite

phase would be the solid state interaction and lead to mullitization on continued heating. Variations in phase evolution regarding (i) the ratio of nonstoichiometric mullite to Al-Si spinel phase formed at 980°C and (ii) residual noncrystalline aluminosilicate phases as noticed apparently in above monophasic gels are due to processing method used in gel synthesis e.g., pH persisted and water/alcohol as hydrolysing agents involved during synthesis beside the choice of two components.

20.2.3 Formation of Noncrystalline Aluminosilicate Phase is a Function of pH on in Monophasic Gels e.g., SHI, SHII, SHIII and SHIV

Chakraborty (2003a, 2004) indicated the absence of silica (A) as a residual phase in monophasic gels heat treated to 980°C. On the other hand, noncrystalline aluminosilicate phases are developed as the accompanying phases in heated products of mullite precursors as shown in Table 20.1. The amount of their formation is assessed from the mullitization growth curve (Fig. 12.6) and it is found to be in the following order.

SHI < SHII < SHIII< SHIV.

pH noted during the gelation process of those gels is as follows.

SHI = 4, SHII = 3 – 2.8, SHIII = 2.8 – 2, SHIV = 6.

These results show that the remainder part of noncrystalline silica rich aluminous phase present during crystallization of noncrystalline aluminosilicate precursor phase at first exotherm of above four precursors is pH related. It is most predominant in the case of SHIV when gelation was done at pH = 6 and it is least in SHI when it was made at pH = 4. With further decrease of pH the formation of noncrystalline silica rich aluminous phase increases. Therefore, the ultimate mullite formation in case of monophasic SH gels may occur at high temperature in multistep stages out of three-four intermediate compounds generated at ~1000°C. For examples,

(i) Residual noncrystalline phase alumina rich silicate (AS2) may undergo solid state reaction with noncrystalline silica rich aluminous phase (AS3) to form o-mullite during continued heating.

(ii) Noncrystalline aluminosilicate phase (AS1) may itself nucleate to crystallize into metastable nonstoichiometric mullite.
(iii) Al-Si spinel phase although present in a minor amount may transform to o-mullite by reacting with AS2 on further heating.
(iv) Noncrystalline aluminosilicate phase (AS2) may react with metastable primary mullite on the course of heating.

Mullite formation out of later phase found to increase with increase of pH maintained during gelation. Since predominance of mullitization in high temperature regions is more in SHIII than in SHII gel than in SHI gel. In conclusion, it is predicted that primary mullite and AS phases were formed at ~ first exotherm in case of SH gels. Quantity of formation of spinel phase follows in the order as

$$\text{SHII} < \text{SHIII} < \text{SHIV}.$$

On further heating, alumina rich AS may react with silica rich AS to develop mullite type phase.

20.2.4 Formation of Noncrystalline Aluminosilicate Phase Based on Water of Hydrolysis Used in Monophasic Gels

Example 1

In gel marked G-104 of Chakraborty and Ghosh (1988): It is thought that during hydrolysis of TEOS, water aids in polymerization of silica micelles and results in generation of large polymeric growth. These bigger silica units possibly hinder o-mullite formation at 980°C and rather develop noncrystalline aluminosilicate phase in large amounts. It was also assumed that the polymerized silicic acid would form both Al-Si spinel phases besides formation of a substantial quantity of later phase. To suppress polymeric growth of silica unit, an attempt was made by this author to remove the water of crystallization part of ANN by repeated drying of it prior to preparation of SHI. However, the result showed that although Al-Si spinel formation is avoided in SHI but residual noncrystalline aluminosilicate phase was still formed in minor quantities. Since mullite formation in the temperature above 1000°C grew gradually and this accounts for mullitization out of this phase. The decrease of DTA exothermic peak intensity with increase of quantity of alcohol was shown by Hoffman et al. (1984). However, the cause of the relationship between ratio of quantity of mullite + Al-Si spinel developed vs. the amount of alcohol used was explained as due to formation of a large volume of gel at the gelation stage. Reasonably, higher amounts of alcohol may lead to liberation of large quantities of noncrystalline aluminosilicate phase with the reduction of mullite phase during heating at exothermic stage as shown here.

Example 2

Three kinds of Al_2O_3-SiO_2 gels out of TEOS and AlOBu with variation of water content, and marked as G-152, G-152 (i) and G-152 (ii) were synthesized by Chakraborty (1996). Three other kinds of gels marked as G-150, Gel-A and G-31 out of TEOS and ANN by varying water contents were also prepared by Chakraborty (1994a). Water affects the growth of silicon hydroxide gel species (see Chapter 7). Various authors use hydrolysis water to different extent which also shows a direct relationship with Al-Si spinel crystallization (Table 20.2).

Table 20.2 Effect of water on the synthesis of mullite precursors by coprecipitation method and their phase evolution

1. Prochazka and Klug (1983) used TEOS, aluminium isopropoxide in cyclohexane and hydrolyzed with water.	Noted spinel phase formation at ~950°C which proceeded mullitization.
2. Suzuki et al. (1990) followed partial hydrolysis method of TEOS in acidic water and mixed with aluminium isopropoxide.	It crystallized to Al-Si spinel at 1000°C.
3. Yoldas and Partlow (1988) synthesized sample C (as water free approach).	It crystallized to mullite with a sharp first exothermic peak.
4. Schneider et al. (1993) synthesized Type I of mullite gels from AlOBU and TEOS (as per Yoldas and Partlow, 1988).	It also formed mullite at the strong 980°C exotherm.
5. Taylor and Holland (1993) synthesized GTW and GNW from TEOS and Al $(OPr)_3$ difference in the content of water content.	Showed differences in phase and exothermic peak exhibitions.

6. Effect of water and pH both: Schneider et al. (1993) synthesized Type III of mullite gels from AlOBu and TEOS. With water and at pH > 10.	It crystallized into a spinel phase with exhibition of reduced 980°C exotherm. A broad second exotherm at ~1250°C due to mullite formation occurred.
7. Hsi et al. (1989) synthesized gel at four different pH values namely 8.3, 9.5, 10.1 and 10.4.	Ratios of pseudoboehmite to barite formation is a function of pH value. Obviously, phase morphology and transformation during calcination were different.
8. Mitachi et al. (1990) synthesized sample Powder A and B in different pH solutions.	Both 980°C exotherm and mullite formation pathways were different.
9. Hirata et al. (1985) studied the effect of pH on coprecipitation out of SiO_2-Al_2O_3 powder from TEOS and aluminium isopropoxide.	Sample A prepared at pH = 8.2 was amorphous. Contrary, sample 2 made at pH >1.9 showed barite initially. It formed a spinel to temperature as high as 1200°C, later developed a large quantity of mullite at ~1300°C.

Example 3

The source of silica also affects the development of 980°C exotherm and mullite formation as shown in Table 20.3. Variations of both two are shown below which are self-explanatory.

Table 20.3 Effect of silica source on development of 980°C exotherm and mullite formation

1. $SiCl_4$ as source: Horte and Weigman (1956) obtained gelatinous precipitates by using ammoniated water out of aluminium and silicon chloride solutions.	Observed a sharp exothermic peak at ~980°C in DTA.
2. $SiCl_4$ as source: Demediwk and Cole (1958) prepared gels by mixing $SiCl_4$ and $AlCl_3.6H_2O$ followed by cohydrolysis with NH_4OH solution.	Showed highest DTA exotherm at the composition of 3:2 mullite.

3. Sodium silicate as source: Ossaka (1961) took sodium silicate potassium aluminium sulphate solution and coprecipitated them with hexamethylenetetramine and water.	DTA showed a sharp 980°C exothermic peak and formed mullite.
4. Sodium silicate as source: Insley and Ewell (1935) used sodium silicate and aluminium sulphate as raw materials and sodium hydroxide as neutralizing agent. Washed gel contained traces of soda.	Al_2O_3:SiO_2 ratios of gels of 1:4, 1:2, 1:1, 2:1 and 4:1 showed the difference in DTA exotherms and phase transformation behaviors.
5. Silica sol as source: Chakraborty and Ghosh (1988) freed soda content by ion exchange technique of silica sol prepared out of sodium silicate solution.	Independency of crystallization behavior of β-cristobalite at 1060°C and α-Al_2O_3 at 1130°C in DTA are lost. On the contrary, a new exotherm was seen at ~980°C due to mullitization alone.
6. Silica sol as source: Schneider et al. (1992) synthesized coprecipitated material (CM) out of sodium aluminate and silica sol and pH = 7 was adjusted by sulphuric acid for coprecipitation.	Spinel phase (minor) with mullite phase (major) are formed at first exotherm at 988°C in DTA. It showed a broad endotherm followed by extensive dehydroxylation of final OH^{-1} groups.
7. $SiCl_4$ as source: Jaymes et al. (1994) dissolved silicon tetrachloride and aluminiun isopropoxide as source components in tetrahydrofuran (THF) and hydrolyzed by ^{17}O enriched water.	Showed mullite at first exotherm only

As a result of variations of (i) pH; (ii) water content; and (iii) source of silica component, the 980°C phase development of different types of mullite gels likely vary. It consists either of mullite as a major phase and Al-Si spinel as minor phase or of reverse order. As water helps formation of Al-Si spinel phase, it is conjectured that the mullite gel would be initially diphasic in character. Secondly, with more and more increase in concentration of water Al-Si spinel formation increases, the resultant gel becomes enriched with more quantity of diphasic gel i.e., a mixture of gels consisting of mono and diphasic character gradually develops. Accordingly, on heating the ratio of XRD intensities of mullite to Al-Si spinel varies during 980°C exotherm at the first stage. Subsequent transformations to mullite at the second stage also show variations during continued heating at >1250°C in the so-called mixed precursors.

20.2.5 Phase Transformation of (SHIII) Gel Out of Six Types of SH Mullite Gels: As a Representative Example

In some previous cases e.g., monophasic gel G-12 prepared by Chakraborty (1988); Polymeric gel synthesized by Rajendran et al. (1990); Nitrate derived gel made by Yoldas (1992); SGM material of Schneider et al. (1992), formation of Al-Si spinel phase (major amount) and o-mullite (minor amount) were noted to crystallize at ~1000°C. Precursor marked MP of Fonesca et al. (1997); Polymeric gel made by Cassidy et al. (1997) by their Nitrate method used ammonia as co precipitating agent and showed the crystallization of Al-Si spinel phase (major amount) together with metastable mullite (minor amount). Formation of these two phases with similar ratio during sharp exotherm at 980°C were also noted by Hirata et al. (1989) and (Haque (2000) vide Table 20.1 out of his monophasic gel SHIII synthesized by using TEOS and ANN as components. Besides formation of two crystalline phases namely Al-Si spinel phase (major) and mullite (minor), formation of much quantity of noncrystalline phase was detected by MAS-NMR studies of several researchers. For example, an ideal case of transformation of SGM precursor is cited here.

At 600°C–800°C, SGM precursor showed an amorphous XRD pattern. ^{29}Si NMR study showed symmetrical ^{29}Si resonance at ~92 ppm. ^{27}Al NMR study showed bands at ~2, 30, and 58–60 ppm, respectively with low tetrahedrally coordinated Al.

Above 900°C Si resonance split into two peaks at –90 and –110 ppm. These are indicative of a separation into silica poorer (~ –90 ppm) and silica richer-domains (~ –110 ppm). XRD showed spinel (major) and of mullite (minor) formations at 900°C. DTA trace showed an exothermic peak at 980°C. Al resonance showed octahedral and tetrahedral resonances, 30 ppm peaks disappeared. Thus amorphous resonance collapses into intermediate reaction type resonance. ^{29}Si resonance peak at 90 and 110 ppm peaks were clumped together at 1000°C. Si resonance also displayed a big shoulder of low intensity at −80 ppm. Octahedral peak shifted to 8 ppm, tetrahedral peak moved to 70 ppm. Therefore, it is concluded that noncrystalline aluminosilicate phase is one of the reaction products of decomposition of mullite precursor. The presence of noncrystalline aluminosilicate phase at ~1000°C should be accounted for. It is very much in the system even in diphasic gel or monophasic gel/precursors which showed only spinel or major spinel phase development at 1000°C. Thus prior to mullitization, there are four considerable phases, (i) Al-Si spinel phase; (ii) alumina rich mullite phase (AS3) as residue; (iii) Silica rich aluminous phase (AS2); and (iv) weakly crystalline mullite type phase. The phase transformation of this gel (SHIII) on the due course of heating consisting of such phases was given in final Eqs. 16.7–16.8.

20.2.6 Phase Transformation of Polymeric Gels Showing Mullitization Reaction in a Single Step at 980°C Exotherm

(A) Crystallization sequence of polymeric gel

Polymeric mullite gel was first made out of TEOS/AlOBu by hydrolysis and cogelation in neutral medium (Yoldas, 1992). Fundamental transformation of it occurred between 950°C–1050°C. On heating, its aluminum environment was pentahedraly coordinated at 500°C. It transformed next at 950°C to 1050°C, when 5-coordinated aluminum sites were eliminated and it changed to two broad peaks comprising 4- and 6-coordinated Al sites nearly attributed to the characteristic of mullite MAS-NMR spectra. Heating to high temperature is required to convert to o-mullite.

(B) Crystallization sequence of Type I precursor

Type I gel as per sample C of Yoldas and Partlow (1988) was synthesized by Schneider et al. (1993) as shown by MAS-NMR is given below. At the first stage between 350°C–900°C, it transformed to noncrystalline aluminosilicate precursor phase which was amorphous in character by XRD. Si resonance showed an intense and sharp symmetric peak at –85 ppm.

At 900°C, Si resonance peak shifted from –85 to –90 i.e., to more negative value at second stage. DTA exhibited a sharp and strong exotherm at 980°C and no peak at 1250°C. At 1000°C, it crystallized to a poorly crystallized metastable mullite phase (abundant quantity) with Al-Si spinel (minor) phase. Correspondingly, ^{29}Si resonance remained almost unaltered at –90 ppm. Second characteristic resonance peak newly appeared at ~110 ppm accounts for noncrystalline silica rich aluminous phase. Heating to high temperature is required in this case also to convert to o-mullite.

(C) Crystallization sequence of GNW

GNW is a polymeric gel synthesized at different pH and water conditions by Taylor and Holland (1993). In the amorphous stage, xerogel contained predominantly pentahedral coordinated Al. Instead of forming spinel phase as intermediary, crystallization of metastable mullite phase took place at 980°C exotherm. Correspondingly, ^{27}Al MAS-NMR spectra of it showed two types of tetrahedral aluminium resonances.

(D) Crystallization sequence of MI

MI is a polymeric gel synthesized (as per Yoldas using pre hydrolyzed TEOS) by Geradin et al. (1994) shows transformation in the following stages.

At the first stage, Si spectra of MI below 900°C, were characteristic of amorphous aluminosilicate structures with random distribution of silicon and aluminum. Local structure was described as $Si(OAl)_n(OSi)_{4-n}$ where n varies between 1 and 4. They noted predominant five coordinated environments just before mullite crystallization.

At the second stage, during its crystallization of their M1 precursor, three types of environments that are characteristic of mullite [Al(IV), Al(IV*), and Al(VI)] were noted but in proportions different from those found in completely crystallized mullite.

At the third stage, the ratios of AlO_4 to AlO_6 groups were also found to increase with heating from 950°C to 1200°C as per the NMR spectrum of MI. In the Si spectrum of M1 heated to 1100°C, the peak at –86 ppm and the shoulders at –90 and –93 ppm correspond to mullite, though the feature was broader than observed for completely crystallized mullite. With increasing temperature to 1200°C features characteristic of mullite get sharper, indicative of more crystalline order. At 1650°C, the chemical shifts at –86.2, –93.7, and a shoulder at –89 ppm were in good agreement with synthesized mullite.

(E) Crystallization sequence of S20 precursors

Instead of using inorganic or organic aluminium salts Yasumori et al. (1990) chose an aluminosilicate compound silicon aluminum ester (SAE) and usual TMOS for synthesis of aluminosilicate xerogel (S20) and studied for its evolution on heating by NMR spectroscopy. The spectrum of it showed a broad peak in the range –70 to –120 ppm. Al spectra showed mostly fourfold coordinated Al. During gradual heating, sixfold coordinated Al at 0 ppm was decreased whereas fourfold coordinated Al at 55 ppm was increased. On heat treatment at ~1000°C, a drastically sharp decrease in Si resonance peak was noted and spectrum became to –111 ppm. The fivefold coordinate Al at 30 ppm which appeared during heating at 950°C disappeared. Precipitation of broad mullite peaks during heating at ~1000°C were observed and the peak for Al ions in fourfold coordination split into two peaks at 47 and 65 ppm. During isothermal heating at 1000°C, the positions of all peaks remained unchanged, while the ratio of octahedral Al ions decreased, the peak heights of the two peaks for tetrahedral Al ions increased and became similar to those in the spectrum of mullite crystal with a tendency opposite to exhibit any metastable phase separation.

(F) Crystallization sequence of (precursor A, C and E)

These precursors of Jaymes et al. (1996) transformed at first exothermic peak temperature to nonstoichiometric mullite and spinel phase (trace) amounts, respectively. Sequences of phase evolution of these monophasic precursors may be similar in nature as those of GNW of Taylor and Holland (1993), MI of Geradin (1994) as shown above.

20.3 Mullite Precursors Synthesized at Basic pH as Second Route of Mullitization and Their Phase Evolution

20.3.1 Mullite Formation Out of Diphasic Gel Synthesized by General Method Using Boehmite Sol Considering Alumina and TEOS, Ludox/Silica Sol as Source

Components

Details of synthesis of diphasic gels are already shown in Chapter 8 and their phase evolution, characterization of intermediates and lastly mullite development processes are discussed in Chapter 16. Readers may go through there. Al_2O_3-SiO_2 gels those developed spinel phases on heating should be termed henceforth as diphasic in character. Accordingly, all these gels mentioned below are characterized as diphasic in nature. General diphasic gel was made by use of pseudoboehmite (based on choice of it as one component). Other gels derived out of the following ways are also diphasic in character. For example, precursor M3 which was made by adding TEOS directly to an aqueous dispersion of boehmite and stirred for 24 h by (Geradin et al., 1994).

20.4 Redesignation of Al_2O_3-SiO_2 Gels Synthesized by Earlier Researchers

There are three kinds of diphasic gels based on the detailed review and critical analysis of phase transformations studies of these by various researchers are given below.

20.4.1 Coprecipitated Gels Prepared by Using ANN and TEOS by NH_4OH at pH = 6 is to be Designated as Diphasic Gel: First Kind

Gels made using ANN as component and coprecipitated by NH_4OH are called as diphasic gel of first kind since aluminum component was not precipitated as either boehmite or bayerite during coprecipitation process using ammonia e.g., RH synthesized by Okada and Otsuka (1986), CP gel made by Chakraborty (1997) are designated as first kind diphasic gel.

(1) Transformation of RH gel: first kind

This diphasic gel of first kind exhibited a small 980°C exotherm in DTA with formation of Al-Si spinel phase as noted by XRD analysis. The later phase started decreasing at ~1100°C and formed mullite. At 1150°C, mullite developed predominantly (Okada et al., 1986). Sequence of phase evolution of RH gel is given in Eqs. 16.7–16.8 in Chapter 16.

(2) Transformation of SHIII and CP6: first kind

At 1000°C, the spinel phase formed in case of CP6 of Haque (2000) may be alumina rich and it is forming along with AS2 and residual AS3 phases. With continued heating it may be assumed that incorporation of Si into the spinel phase continued and the composition of spinel become analogous to composition of 3:2 mullite. Only after that Al-Si spinel transforms to a mullite type phase. Before elucidating the thermal change in this case, it is essential to know the reasons for the three exotherms which are found in the DTA curve of this gel (Fig. 7.20). The exothermic peaks namely 980°C, ~1150°C and 1250°C correspond with those occurring in the DTA trace of a kaolinite clay as shown in (Fig. 7.7 and/or Fig. 2.2) of Chakraborty (1994b) (see Table 20.1).

Thermal transformation sequence of the aluminosilicate gels was found to be analogous to that of kaolinite and this view was proposed by other earlier researchers namely Insley and Ewell (1935) for some of their coprecipitated gels. All these coprecipitated gels gave 980°C exothermic effects and the peaks were of the same character and occurred at the same temperature (see Chapter 2). Furthermore, heat treatment of these mullite gels at constant temperatures just below 925°C resulted in reduced intensity of the heat effect at 980°C which just happened as in kaolin minerals. Besides thermal analysis data, X ray results also showed that both coprecipitated gels and kaolinite were amorphous in the temperature range from about 700°C up to the start of the exothermic peak, when heated at the rate of 6°C/min. Phase changes in both cases are similar as shown by the later author. With these observations, they concluded that dehydrated kaolin and coprecipitated gel of the same composition ($Al_2O_3 \cdot 2SiO_2$) are very similar in internal structure. The reason for this exotherm for metakaolin was explained by them as due to either mullite formation or to Al-Si spinel formation. This result indicates that the thermal transformation phenomenon of CP6 bears a close similarity to the high temperature reaction sequence of kaolinite. Regarding phase transformation of a kaolinitic clay, it is shown by Chakraborty and Ghosh (1991) that kaolinite at 980°C exotherm formed 20–25 wt % Al-Si spinel and ~40 wt % aluminosilicate phase (A) as residue. Besides evaluating the composition of spinel phase, Chakraborty and Ghosh (1987) iterated the fact that noncrystalline aluminosilicate phase invariably present as one of the constituent when their mixed Al_2O_3-SiO_2 gel was heated at ~1000°C. It was also shown that when the above gel was heated, the independent crystallization sequence of alumina component was modified in presence of SiO_2 and vice versa.

Al-Si spinel and noncrystalline aluminosilicate phase exhibited two consecutive exotherms during precise DTA analysis (Fig. 7.20) as shown further by Chakraborty (2008). It is also noted that intensity of mullite increased after completion of the exotherms. In the analogous study, CP6 first formed Al-Si spinel phase at the 980°C exotherm like kaolinite and then transformed to mullite type phase at 1250°C. Furthermore, the mullite growth curve of CP6 (Fig. 12.6) consists of two stages. In the first stage, mullite formation occurred rapidly at 1200°C–1250°C and thereafter in the second stage it increased gradually with rise of temperature. Thus occurrence of the DTA exotherm at 1250°C coincides with the formation of the mullite type phase from the complete elimination of the Al-Si spinel phase as shown by XRD analysis (Fig. 2.2 and/or Fig. 12.6). Thus, mullitization in the case of CP6 is analogous to kaolinite and it follows two reaction paths; (i) Al-Si spinel formed earlier at 980°C, transforms to mullite at 1250°C, constituting path IV; and the exhibition of the new exotherm at 1150°C could be explained as being due to mullitization in the aluminosilicate (A) phase as in path V, similar to kaolinite as was reported earlier. This step proceeded at a comparatively slower rate than path IV. The initial step portion accounts for mullite formation by path IV. (ii) Subsequent gradual rise is due to mullitization from the aluminosilicate (A) phase by nucleation and subsequent crystallization by path III.

Mullite formation in the previous case of gel e.g., SHIII would be similar to CP6 and it would be two-step in nature as like the above case of gel. Only difference lies in nature and quantity of different phases developed at 980°C and on their individual phase transformation or solid state reaction among them.

(3) Transformation of some coprecipitated gel of made by double alkoxide system: first kind

Coprecipitated gels synthesized by Prochazka and Klug (1983); Klug and et al. (1987) and three precursors (A, B and C) synthesized by Yamada and Kimura (1962) which did not contain boehmite initially which is a diphasic precursor of (first kind) and transformed to purely Al-Si spinel phase during first exotherm. The later phase further transformed into mullite during the second exotherm at ~1250°C. The coprecipitated gel of these authors followed purely two steps of transformation processes.

(4) Transformation of coprecipitated gel made at pH < 8

Some precursors were prepared with addition of ammonia such that pH < 8 so that no precipitation of crystalline aluminum hydroxide and or pseudocrystalline boehmite was formed. e.g., Hirata et al. (1985) choose pH = 8.2; and gels made by Croft and Marshall (1967); Mcgee and Wirkus (1972); Perry (1973) are diphasic precursor of (first kind).

(5) Transformation of colloidal gel: first kind

Colloidal gel (CG) synthesized by (Yoldas, 1979; Yoldas and Partlow, 1988; and Pask et al., 1987) behaves analogously to that of diphasic gel. During prehydrolysis, to one mole of AlOBu, an excess quantity (~100 mole) of water was used at 70°C when a thick slurry (which is undesirable) was formed first. Later on, it was peptized with dilute HNO_3 solution to a clear sol. The sol contained large colloids of 10–30 nm size of aluminium hydroxide. To it, TEOS was added, the resulting gel was colloidal in nature and did not show boehmite by XRD at the gel state and followed the course of diphasic character in thermal transformation behavior. Absence of a 980°C exothermic phenomenon was noted. Secondly, spinel formation was reported over a range of temperature and lastly mullite formation took place after a broad exothermic phenomenon at ~1270°C in DTA. This colloidal gel is designated as diphasic of (first kind) in character. By NMR studies the thermal transformation behaviors of colloidal gel heated to 1050°C (prepared earlier by Yoldas, 1992) as shown in Chapter 2 showed two Al resonance peaks characteristic of minor tetrahedral and major octahedral peaks at 64.5 and 7.1 ppm, respectively. The Al environment at AlO_6 peak was much larger than AlO_4 peak. XRD studies of heated gel at 500°C–1050°C showed cubic Al-Si spinel phase only.

(6) Colloidal gel marked CA

It was synthesized by Sanz et al. (1991). It's DTA and phase development process indicate that the gel is diphasic in nature.

(7) Transformation of (HB10 or Type III): first kind

Type III is a coprecipitated gel synthesized by Schneider et al. (1993) likely resembling diphasic gel of first kind as designated by the present author and transformed as per RH gel of Okada et al. (1986). It is amorphous and showed Si resonance maxima at –85 ppm at 350°C became sharper and shifted to –95 ppm at 900°C. At the second stage, it crystallized to spinel phase at the sharp 980°C exotherm. Si resonance showed two maxima due to spinel and third due to phase separated phase. The pattern showed a very broad peak in the range –80, –100 and –110 ppm and interestingly it was noted that these three peaks were clumped together and showed a peak of broad nature. This spinel phase further exhibited another broad exothermic peak of low intensity at 1250°C and formed mullite type phase.

(8) MI precursor: first kind

Similar observation as noted in case of Type III mullite (HB10) precursor was also observed in phase transformation of MI precursor of Geradin et al. (1994). It formed only spinel phase at 1000°C. ^{29}Si spectrum of it showed three peaks namely –80 ppm due to Al-Si spinel phase, –100 ppm due to noncrystalline alumina rich silicate phase and –110 ppm due to noncrystalline silica rich aluminous phase.

(9) Transformation of polymeric gel PB2

XRD patterns synthesized by Sanz et al. (1991) were mostly amorphous up to 900°C. ^{27}Al NMR spectra showed pentahedral peak at ~32 ppm in addition to tetrahedral and octahedral peaks. On heating up to 1000°C, it showed two peaks at ~65 and ~6 ppm which are similar to that of spinel phase. It exhibited two small DTA exotherms, one at 985°C, other at 1250°C those are related to transformation of it to mullite in two steps i.e., it crystallized to mullite via spinel as an intermediate phase.

(10) Transformation of SGM precursor: first kind

Studied by NMR spectroscopic technique, it showed the following sequences of phase developments. Although SGM precursor synthesized by Schneider et al. (1992) was called as polymeric gel, it would be better be re-designated as diphasic gel of first kind. Since it exhibited two exotherms in DTA. It was earlier shown by XRD that spinel phase formation occurred in major quantity, and due to that intensity of tetrahedral Al resonance was found less (as expected for spinel of Chakraborty's spinel model). Al NMR resonance showed larger AlO_6 than AlO_4 at this temperature range. The substantial amount of this spinel phase later on transformed to mullite at 1253°C exotherm. The DTA pattern in the high temperature region for both SGM and kaolinite are similar. Thus, the transformation of the first kind of diphasic gel, during the spinel-to-mullite phase evolution process shows the following observations.

Octahedral and tetrahedral peaks situated at 8 ppm and 70 ppm shifted to 5 ppm, 62 ppm and in addition the spectrum showed a third resonance at 45 due to another AlO_4 peak which are characteristic of typical mullite. Two ^{29}Si resonance peaks at –90 and –110 ppm are also changed. The later resonance peak is due to this noncrystalline aluminosilicate may consist of two types. One may be alumina rich and the other may be rich in silica or amorphous silica containing some alumina. There is a possibility that these two phases may react further during heating to high temperature and nucleates mullite with complete vanish of –110 ppm peaks. Broad peak at 90 ppm was replaced and finally Si resonance displayed peaks near –87, –94 and shoulder near –90 ppm of typical mullite. Development of three sites of Al, one octa, and two tetra, ascribed as T and T* sites mullite structure are also noted.

(11) Crystallization sequence of MI precursors

The synthesis procedure of MI of Huang et al. (1997) was different from usual synthesis process of mixing TEOS and $AlCl_3$ followed by hydrolysis in acidic medium as like as in SH gel making process. These precursors followed phase change in diphasic manner. Precursors obtained by these authors showed spinel formation during 980°C exotherm but did not mention the formation of any noncrystalline aluminosilicate phase along with former phase.

(12) Transformation of coprecipitated gel sample mark B

Sample mark B of Sales and Alarcon (1996) showed spinel formation followed by 980°C exotherm. Accordingly, this sample is diphasic in nature. However, these authors did not predict the formation of any noncrystalline aluminosilicate phase along with the crystallization of spinel phase.

20.4.2 Coprecipitated Gels Prepared by Using ANN and TEOS by NH_4OH at pH = 9 is to be Designated as "in situ Gel" Diphasic Gel: Second Kind

(1) Al-Si spinel forms partly out of γ-alumina (generated by the decomposition and dehydroxylation of boehmite) out of "in situ" gel of the author. As such this diphasic gel synthesized by the author is designated as the second kind of diphasic gel.

(2) **Transformation of Type II (HB13) precursor of Schneider et al. (1993): second kind**

Prehydrolysed TEOS was mixed with AlOBu and finally hydrolysed by using high water content and at high pH = 13. Under this condition, AlOBu was hydrolysed very rapidly and they believed that large colloids with a composition similar to pseudoboehmite were crystallized. The size of it was found to be in the range 100 to 1000 Å as estimated by XRD line broadening study. Thus, Type II precursor is not amorphous as Type I and Type III precursors. However, it shows pseudoboehmite. Thus, it is designated as a diphasic gel of the second kind. ^{29}Si and ^{27}Al NMR spectroscopy and ATEM studies were made to note the sequence of transformation. ^{29}Si resonance of it on heating at 350°C showed two components at −80 and −110 ppm. Interestingly, on continued heating the intensity ratio of the two Si resonance changed i.e., −80 ppm peak increased at the cost of decrease of −110 ppm peak. This event is likely related to bond development. On further heating to 1000°C, a reversal of intensities of Si resonance occurred when the spinel phase was noted by XRD analysis. DTA exhibited no exotherm up to 1000°C. Spinel phase (like γ-alumina) formed after dehydration at the endotherm. Its intensity and sharpness of X-ray peaks were temperature dependent. With increase of temperature, both ordering and growth of spinel crystallites occurred. It was stable up to 1250°C. Mullite formation occurred thereafter.

(3) **Transformation of M3** of **Geradin et al. (1994): second kind**

Gel made by using NH_4OH at pH > 9 e.g., M3 of them should be diphasic in character. Therefore, the course of spinel phase development out of in situ diphasic gel by Chakraborty (1996b) or gel mark Type II by Schneider et al. (1993) are really different from that of or diphasic gel made by Wei and Halloran (1988). As like "in situ" diphasic gel of the author, gel mark Type II of Schneider et al. is also to be called as "in situ" diphasic gel. Because boehmite develops during processing of pH adjustment to 9–13. Whereas boehmite sol itself is taken in case of synthesis of diphasic gel by Wei and Halloran (1988). That may be the reason for easy and rapid development of solid solubility of Si in γ-Al_2O_3 in case of "in situ" diphasic gel and in Type II gel than that occur in diphasic gel made by Wei and Halloran (1988).

20.4.3 Coprecipitated Gels Prepared by Using ANN and TEOS by NH_4OH at pH = 9–14 is to be Designated as Diphasic Gel: Third Kind

(1) **Transformation of diphasic gel made by Hsi et al. (1989): third kind**

Diphasic gels of different characters could be synthesized by monitoring pH of processing—on scrutinizing the results as shown by Hsi et al. (1989) (in their Table 1), it seems that pH has an effect on the formation of gels different in diphasic characters. Other than boehmite, different types of alumina bearing crystalline phases e.g., bayerite, gibbsite are generated with changes of pH. As such, this diphasic gel is called a third kind of diphasic gel. These have a large influence in subsequent mullitization paths. Gel made at low pH formed mullite early in comparison to gel made at high pH. There is a broad exothermic peak before mullitization exotherm at ~1300°C in DTA analysis. This observation is analogous to that observed by Chakraborty (1993, 1996b) in his DTA curve shown for gel marked G-49.

(2) **Transformation of diphasic gel marked MII made by Huang et al. (1997): third kind**

Diphasic gel marked MII synthesized by Huang et al. (1997) also found to be different from the gel marked CP9 so far as reaction series is concerned. They took TEOS and $AlCl_3 \cdot 6H_2O$ as components and gel made at pH = 11. This gel behaves as a diphasic precursor heated to 450°C showed two resonances: one small resonance at –81 ppm, other large resonance at –108 ppm. It formed γ-Al_2O_3 at 960°C. Si spectrum showed a shift from –81 ppm to –83 ppm with more increase in intensity and another usual peak at –108 ppm followed by a shoulder at –100 ppm due to beginning of noncrystalline aluminosilicate phase formation. The peak intensity at –81 ppm increased and gradually shifted to –83 to –91 at 1100°C. A large quantity of noncrystalline aluminosilicate formed with diminution of the peak at –108 ppm due to amorphous phase. Although this observation is found to be similar to some extent to that noted in heated Type II precursor, a difference still there. Basically this diphasic gel contained not only boehmite but it also contained gibbsite and bayerite unlike the sample marked HB13. That may be the reason why noncrystalline aluminosilicate formation was sluggish in nature.

20.4.4 Coprecipitated Gels Prepared Out of Large Silicon Hydroxide Colloids Considering Silica Component as Source is Designated as Diphasic Gel

(1) Transformation of **aged gel** of Chakraborty and Ghosh (1988). Diphasic character of gel will evolve at the cost of forming

monophasic gel when the silica sol used in gel synthesis was aged prior to addition of aluminium sulphate solution to it. Precursors made out of silica sol (made out of aging at different duration of time as reported by Chakraborty and Ghosh (1988). These gels are to be designated as diphasic in nature.

(2) Transformation of **Precursor D** of Jaymes et al. (1996), Considering the silica component, diphasic gel was formed (based on formation of large silicon hydroxide colloids). For example, in Precursor D of Jaymes et al. (1996), a large $Si(OH)_4$ colloid was formed during aging of the silica sol prior to the spray pyrolysis/spray drying process.

20.4.5 Diphasic Gels Considering Alumina as Source Component is Designated as Diphasic Gel

Transformation of Precursor B of Jaymes et al. (1996)

Diphasic gel was formed (based on formation of large aluminum hydroxide colloids) e.g., Precursor B of Jaymes et al. (1996)—Large $Al(OH)_3$ colloids was formed using ammonia at pH = 6 to 8 during its synthesis. Obviously, DTA curve (Fig. 12.1) shown by them for their sample marked B exhibited two exotherms. Although, they reported only one exotherm at ~1260°C, there is one more exotherm roughly at 1200°C before the occurrence of the former exotherm. This verifies the findings that mullitization reaction path via spinel phase formation for CP6 as well as other gels of the above researchers are similar in nature. This result also verifies the previous observation showing two exotherms in the region of the second exotherm by G-150 synthesized by Chakraborty (1994).

20.4.6 Diphasic Gels Considering Use of Large Quantity of Hydrolysis Water

(1) Transformation of G-150 of Chakraborty (1994), G-152(I) of Chakraborty (1996)—some previous authors also noted mullite formation through intermediate spinel phase formation only. It was generally formed (i) in presence of/use of large concentrations of water e.g., G-150 of Chakraborty (1994), G-152(I) of Chakraborty (1996).

(2) Type III of Schneider et al. (1993) was synthesized using large excess of water. SGM gel should be called a colloidal gel which formed below pH = 7 where water was used for hydrolysis of TEOS and AlOBu.

(3) M2 precursor too was synthesized by Geradin et al. (1994) in presence of much quantity of water medium. These gels synthesized by using large excess of water during the gelation process are to be called as diphasic mullite gel.

(4) When both two components are organic and water as hydrolysis medium, for example in cases of precursors made by Prochazka and Klug (1983), Suzuki et al. (1987), Sakurai et al. (1988), GYW of Taylor and Holland (1993), these showed invariably spinel formation before mullitization.

20.4.7 Probability of Formation of Mixed Monophasic and Diphasic Gel

That there occurs every possibility of formation of gels of two characters during the synthesis process of mullite gels. A variation of $Si(OH)_4$ colloidal particles will be generated by polymerization/ later on by condensation process during heating. Similarly, a large variation in the size of $Al(OH)_3$ colloids will occur. Thus, both colloids may be of various sizes which results in formation of a mixture of gels during the synthesis process.

Diphasic gels are synthesized in general by various earlier researchers (based on choice of pseudocrystalline compound as one component). For example, gels made by use of boehmite sol/ TEOS/Ludox/silica fume or sol. It is indicated in previous chapters that all mullite gels synthesized by coprecipitation process in presence of ammonia are diphasic in nature since spinel phase was developed as intermediary. Accordingly, this coprecipitated gel is newly designated as "derived diphasic gel". The following cases are suggested below.

(1) Polymeric precursor of Rajendran et al. (1990) showed spinel phase formation with exhibition of a large peak at ~1000°C. In a obvious reason, this spinel phase converted to mullite and and showed a small exotherm at ~1250°C in DTA.

(2) Colloidal gel marked CG of Li and Thomson (1991) showed 980°C peak and formed a spinel phase. In comparison of the DTA curve with the DXRD curve in their Fig. 6, mullite formation was observed at 1200°C.

(3) Both polymeric gels marked PA and PB showed 980°C exotherm as well as 1250°C exotherms for the crystallization of spinel and mullite respectively. However, the intensity of the first exotherm in two cases are different due to difference in gel preparation conditions. Some precursors showed spinel formation without any 980°C exotherm e.g., colloidal gel of Cassidy (1997).

(4) On careful observation of the DTA curve Fig. 4 of Sales and Alarcon (1996) for their sample marked B, two exotherms in the high temperature region, one at 1200°C and another at 1330°C are noted.

(5) Other examples include Type II gel of Schneider et al. (1993), M3 gel of Geradin et al. (1994) and colloidal gel (even not showing boehmite at the gel state) of Yoldas (1992), Yoldas and Partlow (1988).

Finally, it is concluded that all derived diphasic gels must show intermediary spinel phase.

20.4.8 Evolution Mixed Gel to Mullite Phase: Fourth Kind

Even the phase changes of mixed gel (MG) or gel synthesized by consecutive method (where components of two gels are separately synthesized and then homogenized by mixing followed by stirring) during heat treatment process develops Al-Si spinel phase first and then crystallized to mullite by a few parallel steps (see Chapters 15 and 16). Such kinds of mixed gels are designated as the fourth kind of diphasic gel system.

20.5 Categorization of Diphasic Gels

A critical review on the thermal sequence presented above has shown significant variations. Reaction process of diphasic gel-to-noncrystalline aluminosilicate precursor phase-to-Al-Si spinel-to-mullite shows two significant aspects. Phase transformation sequence up to mullitization constitutes one aspect of research. Crystallization of phases cum densification characteristics of the diphasic gel is another aspect of study. The later phenomenon is more important in diphasic gel systems for obtaining sintered mullite at a comparatively low temperature. Accordingly, diphasic gels show the following four categories.

Category 1, where phase transformation is the main aspect in diphasic gel

For examples, diphasic gels synthesized by Hoffman et al. (1984), Komarneni and Roy (1986), Ismail et al. (1986), Hyatt et al. (1990), Hamano et al. (1985), Hong et al. (1995), Li and Thomson (1991), Sanz et al. (1991), Schneider et al. (1994), Huang et al. (1997), Hyatt et al. (1990), Hulling and Messing (1989), Nietto et al. (1998) fall in this category. The concept of diphasic Al_2O_3-SiO_2 xerogel in three different ways was first described and prepared by Hoffman et al. (1984). A combination of two components used and processing techniques followed by some researchers during synthesis in their diphasic gels are shown below. It's components partially crystallize before mullitization.

(i) The presence of corundum was noted by Chakraborty and Ghosh (1986) in heated diphasic gel made by Hoffman et al. (1984).

(ii) According to Ismail et al. (1986) the diphasic gel formed γ-Al_2O_3 and silica (A) individually by decomposition of boehmite and silica gel respectively. These two phases interacted during continued heating and formed mullite at ~1285 °C as shown by an exotherm in DTA.

(iii) Diphasic gel synthesized by mixing boehmite sol prepared out of aluminium ethoxide and silica sol prepared from methyl silicate crystallized to β-cristobalite before mullitization (Hyatt et al., 1990).

(iv) During XRD studies of heat treated diphasic gel, Hamano et al. (1985) showed cristobalite in the temperature range of 1250°C–1600°C and also showed the crystallization of α-Al_2O_3 in the temperature range 1300°C–1700°C in addition. It was predicted that mullite was formed by the solid state reaction between β-cristobalite and corundum crystallized earlier during the heating process of the gel (Hyatt et al., 1990).

(v) Precursor marked MII synthesized by Huang et al. (1997) using ammonia solution resulted into development of crystalline modifications of alumina component such as bayerite, gibbsite and boehmite and silica (A) developed large cluster at high pH. During this gelation environment, MII finally causes segregation of two components. According to Huang et al. this state is called as nanohomogeneity of Al-Si components. Obviously, alumina and silica phases followed their independent character during the subsequent heat treatment.

Other diphasic gel synthesized by Hyatt et al. (1990); precursor MPP powder of Fonseca et al. (1997) and diphasic gel synthesized by Li and Thomson (1991) pH = 2 showed this type of crystallization behavior.

(vi) Gel B synthesized by Hulling and Messing (1989) preferentially crystallized in the temperature range 1250°C–1270°C into cristobalite, δ-, θ- and α-Al_2O_3. Mullite formation started thereafter on heating.

(vii) Mullite formation of six mullite precursors synthesized by the author presented in the Chapter 12 are really of different characteristics. Accordingly, their mullitization routes are also found to be different as shown and discussed by corroborating the DTA and XRD results with MAS-NMR data and with results of previous authors. There may be single step or two steps or even multi steps of mullitization processes. Question is: why there are differences in their mechanisms? Really, what are the basic differences in their internal structures of the Al_2O_3-SiO_2 precursors? The variation in mechanisms in about six routes are usually dependent upon its synthesis out of precursors made from the choice of two components and their processing techniques and these are discussed elaborately in the Chapter 16.

(viii) Colloidal gel prepared by Cassidy et al. (1997) out of ASB and TEOS did not indicate any thermal event in the region of 900°C associated with initial crystallization of γ-alumina, did not induce crystallization of mullite presumably due to mixing of the constituents in colloidal scale and likely the mullitization exotherm was raised to as high as 1270°C.

(ix) Coprecipitated precursor of Rajendran et al. (1990) contained bayerite and SiO_2 (A) phase. The bayerite component decomposed at ~320°C and thereafter the spinel phase gradually developed at ~1225°C and it was maximum in concentration at 1350°C, it decomposed and formed a large quantity of mullite.

As like as in monophasic gels, processing condition and source of different components of silica and alumina used may be the causes of variations in thermal sequences in diphasic system.

Category 2, where mullite formation by dissolution of γ-/δ-alumina in vitreous matrix in diphasic gels

(i) Synthesizing diphasic gel using TEOS and colloidal pseudoboehmite sol, Wei and Halloran (1988) showed a lamellar feature in the electron micrograph from heated gel. By XRD technique, they designated it was γ-Al_2O_3, as proposed earlier. They further claimed that diphasic gel was an ultra-homogeneous mixture of phases of 7–10 nm in size which crystallized independently up to 1200°C and then reacted directly to nucleate mullite, i.e., they had conjectured beyond what Hoffman et al. has proposed that the two discrete phases reacted only up to 1000°C. They predicted direct solid state reaction between γ-alumina and SiO_2 (A) as responsible for mullite development.

By XRD studies, mullitization processes suggested by these researchers are also shown to be different. Since Wei and Halloran (1988) and Ismail et al. (1986) did not observe cristobalite at all. Reason behind it was not mentioned neither by Wei and Halloran (1988) nor by Ismail et al. (1986). According to Wei and Halloran (1988) nucleation of mullite occurred at the SiO_2/δ-Al_2O_3 surface and transformation was probably controlled by a near-interface diffusion process or by the interfacial reaction itself. The transformation curves showed that it was preceded by a temperature-dependent incubation period where nucleation occurred in a short time. The nucleation density remained constant throughout the rest of the transformation process.

(ii) The mullitization process of diphasic mixtures was explained by Sundaresan and Aksay (1991) in a different way. During heating, alumina particles dissolved into the silica phase at the first stage. Mullite nuclei formed at the second stage when concentration of alumina in silica exceeds the critical nucleation concentration (CNC) value. Mullite nuclei grow at the third stage. Growth rate was found to be time dependent experimentally.

Then, it is highly expected that concentration of δ-Al_2O_3 phase will decrease with time or temperature of heat treatment. However, the said phase disappears in a short temperature interval.

(iii) Formation of a siliceous glassy phase in thermal transformation of gel is not unlikely. Formation of it is more evident in case gel synthesis in acidic medium for example in case of Sonuparlak et al. (1988). Wei and Halloran (1988) too synthesized diphasic gel using colloidal pseudoboehmite, unreacted and partially hydrolysed TEOS and in acidic condition (0.018 N HCL) at ~50°C for 3 days allowing aging for condensation of silica sol. The deliberate use of large $Si(OH)_4$ colloids would assist in glassy silica rich alumina phase during the heating process of gel around 1200°C approx. other than formation of noncrystalline aluminosilicate phase as noted by the author in thermal transformation of in situ diphasic gel.

Category 3, where densification prior to mullitization as a characteristic feature of thermal transformation of diphasic gel

The precursors studied by the following researchers namely (Komarneni et al. (1986), Sonuparlak, 1988 and Pach, 1985), Shinohara et al. (1986), Wu et al. (1993), Mizuno and Sato (1989) and subsequently Sugita et al. (1998) Cassidy et al. (1997), Rajendran et al. (1990)), may be in category 3.

(i) Early densification to the extent of ~97% of theoretical density at ~1300°C was observed by Komarneni et al. (1985). Densification of course varies during heating some diphasic gels.

(ii) Densification of diphasic gels to ~96% of theoretical density is generally achieved before crystallization to mullite (Pach, 1985).

This author synthesized diphasic mullite gel using boehmite (10 nm) and silica sol (12 nm). They observed a narrow difference of temperature ~ 20°C between densification and crystallization processes. After sintering at 1220°C for 5 h, the mullite gel yielded (δ, θ)-Al_2O_3 phases, the relative density was found ~95.8 but mullite did not crystallize. The former phase crystallized on heating 1240°C and above. At that temperature the relative density increased to 99%.

(iii) Densification in diphasic gels prepared out of colloidal boehmite suspensions and TEOS both in case of acidic and in basic condition at 1250°C by Sonuparlak (1988). Accordingly, densification temperatures in both cases vary.

Complete mullitization after densification occurred at 1300°C in basic gel. Whereas spinel phase continued at 1300°C together with partial mullitization in other cases of gel. It was concluded that when intimate mixing was prevented during the gel processing step, the most intermediate favor phase was spinel. It inhibited the initial crystallization of mullite as happened in case of monophasic gel at ~1000°C but allowed viscous phase sintering to achieve earlier densification as low as 1250°C.

In acidic condition, glass development was more predominant as a result rapid densification occurs (Fig. 1A of Sonuparlak, 1988).Contrary, in basic condition glass development is less predominant as a result uniaxial pressing of dried gel is required prior to densification (Fig. 2A of Sonuparlak, 1988).

(iv) Diffusion of SiO_2 into γ-Al_2O_3 spinel phase was slow in acidic condition. As a result, two components may crystallize to corundum and cristobalite on heating with partial crystallization to mullite. Independent transformation of components was noted by Hyatt et al. (1990) and also by the present author (1994) when gelation was done at pH < 2 (see Table 7.1 in Chapter 7). Contrary, diffusion of SiO_2 into γ-Al_2O_3 spinel phase is more in basic condition, consequently Al-Si spinel crystallization occurs easily which transformed fully to mullite at 1285°C exotherm in DTA/or heating directly to 1300°C.

(v) Diphasic gel prepared out of colloidal boehmite suspension and a partially hydrolyzed TEOS solution at pH = 2.5 for ~4 days showed mullitization but it resulted a detrimental effect on the densification process (Shinohara et al., 1986).

Although the above researchers noted densification prior to mullitization. Other authors e.g., Ismail et al. (1986), Li and Thomson (1991), Wu et al (1993), Mizuno and Sato (1989) and Sugita et al. (1998) however noted that it needs high temperature calcinations at >1600°C to complete both mullitization and achievement of highest densification. Synthesizing diphasic gel out of mixed aqueous silica and boehmite (prepared out of γ-Al_2O_3) sols under acidic conditions at pH ~1.8) followed by gelling through solvent evaporation, Ismail et al. (1986) showed that even in this case sintering temperature had to increase to as high as 1650°C for 1.5 h to achieve 98% of its theoretical density to obtain mullite ceramic.

(vi) Wu et al. (1993) synthesized it using sub micrometer alumina powder and (Ludox) colloidal silica sol as precursors and maintained at pH 5–6. Mizuno and Sato (1989) and subsequently Sugita et al. (1998) synthesized it from fume silica coated with aluminum sulfate salt precipitates. However, mullitization occurred in these cases at high temperature.

Category 4, synthesis of in situ diphasic gel and its transformation

Diphasic mullite gel was synthesized by using Ludox and ANN in an ammonia solution where boehmite is being developed "in situ" during the gelification process itself (Chakraborty, 1996b). Comparative mullite formation process of diphasic gel of standard composition vs. silica rich and alumina rich compositions are shown in Chapter 15. Where spinel (γ-Al_2O_3) forms usually at lower temperature than 980°C. DTA shows a broad exotherm over a long range of temperature from 1100°C–1300°C due to formation of Si substituted spinel, and thereafter it further exhibits a broad exotherm at ~1320°C due to crystallization of mullite out of transformation of intermediate Al-Si spinel. In the entire transformation range neither corundum nor cristobalite crystallize out in G-49 and in CP-9 synthesized by Chakraborty (1996) (see Chapter 4).

In conclusion, syntheses and phase development of diphasic mullite gel like monophasic gel also depend upon components chosen and the process of gelation. The categories of formation of monophasic to various kinds of diphasic gels are dependent on hydrolysis cum condensation of TEOS.

A comparison of the evolution process of monophasic gels, coprecipitated gel to diphasic or colloidal gels and SP precursors showing changes of phases are given in schematic diagrams. Three different ways of phase transformation sequence based on XRD intensities of phases formed were shown by Sugita et al. (1998) vide their (Fig. 3).

(i) The phase transformation processes of coprecipitated powder (CP) with sol mixture (SM) and with oxide mixture (OX) were schematically presented in their (Fig. 7.19) and compared by Hamano et al. (1986). Coprecipitated samples on heating first changed into spinel phase from amorphous phase at ~980°C exotherm, and then into mullite at ~1250°C exotherm.

(ii) Five mullite precursors were synthesized by Kumazawa et al. (1991). They noted the changes in crystalline phases formed on heating schematically.

(iii) XRD intensity of mullite formed in mechanically mixed oxides AS, MO and three cases of diphasic gels were shown in a schematic diagram (Fig. 16.5) for the phase development of three cases of diphasic gels (Chakraborty, 2005a). Other examples showed the absence of even partial crystallization of components prior to formation.

Diphasic gel synthesized by Hong et al. (1995) out of $AlCl_3$ and fume silica in ammoniacal condition showed mullitization during the occurrence of a broad exothermic peak at ~1335°C.

Mullitization was also detected at ~1300°C in diphasic gel (DG) by Li and Thomson (1991).

Similar exotherm at ~1285° C in DTA due to crystallization of mullite out of previously formed Al-Si spinel phase by Sanz et al. (1991) in their colloidal gel marked CA was shown.

HB13 of Schneider et al. (1993) on heating first transformed to γ-Al_2O_3 after decomposition of pseudoboehmite. It is conceived that Si spices of amorphous silica gradually diffuse into γ-Al_2O_3 crystals along the crystallite boundaries with rise of temperature as high as 1100°C–1200°C and mullite crystallized thereafter.

Colloidal gel marked CA of Nietto et al. (1998) showed IR absorption peak due to octahedral AlO_6 groups of alumina as low as 500°C and finally the gel crystallized to mullite at ~1285°C as shown in DTA.

20.6 Mullite Formation in AS-γ-Al_2O_3 Microcomposites Behaves as Like as Diphasic Route

The so called microcomposite gel (AS-γ-Al_2O_3) synthesized by Bartsch et al. (1999), and Gel D (derived resembling as microcomposite) prepared by Ivankovic et al. (2003) may behave as diphasic gel (Gel C). These composites exhibit exotherm at 1360°C and at 1361°C (Fig. 2 of Ivankovic et al.), respectively. These are slightly higher than that occur in in situ diphasic at 1300°C. Details of mullite formation in microcomposites are given in Chapters 4 and 15 and summarized in Table 20. 4 below.

Table 20.4 Some mullite precursors synthesized by using γ-Al_2O_3 and theirs phase evolution

1. Ivankovic et al. (2003) studied DTA studies of gels. Gel C and D (purely diphasic in nature).	Since the crystallization pathways are different from monophasic gels. Exhibited exotherms due to mullitization at 1346°C in gel C and at 1361°C in gel D.
2. Bartsch et al. (1999) studied amorphous SiO_2-coated γ-Al_2O_3 nanocomposites.	AS-γA exhibited exotherm at 1360°C attributed to mullite formation.
3. Wu et al. (1993) studied alumina-silica matrices out of 0.5 μm alumina and Ludox as sources.	Observed densification in between 1100°C–1300°C which is much before the crystal phase development of mullite detectable at 1450°C.
4. Sacks et al. (1991) studied AS-α-Al_2O_3 composites	They noted enhanced sintering attributed to transient viscous sintering (TVS).
5. Wang and Sacks (1996) studied α-Al_2O_3/silica microcomposites	Exhibited an endotherm at ~1500°C associated with mullite formation.

Discussion of earlier results

There is a question: whether diffusion of alumina in the siliceous phase controls the mullitization process or dissolution of alumina in siliceous phase is the rate controlling reaction step. Figure 6 of Sacks et al. (1997) shows the DTA plots of both the "74/26" microcomposite powder and a "74/26" mechanical powder mixture of α-alumina and amorphous silica. Both the two powders exhibit endotherms due to mullite formation at 1499°C and 1506°C, respectively, but the difference of peak temperature differs only 7°C. Of course the geometry of the endotherm of the composite sample is apparently more than that for mixture. This study provides clear evidence that the reaction rate during the first stage which is controlled by dissolution of alumina in the siliceous phase is predominant. Similar evidence was also proposed by Huling and Messing (1987), Sundaresan and Aksay (1991) for mullitization of diphasic gel. Second stage of mullitization is composition dependent as noted in three cases of microcomposites. The growth of mullite grains also is probably controlled by chemical inter diffusion within the mullite grains.

(i) Reaction was completed at a much lower temperature in the alumina rich sample compared to a near-stoichiometric sample.
(ii) A silica rich sample also showed enhanced mullitization relative to the near-stoichiometric composition.
(iii) Temperature required to complete the reaction in a silica-rich sample was still considerably higher than in the alumina-rich sample. Question is: why mullitization is faster in case of alumina rich composite has to be ascertained.

Bartsch et al. (1999) claimed that densification of AS-γA nanocomposite and AS-αA compacts completed prior to initial mullite crystallization. However, it is observed by XRD patterns and microstructural studies of heat treated nanocomposites of them that both densification and mullite formation occurred between 1275°C to 1300°C i.e., both processes superimpose or overlap over the said temperature interval. Of course, dilatometric studies of both AS-γA and AS-αA showed a difference of contraction > than 100°C. Density measurement during sintering in dilatometer also yielded a low bulk density of 2.22 g/cc for AS-αA microcomposite as against 2.96 g/cc for AS-γA nanocomposites. Two observations of these authors are strikingly important. One is the persistence of large XRD intensities of γ-Al_2O_3 phase at a temperature as high as 1275°C. Secondly, crystallization of substantial quantities of mullite at 1285°C as noted by sharp increase in XRD intensities of it. Thus, within a narrow temperature range 1275°C to 1285°C i.e., 10°C interval a sudden transition of γ-Al_2O_3 spinel phase to mullite formation occurs. These two events remind us of the analogous transformation of Al-Si spinel phase-to-mullite polymorphic reaction (as discussed by the present author earlier). It is predicted that a part of SiO_2 (A) diffuses into nanocrystalline (~10 nm) γ-Al_2O_3 structure and develops Al-Si spinel phase and that the concept of formation of later phase is extended. Its generation is also possible in nanocomposite systems as usually happens in diphasic gel systems. The incorporation of SiO_2 in γ-Al_2O_3 particles is probably the reason for increment of transformation temperature of nanosized spinel phase as high as 1275°C. Side by side, dissolution of a fraction of γ-Al_2O_3 in to the SiO_2 (A) may result to formation of silica rich glassy phase and generation of the later phase is responsible for early densification. Thus, two parallel reactions are operative in densification as well as mullitization processes in AS-γA nanocomposites.

Due to absence of formation any silica rich glassy phase, it is difficult to sinter the monophasic/SH/Type I/polymeric mullite gel powder to high density in spite of its molecular level homogeneity at <1650°C. Secondly, monophasic gel shows a very high contraction as per TMA study and finally produces a low solid yield on heating. Contrary, diphasic gel processed with sub-micrometer alumina powder or boehmite suspension with either TEOS or Ludox could be sintered at temperature as low as 1300°C with increment of more solid yield and with control able low processing shrinkage.

TEM electron micrograph (Fig. 9) shown by Ivankovic et al. (2003) indicates a sharp boundary between two reacting particles prior to mullitization as shown by Sack et al. (1991). They thought that alumina core coated with amorphous silica core originally. Question is: is it pure alumina core or pure silica core?

In the author's study, the XRD result does not really resemble δ-Al_2O_3 as identified above. Neither a mixture of θ- and α-Al_2O_3

does crystallize nor cristobalite forms when DG72 is heated at high temperature (Fig. 11.5). With this result it is presumed that Si incorporation may take place in the alumina core during heating processes. Secondly the formation of an aluminosilicate glassy phase as noted by the presence of band at 22° 2θ, indicates that Al likely diffuses into the pure silica phase i.e., silica core. This view supports the previous alkali extraction study of Chakraborty (1979). Therefore, the conjecture that the so-called δ- or θ-Al_2O_3 is a Si substituted alumina phase is viable. The increase in intensity of the amorphous band occurs during heating of both DG66 and DG72 (Fig. 15.8) which indicates the development of a more amorphous phase. Moreover, Chakraborty (2003) showed the formation of gehlenite rather than calcium aluminate when the precursor of diphasic mullite gel was heat treated. It was concluded that the precursor was an amorphous aluminosilicate phase which substantiates the present study. Therefore, alumina and silica core as shown in the TEM figure of Ivankovic et al. (2003) are not probably remain as free and independent entities during heating.

20.7 Mullite Precursors Synthesized by Codecomposition Method as Third Route and Their Phase Evolution

Mullitization process in codecomposition process is shown in Chapter 5 and summarized in Table 20.6.

The tentative phase evolution of SP precursor route of transformation to mullite type phase is as follows.

$$\underset{\text{Amorphous precursor phase}}{\text{SP}} \xrightarrow{\sim 970^\circ\text{C}} \underset{\text{[Alumina rich mullite phase(AS1)]}}{3Al_2O_3\ 2SiO_2 \times H_2O} \xrightarrow[\text{(ii) }\sim 980^\circ\text{C exotherm}]{\text{(i) At D } -H_2O} \underset{\text{(Mullite type phase)}}{3Al_2O_3 \cdot 2SiO_2} \tag{20.1}$$

This transformation apparently similar to (i) aluminosilicate hydrate gel synthesized by aqueous sol-gel method marked G-104 of Chakraborty and Ghosh (1988), (ii) in precursors e.g., G-31 where water is mostly avoided during synthesis (see Chapter 7).

Table 20.5 Mixing scale vs. temperature of densification and mullite formation of mixture of components

Gel/composite/oxide mixture category	Scale of mixing	Densification temperature	Cristobalite/mullitization temperature
1. Polymeric gel	May be molecular level	Not known	Mullite only
2. (i) Diphasic gel and (ii) Colloidal mixture	(i) Boehmite sol (nano, 5 to 50 nm) and TEOS or Ludox (ii) Fumed silica, 40 nm and γ-Al_2O_3 nm/isostatic pressing	(i) ρ at 1300°C = 3.06 g/cc (ii) ρ at 1350°C = 3.08 g/cc	(i) XRD revels mullite (ii) XRD revels mullite
3. Microcomposite powder AS-γA	Micro, ~0.2 μm (γ-Al_2O_3) & silica coating on each alumina core = 15–20 nm	1285°C–1300°C	Cristobalite formation occurs between 1425°C–1475°C. Mullitization formation occurs between 1285°C–1300°C.
4. Microcomposite powder AS-αA	(i) Micro, ~0.5 μm (α-alumina). Ludox silica coating on each alumina core = 15–20 nm (ii) ~0.25 μm and TEOS as silica source.	(i) 1100°C–1300°C (ii) ~1300°C	(i) Cristobalite between 1300°C–1400°C. Mullitization between 1575°C–1625°C (ii) Mullitization at ≥1500°C
5. Coarse material – quartz and α-Al_2O_3 (corundum)	Micrometer scale	≥1650°C	XRD revels mullitization at ~1650°C

Sources: (1) Yoldas (1992); (2) Komarneni et al. (1986); and Mizuno and Saito (1989); (3) Bartsch et al. (1999); (4) Sacks et al. (1991); Schmucker et al. (1995); and Wu et al. (1993, 1996); (5) Boccaccini and Ponton (2000).

Table 20.6 Spray pyrolyzed and spray dried processes

1. Kanzaki et al. (1985) synthesized mullite powder by spray pyrolysis method out of ANN and TEOS	DTA showed a very sharp exothermic peak at 970°C with formation of mullite directly.
2. Sato et al. (1991) synthesized it out of aluminium sulfate and TEOS for pyrolysis.	Powder showed spinel formation at 1000°C.
3. Sakurai et al. (1988) spray pyrolyzed a mixture of TEOS and aluminium alkoxide	It showed direct mullitization at 1000°C.
4. Hamano et al. (1986) spray dried (i) an ethereal mixture of ANN and TEOS, (ii) a mixture of aluminium sulfate with TEOS	This spray dried showed a strong 980°C exotherm and formed mullite. (ii) It showed a weak 980°C exotherm.
5. Jaymes and Douy (1992) and Jaymes et al. 1996) spray dried a solution of ANN mixed with hydrolysed TEOS	DSC trace shows strong exotherm at 980°C (240 J/g).

Components used for making monophasic mullite precursor as in SHI and spray pyrolysis powder are the same, but the processing conditions followed in above two processes are different as shown in Table 20.6. Indeed, the mullitization route in those cases shows differently. SP precursor forms mullite type phase directly whereas the others e.g., SHI, G-104 and G-31 crystallize to alumina rich nonstoichiometric mullite. Question is: whether the mechanism of mullite formation in those methods are the same or different?

For example, (i) in spray dried or spray pyrolysis process vs. monophasic gel synthesized by aqueous sol-gel process.

^{29}Si MAS-NMR study of spray dried mullite precursor (Sample C) heated to 1000°C by Jaymes et al. (1996) did not show any resonance peak at –110 ppm. This indicates that no phase separation practically occurred during exothermic events as usually noticed by various researchers in their MAS-NMR studies. So there is no possibility of any type of solid state reactions among noncrystalline aluminosilicate phases with noncrystalline alumina rich mullite and other phases as shown above during further heating to 1150°C–1600°C. In this case, as no segregation occurred during the 980°C exotherm in DTA, it is presumed that this precursor only directly transformed to mullite type phase and may be to the fullest extent at the exotherm itself.

(i) Contrary, ^{29}Si MAS-NMR data of precursors obtained by sol-gel processes using both water soluble components at successive increase of temperatures are not available.

(ii) ^{29}Si NAS-NMR spectra of precursors obtained from same sources such as ANN and TEOS made by homogeneous precipitation using urea by Jaymes et al. (1995) showed that initial gel or precursor marked F is of specific structure (Fig. 8 of them). They conceived that the resonance at –78 ppm is due to three hexa coordinated Al atoms and one -OH group. On heat treatment, the Si spectrum becomes broad and centered at –90 ppm indicates random distribution of Si and Al atoms in the lattice. During crystallization the broad resonance is replaced by characteristic mullite resonances at –86, –90, –94 and –80 ppm and another broad signal at –110 ppm. This points to phase segregation during transformation i.e., noncrystalline aluminosilicate phase formation is obvious. This is in contrast to the crystallization scheme of spray pyrolyzed precursor C.

Another point of contrast is the transformation way of mullite formed at 980°C to high temperature. The poorly crystalline mullite crystallized at exothermic peak temperature is nonstoichiometric and rich in alumina as shown by the author during transformation of noncrystalline aluminosilicate precursor at final dehydroxylation process (Chapter 14). There is no disagreement with this view with Jaymes et al. (1995). But they conceive that SiO_2 (A) is liberated during this transformation by observing –110 ppm broad peak in ^{29}Si resonance. As if mullite precursor fully and completely transforms to alumina rich mullite and SiO_2 (A) at 980°C exotherm. Here lies the difference. Author first showed that final dehydroxylation is inhomogeneous in nature instead of homogeneous dehydroxylation as assumed by various researchers. Migration of both Si^{+4} and Al^{+3} cations are essential for elimination of last trace OH^{-1} groups in the precursor phase. Noncrystalline aluminosilicate mullite precursor phase does not probably exist as a fully anhydrous state at the point D just prior to the occurrence of the first exotherm. Secondly, experimental evidence shows the absence of SiO_2 (A) phase (Chapter 15) or the presence of nonleachable SiO_2. Furthermore, alkali leaching studies as shown in Chapter 15 where in Chakraborty and Ghosh (1987) performed the phase transformation study of silica rich gels in presence $Al(OH)_3$ gel and observed the presence of aluminosilicate amorphous hump (Fig. 15.7).

Jaymes et al. (1995) conjectured alumina rich composition of mullite changes on heating by incorporating amorphous silica. Since silica signal disappears and the spectrum of sample annealed at 1300°C belongs to characteristic o-mullite. Contrary, the author suggests that incorporation of silica comes from residual silica rich noncrystalline aluminosilicate phase (present in Donor region) named as AS2 phase. Thus solid state reaction between alumina rich mullite with AS2 on heating process at ~1300°C is responsible for o-mullite formation. Therefore, obviously the characteristic broad −110 ppm resonance peak belongs to silica rich aluminosilicate phase rather than free amorphous alone.

20.8 Emergence of Composition of Intermediate Mullite and Spinel Phases

20.8.1 Composition of Weakly Crystalline Primary Mullite

Prior to assessing the composition of it, following experimental facts are first enlightened and then discussed.

(1) Based on lattice parameter measurement by XRD method, and also on the observation that a lattice parameter of o-mullite increases linearly from a = 0.754 nm of 3:2 mullite (sinter variety where x = 0.25) to a = 0. 757 nm for 2:1 mullite (melt variety where x = 0.4), Cameron (1977) estimated the composition of mullite.

(2) Existence of interrelation between chemical composition of mullite with absorbance of 1130 cm^{-1} (A_{1130}) and 1170 cm^{-1} (A_{1170}) by Cameron (1977).

(3) On comparing with these previous data of mullite given by Cameron (1977), mol % alumina in mullite which is formed

on heating SH mullite gels at 1000°C was determined (Okada et al., 1986) and (Okada and Otsuka, 1986).

(4) Two peaks namely, 1165 cm^{-1} and 1130 cm^{-1} in between 1100°C to 1200°C vary much with the composition of mullite sensitivity (Suzuki et al., 1990).

(5) The IR absorption band of primary mullite which is formed on heating sample B at ~1000°C/15 h at the first stage shows that the 1130 cm^{-1} band is stronger than the 1165 cm^{-1} band (Sales and Alarcon, 1996).

(6) The composition of the mullite changes gradually from Al-rich mullite to bulk 3:2 mullite as a function of increasing temperature of the firing process of Type I precursor (Ruscher et al., 1996).

Based on these experimental data as shown above and with reference to (Chapters 8, 16 and 19), a logical conclusion has been made as follows.

(i) At the first stage: The weakly crystalline mullite which is crystallized rapidly at the occurrence of at 980°C exotherm should be nonstoichiometric in nature and rich in alumina which is within the solid solution range of mullite in the metastable state. As soon as it formed out of noncrystalline aluminosilicate precursor phase its "*a*" LC value corresponds to 0.761–0.762 nm. It immediately changes to a more ordered state within a short temperature range from 1000°C to 1050°C. Within this 50°C range, change in LC value was noted as follows.

0.761–0.762 — first stage — to — 0.757 nm ——— (1)

(ii) At the second stage: In the plateau region at an interval of 50°C (Figs. 16.1 and 16.7), it's *a* axis value shortens very slightly without changes in its composition. Change in LC value was noted as follows.

0.757 — second stage — to — 0.7568 nm ——— (2)

(iii) At the third stage: It's constitution likely changes by inter diffusion of SiO_2 out of silica rich aluminosilicate phase (AS2) during solid state reaction with nonstoichiometric mullite during the continuous heating process from 1100 to 1300°C. There may be many probabilities of composition of mullite formed in between the range 2:1 to 3:2 composition (Fig. 16.7, see the portion of the curve shown by Cameron (1977). In this stage, the composition of alumina rich nonstoichiometric mullite now really decreases only with heating by the above solid state reaction until the reaction is completed and it reached to the stoichiometric (3:2) stable state at temperature >1400°C. Likely, it's *a* lattice constant value change was noted as follows.

0.7568 — third stage — to — 0.7545 nm ——— (3)
(based on data shown by Okada et al. (1986).

20.8.2 Composition of Weakly Crystalline Spinel Phase

Spinel phase which is weakly crystalline in nature exhibiting only two XRD peaks at 400 and 440 nm may be unsaturated in nature and probably rich in alumina composition as conjectured by Okada et al. (1986) and others. As soon as it formed out of noncrystalline precursor phase at the occurrence of first exotherm, it starts to change its constitution by taking up SiO_2 out of noncrystalline silica rich aluminosilicate phase (AS2) during ongoing rise of temperature. During the incorporation process, when its composition matches the composition of stable 3:2 mullite, this saturated Al-Si spinel structurally transforms from cubic form to orthorhombic form. Mullite formation of the two categories are finally stated below.

1. **Mullite formation in RH category gel (other type of monophasic gel)**
   Details of the mullite formation process is shown in Chapter 16 vide Eqs. 16.6–16.9
2. **Mullite formation in diphasic gel made out of boehmite**
   Details of the mullite formation process is shown in Chapter 16 vide Eq. 16.12

Conclusion

(1) Phase transformations of the different mullite gels or precursors are found to be six different types or categories. Particularly, it is observed that the 980°C phase developments

is a function of the starting materials of silica and alumina components and process techniques of gelation.

(i) In this category, some precursor forms alumina rich nonstoichiometric metastable mullite and exhibit only 980°C exotherm,

(ii) In the second category, precursor forms both metastable form of alumina rich mullite and alumina rich Al-Si spinel phases at the 980°C and thereafter completes mullite formation at usual at 1250°C exotherm and then on further heating at 1600°C. Due to varying ratios of formation of either minor spinel/major mullite or major spinel/minor mullite, two different types of precursors within this category are found.

(iii) In the third category, the precursor forms alumina rich Al-Si spinel only and no mullite phase at 980°C exotherm and thereafter forms mullite type phase at 1250°C exotherm. Lastly, transforms to 3:2 mullite on heating at 1600°C.

(iv) In the fourth category, diphasic gel forms Al-Si spinel phase over a long range of temperature between 1000 to 1300°C through a broad exothermic peak and thereafter crystallizes to mullite type phase at exotherm ~ 1320 °C. Finally, transforms to o-mullite during heating at 1600°C

(v) In the fifth category, both SP/SD precursor rapidly crystallizes to mullite type phase directly by way of decomposition at a sharp 980°C exotherm. It transforms usually to o-mullite during heating at 1600°C

In this way, the author describes six different types of transformations, the sequential stages of their phase evolution processes are narrated in the text above.

(2) On analysis of the whole mullite literature, three routes of phase transformation sequences of mullite precursors emerged out.

Changes in composition of primary mullite to stable mullite and secondly, the changes in composition of intermediary alumina rich spinel phase to Al-Si spinel (cubic mullite) are discussed.

First route:

Mullite formation path of first route/kind is observed in some gels synthesized in acidic conditions e.g., aqueous Al_2O_3-SiO_2 gel, monophasic gel, SH gel, polymeric gel C, molecular gel etc.

Second route:

Similarly, mullite formation path of second route/kind is noted in some precursors prepared out of other different techniques at ammoniacal condition. For example, RH gel, CP gel, colloidal gel, diphasic gels etc.

Third route:

Some SP/SD precursors decompose to mullite in a single step. In all three routes, gelation techniques play an extremely crucial role in phase sequences. Use of water during hydrolysis, basic pH and aging of either sols lead to develop large colloidal particles of either silica or alumina components as a whole, As a result homogeneity is affected and finally diphasicity will prevail both in acidic as well as in basic methods of synthesis of precursors. These are the causes of drawback in precursor synthesis of large sections of earlier researchers. Finally, phase transformations of mullite from different methods vis-a-vis routes are being generalized in this present book.

(3) The mullitization process of three routes consist of various steps.

First route of transformation of monophasic gel to mullite:

Monophasic gel transforms to mullite in three steps. At the first step, formation of alumina rich mullite rapidly takes place to a major extent at ~1000°C exothermic peak temperature by Path I e.g., transformation is noticed in case of (SH) gel. At the second step, slow rise in formation of mullite type phase occurs from 1200°C–1300°C by solid state reactions by Path II. At the third step, mullite type phase transforms to 3:2 mullite at 1300°C to 1600°C by Path III. Other examples. SHI Polymeric gel C, Type I precursor. Other monophasic gels show considerable variation in mullitization steps e.g., transformations are noticed in cases of SHII, polymeric gel (see Chapter 16).

Second route of transformation of diphasic gel to mullite:

Diphasic gels of five types transform to mullite in three steps.

(i) General diphasic gel: It was made by use of pseudoboehmite (based on choice of it as one component) and TEOS. At the first step, two components crystallize independently to corundum and cristobalite (Hyatt et al., 1990) and Hamano et al., 1985). At the second stage mullite was formed possibly by solid state reaction between α-Al_2O_3 and β-cristobalite at high temperature ~1600°C. However, the mode of phase transformation of the above diphasic gel is completely different from those described by Wei and Halloran, Ismail et al. and M3 (Geradin et al., 1994) (see Chapter 4). Transformation of some diphasic gel is also depending on pH of gelation (Sonuparlak, 1988).

(ii) First type of diphasic gel: At the first step, formation of alumina rich spinel phase rapidly takes place to major extent at first exothermic peak temperature at ~1000°C. Mullite type formation occurs by polymorphic transformation of silica saturated Al-Si spinel phase at second exothermic peak temperature at ~1250°C by Path IV e.g., transformations are noticed in cases of RH gel, SHIII. SHIV/CP6 colloidal gel HB 10/Type III. At the second step, a slow rise in formation of 3:2 mullite occurs on heating from 1300°C–1600°C by solid state reactions by Path III.

(iii) Second type of diphasic gel: At the first step, gradual crystallization of Al-Si spinel phase occurs in the temperature range 1000°C to 1300°C and its subsequent polymorphic transformation to mullite type phase at ~1300°C exotherm is observed by Path IV e.g., in case of "in situ" diphasic gel (CP9). At the second step, mullite type phase transforms to 3:2 mullite at 1400°C to 1600°C by Path III. Other examples include Type II (HB13).

(iv) Third type of diphasic gel: The formation of crystalline phases out of alumina components depended on the pH value. At a lower pH value of 8.3, fibrous pseudoboehmite was formed, whereas at pH greater than 9.5, bayerite appeared and it became predominant at pH = 10.4. In the first case, a mixture of pseudoboehmite and $Si(OH)_4$ decomposes and forms SiO_2(A) + γ-Al_2O_3 type spinel phase at 960°C at the first stage. In the second stage, it converts to a mullite type phase on heating at 1250°C.

In the second case, a mixture of bayerite and $Si(OH)_4$ decomposes and forms SiO_2(A) + γ-Al_2O_3. The later phase appeared initially at ~600°C after decomposition of xerogel transformed to spinel like phase on heating at ~1000°C. In comparison a higher temperature is required to achieve transformation of diphasic gel containing bayerite to form intermediate spinel type phase and θ-Al_2O_3 which converts to mullite on heating at 1300°C. Example, gels made by Hsi et al. (1989) and Su Yen et al. (1991).

(v) Fourth type of diphasic gel: MG gels (made by consecutive precipitation method)

At first stage two components $Al(OH)_3$ gel and $Si(OH)_4$ gel react at the temperature range 400°C–900°C and form (i) silica rich alumina phase (A) (ii) SiO_2(A) and γ-alumina.

At the second stage these form (i) Al-Si spinel, (ii) silica incorporated γ-Al_2O_3, (iii) Alumina rich silica phase (A) and (iv) Si incorporated θ-Al_2O_3. At the third stage, inter diffusion takes place between (i) to (iii) phases which result in crystallization to mullite type phase at ~1300°C, at fourth step, Si incorporated θ-Al_2O_3 transformed to mullite and corundum. And then the mullite type phase transforms to 3:2 mullite at >1600°C.

Third route of transformation of spray pyrolyzed powder to mullite:

SD precursor transformed to mullite type phase at first exothermic temperature itself by Path VI. It completes mullitization at the second step on calcination at 1600°C.

(4) Mechanism:

Following results (i) the changes in phase transformations behaviors of the six types of precursors; (ii) the emergence of maximum three routes of phase transformation sequences of mullite; and (iii) various steps of transformations among monophasic mullite gel, five types of diphasic gels and SP/SD precursors are elaborately shown. A preliminary study of the mechanism of mullite formation of different precursors are outlined below.

(i) Formation of mullite by decomposition of precursor phase itself. Monophasic precursor crystallized to nonstoichiometric mullite at first exotherm. SP/SD precursor crystallized to mullite type phase at same exotherm.

(ii) First type of diphasic gel crystallized to mullite type phase at second exotherm via Al-Si spinel phase.

(iii) Second type of diphasic gel crystallized to mullite type phase at fourth exotherm via Al-Si spinel phase.

Besides these three basic transformations, there develops three phase boundary regions during heating of those precursors at dehydroxylation stage. For example, a noncrystalline silica rich phase (AS2) region at the donor side. Two other noncrystalline phases namely (i) alumina rich mullite phase (AS1) with its residual noncrystalline alumina rich mullite phase (AS3) and or (ii) noncrystalline alumina rich spinel phase (AS4) with its residual phase (AS5) in the acceptor side. These noncrystalline phases are the most important sources for solid state reactions in subsequent heating processes of precursors and play a major role in future mullite development. Sketch A (Fig. 7.18) shows a representative example of reaction of two discrete phases of diphasic gel system (see Chapter 7). The possible solid state reactions are shown in Eqs. 16.2–16.9. There lies the differences in their mechanisms of formations in four precursors. Really, there are some basic differences in their internal structures of the Al_2O_3-SiO_2 precursors.The variation in mechanisms in about six methods are usually dependent upon its synthesis of precursors made from the choice of two components and their processing techniques as discussed elaborately in the discussion section. In conclusion, the following parameters are found to be interrelated in precursor-to-mullite transformation series. Source of silica alumina components, techniques of precursor synthesis, kinds of gel/precursor produced, routes of mullite formation favored. Details about the scale of homogeneity vs. mullitization aspects are given in Chapter 11.

References

1. A. Douy, Crystallization of amorphous spray-dried precursors in the Al_2O_3-SiO2 system. *J. Euro. Ceram. Soc.*, **26**, 1447–1454 (2006).
2. A. K. Chakraborty and D. K. Ghosh, Synthesis and 980°C phase development of some mullite gels. *J. Am. Ceram. Soc.*, **71**(11), 978–987 (1988).
3. D. W. Hoffman, R. Roy, and S. Komarneni, Diphasic xerogels, a new class of materials: phases in the system Al_2O_3-SiO_2. *J. Am. Ceram. Soc.*, **67**, 468–471 (1984).
4. K. Okada, Y. Hoshi, and N. Otsuka, Formation reaction of mullite from SiO_2-Al_2O_3 xerogels. *J. Mater. Sci. Lett.*, **5**, 1315–1318 (1986).
5. B. E. Yoldas, Effect of ultrastructure on crystallization of mullite. *J. Mater. Sci.*, **27**(24), 6667–6672 (1992).
6. J. C. Huling and G. L. Messing, Hybrid gels for homoepitactic nucleation of mullite. *J. Am. Ceram. Soc.*, **72**(9), 1725–1729 (1989).
7. J. C. Huling and G. L. Messing, Epitactic nucleation of spinel in aluminosilicate gels and its effect on mullite crystallization. *J. Am. Ceram. Soc.*, **74**(10), 2374–2381(1991).
8. H. Schneider, B. Saruhan, D. Voll, L. Merwin, and A. Sebald, Mullite precursor phases. *J. Euro. Ceram. Soc.*, **11**, 87–94 (1993).
9. M. Haque, Thermal decomposition of some alumino silicate gels. Thesis, University of Kolkata (2000).
10. I. Jaymes, A. Douy, D. Massiot, and J. P. Coutures, Characterization of mono- and diphasic mullite precursor powders prepared by aqueous routes, ^{27}Al and ^{29}Si MAS-NMR spectroscopy. *J. Mater. Sci.*, **31**, 4581–4589 (1996).
11. A. K. Chakraborty, An analysis of the phase evolution of six types of mullite gels. (2008) Unpublished.
12. J. C. Huling and G. I. Messing, Chemistry–crystallization relations in molecular gels. *J. Non-Cryst. Solids*, **147–148**, 213–221 (1992).
12. J. F. MacDowell and G. H. Beall, *Proc. Br. Ceram. Soc.*, **3**, 229–240 (1965).
13. A. K. Chakraborty and S. Das, Al-Si spinel phase formation in diphasic mullite gels. *Ceram. Int.*, **29**, 27–33 (2003a).
14. A. K. Chakraborty, Characterization of monophasic and diphasic mullite precursors by solid state reaction study. *Br. Ceram. Trans.*, **103**(1), 33–36 (2004).

15. A. K. Chakraborty, DTA characterisation of three types of Al_2O_3-SiO_2 gels made from TEOS-Al(OBu)$_3$ mixture with variation of water. *Ceram. Int.*, **22**, 463–469 (1996).

16. A. K. Chakraborty, Effect of pH on 980°C spinel phase-mullite formation of Al_2O_3-SiO_2 gels. *J. Mater. Sci.*, **29**, 1558–1568 (1994a).

17. S. Prochazka and F. J. Klug, Infrared-transparent mullite. *J. Am. Ceram. Soc.*, **66**(12), 874–880 (1983).

18. H. Suzuki, H. Saito, Y. Tomokiyo, and Y. Suyama, Processing of ultrafine mullite through alkoxide route. *Ceram. Trans.*, Vol. 6, edited by S. Somiya, R.F. Davis, and J. A. Pask, *Am. Ceram. Soc.*, Westerville, OH (1990) p. 263.

19. B. E. Yoldas and D. P. Partlow, Formation of mullite and other alumina-based ceramics via hydrolytic poly condensation of alkoxides and resultant ultra- and micro- structural effects. *J. Mater. Sci.*, **23**, 1895–1900 (1988).

20. A. Taylor and D. Holland, The chemical synthesis and characterization sequence of mullite. *J. Non-Cryst. Solids*, **152**, 1–17 (1993).

21. C. S. Hsi, H. Y. Lu, and F. S. Yen, Thermal behavior of aluminosilica xerogels during calcination. *J. Am. Ceram. Soc.*, **72**, 2208–2210 (1989).

22. S. Mitachi, M. Matsuzawa, K. Kaneko, S. Kanzaki, and Y. Tabata, Characterization of SiO_2-Al_2O_3 powders prepared from metal alkoxides. *Ceram. Trans.*, **6**, 275–286 (1990).

23. Y. Hirata, H. Minamizono, and K. Shimada, Property of SiO_2-Al_2O_3 powders prepared from metal alkoxide. *Yogo Kyokai Shi*, **93**(1), 36–54 (1985).

24. C. H. Horte and J. Wiegmann, *Naturwiss*, **43**, 9 (1956).

25. T. Demediuk and W. F. Cole, *Nature*, **181**, 1400 (1958).

26. J. Ossaka, Tetragonal mullite type phase from co precipitated gels. *Nature (London)*, **19**, 1000–1001 (1961).

27. H. Insley and R. H. Ewell, Thermal behavior of the kaolin minerals. *J. Res. Natl. Bur. Stand.*, **14**(5), 615–627 (1935).

28. H. Schneider, I. Merwin, and A. Sebald, Mullite formation from non-crystalline precursors. *J. Mater. Sci.*, **29**, 805–812 (1992).

29. I. Jaymes, A. Douy, P. Florian, D. Massiot, and J. P. Coutures, Mew synthesis of mullite. Structural evolution study by ^{17}O, ^{27}Al and ^{29}Si MAS NMR spectroscopy. *J. Sol-Gel Sci. Technol.*, **2**, 367–370 (1994).

30. S. Rajendran, H. J. Rossell, and J. V. Sanders, Crystallization of a co-precipitated mullite precursor during heat treatment. *J. Mater. Sci.*, **25**, 4462–4471 (1990).

31. A. M. L. M. Fonseca, J. M. F. Ferreira, I. M. M. Salvado, and J. L. Baptista, Mullite based compositions prepared by sol-gel techniques. *J. Sol-Gel Sci. Techno.*, **8**, 403–407 (1997).

32. D. J. Cassidy, J. L. Woolerey, J. R. Bartlett, and B. Ben-Nissan, The effect of precursor chemistry on the crystallization and densification of sol-gel derived mullite gels and powders. *J. Sol-Gel Sci. Technol.*, **10**, 19–30 (1997).

33. Y. Hirata, K. Sakeda, Y. Matsushita, K. Shimada, and Y. Ishihara, Characterization and sintering behavior of alkoxide-derived aluminosilicate powders. *J. Am. Ceram. Soc.*, **72**(6), 995–1002 (1989).

34. C. Gerardin, S. Sundaresan, J. Benziger, and A. Navrotsky, Structural investigation and energetics of mullite formation from sol-gel precursors. *Chem. Mater.*, **6**, 160– 170 (1994).

79. A. Yasumori, M. Iwasaki, H. Kawazoe, M, Yamane, and Y. Nakamura, Nuclear magnetic resonance study of the structure of aluminosilicate gel and glass. *Phys. Chem. Glasses*, **31**(1), 1–9 (1990).

35. A. K. Chakraborty, Further studies on mullitization of diphasic gel. *Trans. Ind. Ceram. Soc.*, **56**(1), 9–15 (1997).

36. A. K. Chakraborty, Role of hydrolysis water-alcohol mixture on mullitization of Al_2O_3-SiO_2 monophasic gels. *J. Mater. Sci.*, **29**, 6131–6138 (1994b).

37. A. K. Chakraborty and D. K. Ghosh, Kaolinite-mullite reaction series. The development and significance of a binary aluminosilicate phase. *J. Am. Ceram. Soc.*, **74**(6), 1401–1406 (1991).

38. A. K. Chakraborty and D. K. Ghosh, Crystallization behaviors of Al_2O_3 in presence of SiO_2. *J. Am. Ceram. Soc.*, **70**(3), C-45–C-48 (1987).

39. A. K. Chakraborty, An analysis of the phase evolution of six types of mullite gels. Unpublished (2008).

40. F. J. Klug, S. Prochajka, and R. H. Doremus, Aluminosilicate phase diagram in mullite region. *J. Am. Ceram. Soc.*, **70**, 750–759 (1987).

41. H. Yamada and S. Kimura, Studies on co precipitates of alumina and silica gels and its transformations at higher temperatures. *Yogo Kyokai Shi*, **70**, 87–93 (1962).

42. D. Croft and W. W. Marshall, A novel synthesis of alumino-silicates and similar materials. *Trans. Br. Ceram. Soc.*, **64**(3), 121–126 (1967).

43. T. D. McGee and C. D. WirKus, Mullitization of aluminum-silicate gels. *Am. Ceram. Soc. Bull.*, **51**, 577–581 (1972).

44. G. S. Perry, Microwave dielectric properties of mullite. *Trans. J. Br. Ceram. Soc.*, **72**, 279–283 (1973).

45. B. E. Yoldas, Monolithic glass formation by chemical polymerization. *J. Mater. Sci.*, **14**, 1843–1849 (1979).

46. J. A. Pask, X. W. Zhang, A. P. Tomsia, and B. E. Yoldas, Effect of sol-gel mixing on mullite microstructure and phase equilibria in the α-Al_2O_3-SiO_2 system. *J. Am. Ceram. Soc.*, **70**(10), 704–707 (1987).

47. J. Sanz, I. Sobrados, A. L. Cavalieri, P. Pena, S. de Aza, and J. S. Moya, Structural changes induced on mullite precursors by thermal treatment: a ^{27}Al MAS-NMR investigation. *J. Am. Ceram. Soc.*, **74**(10), 2398–2403 (1991).

48. Y. X. Huang, A. M. R. Senos, J. Rocha, and J. L. Baptista, Gel formation in mullite precursors obtained via tetraethylorthosilicate (TEOS) prehydrolysis. *J. Mater. Sci.*, **32**, 105–110 (1997).

49. M. Sales and J. Alarcon, Synthesis and phase transformations of mullite obtained from SiO_2-Al_2O_3 gels. *J. Euro. Ceram. Soc.*, **16**, 781–789 (1996).

50. A. K. Chakraborty, New data on thermal analysis of diphasic mullite gels. *J. Therm. Anal.*, **46**, 1413–1419 (1996b).

51. W.-C. Wei and J. W. Halloran, Phase transformation of diphasic aluminosilicate gels. *J. Am. Ceram. Soc.*, **71**(3), 166–172 (1988).

52. A. K. Chakraborty, Intermediate Si-Al spinel formation in phase transformation of diphasic mullite gel. *J. Mater. Sci.*, **28**, 3839–3844 (1993).

53. O. Sakurai, N. Mizutani, and M. Kato, Preparation of mullite powders from metal alkoxides by ultrasonic spray pyrolysis. *J. Ceram. Soc. Jpn.*, **96**, 639–645 (1988).

54. D. X. Li and W. J. Thomson, Effects of hydrolysis on the kinetics of high temperature transformations in aluminosilicate gels. *J. Am. Ceram. Soc.*, **74**, 574–578 (1991).

55. S. Komarneni and R. Roy, Solid-state ^{27}Al and ^{29}Si magic-angle spinning NMR of aluminosilicate gels. *J. Am. Ceram. Soc.*, **69**(3), C-42–C-44 (1986).

56. S. Komerneni and C. Ruscher, Single-phase and diphasic aerogels and xerogels of mullite: preparation and characterization. *J. Euro. Ceram. Soc.*, **16**, 143–147 (1996).

57. M. G. M. U. Ismail, Z. Nakai, K. Minegishi, and S. Somiya, Synthesis of mullite powder and its characteristics. *Int. J. High Technol. Ceram.*, **2**, 123–134 (1986).

58. M. J. Hyatt and N. P. Bansal, Phase transformations in xerogels of mullite composition. *J. Mater. Sci.*, **25**, 2815–2821 (1990).

59. K. Hamano, Z. Nakagawa, G. Cun-Ji, and T. Sato, *Mullite*, edited by S. Somiya (Uchida Rokakuho Publishing Co., Tokyo, Japan) 1985, p. 37.

60. C. S. Hong, P. Ravindranathan, D. K. Agrawal, and R. Roy, Synthesis and sintering of mullite powders by the decomposition/combustion of aluminium nitrate- amorphous fumed silica-urea mixtures. *J. Mater. Sci. Lett.*, **13**, 1072–1075 (1994).

61. H. Schneider, D. Voll, B. Saruhan, J. Sanz, G. Schrader, C. Ruscher, and A. Mosset, Synthesis and structural characterization of non-crystalline mullite precursor. *J. Non-Cryst. Solids*, **78**, 262–271 (1994).

62. J. C. Huling and G. L. Messing, In *Ceram. Trans.*, Mullite and mullite matrix composites, edited by S. Somiya, R. F. Davis, and J. A. Pask, *Am. Ceram. Soc.*, **6**, 221 (1990).

63. M. I. Nieto, G. Urretavizcaya, A. L. Cavalieri, and P. Rana, Structural changes in colloidal and polymeric aluminosilicate gels with mullite composition. *Br. Ceram. Trans.*, **97**(1), 17–23 (1998).

64. A. K. Chakraborty and D. K. Ghosh, Comment on "diphasic xerogels, a new class of materials phases in the system Al_2O_3-SiO_2. *J. Am. Ceram. Soc.*, **69**(8), C-202–C-203 (1986). Reply by S. Komarneni and R. Roy, ibid, **69**(8), C-204 (1986).

65. S. Sundaresan and I. A. Aksay, Mullitization of diphasic aluminosilicate gels. *J. Am. Ceram. Soc.*, **74**(10), 2388–2392 (1991).

66. B. Sonuparlak, Sol-gel processing of infrared transparent mullite. *Adv. Ceram. Mater.*, **3**(3), 263–267 (1988).

67. S. Komarneni, R. Roy, C. A. Fyfe, and G. J. Kennedy, Preliminary characterization of gel precursors and their high- temperature products by ^{27}Al magic-angle spinning NMR. *J. Am. Ceram. Soc.*, **68**(9), C-243–C-245 (1985).

68. L. Pach, A. Iratni, Z. Hrabe, S. Svetik, and S. Komarneni, Sintering and crystallization of mullite in diphasic gels. *J. Mater. Sci.*, **30**, 5490–5494 (1995).

69. N. Shinohara, D. M. Dabbs, and I. A. Aksay, Infrared transparent mullite through densification of monolithic gels at 1250°C. *Proc SPIE*, **683**, 19–24 (1986).

70. J. Wu, M. Chen, F. R. Jons, and P. F. James, Mullite and alumina-silica matrices for composites by modified sol-gel processing. *J. Non-Cryst. Solids*, **162**, 197–200 (1993).

71. M. Mizuno and H. Saito, Preparation of highly pure fine mullite powder. *J. Am. Ceram. Soc.*, **72**, 377–382 (1989).

72. S. Sugita, Sueyoshi and C. A. Contreras Soto, Fine pure mullite powder by homogeneous precipitation. *J. Euro. Ceram. Soc.*, **18**, 1145–1152 (1998).

73. T. Kumazawa, S. Ohta, S. Kanzaki, and H. Tabata, Influence of powder characteristics on microstructural and mechanical properties of mullite ceramics (74 wt. % Al_2O_3). *J. Jpn. Ceram. Soc.*, **99**, 1228–1233 (1991).

74. A. K. Chakraborty, Reinvestigation of Al-Si spinel phase in diphasic Al_2O_3–SiO_2 gels. *J. Am. Ceram. Soc.*, **88**(1), 134–140 (2005a).

75. M. Bartsch, B. Saruhan, M. Schmucker, and H. Schneider, Novel low-temperature processing route of dense mullite ceramics by reaction sintering of amorphous SiO_2-coated γ-Al_2O_3 particle nanocomposites. *J. Am. Ceram. Soc.*, **82**(6), 1388–1392 (1999).

76. H. Ivankovic, E. Tkalcec, R. Nass, and H. Schmidt, Correlation of the precursor type with densification behavior and microstructure of sintering mullite ceramics. *J. Euro. Ceram. Soc.*, **23**, 283–292 (2003).

77. M. D. Sacks, N. Bozkurt, and G. W. Scheiffele, Fabrication of mullite and mullite-matrix composites by transient viscous sintering of composite powders. *J. Am. Ceram. Soc.*, **74**(10), 2428–2437 (1991).

78. K. Wang and M. D. Sacks, Mullite formation by endothermic reaction of α-Al_2O_3/SiO_2 micro-composite particles. *J. Am. Ceram. Soc.*, **79**(1), 12–16 (1996).

79. M. D. Sacks, N. Bozkurt, and G. W. Scheiffele, Effect of composition on mullitization behavior of α-alumina/silica micro composite powders. *J. Am. Ceram. Soc.*, **80**(3), 663–672 (1997).

80. M. Schmucker, H. Schneider, M. Poorteman, F. Cambier, and R. J. Meinhold, *J. Euro. Ceram. Soc.*, **15**, 1201 (1995).

81. A. R. Boccaccini and C. B. Ponton, Colloidal processing of mullite ceramics: are silica-alumina "composite" precursor particles essential? *J. Mater. Sci. Lett.*, **19**, 1687–1688 (2000).

82. S. Kanzaki and H. Tabata, Sintering and mechanical properties of stoichiometric mullite. *J. Am. Ceram. Soc.*, **68**(1), C-6–C-7 (1985).

83. Y. Sato, T. Takei, S. Hayashi, A. Yasumori, and K. Okada, *J. Am. Ceram. Soc.*, **81**, 2197 (1998).

84. K. Hamano, T. Sato, and Z. Nakagawa, Properties of mullite prepared by co-precipitation and microstructure of fired bodies. *Yogo Kyokai Shi*, **94**(8), 818–822 (1986).

85. I. Jaymes and A. Douy, Homogeneous mullite-forming powders from spray-drying aqueous solutions. *J. Am. Ceram. Soc.*, **75**(11), 3154–3156 (1992).

86. I. Jaymes, A. Douy, and D. Massiot, Synthesis of a mullite precursor from aluminum nitrate and tetraethoxysilane via aqueous homogeneous precipitation: an ^{27}Al and ^{29}Si liquid- and solid-state NMR spectroscopic study. *J. Am. Ceram. Soc.*, **78**(10), 2648–2654 (1995).

87. C. H. Ruscher, G. Schrader, and M. Gotte, Infra-red spectroscopic investigation in the mullite field of composition: $Al_2(Al_{2+2x}Si_{2-2x})O_{10-x}$ with 0.55 > x > 0.25. *J. Euro. Ceram. Soc.*, **16**, 169–175 (1996).

88. K. Okada and N. Otsuka, Characterization of the spinel phase from SiO_2-Al_2O_3 xerogels and the formation process of mullite. *J. Am. Ceram. Soc.*, **69**(9), 652–656 (1986).

Problems

Question 1. What is the meaning of solid state reaction of γ-Al_2O_3 with SiO_2 (A) and solid state reaction of Al-Si spinel (low SiO_2 content) with SiO_2 (A) on heating?

During the continued heating process, Si will diffuse into the weak phase. How long will this incorporation continue?

Will incorporation continue till the quantity of SiO_2 into the Al-Si spinel or γ-Al_2O_3 structure is fully saturated and corresponds to composition equivalent to 3:2 composition of mullite?

Question 2. Okada et al. (1986) determined the composition of the Al-Si spinel phase formed at ~1000°C.

This spinel phase exists up to ~1250°C prior to forming mullite. Is it necessary to estimate the silica content of spinel phase in the range between 1000°C–1250°C in order to obtain the final composition of silica saturated Al-Si spinel phase before its transformation to mullite?

Question 3. Why mullite formation in two cases namely mullite gel synthesized out of TEOS and ANN at pH 6 to 7 and another gel made using same reagents but at pH > 11 using ammonia as co precipitating agent are different?

Is it due to the colloidal sizes of $Al(OH)_3$ particles and its crystalline phase obtained during coprecipitation processes at two different pH?

It is assumed that the size of aluminium hydroxide colloids formed at low pH is small (measurement data should be obtained) and as a result penetration of silica is more easier during heating such mullite gel. In comparison, at higher pH aluminium hydroxides missiles crystallizes into pseudoboehmite crystals. Consequently, it dehydrates and forms amorphous alumina partially at the initial stage of heating of mullite gel. Likely, high temperature is needed to incorporate silica into amorphous alumina and into γ-Al_2O_3 structure to develop fully the Al-Si spinel phase. Why does mullite formation in the first case occur at 1250°C, whereas the same crystallization phenomenon will occur at 1300°C in the second case?

Question 4. Why is viscous glass formation a preferred reaction than solid state reaction between α-alumina and amorphous silica to nucleate mullite at early temperature at ~1300°C?

Chapter 21

Critical Analysis of Classification Scheme of Mullite Gels

21.1 Introduction

Mullite precursors synthesized by various authors at varying processing conditions as shown in previous chapters lead to develop gels of different characters. For example,

(1) Same source of components (ANN and TEOS) produce two different types of monophasic mullite gels (Hoffman et al., 1984; Okada et al., 1986 and Chakraborty and Ghosh, 1988). Same source of other components (AlOBu and TEOS) produce two other different types of gels namely polymeric gel and colloidal gel (Yoldas and Partlow, 1988 and Yoldas, 1992) (see Chapters 2 and 3).
(2) Using ANN and TEOS by changing both pH and water of processing three types of mullite gels are produced (Schneider et al., 1993; Schmucker et al., 1999) (see Chapter 3).
(3) Using ANN and TEOS and by changing processing, four kinds of mullite precursors are synthesized (Jaymes et al., 1996) (see Chapter 12).

Mullite Formations: Analysis and Applications
Akshoy Kumar Chakraborty

ISBN 978-981-4877-05-3 (Hardcover), 978-1-003-03167-3 (eBook)
www.jennystanford.com

(4) Interestingly, using ANN and TEOS and by changing processing, six kinds of mullite precursors are synthesized (Chakraborty, 2008) (see Chapter 12).

(5) Precursor character also changes with use of silicon hydroxide of large colloids or aged gels considering silica component as source or with large aluminum hydroxide colloidal particles during gelation either in aqueous method, in monophasic gel method or in SD or SP (spray pyrolyzed) techniques (see Chapter 16).

Thus, with variations in both sources of components and technique of precursor synthesis., different types or kinds of mullite gels could be prepared. The number and types of precursor synthesized by various authors in detail are summarized in earlier chapters and thermal evolution of these mullite gels are consequently shown to vary to a great extent as elaborately discussed in Chapters 12, 15 and 16.

During the classification process of these mullite precursors, various researchers named the mullite gel as per the same two basic points viz. (i) the components used and or (ii) by processing conditions maintained as highlighted above. For example, (A) monophasic gel; (B) polymeric gel; (C) nitrate gel; (D) coprecipitated gel; (E) colloidal gel; (E) diphasic gel.

It is observed that parameters namely choice of component selected, hydrolysis method and or coprecipitation technique and code composition process followed influence much of the internal structure of the precursor which in turn display a significant role in crystallization chemistry during heating. It is felt necessary to classify different mullite precursors synthesized out of various sources.

21.2 Old Basis of Classification of Different Types of Mullite Gels

The basis of classifications of the said Al_2O_3-SiO_2 gels by the various researchers are cited in chronological ways.

(1) Monophasic vs. diphasic gel: Phase evolution behaviour is the criteria for classification of two types of nano structurally different gels was considered by Hoffman et al. (1984). Single phase xerogel exhibits 980°C exotherm in DTA and forms mullite and Al-Si spinel phase. Whereas diphasic xerogels show exactly different behaviour e.g., DTA exhibits a very broad exotherm over the entire region from about 700°C–1250°C and there after forms mullite.

(2) Slow hydrolysis vs. rapid hydrolysis technique: The method of gel synthesis i.e., how slowly or rapidly gelation/precipitation occur i.e., whether hydrolysis of TEOS has been done in acidic or in alkaline medium and classified them as SH and RH gel by Okada et al. (1986).

(3) Polymeric vs. colloidal gel: The choice of starting materials was considered by Yoldas (1992) e.g., whether TEOS and Al(OBu) mixture are hydrolysed and to create aluminum hydroxy bonds which thereafter reacts with the alkoxy bond of Si $(OEt)_4$ as such or with partially hydrolysed silicon ester and finally allow gelatin to form polymeric gel. In another case, Al(OBu) is first hydrolysed to form colloidal AlO(OH) and then reacted with Si $(OEt)_4$ with creation of chemically bonded silica.

    The synthesis of two sol-gel powders are classified as polymeric and colloidal by Yoldas and Partlow (1988). They made this distinction with a view that ideal molecular mixing or homogeneity wherein oxygen ions are bridging oxygens between Al and Si cations, corresponding to complete polymerization has been achieved in polymeric gel. Whereas in colloidal gel, boehmite particles reacted with TEOS with formation of = Si-O-[AlO(OH)$(n - 1)/n$]n linkages.

(4) Gels synthesized at various processing conditions: Preparation condition/i.e., pH and water/alcohol mixture or gel homogeneity used as the basis of classification of gels tentatively as shown in Table 21.1 by Schneidar et al. (1994). Spraying followed by pyrolysis is another technique used for mullite precursor synthesis.

Table 21.1 Classification of mullite precursors from various sources (Schneider et al., 1994) gel homogeneity

	High	Low	Medium	References
1.	Type I precursor	Type II precursor	Type III precursor	Schneider et al. (1993)
2.	Single phase gel	Diphasic gel	—	Hoffman et al. (1984)
3.	SH gels slow hydrolysis	NM-gel no mixing	RH-gel rapid hydrolysis	Okada et al. (1986)
4.	Polymeric gel	Colloidal	—	Yoldas (1990)
5.	Sample spray-pyrolized	Sample CA colloidal	Sample PB2 polymeric	SP Sanz et al. (1991)

Crystallization paths of amorphous mullite precursor powders (bulk composition close to 3:2 mullite) synthesized out of various components and in different processing conditions depend upon the scale of homogeneity of the primary amorphous phases and these vary considerably.

(1) Single phase gel of Hoffman et al. (1984); slow hydrolysis gel of Okada et al. (1986); Type I precursor of Schneider et al. (1993) are not most homogeneous gels/precursors, since minor amount of Al-Si spinel phase was developed during the occurrence of first exothermic peak. On considering the proposed scale, above such gels were designated as "homogeneous" in Chapter 11.

(2) Colloidal gel of Yoldas (1992) is a "low homogeneous" gel, contrary polymeric gel is not such type highly homogeneous as shown above by Schneider et al. (1993) in Table 21.1.

(3) Sample PB2 polymeric is nearly the same character as RH gel. Both are other kinds of monophasic gel and these are assumed as "less homogeneous".

(4) Deviation from normal gelation procedure of SH gels and polymeric gel are also observed in the literature. Accordingly, a lot of variations in their phase transformation are noted among groups of gels and a few of these are mentioned below.

(i) Nitrate gel

Single phase gel/SH gel of Okada et al./SHII of Haque (2000) are basically nitrate gel. These crystallized to mullite as a major phase at the occurrence of the first exothermic peak. Only a minor quantity of Al-Si spinel phase is shown to be associated with mullite phase at the same exotherm. Deviated nitrate gel is nitrate gel of Cassidy (1977) and nitrate gel of Yoldas (1992). All these gels showed crystallization of major spinel phase when heated at 1050°C contrary to basic nitrate gel synthesized by Okada et al. (1986).

(ii) Polymeric gel

Yoldas (1990) first named polymeric gel out of (TEOS and ALOBU) which develop major mullite and minor spinel phase at first exothermic peak temperature and further exhibits second exotherm for transformation of spinel at 1250°C. Deviated polymeric gels usually are polymeric gel made by Rajendran et al. (1990); PB2 gel made by Sanz et al. (1991). PA and PB gels are both polymeric gels as per Nieto et al. (1998). However, these polymeric gels exhibited different thermal behaviours. Similarly, different behaviour is also noted in the phase evolution process of Polymeric gel of Cassidy et al. (1977).

21.3 New Classification Scheme, Deviation from Normal Gelification Procedure of SH Gels, Colloidal Gels and Others

Now the question is: what should be the proper criteria for classification of alumina silica precursor powder? It is noted that precursors show variation in their 980°C phase developments and probably there lies the main difference. It would be worthwhile to consider the following points prior to make the classification. For example,

(1) Nature of the as prepared gel or precursor i.e., whether it is amorphous or crystalline.

(2) How many exothermic peaks the gel exhibits and nature of exotherms. There are three possibilities based on experimental data of DTA runs. Probably,

(i) It may exhibit only one sharp peak at ~980°C.

(ii) It may exhibit two exothermic peaks, one comparatively less sharp peak at ~980°C and a broad and short peak at ~1250°C.

(iii) It may exhibit two exothermic peaks, one broad peak over a long range of temperature at ~1260°C following a short peak at ~1300°C.

(3) Nature of crystalline phase or phases forms at 980°C and 1250°C or 1300°C exothermic temperatures.

(4) Probably, QXRD values of the crystalline phases develop at the aforesaid exothermic peaks may be noted and these data are helpful in assessing the path of thermal evolution.

(5) Besides, mullite formation processes occur out of (using two component particles of) in case of fused oxide mixtures (mm or μ range), in case of oxide mixtures (μm range), in case of microcomposite powders (nm range) are also to be mentioned for comparison to the phase evolution processes for alumina-silica mullite gels system.

Accordingly, thermal sequences of five-six types of gels and precursors are described in Chapters 2–5 and 12 and subsequently evaluated in Chapter 16 will be taken into account.

The individual finding of a precursor with those of established thermal sequences given by various earlier authors in the conquest of five points as mentioned above are compared.

Considering five points as suggested above and on the basis of analysis of five to six types of mullite precursors as mentioned earlier and their comparisons of thermal changes, a newer classification scheme is globally presented in Table 21.2 with few examples.

Table 21.2 Different classes of mullitization reactions based on sources of components

Class 1 Mullite formation during heat treatment of fused-oxides mixtures (academic interest)	
Class 1.1	
Fused-oxides mixtures	
Davis and Pask (1972)	Single crystal α-alumina (sapphire)/fused silica, cristobalite.
Aksay and Pask (1975)	Sapphire/aluminosilicate glasses

Nurishi and Pask (1982)	α-alumina/fused silica
Class 1.2 Mullite formation during heat treatment of oxides mixtures (industrial process)	
Oxides mixtures	
Rana et al. (1982)	Alcoa A-14 α-alumina powder/quartz powder (Ottawa flour)/cristobalite
Schmucker et al. (1994)	Quartz/α-alumina
Saruhan et al. (1993) Saruhan et al. (1994) Saruhan et al. (1996)	-Reaction sintered mullite (α-alumina/ amorphous silica) sample named as GK. Precursor seeded material (sample mark-PS). It is a mixture of 57.4% α-alumina, 22.6% powder and 20% mullite precursor. - Cristobalite/α-alumina - Amorphous silica/α-alumina
Nurishi and Pask (1982)	Alcoa XA-16 reactive α-alumina/Corning 7940 fused silica glass.
Sacks and Pask (1982)	Powder mixture of A-14 Al_2O_3/Silica flour
Sacks et al. (1997)	TEOS – to produce silica particles, average dia. 160 nm/corundum (180 nm) are separately mixed

Class 2 Mullite formation during heat treatment of microcomposite powder (commercial process)	
Class 2.1	
Microcomposite powder (precursor IV). Assumed as "more heterogeneous"	
Sacks et al. (1991)	TEOS/corundum (~0.2 μm) – microcomposite powder
Wang et al. (1992)	TEOS/corundum (~0.75 μm)
Wu et al. (1993)	Ludox and α-alumina
Wang and Sacks (1996)	TEOS/corundum (~0.18 μm
Bartsch et al. (1999)	(i) γ-alumina nano size/TEOS (ii) α-alumina sub micrometer size /TEOS
Ivankovic et al. (2003)	γ-alumina/TEOS

Class 2.2 Mullite formation during heat treatment of colloidal gel (commercial process)	
Colloidal gel (precursor IV). Assumed as "heterogeneous"	
Ghate et al. (1973)	Fumed alumina (γ-Al_2O_3, surface area-100 m^2/g) and Ludox (SiO_2-31 wt %, particle dia. -140 Å)
Sacks and Pask (1982)	Aq. suspension of alumina (γ-Al_2O_3) and silica (Ludox AS)
Metcalfe and Sant (1975)	Consecutive precipitation

Class 2.3 Mullite formation during heat treatment of diphasic gel/ colloidal gel	
Diphasic gel (precursor IV). Assumed as "heterogeneous"	
"In situ" diphasic gel, Chakraborty (1979)	ANN/TEOS
Diphasic gel made by Sonuparlak (1988)	Boehmite/TEOS both in acidic and basic condition
Wei and Halloran (1988)	Boehmite/TEOS in acidic condition
Derived diphasic gel made by Hsi et al. (1989)	ANN and TEOS in absolute alcohol at four different pH values namely 8.3, 9.5, 10.1 and 10.4
Ivankovic et al. (2003)	Pseudo boehmite/TEOS
Yoldas (1992)	Colloidal gel

Class 3	
Class 3.1 Mullite formation during heat treatment of coprecipitated gel	
Coprecipitated gel (precursor III). Assumed as "less homogeneous"	
Yamada and Kimura (1962)	Gel samples A, B, C
Okada et al. (1986) Low and McPherson (1989)	–RH gel –Aluminosilicate gels
Hamano et al. (1986)	Coprecipitated mullite gel (CP)
Li and Thomson (1991)	Colloidal gel marked CG

Kanka and Schneider (1994)	Coprecipitated mullite gel
Dafader et al. (1998a)	Coprecipitated mullite gel (CP6)

Class 3.2 Mullite formation during heat treatment of monophasic gel	
Monophasic gel SH gel/polymeric gel molecular gel. Assumed as "homogeneous"	
Hoffman et al. (1984)	Monophasic phase gel
Okada et al. (1986)	SH gel
Yoldas (1992)	Polymeric gel
Huling and Messing (1991)	100 M gel
Jaymes et al. (1996)	Sample A, chelated precursor
Geradin et al. (1994)	M1 sample
Orefice et al. (1997)	Polymeric gel

Class 3.3 Mullite formation during heat treatment of mullite gel of water free approach	
Polymeric gel/Type I precursor. Assumed as "most homogeneous"	
Yoldas and Partlow (1988)	Sample C
Chakraborty and Ghosh (1988)	Sample G 31
Schneider et al. (1993)	Type I
Taylor and Holland (1993)	GNW

Class 3.4 Mullite formation during heat treatment of aqueous gel mullite gel	
Assumed as "most homogeneous" synthesized by	
Insley and Ewell (1935)	Sodium silicate and aluminium sulphate
Chakraborty and Ghosh (1988)	Silica sol and aluminium sulphate

Class 4 Mullite formation during heat treatment of sprayed precursor	
Sprayed precursor (precursor I). Assumed as "most homogeneous"	
(i) Spray pyrolyzed powder first synthesized by	
Kanzaki et al. (1985)	Used ANN and TEOS
(ii) Spray dried powder	
Douy (2006)	Hydrolysed TEOS and ANN
Chakraborty (2008)	Used ANN and TEOS

Classes of Precursors

(1) Mullite gel/precursor synthesized in aqueous gel technique is classified as (Class I precursor).

(2) Precursor synthesized by a water free approach is to be classified as (Class I precursor).

(3) The precursors those have been synthesized using different combination of raw materials under varying processing conditions in different names such as (1) monophasic xerogel; (2) polymeric gel; (3) slow hydrolysis gel (SH); Type II precursor are noncrystalline up to the commencement of 980°C exotherm and these gels form major nonstoichiometric mullite and minor amount of Al-Si spinel phase with exhibition of sharp and pronounced 980°C exotherm in DTA is to be designated as Class II precursor.

(4) Monophasic gels e.g., RH gel (pH maintained during gelation is not known from Okada et al.'s work, which may be less than 10); G-150 of Chakraborty and Ghosh (1988); Colloidal gel of Li and Thomson (1991); Type III of Schneider et al. (1993); Coprecipitated gels e.g., gels made by Yamada and Kimura (1962), CP6 gels made by Dafather (1988a) are amorphous in character although they have been synthesized by using ammonia and remain amorphous till the occurrence of first exotherm i.e., these gels are monophasic in nature. But unlike the Class II precursor, they form only Al-Si spinel phase with comparatively less pronounced 980°C exotherm in DTA to be designated as Class III precursor. These kind of

gel or precursor on further DTA analysis exhibits a second exotherms at 1250°C with formation of mullite followed by complete transformation of Al-Si spinel phase

(5) Precursors synthesized by a large section of mullite researchers show a varying ratio of mullite to Al-Si spinel phase at 980°C. These are to be designated as Class III precursor.

(6) Diphasic gel; G-49 i.e., xerogels prepared at high pH; Type II; CP10. show boehmite phase at room temperature, does not exhibit 980°C exotherm but forms Al-Si spinel phase much below this temperature and exists above 1250°C and crystallizes to mullite at 1300°C exotherm. Such gels are to be designated as Class IV precursor.

(7) Precursors synthesized using ammonia as precipitating agents generally crystallized Al-Si spinel phase with mullite and thus developed diphasicity as a character of gel. These coprecipitated gels are to be designated as Class IV precursor.

(8) Precursors powder e.g., SP i.e., spray pyrolyzed; SP i.e., spray dried; G-31 are purely amorphous in character. These precursors exhibit the most pronounced 980°C exotherm with the formation of only predominant nonstoichiometric mullites. DTA study shows no further exotherm. Such precursors are to be designated as Class I precursor.

References

1. D. W. Hoffman, R. Roy, and S. Komarneni, Diphasic xerogels, a new class of materials: phases in the system Al_2O_3-SiO_2. *J. Am. Ceram. Soc.*, **67**, 468–471 (1984).

2. K. Okada and N. Otsuka, Characterization of the spinel phase from SiO_2-Al_2O_3 xerogels and the formation process of mullite. *J. Am. Ceram. Soc.*, **69**(9), 652–656 (1986).

3. A. K. Chakraborty and D. K. Ghosh, Synthesis and 980°C phase development of some mullite gels. *J. Am. Ceram. Soc.*, **71**(11), 978–987 (1988).

4. B. E. Yoldas and D. P. Partlow, Formation of mullite and other alumina-based ceramics via hydrolytic polycondensation of alkoxides and resultant ultra- and micro-structural effects. *J. Mater. Sci.*, **23**, 1895–1900 (1988).

5. B. E. Yoldas, Effect of ultrastructure on crystallization of mullite. *J. Mater. Sci.*, **27**(24), 6667–6672 (1992).
6. H. Schneider, B. Saruhan, D. Voll, L. Merwin, and A. Sebald, Mullite precursor phases. *J. Euro. Ceram. Soc.*, **11**, 87–94 (1993).
7. M. Schmucker and H. Schneider, Structural development of single phase (Type I) mullite gels. *J. Sol-Gel Sci. Technol.*, **15**, 191–199 (1999).
8. I. Jaymes, A. Douy, D. Massiot, and J. P. Coutures, Characterization of mono- and diphasic mullite precursor powders prepared by aqueous routes, ^{27}Al and ^{29}Si MAS-NMR spectroscopy. *J. Mater. Sci.*, **31**, 4581–4589 (1996).
9. A. K. Chakraborty, An analysis of the phase evolution of six types of mullite gels. Unpublished data (2008).
10. H. Schneider, D. Voll, B. Saruhan, J. Sanz, G. Schrader, C. Ruscher, and A. Mosset, Synthesis and structural characterization of non-crystalline mullite precursor. *J. Non-Cryst. Solids*, **78**, 262–271 (1994).
11. J. Sanz, I. Sobrados, A. L. Cavalieri, P. Pena, S. de Aza, and J. S. Moya, Structural changes induced on mullite precursors by thermal treatment: a ^{27}Al MAS-NMR investigation. *J. Am. Ceram. Soc.*, **74**(10), 2398–2403 (1991).
12. M. Haque, Thesis, University of Kolkata (2000).
13. D. J. Cassidy, J. L. Woolerey, J. R. Bartlett, and B. Ben-Nissan, The effect of precursor chemistry on the crystallization and densification of sol-gel derived mullite gels and powders. *J. Sol-Gel Sci. Technol.*, **10**, 19–30 (1997).
14. S. Rajendran, H. J. Rossell, and J. V. Sanders, Crystallization of a co-precipitated mullite precursor during heat treatment. *J. Mater. Sci.*, **25**, 4462–4471 (1990).
15. M. I. Nieto, G. Urretavizcaya, A. L. Cavalieri, and P. Rana, Structural changes in colloidal and polymeric aluminosilicate gels with mullite composition. *Br. Ceram. Trans.*, **97**(1), 17–23 (1998).
16. R. F. Davis and J. A. Pask, Diffusion and reaction studies in the studies in the system Al_2O_3-SiO_2. *J. Am. Ceram. Soc.*, **55**(10), 525–531 (1972).
17. I. A. Aksay and J. A. Pask, Stable and metastable equilibria in the system SiO_2-Al_2O_3. *J. Am. Ceram. Soc.*, **58**(11–12), 507–512 (1975).
18. Y. Nurishi and J. A. Pask, Sintering of α-Al_2O_3-amorphous silica compacts. *Ceram. Int.*, **8**, 57–59 (1982).
19. A. P. S. Rana, O. Aiko, and J. A. Pask, Sintering of α-Al_2O_3/quartz, and α-Al_2O_3/cristobalite related to mullite formation. *Ceram. Int.*, **8**, 151–157 (1982).

20. M. Schmucker, W. Albers, and H. Schneider, Mullite formation by reaction sintering of quartz and α-Al_2O_3 – a TEM study. *J. Euro. Ceram. Soc.*, **14**, 511–515 (1994).

21. B. Saruhan, W. Albers, and H. Schneider, Improved densification by seeding of reaction sintered mullite with sol-gel derived precursor. *J. Mater. Sci. Lett.*, **12**, 1812–1814 (1993).

22. B. Saruhan, Solid solution range of mullite up to 1800°C & microstructural development of ceramics. *J. Mater. Sci.*, **29**, 3261–3268 (1994).

23. B. Saruhan, W. Albers, H. Schneider, and W. A. Kaysser, Reaction sintering mechanism of mullite in system cristobalite/α-Al_2O_3 and amorphous SiO_2/α-Al_2O_3. *J. Euro. Ceram. Soc.*, **16**, 1075–1081 (1996).

24. M. D. Sacks and J. A. Pask, Sintering of mullite containing materials: I, effect of composition. *J. Am. Ceram. Soc.*, **65**, 65–70 (1982a).

25. M. D. Sacks and J. A. Pask, Sintering of mullite containing materials: I, effect of agglomeration. *J. Am. Ceram. Soc.*, **65**, 70–77 (1982b).

26. M. D. Sacks, K. Wang, G. W. Scheiffele, and N. Bozkurt, Effect of composition of mullitization behavior of α-alumina/silica microcomposite powders. *J. Am. Ceram. Soc.*, **80**(3), 663–672 (1997).

27. M. D. Sacks, N. Bozkurt, and G. W. Scheiffele, Fabrication of mullite and mullite – matrix composites by transient viscous sintering of composite powders. *J. Am. Ceram. Soc.*, **74**(10), 2428–2437 (1991).

28. J. G. Wang, C. B. Ponton, and P. M. Marquis, Effects of green density on crystallization and mullitization in the transiently sintered mullite. *J. Am. Ceram. Soc.*, **75**(12), 3457–3461 (1992).

29. J. Wu, M. Chen, F. R. Jones, and P. F. James, Mullite and alumina-silica matrices for composites by modified sol-gel processing. *J. Non-Cryst. Solids*, **162**, 197–200 (1993).

30. K. Wang and M. D. Sacks, Mullite formation by endothermic reaction of α-Al_2O_3/SiO_2 micro-composite particles. *J. Am. Ceram. Soc.*, **79**(1), 12–16 (1996).

31. M. Bartsch, B. Saruhan, M. Schmucker, and H. Schneider, Novel low-temperature processing route of dense mullite ceramics by reaction sintering of amorphous SiO_2-coated γ-Al_2O_3 particle nanocomposites. *J. Am. Ceram. Soc.*, **82**(6), 1388–1392 (1999).

32. H. Ivankovic, E. Tkalcec, R. Nass, and H. Schmidt, Correlation of the precursor type with densification behavior and microstructure of sintered mullite ceramics. *J. Euro. Ceram. Soc.*, **23**, 283–292 (2003).

33. B. B. Ghate, D. P. H. Hasselman, and R. M. Spriggs, Synthesis and characterization of high purity, fine grained mullite. *Ceram. Bull.*, **52**(9), 670–672 (1973).

34. B. L. Metcalfe and J. H. Sant, The synthesis, microstructure and physical properties of high purity mullite. *Trans. J. Br. Ceram. Soc.*, **74**, 193–201 (1975).

35. A. K. Chakraborty, Formation of silicon-aluminium spinel. *J. Am. Ceram. Soc.*, **62**, 120 (1979).

36. B. Sonuparlak, Sol-gel processing of infrared transparent mullite. *Adv. Ceram. Mater.*, **3**(3), 263–267 (1988).

37. W.-C. Wei and J. W. Halloran, Phase transformation of diphasic aluminosilicate gels. *J. Am. Ceram. Soc.*, **71**(3), 166–172 (1988).

38. C. S. Hsi, F. S. Yen, and Y. H. Chang, Characterization of co-precipitated Al_2O_3-SiO_2 gels. *J. Mater. Sci.*, **24**, 2041–2046 (1989).

39. H. Yamada and S. Kimura, Studies on co-precipitates of alumina and silica gels and its transformations at higher temperatures. *Yogo Kyokai Shi*, **70**, 87–93 (1962).

40. M. Low and R. McPherson, The structure and composition of Al- Si spinel. *J. Mater. Sci.*, **7**, 1196–1198 (1989).

41. K. Hamano, T. Sato, and Z. Nakagawa, Properties of mullite prepared by co-precipitation and microstructure of fired bodies. *Yogo Kyokai Shi*, **94**(8), 818–822 (1986).

42. D. X. Li and W. J. Thomson, Effects of hydrolysis on the kinetics of high temperature transformations in aluminosilicate gels. *J. Am. Ceram. Soc.*, **74**, 574–578 (1991).

43. B. Kanka and H. Schneider, Sintering mechanism 7 development of co-precipitated mullite. *J. Mater. Sci.*, (1993).

44. B. Kanka and H. Schneider, Sintering mechanical and microstructural development of co-precipitated mullite. *J. Mater. Sci.*, **29**, 1239–1249 (1994).

45. M. H. Dafader, S. Das, and A. K. Chakraborty, Effect of water used during gelation on the phase transformation of Al_2O_3 gel. *Trans. Ind. Ceram. Soc.*, **57**(5), 121–126 (1998b).

46. J. C. Huling and G. L. Messing, Epitactic nucleation of spinel in aluminosilicate gels and its effect on mullite crystallization. *J. Am. Ceram. Soc.*, **74**(10), 2374–2381 (1991).

47. C. Gerardin, S. Sundaresan, J. Benziger, and A. Navrotsky, Structural investigation and energetics of mullite formation from sol-gel precursors. *Chem. Mater.*, **6**, 160–170 (1994).

48. R.-L. Orefice and W.-L. Vasconcelos, Sol-gel transition and structural evolution on multicomponent gels derived from the alumina-silica system. *J. Sol-Gel Sci. Technol.*, **9**, 239–249 (1997).

49. A. Taylor and D. Holland, The chemical synthesis and characterization sequence of mullite. *J. Non-Cryst. Solids*, **152**, 1–17 (1993).

50. H. Insley and R. H. Ewell, Thermal behavior of the kaolin minerals. *J. Res. Natl. Bur. Stand.*, **14**(5), 615–627 (1935).

51. S. Kanzaki and H. Tabata, Sintering and mechanical properties of stoichiometric mullite. *J. Am. Ceram. Soc.*, **68**(1), C-6–C-7 (1985).

52. A. Douy, Crystallization of amorphous spray-dried precursors in the Al_2O_3-SiO_2 system. *J. Euro. Ceram. Soc.*, **26**, 1447–1454 (2006).

Author Index

Aelion et al. (1950) 292–293, 297
Agrell and Smith (1960) 604, 606, 806
Aksay and Pask (1975) 2, 582, 582, 830
Aksay and Wiederhorn (1991) 7
Aksay et al. (1991) 7, 9
Al-Jarsha et al. (1985) 27
Angel and Prewitt (1986) 3, 609, 730
Anilkumar et al. (1997) 584
Aoki et al. (1992) 346
Aramaki and Roy (1962) 2, 604, 606–607, 733, 745

Ban and Okada (1992) 732
Ban and Okada (1993) 271, 754
Ban et al. (1996) 732
Baranwal et al. (2001) 99, 613
Bartsch et al. (1999) 83, 801, 805, 831
Bechtold (1968) 311
Bhattcharya et al. (1996) 96, 304, 354
Boccaccini and Ponton (2000) 805
Bowen and Greig (1924) 2, 4, 604
Bradley and Roussin (1932) 604
Brindley and Nakahira (1959) 534
Brinker et al. (1982) 291–292, 320
Brinker et al. (1984) 293, 296, 311
Burnham (1963) 3
Burnham (1964) 3
Bye and Simpkin (1974) 685
Cameron (1977) 212, 214–216, 231–232, 604, 607, 611, 615, 634, 638, 731–732, 744–745, 748, 808, 810
Carman (1940) 284
Cassidy et al. (1997) 382, 825, 836
Chakraborty (1979) 79, 513, 516–517, 520, 530, 535, 544, 547, 549, 569, 591, 626, 628, 639, 657, 665, 715, 804, 832
Chakraborty (1993) 486, 544, 550, 657, 703
Chakraborty (1994a) 24, 31, 37, 140–142, 148, 157, 169–170, 314, 319–320, 322, 343, 345, 466, 775
Chakraborty (1994b) 24, 162, 286
Chakraborty (1996a) 55, 112, 175, 349
Chakraborty (1996b) 473, 483, 489, 681
Chakraborty (1997) 483, 489, 570, 585
Chakraborty (2000) 526
Chakraborty (2003a) 485, 570, 657, 670, 739, 773
Chakraborty (2003b) 189, 506, 657, 702, 773
Chakraborty (2004) 438, 508, 570, 773
Chakraborty (2005a) 243, 495, 566, 587
Chakraborty (2005b) 616, 637
Chakraborty (2005c) 617, 736, 756

Chakraborty (2006) 245, 498, 533, 600, 677, 738
Chakraborty (2008a) 440, 501, 585, 675, 681
Chakraborty (2008b) 299, 460
Chakraborty and Das (2003a) 526, 570
Chakraborty and Das (2003b) 525
Chakraborty and Ghosh (1978) 656
Chakraborty and Ghosh (1979) 569
Chakraborty and Ghosh (1986) 27, 582, 794
Chakraborty and Ghosh (1987) 58, 530, 653, 784, 808
Chakraborty and Ghosh (1988) 24, 28, 37, 47, 58, 62, 139, 145, 162, 283, 287, 304, 342, 344, 349–350, 352, 371, 377, 411, 421, 473, 525, 541–542, 664, 774, 777, 790–791, 804, 833–834
Chakraborty and Ghosh (1991) 784
Chakraborty and Macnenzie (2006) 591
Chakraborty, unpublished (2008) 112, 200, 271, 389, 440, 466, 573, 585, 657, 659, 735, 769, 772, 834
Charles (1949) 23
Chen and Vilminot (1994) 118
Cho et al. (1998) 7
Chung et al. (1992) 98
Clark and White (1950) 685
Cloos et al. (1969) 22, 36
Colomban (1989) 61, 109, 438, 443, 465, 473, 483
Colomban (1991) 465
Colomban (1993) 474
Colomban and Mazerolles (1990) 126, 466, 473
Colomban Vendange (1992) 475
Croft and Marshall (1967) 45, 343, 541

Dafader et al. (1998a) 152, 833
Davis and Pask (1971) 7, 830
de Boer et al. (1960) 22
Dekeyser (1965) 576, 581, 604
Demediuk and Cole (1958) 43, 343, 344, 541
Demediuk and Cole (1960) 44
Dessalces et al. (1992) 22
Douy (1991) 145, 338,
Douy (2006) 198, 336, 344, 355, 463, 466, 706, 768, 834
Douy and Odier (1989) 113
Durovic (1962) 3
Durovic (1969) 3

Engelhadt and Michel (1987) 235, 687, 740

Fahrenholtz et al. (1991) 185, 197
Fenstermacher and Hummel (1961) 7
Fischer et al. (1996) 599, 617
Fisher et al. (1993) 617
Fonseca et al. (1997) 110, 173, 349
Foster (1959) 730
Freund (1967) 446
Fukuoka et al. (1993) 144, 352, 372, 423, 427, 443, 664

Gani and McPharson (1977) 98
Garvey (1980) 341
Ge et al. (1992) 110, 148
Gelsdorf et al. (1958) 605
Geradin et al. (1994) 54, 110, 210, 262, 341, 382, 471, 613, 627, 833
Ghate et al. (1973) 80, 129, 832
Gossink (1975) 291
Grofesik and Vago (1961) 47, 383

Guo et al. (1986) 101

Hamano et al. (1985) 50, 64, 71, 85, 160, 317, 319, 323, 483–484, 543, 562, 566, 583, 794, 813
Hamano et al. (1986) 64, 71, 77, 85, 93, 109, 160, 187, 388, 483, 543, 562, 581, 800, 806, 832
Haque (2000) 112, 200, 271, 305, 389, 769, 825
Heinrich and Raether (1992) 117, 181, 343, 347, 351
Heinrich et al. (1991) 117
Hirata et al. (1989) 35, 38, 222, 475, 535, 562, 609, 611, 618, 635, 637, 702, 733, 737, 743, 749–750
Hirata et al. (Jan., 1985a) 35, 50, 156, 160, 217, 286, 340, 483, 536, 543, 562, 609, 611, 635, 733, 754
Hirata et al. (Sept., 1985b) 35, 218, 483, 535, 551, 635, 733
Hoffman et al. (1984) 27, 71, 161, 347, 373, 541, 566, 627, 825, 833
Hong et al. (1995) 800
Horiuchi et al. (1999) 708
Horte and Wiegmann (1956) 43, 343
Hsi et al. (1989) 48, 153, 193, 599, 832
Huang et al. (1997) 48, 111, 150, 249, 342, 430, 471, 599
Huizhong et al. (2003) 127
Huling and Messing (1989) 74, 109, 122
Huling and Messing (1990) 74, 109, 188, 338, 677
Huling and Messing (1991) 338, 538, 540, 833
Hwang et al. (2000) 664
Hyatt and Bansal (1990) 30, 76, 188, 220, 627

Iller (1963) 677, 720
Iller (1964) 668
Iller (1979) 143, 294, 322, 459
Inoue et al. (1996) 96, 354
Insley and Ewell (1935) 23, 128, 168, 420, 541, 833
Irwin et al. (1987) 297
Irwin et al. (1988) 297, 424
Ismail et al. (1986) 72, 568
Ismail et al. (1987) 72
Ivankovic et al. (2003) 374, 473, 538, 564, 585, 652, 681, 801, 831, 832
Iwahiro et al. (2001) 127

Janackovic et al. (1996) 98
Jaymes and Douy (1992) 93, 113, 198, 354, 472
Jaymes and Douy (1995) 113, 199
Jaymes et al. (1994) 48, 240, 336, 346
Jaymes et al. (1995) 56,94, 199, 242, 348, 354, 358
Jaymes et al. (1996) 246, 265, 307, 338, 344, 347, 373, 429, 470, 655, 686, 833
Jin et al. (2002) 196, 226, 521
Jones and Fischbach (1988) 302
Jones et al. (1984) 10

Kamijo et al. (1994) 431
Kanka and Schneider (1994) 833
Kansal and Laine (1997) 127
Kanzaki et al. (1985) 91, 198, 358, 542, 834
Kanzaki et al. (1990) 626
Keefer (1984) 297, 320, 459
Kim et al. (2003) 46, 144
Klaussen (1990) 75, 188
Klaussen (1992) 569
Klein and Garvey (1980) 341

Klug et al. (1987) 2, 616, 755
Klug et al. (1990) 2
Komarneni and Roy (1986) 241, 304, 430, 805
Komarneni et al. (1985) 240, 431
Komerneni and Ruscher (1996) 17
Konard et al. (1867) 290
Konard et al. (1929) 290, 306
Kruit (1952) 286
Kubota and Takagi (1986) 92, 286
Kumazawa et al. (1990) 128, 153, 385
Kumazawa et al. (1991) 140, 181, 385, 614

Lee et al. (1992) 702
Lee et al. (2002) 46, 319
Lee et al. (2003) 46, 319
Lee et al. (2004) 46, 319
Leipold and Sibod (1982) 10
Leonard (1964) 438
Leonard et al. (1971) 23, 309, 433
Lewis (1991) 98
Li and Thomson (1991) 30, 146, 161, 308, 342, 598, 832
Lippmaa (1980) 357
Lopez et al. (1986) 10
Lopez et al. (1989) 10
Low and McPherson (1989) 51, 533, 604, 614, 748, 832

MacDowell and Beall (1969) 772
Mackenzie (1972) 212
Mackenzie et al. (1996) 439, 444, 464, 475
Majumder and Walch (1963) 604, 606
Marcilly et al. (1970) 114
Mazdiyasni and Brown (1972) 49, 317, 340
McAteee and Milligan (1950) 79
McGee and Wirkus (1972) 75
McHale et al. (1997) 530, 707
McMillan and Piriou (1982) 212
Mehrotra and Pant (1963) 122
Meng and Ruggins (1983) 10, 126
Merwin et al. (1991) 270
Metcalfe and Sant (1975) 128, 832
Milliken et al. (1950) 22
Mitachi et al. (1990) 109, 157, 317
Mizukami et al. (1990) 117
Mizukami et al. (1997) 117
Mizuno and Saito (1989) 101
Mizuno et al. (1990) 193 286, 345
Moora (1992) 98, 102
Morikawa et al. (1982) 442

Nakajima (1975) 3
Nieto et al. (1998) 111, 220, 347, 473
Nishu et al. (1989) 144, 239, 286, 345
Nogami and Moriya (1980) 288
Nurishi and Pask (1982) 831

Oblad et al. (1951) 22
Ocana et al. (1993) 98
Ogihara et al. (1994) 102
Okada (1993) 7
Okada (1996) 170–171
Okada and Otsuka (1986b) 27, 47, 152, 215, 338, 339, 482, 517, 524, 528, 534, 565, 573, 609, 613, 685, 735, 746, 831, 833
Okada and Otsuka (1987) 482, 731-32, 738
Okada and Otsuka (1990) 168, 344, 751
Okada et al. (1991) 7, 129, 287, 345
Okada, Hoshi, and Otsuka (1986a) 214–215, 230–232, 564, 809
Okuno et al. (1997) 441, 441-443
Orefice and Vasconcelos (1997) 61, 126, 184, 226, 696, 833

Ossaka (1961) 37, 44, 140, 745

Pach et al. (1995) 78
Padmaja et al. (1998) 195
Pask et al. (1987) 308, 317, 339, 341, 358, 598
Paulick et al. (1987) 158
Percival et al. (1974) 212
Perry (1973) 9, 128
Pope and Mackenzie (1886) 294
Pouxviel et al. (1986) 122
Prochazka and Klug (1983) 49, 168, 317, 340, 609

Rajendran et al. (1990) 159, 193, 347
Rana et al. (1982) 831
Reynen and Faizullah (1974) 126
Risbud (1987) 236
Rooksby and Partridge (1939) 604
Roy (1956) 8
Roy and Osborn (1954) 8
Roy et al. (1977) 98
Ruscher et al. (1996) 216

Saalfeld and Guse (1981) 3
Sacks and Pask (1982a) 80, 336, 831
Sacks and Pask (1982b) 80, 832
Sacks et al. (1990) 129
Sacks et al. (1991) 81, 801, 805, 831
Sacks et al. (1995) 81
Sacks et al. (1997) 81, 831
Sadanaga et al. (1962) 3
Saha and Pramanik (1994) 32
Sakka and Kamiya (1982) 289
Sakurai et al. (1988) 51, 93, 318
Sales and Alarcon (1996) 55, 222, 655, 750, 754
Sanz et al. (1991) 182, 264, 358, 387, 473
Saruhan et al. (1993) 831
Saruhan et al. (1994) 831
Saruhan et al. (1996) 831
Sato et al. (1998) 92, 677, 717
Schaefer et al. (1984) 291
Schmidt et al. (1984) 291, 295, 316
Schmucker and Schneider (1999) 315, 397, 443, 448, 476, 612, 825
Schmucker et al. (1994) 831
Schneider and Komarneni (2005) 7, 10, 36
Schneider and Rymon-Lipinski (1988) 612, 636, 732, 745
Schneider et al. (1992) 25, 52, 139, 236, 247, 372, 429, 463, 466, 471, 541, 564, 654, 745
Schneider et al. (1993) 11, 53, 112, 153, 159, 180, 225, 259, 314, 316, 338, 346, 350, 358, 463, 541, 564, 655, 691, 745, 825, 833
Schneider et al. (1994) 433, 437, 444
Schneider et al. (1994) 518
Schneider, Fischer, and Voll (1993) 11, 613
Schneider, Okada, and Pask (1994) 6
Scholze (1955) 4
Sen and Thiagarajan (1988) 373, 625
Shiga et al. (1991) 56, 319
Shinohara et al. (1986) 73
Sin et al. (2001) 149
Sinko et al. (2001) 311
Skorodumora (1989) 28
Somiya and Hirata (1991) 7
Somiya et al. (1985) 50, 317
Sonuparlak (1988) 73, 109, 187, 832
Staley and Brindley (1969) 581
Su Yen et al. (1991) 189, 195
Sugita et al. (1998) 195, 537

Sundaresan and Aksay (1991) 82, 335, 568
Suttor et al. (1997) 127
Suzuki et al. (1990) 52, 168, 219, 317, 341, 524

Tamele (1950) 8, 22, 663
Taylor (1928) 3
Taylor (1932) 606
Taylor and Holland (1993) 53, 110, 168, 254, 317, 339, 348, 352, 833
Thim et al. (2001) 25, 151
Tkalcec et al. (1999) 628
Tokonami et al. (1980) 4
Tromel et al. (1958) 606
Tsuchiya et al. (1989) 307
Tummala (1991) 8

Uhlmann et al. (1984) 296

Voll et al. (1998) 443, 462, 464, 473, 613

Wakao and Hibino (1962) 677
Wang and Sacks (1996) 538, 801, 831
Wang and Thomson (1992) 831
Wang and Thomson (1995) 532, 732, 572, 744
Wei and Halloran (1988) 76, 188, 533, 566, 568, 582, 690, 832
West and Gray (1958) 102, 114
Wheat (1977) 8
Wheat et al. (1979) 101, 324
Wilson (1979) 489
Wu et al. (1993) 79, 831
Wu et al. (1996) 83
Wyckoff, Greig, and Bowen (1926) 2

Xue and Chem (1992) 686

Yamada and Kimura (1962) 55, 348, 421, 534, 536, 572, 604, 613, 832
Yamane et al. (1978) 289
Yasumori et al. (1990) 253, 304, 476
Yoldas (1975) 57, 298
Yoldas (1976) 57, 599, 677, 707
Yoldas (1977) 33
Yoldas (1979) 33
Yoldas (1980) 33, 57, 123, 312
Yoldas (1992) 34, 257, 312, 315, 316, 337, 341, 346, 347, 377, 473, 825, 833
Yoldas and Partlow (1988) 11, 34, 183 296, 308, 341, 348, 352, 469, 473, 542, 833

Subject Index

^{27}Al and ^{29}Si MAS-NMR spectra of heat treated SGM mullite precursor 268
^{27}Al and ^{29}Si spectra of heat treated CM precursor shows two steps of crystallization 236
^{27}Al spectra of gel between 450°C–800°C shows 52 ppm peak of $[AlO_4]$ groups 425
Absence of –110 nm Si resonance 358
Absence of cristobalite formation on heating diphasic gel 76
Absence of exotherm in freeze drying and SD process 101
Absence of free SiO_2(A) in heated SH, RH, polymeric, diphasic, and SP precursors 570
Absence of phase separation in precursor C during heating process 269
Acidic gels vs. basic gel and their homogeneous in contrast to heterogeneous nature 283
Activation energy data of mullite crystallization 628
Aerosol as a source 98
Aerosol decomposition 101
Agglomerated mullite powder developed at 1350°C 81
Aggregation of $Si(OC_2H_5)_4$ 289
Aging affect the heterogeneity of the powder 198
Aging of silica sol reduces intensity of exotherm of SD powder 93
Al cation substitute Si in the tetra position of silica–alumina gel at 500°C 434
Al(IV) NMR resonance precursors increases with heating at 800°C 423
Alkali leaching of heated gel performed under direct boiling condition 517
Alkali leaching process is more drastic prior to IR studies 521
Alkali leaching process is not as per schedule 521
Alkali leaching shows siliceous rich and aluminous rich phases at dehydroxylation 512
All four gels synthesized at basic pH ranges develops diphasic character 189
Al-O distance on heating gel is large, varies between 2.1 and 2.5 Å 433
Al-Si spinel and silica incorporated θ-Al_2O_3 formed on heating mixed gels 677
Al-Si spinel polymorphically transforms to mullite type phase at ~1300°C exotherm 700
Al-Si spinel to mullite formation is interrelated 536
Alumina promotes crystallization o SiO_2(A) to crystallization of cristobalite 61
Alumina rich amorphous phase transformed predominantly to metastable mullite 772
Alumina sol out of fume γ-alumina 80
Aluminate structure and its stabilization 22

Aluminium alkoxide and TEOS are used in synthesis of SP powder 93
Aluminium alkoxide fume silica, silicic acid sol &TMOS are used in synthesis of SP 98
Aluminium sulphate and TEOS are used in synthesis of SP powder 92
Aluminosilicate bond formation occurs on heating CM, Type I, II, III gels 429
Aluminosilicate phase separates (i) alumina rich (–89 ppm), (ii) silica rich phases (–110 ppm) 471
Ammonia changes solubility of Al and hydrolyse Si components 43
Amorphous alumina silicate structure below 900°C 264
Amorphous citrate process 114
Amorphous structure studied by XAFS and XANES 431
Amount of glassy phase changes with the wt % of Al_2O_3 in the gels 533
Amount of glassy phase changes with the wt % of Al_2O_3 in the gels 699
Amount of noncrystalline alumina silicate phase ~21 wt % 528
Analogous thermal change of clays and co-precipitated gels studied by DTA 23–32
Analyzing whole mullite literature, three routes of phase evolution mullite emerge out 811
ANN and TEOS are dried in synthesis of SD powder 93
ANN and TEOS are used in synthesis of SP powder 92
Apparent solid state reaction of γ-/δ-alumina and silica phases 568
Appearance and disappearance of spinel phase to mullite transformation 704
Aq. mullite gel at pH ≤ 1 neither exhibits 980°C exotherm nor forms Al-Si spinel 140
Attempted to minimize difference in hydrolysis rate of Si and Al alkoxides 52

Based on appearance, existence and disappearance of spinel, 3:2 mullite phase is detected 714
Boehmite is developed "in situ" during gelification process 79
Breadth of 0.137 nm peak of Al-Si spinel changes on 1000°C–1250°C 500.
Broad exotherm 800°C–1300°C corresponds to crystallization of Al-Si spinel 486
Broad resonance at –103.3 ppm due to disordered Si(Al) environment 430

Calculated I_{spinel}/I_{CaF2} lies above theoretical curve 526
Cause of favourable formation of spinel phase 538
Cause of variation in phase transformation behaviours of diphasic gels 187
Cell dimension of mullite vary linearly in the range 60–71 mol % alumina 600
Cell dimensions of mullite 604
Change in LC with heating diphasic gel >1300°C is due to increase in crystallinity 756

Changes in a-, b-, c- LC values of mullite during heating SH and RH gels 610
Changes in composition of noncrystalline phase on heating 675
Changes in IR spectra in the composition range 3:2 to 2:1 mullite 213
Changes in IR spectra of first step of transformation 230
Changes in IR spectra of second step of transformation 230
Changes in IR spectra of third step of transformation beyond 1400°C 231
Changes in relative IR resolution of Al(4)/Al(6) during heating precursor 231
Changes in the ratio AlO_4/AlO_6 Al resonance of CM during heating 267
Changes in the ratio of AlO_6 to AlO_4 NMR resonance SGM etc with temperature 618
Characteristic –110 ppm resonance is obvious to silica rich aluminosilicate phase 808
Characteristic mullite peaks (AlIV, AlIV*, and AlVI) at first exotherm at 950°C are different 262
Characterization of –110 ppm Si MAS NMR peak 686
Chelating agents to control rate of hydrolysis of Al^{+3} 121
Chelation of alkoxide with acetyl acetone 117
Chelation of AlOBu with citric acid in presence of TESPA 114
Chelation of AlOBu with diketone 181
Chemical composition approaches 3:2 mullite when $A_{1170} > A_{1130}$ 215
Chemical homogeneity is a question 478
Chemical homogeneity is very difficult to preserve and problematic 94
Chemisorption of aluminium hydroxide on silicon hydroxide gel surface 21
Choice of with or without de hydroxylated boehmite as source 195
Classification of different mullite precursors synthesized out of various sources 826
Classification of five species of aluminosilicates 235
Classification of six dried precursors 445
Coalescence of silica colloids with aging time and its effect on DTA 169
Coated precursor 81
Coexistence of γ-Al_2O_3 type spinel and mullite type phase 52
Colloidal gel (CG) prepared with excess water designated as first kind of diphasic gel 785
Colloidal gel shows both tetra and octa peaks at 64.5 and 7.1 ppm at 500°C 259
Colloidal gel shows mullitization by reverse change in ratio of Al(VI)/Al(IV) at exotherm at 1300 C 259
Colloidal process scale of synthesis of mullite powder 128
Colloidal silica spherulites 289
Combustion of chlorides of two components 98
Comparative phase evolution of polymeric vs. colloidal gels by NMR spectra 257

Comparative phase evolution of polymeric, CP and diphasic gels by NMR spectra 262
Comparative phase evolution of Type I, II, and III by NMR spectra 259
Comparative phase transformation of CM vs. SGM by NMR spectroscopy 247
Comparative phase transformation of GNW vs. GYW by NMR spectroscopy 254
Comparative phase transformation of MI vs. MII by NMR spectroscopy 249
Comparison of phase transformation of three types with that of kaolinite 162
Complete mullitization in a single step 97
Composition belonging to 3/1 = Al/Si ratio forms only mullite at 980°C 533
Composition of Al-Si spinel remains constant irrespective of composition of mixed gels 689
Composition of noncrystalline alumina silicate by solid state reaction 439
Composition of noncrystalline phase corresponds to 3:3 mullite 509
Composition of spinel by alkali leaching with TEM and EDX 693
Condensation of silanols form polymeric linkages, growth and set to gel 292
Condition of hydrolysis include amount of water, catalyst and temperature 288
Condition of hydrolysis: amount of water/solvent, velocity of hydrolysis process and pH 83
Contour of absorption band between A_{1100} to A_{1200} changes 215
Control the size of primary particles 343
Controversy over composition of spinel phase 482
Coordination environment of Al-O decreases to 4–5 on heating gel at 800°C 433
Coprecipitation in presence of amine 44
Correlation between decrease of XRD band of amorphous phase to spinel formation 673
Correlation between exothermicity of two gels with mullite formation 173
Correlation between mixing scale-mullitization temperature 336
Correlative effects, DTA at 1285°C and expansive effect in dilatometry 265
CP at 1780°C showed smallest amount glass phase than SM and OM 389
CP gel is designated newly as "derived diphasic gel" 792
CP gel made at pH < 8 boehmite was not formed. designated as first kind diphasic gel 785
CP gel of made by double alkoxide system designated as first kind diphasic gel 785
CP intensity decreased to 39% indicates removal silanol protons at 650°C 444
CP powder 182

CP prepared out of large quantity of water of hydrolysis designated diphasic gel 791
CP prepared out of large silicon hydroxide colloids designated diphasic gel 790
Criteria of attaining highest homogeneity -448
Criteria of stability of noncrystalline phase 446
Critical point of drying condition 173
Crystal growth is due to long range diffusion on continued heating 566
Crystallization at first exotherm shows an abrupt shift of IR frequency 218
Crystallization behaviour of alumina in presence of silica 58
Crystallization of noncrystalline phase 680
Crystallization of Al-Si spinel even using partially hydrolysed TEOS 52
Crystallization of Al-Si spinel from noncrystalline aluminosilicate shows more Al(VI) 738
Crystallization of into cristobalite and corundum prior to mullitization 569
Crystallization of noncrystalline alumina silicate phase on heating 664
Crystallization of transitional alumina phase during heating diphasic gel 188

2D chain of Si-O-Si network 306
3D cross linked chain formation 311
3D cross linked cyclic chain condensate formation 312
3D network structure, chain like or spherical 289
D mullite to S mullite 605
D(r) and G(r) precursors at different temperatures 441
Decrease in intensity of amorphous band is correlated Al-Si spinel 688
Decrease in intensity of silica (A) with increase of the peak of spinel phase 504
Decrease in peak width with calcination temperature 620, 623, 625
Decrease of reactivity of aluminium alkoxides 343
Degree of homogeneity decreased in the order of S < Ha < Hb < C powders 386
Degree of polymerization is a factor 287, 290
Dehydration of mix organic compounds of Al and Si 101
Densification before mullitization 78, 187
Determination of analytical water content by MEA and spectroscopy 463
Determination of residual water content by TG and DTG 464
Determination of silica content of spinel phase by EDX analysis 690
Development mullite precursors of different characteristics 107
Development of mullite gels of different characters 825
Development of LC values on variation of *x* 212
Development of noncrystalline aluminosilicate precursor phase by –95 ppm 245

Development of Si-O-Al linkages on heating both mono & diphasic gels 439
Development of Si-O-Si linkages in the form of chain or branches on heating 430
Deviation of X-ray reflections in normal mullite to mullite type phase 604
Difference in condensation of volatiles 99
Differences in evolution behaviour of polymeric vs. colloidal gel 35
Different angles and approaches are studied to solve problems of spinel 544
Different techniques of study noncrystalline mullite precursor phase 420
Difficulties to synthesize homogeneous mullite precursors 358
Diffusion of silica into alumina structure 86
Diffusion process of Si/Al on heating all four gels 427
Diphasic gel also shows phase separation to –89 and –109 ppm resonances 273
Diphasic gel showed a broad exotherm between 1000°C–1300°C and a sharp exotherm at 1320°C 201
Diphasic gel shows most pronounced broad Si resonance at –90 ppm 272
Diphasic gel up to composition of mullite forms gehlenite 506
Diphasic gels show four categories and show varying aspects 794
Diphasic gels synthesized by monitoring pH 9–14 designated as third kind of diphasic gel 789
Direct formation mullite and spinel by monophasic, polymeric & CP gels 560–562
Direct mullite formation out of mullite precursor by aq. gel method 560
Discrete silica and alumina sols 71
Dissolution of γ/δ alumina in vitreous phase 566
Dissolution of alumina in the siliceous phase is predominant 802
Dissolution velocity of alumina into the SiO_2 matrix may be the rate controlling step 112
Doping of active alumina by silica from 6–20 wt % 57
Drastic alkali leaching study 517
Dried and calcined SH gel form gehlenite indicate bonded hypothesis 504–508
DTA events, phase evolution vary with components & techniques of gelation 584
DTA analysis of precursors of nano scale mixing 336
DTA analysis of silica gel, alumina (A), aluminium sulphate 25
DTA and XRD intensity increase with pH and maximum at pH ~4.5 140
DTA behavior of monolithic alumina at various amounts of doping silica 711
DTA peak height of first exotherm 626
DXRD recording of intensities of mullite 31

Effect if size of silica 197

Effect of Al/Si ratio on development process of diphasic gel 597
Effect of ANN vs. aluminum sulfate as source in synthesis of SP or SD precursors 101
Effect of hydrolysis by ^{17}O enriched water 49
Effect of pH on coprecipitation 186
Effect of rapid hydrolysis of AlOBu or Al(iPro) 345
Effect of reduction of chemical encounter rate with water 183
Effect of silica doping on the occurrence of DTA peak of alumina 57
Effect of source of components processing variables 368
Effect of water of hydrolysis on thermal change of monophasic gal 161
Effect of water on hydrolysis of components 85
Effect of water, ammonia sequential hydrolysis-mixing on phase evolution 181
Elimination of penta with 42 ppm and 2 ppm resonances of heated SP at 1000°C 264
Elimination of spinel phase by adopting SP process 542
End point of SiO_2-γ-Al_2O_3 solid solution should tested on three criteria 713
Endothermic dip marked D 465, 466
Engineered coprecipitation technique by raising the pH using urea 95
Enriched Al-Si spinel in RH etc. changes to mullite at second & third exotherm by Part IV & Part V 748
Enthalpy of crystallization data of mullite out of different precursors 625
Essential condition for obtaining pure mullite at 1000°C 169
Estimated value of spinel is greater than percent alumina content 526
Evaporative decomposition 98
Evidences of noncrystalline alumina silicate phase on heating BS gel 427
Evolution of aqueous mullite gel 139
Evolution of diphasic mullite gel 186
Evolution of diphasic mullite gel prepared at different pH 187
Evolution of monophasic mullite gel prepared at different water 161
Evolution of monophasic mullite gel prepared at different pH 145
Evolution of polymeric mullite gel 175
Evolution of polymeric mullite gel prepared at different temperature 184
Evolution of polymeric mullite gel prepared at different water 183
Evolution of polymeric mullite gel prepared on washing 185
Evolution of polymeric mullite gel prepared at different pH 182
Evolution of SD & SP mullite precursor 197
Evolution of six mullite precursors prepared out of two fixed sources 199
Existence of –110 ppm Si resonance even heating precursor B at 1200°C 266

Existence of OH groups vs. slow dehydroxylation 444
Existence of XRD peaks related to alumina > than 1200°C in diphasic gel (M3) 381
Extensive cross-linking or highly condensed species 296
Extrapolation of a- and b- lattice parameter curves 600

First category of diphasic gel deals with phase transformation 794
First coherent part of transformation of diphasic gel 666
First rapid step of mullitization between 900°C–1000°C occurs by Path I 746
First stage of transformation at 980°C exotherm due to spinel and mullite formation 144
Five general methods of mullite synthesis 10
Flame spray pyrolysis (FSP) 98.99
Formation a ratio of mullite and spinel by some colloidal other gels 563
Formation glassy alumina silicate matrix 191
Formation of 3D silica frame work structure 171
Formation of Al-Si spinel (minor) with primary mullite is invariable in three methods 742
Formation of Al-Si Spinel is pH dependent 692
Formation of Al-Si spinel phase of composition 3:2 at 28 wt % silica addition 58
Formation of aluminosilicate phase by condensation based measurement of T–O, T–(O2) distances 442
Formation of amorphous band of noncrystalline silica rich aluminous phase (AS1) 653
Formation of bi modular polymeric gel phases 315
Formation of heterogeneous –Al–O–Si– matrix network structure 435
Formation of mixture mono and diphasic gel 195
Formation of noncrystalline aluminosilicate as per order SHI < SHII < SHIII < SHIV 775
Formation of precursor of two types 54
Formation of silica rich alumina phase 689
Formation of spinel phase with longer aging time 168
Formation of white gel during adding water drop wise to mixed alkoxides 52
Formation θ-alumina in addition to Al-Si spinel in mixed gel 499
Four different characters of mullite gels at varying processing conditions 112
Four classes of precursors are presented in details 834
Four controlling parameters of hydrolysis 342
Four different types of Al_2O_3-SiO_2 gels vs. thermal reaction paths 195
Four factors influence hydrolysis two components 369
Four fold peaks of Al at 47 and 65 ppm of SAE and forms mullite 253
Four phase equilibrium curves 2
Four processing variables lead to two class of precursors as per DSC 373
Four sources of aluminium components 426

Four stages of hydrolysis 287
Four steps of changes in LC curve of SH gel compared with mullite formation curve 736
Four steps of changes in quantitative value of mullite over a range of temperature 746
Four variables affecting crystallization sequences of four mullite gels 382
Four varieties of inorganic alumina sols for mullite gel 96
Four views of mullite formation in diphasic gel 568
Fourth category deals in situ diphasic gel & its phase transformation 799
Fourth step of mullitization between 1100°C–1300°C occurs by Path III 746
Freeze dried gel 100
Frequency shift of the Al–O "stretching mode" 229
Fume alumina accepts silica by diffusion 80
Fundamental transformation between 950°C–1050°C 242

Gel prepared at highly acidic condition (pH ≤ 1) 31
Gel prepared out of large aluminum hydroxide colloids designated diphasic gel 790
Gel synthesis by solution-precipitation technique 79
Gelation time correlated with mullite formation 31
Gelation time was correlated with water content 182
Gels at 1200°C XRD data of mullite does not conform IR observations 754
Global presentation of newer classification scheme 830
Growth curve was reported to be parabolic for oxide mixture 582
Growth curve in OM & diphasic gel are dissimilar 582
Growth curve of mullite out of six precursors 573
Growth curves of SHIII, SHIV & diphasic gel lie below curve of SHII 581
Growth of mullite grains is controlled by chemical inter diffusion within mullite 802
Growth of mullite in (AS), MO compared with diphasic gels 567
Growth of mullite in diphasic gel is analogous to that of kaolinite 585
Growth of mullite in diphasic gel system was sigmoidal in nature 582

Heat evolution of various gels at 980°C exotherm 627
Heat of reaction of TEOS and ANN 302
Heating rate favors crystallization of spinel or mullite due to difference in kinetics 715
Height of sharp exotherm is found highest in STD sample 92
Hetero coagulation 74
Higher addition of alumina decreases cristobalite formation 61
Higher amount of alcohol lead to large quantity of noncrystalline aluminosilicate 774
Highest crystallization energy 344

Highest enthalpy of crystallization in SD corresponds to ~60 mol % Al_2O_3 706
Highest intensity of mullite without crystallization of components in SD is 40–30 mol % silica 712
Highly cross linked polymers in base catalysed TEOS 319
Homogeneity affect microstructure development of diphasic gel 188
Homogeneity of gel correlated with time/temperature of mullitization 74.
Homogeneity of GNW vs. GYW 256
Homogeneous co-polymerization is rather a difficult task 91
Hydrolysis in presence vs. absence of water 53
Hydrolysis of alkoxide groups to develop silanols 292
Hydrolysis water shows a direct relationship with Al-Si spinel crystallization 775
Hydrolysis with less ethanol vs. more ethanol 148
Hydroxyl alumina silicate sulphate 284
Hydroxyl groups a part of noncrystalline aluminosilicate structure 439

Identification of spinel phase a problem due to θ-alumina 669
In situ diphasic gel, HB13, M3 designated as second kind of diphasic gel 789
In situ diphasic gel, HB13, M3 designated as second kind of diphasic gel 789
Incomplete extraction of siliceous phase prior to EDS study 517
Incorporation leads to diminution of XRD peaks of θ-alumina 670
Incorporation of silica in spinel structure forms Al-Si spinel phase etc. 504
Incorporation phenomenon 87
Increase in ratio of Al(IV)/Al(VI) is contrary to formation of 2:1 mullite at 980°C exotherm 739
Increase in network formation & decrease in bridging angles on condensation 442
Independent crystallization of components 78
Independent partial crystallization of two components of diphasic gels 187
Indirect evidences of presence of silica in spinel phase derived out of mono and diphasic gels 659
Influence of pH on the mixing scale of the components on mullitization 48
Influence of water of hydrolysis 346
Inhibition of nucleation of α-alumina 686
Inhomogeneous dehydroxylation 468
Intensity of exothermic effect with Si-O-Al in the precursor 337
Intensity of 980°C exotherm directly proportional to quantity of mullite 140
Intensity of spinel phase increases with heating 499
Intensity of spinel phase increases with heating in three diphasic gels 504

Intensity of the amorph. band at 22°2θ should decrease if spinel forms 681
Inter diffusions between AS3, alumina rich Al-Si spinel & AS2 occur at third & fourth steps 747
Inter relationship between use of water with Al-Si spinel formation 167
Interaction between TEOS and melted ANN 305
Interaction of aq. solution of silicon acetate and ANN 126
Intermediates during hydrolysis process 295
Interrelationship between heating temperature with Al(IV) 443
Introduction of Si (small amount) forms γ-Al_2O_3 and then δ-Al_2O_3 61
IR frequency data of ideal mullite 212
IR intensity ratio A_{1130}/A_{1170} a function of a LC value of mullite composition 217
IR peak at 1070 cm^{-1} corresponds to Al-Si Spinel 696
IR result used to interpret mullitization process 216
IR study indicates the introduction of Al into glass network 421
Isolation of Al-Si spinel and noncrystalline aluminosilicate 509

Lack of reflection splitting of the doublets 73
Large problems of gel-mullite transformation processes data to others 392
Lattice constant value *a* showed the greatest composition dependence 611
Lattice parameters of diphasic gels of varying compo. fired at different temp. 622
LC of poorly vs. well crystallized mullite 598
LC of spinel phase a function of composition 534
Level of homogeneity vs. intensity of 980°C exotherm 379
Limitations of distinguishing γ-alumina from Al-Si Spinel phase 661
Line profile of IR spectra in the range 1100–1200 cm^{-1} a function of alumina content 213
Linear relationship between heating temperature to (AlIV/AlIV+AlVI) 372
Logical conclusion is made on composition of weakly crystalline primary mullite 809

M1 & M2 showed differently in exhibition in DTA peaks vs. mullite/spinel ratio 382
Major jump of mullitization at 1300°C coincides with DTA peak 76
Many predictions prevail on thermal changes of gels among researchers 399
Matching copolymerization rate by urea 345
Maximum limit of silica addition to γ-alumina spinel assigned as 40 mol % (~28 wt %) 714
Maximum solubility of silica in θ-Al_2O_3 is considered as ~6 wt % silica 707
Measurement of amorphous hump by a software programme 654

Measurement of amorphous hump manually 654
Measurement of changes in XRD intensity of amorphous band 440
Mechanism of condensation reaction 293
Mechanism of the reaction between silicic acid sol and Al^{+3} 285
Metal alkoxides as source components and their coagulation 33
Methods of characterization of the nature of t-mullite formed at 980°C 731
MI and MII both develops spinel phase at second stage 252
Minimization to avoid local enrichment of alumina 182
Minimization of selective hydrolysis 45
Minimization of water addition during processing 542
Missing of few peaks & absence of splitting of Miller indices of 120, 240, 041, and 250 756
Mixed gel synthesized by consecutive method is designated as fourth kind 792
Mixing scale of components (Table 18.2) controls course of mullitization 693
Mixture of Al-Si spinel & Al-rich spinel formed at 1000°C is 70–80 mol % alumina 717
Mixtures of polymeric gel particles of different in chains lengths and clusters 316
Mode of development of Al-Si spinel phase formed on heating G-173 704
Modification of hydrolysis-polymerization of metal ions by complexing agents 117
Modifying AlOBu with ethyl acetoacetate 118
Monitoring of condensation process by ^{29}Si NMR 297
Mono dispersed spherical fine mullite exhibit 980°C exotherm 102
Monophasic gel prepared at acidic vs. basic pH affect mullite formation 148
Monophasic gel prepared at purely basic pH develops diphasicity 153
Monophasic mullite gel prepared by polymeric vs. colloidal conditions 175
Most homogeneous precursor may exhibit strongest exotherm 335
Most homogeneous precursor shows high ratio of AlO_4 groups 355
Most important parameter : solvent used and time of precipitation 95
Mullite phase formation was major in CM than that of same in SGM at 1000°C 249
Mullite exists as a two-component solid solution, Al_2O_3/SiO_2 = 3/2~2/1 605
Mullite formation out of two aluminosilicate phases 659
Mullite formation out of two aluminosilicate phases 659
Mullite formation by dissolution-precipitation reactions in AS-γA 82
Mullite formation by nucleation and growth within the siliceous phase 82
Mullite formation by solid state reaction of alumina & siliceous phase by inter diffusion 82

Mullite formation in SP corroborate DTA 980°C exotherm & expansive effect in TMA 387
Mullite formation in stoichiometric SH gel never riches full conversion at 1000°C 735
Mullite formed by inter diffusion among two phase separated phases 564
Mullite formed by thermal decomposition of spinel phase at second exotherm 572
Mullite precursor out of alumina-filled siloxanes 127
Mullite precursors by spray pyrolysis & spray dried (SP and SD) techniques 197
Mullite precursors made by use of organic substances 114
Mullite products for various technological applications 7
Mullite type phase 37
Mullitization curve of diphasic gel predicts two paths like kaolinite 686
Mullitization in OM, alumina-silica compact are diffusion controlled 582
Mullitization in OM, alumina-silica compact are diffusion controlled 582
Mullitization process of three routes consist of various steps 812
Mullitization processes vs. changes of IR data (i) $A_{1130} > A_{1170}$, (ii) $A_{1130} \cong A_{1170}$, (iii) $A_{1130} < A_{1170}$ 737
Mullitization temp. of four increased as per order S < Ha < Hb < C 386

Narrowing of Raman spectra during evolution process 228
Narrowing of the SiO_4 mode at 970 cm^{-1} (above 1000°C) and AlO_4 (above 1300°C) 229
Nature and composition of 980°C heated mullite phase-two possibilities 665
Nature of 980°C heated spinel phase-three possibilities 664
Nature of phase forms varies at four exotherms e.g., 980°C, 1150°C, 1250°C and 1300°C in DTA 729
Noncrystalline aluminosilicate phase crystallizes to spinel phase in two ways 684
Noncrystalline aluminosilicate phase itself crystallizes to Al-Si spinel 688
Noncrystalline aluminosilicate phase vs. initial composition of gel 527
Noncrystalline aluminosilicate phase formation shows resonance at –104 ppm 272
Noncrystalline mullite precursor studied by four physicochemical methods 649
Noncrystalline silica rich alumina phase decreased with decreasing silica content 245
Nonstoichiometric alumina rich mullite 469
Nucleation is due to short range migration of diffusing species 566

Objectives of mullite research Book 14
OCM heated at 1100°C//20 h showed ring pattern of γ-Al_2O_3 spinel 125
Octa. Al in GYW is significant at 980°C 257

Octahedral peaks at 600 & 550 cm^{-1}, tetrahedral peak at 740 cm^{-1} become distinct at 1400°C 755.
Old method of classification of mullite precursors from various sources 828
One octa and two tetrahedral ascribed as T and T* sites of Al in mullite at ~1000°C 243
Origin of negative lattice charge 23
Oxidation of organometallic compounds 98
Oxide one pot synthesis process 127

Parameters guide hydrolysis & condensation of TEOS 287
Peak at 0.137 nm is a mixture of Al-Si spinel & Si incorporated θ-Al_2O_3 500
Peak at –110 ppm of precursor B increases largely 266
Penta Al in GNW is predominant at 980°C exotherm 257
Percent of leached silica and alumina of diphasic gels vs. heating temperature 652
Persistence of optical clarity of OCM even on heating up to 1480°C 124
pH during chemical precipitation affect boehmite or bayerite formation 188
Phase changes of co precipitated gels of aq. solutions by DTA and XRD studies 43
Phase changes of diphasic gels by DTA and XRD studies 71
Phase evolution of codecomposed powders by DTA and XRD studies 91
Phase evolution behavior of precursors by sol-gel is absolutely different from SD 710
Phase evolution of aqueous and monophasic gels by DTA and XRD studies 23
Phase evolution of aqueous and monophasic gels by DTA and XRD studies 23
Phase evolution of co precipitated gels out of double alkoxides by DTA & XRD 55
Phase evolution of mixed mullite gels by ^{29}Si NMR spectroscopy 244
Phase evolution of polymeric gels by DTA and XRD studies 32
Phase evolution of RH gels by DTA and XRD studies 47
Phase evolution of SD/SP precursors by NMR spectra 264
Phase evolution of six precursors correlated by DTA, XRD, IR, NMR etc. 770
Phase evolution process on heating was influenced greatly by physical structure 186
Phase evolution process on heating was influenced greatly by physical structure 186
Phase evolution two gels made at excess water (A) or in ammonia (B) 374
Phase separation related to two peaks at –95 ppm and –110 ppm 245
Phase transformation of SHIII on heating consisting four phases 779
Phase transformation of aq. mullite gels by using NMR spectroscopy 236

Phase transformation of coprecipitated mullite gels by NMR spectroscopy 242
Phase transformation of diphasic gels by NMR spectroscopy 243
Phase transformation of mechanically mixed oxides shows θ, α alumina 494, 497
Phase transformation of mixed gel at ~1000°C 669
Phase transformation of polymeric and colloidal mullite gels by NMR spectroscopy 241
Phase transformation silica rich Al_2O_3–SiO_2 gels 61
Pilot scale synthesis of mullite powder 128
Plot of converging value of lattice constant a/b vs. c/b in Smith's chart 614
Plot of a-spacing varies with composition 72–77 wt % alumina 609
Polymeric gel shows major penta coordinated Al sites 258
Polymeric gel shows mullitization by drastic change of penta during exotherm at 1050°C 258
Polymeric silicic acid units play a key role in phase evolution 143
Polymeric silicic acid units play a key role in phase evolution 143
Polymerization with acrylamide monomers 120
Polymerized silica solution is to be avoided in SP process 542
Polymorphic transformation of Al-Si spinel phase to mullite at third exotherm 569
Portion of curve in 2–10 wt % SiO_2 analogous to portion of first stage of Yoldas 710
Possibility of cluster formation 319
Possibility of formation of mixed gels of mono and diphasic characters 792
Possible steps of evolution of six mullite precursors to be discussed 450
Precursor A still shows phase separation due to appearance of –110 peak 268
Precursor D shows sharper resonance at –110 ppm persisted at 1400°C 267
Precursor structure of SP & CA at 1000°C are different as per ^{27}Al MAS-NMR 388
Precursor synthesis at hydro thermal condition 50
Precursor synthesis out of double alkoxide in ammonia 49
Precursor when heated, the T-O bond distance was found large 441
Primary mullite vs. well crystallized mullite 614
Probable solutions of problems of gel-mullite reaction series 402
Problems of inhomogeneity in Al_2O_3-SiO_2 precursor system 459
Processing in presence of organic gel & in absence of organic gel with urea 113
Processing of aqueous spray-drying of aged and without aged silica source 113
Progressive changes of octa to tetra ratio of Al on heating 244

Progressive ordering of local structure by Raman spectroscopy 438
Proposed gel model formed out of aq. sol-gel method 286
Pseudo tetragonal metric 745

Quantity of amorphous phase dependent on pH during gelation 194
Quantity of formation of spinel phase follows as order SHII < SHIII < SHIV 774
Quantity of residual non crystalline alumina rich mullite phase (AS3) changes in precursors 744
Quantity of SiO_2 more than 24 wt % in solid solution with γ-Al_2O_3 is feasible 702
Question of inhomogeneity during hydrolysis 56
QXRD data of Al-Si spinel phase is ~79–80 wt % 523, 528

Raman frequency data of mullite 213
Raman spectra of mullite observed bands at 320 cm^{-1}, 430 cm^{-1} and 970 cm^{-1} 229
Rapid drying vs. slow drying affect phase transition 174
Rapid growth Al-Si spinel occurs between 800°C~1250°C 490, 492
Rapidity of mullitization at first exotherm follows as SD > SHI > SHII 580
Ratio "Al(IV)/Al(VI)" increases from 1000°C to 1300°C of mullite gel 240
Ratio of tetra to octahedral Al in diphasic gel at 1010°C is 0.45 241
Ratio of tetra to octahedral Al in monophasic gel at 1075°C is 0.48 241
Ratios of Al-Si spinel/mullite of SHII and SHIII at ~1000°C 580
Raw SH gels forms $C_{12}A_7$ concludes mixture hypotheses 503
Reaction at points of contact (interface) of SiO_2/γ-Al_2O_3 particles 79
Reaction limited cluster–cluster aggregation 181
Reaction taking place at first stage of transformation of mullite precursor 419
Reactive aluminium species is of paramount importance 97
Reactivity difference between silicon and aluminium alkoxide 181
Reason of low sintering temperature 79
Reason of relative ratio of SiO_4 tetra (A_{1130}) and AlO_4 (A_{1170}) vibrations changes on heating 737
Reasons of noncrystallization of the components of diphase gel 86
Relation bet. composition of spinel coincides with starting materials 530
Relation between first exotherm with mullite formation 337
Relationship between a-parameter with composition of mullite 606
Relative ratio of A_{1165} to A_{1130} changes in mullitization stage 215
Relative ratio of Al (IV)/Al (VI) NMR of mullite at 980°C are different to well crystallized state 737

Relief of strain present in mullite at incipient formation stage to >1300°C 756
Residual non crystalline alumina rich mullite phase (AS3) 469
Residual non crystalline alumina rich silicate after Al-Si spinel/ mullite (AS2) 653
Retardation of transformation of alumina component by glassy phase 685
RH gel forms unsaturated Al-Si spinel at 980°C exotherm in DTA 699
RH gel, CP gel SHIII designated as first kind diphasic gel 783
Role of bonded protons its consequences on removal 465
Role of organic silica affects spinel to mullite development 541
Role of acidic pH on nature of exotherm and thermal change of monophasic gal 145
Role of alumina source in phase transformation of C & D mullite gels 374
Role of hydrolysing catalysts 148
Role of mono silicic vs. fume silica on thermal change monophasic gal 144
Role of pH and agent like ammonia, urea amine on homogeneity 159
Role of pH on phase transformation behavior diphasic gel 187
Role of reactivity of alumina component during gelation 195
Role of reactivity of silica particles of varying sizes 225
^{29}Si spectra of first type of a monophasic gel heated to 1000°C is reanalysed 747
^{29}Si peak maximum at –90 ppm in heated precursors A, C, E & B bond formation 429
^{29}Si resonance at ~ –80 ppm due to silicon in an Al_2O_3 rich noncrystalline phase 430
^{29}Si resonance of Al-Si spinel phase at –89 ppm shifted to –86.6 ppm of mullite 273
^{29}Si spectra of CM, mullite does not attain stability between 1100°C–1200°C 236
Same sources component (AIP, TEOS) at varying conditions of water 110
Same sources component (AIP, TEOS) hydrolysis with water or without water 110
Same sources component (AIP, TMOS) at hydrolysis with H_2O vs. H_2O and NH_4OH 110
Same sources component (AlCl3, TEOS) at variation of pH from acidic to basic range 110
Same sources component (AlOBu, TEOS) at pH and water of processing produce three kinds of mullite gels 111
Same sources component (AlOBu, TEOS) by changing both pH and water of processing 111
Same sources component (ANN, TEOS) at four different processing 112
Same sources component (ANN, TEOS) at three different processing 112
Same sources component (ANN, TEOS) at differ in pH form gels of differ in thermal characters 111

Same sources component (ANN, TEOS) at six processing parameter 112
Same sources component (ANN, TEOS) in presence of three catalysts 110
Same sources component (boehmite, silica sol) at two different processing 109
Same sources component (boehmite, TEOS) at acidic vs. basic conditions 109
Same sources of components (AIP, TEOS) at varying coprecipitation techniques 109
Same sources of components (AIP, TMOS) coprecipitated in two different hydrolysis conditions 109
Same sources of components (ANN, TEOS) at two different pH values 107
Same sources of components at two different pH values/ methods 109
Same sources show variations DTA & mullitization of 6 precursors synthesis 392
Scale of homogeneity 356
Schematic mullite formation in the three cases of fired diphasic gels 588
SD (16 wt %) SiO_2 crystallizes to mullite directly & not spinel solid solution~1273°K 708
Second category diphasic gel deals densification and microstructure 797
Second nucleation process of transformation of diphasic gel 666
Second stage of transformation at 1200°C exo. due to final mullite formation 144
Second step of mullitization ceases between 900°C–1100°C, width & *a* LC are constant 746
Selection of low heating temperature below 1300°C prior to MAS-NMR 518
Self-condensation of alumina 340
Semi quantitative cross polarization intensity heated precursor 439
Semi quantitative estimation of mullite 6 precursors (SHI to SP) 390
Semi quantitative value of aluminosilicate phases 657
Sequence of phase evolution of A, C, E similar in nature as GNW & MI 781
Sequence of phase transformation pure and silica doped alumina gel 61
Sequential hydrolysis 128
Seven steps transformation of diphasic gel for clarity 595
Several approaches to achieve homogeneity 357
SGM, Type III, M1, MI are designated as first kind of diphasic gel 788
Sharp 980°C exotherm pf SP corroborates with expansive effect in thermo mechanical analyser 264
Sharp decrease in surface of SP powder during mullitization 92
Shifting of absorption band of silica by incorporation of alumina 217
SHIII exhibited three exotherms 201
SHIV shows profound segregation at 1000°C 272

Short distance RED function ($R < 6$ Å) of Type I and III (oi) changes on heating 435
Silica coated γ-alumina, and or α-alumina 83
Silica doped alumina gel forms pseudo boehmite to α-alumina 61
Silica doped alumina gel forms pseudo boehmite to α-alumina 61
Silica hinders crystallization of alumina 58
Silica in aq. form plays a major role in phase transformation path 540
Silica incorporated γ-Al_2O_3 changes to silica incorporated θ-Al_2O_3 at 1200°C 675
Silica rich aluminous phase (AS2) 469
Silica sol should be primary (–Si-OH) species instead of polymerized units 143
Silicon tetrachloride as like aq. silica sol an effective source 344
Si-O-Al bond development in spray pyrolysis method 322
Six different characters of mullite gels at varying processing conditions 113
Six mullite precursors 368
Six mullitization paths 354
Size and strain values of mullite crystallites vs. heat temperature 619
Size of silica sol particles 345
Slow hydrolysis (SH) gel 28
Slower hydrolysis rate may tend to produce better mixing of components 147
Solid state interaction between AS3 and AS2 with AS1 phases on further heating 751
Solid state reaction by diffusion process of Si and or Al atoms on heating 431
Solid state reaction of spinel crystallized at first exotherm with SiO_2 (A) at second exotherm 565
Some characterization methods of Al-Si Spinel phase 662
Some models of gel precursors 461
Some proposals on early & final mullitization 564
Source of silica also affect the development of 980°C exotherm and mullite formation 776
SP precursor shows predominant Al(V) resonance at 400°C 264
Spherical hollow particles formed during spray pyrolysis 92
Spinel is the favoured phase than mullite in some SP powder at 1000°C 98
Spinel phase formation was major in SGM than that of same in CM at 1000°C 249
Spinel is a polymorph of mullite 665
Spinel like phase is characterized by the 350 cm^{-1} shoulder in Raman spectra 229
spinel phase (γ-Al_2O_3 type) 37
Spinel phase formation is inhibitable in al monophasic to diphasic gels 481
Spinel phase formation prior to mullitization 49
Splitting of peaks with Miller indices of 615
Splitting of Si spectra of CM occurs during crystallization processes at >1400°C 237
Stabilization of δ-alumina 57

Starting silica sol controls two forms of mullite 142
States of first category of mullite precursor (hydrolysis cum gelation) 460
States of second category of mullite precursor (hydrolysis cum precipitation) 461
States of third category of mullite precursor (spray dried or pyrolysis) 461
Static heating reduces intensity of first exotherm 420
Steps of phase transformation of diphasic gel 590
Stoichiometric mullite phase is $[Al_2^{vi}(Si_{2-2x}Al_{2+2x})^{iv}O_{10-x}$, x = 0.17 to 0.59 607
Structural changes of 3:2 mullite (weakly to well variety) is cause change of a LC value 745
Structural model of mullite 3
Structural transformation of Al-Si spinel phase by MAS-NMR studies 590
Structure of the amorph. GYW and GNW xerogel played a central role 255
Suggestion of six proposals of mullite formation 559
Summarized studies of six precursors explain exothermicity 371
Summarized thermal transformation of five mullite gels 139
Summary of comparative phase transformations vis-à-vis mullite gels 403
Super critical fluid drying process 127
Suppression of crystallization of α-alumina is ~200°K by silica doping 708
Surface area, pore structure of alumina silicate gel 186
Symmetric vs. asymmetric shape of MAS-NMR peaks four precursors 427
Synthesis of monophasic gels out of water soluble components 25
Synthesis of monophasic out of ANN-TEOS 27
Synthesis of mullite by hybrid gel methods 122
Synthesis of mullite gels at varying hydrolysis conditions 30
Synthesis of precursors of three types 48, 54
Synthesis of three monophasic gels prepared in different water contents 161
Synthesis process of colloidal gel 34
Synthesis process of polymeric gel 35

Techniques used vs. six different types of phase transformation paths 370
Temperature during hydrolysis affect character of mullite gel 184
Temperature of gelation in the range 25°C–40°C–50°C–60°C and aging for 1D to 5D 172
Temperature scale of mullite crystallization 353
TEOS-AC, TEOS-ANN,TEOS-AF show Si-O-Al bond by NMR 427
Tetra of Al resonance of CM is more than that of same in SGM at 1000°C 249
The a (cell edge) vs. V (cell volume) plot for mullite shows linearity 607

The *a* lattice parameter of o-mullite increases linearly from 3:2 to 2:1 mullite 212
The *a*–*X* plot of mullite from various sources also shows linearity 607
The broad peak in precursor B is due to binary aluminosilicate phase 268
Thermal analysis of aq. CP, (CM) vs. gel out of alkoxides (SGM) 372
Thermal behaviour of following six (6) representative mullite precursors 199
Thermal changes gels are monitored by seven techniques starting from TGA etc. 384
Thermal changes M1 prepared at acidic pH vs. M2 with much water 380
Thermal transformation of Type I, Type II and Type III are different 292
Third category diphasic gel deals densification before mullitization 797
Third slow step of mullitization between 1100°C–1300°C occurs by Path III 746
Three different characters of mullite gels at varying processing conditions 111
Three different characters of mullite gels at varying processing conditions 111
Three different sets of conditions of diphasic gel 657
Three different types of mullite gels were synthesized 183
Three kinds mullite gels out of organic sources with variation hydrolysis water 175
Three major problems of identification of weakly crystalline 980°C formed mullite 734
Three phase boundary regions on heating of precursors at dehydroxylation stage 815
Time of hydrolysis 347
Time of aging shows inverse relationship with mullite development 93
Time variation of b- vs. ordering of lattice 606
Time variations of a- and c- vs. change of composition of mullite 606
Transformation of Al-Si spinel to mullite exhibit exotherm 1320°C 487
Transformation of diphasic gel at acidic vs. basic condition 77
Transformation of Gel→Glass→Spinel→ Mullite is continuous 229
Transformation of silica incorporated γ-alumina 667
Transformation stages of mixed Al-Si spinel and θ-alumina heating 600
Transparent alumina gel 57
Transparent monolithic xerogel 126
Transparent mullite coating film 168
Triplex nature of the spectrum 243
True diphasic route of crystallization 187
Two co precipitated mullite powder (A and B) 109
Two criteria of formation of cristobalite and corundum are considered in solid solution 700

Two different characters of mullite gels 107
Two parallel reactions operate densification & mullitization processes in AS-γA 803
Two stage mullite formation 31
Two steps of densification 351
Two steps of mullite formation in monophasic SH gel 564
Two types aluminosilicate phases 656
Type I showed major Si resonance at –90 ppm and a minor at –110 at 1000°C 261
Type II changes ratio of Si resonance –80 ppm at –110 on heating 350°C–1000°C 261
Type III showed broad Si resonance at –90 ppm and merged with others at 1000°C 261

Unidirectional short chain formation 308
Uniform nucleation of mullite phase within alumina grain 690
Unstable noncrystalline alumina rich silica phase (ASI) 469
Upper limit of silica solubility near 60 mol % alumina, SD precursor shows highest enthalpy 717
Use of alumina siloxane, silicon ethyl ester as source 122
Use of pre hydrolysis of TEOS 48, 55, 341
Use of urea as co precipitating agent 55, 56

Variation in high temp DTA runs 201
Variation in quantity of aluminosilicate phase 658
Variation of character of gels depend upon five variables 173
Variation of LC of mullite vs. chemical composition of diphasic gel 616
Variation of LC of mullite vs. degree of crystallization 607
Various physicochemical techniques used 366
Various sources of components 366
Various views on composition of spinel in different diphasic gels 484
Various ways of synthesis of optically clear mullite ceramic (OCM) 123
Varying degree of spinel formation vs. mullite formation 146

Washing treatment of gel affect surface area and magnitude of exotherm 185
Water invariably developed spinel phase as intermediary 167
Water affects formation of spinel and noncrystalline aluminosilicate phase 178
Water free approach 314
What controls the development of spinel phase? 538
When extent of spinel formation would be maximum 530
Which composition is the end point of γ-Al_2O_3–SiO_2 solid solution? 533

Wide distribution of monomers, dimer, chain 296
Wide variety of ceramic and glass products 4

X-ray amorphous hump vs. temperature were recalculated 585
X-ray peak width and a LC of SH gel are very high at 900°C 746
XRD intensity area is greater than γ-alumina crystallization 525
XRD intensity of amorphous band increases with heat treatment 503
XRD peak width & *a* LC are shortened & attained 3:2 mullite between 1300°C–1600°C 746
XRD peak width & *a* LC are shortened between 1150°C–1300°C 746
XRD peaks of primary mullite are broadened at 1000°C 756

Zeolite phase formation is evident when leaching is extended 514

For Product Safety Concerns and Information please contact our EU representative GPSR@taylorandfrancis.com Taylor & Francis Verlag GmbH, Kaufingerstraße 24, 80331 München, Germany

Batch number: 10726503

Printed by Printforce, the Netherlands